T0342344

**Digital Speech Transmission and Enhancement**

# Digital Speech Transmission and Enhancement

*Second edition*

*Peter Vary*
Institute of Communication Systems
RWTH Aachen University
Aachen, Germany

*Rainer Martin*
Institute of Communication Acoustics
Ruhr-Universität Bochum
Bochum, Germany

**Library of Congress Cataloging-in-Publication Data applied for**

Hardback: 9781119060963
ePdf: 9781119060994
ePub: 9781119060987

Cover Design: Wiley
Cover Image: © BAIVECTOR/Shutterstock

Set in 9.5/12.5pt STIXTwoText by Straive, Chennai, India
Printed and bound by CPI Group (UK) Ltd, Croydon, CR0 4YY

C9781119060963_201123

# Contents

# Preface

Digital processing, storage, and transmission of speech signals have gained great practical importance. The main areas of application are digital mobile radio, audio-visual conferencing, acoustic human–machine communication, and hearing aids. In fact, these applications are the driving forces behind many scientific and technological developments in this field. A specific feature of these application areas is that theory and implementation are closely linked; there is a seamless transition from theory and algorithms to system simulations using general-purpose computers and to implementations on embedded processors.

This book has been written for engineers and engineering students specializing in speech and audio processing. It summarizes fundamental theory and recent developments in the broad field of digital speech transmission and enhancement and includes joint research of the authors and their PhD students. This book is being used in graduate courses at RWTH Aachen University and Ruhr-Universität Bochum and other universities.

This second edition also reflects progress in digital speech transmission and enhancement since the publication of the first edition [Vary, Martin 2006]. In this respect, new speech coding standards have been included, such as the Enhanced Voice Services (EVS) codec. Throughout this book, the term *enhancement* comprises besides noise reduction also the topics of error concealment, artificial bandwidth extension, echo cancellation, and the new topic of near-end listening enhancement.

Furthermore, summaries of essential tools such as spectral analysis, digital filter banks, including the so-called filter bank equalizer, as well as stochastic signal processing and estimation theory are provided. Recent trends of applying machine learning techniques in speech signal processing are addressed.

As a supplement to the first and second edition, the companion book *Advances in Digital Speech Transmission* [Martin et al. 2008] should be mentioned that covers specific topics in Speech Quality Assessment, Acoustic Signal Processing, Speech Coding, Joint Source-Channel Coding, and Speech Processing in Hearing Instruments and Human–Machine Interfaces.

Furthermore, the reader will find supplementary information, publications, programs, and audio samples, the Aachen databases (single and multichannel room impulse responses, active noise cancellation impulse responses), and a database of simulated room impulse responses for acoustic sensor networks on the following web sites:

http://www.iks.rwth-aachen.de
http://www.rub.de/ika

The scope of the individual subjects treated in the book chapters exceeds that of graduate lectures; recent research results, standards, problems of realization, and applications have been included, as well as many suggestions for further reading. The reader should be familiar with the fundamentals of digital signal processing and statistical signal processing.

The authors are grateful to all current and former members of their groups and students who contributed to the book through research results, discussions, or editorial work. In particular, we like to thank Dr.-Ing. Christiane Antweiler, Dr.-Ing. Colin Breithaupt, Prof. Gerald Enzner, Prof. Tim Fingscheidt, Prof. Timo Gerkmann, Prof. Peter Jax, Dr.-Ing. Heiner Löllmann, Prof. Nilesh Madhu, Dr.-Ing. Anil Nagathil, Dr.-Ing. Markus Niermann, Dr.-Ing. Bastian Sauert, and Dr.-Ing. Thomas Schlien for fruitful discussions and valuable contributions. Furthermore, we would especially like to thank Dr.-Ing. Christiane Antweiler, for her tireless support to this project, and Horst Krott and Dipl.-Geogr. Julia Ringeis for preparing most of the diagrams.

Finally, we would like to express our sincere thanks to the managing editors and staff of John Wiley & Sons for their kind and patient assistance.

Aachen and Bochum                                             *Peter Vary and Rainer Martin*
October 2023

# References

Martin, R.; Heute, U.; Antweiler, C. (2008). *Advances in Digital Speech Transmission*, John Wiley & Sons.

Vary, P.; Martin, R. (2006). *Digital Speech Transmission – Enhancement, Coding and Error Concealment*, John Wiley & Sons.

# 1

## Introduction

Language is the most essential means of human communication. It is used in two modes: as spoken language (*speech communication*) and as written language (*textual communication*). In our modern information society both modes are greatly enhanced by technical systems and devices. E-mail, short messaging, and the worldwide web have revolutionized textual communication, while

- digital cellular radio systems,
- audio–visual conference systems,
- acoustic human–machine communication, and
- digital hearing aids

have significantly expanded the possibilities and convenience of speech and audio–visual communication.

*Digital processing and enhancement of speech signals* for the purpose of transmission (or storage) is a branch of information technology and an engineering science which draws on various other disciplines, such as physiology, phonetics, linguistics, acoustics, and psychoacoustics. It is this multidisciplinary aspect which makes digital speech processing a challenging as well as rewarding task.

The goal of this book is a comprehensive discussion of fundamental issues, standards, and trends in speech communication technology. Speech communication technology helps to mitigate a number of physical constraints and technological limitations, most notably

- bandwidth limitations of the telephone channel,
- shortage of radio frequencies,
- acoustic background noise at the near-end (receiving side),
- acoustic background noise at the far-end (transmitting side),
- (residual) transmission errors and packet losses caused by the transmission channel,
- interfering acoustic echo signals from loudspeaker(s).

The enormous advances in signal processing technology have contributed to the success of speech signal processing. At present, integrated digital signal processors allow economic real-time implementations of complex algorithms, which require several thousand operations per speech sample. For this reason, advanced speech signal processing functions can be implemented in cellular phones and audio–visual terminals, as illustrated in Figure 1.1.

*Digital Speech Transmission and Enhancement*, Second Edition. Peter Vary and Rainer Martin.
© 2024 John Wiley & Sons Ltd. Published 2024 by John Wiley & Sons Ltd.

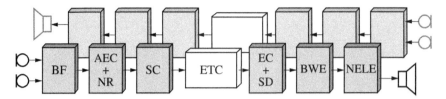

**Figure 1.1** Speech signal processing in a handsfree cellular terminal. BF: beamforming, AEC: acoustic echo cancellation, NR: noise reduction, SC: speech coding, ETC: equivalent transmission channel, EC: error concealment, SD: speech decoding, BWE: bandwidth extension, and NELE: near-end listening enhancement.

The handsfree terminal in Figure 1.1 facilitates communication via microphones and loudspeakers. Handsfree telephone devices are installed in motor vehicles in order to enhance road safety and to increase convenience in general.

At the far end of the transmission system, three different pre-processing steps are taken to improve communication in the presence of ambient noise and loudspeaker signals. In the first step, two or more microphones are used to enhance the near-end speech signal by **beamforming (BF)**. Specific characteristics of the interference, such as the spatial distribution of the sound sources and the statistics of the spatial sound field, are exploited.

**Acoustic echoes** occur when the far-end signal leaks at the near-end from the loudspeaker of the handsfree set into the microphone(s) via the acoustic path. As a consequence, the far-end speakers will hear their own voice delayed by twice the signal propagation time of the telephone network. Therefore, in a second step, the acoustic echo must be compensated by an adaptive digital filter, the **acoustic echo canceller (AEC)**.

The third module of the pre-processing chain is **noise reduction (NR)** aiming at an improvement of speech quality prior to coding and transmission. Single-channel NR systems rely on spectral modifications and are most effective for short-term stationary noise.

**Speech coding (SC)**, **error concealment (EC)**, and **speech decoding (SD)** facilitate the efficient use of the transmission channel. SC algorithms for cellular communications with typical bit rates between 4 and 24 bit/s are explicitly based upon a model of speech production and exploit properties of the hearing mechanism.

At the receiving side of the transmission system, speech quality is ensured by means of error correction (channel decoding), which is not within the scope of this book. In Figure 1.1, the (inner) channel coding/decoding as well as modulation/demodulation and transmission over the physical channel are modeled as an **equivalent transmission channel (ETC)**. In spite of channel coding, quite frequently residual errors remain. The negative auditive effects of these errors can be mitigated by **error concealment (EC)** techniques. In many cases, these effects can be reduced by exploiting both residual source redundancy and information about the instantaneous quality of the transmission channel.

Finally, the decoded signal might be subjected to artificial **bandwidth extension (BWE)** which expands narrowband (0.3–3.4 kHz) to wideband (0.05–7.0 kHz) speech or wideband speech to super wideband (0.05–14.0 kHz). With the introduction of true wideband and super wideband speech audio coding into telephone networks, this step will be of significant importance as, for a long transition period, narrowband and wideband speech terminals will coexist.

At the receiving end (near-end), the perception of the decoded (and eventually band-width expanded) speech signal might be disturbed by acoustic background noise. The task of the last module in the transmission chain is to improve intelligibility or at least to reduce the listening effort. The received speech signal is modified, taking the near-end background noise into account, which can be captured with a microphone. This method is called **near-end listening enhancement (NELE)**.

Some of these processing functions find also applications in audio–visual conferencing devices and digital hearing aids.

The book is organized as follows. The first part *fundamentals* (Chapters 2–5) deals with models of speech production and hearing, spectral transformations, filter banks, and stochastic processes.

The second part *speech coding* (Chapters 6–8) covers quantization, differential waveform coding and especially the concepts of code excited linear prediction (CELP) are discussed. Finally, some of the most relevant speech codec standards are presented. Recent developments such as the *Adaptive Multi-Rate, AMR* codec, or the *Enhanced Voice Services* (EVS) codec for cellular and IP communication are described.

The third part *speech enhancement* (Chapters 9–15) is concerned with error concealment, bandwidth extension, near-end listening enhancement, single and dual-channel noise and reverberation reduction, acoustic echo cancellation, and beamforming.

# 2

# Models of Speech Production and Hearing

Digital speech communication systems are largely based on knowledge of speech production, hearing, and perception. In this chapter, we will discuss some fundamental aspects in so far as they are of importance for optimizing speech-processing algorithms such as speech coding, speech enhancement, or feature extraction for automatic speech recognition.

In particular, we will study the mechanism of speech production and the typical characteristics of speech signals. The digital speech production model will be derived from acoustical and physical considerations. The resulting *all-pole model of the vocal tract* is the key element of most of the current speech-coding algorithms and standards.

Furthermore, we will provide insights into the human auditory system and we will focus on perceptual fundamentals which can be exploited to improve the quality and the effectiveness of speech-processing algorithms to be discussed in later chapters. With respect to perception, the main aspects to be considered in digital speech transmission are the *masking effect* and the spectral resolution of the auditory system.

As a detailed discussion of the acoustic theory of speech production, phonetics, psychoacoustics, and perception is beyond the scope of this book, the reader is referred to the literature (e.g., [Fant 1970], [Flanagan 1972], [Rabiner, Schafer 1978], [Picket 1980], and [Zwicker, Fastl 2007]).

## 2.1 Sound Waves

Sound is a mechanical vibration that propagates through matter in the form of waves. Sound waves may be described in terms of a sound pressure field $p(\mathbf{r}, t)$ and a sound velocity vector field $\mathbf{u}(\mathbf{r}, t)$, which are both functions of a spatial co-ordinate vector $\mathbf{r}$ and time $t$. While the sound pressure characterizes the density variations (we do not consider the DC component, also known as atmospheric pressure), the sound velocity describes the velocity of dislocation of the physical particles of the medium which carries the waves. This velocity is different from the speed $c$ of the traveling sound wave.

In the context of our applications, i.e., sound waves in air, sound pressure $p(\mathbf{r}, t)$ and resulting density variations $\rho(\mathbf{r}, t)$ are related by

$$p(\mathbf{r}, t) = c^2 \rho(\mathbf{r}, t) \tag{2.1}$$

*Digital Speech Transmission and Enhancement*, Second Edition. Peter Vary and Rainer Martin.
© 2024 John Wiley & Sons Ltd. Published 2024 by John Wiley & Sons Ltd.

and also the relation between $p(\mathbf{r}, t)$ and $\mathbf{u}(\mathbf{r}, t)$ may be linearized. Then, in the general case of three spatial dimensions these two quantities are related via differential operators in an infinitesimally small volume of air particles as

$$\text{grad}\,(p(\mathbf{r}, t)) = -\rho_0 \frac{\partial \mathbf{u}(\mathbf{r}, t)}{\partial t} \quad \text{and} \quad \text{div}\,(\mathbf{u}(\mathbf{r}, t)) = -\frac{1}{\rho_0 c^2} \frac{\partial p(\mathbf{r}, t)}{\partial t}, \tag{2.2}$$

where $c$ and $\rho_0$ are the speed of sound and the density at rest, respectively. These equations, also known as Euler's equation and continuity equation [Xiang, Blauert 2021], may be combined into the wave equation

$$\Delta p(\mathbf{r}, t) = \frac{1}{c^2} \frac{\partial^2 p(\mathbf{r}, t)}{\partial t^2}, \tag{2.3}$$

where the Laplace operator $\Delta p(\mathbf{r}, t)$ in Cartesian coordinates $\mathbf{r} = (x, y, z)$ is

$$\Delta p(\mathbf{r}, t) = \frac{\partial^2 p(\mathbf{r}, t)}{\partial x^2} + \frac{\partial^2 p(\mathbf{r}, t)}{\partial y^2} + \frac{\partial^2 p(\mathbf{r}, t)}{\partial z^2}. \tag{2.4}$$

A solution of the wave equation (2.3) is plane waves which feature surfaces of constant sound pressure propagating in a given spatial direction. A harmonic plane wave of angular frequency $\omega$ which propagates in positive $x$ direction or negative $x$ direction may be written in complex notation as

$$p_f(x, t) = \hat{p}_f e^{j(\omega t - \tilde{\beta} x)} \quad \text{and} \quad p_b(x, t) = \hat{p}_b e^{j(\omega t + \tilde{\beta} x)} \tag{2.5}$$

where $\tilde{\beta} = \omega/c = 2\pi/\lambda$ is the *wave number*, $\lambda$ is the *wavelength*, and $\hat{p}_f, \hat{p}_b$ are the (possibly complex-valued) *amplitudes*. Using (2.2), the $x$ component of the sound velocity is then given by

$$u_x(x, t) = \frac{1}{\rho_0 c} p(x, t). \tag{2.6}$$

Thus, for a plane wave, the sound velocity is proportional to the sound pressure.

In our applications, waves which have a constant sound pressure on concentrical spheres are also of interest. Indeed, the wave equation (2.3) delivers a solution for the *spherical wave* which propagates in radial direction $r$ as

$$p(r, t) = \frac{1}{r} f(r - ct), \tag{2.7}$$

where $f$ is the propagating waveform. The amplitude of the sound wave diminishes with increasing distance from the source. We may then use the abstraction of a *point source* to explain the generation of such *spherical waves*.

An ideal point source may be represented by its source strength $v_0(t)$ [Xiang, Blauert 2021]. Furthermore, with (2.2) we have

$$\frac{\partial u_r(r, t)}{\partial t} = -\frac{1}{\rho_0} \frac{\partial p(r, t)}{\partial r} = \frac{1}{\rho_0} \left( \frac{f(r - ct)}{r^2} - \frac{1}{r} \frac{df(r - ct)}{dr} \right). \tag{2.8}$$

Then, the radial component of the velocity vector may be integrated over a sphere of radius $r$ to yield $v_0(t) \approx 4\pi r^2 u_r(r, t)$. For $r \to 0$, the second term on the right-hand side of (2.8) is smaller than the first. Therefore, for an infinitesimally small sphere, we find with (2.8)

$$\frac{dv_0(t)}{dt} \Big|\, r \to 0 \approx \frac{4\pi}{\rho_0} f(r - ct) \tag{2.9}$$

and, with (2.7), for any $r$

$$p(r, t) = \frac{\rho_0}{4\pi r} \frac{dv_0(t - r/c)}{dt},$$  (2.10)

which characterizes, again, a spherical wave. The sound pressure is inversely proportional to the radial distance $r$ from the point source. For a harmonic excitation

$$v_0(t) = \hat{v}e^{j\omega t}$$  (2.11)

we find the sound pressure

$$p(r, t) = \frac{j\omega \rho_0 \hat{v}e^{j\omega(t - r/c)}}{4\pi r} = \frac{j\omega \rho_0 \hat{v}e^{j(\omega t - \tilde{\beta} r)}}{4\pi r}$$  (2.12)

and hence, with (2.8) and an integration with respect to time, the sound velocity

$$u_r(r, t) = \frac{p(r, t)}{\rho_0 c} \left(1 + \frac{1}{j\tilde{\beta} r}\right).$$  (2.13)

Clearly, (2.12) and (2.13) satisfy (2.8). Because of the second term in the parentheses in (2.13), sound pressure and sound velocity are not in phase. Depending on the distance of the observation point to the point source, the behavior of the wave is distinctly different. When the second term cannot be neglected, the observation point is in the *nearfield* of the source. For $\tilde{\beta} r \gg 1$, the observation point is in the *farfield*. The transition from the nearfield to the farfield depends on the wave number $\tilde{\beta}$ and, as such, on the wavelength or the frequency of the harmonic excitation.

## 2.2   Organs of Speech Production

The production of speech sounds involves the manipulation of an airstream. The acoustic representation of speech is a sound pressure wave originating from the physiological speech production system. A simplified schematic of the human speech organs is given in Figure 2.1. The main components and their functions are:

- lungs:                      the energy generator,
- trachea:                    for energy transport,
- larynx with vocal cords:    the signal generator, and
- vocal tract with pharynx,
  oral and nasal cavities:    the acoustic filter.

By contraction, the *lungs* produce an airflow which is modulated by the *larynx*, processed by the *vocal tract*, and radiated via the lips and the nostrils. The *larynx* provides several biological and sound production functions. In the context of speech production, its purpose is to control the stream of air that enters the vocal tract via the *vocal cords*.

Speech sounds are produced by means of various mechanisms. *Voiced sounds* are produced when the airflow is interrupted periodically by the movements (vibration) of the vocal cords (see Figure 2.2). This self-sustained oscillation, i.e., the repeated opening and closing of the vocal cords, can be explained by the so-called *Bernoulli effect* as in fluid

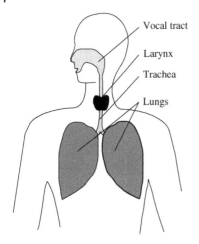
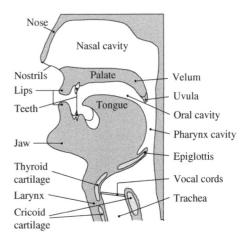

**Figure 2.1** Organs of speech production.

dynamics: as airflow velocity increases, local pressure decreases. At the beginning of each cycle, the area between the vocal cords, which is called the *glottis*, is almost closed by means of appropriate tension of the vocal cords. Then an increased air pressure builds up below the glottis, forcing the vocal cords to open. As the vocal cords diverge, the velocity of the air flowing through the glottis increases steadily, which causes a drop in the local pressure. Then, the vocal cords snap back to their initial position and the next cycle can start if the airflow from the lungs and the tension of the vocal cords are sustained. Due to the abrupt periodic interruptions of the glottal airflow, as schematically illustrated in Figure 2.2, the resulting excitation (pressure wave) of the vocal tract has a fundamental frequency of $f_0 = 1/T_0$ and has a large number of harmonics. These are spectrally shaped according to the frequency response of the acoustic vocal tract. The duration $T_0$ of a single cycle is called the *pitch period*.

*Unvoiced sounds* are generated by a constriction at the open glottis or along the vocal tract causing a nonperiodic turbulent airflow.

*Plosive sounds* (also known as stops) are caused by building up the air pressure behind a complete constriction somewhere in the vocal tract, followed by a sudden opening. The released airflow may create a voiced or an unvoiced sound or even a mixture of both, depending on the actual constellation of the articulators.

The *vocal tract* can be subdivided into three sections: the pharynx, the oral cavity, and the nasal cavity. As the entrance to the nasal cavity can be closed by the velum, a distinction is often made in the literature between the nasal tract (from velum to nostrils) and the other

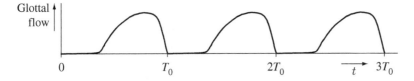

**Figure 2.2** Glottal airflow during voiced sounds.

two sections (from trachea to lips, including the pharynx cavity). In this chapter, we will define the *vocal tract* as a variable acoustic resonator including the nasal cavity with the velum either open or closed, depending on the specific sound to be produced. From the engineering point of view, the resonance frequencies are varied by changing the size and the shape of the vocal tract using different constellations and movements of the *articulators*, i.e., tongue, teeth, lips, velum, lower jaw, etc. Thus, humans can produce a variety of different sounds based on different vocal tract constellations and different acoustic excitations.

Finally, the acoustic waves carrying speech sounds are radiated via the mouth and head. In a first approximation, we may model the radiating head as a spherical source in free space. The (complex-valued) acoustic load $Z_L(r_H)$ at the lips may then be approximated by the radiation load of a spherical source of radius $r_H$ where $r_H$, represents the head radius. Following (2.13), this load exhibits a high-pass characteristic,

$$Z_L(r_H) = \frac{\hat{p}(r_H)}{\hat{u}_r(r_H)} \sim \frac{j\omega\, r_H/c}{1 + j\omega\, r_H/c} = \frac{\left(\omega\, r_H/c\right)^2}{1 + \left(\omega\, r_H/c\right)^2} + j\frac{\omega\, r_H/c}{1 + \left(\omega\, r_H/c\right)^2} \qquad (2.14)$$

where $\omega = 2\pi f$ denotes angular frequency and $c$ is the *speed of sound*. This model suggests an acoustic "short circuit" at very low frequencies, i.e., little acoustic radiation at low frequencies, which is also supported by measurements [Flanagan 1960]. For an assumed head radius of $r_H = 8.5$ cm and $c = 343$ m/s, the 3-dB cutoff frequency $f_c = c/(2\pi r_H)$ is about 640 Hz.

## 2.3 Characteristics of Speech Signals

Most languages can be described as a set of elementary linguistic units, which are called *phonemes*. A phoneme is defined as the smallest unit which differentiates the meaning of two words in one language. The acoustic representation associated with a phoneme is called a *phone*. American English, for instance, consists of about 42 phonemes, which are subdivided into four classes:

*Vowels* are voiced and belong to the speech sounds with the largest energy. They exhibit a quasi periodic time structure caused by oscillation of the vocal cords. The duration varies from 40 to 400 ms. Vowels can be distinguished by the time-varying resonance characteristics of the vocal tract. The resonance frequencies are also called *formant frequencies*. Examples: /a/ as in "father" and /i/ as in "eve."

*Diphthongs* involve a gliding transition of the articulators from one vowel to another vowel. Examples: /oU/ as in "boat" and /ju/ as in "you."

*Approximants* are a group of voiced phonemes for which the airstream escapes through a relatively narrow aperture in the vocal tract. They can, thus, be regarded as intermediate between vowels and consonants [Gimson, Cruttenden 1994]. Examples: /w/ in "wet" and /r/ in "ran."

*Consonants* are produced with stronger constriction of the vocal tract than vowels. All kinds of excitation can be observed. Consonants are subdivided into *nasals, stops, fricatives, aspirates,* and *affricatives*. Examples of these five subclasses: /m/ as in "more," /t/ as in "tea," /f/ as in "free," /h/ as in "hold," and /tʃ/ as in "chase."

Each of these classes may be further divided into subclasses, which are related to the interaction of the articulators within the vocal tract. The phonemes can further be classified as either *continuant* (excitation of a more or less non time-varying vocal tract) or *non continuant* (rapid vocal tract changes). The class of continuant sounds consists of vowels and fricatives (voiced and unvoiced). The non continuant sounds are represented by diphthongs, semivowels, stops, and affricates.

For the purpose of speech-signal processing, specific articulatory and phonetic aspects are not as important as the typical characteristics of the waveforms, namely, the basic categories:

- voiced,
- unvoiced,
- mixed voiced/unvoiced,
- plosive, and
- silence.

Voiced sounds are characterized by their fundamental frequency, i.e., the frequency of vibration of the vocal cords, and by the specific pattern of amplitudes of the spectral harmonics.

In the speech signal processing literature, the fundamental frequency is often called *pitch* and the respective period is called *pitch period*. It should be noted, however, that in psychoacoustics the term pitch is used differently, i.e., for the perceived fundamental frequency of a sound, whether or not that frequency is actually present in the waveform (e.g., [Deller Jr. et al. 2000]). The fundamental frequency of young men ranges from 85 to 155 Hz and that of young women from 165 to 255 Hz [Fitch, Holbrook 1970]. Fundamental frequency, also in combination with vocal tract length, is indicative of sex, age, and size of the speaker [Smith, Patterson 2005].

*Unvoiced sounds* are determined mainly by their characteristic spectral envelopes. Voiced and unvoiced excitation do not exclude each other. They may occur simultaneously, e.g., in fricative sounds.

The distinctive feature of *plosive sounds* is the dynamically transient change of the vocal tract. Immediately before the transition, a total constriction in the vocal tract stops sound radiation from the lips for a short period. There might be a small amount of low-frequency components radiated through the throat. Then, the sudden change with release of the constriction produces a plosive burst.

Some typical speech waveforms are shown in Figure 2.3.

## 2.4   Model of Speech Production

The purpose of developing a model of speech production is not to obtain an accurate description of the anatomy and physiology of human speech production but rather to achieve a simplifying mathematical representation for reproducing the essential characteristics of speech signals.

In analogy to the organs of human speech production as discussed in Section 2.2, it seems reasonable to design a parametric two-stage model consisting of an *excitation source* and a

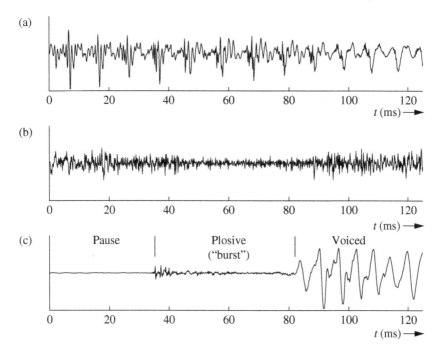

**Figure 2.3** Characteristic waveforms of speech signals: (a) Voiced (vowel with transition to voiced consonant); (b) Unvoiced (fricative); (c) Transition: pause–plosive–vowel.

*vocal tract filter*, see also [Rabiner, Schafer 1978], [Parsons 1986], [Quatieri 2001], [Deller Jr. et al. 2000]. The resulting digital *source-filter model*, as illustrated in Figure 2.4, will be derived below.

The model consists of two components:

- the *excitation source* featuring mainly the influence of the lungs and the vocal cords (voiced, unvoiced, mixed) and
- the *time-varying digital vocal tract filter* approximating the behavior of the vocal tract (spectral envelope and dynamic transitions).

In the first and simple model, the *excitation generator* only has to deliver either white noise or a periodic sequence of *pitch pulses* for synthesizing unvoiced and voiced sounds, respectively, whereas the vocal tract is modeled as a time-varying discrete-time filter.

**Figure 2.4** Digital source-filter model.

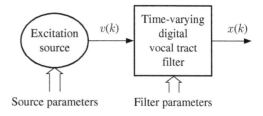

### 2.4.1 Acoustic Tube Model of the Vocal Tract

The digital source-filter model of Figure 2.4, especially the vocal tract filter, will be derived from the physics of sound propagation inside an acoustic tube. To estimate the necessary filter order, we start with the extremely simplifying physical model of Figure 2.5. According to this simplistic model, the pharynx and oral cavities are represented by a lossless tube with constant cross section and the nasal cavity by a second tube which can be closed by the velum. The length of $L = 17$ cm corresponds to the average length of the vocal tract of a male adult. The tube is (almost) closed at the glottis side and open at the lips.

In the case of a non nasal sound, the velum is closed. Then, the wavelength $\lambda_i$ of each resonance frequency of the main tube from the glottis to the lips fulfills the standing wave condition

$$(2i - 1) \cdot \frac{\lambda_i}{4} = L; \quad i = 1, 2, 3, \dots . \tag{2.15}$$

For $L = 17$ cm, we compute the resonance frequencies

$$f_i = \frac{c}{\lambda_i} = (2i - 1)\frac{c}{4L} \in \{500, 1500, 2500, 3500, \dots \} \text{ Hz}, \tag{2.16}$$

where the speed of sound is given by $c = 340$ m/s.

Taking (2.16) into account as well as the fact that the conventional *narrowband telephone* (NB) service has a frequency range of about 200–3400 Hz, and that the *wideband telephone* (WB) service covers a frequency range from 50 to 7000 Hz, we have to consider only four (NB) and eight resonances (WB) of the vocal tract model, respectively. As the acoustical bandwidth of speech is wider than 3400 Hz and even wider than 7000 Hz, lowpass filtering with a finite transition width from pass band to stop band is required as part of analog-to-digital conversion. Thus, the sampling rate for telephone speech is either 8 kHz (NB) or 16 kHz (WB), and the overall filter order for synthesizing telephone speech is roughly only $n = 8$ or $n = 16$. Each resonance frequency corresponds to a pole-pair or second-order filter section. As a rule of thumb, we can state the need for *"one resonance per kHz."*

In the second step, we improve our acoustic tube model, as shown in Figure 2.6. For simplicity, the nasal cavity is not considered (velum is closed). The cylindrical lossless tube

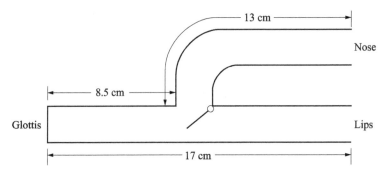

**Figure 2.5**  Simplified physical model of the vocal tract.

of Figure 2.5 is replaced by a tube with a continuous area function $A(x)$. This area function can be approximated by a stepwise constant contour corresponding to the concatenation of $n$ short cylindrical tubes of length $\Delta x = L/n$ with cross-section areas $A_i$, $i = 1, 2, \ldots, n$, as shown in Figures 2.6 and 2.7.

The generation of sound is related to the vibration and perturbation of air particles. When describing sound propagation through the concatenated tube segments, we have to deal with the *particle velocity* $u(x, t)$ (unit m/s) and the *sound pressure*, strictly speaking the pressure fluctuations $p(x, t)$ (unit N/m²) about the ambient or average (atmospheric) pressure. There is a vast amount of literature on relations between the velocity and the pressure (e.g., [Fant 1970]) inside a tube segment, taking different degrees of simplifying assumptions into account.

Within the scope of this book, we are interested in the derivation of the discrete-time filter structure, which is widely used for digital speech synthesis and speech coding. It will be outlined here that this filter can be derived under simplifying assumptions from the acoustic vocal tract model of Figure 2.7.

If we make the usual assumptions of *lossless tube segments* aligned with the $x$ axis, constant area $A$ in both time and space, and lossless plane wave propagation, the relation

**Figure 2.6** The lossless vocal tract tube model (neglecting the nasal cavity). (a) Continuous area function $A(x)$. (b) Stepwise approximation of $A(x)$.

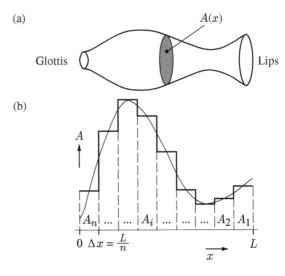

**Figure 2.7** Representation of the vocal tract by a concatenation of uniform lossless acoustic tube segments each of length $\Delta x = L/n$.

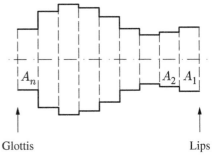

between sound pressure $p(x, t)$ and sound velocity $u(x, t)$ may be stated as (see (2.2))

$$-\frac{\partial p(x, t)}{\partial x} = \rho_0 \cdot \frac{\partial u(x, t)}{\partial t} \tag{2.17a}$$

$$-\frac{\partial u(x, t)}{\partial x} = \frac{1}{\rho_0 c^2} \cdot \frac{\partial p(x, t)}{\partial t}. \tag{2.17b}$$

With the *volume velocity* $v(x, t)$ as the product of the particle velocity in $x$ direction $u(x, t)$ and the cross-sectional area $A$,

$$v(x, \ t) = u(x, \ t) \cdot A, \tag{2.18}$$

we modify (2.17a) and (2.17b) as follows:

$$-\frac{\partial p(x, t)}{\partial x} = \frac{\rho_0}{A} \cdot \frac{\partial v(x, t)}{\partial t} \tag{2.19a}$$

$$-\frac{\partial v(x, t)}{\partial x} = \frac{A}{\rho_0 c^2} \cdot \frac{\partial p(x, t)}{\partial t}. \tag{2.19b}$$

The sound pressure $p(x, t)$ and volume velocity $v(x, t)$ are dependent on time $t$ and spatial coordinate $x$. Next, the combination of Eqs. (2.19a) and (2.19b) leads to a partial differential equation of second order, the *wave equation*, e.g., for the sound pressure $p(x, t)$ in one spatial coordinate $x$ (see (2.3)),

$$\frac{\partial^2 p(x, t)}{\partial x^2} = \frac{1}{c^2} \cdot \frac{\partial^2 p(x, t)}{\partial t^2}. \tag{2.20}$$

The wave equation (2.20) can be solved by a superposition of forward and backward traveling sound pressure waves $f(t)$ and $b(t)$, respectively, moving at velocities $c$.

In the literature, we find two alternative but equivalent solution approaches which differ in defining $f(t)$ and $b(t)$ either as sound pressure (approach A) or as volume velocity waves (approach B).

**Approach A: Sound pressure waves $f(t)$ and $b(t)$**
Here, $f(t)$ and $b(t)$ represent sound pressure waves. Therefore,

$$p(x, \ t) = f\left(t - \frac{x}{c}\right) + b\left(t + \frac{x}{c}\right) \tag{2.21a}$$

$$v(x, \ t) = \frac{1}{Z}\left[f\left(t - \frac{x}{c}\right) - b\left(t + \frac{x}{c}\right)\right], \tag{2.21b}$$

where

$$Z = \frac{\rho_0 \cdot c}{A} \tag{2.22}$$

represents an *acoustic impedance*.

**Approach B: Volume velocity waves $f(t)$ and $b(t)$**
Alternatively, we may define

$$p(x, \ t) = Z\left[f\left(t - \frac{x}{c}\right) + b\left(t + \frac{x}{c}\right)\right] \tag{2.23a}$$

$$v(x, \ t) = f\left(t - \frac{x}{c}\right) - b\left(t + \frac{x}{c}\right). \tag{2.23b}$$

**Figure 2.8**   Forward and backward traveling waves
at the junctions of tube segments $i$ and $i + 1$.

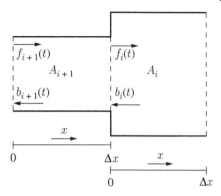

While approach A is mostly used in the field of acoustics (e.g., [Fant 1970], [Xiang, Blauert 2021]), approach B is often used in digital signal processing and coding (e.g., [Rabiner, Schafer 1978], [Vary, Martin 2006], see also Chapter 6, Section 6.4).

The solution approaches A and B can be applied to each of the tube segments, where boundary conditions exist at the junctions of adjacent segments. In each segment, we use a local spatial coordinate $x$, ranging from 0 to $\Delta x$ as shown in Figure 2.8, and consider the junction of segments $i$ and $i + 1$.

Pressure and volume velocity must be continuous both in time and in space. Therefore, the following constraints have to be fulfilled at the junction:

$$p_{i+1}(x = \Delta x,\ t) = p_i(x = 0,\ t) \tag{2.24a}$$

$$v_{i+1}(x = \Delta x,\ t) = v_i(x = 0,\ t). \tag{2.24b}$$

First, approach A will be evaluated here. The differences between solutions A and B will be discussed thereafter.

Using the solution approach A (2.21a) and (2.21b) with the boundary conditions (2.24a), (2.24b) and introducing the notation $\tau = \Delta x/c$ for the propagation time through one tube segment, the boundary conditions can be formulated in terms of the forward and the backward traveling waves:

$$f_{i+1}(t - \tau) + b_{i+1}(t + \tau) = f_i(t) + b_i(t) \tag{2.25a}$$

$$\frac{1}{Z_{i+1}}[f_{i+1}(t - \tau) - b_{i+1}(t + \tau)] = \frac{1}{Z_i}[f_i(t) - b_i(t)]. \tag{2.25b}$$

Note that we have eliminated the space coordinate as we are now considering the forward and backward traveling waves at the junction only. We are interested in how the forward wave $f_{i+1}$ arriving at the junction from the left and the backward wave $b_i$ arriving from the right are combined to form the waves $f_i$ and $b_{i+1}$ originating from the junction.

Thus, we have to solve the boundary conditions (2.25a) and (2.25b) with respect to the following input–output relations:

$$f_i(t) = F_1(f_{i+1}(t),\ b_i(t)) \tag{2.26a}$$

$$b_{i+1}(t) = F_2(f_{i+1}(t),\ b_i(t)). \tag{2.26b}$$

After some elementary algebraic operations, we obtain

$$f_i(t) = (1 + r_i) \cdot f_{i+1}(t - \tau) - r_i \cdot b_i(t) \tag{2.27a}$$

$$b_{i+1}(t) = r_i \cdot f_{i+1}(t - 2\tau) + (1 - r_i) \cdot b_i(t - \tau), \tag{2.27b}$$

where

$$r_i = \frac{Z_i - Z_{i+1}}{Z_i + Z_{i+1}} = \frac{A_{i+1} - A_i}{A_{i+1} + A_i} = \frac{A_{i+1}/A_i - 1}{A_{i+1}/A_i + 1} \tag{2.28}$$

is called the *reflection coefficient*. Note that the reflection coefficient $r_i$ can take positive and negative values depending on the relative sizes of the areas $A_i$ and $A_{i+1}$. From $A_i > 0$, it readily follows that $-1 \leq r_i \leq +1$.

The solution (2.27a,b) is called the *Kelly–Lochbaum* equations [Kelly, Lochbaum 1962]. Kelly and Lochbaum used this structure, illustrated in Figure 2.9a, for the generation of synthetic speech in 1962.

The forward traveling wave $f_i(t)$ into segment $i$, which *arises* at $\Delta x = 0$ from reflections at the junction, consists of two components:

- a portion of the backward traveling wave $b_i(t)$, which is partly reflected at the junction with weight $-r_i$ and
- a portion of the forward traveling wave $f_{i+1}(t)$, which is delayed by $\tau$ due to the segment length $\Delta x$ and partly propagated into segment $i$ with weight $1 + r_i$.

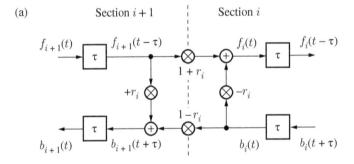

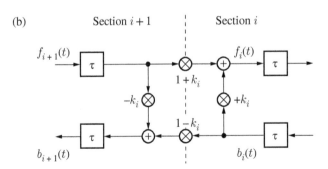

**Figure 2.9** Signal model of the junction between two adjacent tube sections $i + 1$ and $i$. (a) Solution A (2.27a,b) for the sound pressure. (b) Solution B (2.29a,b) for the volume velocity.

The backward traveling wave $b_{i+1}(t)$ arriving in segment $i+1$ at $\Delta x = 0$ also consists of two components:

- a portion of the backward traveling wave $b_i(t)$, which is partly propagated at the junction with weight $1 - r_i$, and delayed by $\tau$ and
- a portion of the forward traveling wave $f_{i+1}(t)$, which is partly reflected at the junction between $i+1$ and $i$ with weight $+r_i$ and delayed by $2\tau$ due to traveling back and forth through segment $i$.

Next, we use the volume velocity approach B (2.23a) and (2.23b) with the boundary conditions (2.24a), (2.24b) and obtain after some algebraic operations the following solution

$$f_i(t) = (1 + k_i) \cdot f_{i+1}(t - \tau) + k_i \cdot b_i(t) \tag{2.29a}$$

$$b_{i+1}(t) = -k_i \cdot f_{i+1}(t - 2\tau) + (1 - k_i) \cdot b_i(t - \tau), \tag{2.29b}$$

where

$$k_i = \frac{Z_{i+1} - Z_i}{Z_{i+1} + Z_i} = \frac{A_i - A_{i+1}}{A_i + A_{i+1}} = \frac{1 - A_{i+1}/A_i}{1 + A_{i+1}/A_i} \tag{2.30}$$

is also called the *reflection coefficient*.

The relation between the two definitions of the reflection coefficient is given by

$$k_i = -r_i \tag{2.31a}$$

$$1 \pm k_i = 1 \mp r_i. \tag{2.31b}$$

The comparison of solutions A and B is given in Figure 2.9a,b.

Finally, we illustrate for the sound pressure approach A the complete filter model in Figure 2.10. Special considerations are needed at the terminating glottis and at the lips. We may model the free space beyond the lips as an additional tube with index $i = 0$, with an infinite area $A_0$, and with infinite length. Thus, the first reflection coefficient becomes

$$r_0 = \frac{A_1 - A_0}{A_1 + A_0} = -1, \tag{2.32}$$

which is characteristic of an ideal soft termination.

Since a backward traveling wave does not exist, we have $b_0(t) = 0$. Furthermore, the Kelly–Lochbaum model suggests no sound radiation at the lips as the term $1 + r_0$ is zero and thus $f_0(t) = 0$. This is quite in line with our previous acoustic considerations: the lossless tube model leads to a reflection of the forward-moving acoustic wave at the lips. While little radiation is observed at low frequencies in practice (see (2.14)), sound waves are radiated from the mouth, indicating limitations of the above acoustic model. Thus, in general we may assume that $r_0$ is slightly larger than $-1$ and frequency dependent. In a practical implementation of this network, we may disregard the final factor $1 + r_0$ in Figure 2.10 and consider the output signal $f_1(t - \tau)$ instead.

At the glottis, the left side of section $n$ is almost closed, such that the backward traveling wave $b_n(t)$ is reflected $-r_n \approx +1$ according to an almost hard acoustic termination. We thus model the glottal source as a terminating segment which generates a forward-traveling wave $f_n(t)$.

(a)

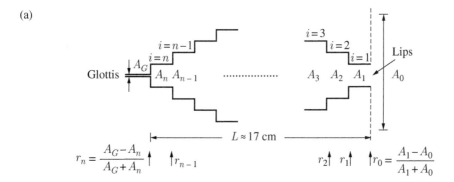

(b)

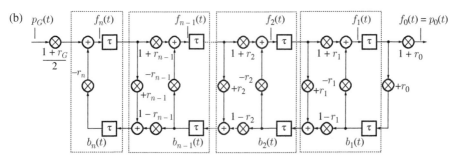

**Figure 2.10** Concatenation of multiple tubes of uniform cross sections and equivalent continuous-time filter structure (approach A, sound pressure). (a) Acoustical tube model. (b) Block diagram.

To compute the generated forward wave, we make use of an analogy between the concatenation of acoustic tube segments and the serial connection of homogeneous electrical line segments (e.g., [Xiang, Blauert 2021]). Basically, the mathematical description by the differential equations (2.20), (2.21a), and (2.21b) is the same, where the sound pressure $p_i(t)$ is corresponding to voltage and the volume velocity $v_i(t)$ to electric current.

Then, the glottal excitation source is a current source of strength $v_G(t)$ in parallel with impedance $Z_G$ and connected to the acoustic impedance $Z_n$ of the $n$-th segment, as shown in Figure 2.11. Thus the "input current" $v_n(0, t)$ to the segment $n$ is given by

$$v_n(0, t) = v_G(t) - \frac{p_n(0, t)}{Z_G}. \tag{2.33}$$

Using this relation as the boundary constraint and the solution (2.21a) and (2.21b) for segment $n - 1$ with $x = 0$, i.e.,

$$v_n(0, t) = \frac{1}{Z_n}[f_n(t) - b_n(t)] \quad \text{and} \quad p_n(0, t) = f_n(t) + b_n(t), \tag{2.34}$$

we derive from (2.33) and (2.34) with a few algebraic steps the equation

$$f_n(t) = v_G(t)Z_G \cdot \frac{Z_n}{Z_n + Z_G} + b_n(t) \cdot \frac{Z_G - Z_n}{Z_n + Z_G} \tag{2.35a}$$

$$= p_G(t) \cdot \frac{1 + r_G}{2} - b_n(t) \cdot r_n \tag{2.35b}$$

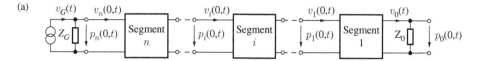

**Figure 2.11** Equivalent electric circuit of the vocal tract using electro-acoustic analogies. (a) Concatenation of transmission line segments. (b) Interface between the glottal volume velocity source and tube segment $n$. Note: Sound pressure $p_i(t)$ is corresponding to voltage and volume velocity $v_i(t)$ to electric current.

with the (load-independent) sound pressure signal $p_G(t) = v_G(t)Z_G$ (open circuit voltage of the current source) and

$$r_n = \frac{Z_n - Z_G}{Z_n + Z_G}. \tag{2.35c}$$

These *electro-acoustic* analogies can also be used for further refinements of the model, in particular in modeling the glottal excitation and the load effects from lip radiation. For more detailed studies, the reader is referred to the literature (e.g., [Rabiner, Schafer 1978]).

### 2.4.2 Discrete Time All-Pole Model of the Vocal Tract

So far we have derived a solution for the propagation of sound in concatenated lossless tube segments. We obtained a functional relation between the sound pressure of the forward-moving wave $p_0(t)$ at the lips and the input pressure signal $p_G(t)$ at the glottis as described by the block diagram of Figure 2.10b. From this representation, an equivalent discrete-time filter model can easily be derived.

The structure in Figure 2.10b represents an analog network consisting of delay elements, multipliers, and adders, which in the case of constant coefficients $r_G$, and $r_i$, $i = 0, 1, 2, \dots, n$, can be interpreted as a linear time-invariant system (LTI system). Any LTI system may be characterized by its impulse response or its corresponding frequency response. Obviously, the impulse response of the system in Figure 2.10 is discrete in time, as the internal signals $f_i(t)$ and $b_i(t)$, $i = 0, 1, 2, \dots, n$ are delayed by multiples of the basic one-way delay $\tau$ caused by each tube segment.

If we apply a Dirac impulse $\delta(t)$ as stimulus signal to the system input, i.e., $p_G(t) = \delta(t)$, we observe a response $p_0(t)$ at the output which is the impulse response $h_0(t)$. Although time $t$ is still a continuous variable, it is obvious that the output takes non zero values only at multiples of the basic delay time $\tau$. The first non zero value occurs at time $t = n \cdot \tau$, the propagation time through the concatenated tube segments. Due to the feedback structure, the output signal is a sequence of weighted Dirac impulses with a minimum distance between the pulses of $2\tau$. More specifically, the system response can be written as:

$$p_0(t) = h_0(t) = \sum_{\kappa=0}^{\infty} \delta(t - n\tau - 2\kappa\tau) \cdot h(\kappa), \tag{2.36}$$

(a)
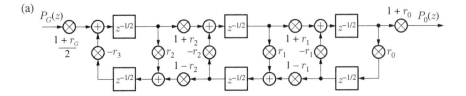

(b)
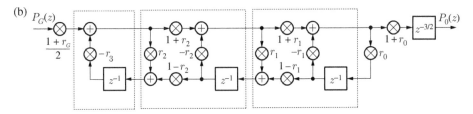

**Figure 2.12** Discrete-time models of the vocal tract (example $n = 3$). (a) Structure derived from equations (2.27a,b); $z^{-1/2}$ corresponds to a delay by $\tau$. (b) Modified structure with delay elements $z^{-1}$ corresponding to $T = 2\tau$ and an additional fractional output delay.

where the values $h(\kappa)$ denote the weights and the first weight is given by

$$h(0) = \frac{1 + r_G}{2} \cdot \prod_{i=0}^{n-1}(1 + r_i). \tag{2.37}$$

Thus, the impulse response of the tube model is already discrete in time, and the signal-flow graph of the equivalent digital filter may be considered as a one-to-one translation, as shown in Figure 2.12-a for the example of a lossless tube model with $n = 3$. Note that in Figure 2.12 we have introduced the $z$-transforms $P_G(z)$ and $P_0(z)$ of the sequences $p_G(t = \kappa \cdot T)$ and $p_0(t = \kappa \cdot T)$, with $T = 2 \cdot \tau$, respectively. Therefore, a delay by $\tau$ is denoted in Figure 2.12a by $z^{-1/2}$. The time discretization $\tau$ of the block diagram is intimately connected with the spatial discretization $\Delta x$ of the tube model. However, the effective time discretization (sampling interval) of the output signal $p_0$ is $T = 2\tau$. Thus, the discrete-time filter would have to be operated at twice the sampling rate, i.e.,

$$\frac{1}{\tau} = \frac{2}{T} = 2 \cdot f_s. \tag{2.38}$$

By systematically shifting the delay elements $z^{-1/2}$ across the nodes, the equivalent signal-flow graph of Figure 2.12b can be derived. This structure contains only internal delays of $T = 2\tau$ corresponding to $z^{-1}$ and one additional factional output delay of $n/2 \cdot \tau = 3/2 \cdot \tau$ corresponding to $z^{-3/2}$. If the number $n$ of segments is not even, the delay by $n/2 \cdot \tau$ would imply an interpolation by half a sampling interval. In any case, the fractional output delay will not have any influence on the quality of the synthesized speech. Therefore, this delay can be omitted in the implementation and the filter can be operated at the sampling rate $f_s = 1/T$.

The structure of Figure 2.12b (without the output delay) is used for speech synthesis and model-based speech coding. This special structure is called the *ladder structure*.

It is known from network theory that there are filter structures having the same transfer function but different signal-flow graphs. They may differ, for instance, with respect to

(a)

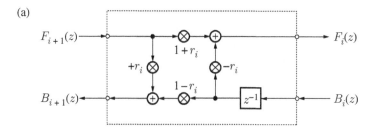

(b)

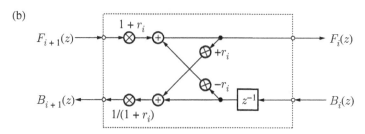

**Figure 2.13** Equivalent filter sections. (a) Ladder structure. (b) Lattice structure. Note: In comparison to Figure 2.10, $B_i(z)$ and $F_i(z)$, $i = 1, 2, ...n$ differ here by delays. However, the magnitude frequency response of the filters are the same.

the number of arithmetic operations needed to calculate one output sample and/or with respect to numerical precision of the impulse response in the case of implementation with finite-precision arithmetic (fixed point arithmetic).

Two equivalent structures are shown in Figure 2.13a,b, the *ladder structure* as derived above and the *lattice structure*. Their equivalence can readily be checked by the analysis of the two block diagrams:

$$F_i(z) = (1 + r_i) \cdot F_{i+1}(z) - r_i \cdot B_i(z) \cdot z^{-1} \qquad \text{Figure 2.13a,b} \qquad (2.39a)$$

$$B_{i+1}(z) = r_i \cdot F_{i+1}(z) + (1 - r_i) \cdot B_i(z) \cdot z^{-1} \qquad \text{Figure 13.a} \qquad (2.39b)$$

$$= \frac{1 + r_i}{1 + r_i} r_i \cdot F_{i+1}(z) + \frac{1 - r_i^2}{1 + r_i} \cdot B_i(z) \cdot z^{-1}. \qquad \text{Figure 13.b} \qquad (2.39c)$$

The *lattice structure* also requires four multiplications and two additions per section. However, the pre and post scaling factors $1 + r_i$ and $1/(1 + r_i)$ can be shifted and merged along the cascade of filter sections (cf. Figure 2.14a,b) in such a way that they can be combined to a single scaling or gain factor

$$g_r = \prod_{i=0}^{n-1} (1 + r_i), \qquad (2.40)$$

at the filter output cascade, as shown in Figure 2.14c.

Now we need, as shown in Figure 2.14c for the filtering operation, only two multiply-and-add operations (MAC operation) per section $i = 1, 2, ..n - 1$, one MAC operation each for section $i = 0$ (reflection $r_0$) and section $i = n$ (glottis reflection gain $-r_n$), one input scaling operation by $\frac{1+r_G}{2}$, $n - 1$ MAC operations for the calculation of $g_r$, and one scaling multiplication by $g_r$ at the output, i.e., a total number of $3(n-1)+4$ operations per output sample $p_0(\kappa \cdot T)$, where we use a basic MAC operation as available in signal processors.

In comparison to the ladder structure, we have reduced the computational complexity by a factor of almost 2. The ladder and the lattice structure are stable as long as the reflection coefficients are limited to the range

$$-1 < r_i < +1. \tag{2.41}$$

This stability can be guaranteed due to the underlying physical tube model of the vocal tract.

The input and output signals of the 2-multiplier-lattice elements of Figure 2.14c are different from those of the original lattice structure of Figure 2.14a by the corresponding scaling factors. Therefore, the $z$-transforms of the intermediate signals of the lattice sections are denoted by $\tilde{F}_i(z)$ and $\tilde{B}_i(z)$.

The lattice structure of Figure 2.14 can be translated into the direct-form structure. For reasons of simplicity, we use the notation $\tilde{F}_i$ instead of $\tilde{F}_i(z)$, etc., and rewrite the input–output relation of the 2-multiplier-lattice element in Figure 2.14b,c as

$$\tilde{F}_i = \tilde{F}_{i+1} - r_i \cdot \tilde{B}_i \cdot z^{-1} \tag{2.42a}$$

$$\tilde{B}_{i+1} = r_i \cdot \tilde{F}_i + \tilde{B}_i \cdot z^{-1}. \tag{2.42b}$$

From (2.42a) and (2.42b), we derive the following vector–matrix representation

$$\begin{pmatrix} \tilde{F}_{i+1} \\ \tilde{B}_{i+1} \end{pmatrix} = \begin{pmatrix} 1 & r_i \cdot z^{-1} \\ r_i & z^{-1} \end{pmatrix} \cdot \begin{pmatrix} \tilde{F}_i \\ \tilde{B}_i \end{pmatrix}. \tag{2.43}$$

(a)
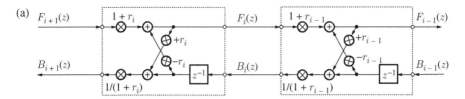

(b)
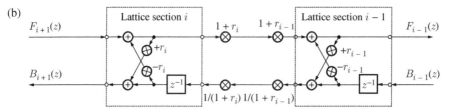

(c)
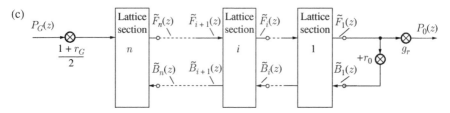

**Figure 2.14** Vocal tract model as a cascade of lattice filter sections (Approach A). (a) Lattice sections with four multipliers. (b) Lattice section $i$ with shifted scaling multipliers. (c) Lattice sections with two multipliers and output gain scaling by $g_r$.

Furthermore, we have constraints at the lips

$$P_0 = \tilde{F}_1 \cdot g_r \tag{2.44a}$$

$$\tilde{B}_0 = 0 \tag{2.44b}$$

$$\begin{pmatrix} \tilde{F}_1 \\ \tilde{B}_1 \end{pmatrix} = \begin{pmatrix} 1 & 0 \\ r_0 & 0 \end{pmatrix} \cdot \begin{pmatrix} P_0 \\ 0 \end{pmatrix} \cdot \frac{1}{g_r} \tag{2.44c}$$

$$= \begin{pmatrix} 1 & r_0 \cdot z^{-1} \\ r_0 & z^{-1} \end{pmatrix} \cdot \begin{pmatrix} 1 \\ 0 \end{pmatrix} \cdot \frac{P_0}{g_r} \tag{2.44b}$$

and at the glottis

$$\tilde{F}_n = P_G \cdot \frac{1 + r_G}{2} - r_n \cdot \tilde{B}_n \cdot z^{-1}, \tag{2.45a}$$

i.e.,

$$P_G = \frac{2}{1 + r_G} \cdot [\tilde{F}_n + r_n \cdot \tilde{B}_n \cdot z^{-1}] \tag{2.45b}$$

$$= \frac{2}{1 + r_G} \cdot (1 \ r_n \cdot z^{-1}) \cdot \begin{pmatrix} \tilde{F}_n \\ \tilde{B}_n \end{pmatrix}, \tag{2.45c}$$

where $(1 \ r_n \cdot z^{-1})$ denotes a row vector with two elements. From (2.43), (2.44c), and (2.45c), we finally derive the following expression

$$P_G(z) = \frac{2}{1 + r_G} \cdot (1 \ r_n \cdot z^{-1}) \prod_{i=0}^{n-1} \begin{pmatrix} 1 & r_i \cdot z^{-1} \\ r_i & z^{-1} \end{pmatrix} \begin{pmatrix} 1 \\ 0 \end{pmatrix} \cdot \frac{P_0(z)}{g_r}$$

$$= \frac{2}{1 + r_G} \cdot \frac{1}{g_r} \cdot D(z) \cdot P_0(z), \tag{2.46}$$

where $D(z)$ constitutes a polynomial in $z^{-1}$ and may be rewritten as

$$D(z) = 1 - \sum_{i=0}^{n} c_i \cdot z^{-i} = 1 - C(z). \tag{2.47}$$

By inversion of (2.46), we find with (2.40) the frequency response $H(z)$ and its dependency on $z$

$$H(z) = \frac{P_0(z)}{P_G(z)} = \frac{1 + r_G}{2} \cdot \frac{g_r}{D(z)} \tag{2.48a}$$

$$= \frac{1 + r_G}{2} \cdot \frac{\prod_{i=0}^{n-1}(1 + r_i)}{1 - \sum_{i=0}^{n} c_i \cdot z^{-i}} \tag{2.48b}$$

$$= g_0 \cdot \frac{1}{1 - \sum_{i=0}^{n} c_i \cdot z^{-i}}. \tag{2.48c}$$

The direct-form vocal tract transfer function $H(z)$ has $n$ poles and zeros only in the origin of the z-plane. This type of filter is called an *all-pole filter*. The corresponding direct-form implementation is illustrated in Figure 2.15a.

(a)

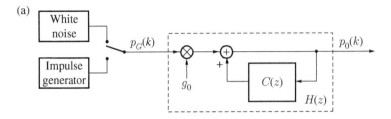

(b)

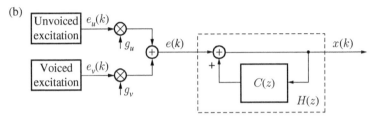

**Figure 2.15** All-pole vocal tract model. (a) Vocoder with voiced or unvoiced excitation $p_G(k)$. (b) Improved vocoder with mixed unvoiced/voiced excitation $e(k)$.

For producing purely unvoiced sounds, a noise-like excitation signal $p_G(k)$ is required, while for producing purely voiced sounds, a sequence of pitch pulses according to Figure 2.2 is needed. However, it should be noted that the model of Figure 2.15a is too simplistic as natural speech sounds mostly consist of a variable mixture of voiced and unvoiced components. This is taken into consideration in speech coding by replacing the source switch in Figure 2.15a by an adder for generating a weighted mixture of an unvoiced and a voiced excitation component $e_u(k)$ and $e_v(k)$, respectively, according to

$$e(k) = g_u \cdot e_u(k) + g_v \cdot e_v(k). \tag{2.49}$$

The *gain factors* $g_u \geq 0$ and $g_v \geq 0$ control the amount of voicing. The corresponding block diagram, which is widely used in standardized speech codecs (see Chapter 8), is shown in Figure 2.15b.

The periodic voiced excitation signal $e_v(k)$ can be generated, as illustrated in Figure 2.16 where

$$\delta(k) = \begin{cases} 1 & k = 0 \\ 0 & k \neq 0 \end{cases} \tag{2.50}$$

denotes the unit impulse. The output signal is given by

$$e_v(k) = \sum_\lambda g(k - \lambda \cdot N_0) \tag{2.51}$$

with the glottis impulse response $g(k)$ and the pitch period $N_0$. The impulse response $g(k)$ determines the shape of a single pitch cycle. In practical implementations of speech codecs,

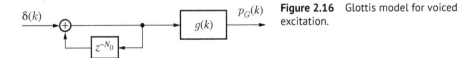

**Figure 2.16** Glottis model for voiced excitation.

usually the glottis impulse response is merged with the impulse response $h(k)$ of the vocal tract filter.

Finally, it should be mentioned that the parameters of the vocal tract transfer function $H(z)$ can directly be calculated from the auto-correlation function of the speech signal, e.g., using the Levinson–Durbin algorithm (see also Chapter 6, Section 6.4) which delivers both the reflection coefficients $k_i = -r_i$ as well as the direct form filter coefficients $c_i, i = 1, 2, ..n$. Furthermore, reflection coefficients can be converted by iterative calculations into direct form coefficients and vice versa (see Section 6.4.1.3).

## 2.5  Anatomy of Hearing

The peripheral hearing organ is divided into three sections (e.g., [Hudde 2005]):

- outer ear,
- middle ear, and
- inner ear,

as illustrated in Figure 2.17.

The *outer ear* consists of the pinna, the outer ear canal, and the ear drum. The pinna protects the opening and contributes, in combination with head and shoulders, to the directivity of the ear. The outer ear canal is a nearly uniform tube, closed at the inner end, with a length up to 3 cm and a diameter of about 0.7 cm. This tube transmits the sound to the ear

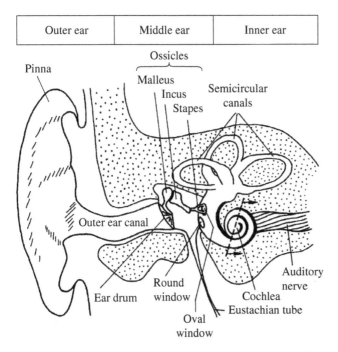

**Figure 2.17**  Schematic drawing of the ear. Source: [Zwicker, Fastl 2007] with permission of Springer Nature.

drum. Within the frequency range of speech, it has a single resonance frequency between 3 kHz and 4 kHz, which results in an increased sensitivity of the ear in this frequency range. The ear drum is a stiff, conical membrane, which vibrates because of the forces of the oscillating air particles. In the ear canal, sound waves travel predominantly as plane waves.

The *middle ear* is an air-filled cavity which is connected to the outer ear via the ear drum and to the inner ear via the round window and the oval window. It contains three tiny bones, the ossicles, which provide the acoustic coupling between the ear drum and the oval window. The ossicles perform a mechanical impedance transformation by a factor of about 15 from the airborne sound to the fluids of the inner ear commensurate with the area ratio of the ear drum and the oval window. Consequently, oscillations of the air particles by small forces and large displacements are transformed into large forces and small displacements.

Additionally, the middle ear is connected to the upper throat via the Eustachian tube, which is opened briefly during swallowing, to equalize the air pressure in the middle ear to that of the environment. This is necessary to adjust the resting point of the ear drum and the working point of the middle ear ossicles.

The *inner ear* consists of the organs for the sense of equilibrium and orientation (vestibular system) with the semicircular canals (which are not involved in the hearing process) and the *cochlea* with the round and the oval windows. The cochlea contains the *basilar membrane* with the *organ of Corti*, which converts mechanical vibrations into impulses in the auditory nerve. The cochlea is formed like a snail-shell with about 2.5 turns. Figure 2.18

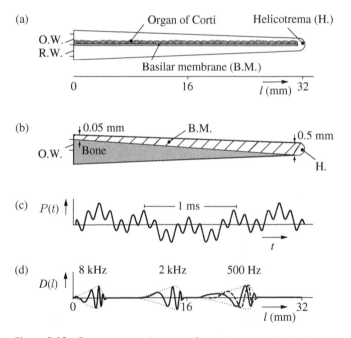

**Figure 2.18** Frequency-to-place transformation along the basilar membrane from the base (left) to the apex (right) of the cochlea. (a) Top-down view into the uncoiled cochlea. (b) Side view. (c) Triple-tone audio signal (sound pressure, 500 Hz, 2000 Hz, 8000 Hz). (d) Displacement $D(l)$ by excitation (traveling waves and envelopes). O.W., Oval Window; R.W., Round Window. Source: Adapted from [Zwicker, Fastl 2007].

illustrates the schematic view of the uncoiled cochlea, which has an average length of about 32 mm. The movements of the air particles are transmitted by the *ossicles* through the oval window (O.W.) to the incompressible fluids of the cochlea, which results in a traveling wave on the basilar membrane (B.M.). The cochlea is separated by Reissner's membrane and the basilar membrane into three parallel canals, the *scala vestibuli*, the *scala media*, and the *scala tympani*. The scala vestibuli and the scala tympani are connected at the far end (apex) by an opening, the *helicotrema* (Figure 2.18a).

The organ of Corti, situated on the basilar membrane, supports about 3600 inner hair cells and about 11 000 outer hair cells. While the inner hair cells pass information predominantly via the spiral ganglion to the auditory nerve and the brain (via *afferent* fibers), the outer hair cells receive information from the brain (*efferent* fibers). The outer hair cells are thus involved in the regulation of the sensitivity of the cochlea. A loss of the outer hair cells leads typically to an elevated hearing threshold, i.e., a hearing loss.

Through the helicotrema and the round window (R.W.), pressure compensation of the traveling fluid waves is performed.

The B.M. is about 0.05 mm wide at the oval window and about 0.5 mm wide at the helicotrema (see Figure 2.18b) and thus features varying mechanical properties. The B.M. performs a transformation of sound frequency to place by means of the traveling wave mechanism. High frequencies stimulate the membrane and thus the hair cells of the organ of Corti near the oval window, while the resonances for low frequencies are near to the helicotrema. This transformation corresponds to a spectral analysis using a non-uniform filter bank.

A schematic drawing of the frequency-to-place transformation is shown in Figure 2.18c,d for an audio signal (sound pressure $p(t)$), consisting of three sinusoidal tones of 500, 2000, and 8000 Hz. The three signal components cause vibrations on the B.M. such that the envelope of the dynamical displacement $D(l)$ reaches local maxima at distinct locations each of which is linked to the *characteristic frequency* of a sinusoidal tone. The envelope of the displacements is quite steep toward the helicotrema, while in the direction of the oval window, a flat descent can be observed. This behavior determines the characteristics of the masking effect to be discussed in Section 2.6.3. In general, the frequency-to-place mapping is well described by Greenwood's function [Greenwood 1990]

$$f = A \left( 10^{ax} - k \right) \tag{2.52}$$

where $x$ describes the distance from apex and the constants are chosen to $A = 165.4$ Hz, $k = 0.88$, and $a = 0.06/$mm for the human ear.

## 2.6 Psychoacoustic Properties of the Auditory System

### 2.6.1 Hearing and Loudness

Speech is carried by small temporal variations of the sound pressure $p(t)$. The physical unit of the sound pressure is the pascal (Pa), where the relevant range for hearing covers more than seven decades from the threshold of hearing at about $20 \cdot 10^{-6}$ Pa to the threshold of pain at $10^2$ Pa. The normalized root mean square (rms) *sound pressure* $\bar{p}/p_0$ and the

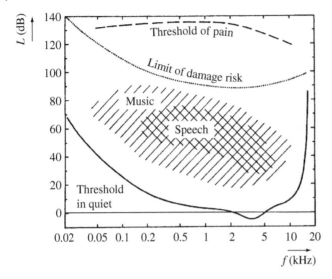

**Figure 2.19** Typical frequency-SPL regions for speech and music, lower bounded by the hearing threshold in quiet and upper bounded by the threshold of pain. Source: Adapted from [Zwicker, Fastl 2007].

normalized *sound intensity* $\bar{I}/I_0$ are measured on logarithmic scales leading to the *sound pressure level* (SPL) and the *intensity level*,

$$L = 20 \cdot \log_{10}\left(\frac{\bar{p}}{p_0}\right), \quad L_I = 10 \cdot \log_{10}\left(\frac{\bar{I}}{I_0}\right), \tag{2.53}$$

respectively. The level $L$ is given in decibels (dB), with the standardized reference values for sound pressure $p_0 = 20\,\mu\text{Pa}$ and for sound intensity $I_0 = 10^{-12}\,\text{W/m}^2$.

The audible ranges of frequency and SPL can be visualized in the so-called *hearing area*, as shown in Figure 2.19. The abscissa represents the frequency $f$ on a logarithmic scale, the ordinate the sound pressure level $L$.

The threshold in quiet is dependent on frequency. The highest sensitivity between 3 and 4 kHz is due to the resonance frequency of the outer ear canal as mentioned above.

The dotted line indicates the limit of damage risk for an "average person," which strongly depends on the duration of exposure and on frequency, taking the smallest values in the range between 1 and 5 kHz. The areas of speech and music sounds are indicated by hatched patterns.

In general, the perceived *loudness level* $L_L$ is a function of both frequency $f$ and sound pressure level $L$ as shown in Figure 2.20.

Contours of constant subjective loudness can be found by auditive comparison of a sinusoidal test tone at different frequencies and amplitudes with a sinusoidal reference at 1 kHz. At the reference frequency, the loudness level is identical to the sound pressure level. The loudness level $L_L$ of a test tone is the same as the loudness level of a reference 1 kHz tone with sound pressure level $L = L_L$. The loudness level $L_L$ is given in the pseudo-unit *phon*.

In psychoacoustics, the perceived loudness, which is defined in the unit *sone*, is often evaluated by auditive tests. The relation between loudness $N$ in sone and loudness level $L_L$

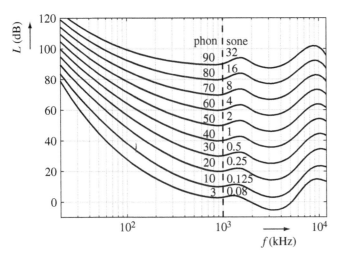

**Figure 2.20** Contours of equal loudness level $L_L$ for sinusoidal tones. Source: Adapted from [ISO226 2003].

in phon can be approximated by

$$N \approx 2^{(L_L-40)/10}.$$ (2.54)

A perceived loudness of $N = 1$ sone corresponds to a loudness level of $L_L = 40$ phon.

### 2.6.2 Spectral Resolution

The spectral resolution of the ear is related to the frequency-to-place transformation on the B.M. The frequency resolution can be analyzed by specific auditive tests.

The just noticeable variations in frequency can be measured as follows. Sinusoidal frequency modulation with a modulation frequency of about 4 Hz and different values of the modulation index and thus different frequency deviations $\Delta f$ is applied to a "carrier sine wave" of frequency $f_0$. The test persons are asked to detect the difference between the unmodulated carrier signal with frequency $f_0$ and the modulated tone. As the frequency-modulated signal has two main spectral components at $f_0 \pm \Delta f$, the effective variation in frequency is $2 \cdot \Delta f$.

At frequencies $f_0 < 500$ Hz, a constant, just noticeable difference of $2 \cdot \Delta f = 3.6$ Hz can be measured. Above 500 Hz, this value increases proportionally according to

$$2 \cdot \Delta f \approx 0.007 \cdot f_0.$$ (2.55)

We can distinguish very small frequency changes of about 0.7% in this range. Thus, in the frequency range up to 16 kHz, about 640 frequency steps can be distinguished.

Another important aspect is the effective spectral resolution, which is also linked to loudness perception. By using test signals either consisting of several sinusoidal tones closely spaced in frequency or consisting of bandpass noise with adjustable bandwidth, it can be shown that the ear integrates the excitation over a certain frequency interval. By means of listening experiments, 24 intervals can be identified in the audible frequency range up to

**Table 2.1** Critical bands.

| $b$/Bark | $f_c$/Hz | $\Delta f_b$/Hz | $b$/Bark | $f_c$/Hz | $\Delta f_b$/Hz |
|---|---|---|---|---|---|
| 0.5 | 50 | 100 | 12.5 | 1850 | 280 |
| 1.5 | 150 | 100 | 13.5 | 2150 | 320 |
| 2.5 | 250 | 100 | 14.5 | 2500 | 380 |
| 3.5 | 350 | 100 | 15.5 | 2900 | 450 |
| 4.5 | 450 | 110 | 16.5 | 3400 | 550 |
| 5.5 | 570 | 120 | 17.5 | 4000 | 700 |
| 6.5 | 700 | 140 | 18.5 | 4800 | 900 |
| 7.5 | 840 | 150 | 19.5 | 5800 | 1100 |
| 8.5 | 1000 | 160 | 20.5 | 7000 | 1300 |
| 9.5 | 1170 | 190 | 21.5 | 8500 | 1800 |
| 10.5 | 1370 | 210 | 22.5 | 10 500 | 2500 |
| 11.5 | 1600 | 240 | 23.5 | 13 500 | 3500 |

$b$: critical-band rate; $f_c$: center frequency; $\Delta f_b$: bandwidth.
Source: Adapted from [Zwicker 1961].

16 kHz, which are called the *critical bands* [Zwicker 1961]. In technical terms, they can be described as a filter bank with non-uniform spectral resolution. There is a strong correlation between the critical bands and the excitation patterns of the B.M. If the B.M. of length $d = 32$ mm is divided into 24 uniform intervals of length $d_B = d/24 = 4/3$ mm, each segment is equivalent to a critical band. The critical bands do not correspond to 24 discrete filters, but they quantify the effective frequency-dependent frequency resolution of loudness perception. The uniform subdivision of the basilar membrane is widely known as the *Bark scale* in honor of the famous physicist H.G. Barkhausen. Table 2.1 lists the Bark number $b$ (*critical-band rate*), the center frequency $f_c$, and the bandwidth $\Delta f_b$ of each critical band. Within the narrow-band telephone frequency band there are 16 critical bands, while wide-band telephony covers 21 critical bands.

The Bark scale in Table 2.1 is approximately related to linear frequency $f$ as

$$\frac{b}{\text{Bark}} = 13 \arctan\left(0.76\frac{f}{\text{kHz}}\right) + 3.5 \arctan\left(\frac{f}{7.5\ \text{kHz}}\right)^2 \tag{2.56}$$

and the bandwidth of a Bark band at any particular center frequency $f_c$ as

$$\frac{\Delta f_b}{\text{Hz}} = 25 + 75 \cdot \left[1 + 1.4\left(\frac{f_c}{\text{kHz}}\right)^2\right]^{0.69}. \tag{2.57}$$

In more precise measurements of auditory filters using the *notched-noise method*, the loudness of a tone is measured as a function of a frequency gap in masking noise around the test tone [Patterson 1976]. The resulting masking and loudness effects are then described via the concept of *equivalent rectangular bandwidth* (ERB) [Moore, Glasberg 1996], which leads to bands which are narrower than Bark bands, especially at low frequencies. In analogy to the Bark scale, these bands define an ERB scale. The *ERB rate*, i.e., the ERB units below a given frequency $f$, may be approximated by

$$\frac{z(f)}{\text{ERB-units}} = 21.4 \log_{10}\left(1 + 4.37\frac{f}{\text{kHz}}\right), \tag{2.58}$$

while the ERB of the auditory filter at any center frequency $f_c$ is described by

$$\frac{\text{ERB}(f_c)}{\text{Hz}} = 24.7\left(1 + 4.37\frac{f_c}{\text{kHz}}\right). \tag{2.59}$$

The critical-band (or ERB) feature and the masking effect resulting from it can be exploited to improve the performance and to reduce the complexity of speech-signal-processing algorithms such as speech recognition, speech coding, or speech enhancement.

### 2.6.3 Masking

Masking occurs in many everyday situations, where a dominant sound renders a weaker sound inaudible. One example is music at "disco level," which might completely mask the ringing of a mobile phone. In that situation, the music signal is called the *masker* and the ringing tone is called the *test tone*.

The masking effect can easily be demonstrated by using a narrow bandpass noise or a sinusoidal tone with fixed frequency $f_M = f_c$ as masker and a sinusoidal tone with variable frequency $f_T$ or a fixed narrowband noise as test tone. Narrowband means that in the masking experiment the bandwidth of the noise (being either masker or test signal) does not exceed the critical bandwidth at this frequency.

One example is given in Figure 2.21 for a narrowband noise as masker centered at $f_M = 1$ kHz, having a bandwidth of $\Delta f_M = 160$ Hz and a fixed sound pressure level $L_M$. For any frequency $f_T$ of the test tone, the sinusoidal tone is masked by the noise as long as

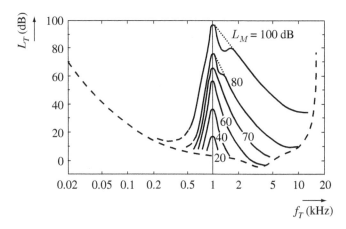

**Figure 2.21** Level $L_T$ of a sinusoidal test tone of frequency $f_T$ just masked by narrowband noise with level $L_M$ and with critical bandwidth, centered at $f_M = f_c = 1$ kHz. The dashed line indicates the threshold of hearing whereas the dips for $L_M = 80$ dB and $L_M = 100$ dB result from audible non-linear interactions between the tone and masking narrow-band noise. For these levels, the level of the test tone must be further raised to become just audible as indicated by the dotted lines. Source: Adapted from [Zwicker, Fastl 2007].

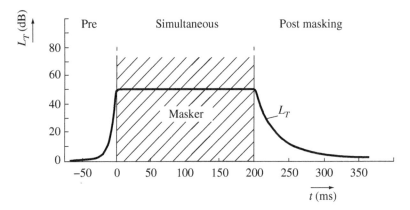

**Figure 2.22**  Pre and post masking: necessary level $L_T$ for the audibility of a sinusoidal burst (test tone) masked by wideband noise. Source: Adapted from [Zwicker, Fastl 2007].

its level $L_T$ is below the masking threshold. The masking curves are not symmetric with frequency $f_T$. They are steeper toward lower frequencies than toward higher frequencies. This follows, as discussed in Section 2.5, from the shape of the traveling wave envelope on the basilar membrane (see also Figure 2.18). The dips in the curves at masker levels $L_M$ of 80 dB and 100 dB are caused by non-linear effects of the hearing system.

As mentioned earlier, noise can also be masked by a sinusoidal signal. Furthermore, the masking threshold can be evaluated for complex test sounds such as speech signals. This is widely used in speech and audio coding to mask quantization noise, i.e., to improve the subjective quality of the decoded signal at a given bit rate.

The masking effect discussed so far is called masking in the frequency domain or *simultaneous masking* as the masker and the test signal are present at the same time.

Apart from this, *non-simultaneous masking* can be observed in the time domain. When the test sound is presented either before the masker is switched on or after the masker is switched off, the onset of the test tone is inaudible, resulting in *pre masking* or *post masking*, respectively. The effect of pre masking is not very strong but post masking is quite pronounced. The latter can be explained by the fact that the B.M. and the organ of Corti need some time to recover the threshold in quiet after the masker has been switched off. The principle of masking with wideband noise as masker and sinusoidal bursts as test tone is illustrated in Figure 2.22.

Pre and post masking are exploited in speech and audio processing, such as frequency-domain-based coding, to hide pre and post echoes originating from non perfect reconstruction.

### 2.6.4  Spatial Hearing

Equipped with two ears, the human auditory system has the ability of spatial hearing and thus provides information about the spatial location of sound events. Spatial hearing relies on two basic auditory cues, i.e.,

- the inter-aural time difference (ITD), and
- the inter-aural level difference (ILD).

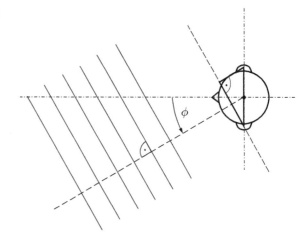

**Figure 2.23** Plane wave impinging on a human head. The direction of arrival is defined by the azimuth $\phi$.

As illustrated in Figure 2.23, these cues arise as a result of diffraction and scattering when a sound wave (e.g., a plane wave) impinges on the human head. While the signals at the left and the right ears will be very similar for a source impinging from the frontal or the rear direction, any other direction leads to delay (or phase) differences and level differences. The maximum ITD of approx. 0.7 ms is observed for an azimuth of about ±90°. At low frequencies, the acoustic wave will be bend around the head but may lead to destructive interference at certain locations. At higher frequencies, level differences occur as a result of reflections and scattering, leading to an effective sound pressure increase at the ear directed toward the source (ipsi-lateral ear) and a shadowing of the ear on the other side (contra-lateral ear).

In the case of high-frequency sinusoidal tones, the level differences at the two ears dominate the location estimate. In the case of low-frequency tones, both time and level differences support directional hearing, where the latter can be small due to diffraction. For broadband signals, the inter-aural delay differences of the (low-frequency) envelopes are also evaluated in this process. Spatial hearing is discussed in depth in [Blauert 1997].

### 2.6.4.1 Head-Related Impulse Responses and Transfer Functions

In an acoustic scenario with a listener in the far-field of a single source, the ITD and ILD cues are well represented by

- the *head-related impulse response* (HRIR) or
- the *head-related transfer function* (HRTF)

of the left and the right ears. To measure HRIRs, miniature microphones are placed at the entrance of the (sealed) ear canal and the signals are recorded in an anechoic room. Thus, the ear canal does not play a role. The head-related coordinate system characterizing directions of incidence via the azimuth and elevation is shown in Figure 2.24. Figure 2.24 also depicts the symmetry planes which are denoted as horizontal plane, frontal plane, and median plane.

The measurement signal typically is one of white noise, maximum-length sequences, or chirp sequences. The HRIRs for all azimuth and elevations may be either measured in a stop-and-go fashion for each direction separately or via a continuous acquisition method

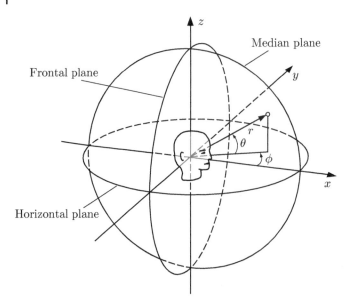

**Figure 2.24** The head-related coordinate system. Source: Adapted from [Blauert 1997].

[Enzner 2008], [Enzner et al. 2013], [Enzner et al. 2022]. In the latter case, an array of loud-speakers covers the desired set of elevations and emits the measurement signals, while the person or device under test is rotated and the received signals are continuously recorded. Using adaptive filters, the impulse responses may then be extracted from the continuous measurement signal.

To eliminate the effects of the loudspeaker when measuring HRIRs, the measurement is often normalized on the sound field at a reference point. Here, several options are available:

- free-field HRIRs are normalized on the sound field at the position of the center of head (with no head in place),
- monaural HRIRs are normalized on the HRIRs of a specific direction, most often the frontal direction,
- interaural HRIRs are normalized on the HRIR of the contra-lateral ear.

A set of exemplary free-field head-related impulse responses, measured in the horizontal plane, is shown in Figure 2.25 for the left and the right ears. The impact of the angle of incidence on the maximum amplitudes and the lag between the left and the right ear can be clearly observed. Note that the HRIRs do not contain binaural (i.e., ILD and ITD) cues for 0° and 180°. In this case, spectral (pinna) cues allow the detection of the front and back direction, albeit with errors (also known as front-back confusion). Given the HRIRs of the left and the right ears, a monaural sound source may then be rendered at any direction via a convolution of the source signal with these HRIRs.

### 2.6.4.2 Law of The First Wavefront
In a reverberant environment, the directional perception is determined by the direction of the first incident wavefront (also known as the precedence effect [Wallach et al. 1949]).

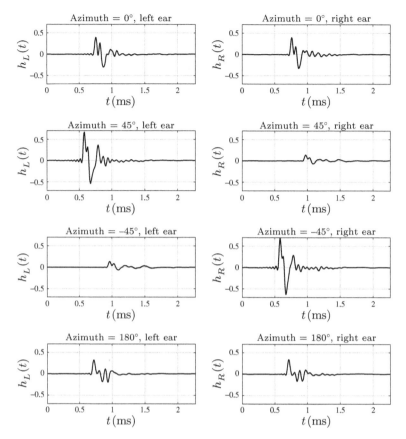

**Figure 2.25** Examples of measured HRIRs for azimuths $\phi = 0, \pm 45, 180$ degrees in the horizontal plane. Left column: left ear, right column: right ear.

A selective processing mechanism in the auditory system assigns less importance to the echoes when they arrive between 1 and 50 ms after the first wavefront, thus facilitating sound localization in reverberant rooms [Blauert 1997].

## References

Blauert, J. (1997). *Spatial Hearing - The Psychoacoustics of Human Sound Localization*, revised edn, The MIT Press, Boston, Massachusetts.

Deller Jr., J. R.; Proakis, J. G.; Hansen, J. H. L. (2000). *Discrete-Time Processing of Speech Signals*, 2nd edn, IEEE Press, New York.

Enzner, G. (2008). Analysis and Optimal Control of LMS-Type Adaptive Filtering for Continuous-Azimuth Acquisition of Head Related Impulse Responses, *2008 IEEE International Conference on Acoustics, Speech and Signal Processing*, pp. 393–396.

Enzner, G.; Antweiler, C.; Spors, S. (2013). Trends in Acquisition of Individual Head-Related Transfer Functions, *in* J. Blauert (ed.), *The Technology of Binaural Listening. Modern Acoustics and Signal Processing*, Springer-Verlag, Berlin, Heidelberg, pp. 57–92.

Enzner, G.; Urbanietz, C.; Martin, R. (2022). Optimized Learning of Spatial-Fourier Representations from Fast HRIR Recordings, *30th European Signal Processing Conference (EUSIPCO)*, pp. 304–308.

Fant, G. (1970). *Acoustic Theory of Speech Production*, 2nd edn, Mouton, The Hague.

Fitch, J.; Holbrook, A. (1970). Modal Vocal Fundamental Frequency of Young Adults, *Archives of Otolaryngology*, vol. 92 no. 4, pp. 379–382.

Flanagan, J. L. (1960). Analog Measurements of Sound Radiation from the Mouth, *Journal of the Acoustical Society of America*, vol. 32, no. 12, pp. 1613–1620.

Flanagan, J. L. (1972). *Speech Analysis, Synthesis, and Perception*, 2nd edn, Springer-Verlag, Berlin.

Gimson, A. C.; Cruttenden, A. (1994). *Gimson's Pronunciation of English*, 5th revised edn, Arnold, London.

Greenwood, D. (1990). A Cochlear Frequency-Position Function for Several Species - 29 Years Later, *Journal of the Acoustical Society of America*, vol. 87, no. 6, pp. 2592–2605.

Hudde, H. (2005). A Functional View on the Peripheral Human Hearing Organ, *in* J. Blauert (ed.), *Communication Acoustics*, Springer-Verlag, Berlin, pp. 47–74.

ISO226 (2003). Acoustics – Normal Equal-Loudness-Level Contours, ISO 226:2003(E), International Standardization Organization. Geneva, Switzerland.

Kelly, J. L.; Lochbaum, C. C. (1962). Speech Synthesis, *Proceedings of Fourth International Congress on Acoustics*, Copenhagen, Denmark, pp. 1–4.

Moore, B.; Glasberg, B. (1996). A Revision of Zwicker's Loudness Model, *Acta Acustica United with Acustica*, vol. 82, pp. 335–345.

Parsons, T. W. (1986). *Voice and Speech Processing*, McGraw-Hill, New York.

Patterson, R. (1976). Auditory Filter Shapes Derived with Noise Stimuli, *Journal of the Acoustical Society of America*, vol. 59, no. 3, pp. 640–654.

Picket, J. M. (1980). *The Sounds of Speech Communication*, Pro-Ed, Inc., Austin, Texas.

Quatieri, T. F. (2001). *Discrete-Time Speech Signal Processing*, Prentice Hall, Englewood Cliffs, New Jersey.

Rabiner, L.; Schafer, R. (1978). *Digital Processing of Speech Signals*, Prentice-Hall International.

Smith, D. R. R.; Patterson, R. D. (2005). The Interaction of Glottal-Pulse Rate and Vocal-Tract Length in Judgements of Speaker Size, Sex, and Age, *Journal of the Acoustical Society of America*, vol. 118, no. 5, pp. 3177–3186.

Vary, P.; Martin, R. (2006). *Digital Speech Transmission - Enhancement, Coding and Error Concealment*, 1st edn, John Wiley & Sons Ltd., England.

Wallach, H.; Newman, E. B.; Rosenzweig, M. R. (1949). The Precedence Effect in Sound Localization, *American Journal of Psychology*, vol. 62, no. 3, pp. 315–336.

Xiang, N.; Blauert, J. (2021). *Acoustic for Engineers – Troy Lectures*, 3rd edn, Springer-Verlag, Berlin, Heidelberg.

Zwicker, E. (1961). Subdivision of the Audible Frequency Range into Critical Bands (Frequenzgruppen), *Journal of the Acoustical Society of America*, vol. 33, no. 2, p. 248.

Zwicker, E.; Fastl, H. (2007). *Psychoacoustics*, 3rd edn, Springer-Verlag, Berlin, Germany.

# 3

# Spectral Transformations

Spectral transformations are key to many speech-processing algorithms. The purpose of a spectral transformation is to represent a signal in a domain where certain signal properties are better accessible or a specific processing task is more efficiently accomplished. In this chapter, we will summarize the definitions and properties of the Fourier transform (FT) for continuous and discrete time signals as well as the discrete Fourier transform (DFT), its fast realizations, and the $z$-transform. No extensive derivations will be given. For more elaborate expositions of this material the reader is referred to the many excellent textbooks on digital signal processing. The chapter concludes with an introduction into the real and the complex cepstrum and applications.

## 3.1 Fourier Transform of Continuous Signals

The *FT* provides an analysis of signals in terms of its spectral components. It relates a continuous (time) domain signal $x_a(t)$ of, in general, infinite support to its Fourier domain representation $X_a(j\omega)$

$$X_a(j\omega) = \int_{-\infty}^{\infty} x_a(t)e^{-j\omega t}dt \qquad (3.1)$$

with $\omega = 2\pi f$ denoting the radian frequency. The inverse transformation is given by

$$x_a(t) = \frac{1}{2\pi} \int_{-\infty}^{\infty} X_a(j\omega)e^{j\omega t}d\omega. \qquad (3.2)$$

The FT converges in the mean square if

$$\int_{-\infty}^{\infty} |x_a(t)|^2 dt < \infty. \qquad (3.3)$$

This condition includes all signals of finite amplitude and duration. When $x(t)$ is absolutely summable and other conditions, known as the Dirichlet conditions, e.g., [Oppenheim et al. 1996], are met, the inverse FT reconstructs a signal, which is identical to the original signal

*Digital Speech Transmission and Enhancement*, Second Edition. Peter Vary and Rainer Martin.
© 2024 John Wiley & Sons Ltd. Published 2024 by John Wiley & Sons Ltd.

**Table 3.1** Properties of the FT.

| Property | Time domain | Frequency domain |
|---|---|---|
| Definition | $x(t) = \dfrac{1}{2\pi}\displaystyle\int_{-\infty}^{\infty} X(j\omega)\,e^{j\omega t}\,d\omega$ | $X(j\omega) = \displaystyle\int_{-\infty}^{\infty} x(t)\,e^{-j\omega t}\,dt$ |
| Linearity | $a\,x_1(t) + b\,x_2(t)$ | $a\,X_1(j\omega) + b\,X_2(j\omega)$ |
| Conjugation | $x^*(t)$ | $X^*(-j\omega)$ |
| Symmetry | $x(t)$ is a real-valued signal | $X(j\omega) = X^*(-j\omega)$ |
| Even part | $x_e(t) = 0.5\,(x(t) + x(-t))$ | $\mathrm{Re}\{X(j\omega)\}$ |
| Odd part | $x_o(t) = 0.5\,(x(t) - x(-t))$ | $j\,\mathrm{Im}\{X(j\omega)\}$ |
| Convolution | $x_1(t) * x_2(t)$ | $X_1(j\omega)\cdot X_2(j\omega)$ |
| Time shift | $x(t - t_0)$ | $e^{-j\omega t_0} X(j\omega)$ |
| Modulation | $x(t)\,e^{j\omega_M t}$ | $X(j(\omega - \omega_M))$ |
| Scaling | $x(at),\ a \in \mathbb{R},\ a \neq 0$ | $\dfrac{1}{|a|} X\left(j\dfrac{\omega}{a}\right)$ |
| Parseval's theorem | $\displaystyle\int_{-\infty}^{\infty} x(t)y^*(t)\,dt = \dfrac{1}{2\pi}\displaystyle\int_{-\infty}^{\infty} X(j\omega)Y^*(j\omega)\,d\omega$ | |

except for a finite number of discontinuities. $X_a(j\omega)$ is, in general, a continuous and non periodic function of frequency. The FT has a number of well-known properties, which are summarized in Table 3.1.

## 3.2 Fourier Transform of Discrete Signals

The *Fourier transform of discrete signals* (FTDS)[1] is derived using the representation of sampled signals as pulse trains

$$x_s(t) = \sum_{k=-\infty}^{\infty} x_a(kT)\,\delta_a(t - kT) \tag{3.4}$$

and the sifting property of the Dirac impulse[2]

$$\int_{-\infty}^{\infty} \delta_a(t - t_0) f(t)\,dt = f(t_0). \tag{3.5}$$

---

1 The FTDS is also known as the *discrete time Fourier transform* (DTFT), e.g., [Oppenheim et al. 1996]. We prefer the more general terminology.
2 We will denote the continuous time Dirac impulse by $\delta_a(t)$ and use $\delta(k)$ for the discrete unit impulse (3.8).

For a sampling period of $T = \dfrac{1}{f_s}$ we obtain from the continuous time FT (3.1)

$$X_s(j\omega) = \int_{-\infty}^{\infty} x_s(t)\, e^{-j\omega t}\, dt = \sum_{k=-\infty}^{\infty} x_a(kT) \int_{-\infty}^{\infty} \delta_a(t - kT)\, e^{-j\omega t}\, dt$$

$$= \sum_{k=-\infty}^{\infty} x_a(kT)\, e^{-j\omega kT}.$$

$X_s(j\omega)$ is a continuous and periodic function of the radian frequency $\omega$ and hence also of frequency $f$. To see this we note that the complex phasor

$$e^{-j\omega kT} = e^{-j2\pi fkT} = \cos(2\pi fkT) - j\sin(2\pi fkT)$$

is periodic in $f$ with period $f = \dfrac{1}{T} = f_s$. Therefore, we have

$$X_s\left(j\left(\omega + \frac{2\pi\ell}{T}\right)\right) = X_s(j\omega) \text{ for any } \ell \in \mathbb{Z}.$$

To facilitate the treatment of sampled signals in the Fourier domain we normalize the frequency variable $f$ on the sampling rate $f_s$ and introduce the *normalized radian frequency* $\Omega = \omega T = 2\pi fT = 2\pi\dfrac{f}{f_s}$. We then obtain the FTDS and the inverse transform with $x(k) = x_a(kT)$ as

$$X(e^{j\Omega}) = \sum_{k=-\infty}^{\infty} x(k)\, e^{-j\Omega k} \quad \text{and} \quad x(k) = \frac{1}{2\pi} \int_{-\pi}^{\pi} X(e^{j\Omega})\, e^{j\Omega k}\, d\Omega. \tag{3.6}$$

Note that the inverse transform is evaluated over one period of the spectrum only. $X(e^{j\Omega})$ is a complex quantity and may be written in terms of its real and imaginary parts

$$X(e^{j\Omega}) = \text{Re}\{X(e^{j\Omega})\} + j\,\text{Im}\{X(e^{j\Omega})\} = X_R(e^{j\Omega}) + jX_I(e^{j\Omega})$$

or in terms of its magnitude and phase

$$X(e^{j\Omega}) = |X(e^{j\Omega})|\, e^{j\phi(\Omega)}.$$

$|X(e^{j\Omega})|$ is called the amplitude spectrum and $\phi(\Omega)$ is called the phase spectrum. The principal value of the phase is denoted by $\arg\{X(e^{j\Omega})\} \in \{-\pi, \pi\}$. Frequently, we will also use the logarithm of the amplitude spectrum $20\log_{10}(|X(e^{j\Omega})|)$. The properties of the FTDS are summarized in Table 3.2.

As an example we compute the FTDS of a discrete rectangular pulse

$$x(k) = \sum_{\ell=-(N-1)/2}^{(N-1)/2} \delta(k - \ell) \tag{3.7}$$

where $N$ is odd and $\delta(k)$ is the discrete unit impulse sequence, i.e.,

$$\delta(k) = \begin{cases} 1 & k = 0 \\ 0 & k \neq 0. \end{cases} \tag{3.8}$$

**Table 3.2** Properties of the FTDS.

| Property | Time domain | Frequency domain |
|---|---|---|
| Definition | $x(k) = \dfrac{1}{2\pi} \displaystyle\int_{-\pi}^{\pi} X(e^{j\Omega}) e^{j\Omega k} d\Omega$ | $X(e^{j\Omega}) = \displaystyle\sum_{k=-\infty}^{\infty} x(k) e^{-j\Omega k}$ |
| Linearity | $a x_1(k) + b x_2(k)$ | $a X_1(e^{j\Omega}) + b X_2(e^{j\Omega})$ |
| Conjugation | $x^*(k)$ | $X^*(e^{-j\Omega})$ |
| Symmetry | $x(k)$ is a real-valued signal | $X(e^{j\Omega}) = X^*(e^{-j\Omega})$ |
| Even part | $x_e(k) = 0.5\,(x(k) + x(-k))$ | $\mathrm{Re}\{X(e^{j\Omega})\}$ |
| Odd part | $x_o(k) = 0.5\,(x(k) - x(-k))$ | $j\,\mathrm{Im}\{X(e^{j\Omega})\}$ |
| Convolution | $x_1(k) * x_2(k)$ | $X_1(e^{j\Omega}) \cdot X_2(e^{j\Omega})$ |
| Time shift | $x(k - k_0)$ | $e^{-j\Omega k_0} X(e^{j\Omega})$ |
| Modulation | $x(k)\, e^{j\Omega_M k}$ | $X(e^{j(\Omega - \Omega_M)})$ |
| Parseval's theorem | $\displaystyle\sum_{k=-\infty}^{\infty} x(k) y^*(k) = \dfrac{1}{2\pi} \displaystyle\int_{-\pi}^{\pi} X(e^{j\Omega}) Y^*(e^{j\Omega}) d\Omega$ | |

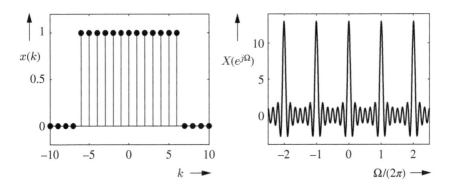

**Figure 3.1** FT of a discrete rectangular pulse ($N = 13$).

With

$$X(e^{j\Omega}) = \sum_{k=-(N-1)/2}^{(N-1)/2} e^{-j\Omega k} = e^{j\Omega(N-1)/2} \sum_{k=0}^{N-1} e^{-j\Omega k} = \frac{e^{j\Omega N/2} - e^{-j\Omega N/2}}{e^{j\Omega/2} - e^{-j\Omega/2}}$$

we obtain

$$X(e^{j\Omega}) = \frac{\sin(\Omega N/2)}{\sin(\Omega/2)}. \tag{3.9}$$

For $N = 13$, the sequence $x(k)$ and the FTDS $X(e^{j\Omega})$ are plotted in Figure 3.1. For the above example, $X(e^{j\Omega})$ is a real-valued function.

## 3.3 Linear Shift Invariant Systems

Systems map one or more input signals onto one or more output signals. In the case of a single input and a single output we have

$$y(k) = T\{x(k)\}, \tag{3.10}$$

where $T\{\cdot\}$ defines the mapping from any input sequence $x(k)$ to the corresponding output sequence $y(k)$.

A system is *linear* if and only if for any two signals $x_1(k)$ and $x_2(k)$ and any $a, b \in \mathbb{C}$ we have

$$T\{a x_1(k) + b x_2(k)\} = a\, T\{x_1(k)\} + b\, T\{x_2(k)\}. \tag{3.11}$$

A system is *memoryless* when $y(k = k_0)$ depends only on $x(k = k_0)$ for any $k_0$.

A system is *shift invariant* if for any $k_0$

$$y(k - k_0) = T\{x(k - k_0)\}. \tag{3.12}$$

A system that is linear and shift invariant is called a *linear shift invariant* (LSI) system.

An LSI system is defined by a mapping $T_{\mathrm{LSI}}\{\cdot\}$, as shown in Figure 3.2. It may also be characterized by its impulse response $h(k) = T_{\mathrm{LSI}}\{\delta(k)\}$ if we assume that the system is at rest before any signal is applied. The latter condition implies that initially all system state variables are zero.

Using the linearity and the shift invariance, we obtain

$$y(k) = T_{\mathrm{LSI}}\{x(k)\}$$

$$= T_{\mathrm{LSI}}\left\{\sum_{l=-\infty}^{\infty} x(l)\,\delta(k-l)\right\} = \sum_{l=-\infty}^{\infty} x(l)\, T_{\mathrm{LSI}}\{\delta(k-l)\}$$

$$= \sum_{l=-\infty}^{\infty} x(l)\, h(k-l) = x(k) * h(k).$$

This relation is known as *convolution* and denoted by $*$:

$$y(k) = x(k) * h(k) = \sum_{l=-\infty}^{\infty} x(l)\, h(k-l)$$

$$= \sum_{l=-\infty}^{\infty} h(l)\, x(k-l) = h(k) * x(k). \tag{3.13}$$

An especially useful class of systems is specified via a *difference equation* with constant coefficients $a_\mu$ and $b_\nu$

$$y(k) = b_0 x(k) + b_1 x(k-1) + \ldots + b_N x(k-N)$$
$$+ a_1 y(k-1) + a_2 y(k-2) + \ldots + a_M y(k-M), \tag{3.14}$$

**Figure 3.2** LSI system.

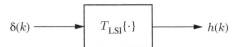

$\delta(k) \longrightarrow \boxed{T_{\mathrm{LSI}}\{\cdot\}} \longrightarrow h(k)$

where input and output samples with index $l \le k$ contribute to the current output sample $y(k)$. This equation may be cast in a more compact form

$$\sum_{\mu=0}^{M} \tilde{a}_\mu y(k-\mu) = \sum_{v=0}^{N} b_v x(k-v) \quad \tilde{a}_0 = 1 \text{ and } \tilde{a}_\mu = -a_\mu \quad \forall \mu > 0. \tag{3.15}$$

The maximum of $N$ and $M$ is called the order of the system. Additionally, we must specify $M$ initial conditions to obtain a unique solution. In most practical cases, however, we assume that the system is causal and that it is at rest (zero state variables) before an input signal is applied.

### 3.3.1 Frequency Response of LSI Systems

The non causal complex exponential $x(k) = e^{j(\Omega_0 k + \phi_0)}$, $-\infty < k < \infty$, is an eigenfunction of an LSI system, i.e.,

$$y(k) = \sum_{\ell=-\infty}^{\infty} h(\ell) e^{j(\Omega_0(k-\ell)+\phi_0)} = e^{j(\Omega_0 k + \phi_0)} \sum_{\ell=-\infty}^{\infty} h(\ell) e^{-j\Omega_0 \ell} \tag{3.16}$$

$$= e^{j(\Omega_0 k + \phi_0)} H(e^{j\Omega_0}), \tag{3.17}$$

where the *frequency response*

$$H(e^{j\Omega}) = \sum_{k=-\infty}^{\infty} h(k) e^{-j\Omega k}$$

is the FT of the discrete impulse response $h(k)$ and

$$H(e^{j\Omega}) = \text{Re}\{H(e^{j\Omega})\} + j \text{Im}\{H(e^{j\Omega})\} = |H(e^{j\Omega})| e^{j\phi(\Omega)}.$$

## 3.4 The z-transform

The FTDS does not converge for signals, such as the unit step $u(k)$, which is defined as

$$u(k) = \begin{cases} 1 & k \ge 0 \\ 0 & k < 0. \end{cases} \tag{3.18}$$

Therefore, we extend the FTDS into the complex plane by substituting

$$j\Omega = j\omega T \to sT \quad \text{with} \quad s = \alpha + j\omega.$$

To avoid the periodic repetition of the resulting spectral function along the imaginary axis, it is common practice to map the left half plane into the unit circle and to map periodic repetitions onto each other. This is achieved by introducing the complex variable $z = e^{sT} = e^{\alpha T} e^{j\omega T} = r e^{j\Omega}$ with $r = e^{\alpha T}$. The *two-sided z-transform* of a discrete sequence $x(k)$ is then given by

$$\mathcal{Z}\{x(k)\} = X(z) = \sum_{k=-\infty}^{\infty} x(k) z^{-k}. \tag{3.19}$$

Alternatively, we could use the Laplace transform [Oppenheim et al. 1996] of a sampled signal $x_s(t)$ as in (3.4) and obtain

$$\int_{-\infty}^{\infty} x_s(t) e^{-st} dt = \sum_{k=-\infty}^{\infty} x_a(kT) \int_{-\infty}^{\infty} \delta_a(t - kT) e^{-st} dt = \sum_{k=-\infty}^{\infty} x(k) e^{-sTk},$$

which, with the above mapping $z = e^{sT}$, results in the definition (3.19) of the z-transform.

The z-transform converges for a given $z$ when

$$|X(z)| = \left| \sum_{k=-\infty}^{\infty} x(k) z^{-k} \right|$$

attains a finite value, i.e.,

$$\left| \sum_{k=-\infty}^{\infty} x(k) z^{-k} \right| < \infty.$$

Since

$$X(z) = \sum_{k=-\infty}^{\infty} x(k) z^{-k} = \sum_{k=-\infty}^{\infty} x(k) e^{-\alpha Tk} e^{-j\omega Tk} = \sum_{k=-\infty}^{\infty} x(k) r^{-k} e^{-j\omega Tk}$$

the z-transform converges when

$$\sum_{k=-\infty}^{\infty} |x(k) r^{-k}| < \infty. \tag{3.20}$$

The region of the complex plane for which the above condition holds is known as the region of convergence (ROC). The ROC depends on $r = |z|$ since within the ROC we have

$$\sum_{k=-\infty}^{\infty} \left| x(k) z^{-k} \right| = \sum_{k=-\infty}^{\infty} |x(k)| \left| z^{-k} \right| = \sum_{k=-\infty}^{\infty} |x(k)| r^{-k} < \infty. \tag{3.21}$$

Note that the z-transform of a sequence with finite support with $|x(k)| < \infty$ does converge in the whole complex plane except for $z = 0$. Other properties of the z-transform are summarized in Table 3.3. The last column of Table 3.3 specifies the ROC. For example, when the ROC of $X_1(z)$ is $R_{x_1}$ and the ROC of $X_2(z)$ is $R_{x_2}$ then the ROC of $aX_1(z) + bX_2(z)$ is at least equal to the intersection of both ROCs. Since poles of the individual z-transforms may cancel, it can be larger than the intersection of $R_{x_1}$ and $R_{x_2}$. Finally, we state the initial value theorem: when $x(k)$ is causal we have $x(0) = \lim_{z \to \infty} X(z)$. The proofs for this and the theorems in Table 3.3 can be found, for instance, in [Oppenheim et al. 1999].

### 3.4.1 Relation to Fourier Transform

In general, the z-transform is equivalent to the FT of $x(k) r^{-k}$. Moreover, for $z = e^{j\Omega}$ we obtain

$$X(e^{j\Omega}) = \sum_{k=-\infty}^{\infty} x(k) e^{-j\Omega k},$$

which is the FT of a discrete signal $x(k)$, provided that the unit circle is within the ROC.

**Table 3.3** Properties of the *z*-transform.

| Properties | Time domain | z-domain | ROC |
|---|---|---|---|
| Definition | $x(k) = \oint_C \dfrac{X(z)}{2\pi j} z^{k-1} dz$ | $X(z) = \displaystyle\sum_{k=-\infty}^{\infty} x(k)\, z^{-k}$ | $R_x$ |
| Linearity | $a\, x_1(k) + b\, x_2(k)$ | $a\, X_1(z) + b\, X_2(z)$ | Contains at least $R_{x_1} \cap R_{x_2}$; it may be larger, when poles and zeros cancel |
| Conjugation | $x^*(k)$ | $X^*(z^*)$ | $R_x$ |
| Time reversal | $x(-k)$ | $X(z^{-1})$ | $1/R_x$ |
| Convolution | $x_1(k) * x_2(k)$ | $X_1(z) \cdot X_2(z)$ | Contains at least $R_{x_1} \cap R_{x_2}$; it may be larger, when poles and zeros cancel |
| Time shifting | $x(k + k_0)$ | $z^{k_0} X(z)$ | $R_x$ (it may not contain $z = 0$ or $z = \infty$) |

The *z*-transform may be interpreted as a Laurent series expansion of $X(z)$. $X(z)$ is an analytic (holomorphic, regular) function within the ROC. Therefore, the *z*-transform and all its derivatives are continuous functions in the ROC. More details on complex analysis can be found, for instance, in [Churchill, Brown 1990].

We may also define a *one-sided z-transform*

$$\mathcal{Z}_1\{x(k)\} = \sum_{k=0}^{\infty} x(k)\, z^{-k}$$

that is for causal signals identical to the two-sided transform.

### 3.4.2 Properties of the ROC

For a *z*-transform, which is a rational function of *z*, we summarize the following properties of the ROC [Oppenheim, Schafer 1975]:

- The ROC is a ring or a disk centered around the origin. It cannot contain any poles and must be a connected region.
- If $x(k)$ is a right-sided sequence, the ROC extends outward from the outermost pole to $|z| < \infty$.
- If $x(k)$ is a left-sided sequence, the ROC extends from the innermost pole to $z = 0$.
- The FT of $x(k)$ converges absolutely if and only if the ROC includes the unit circle.

### 3.4.3 Inverse *z*-Transform

The inverse *z*-transform is based on the Cauchy integral theorem

$$\frac{1}{2\pi j} \oint_C z^{k-1} dz = \begin{cases} 1 & k = 0 \\ 0 & k \neq 0 \end{cases} \tag{3.22}$$

and may be written as

$$x(k) = \frac{1}{2\pi j} \oint_C X(z) z^{k-1} dz. \tag{3.23}$$

To prove the inverse transform we note that

$$x(k) = \frac{1}{2\pi j} \oint_C X(z) z^{k-1} dz = \frac{1}{2\pi j} \oint_C \sum_{m=-\infty}^{\infty} x(m) z^{-m} z^{k-1} dz$$

$$= \sum_{m=-\infty}^{\infty} x(m) \frac{1}{2\pi j} \oint_C z^{k-m-1} dz = x(k).$$

In general, the evaluation of the above integral is cumbersome. For rational $z$-transforms methods based on

- long division,
- partial fraction expansions, and
- table look-up

are much more convenient. A description of these methods can be found, for instance, in [Proakis, Manolakis 1992] and [Oppenheim et al. 1999]. A selection of transform pairs is given in Table 3.4.

**Table 3.4**  Selected $z$-transform pairs.

| Sequence | $z$-Transform | ROC |
|---|---|---|
| $\delta(k)$ | $1$ | All $z$ |
| $\delta(k - k_0)$, $k_0 > 0$ | $z^{-k_0}$ | All $z$ except 0 |
| $\delta(k - k_0)$, $k_0 < 0$ | $z^{-k_0}$ | All $z$ except $\infty$ |
| $u(k)$ (see (3.18)) | $\dfrac{1}{1 - z^{-1}}$ | $|z| > 1$ |
| $a^k u(k)$ | $\dfrac{1}{1 - az^{-1}}$ | $|z| > |a|$ |
| $\cos(\Omega_0 k)u(k)$ | $\dfrac{1 - \cos(\Omega_0)z^{-1}}{1 - 2\cos(\Omega_0)z^{-1} + z^{-2}}$ | $|z| > 1$ |
| $\sin(\Omega_0 k)u(k)$ | $\dfrac{\sin(\Omega_0)z^{-1}}{1 - 2\cos(\Omega_0)z^{-1} + z^{-2}}$ | $|z| > 1$ |
| $r^k \cos(\Omega_0 k)u(k)$ | $\dfrac{1 - r\cos(\Omega_0)z^{-1}}{1 - 2r\cos(\Omega_0)z^{-1} + r^2 z^{-2}}$ | $|z| > r$ |
| $r^k \sin(\Omega_0 k)u(k)$ | $\dfrac{r\sin(\Omega_0)z^{-1}}{1 - 2r\cos(\Omega_0)z^{-1} + r^2 z^{-2}}$ | $|z| > r$ |

### 3.4.4 z-Transform Analysis of LSI Systems

When the system states are initially zero, the $z$-transform of the output signal $y(k)$ of an LSI system $T_{\text{LSI}}\{\cdot\}$ is given by

$$Y(z) = H(z)X(z), \tag{3.24}$$

where $X(z)$ and $H(z)$ are the $z$-transforms of the input signal $x(k)$ and the impulse response $h(k)$, respectively. This a direct consequence of the convolution theorem as stated in Table 3.3. We call $H(z)$ the system response (or transfer function) of the system $T_{\text{LSI}}$ (see also Figure 3.3).

An especially important class of LSI systems is characterized by a rational system response $H(z)$ with constant coefficients

$$H(z) = \frac{B(z)}{\tilde{A}(z)} = \frac{\displaystyle\sum_{v=0}^{N} b_v z^{-v}}{\displaystyle\sum_{\mu=0}^{M} \tilde{a}_\mu z^{-\mu}} = \frac{\displaystyle\sum_{v=0}^{N} b_v z^{-v}}{1 - \displaystyle\sum_{\mu=1}^{M} a_\mu z^{-\mu}} = b_0 \frac{\displaystyle\prod_{v=1}^{N}(1 - z_{0,v} z^{-1})}{\displaystyle\prod_{\mu=1}^{M}(1 - z_{\infty,\mu} z^{-1})},$$

where we have set $\tilde{a}_0 = 1$ and $\tilde{a}_k = -a_k$. We distinguish the following cases:

- General recursive (zeros and poles):

$$H(z) = \frac{B(z)}{1 - A(z)} = \frac{\displaystyle\sum_{v=0}^{N} b_v z^{-v}}{1 - \displaystyle\sum_{\mu=1}^{M} a_\mu z^{-\mu}} = \frac{z^M}{z^N} \frac{\displaystyle\sum_{v=0}^{N} b_v z^{N-v}}{z^M - \displaystyle\sum_{\mu=1}^{M} a_\mu z^{M-\mu}}.$$

- Non recursive (all zeros):

$$H(z) = B(z) = \sum_{v=0}^{N} b_v z^{-v} = \frac{\displaystyle\sum_{v=0}^{N} b_v z^{N-v}}{z^N}.$$

- Purely recursive (all poles):

$$H(z) = \frac{b_0}{1 - A(z)} = \frac{b_0}{1 - \displaystyle\sum_{\mu=1}^{M} a_\mu z^{-\mu}} = \frac{b_0 z^M}{z^M - \displaystyle\sum_{\mu=1}^{M} a_\mu z^{M-\mu}}.$$

Note that the frequently used terminology ("all-zeros" or "all-poles") does not account for zeros and poles at $z = 0$. Furthermore:

- $H(z)$ does not uniquely specify the impulse response of the system. We must also specify an ROC.

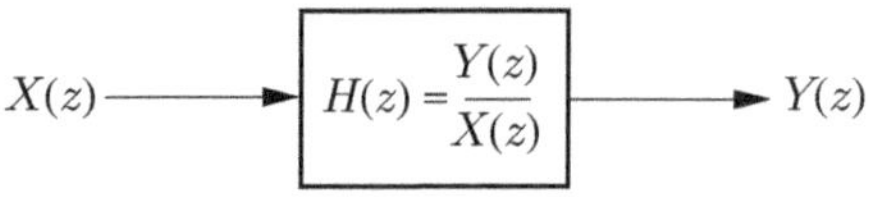

$$X(z) \longrightarrow \boxed{H(z) = \frac{Y(z)}{X(z)}} \longrightarrow Y(z)$$

**Figure 3.3** Input–output relation of LSI systems in the z-domain.

- If the system is stable, the impulse response must be absolutely summable. Therefore, the ROC must include the unit circle.
- If the system is causal, the ROC must extend from the outermost pole outward.
- For a stable and causal system, we must require that all poles lie within the unit circle.
- For a stable and invertible system, we must require that all poles and all zeros are within the unit circle. Systems that meet this condition are also called *minimum-phase* systems.

## 3.5 The Discrete Fourier Transform

The FTDS is not directly suited for numerical computations since it requires in principle an input signal of infinite support and delivers a continuous spectral function $X(e^{j\Omega})$.

By contrast, the DFT computes a finite set of discrete Fourier coefficients $X_\mu$ from a finite number of signal samples. The DFT coefficients represent the spectrum of the input signal at equally spaced points on the frequency axis. However, when the support of the signal is larger than the transformation length, the DFT coefficients are not identical to the FT of the complete signal at these frequencies. Nevertheless, the DFT is an indispensable tool for numerical harmonic analysis of signals of finite or infinite support.

The spectral coefficients $X_\mu$ of the DFT are computed via the finite sum

$$X_\mu = \sum_{k=0}^{M-1} x(k) e^{-j\frac{2\pi \mu k}{M}}, \quad \mu = 0, \ldots, M - 1. \tag{3.25}$$

The $M$ signal samples are recovered by the inverse relationship

$$x(k) = \frac{1}{M} \sum_{\mu=0}^{M-1} X_\mu e^{j\frac{2\pi \mu k}{M}}, \quad k = 0, \ldots, M - 1. \tag{3.26}$$

The DFT coefficients $X_\mu$ are periodical, i.e., $X_{\mu+\lambda M} = X_\mu$ for $\lambda \in \mathbb{Z}$. The same is true for the signal $x(k)$ reconstructed from $M$ complex DFT coefficients, since, using (3.26), we have $x(k + \lambda M) = x(k)$. Therefore, we may extend the $M$ samples of $x(k)$, $k = 0, \ldots, M - 1$, into an $M$-periodic signal $x_{\tilde{M}}(k)$ where $x_{\tilde{M}}(k) = x([k]_{\mod M})$. We then have

$$x_{\tilde{M}}(k) = \frac{1}{M} \sum_{\mu=0}^{M-1} X_\mu e^{j\frac{2\pi \mu}{M} k} \tag{3.27}$$

for any $k$.

The coefficients $X_\mu$ of the DFT are spaced by $\Delta\Omega = \frac{2\pi}{M}$ on the normalized frequency axis $\Omega$ such that we obtain normalized center frequencies $\Omega_\mu = \frac{2\pi\mu}{M}$ for $\mu \in \{0, \ldots, M - 1\}$. When the signal samples $x(k)$ are generated by means of sampling a continuous signal $x_a(t)$ with sampling period $T = \frac{1}{f_s}$, the coefficients of the DFT are spaced by $\Delta f = \frac{f_s}{M}$, i.e., $f_\mu = \frac{\mu f_s}{M}$.

When the signal $x(k)$ is real valued, the DFT coefficients have a number of symmetry properties, which can be exploited to reduce the complexity of frequency domain algorithms. These and other properties of the DFT are summarized in Table 3.5. For a real input sequence the symmetry properties of the DFT are illustrated in Figure 3.4. While

**Table 3.5** Properties of the DFT.

| Property | Time domain | Frequency domain |
|---|---|---|
| Definition | $x(k) = \dfrac{1}{M} \displaystyle\sum_{\mu=0}^{M-1} X_\mu e^{j2\pi \frac{\mu k}{M}}$ | $X_\mu = \displaystyle\sum_{k=0}^{M-1} x(k) e^{-j2\pi \frac{\mu k}{M}}$ |
| | $k = 0 \ldots M - 1$ | $\mu = 0 \ldots M - 1$ |
| Linearity | $a\,x(k) + b\,y(k)$ | $a\,X_\mu + b\,Y_\mu$ |
| Symmetry | $x(k)$ is real valued | $X_\mu = X^*_{[-\mu]_{\mathrm{mod}\,M}}$ |
| Convolution | $\displaystyle\sum_{\ell=0}^{M-1} x(\ell)\,y([k - \ell]_{\mathrm{mod}\,M})$ | $X_\mu Y_\mu$ |
| Multiplication | $x(k)\,y(k)$ | $\dfrac{1}{M} \displaystyle\sum_{\ell=0}^{M-1} X_\ell\, Y_{[\mu-\ell]_{\mathrm{mod}\,M}}$ |
| Delay | $x([k + k_0]_{\mathrm{mod}\,M})$ | $e^{+j2\pi \frac{\mu k_0}{M}} X_\mu$ |
| Modulation | $x(k)\,e^{-j2\pi \frac{k\mu_0}{M}}$ | $X_{[\mu+\mu_0]_{\mathrm{mod}\,M}}$ |
| Parseval's theorem | $\displaystyle\sum_{k=0}^{M-1} x(k)y^*(k) = \dfrac{1}{M} \displaystyle\sum_{\mu=0}^{M-1} X_\mu Y^*_\mu$ | |

the real part and the magnitude are even symmetric, the imaginary part and the phase are odd symmetric.

In what follows, we will discuss some of the properties in more detail.

### 3.5.1 Linear and Cyclic Convolution

The multiplication of two sequences of DFT coefficients, $X_\mu Y_\mu$, corresponds to a *cyclic convolution* of the corresponding length-$M$ segments of time domain sequences $x(k)$ and $y(k)$. The cyclic convolution may be written as

$$\sum_{\ell=0}^{M-1} x(\ell)\,y([k - \ell]_{\mathrm{mod}\,M}) = \sum_{\ell=0}^{M-1} x(\ell)\,y_{\tilde{M}}(k - \ell)$$

$$= \left[x(k)\,R_M(k)\right] * y_{\tilde{M}}(k) \tag{3.28a}$$

or

$$\sum_{\ell=0}^{M-1} y(\ell)\,x([k - \ell]_{\mathrm{mod}\,M}) = \sum_{\ell=0}^{M-1} y(\ell)\,x_{\tilde{M}}(k - \ell)$$

$$= \left[y(k)\,R_M(k)\right] * x_{\tilde{M}}(k) \tag{3.28b}$$

with $\left[x(k)\,R_M(k)\right] * y_{\tilde{M}}(k) = \left[y(k)\,R_M(k)\right] * x_{\tilde{M}}(k)$ where $*$ denotes the aperiodic (linear) convolution and $R_M(k)$ denotes a rectangular window

$$R_M(k) = \begin{cases} 1 & 0 \le k \le M - 1 \\ 0 & \text{otherwise.} \end{cases} \tag{3.29}$$

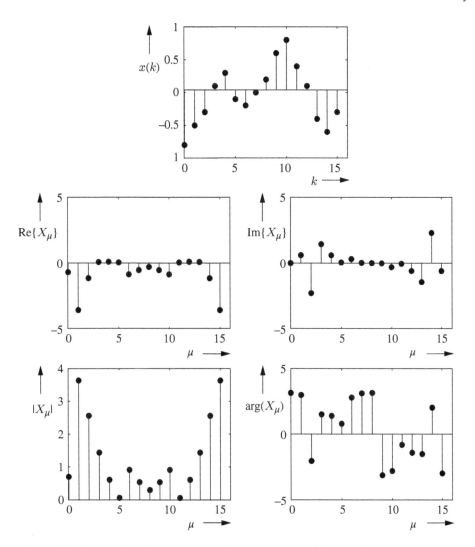

**Figure 3.4** Real part, imaginary part, magnitude, and phase of DFT coefficients of a real-valued sequence $x(k)$.

To see this, we write the inverse DFT (IDFT) of $X_\mu Y_\mu$ as

$$
\text{IDFT}\left\{X_\mu Y_\mu\right\} = \frac{1}{M}\sum_{\mu=0}^{M-1} X_\mu Y_\mu\, e^{j\frac{2\pi k}{M}\mu}
$$

$$
= \frac{1}{M}\sum_{\mu=0}^{M-1}\sum_{\ell=0}^{M-1} x(\ell)\, e^{-j\frac{2\pi\ell}{M}\mu}\, Y_\mu\, e^{j\frac{2\pi k}{M}\mu}
$$

$$
= \sum_{\ell=0}^{M-1} x(\ell)\frac{1}{M}\sum_{\mu=0}^{M-1} Y_\mu\, e^{j\frac{2\pi(k-\ell)}{M}\mu}
$$

$$
= \sum_{\ell=0}^{M-1} x(\ell)\, y_{\tilde{M}}(k-\ell),
$$

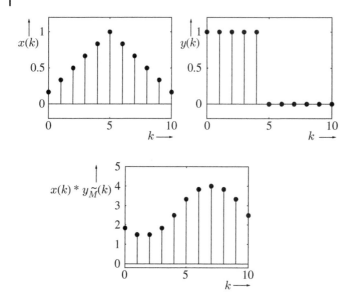

**Figure 3.5** Cyclic convolution of two sequences $x(k)$ and $y(k)$ with $M = 11$.

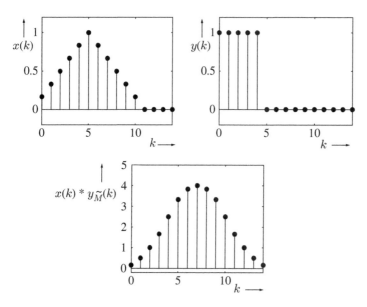

**Figure 3.6** Cyclic convolution of two sequences $x(k)$ and $y(k)$ with $M = 15$. The DFT length is chosen such that a linear convolution is obtained.

while the version in (3.28b) is shown by a simple exchange of variables. The cyclic convolution is therefore equivalent to an aperiodic convolution of $M$ samples of one sequence with the $M$-periodic extension of the other sequence.

The cyclic convolution of two sequences is illustrated in Figures 3.5 and 3.6, where the result in Figure 3.6 corresponds to a linear convolution.

### 3.5.2 The DFT of Windowed Sequences

Quite frequently, the DFT is applied to signals $x(k)$, which extend beyond the length $M$ of the DFT. In this case, the DFT uses only $M$ samples of the signal, which may be extracted via a multiplication of $x(k)$ with a window sequence $w(k)$. The application of a window of length $M$ to the time domain signal corresponds to a convolution of the FTDS of the complete signal with the FTDS of the window function in the spectral domain. After applying a window $w(k)$ to sequence $x(k)$, the DFT coefficients $X_\mu$ are equal to the spectrum of the windowed sequence $w(k)x(k)$ at the discrete frequencies $\Omega_\mu = \frac{2\pi\mu}{M}$. For $\mu = 0, \ldots, M-1$, we may write this as

$$X_\mu = \left[X(e^{j\Omega}) * W(e^{j\Omega})\right]_{\Omega=\Omega_\mu} = \sum_{k=0}^{M-1} w(k)x(k)e^{-j\frac{2\pi\mu k}{M}}. \tag{3.30}$$

We find that the DFT spectrum is a sampled version of the spectrum $X(e^{j\Omega}) * W(e^{j\Omega})$ where $W(e^{j\Omega})$ is the FTDS of the window function $w(k)$. The spread of the spectrum due to this convolution in the frequency domain is also known as *spectral leakage* and illustrated in Figure 3.7. Spectral leakage may be reduced

- by increasing the DFT length $M$ or
- by using a tapered window $w(k)$.

Compared to the rectangular window, tapered windows possess a wider main lobe in their frequency response (Figure 3.8) and thus lead to reduced frequency resolution. Some of the

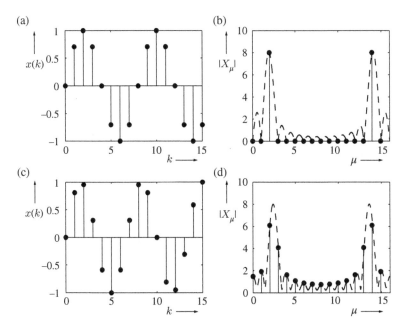

**Figure 3.7** Finite segments of a sinusoid (a,c) and their DFT (b,d) and FTDS (dashed). (a,b) Integer multiple of period is equal to the DFT length $M$. (c,d) Integer multiple of period is not equal to the DFT length $M$.

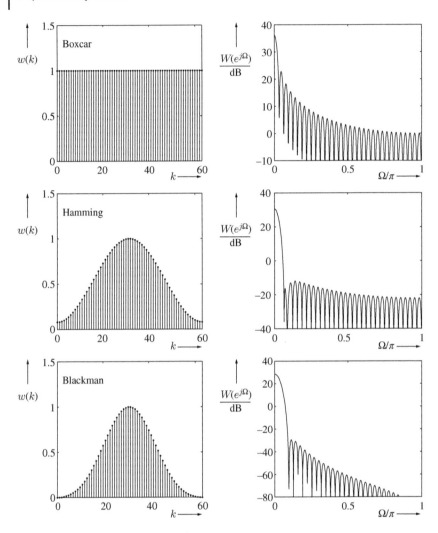

**Figure 3.8** Frequently used window functions and their amplitude spectra for $M = 61$.

frequently used window functions may be written as

$$w(k) = a - (1 - a) \cos\left(\frac{2\pi}{M-1}k\right), \quad k = 0\ldots M - 1, \tag{3.31}$$

with     $a = 1$        for the rectangular (boxcar) window,
             $a = 0.54$    for the Hamming window, and
             $a = 0.5$      for the Hann window.

To compute the frequency response, we rewrite $w(k)$ as

$$w(k) = a - \frac{1-a}{2}\left(e^{j\frac{2\pi}{M-1}k} + e^{-j\frac{2\pi}{M-1}k}\right), \tag{3.32}$$

and multiply (3.32) with the appropriate rectangular window. Then, using the FTDS and the modulation theorem, we obtain the frequency response of this window function as

$$
W(e^{j\Omega}) = e^{-j\frac{M-1}{2}\Omega}\left[a\frac{\sin\left(\frac{M}{2}\Omega\right)}{\sin\left(\frac{1}{2}\Omega\right)}\right.
$$

$$
\left. + \frac{1-a}{2}\left(\frac{\sin\left(\frac{M}{2}\left(\Omega - \frac{2\pi}{M-1}\right)\right)}{\sin\left(\frac{1}{2}\left(\Omega - \frac{2\pi}{M-1}\right)\right)} + \frac{\sin\left(\frac{M}{2}\left(\Omega + \frac{2\pi}{M-1}\right)\right)}{\sin\left(\frac{1}{2}\left(\Omega + \frac{2\pi}{M-1}\right)\right)}\right)\right].
$$

Other well-known windows are the Blackman, the Tukey, and the Kaiser window, e.g., [Oppenheim et al. 1999]. The Blackman window is quite similar to the Hann and Hamming windows, but it has one additional cosine term to further reduce the ripple ratio. Figure 3.8 illustrates the trade-off between the width of the main lobe and the side-lobe attenuation while Figure 3.9 exemplifies the reduction of spectral leakage when a Hamming window is used.

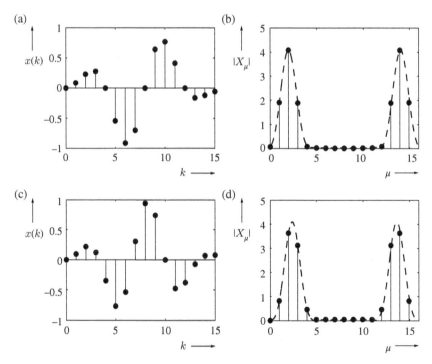

**Figure 3.9** DFT and FTDS (dashed) of Hamming windowed sinusoids. (a,b) Integer multiple of period is equal to DFT length. (c,d) Integer multiple of period is not equal to DFT length. In both cases, spectral leakage is significantly reduced.

### 3.5.3  Spectral Resolution and Zero Padding

The DFT length $M$ is identical to the number of discrete bins in the frequency domain. These bins are spaced on the normalized frequency axis according to

$$\Delta\Omega = \frac{2\pi}{M}. \qquad (3.33)$$

However, the spacing between frequency bins must not be confused with the spectral resolution of the DFT. Spectral resolution may be defined via the capability of the DFT to separate closely spaced sinusoids, as illustrated in Figure 3.10 for the boxcar window. In general, the spectral resolution depends on the number of samples of the transformed sequence $x(k)$ and the window function $w(k)$. It is inversely related to the 3-dB bandwidth of the main lobe of the amplitude spectrum of the window function. The minimum bandwidth of $4\pi/M$ is obtained for the boxcar window. Tapered windows reduce the spectral resolution but also the spectral leakage. The spectral resolution of the DFT may be increased by increasing the length $M$ of the data window, i.e., using more signal samples to compute the DFT coefficients.

The technique known as *zero padding* increases the number of frequency bins in the DFT domain. It does not increase the spectral resolution in the sense that closely spaced sinusoids can be better separated. Obviously, appending zeros to a segment of a signal, where the signal is in general much longer than the DFT length, does not add information about

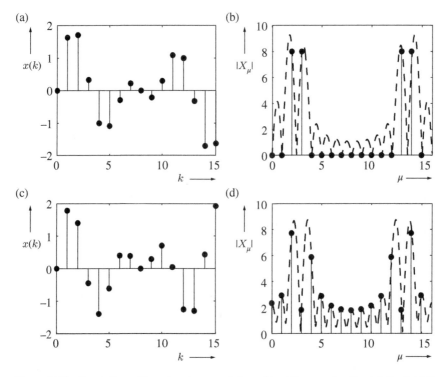

**Figure 3.10** Resolution of two closely spaced sinusoids at frequencies $\Omega_1$ and $\Omega_2$. (a,b) $\Omega_1 = 2\pi/8$, $\Omega_2 = 5\pi/16$. (c,d) $\Omega_1 = 2.2\pi/8$, $\Omega_2 = 2.2\pi/8 + 2\pi/16$.

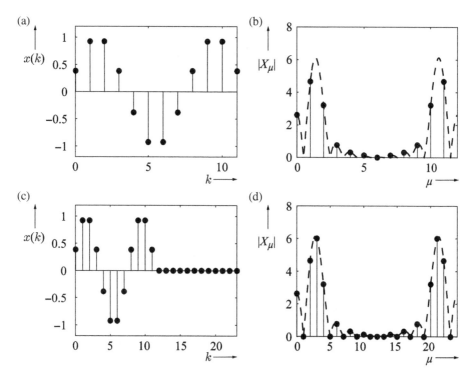

**Figure 3.11** Zero padding a sequence $x(k)$. Sequence (a) and DFT coefficients (b) without zero padding. Sequence (c) and DFT coefficients (d) after zero padding. The dashed line indicates the FTDS spectrum of the sequence, which is the same for both cases.

this signal. This is illustrated in Figure 3.11. However, for finite length signals, the DFT and zero padding allow us to compute the FTDS without error for any number of frequency bins. This is useful, for example, when the frequency response of a *finite impulse response* (FIR) filter must be computed with high resolution.

### 3.5.4 The Spectrogram

The above considerations are important for the analysis of speech samples via time-frequency representations and thus the computation of the *spectrogram*. The spectrogram implements the time-frequency analysis via the short-time Fourier transform (STFT) using a "sliding window" approach. For an input signal $x(k)$, an analysis window $w_A(k)$ and the $\lambda$-th signal frame we obtain

$$X_\mu(\lambda) = \sum_{\ell=0}^{M-1} w_A(\ell) x(\lambda r + \ell) e^{-j\frac{2\pi\mu\ell}{M}}, \qquad (3.34)$$

where $M$ is the length of the DFT, $r$ is the frame shift, and $\lambda$ and $\mu$ are the frame and frequency bin indices. The window function $w_A(k)$ is defined to have non zero samples in the

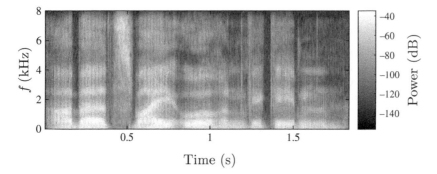

**Figure 3.12** Wide-band spectrogram of a male speech sample. Parameters: $f_s = 16$ kHz, Hann window of length 96 and overlap 64, DFT length $M = 1024$.

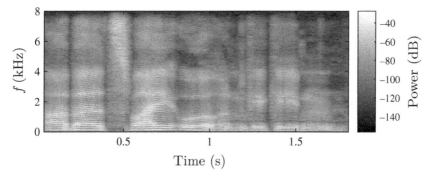

**Figure 3.13** Narrow-band spectrogram of a male speech sample. Parameters: $f_s = 16$ kHz, Hann window of length 1024 and overlap 992, DFT length $M = 1024$.

interval $k = 0 \ldots M - 1$ only. Then, the spectrogram is given by

$$S_\mu(\lambda) = 20 \log_{10}(|X_\mu(\lambda)|) \qquad \lambda \in \mathbb{Z}, \, \mu \in \{0, \ldots, M - 1\}. \tag{3.35}$$

In the analysis of speech signals, it is instructive to consider different settings for the temporal and spectral resolution:

- The *wide-band spectrogram* features a high temporal resolution and low spectral resolution. An example is shown in Figure 3.12. While the frequency resolution is sufficient to resolve resonances of the vocal tract, glottal pulses are resolved in the temporal fine structure.
- The *narrow-band spectrogram* provide a high spectral resolution but a low temporal resolution. An example is shown in Figure 3.13. Here, the glottal excitation is visible in terms of harmonic components along the frequency dimension while formants and fast temporal variations are less prominently displayed.

### 3.5.5 Fast Computation of the DFT: The FFT

*Fast Fourier transform* (FFT) algorithms were used by C.F. Gauß in the 19th century, temporarily forgotten, and rediscovered [Cooley, Tukey 1965] when digital computers emerged

[Cooley 1992]. FFT algorithms provide for an exact computation of the DFT, however, at a significantly reduced complexity.

In fact, the FFT is not a single algorithm but a family of different algorithms which rely on the principle of "divide and conquer" and symmetry considerations. The basic idea of the FFT algorithm is to divide the overall DFT computation into smaller subtasks, compute these subtasks, and then recombine the results. FFT algorithms can be classified with respect to

- their design procedure (*decimation-in-time, decimation-in-frequency*)
- their radix (2, 3, 4, …)
- memory requirements (in-place vs. not in-place)
- addressing schemes (bit-reversal vs. linear).

It should be pointed out that FFT algorithms do not exist only for DFT lengths $M = 2^p$ (powers of two) although these constitute the most widespread form of the FFT. Efficient algorithms are available for other DFT lengths $M \neq 2^p$ as well, e.g., [Oppenheim, Schafer 1975]. In what follows we demonstrate the basic idea for a radix-2 decimation-in-time algorithm [Cochran et al. 1967].

### 3.5.6 Radix-2 Decimation-in-Time FFT

The *decimation-in-time* algorithm splits the sequence of spectral coefficients

$$X_\mu = \sum_{k=0}^{M-1} x(k) e^{-j\frac{2\pi\mu k}{M}}, \quad \mu = 0, \ldots, M - 1,$$

where $M = 2^p$, into an even-indexed and odd-indexed subsequence

$$X_\mu = \sum_{k=0}^{M/2-1} x(2k) e^{-j\frac{4\pi\mu k}{M}} + \sum_{k=0}^{M/2-1} x(2k+1) e^{-j\frac{2\pi\mu(2k+1)}{M}},$$

which may be rewritten as two DFTs of length $M/2$

$$X_\mu = \sum_{k=0}^{M/2-1} x(2k) e^{-j\frac{2\pi\mu k}{M/2}} + e^{-j\frac{2\pi\mu}{M}} \sum_{k=0}^{M/2-1} x(2k+1) e^{-j\frac{2\pi\mu k}{M/2}}.$$

While the computational effort of the length-$M$ DFT is $M^2$ complex multiplications and additions, the effort is now reduced to $2\left(\frac{M}{2}\right)^2 + M = \frac{M^2}{2} + M$ multiplications and additions.

To prepare for the next decomposition we set $M_2 = M/2$, $x_{2e}(k) = x(2k)$, $x_{2o}(k) = x(2k + 1)$, and define a complex phasor $W_M = e^{-j\frac{2\pi}{M}}$. We then obtain

$$X_\mu = \sum_{k=0}^{M_2-1} x_{2e}(k) e^{-j\frac{2\pi\mu k}{M_2}} + W_M^\mu \sum_{k=0}^{M_2-1} x_{2o}(k) e^{-j\frac{2\pi\mu k}{M_2}}, \tag{3.36}$$

a signal-flow graph of which is shown in Figure 3.14. Two DFTs of length $M_2 = M/2$ are combined by means of the "twiddle factors" $W_M^\mu$, $\mu = 0 \ldots M - 1$. Since these DFTs of length

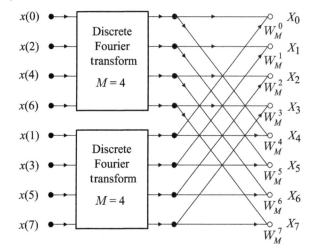

**Figure 3.14** Signal-flow graph after the first decimation step ($M = 8$).

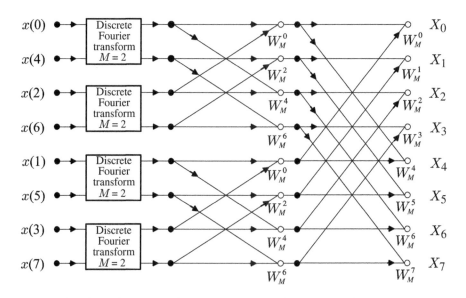

**Figure 3.15** Signal-flow graph after the second decimation step ($M = 8$).

$M_2$ are $M_2$-periodic, we compute them for $\mu = 0 \ldots M_2 - 1$ only and reuse the results for $\mu = M_2 \ldots M - 1$.

In the next decomposition step we split each of the two sequences $x_{2e}(k)$ and $x_{2o}(k)$ into two subsequences of length $M/4$. After this decomposition, which is shown in Figure 3.15, the computational effort is reduced to

$$2\left(2\left(\frac{M}{4}\right)^2 + \frac{M}{2}\right) + M = \frac{M^2}{4} + 2M,$$

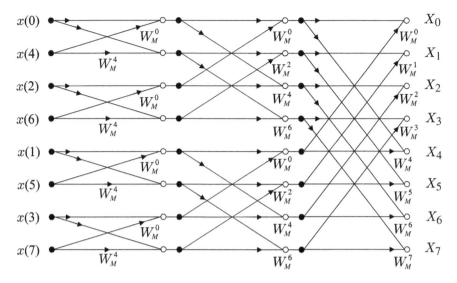

**Figure 3.16** Signal-flow graph after the third decimation step ($M = 8$).

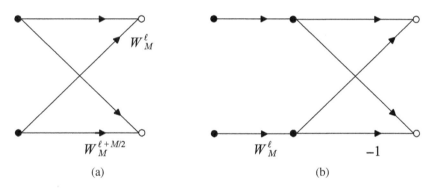

(a)          (b)

**Figure 3.17** The basic computational element of the radix-2 FFT: (a) The butterfly; (b) An efficient implementation.

complex multiplications and additions. After $p - 1$ decimation steps we have $2^{(p-1)}$ sequences of length 2. Hence, the computational effort is

$$\frac{M^2}{2^p} + (p - 1)M = M + (p - 1)M = pM = p\,2^p = M\log_2(M) \tag{3.37}$$

complex multiplications and additions. The signal-flow graph of the final result is shown in Figure 3.16. For example, when the DFT length is $M = 1024$, we obtain an extreme complexity reduction of $\frac{M\log_2(M)}{M^2} \approx 0.01$.

The regular structure which we find at all stages of the algorithm is called a "butterfly." The left plot and the right plot in Figure 3.17 show the butterfly as it was used in the preceding development and in a form which requires only one complex multiplication, respectively. The latter makes use of the relation $W_M^{\ell+M/2} = -W_M^{\ell}$. Without further optimizations, the computational complexity of the FFT algorithm is therefore

$M \log_2(M)$   complex additions and

$$\frac{M}{2} \log_2(M) \text{ complex multiplications,} \qquad (3.38)$$

where one complex addition requires two real-valued additions, and one complex multiplication needs two real-valued additions and four real-valued multiplications.

As a result of the repeated decimations, the input sequence $x(k)$ is not in its natural order. However, the elements of the scrambled input sequence can be addressed using the bit-reversal addressing mode that is supported on most digital signal processors (DSPs). Bit-reversal addressing reads the address bits of each input element in reverse order. For example, the address of the second and the seventh element of the sequence 0 0 1 and 1 1 0, respectively, are read in reverse order as 1 0 0 and 0 1 1. The latter addresses correspond to the actual position of these elements in the scrambled sequence.

## 3.6   Fast Convolution

The use of the convolution theorem of the DFT presents an attractive alternative to the direct computation of the convolution sum. In conjunction with the FFT this method is called *fast convolution* because of its lower number of operations as compared to the direct time-domain convolution.

The linear convolution of two complex-valued sequences of length $N$ requires two FFTs and one inverse fast Fourier transform (IFFT) of length $M = 2N - 1$ and $M$ complex multiplications. If the computational effort for one FFT or IFFT is $KM \log_2(M)$, where $K$ depends on the algorithm and the computer hardware, we then have

$$3KM \log_2(M) + M$$

complex operations for a fast convolution. This compares favourably to $N^2$ operations for the direct computation of the convolution sum when $N$ is large. For example, for $N = 1024$ and $K = 1$ we need 69 594 complex operations for the fast convolution compared to 1 048 576 complex operations for the direct computation.

### 3.6.1   Fast Convolution of Long Sequences

Another, even more interesting case is the convolution of a very long sequence $x(k)$, e.g., a speech signal, with a relatively short impulse response $h(k)$ of an FIR filter with $N_h$ non zero taps. This convolution may be performed in segments using a DFT of length $M > N_h$. In this case, the impulse response of the filter is transformed only once, leading to additional savings.

There are two basic methods (and numerous variations thereof) for the fast convolution of long sequences, which are known as the *overlap-add (OLA)* and *overlap-save* techniques.

The OLA method uses non overlapping segments of the input sequence and adds partial results to reconstruct the output sequence. In contrast, the overlap-save method uses overlapping segments of the input sequence and reconstructs the output sequence without overlapping the results of the partial convolutions. In what follows we will briefly illustrate these two techniques.

### 3.6.2 Fast Convolution by Overlap-Add

The OLA technique segments the input signal $x(k)$ into non overlapping shorter segments $x_1(k), x_2(k), \ldots$ of length $N_x < M$, with

$$
x_\ell(k) = \begin{cases} x\left(k + (\ell - 1)N_x\right) & 0 \le k < N_x \\ 0 & \text{elsewhere.} \end{cases}
\tag{3.39}
$$

Each of these segments is convolved with the impulse response $h(k)$. The results of these convolutions, $y_1(k), y_2(k), \ldots$, are then overlap-added to yield the final result

$$
\begin{aligned}
y(k) &= \sum_{v=0}^{N_h-1} h(v) x(k - v) \\
&= \sum_{v=0}^{N_h-1} h(v) \sum_\ell x_\ell\left(k - (\ell - 1)N_x - v\right) \\
&= \sum_\ell \underbrace{\sum_{v=0}^{N_h-1} h(v) x_\ell\left(k - (\ell - 1)N_x - v\right)}_{=y_\ell\left(k-(\ell-1)N_x\right)}.
\end{aligned}
$$

With appropriate zero padding the convolutions of the shorter segments with the impulse response $h(k)$ may be performed in the frequency domain. The OLA method requires that the convolution is linear.

Therefore, we must choose $N_h + N_x - 1 \le M$. This is illustrated in Figure 3.18, where the top part of the graph shows the linear convolution of the full sequence while the lower parts depicts the partial convolutions of zero-padded shorter segments.

### 3.6.3 Fast Convolution by Overlap-Save

For the overlap-save procedure we use segments $x_1(k), x_2(k), \ldots$ of length $N_x = M > N_h$, which contain $N_h - 1$ ("saved") samples of the previous segment and $M - N_h + 1$ new samples,

$$
x_\ell(k) = \begin{cases} x\left(k + (\ell - 1)M - \ell(N_h - 1)\right) & 0 \le k < M \\ 0 & \text{elsewhere.} \end{cases}
\tag{3.40}
$$

After padding the impulse response with $M - N_h$ zeros the cyclic convolution of these segments with impulse response $h(k)$ yields $N_h - 1$ invalid samples and $M - N_h + 1$ valid

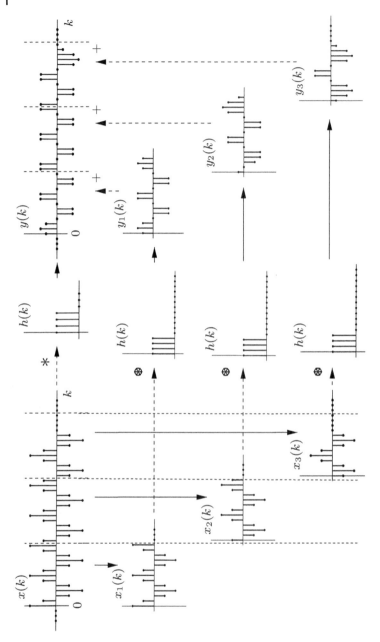

**Figure 3.18** Illustration of the overlap-add technique. $N_x = 13$, $N_h = 4$, and $M = 16$. Symbols $*$ and $\circledast$ denote the linear and the fast convolution, respectively.

samples. The latter are concatenated with the valid samples of the previously computed segment. In this way, all of the output signal is constructed. The overlap-save procedure is illustrated in Figure 3.19. The invalid samples of the output sequences $y_1(k)$, $y_2(k)$, ... are marked x and are discarded. Only the valid samples (marked $\bullet$) are concatenated.

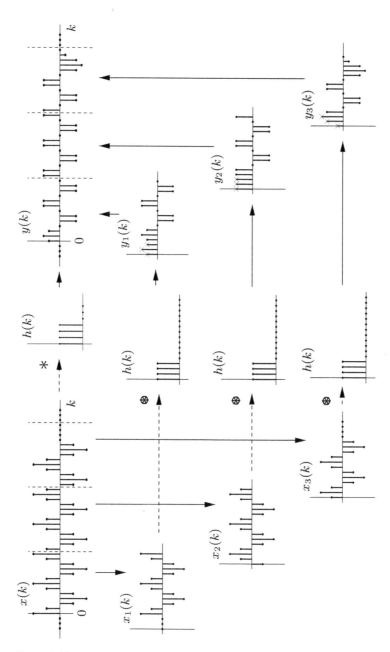

**Figure 3.19** Illustration of the overlap-save technique. $N_x = M = 16$ and $N_h = 4$. Samples marked x are discarded. Symbols $*$ and $\circledast$ denote the linear and the fast convolution, respectively.

## 3.7 Analysis–Modification–Synthesis Systems

While fast convolution implements a linear, time-invariant filter, we now turn to the more general case of time-varying and possibly non-linear filter operations. We will outline the widely used approach for time-variant filtering based on the sliding window DFT, also known as *STFT* analysis–modification–synthesis. The design of these systems should minimize signal distortions both in the presence or the absence of spectral modifications in between the analysis and synthesis stages. This design task has been treated, for example, in [Allen et al. 1977], [Griffin, Lim 1984], and [Cappé 1995].

The analysis stage applies a short-time Fourier analysis to the input signal $x(k)$ by computing the DFT of overlapping windowed frames,

$$X_\mu(\lambda) = \sum_{\ell=0}^{M-1} w_A(\ell) x(\lambda r + \ell) e^{-j\frac{2\pi\mu\ell}{M}}. \tag{3.41}$$

Here, $r$ denotes the frame shift, $\lambda \in \mathbb{Z}$ is the frame index, $\mu \in \{0, 1, \ldots, M-1\}$ is the frequency bin index, which is related to the normalized center frequency $\Omega_\mu = 2\pi\mu/M$, and $w_A(\ell)$ denotes the analysis window.

Given the modified Fourier coefficients $Y_\mu(\lambda) = f(X_\mu(\lambda))$, where $f$ represents any real-valued or complex-valued function, the overlap-add (OLA) synthesis of the modified signal $y(k)$ is then given by

$$y(k) = \sum_{\lambda=-\infty}^{\infty} w_S(k - \lambda r) \frac{1}{M} \sum_{\mu=0}^{M-1} Y_\mu(\lambda) e^{j\frac{2\pi\mu(k-\lambda r)}{M}}, \tag{3.42}$$

where $w_S(k)$ denotes the synthesis window function. The window functions $w_A(k)$ and $w_S(k)$ are defined such that they have non zero samples for $k = 0, \ldots, M-1$ only. This STFT analysis and OLA synthesis process is illustrated in Figure 3.20.

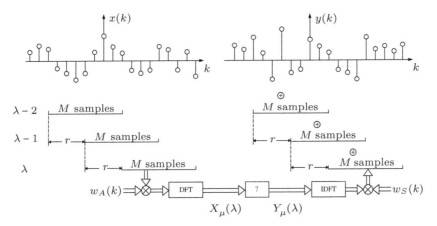

**Figure 3.20** DFT-based analysis–modification–synthesis framework. The block marked with symbol ? implements the spectral modification which is not further specified.

The analysis–synthesis system must balance conflicting requirements of sufficient spectral resolution, little spectral leakage, smooth transitions between signal frames, low latency, and low complexity.

In case of no spectral modifications, i.e., $Y_\mu(\lambda) = X_\mu(\lambda)$, *perfect reconstruction* is achieved when the product of the analysis and synthesis window functions satisfy the *constant-OLA* constraint

$$\sum_{\lambda=-\infty}^{\infty} w_A(k - \lambda r) w_S(k - \lambda r) = 1 \quad \forall k \in \mathbb{Z}. \tag{3.43}$$

To this end, we may use a Hann window with a 50% overlap ($r/M = 0.5$) or a Hamming window with a 75% overlap ($r/M = 0.25$) for spectral analysis, and a rectangular window for synthesis. However, in case of spectral modifications a rectangular synthesis window is not optimal and will lead to artifacts and discontinuities at the frame boundaries. In fact, the (in general) unconstrained modification of $X_\mu(\lambda)$ may lead to a complex-valued spectral representation $Y_\mu(\lambda)$ that could not be computed from any real-valued time domain signal through STFT analysis. In order to compute an artifact-free modified output signal *spectrogram consistency* should therefore be enforced. A spectral representation is called consistent when it can be computed from a real-valued signal via the STFT. Therefore, the objective is to approximate the modified STFT $Y_\mu(\lambda)$ with a consistent spectral representation. The least-squares error between the modified and its consistent spectral representation is minimized when the synthesis window in the OLA process is a normalized version of the analysis window, i.e. $w_S(k) = w_A(k)/\sum_{k=-\infty}^{\infty} \left( w_A(k) \right)^2$ [Griffin, Lim 1984]. This choice will also satisfy the perfect reconstruction condition in (3.43).

A common choice is to use the half-overlapping ($r = M/2$) and appropriately normalized square-root Hann window

$$w_A(k) = w_S(k) = \sqrt{0.5 \left( 1 - \cos\left( \frac{2\pi k}{M} \right) \right)} \quad k \in \{0, \dots, M-1\}. \tag{3.44}$$

When only a modified spectral magnitude is given, consistency may be achieved with iterative phase reconstruction algorithms [Griffin, Lim 1984]. In each iteration, the STFT is computed from the reconstructed time domain signal of the previous iteration. Then, the magnitude of the recomputed STFT is replaced by the given modified amplitude $|Y_\mu(\lambda)|$ and combined with the recomputed phase. Finally, the OLA resynthesis generates a new signal estimate in the time domain for the next iteration. Note that this algorithm also helps to reduce inconsistancies between the magnitude and the phase spectra and can be used to recover the phase when only the magnitude is given.

Improvements to this procedure have been published in several works: improved magnitude updates with faster convergence were demonstrated in [Perraudin et al. 2013], an improved implementation for real-time operation has been proposed in [Gnann and Spiertz, 2010], zero-padded STFT representations in [Wakabayashi, Ono 2019], or an DNN-based iterative approach in [Masuyama et al. 2021]. We note that in the above

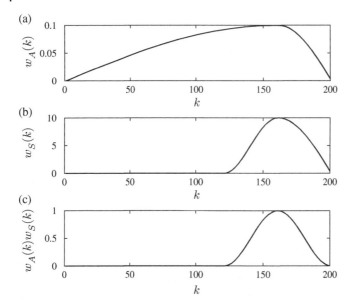

**Figure 3.21** Analysis window $w_A(k)$ of length 200 (a) and synthesis window $w_S(k)$ of length 80 (b) for low-latency spectral analysis–synthesis. The resulting product window is a Hann window of length 80 (c). The frame shift is $r = 40$, the resulting latency corresponds to 80 samples.

analysis–synthesis approach the overall algorithmic latency is governed by the length of the synthesis window. Thus, to reduce the latency we may combine a relatively long analysis window $w_A(k)$ with a shortened synthesis window $w_S(k)$ [Mauler, Martin 2007]. Both windows are designed to satisfy the perfect reconstruction constraint (3.43) leading to non symmetric analysis and synthesis windows. A typical example is shown in Figure 3.21 where the product of both windows corresponds to a Hann window.

## 3.8 Cepstral Analysis

The convolution of two discrete time sequences corresponds to a multiplication of their spectra. It turns out that the logarithm of the spectrum and its inverse transform are also very useful as the logarithm of the product corresponds to a sum of the logarithms. The inverse transform of the logarithm of the spectrum is called the *cepstrum*. Along with other useful vocabulary, the term cepstrum had been coined in [Bogert et al. 1963], where also "quefrency" has been introduced to denote the independent variable of the cepstrum.

The processing related to the computation of the cepstrum is a special case of the more general concept of homomorphic processing [Oppenheim, Schafer 1975]. Cepstral analysis is also useful in the context of stochastic signals, especially when the power spectrum of these signals is obtained from an autoregressive model. It has been used in speech coding for the quantization of the spectral envelope [Hagen 1994], and in speech recognition for the computation of spectral features [Davis, Mermelstein 1980], [Jankowski et al. 1995]. In this section, we briefly summarize some properties and applications of the cepstrum.

### 3.8.1 Complex Cepstrum

The *complex cepstrum* $x_{cc}(q)$ of a sequence $x(k)$ with $X(z) = \mathcal{Z}\{x(k)\}$ is defined as the sequence whose z-transform yields the logarithm of $X(z)$,

$$\sum_{q=-\infty}^{\infty} x_{cc}(q)z^{-q} = \ln(X(z)) = X_{cc}(z), \tag{3.45}$$

where we use the natural complex logarithm $\ln(X(z)) = \log_e(X(z))$. The variable $q$ is denoted as *quefrency*, [Bogert et al. 1963], [Oppenheim, Schafer 1975]. The logarithm of the z-transform may be written in terms of magnitude and phase of $X(z)$ as

$$X_{cc}(z) = \ln(X(z)) = \ln(|X(z)|\, e^{j\phi(z)}) = \ln(|X(z)|) + j\phi(z). \tag{3.46}$$

We further assume that $X_{cc}(z) = \ln(X(z))$ is a valid z-transform and thus converges in some region of the complex z-plane. For practical reasons we also require that $x(k)$ and $x_{cc}(q)$ are real-valued and stable sequences. Then, both $X(z)$ and $X_{cc}(z)$ converge on the unit circle and $\ln(|X(e^{j\Omega})|)$ and $\phi(\Omega)$ are even and odd functions of $\Omega$, respectively. In this case, we may obtain $x_{cc}(q)$ from an inverse FTDS

$$x_{cc}(q) = \frac{1}{2\pi} \int_{-\pi}^{\pi} \left[\ln(|X(e^{j\Omega})|) + j\phi(\Omega)\right] e^{j\Omega q}\, d\Omega. \tag{3.47}$$

Since singularities of $\ln(X(z))$ are found at the poles and the zeros of $X(z)$, a stable and causal sequence $x_{cc}(q)$ is only obtained when both poles and zeros are within the unit circle. As a consequence, $x(k)$ is a minimum-phase sequence if and only if its complex cepstrum is causal [Oppenheim et al. 1999].

The definition of the complex logarithm in (3.45) is not at all trivial as $X_{cc}(z)$ must be an analytic and hence continuous function in the ROC. Since the imaginary part of $X_{cc}(z)$ is identical to the phase of $X(z)$, we cannot substitute the principal value $-\pi < \arg\{X(z)\} \leq \pi$ for the phase in general. The complex logarithm must therefore be defined such that it is invertible and analytic in the ROC [Oppenheim, Schafer 1975].

A recursive relation (see 3.53) linking $x(k)$ and $x_{cc}(q)$ can be derived as follows. First we consider the derivative of $X_{cc}(z)$ in the z-domain

$$X'_{cc}(z) = \frac{dX_{cc}(z)}{dz} = \frac{1}{X(z)} \cdot \frac{dX(z)}{dz}, \tag{3.48}$$

i.e.,

$$X'_{cc}(z) = \frac{X'(z)}{X(z)}. \tag{3.49}$$

Both $X'(z)$ and $X'_{cc}(z)$ can be derived from the z-transforms of $x(k)$ and $x_{cc}(q)$, respectively

$$X'(z) = \frac{d}{dz}\left[\sum_{k=-\infty}^{\infty} x(k)z^{-k}\right] \qquad (3.50a)$$

$$= \sum_{k=-\infty}^{\infty}\left[-k \cdot x(k)z^{-(k+1)}\right], \qquad (3.50b)$$

i.e., we have

$$z X'(z) = -\sum_{k=-\infty}^{\infty} k \cdot x(k)z^{-k}, \qquad (3.51a)$$

$$z X'_{cc}(z) = -\sum_{q=-\infty}^{\infty} q \cdot x_{cc}(q)z^{-q}. \qquad (3.51b)$$

With (3.49) we find

$$z X'(z) = z X'_{cc}(z) \cdot X(z). \qquad (3.52)$$

The inverse z-transformation of both sides of (3.52) delivers

$$k \cdot x(k) = \sum_{q=-\infty}^{\infty} q \cdot x_{cc}(q) \cdot x(k - q), \qquad (3.53)$$

taking into account that the z-domain product in (3.52) corresponds to a convolution in (3.53).

Next we assume a causal and real-valued sequence $x(k)$ for $k \neq 0$ and divide both sides of (3.53) by $k$

$$x(k) = \sum_{q=0}^{\infty} \frac{q}{k} x_{cc}(q) x(k - q), \quad k \neq 0. \qquad (3.54)$$

When $x(k)$ is a minimum-phase sequence with $x(k) = 0$ for $k < 0$ and $X(z)$ is a rational function, we have $x_{cc}(q) = 0$ for $q < 0$. For a minimum-phase sequence, the complex cepstrum may thus be computed recursively

$$x_{cc}(q) = \begin{cases} 0 & q < 0 \\ \ln(x(0)) & q = 0 \\ \dfrac{x(q)}{x(0)} - \displaystyle\sum_{\ell=0}^{q-1} \dfrac{\ell}{q} x_{cc}(\ell)\dfrac{x(q-\ell)}{x(0)} & q > 0. \end{cases} \qquad (3.55)$$

The value of $x_{cc}(0) = \ln(x(0))$ can easily be proven by using the initial value theorem of the z-transform:

$$\lim_{z\to\infty} X(z) = \lim_{z\to\infty}\sum_{k=0}^{\infty} x(k)z^{-k} = x(0) \qquad (3.56a)$$

$$\lim_{z\to\infty} X_{cc}(z) = \lim_{z\to\infty} [\ln(X(z))] = \ln(x(0)). \qquad (3.56b)$$

As an example, we consider the cepstral analysis of an all-pole spectrum model

$$H(e^{j\Omega}) = \frac{1}{1 - A(e^{j\Omega})} \qquad (3.57)$$

as used in speech coding (see Chapter 8). The complex cepstrum is given by

$$H_{cc}(e^{j\Omega}) = \ln\left(H(e^{j\Omega})\right) \tag{3.58a}$$

$$= -\ln(1 - A(e^{j\Omega})) \tag{3.58b}$$

$$= -\ln\left(G(e^{j\Omega})\right), \tag{3.58c}$$

and we need to consider only the LP analysis filter

$$G(e^{j\Omega}) = 1 - A(e^{j\Omega})$$

$$= 1 - \sum_{k=1}^{n} a_k e^{-j\Omega k},$$

with the finite length impulse response

$$g(k) = 1, -a_1, -a_2, \ldots - a_n, 0, 0, \ldots \tag{3.59}$$

Thus, the time domain samples $g(k)$, respectively the sequence of LP coefficients $a_k$ can be converted into cepstral coefficients $g_{cc}(q)$ using the recursive relation (3.55) with $g(k = 0) = 1$

$$g_{cc}(q) = a_q + \sum_{\ell=1}^{q-1} \frac{\ell}{q} g_{cc}(\ell) a_{q-\ell} = a_q + \sum_{\ell=1}^{q-1} \frac{q-\ell}{q} g_{cc}(q-\ell) a_\ell, \quad q \geq 1, \tag{3.60}$$

with $g_{cc}(q = 0) = \ln(g(k = 0)) = \ln(1) = 0$ and $g_{cc}(q) = 0$ for $q < 0$. This recursion between LP coefficients $a_q$ and cepstral coefficients $g_{cc}(q)$ can be used in both directions. By definition, the summations in (3.60) yield zero for $q = 1$.

### 3.8.2 Real Cepstrum

Because of the difficulties in the definition and evaluation of the complex logarithm, the *real cepstrum*, or *cepstrum* for short, is preferred in many applications. Also, for minimum-phase signals the real cepstrum captures all information about the signal since for these signals the phase is determined by the magnitude spectrum [Oppenheim, Schafer 1975]. The real cepstrum is obtained as inverse FTDS of the real part of $X_{cc}(e^{j\Omega})$ as defined in (3.46)

$$c_x(q) = \frac{1}{2\pi} \int_{-\pi}^{\pi} \ln(|X(e^{j\Omega})|) e^{j\Omega q} \, d\Omega, \tag{3.61}$$

where we assume that all singularities of $\ln(|X(z)|)$ are within the unit circle. This implies that $x(k)$ as well as $x_{cc}(q)$ are causal, minimum-phase sequences. It therefore suffices to evaluate the real part of $X_{cc}(z)$, which is $\ln(|X(z)|)$. Note that

$$c_x(0) = \frac{1}{2\pi} \int_{-\pi}^{\pi} \ln\left(\left|X(e^{j\Omega})\right|\right) d\Omega, \tag{3.62}$$

evaluates for an all-pole signal model $X(e^{j\Omega}) = \sigma/(1 - A(e^{j\Omega}))$ to $c_x(0) = \ln(\sigma)$ [Markel, Gray 1976].

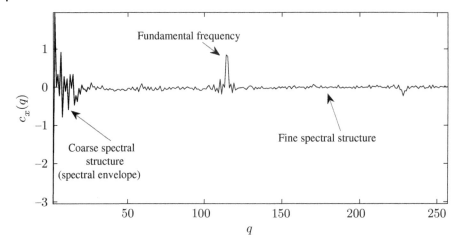

**Figure 3.22** The real cepstrum and its interpretation.

For minimum-phase sequences we have

$$
x_{cc}(q) = \begin{cases} 0 & q < 0 \\ c_x(q) & q = 0 \\ 2c_x(q) & q > 0. \end{cases} \tag{3.63}
$$

For real-valued $x(k)$ the log-magnitude spectrum $\ln(|X(e^{j\Omega})|)$ is a real and even function of $\Omega$. Since $\ln\left(|X(e^{j\Omega})|^2\right) = \ln\left(X(e^{j\Omega})\right) + \ln\left(X^*(e^{j\Omega})\right)$, the real cepstrum $c_x(q)$ is also even and may also be obtained as the conjugate-symmetric part of $x_{cc}(q)$,

$$
c_x(q) = \frac{x_{cc}(q) + x_{cc}^*(-q)}{2}. \tag{3.64}
$$

The interpretation of the real cepstrum is quite intuitive since it describes the Fourier components of the familiar log-magnitude spectrum. It clearly separates coarse spectral features (envelope) at low quefrencies, contributions related to glottal excitation visible as peaks, and the somewhat random spectral fine structure at higher quefrencies. Figure 3.22 depicts an exemplary real cepstrum of a voiced speech sound.

### 3.8.3 Applications of the Cepstrum

#### 3.8.3.1 Construction of Minimum-Phase Sequences
The above relations may be used to construct a minimum-phase sequence from a given non minimum-phase sequence $x(k)$. Any rational $z$-transform $X(z)$ of a stable sequence $x(k)$ may be decomposed into

$$
X(z) = X_{\min}(z)X_{AP}(z),
$$

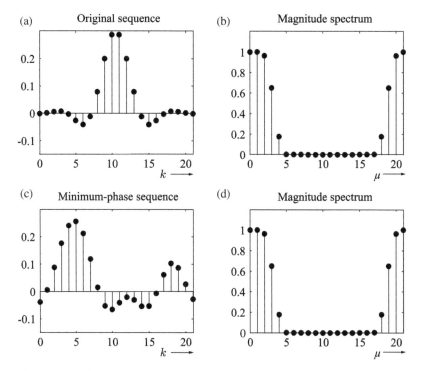

**Figure 3.23** Linear-phase sequence (a,b) and corresponding minimum-phase sequence (c,d). Both sequences have the same magnitude spectrum.

where $X_{min}(z)$ and $X_{AP}(z)$ correspond to a minimum-phase and an allpass sequence, respectively. In general, the ROC of $X(z)$ is an annular region in the $z$-plane, which contains the unit circle. The corresponding complex cepstrum is a two-sided sequence. With

$$\ln(X(z)) = \ln(X_{min}(z)) + \ln(X_{AP}(z)),$$

we obtain the minimum-phase sequence by extracting the causal part of the complex cepstrum. An example is shown in Figure 3.23.

### 3.8.3.2 Deconvolution by Cepstral Mean Subtraction

When a speech signal is recorded in a reverberant acoustic environment, it may be written as the convolution of the original, unreverberated speech signal $s(k)$ with the impulse response $h(k)$ of this environment,

$$x(k) = h(k) * s(k) \quad \Leftrightarrow \quad X(e^{j\Omega}) = H(e^{j\Omega}) S(e^{j\Omega}).$$

A similar situation occurs when a speech signal is recorded via a telephone line. In this case, network echoes contribute to what is called *convolutive noise* in the speech-recognition community. The use of the real cepstrum obviously leads to

$$\ln\left(|X(e^{j\Omega})|\right) = \ln\left(|H(e^{j\Omega})|\right) + \ln\left(|S(e^{j\Omega})|\right) \Leftrightarrow c_x(q) = c_h(q) + c_s(q)$$

or, applying the DFT, to the short-term relation

$$\ln\left(|X_\mu(\lambda)|\right) \approx \ln\left(|H_\mu(\lambda)|\right) + \ln\left(|S_\mu(\lambda)|\right), \tag{3.65}$$

provided that circular convolutive effects are negligibly small. $\mu$ denotes the frequency bin and $\lambda$ the frame index, respectively. The deconvolution might be now achieved by subtracting $\ln\left(|H_\mu(\lambda)|\right)$ from $\ln\left(|X_\mu(\lambda)|\right)$ [Oppenheim, Schafer 1968], [Atal 1974]. When the impulse response is constant over time, an estimate of $\ln\left(|H_\mu(\lambda)|\right)$ is given by the expectation of $\ln\left(|X_\mu(\lambda)|\right) - \ln\left(|S_\mu(\lambda)|\right)$. In a practical realization of this method, we approximate the expectation of $\ln\left(|X_\mu(\lambda)|\right)$ by a time average and compute $\mathrm{E}\left\{\ln\left(|S_\mu(\lambda)|\right)\right\}$ from undisturbed speech data. We then obtain

$$\ln\left(|\widehat{S_\mu(\lambda)}|\right) \approx \ln\left(|X_\mu(\lambda)|\right) - \left(\frac{1}{N}\sum_{i=\lambda-N+1}^{\lambda}\ln\left(|X_\mu(i)|\right) - \mathrm{E}\left\{\ln\left(|S_\mu(\lambda)|\right)\right\}\right).$$

Note that the cepstral coefficients of zero mean Gaussian signals are not zero mean [Stockham et al. 1975], [Ephraim, Rahim 1999], [Gerkmann, Martin 2009]. This technique and its variations [Stockham et al. 1975], [Acero, Huang 1995], [Rahim et al. 1996] are frequently used in speech recognition. They are less successful in mitigating acoustic echoes, as the acoustic echo path is typically of high order and not stationary.

### 3.8.3.3  Computation of the Spectral Distortion Measure

When we are given two magnitude squared spectra $|H(e^{j\Omega})|^2$ and $|\hat{H}(e^{j\Omega})|^2$, where the latter may be an approximation of the former, the *total log spectral distance* (SD) of the two spectra is defined as

$$SD = \sqrt{\frac{1}{2\pi}\int_{-\pi}^{\pi}\left[20\log_{10}|H(e^{j\Omega})| - 20\log_{10}|\hat{H}(e^{j\Omega})|\right]^2 d\Omega}. \tag{3.66}$$

This distance (or distortion) measure is used, for instance, for the quantization of the spectral envelopes of speech signals [Hagen 1994]. The distortion measure may be computed using cepstral coefficients [Markel, Gray 1976],

$$SD = 20\log_{10}(e)\sqrt{\sum_{q=-\infty}^{\infty}\left[c_h(q) - \hat{c}_h(q)\right]^2}.$$

To prove this relation we use $\log_{10}(x) = \ln(x)\log_{10}(e)$ and note that

$$SD = \sqrt{\frac{1}{2\pi}\int_{-\pi}^{\pi}\left[20\log_{10}|H(e^{j\Omega})| - 20\log_{10}|\hat{H}(e^{j\Omega})|\right]^2 d\Omega}$$

$$= 20\log_{10}(e)\sqrt{\frac{1}{2\pi}\int_{-\pi}^{\pi}\left[\sum_{q=-\infty}^{\infty}c_h(q)e^{-jq\Omega} - \sum_{q=-\infty}^{\infty}\hat{c}_h(q)e^{-jq\Omega}\right]^2 d\Omega}.$$

Parseval's theorem (see Table 3.2)

$$
\frac{1}{2\pi} \int_{-\pi}^{\pi} \left( \sum_{q=-\infty}^{\infty} \left( c_h(q) - \hat{c}_h(q) \right) e^{-jq\Omega} \right) \left( \sum_{q=-\infty}^{\infty} \left( c_h(q) - \hat{c}_h(q) \right) e^{-jq\Omega} \right)^* d\Omega
$$

$$
= \sum_{q=-\infty}^{\infty} \left( c_h(q) - \hat{c}_h(q) \right) \left( c_h(q) - \hat{c}_h(q) \right)^*
$$

yields

$$
SD = 20 \log_{10}(e) \sqrt{ \frac{1}{2\pi} \int_{-\pi}^{\pi} \left| \sum_{q=-\infty}^{\infty} \left( c_h(q) - \hat{c}_h(q) \right) e^{-jq\Omega} \right|^2 d\Omega }
$$

$$
= 20 \log_{10}(e) \sqrt{ \sum_{q=-\infty}^{\infty} \left( c_h(q) - \hat{c}_h(q) \right)^2 } . \tag{3.67}
$$

For $c_h(0) = 0$ and with $c_h(-q) = c_h(q)$, (3.67) may be simplified,

$$
SD = 20 \log_{10}(e) \sqrt{2} \sqrt{ \sum_{q=1}^{\infty} \left( c_h(q) - \hat{c}_h(q) \right)^2 }. \tag{3.68}
$$

For minimum-phase signals with

$$
\int_{-\pi}^{\pi} \ln |H(e^{j\Omega})| d\Omega = \int_{-\pi}^{\pi} \ln |\hat{H}(e^{j\Omega})| d\Omega = 0, \tag{3.69}
$$

we may also write

$$
SD = 10 \log_{10}(e) \sqrt{2} \sqrt{ \sum_{q=1}^{\infty} \left( h_{cc}(q) - \hat{h}_{cc}(q) \right)^2 }, \tag{3.70}
$$

where $h_{cc}(q)$ and $\hat{h}_{cc}(q)$ are the corresponding complex cepstra.

This spectral distortion measure [Quackenbush et al. 1988] is widely used to assess the performance of vector quantizers in speech coding [Kleijn, Paliwal 1995] as well as the quality of speech-enhancement algorithms, e.g., [Gustafsson et al. 2002] and [Cohen 2004].

### 3.8.3.4  Fundamental Frequency Estimation

Due to the separation of envelope-related and excitation-related speech features, the cepstrum is also quite useful for the estimation of the fundamental frequency [Noll 1967]. Figure 3.24 depicts an example of a waveform, the corresponding log-magnitude spectrum, and the cepstrum. The peak related to the fundamental frequency at around $q = q_0 = 43$ is clearly observed, as well as the less prominent peak of its first *rahmonic* at $2q_0$.

When the DFT/IDFT is used to compute the real cepstrum,

$$
c_x(q) = \frac{1}{M} \sum_{\mu=0}^{M-1} \ln \left( |X_\mu| \right) e^{j \frac{2\pi \mu q}{M}}, q = 0, \ldots, M/2 - 1, \tag{3.71}
$$

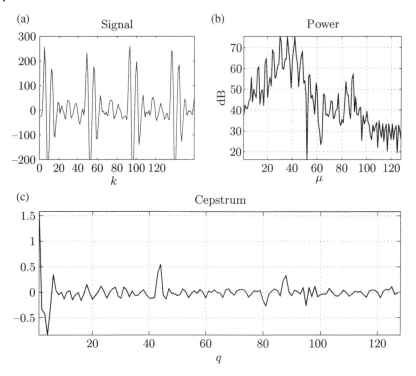

**Figure 3.24** A segment of voiced speech (a), its DFT-based log-amplitude spectrum in dB (b), and the corresponding real cepstrum (c). The sampling rate is $f_s = 8$ kHz. At about $q = 43$, we observe a peak in the cepstrum that results from the inverse transform of the harmonic series describing the glottal excitation.

we note that according to (3.71) a constructive summation of DFT bins with index $\mu$ is achieved for values $q$ that obey to $\mu q_\ell = (\ell + 1)M$ for $\ell = 0,1,2,3,\ldots$, since the corresponding complex exponential functions take the value

$$e^{j\frac{2\pi\mu q}{M}} = e^{j\frac{2\pi(\ell+1)M}{M}} = 1. \tag{3.72}$$

In terms of frequencies we then have

$$\frac{\mu f_s}{M}q_\ell = (\ell+1)\frac{Mf_s}{M} = (\ell+1)f_s . \tag{3.73}$$

For $\ell = 0$ we obtain with the corresponding fundamental frequency at DFT bin $\mu_0$, i.e. $f_0 = \mu_0 f_s/M$ the simple relation $f_0 q_0 = f_s$, and the corresponding cepstral bin as

$$q_0 = \text{round}\left(\frac{f_s}{f_0}\right). \tag{3.74}$$

Multiples of $f_0$ are computed for $\ell = 1,2,3,\ldots$ and constitute the rahmonics $q_\ell = (\ell + 1)q_0$ in the cepstral domain. Surprisingly, the mapping from $f_0$ to $q_0$ does not depend on the DFT length $M$. For the example in Figure 3.24 we find, with $q_0 = 43$ and $f_s = 8$ kHz, the fundamental frequency $f_0 \approx 186$ Hz.

Finally, we note that the cepstrum may be computed for a succession of signal frames and displayed in a similar fashion as the spectrogram. This yields the *cepstrogram*, an example

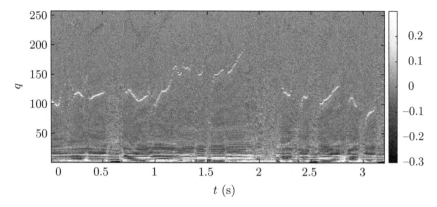

**Figure 3.25** Cepstrogram of a male speech sample. The evolution of the cepstral peak corresponding to fundamental frequency in the range $80 \leq q \leq 200$ can clearly be observed.

of which is shown in Figure 3.25. The temporal evolution of $q_0$ and thus the fundamental frequency can be clearly observed.

## References

Acero, A.; Huang, X. (1995). Augmented Cepstral Normalization for Robust Speech Recognition, *IEEE International Workshop on Automatic Speech Recognition and Understanding*, pp. 147–148.

Allen, J.; Berkley, D.; Blauert, J. (1977). Multimicrophone Signal-Processing Technique to Remove Room Reverberation from Speech Signals, *Journal of the Acoustical Society of America*, vol. 62, no. 4, pp. 912–915.

Atal, B. S. (1974). Effectiveness of Linear Prediction Characteristics of the Speech Wave for Automatic Speaker Identification and Verification, *Journal of the Acoustical Society of America*, vol. 55, no. 6, pp. 1304–1312.

Bogert, B. P.; Healy, M. J. R.; Tukey, J. W. (1963). The Quefrency Alanysis of Time Series for Echoes: Cepstrum, Pseudo-Autocovariance, Cross-Cepstrum, and Saphe Cracking, *Proceedings of the Symposium Time Series Analysis*, pp. 209–243.

Cappé, O. (1995). Evaluation of Short-Time Spectral Attenuation Techniques for the Restoration of Musical Recordings, *IEEE Transactions on Speech and Audio Processing*, vol. 3, no. 1, pp. 84–93.

Churchill, R. V.; Brown, J. W. (1990). *Introduction to Complex Variables and Applications*, 5th edn, MacGraw-Hill, New York.

Cochran, W. T.; Cooley, J. W.; Favin, D. L.; Helms, H. D.; Kaenel, R. A.; Lang, W. W.; Maling, G. C.; Nelson, D. E.; Rader, C. M.; Welch, P. D. (1967). What is the Fast Fourier Transform?, *IEEE Transactions on Audio and Electroacoustics*, vol. AU-15, pp. 45–55.

Cohen, I. (2004). Speech Enhancement Using a Noncausal A Priori SNR Estimator, *Signal Processing Letters*, vol. 11, no. 9, pp. 725–728.

Cooley, J. (1992). How the FFT Gained Acceptance, *IEEE Signal Processing Magazine*, vol. 9, pp. 10–13.

Cooley, J.; Tukey, J. (1965). An Algorithm for the Machine Calculation of Complex Fourier Series, *Mathematics of Computation*, vol. 19, p. 297.

Davis, S.; Mermelstein, P. (1980). Comparison of Parametric Representations for Monosyllabic Word Recognition in Continuously Spoken Sentences, *IEEE Transactions on Acoustics, Speech and Signal Processing*, vol. 28, no. 4, pp. 357–366.

Ephraim, Y.; Rahim, M. (1999). On Second Order Statistics and Linear Estimation of Cepstral Coefficients, *IEEE Transactions on Speech and Audio Processing*, vol. 7, no. 2, pp. 162–176.

Gerkmann, T.; Martin, R. (2009). On the Statistics of Spectral Amplitudes After Variance Reduction by Temporal Cepstrum Smoothing and Cepstral Nulling, *IEEE Transactions on Signal Processing*, vol. 57, no. 11, pp. 4165–4174.

Griffin, D. W.; Lim, J. S. (1984). Signal Estimation from Modified Short-Time Fourier Transform, *IEEE Transactions on Acoustics, Speech and Signal Processing*, vol. 32, no. 2, pp. 236–243.

Gustafsson, S.; Martin, R.; Jax, P.; Vary, P. (2002). A Psychoacoustic Approach to Combined Acoustic Echo Cancellation and Noise Reduction, *IEEE Transactions on Speech and Audio Processing*, vol. 10, no. 5, pp. 245–256.

Hagen, R. (1994). Spectral Quantization of Cepstral Coefficients, *Proceedings of the IEEE International Conference on Acoustics, Speech, and Signal Processing (ICASSP)*, Adelaide, pp. 509–512.

Jankowski, C.; Vo, H.-D.; Lippmann, R. (1995). A Comparison of Signal Processing Front Ends for Automatic Word Recognition, *IEEE Transactions on Speech and Audio Processing*, vol. 3, no. 4, pp. 286–293.

Kleijn, W. B.; Paliwal, K. K. (eds.) (1995). *Speech Coding and Synthesis*, Elsevier, Amsterdam.

Markel, J. D.; Gray, A. H. (1976). *Linear Prediction of Speech*, Springer-Verlag, Berlin, Heidelberg, New York.

Masuyama, Y.; Yatabe, K.; Koizumi, Y.; Oikawa, Y.; Harada, N. (2021). Deep Griffin–Lim Iteration: Trainable Iterative Phase Reconstruction Using Neural Network, *IEEE Journal of Selected Topics in Signal Processing*, vol. 15, no. 1, pp. 37–50.

Mauler, D.; Martin, R. (2007). A Low Delay, Variable Resolution, Perfect Reconstruction Spectral Analysis-Synthesis System for Speech Enhancement, *2007 15th European Signal Processing Conference*, pp. 222–226.

Noll, A. M. (1967). Cepstrum pitch determination, *The Journal of the Acoustical Society of America*, vol. 41, no. 2, pp. 293–309.

Oppenheim, A.; Schafer, R. (1968). Homomorphic Analysis of Speech, *IEEE Transactions on Audio and Electroacoustics*, vol. AU-16, pp. 221–226.

Oppenheim, A.; Schafer, R. (1975). *Digital Signal Processing*, Prentice Hall, Englewood Cliffs, New Jersey.

Oppenheim, A.; Willsky, A.; Nawab, H. (1996). *Signals and Systems*, Prentice Hall, Upper Saddle River, New Jersey.

Oppenheim, A.; Schafer, R.; Buck, J. (1999). *Discrete-Time Signal Processing*, 2nd edn, Prentice Hall, Englewood Cliffs, New Jersey.

Perraudin, N.; Balazs, P.; Søndergaard, P. L. (2013). A Fast Griffin–Lim Algorithm, *2013 IEEE Workshop on Applications of Signal Processing to Audio and Acoustics*, pp. 1–4.

Proakis, J. G.; Manolakis, D. G. (1992). *Digital Signal Processing: Principles, Algorithms and Applications*, 2nd edn, Macmillan, New York.

Quackenbush, S. R.; Barnwell III, T. P.; Clements, M. A. (1988). *Objective Measures of Speech Quality*, Prentice Hall, Englewood Cliffs, New Jersey.

Rahim, M.; Juang, B.-H.; Chou, W.; Buhrke, E. (1996). Signal Condition Techniques for Robust Speech Recognition, *Signal Processing Letters*, vol. 3, no. 4, pp. 107–109.

Stockham, T.; Cannon, T.; Ingebretsen, R. (1975). Blind Deconvolution Through Digital Signal Processing, *Proceedings of the IEEE*, vol. 63, no. 4, pp. 678–692.

Wakabayashi, Y.; Ono, N. (2019). Griffin–Lim Phase Reconstruction Using Short-Time Fourier Transform with Zero-Padded Frame Analysis, *2019 Asia-Pacific Signal and Information Processing Association Annual Summit and Conference (APSIPA ASC)*, pp. 1863–1867.

# 4

# Filter Banks for Spectral Analysis and Synthesis

## 4.1 Spectral Analysis Using Narrowband Filters

In contrast to the transform-based methods of the previous chapter we will now discuss spectral analysis using narrowband filters. Two equivalent measurement procedures are illustrated in Figure 4.1. We are interested in the temporal evolution of a signal $x(k)$ at a certain frequency $\Omega_\mu$. The first approach (see Figure 4.1a) consists of three steps: narrow bandpass (BP) filtering, spectral shifting by $\Omega_\mu$, and sampling rate decimation by $r$.

The intermediate BP signal $v_\mu(k)$, which is centered at $\Omega_\mu$, is shifted to the baseband by complex demodulation, i.e., by multiplication with $e^{-j\Omega_\mu k}$. The resulting complex-valued baseband signal $\bar{x}_\mu(k)$ is called the *sub-band signal*. This signal gives information about the signal $x(k)$ at frequency $\Omega_\mu$ and time instant $k$. The resolution in time and frequency is determined by the impulse response $h_\mu^{BP}(k)$ and the frequency response $H_\mu^{BP}(e^{j\Omega})$ of the bandpass, respectively.

The same result can be obtained by applying first modulation and then feeding the modulated signal to an equivalent lowpass filter with frequency response $H$, as shown in Figure 4.1b. The intermediate output signals $\bar{x}_\mu(k)$ of both systems are identical if the frequency responses of the BP and the lowpass filters satisfy the following relation:

$$H_\mu^{BP}(e^{j\Omega}) = H\left(e^{j(\Omega-\Omega_\mu)}\right). \tag{4.1a}$$

According to the modulation theorem of the Fourier transform the corresponding time domain impulse responses are related by

$$h_\mu^{BP}(k) = h(k)\, e^{j\Omega_\mu k}. \tag{4.1b}$$

The signal $\bar{x}_\mu(k)$ in Figure 4.1a can be formulated explicitly as

$$\bar{x}_\mu(k) = e^{-j\Omega_\mu k} \sum_{\kappa=-\infty}^{\infty} x(\kappa)\, h_\mu^{BP}(k-\kappa) \tag{4.2a}$$

$$= e^{-j\Omega_\mu k} \sum_{\kappa=-\infty}^{\infty} x(\kappa)\, h(k-\kappa)\, e^{j\Omega_\mu(k-\kappa)} \tag{4.2b}$$

$$= \sum_{\kappa=-\infty}^{\infty} x(\kappa)\, e^{-j\Omega_\mu \kappa}\, h(k-\kappa) \tag{4.2c}$$

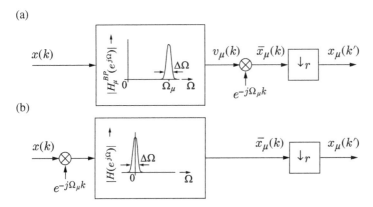

**Figure 4.1** Spectral analysis by (a) Bandpass filtering and subsequent complex demodulation; (b) Complex demodulation and subsequent lowpass filtering.

where (4.2c) describes the lowpass approach of Figure 4.1-b. If the complex samples $\overline{x}_\mu(k)$ are calculated at frequencies

$$\Omega_\mu = \Delta\Omega \cdot \mu, \quad \mu = 0, 1, 2, \ldots, M - 1$$

and if the whole frequency range $0 \leq \Omega \leq 2\pi$ is covered, then each set of $M$ samples $\overline{x}_\mu(k)$, $\mu = 0, 1, 2, \ldots, M - 1$, at any time instant $k$ is called the *short-term spectrum*.

The complex samples $\overline{x}_\mu(k)$ may be represented either by their real and imaginary parts or by (short-term) magnitude and phase

$$\overline{x}_\mu(k) = \overline{x}_{Re,\mu}(k) + j \cdot \overline{x}_{Im,\mu}(k) \tag{4.3a}$$
$$= |\overline{x}_\mu(k)| \cdot e^{j\varphi_\mu(k)}. \tag{4.3b}$$

If the filter is causal and stable, the impulse response $h(k)$ is zero for $k < 0$, and $|h(k)|$ will decay with time $k$. The time dependency of the short-term spectrum is due to the position $k$ of the time-reversed impulse response $h(k - \kappa)$, which acts as a sliding window. As shown examplarily in Figure 4.2, the impulse response $h(k - \kappa)$ weights the most recent signal samples $x(\kappa)$ up to the observation time instant $k$.

Because of the decay of $|h(k)|$, the older parts of the signal have less influence than the most recent samples. Furthermore, the sampling rate of the sub-band signals $\overline{x}_\mu(k)$ can be decimated due to the narrow bandwidth $\Delta\Omega$ of the frequency selective filter $H$. By decimating the sampling rate by $r$, we obtain the output sequence

$$x_\mu(k') = \overline{x}_\mu(k = k' \cdot r). \tag{4.4}$$

The bandwidth of the signals $\overline{x}_\mu(k)$ is limited to the passband width $\Delta\Omega$ of the filter. Usually $\Delta\Omega$ is markedly smaller than $\pi$. If the stopband attenuation of the selective filter is sufficiently high, the output signal $\overline{x}_\mu(k)$ can be calculated at the reduced rate

$$f_s' = f_s \frac{\Delta\Omega}{2\pi}. \tag{4.5}$$

Due to the computational complexity of any further processing of $\overline{x}_\mu(k)$, the sampling rate should be reduced – according to the sampling theorem – as much as possible by an integer

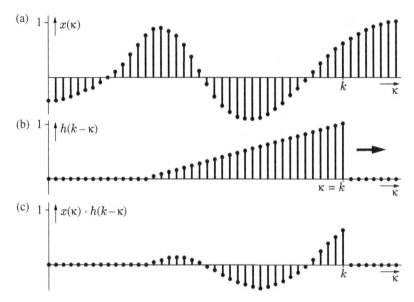

**Figure 4.2** (a) Signal $x(\kappa)$. (b) Impulse response $h(k - \kappa)$. (c) Weighted signal history $x(\kappa) \cdot h(k - \kappa)$.

factor $r > 1$ with

$$r \le r_{max} = \frac{2\pi}{\Delta\Omega} \; ; \; r \in \mathbb{N}. \tag{4.6}$$

In the extreme case with $r = r_{max}$, which is called *critical decimation*, aliasing can only be avoided by using ideal lowpass or BP filters. With non ideal filters the decimation factor $r$ has to be chosen somewhat smaller.

For ease of analytical description a special version $\tilde{x}_\mu(k)$ of the decimated sequence is introduced with the original sampling rate $f_s$ but with $r - 1$ zeros filled in between the decimated samples $x_\mu(k')$. The relations between these different sequences are illustrated in Figure 4.3.

The following notations are used throughout this chapter for decimated and upsampled versions of any sequence:

- $k$, $\kappa$: time indices at original sampling rate $f_s$
- $k'$, $\kappa'$: time indices at reduced sampling rate $f_s' = f_s/r$
- decimated sequence at sampling rate $f_s'$ *without* intermediate zero samples

$$x_\mu(k') = \bar{x}_\mu(k = k' \cdot r) \tag{4.7}$$

- upsampled sequence at sampling rate $f_s$ *with* intermediate zero samples

$$\tilde{x}_\mu(k) = \begin{cases} \bar{x}_\mu(k) & \text{if } k = k' \cdot r, \quad k' = 0, \pm 1, \pm 2, \dots \\ 0 & \text{else.} \end{cases} \tag{4.8}$$

The process of sampling rate reduction by $r$ with subsequent upsampling by $r$ (zero filling) can be described analytically by multiplying $\bar{x}_\mu(k)$ with a periodic pulse sequence $p^{(r)}(k)$

(a)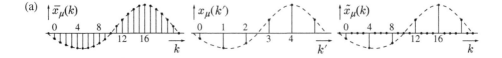

(b)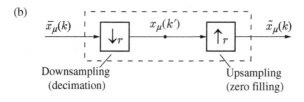

Downsampling (decimation)

Upsampling (zero filling)

(c)

**Figure 4.3** Decimation and zero filling for $r = 4$ with $p^{(r)}(k) = \sum\limits_{k'=-\infty}^{\infty} \delta(k - k'r)$. (a) Signals. (b) Block diagram. (c) Equivalent model.

as shown in Figure 4.3,

$$\tilde{x}_\mu(k) = \bar{x}_\mu(k) \cdot p^{(r)}(k) \tag{4.9}$$

$$p^{(r)}(k) = \sum_{k'=-\infty}^{\infty} \delta(k - k'r). \tag{4.10}$$

The sequence $p^{(r)}(k)$ can be expressed as:

$$p^{(r)}(k) = \frac{1}{r} \sum_{i=0}^{r-1} e^{jki\frac{2\pi}{r}}, \tag{4.11}$$

as the complex exponential is periodic in $i$. According to the modulation theorem, we get

$$\tilde{X}_\mu(e^{j\Omega}) = \frac{1}{r} \sum_{i=0}^{r-1} \bar{X}_\mu \left( e^{j(\Omega - \frac{2\pi}{r}i)} \right). \tag{4.12}$$

An example with $r = 4$ is illustrated in Figure 4.4.

In (4.8) the decimation grid is aligned to the origin of the time axis, i.e., with $k = 0$. In some cases it will be necessary that upsampled sequences have a constant time shift by $k_0$ so that $k = 0$ is not a decimation instant. In any case, upsampled sequences with or without time shift but with intermediate zero samples will be marked by "~".

Note that a displacement of the decimation grid by $k_0$ samples, which will be needed in the context of *polyphase network filter banks* (see Section 4.2), can be described as follows:

$$p^{(r)}(k - k_0) = \frac{1}{r} \sum_{i=0}^{r-1} e^{j\frac{2\pi}{r}(k-k_0)i} \tag{4.13a}$$

$$= \begin{cases} 1 & \text{if } k = k' \cdot r + k_0, \quad k' = 0, \pm 1, \pm 2, \ldots \\ 0 & \text{else.} \end{cases} \tag{4.13b}$$

(a)

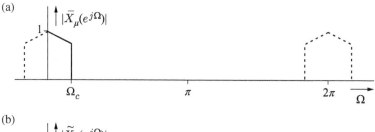

(b)

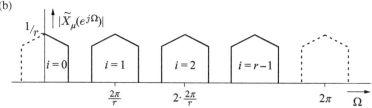

**Figure 4.4** Spectra before and after decimation (with zero filling) for $r = 4$.

As we have kept the zero samples in the time domain, the spectrum is periodic with $2\pi/r$. If these zeros are deleted, we have to consider the normalized frequency $\Omega' = r \cdot \Omega$ (period $2\pi$) corresponding to the reduced sampling rate $f_s' = f_s/r$.

### 4.1.1 Short-Term Spectral Analyzer

The behavior of the system of Figure 4.1 can be described in the frequency domain by considering the Fourier transforms of the signals $x(k), \bar{x}_\mu(k)$, and $\tilde{x}_\mu(k)$. For reasons of simplicity, we assume an ideal BP with center frequency $\Omega_\mu$. The spectral relations are explained exemplarily by Figure 4.5. In this example, a passband width of

$$\Delta\Omega = \frac{2\pi}{16}$$

and a center frequency of

$$\Omega_\mu = \Omega_3 = \frac{2\pi}{16} \cdot 3$$

are considered; $\Omega_\mu$ is an integer multiple of $2\pi/r$. Then the frequency shift of the BP signal $v_\mu(k)$ by $\Omega_\mu$ is not explicitly necessary, as the decimation process of (4.12) implicitly produces the required component in the baseband. This effect can easily be explained in the time domain, if we take into consideration that the complex exponential with frequency

$$\Omega_\mu = \frac{2\pi}{M}\mu$$

is periodic with length $M/\mu$, where $M$ is an integer multiple of $r$ according to

$$M = m \cdot r; \quad m \in \mathbb{N}.$$

Then the decimation process delivers the samples (see also Figure 4.1a)

$$x_\mu(k') = \bar{x}_\mu(k = k'r) = v_\mu(k'r)\, e^{-j\frac{2\pi}{m \cdot r}\mu \cdot k' \cdot r} \tag{4.14a}$$

$$= v_\mu(k'r)\, e^{-j\frac{2\pi}{m}\mu \cdot k'}. \tag{4.14b}$$

(a)

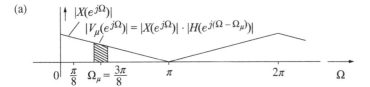

(b)

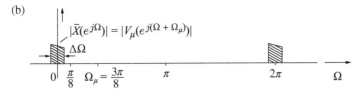

(c)

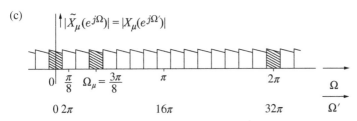

**Figure 4.5** Spectral relations of the short-term spectrum analyzer of Figure 4.1a. (a) Selection of a narrow band by BP filtering. (b) Frequency shift of the filter output by complex demodulation with $e^{-j\Omega_\mu k}$. (c) Periodic spectrum after critical decimation by $r = M = 16$. The normalized frequency axis $f'_s = \frac{f_s}{r} = \frac{f_s}{16}$ is denoted by $\Omega' = r\Omega$. Source: Adapted from [Vary, Wackersreuther 1983].

Two cases are of special interest:

**(a) Critical Decimation**

$$r = M, \qquad \text{i.e., } m = 1.$$

Equation (4.14b) results in

$$x_\mu(k') = v_\mu(k'r).$$

In this case, we can omit the demodulation process in Figure 4.1a, i.e., the multiplication by $e^{-j\Omega_\mu k}$, and apply the decimation process immediately to $v_\mu(k)$. This procedure is called *integer-band sampling*.

**(b) Half-Critical Decimation**

Due to non ideal filter characteristics a decimation factor of

$$r = M/2, \qquad \text{i.e., } m = 2,$$

is often chosen and (4.14b) results in

$$x_\mu(k') = v_\mu(k'r)\, e^{-j\pi\mu \cdot k'} = v_\mu(k'r) \cdot (-1)^{\mu \cdot k'}. \tag{4.15}$$

For channels with even-frequency index $\mu$ we get the same result as before. Explicit demodulation (frequency shift) of $v_\mu(k)$ is not necessary, the decimation process can be applied to $v_\mu(k)$. However, if the channel index $\mu$ is not even, the decimated samples $v_\mu(k'r)$ have to be multiplied by $(-1)^{k'}$ to obtain the same result as in the case

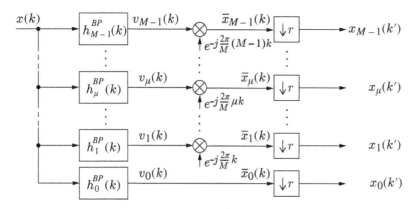

**Figure 4.6**  Short-term spectral analyzer with $M$ parallel channels and decimation by $r$.

with explicit demodulation. At the reduced sampling rate $f'_s = f_s/r$ this corresponds to a frequency shift by $\pi$, and the lowpass frequency spectra of channels having odd indices are mirrored. This should be taken into consideration if we process the decimated samples $v_\mu(k'r)$ without the correction (4.15).

So far we have discussed the measurement of a single component $x_\mu(k')$ (fixed $\mu$) of the short-term spectrum. If we are interested in the complete spectrum we need $M$ parallel filters with different center frequencies $\Omega_\mu = 2\pi\mu/M$, $\mu = 0, 1, \dots, M-1$, to achieve a uniform spectral resolution with

$$\Delta\Omega = \frac{2\pi}{M}.$$

According to (4.1), the impulse responses of these BP filters are modulated versions of the lowpass impulse response $h(k)$, which is called the *prototype impulse response*. The block diagram of the complete short-term spectrum analyzer (*analysis filter bank*) with $M$ BP impulse responses

$$h_\mu^{BP}(k) = h(k)\, e^{j\frac{2\pi}{M}\mu k}, \quad \mu = 0, 1, \dots, M-1 \tag{4.16}$$

is shown in Figure 4.6. In what follows, this block diagram will serve as a *reference model*, although this implementation is obviously sub-optimal with respect to the computational effort:

- most of the filter output samples $v_\mu(k)$ are discarded by decimation
- the impulse responses $h_\mu^{BP}$ and $h_{M-\mu}^{BP}$ are complex conjugates in pairs.

If the BPs are FIR (*finite impulse response*) filters, the computational effort can significantly be reduced by calculating output samples only at the decimated rate. Furthermore, if the input signal $x(k)$ is real-valued, the output samples $x_\mu(k')$ and $x_{M-\mu}(k')$ are complex conjugates of each other. We only need to calculate the samples for $\mu = 0, 1, \dots, M/2$. We will see later on that the computational complexity can be further reduced by using the FFT algorithm in combination with any FIR or IIR (*infinite impulse response*) prototype impulse response $h(k)$ (see Section 4.2).

Nevertheless, this reference system is well suited for studying the filter design issues.

### 4.1.2 Prototype Filter Design for the Analysis Filter Bank

First of all we have to specify the desired spectral resolution, i.e., the number $M$ of channels, the passband width $\Delta\Omega$, and the stopband attenuation. We can use any filter design method such as [Dehner 1979], [Parks et al. 1979], [Parks, Burrus 1987], and [Schüssler 1994], as available in MATLAB® to approximate the desired frequency response $H(z = e^{j\Omega})$ according to an error criterion (e.g., "minimum mean square," "min-max"-, "equi-ripple"- or "Chebyshev"-behavior).

However, the overall frequency response of the filter bank has to be taken into account as a second design criterion.

A reasonable design constraint is that the overall magnitude response should be flat, or that in case of a linear-phase prototype filter with impulse response $h(k)$ the overall impulse response $h_A(k)$ of the analysis filter bank should be a mere delay of $k_0$ samples:

$$h_A(k) = \sum_{\mu=0}^{M-1} h_\mu^{BP}(k) \overset{!}{=} \delta(k - k_0). \tag{4.17a}$$

By inserting (4.16) we get

$$h_A(k) = \sum_{\mu=0}^{M-1} h_\mu^{BP}(k) \tag{4.17b}$$

$$= \sum_{\mu=0}^{M-1} h(k)\, e^{j\frac{2\pi}{M}\mu k} \tag{4.17c}$$

$$= h(k) \cdot \sum_{\mu=0}^{M-1} e^{j\frac{2\pi}{M}\mu k} \tag{4.17d}$$

$$= h(k) \cdot M \cdot p^{(M)}(k) \tag{4.17e}$$

with

$$p^{(M)}(k) = \frac{1}{M} \sum_{\mu=0}^{M-1} e^{j\frac{2\pi}{M}\mu k}$$

$$= \begin{cases} 1 & \text{if } k = \lambda M,\ \lambda = 0, \pm 1, \pm 2, \ldots \\ 0 & \text{else.} \end{cases} \tag{4.17f}$$

As the BP responses are modulated versions of the prototype impulse response $h(k)$, the effective overall impulse response $h_A(k)$ has non zero samples only at $k = \lambda M$. Therefore, an ideal overall response can be obtained if the prototype lowpass filter with $k_0 = \lambda_0 \cdot M$ satisfies the condition

$$h(\lambda M) = \begin{cases} 1/M & \lambda = \lambda_0 \\ 0 & \lambda \neq \lambda_0. \end{cases} \tag{4.18}$$

These filters are also known as $M$-th band filters. The prototype impulse response has equidistant zeros and a non zero sample at time instant $k_0 = \lambda_0 \cdot M$. The samples $h(k)$ with

$k \neq \lambda \cdot M$ have no direct influence on the effective overall frequency response. They can be designed to optimize the frequency selectivity of the filter bank.

The design criterion (4.18) can easily be met if we use the "modified Fourier approximation" method, e.g., [Mitra 1998]. The prototype impulse response of odd length $L$, e.g., $L = 4 \cdot M + 1$, is obtained by multiplying the non causal impulse response $h_{LP}(k)$ of the ideal lowpass filter with cutoff frequency $\Omega_c = 2\pi/M$ by any window $w(k)$ of finite length $L$ centered symmetrically around $k = 0$. Finally, the impulse response is made causal by delaying the product $h_{LP}(k) \cdot w(k)$ by $k_0 = (L-1)/2$ samples:

$$h(k) = h_{LP}(k - k_0) \cdot w(k - k_0).$$

A design example is illustrated in Figure 4.7.

If the impulse response $h(k)$ of the prototype lowpass filter of order $4M$ (length $L = 4M + 1$) follows (4.18), the effective overall impulse response $h_A(k)$ corresponds to a pure delay of $k_0 = 2M$ samples.

### 4.1.3 Short-Term Spectral Synthesizer

In some applications, such as noise suppression, transform, or sub-band coding, we need to recover a time domain signal $y(k)$ from the samples $x_\mu(k')$ of the short-term spectrum. For any frequency index $\mu$, the samples $x_\mu(k')$ are to be considered as decimated sub-band signals associated with center frequency $\Omega_\mu$. If we assume a decimation factor $r = M/m$, the sub-band signals $x_\mu(k')$ can be represented as lowpass signals at the reduced sampling rate

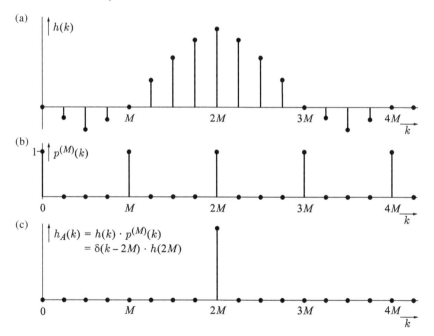

**Figure 4.7** Design criterion for a perfect overall frequency response $h_A(k)$. (a) Prototype impulse response $h(k) = h_{LP}(k - k_0) \cdot w(k - k_0)$: $L = 4 \cdot M + 1$, $M = 4$, Hamming window $w(k)$. (b) Decimation function $p^{(M)}(k)$. (c) Effective overall impulse response $h_A(k)$.

$f'_s = f_s/r$. As indicated in Figure 4.8, the process of reconstruction or re-synthesis consists of the following steps:

1. Up-sampling from $f'_s$ to $f_s = f'_s \cdot r$ by filling in $r - 1$ zeros in between adjacent samples of $x_\mu(k')$.
2. Interpolation by applying a filter with impulse response $g(k)$.
3. Frequency shift of the interpolated lowpass signal $\bar{y}_\mu(k)$ by $\Omega_\mu$ (modulation).
4. Superposition of the interpolated BP signals $y_\mu(k)$.

The interpolation is achieved by applying the samples $\tilde{x}_\mu(k)$ to a lowpass filter with (two-sided) passband width $\Delta\Omega$ and impulse response $g(k)$. The spectral relations are described by Figure 4.9 (in case of an ideal lowpass $g(k)$).

### 4.1.4 Short-Term Spectral Analysis and Synthesis

If the sub-band signals are not modified at all, the signal $x(k)$ can be reconstructed perfectly at the output. As some algorithmic delay caused by filter operations cannot be avoided, it should be a delayed version of the input signal

$$y(k) \overset{!}{=} x(k - k_0).$$

The overall model of the spectral analysis–synthesis system without any modification of the sub-band signals is shown in Figure 4.10a. The more detailed sub-channel model for the contribution $y_\mu(k)$ to the reconstructed output signal $y(k)$ is illustrated in Figure 4.10b. With the spectral representation of the intermediate signal $\bar{x}_\mu(k)$

$$\bar{X}_\mu(e^{j\Omega}) = X(e^{j(\Omega + \frac{2\pi}{M}\mu)}) \cdot H_\mu^{BP}(e^{j(\Omega + \frac{2\pi}{M}\mu)}) \tag{4.19a}$$

$$= X(e^{j(\Omega + \frac{2\pi}{M}\mu)}) \cdot H(e^{j\Omega}) \tag{4.19b}$$

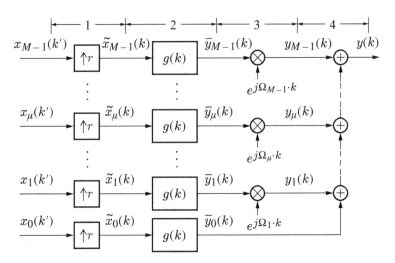

**Figure 4.8** Short-term spectral synthesizer with $M$ parallel channels.

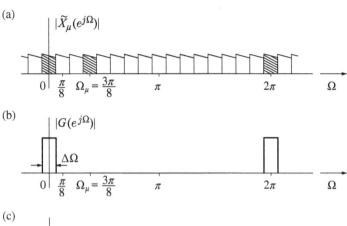

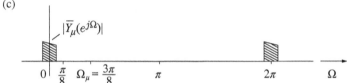

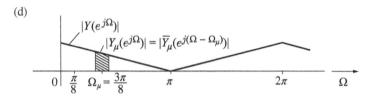

**Figure 4.9**  Spectral relations of short-term spectral synthesis. (a) Magnitude spectrum of sub-band signal $\tilde{x}_\mu(k)$. (b) Magnitude frequency response of the interpolator $g(k)$. (c) Magnitude spectrum of the interpolated sub-band signal $\bar{y}_\mu(k)$. (d) $y(k)$: superposition of the frequency-shifted signals $y_\mu(k)$.

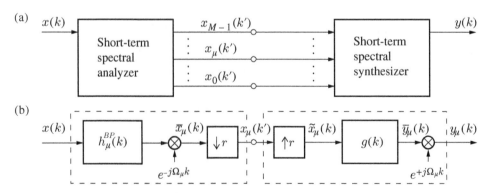

**Figure 4.10**  Reference model of the analysis-synthesis system. (a) Overall system. (b) Sub-channel model.

we get the baseband contribution $\overline{Y}_\mu(e^{j\Omega})$ of channel $\mu$ under the assumptions that

- the decimation factor $r$ is chosen according to the bandwidth $\Delta\Omega$ of the prototype lowpass filter
- the stopband attenuation of $h(k)$ is sufficiently high so that any aliasing due to the decimation process can be neglected
- the interpolation filter $g(k)$ is designed properly, i.e., the spectral repetitions due to the zero filling process are sufficiently suppressed

as follows (see also, e.g., [Vaidyanathan 1993b]):

$$\overline{Y}_\mu(e^{j\Omega}) = \tilde{X}_\mu(e^{j\Omega}) \cdot G(e^{j\Omega}) \tag{4.20a}$$

$$= \frac{1}{r} \cdot X(e^{j(\Omega + \frac{2\pi}{M}\mu)}) \cdot H(e^{j\Omega}) \cdot G(e^{j\Omega}) \tag{4.20b}$$

and finally the spectrum of the output signal (cf. Figures 4.9 and 4.10)

$$Y(e^{j\Omega}) = \sum_{\mu=0}^{M-1} Y_\mu(e^{j\Omega}) \tag{4.21a}$$

$$= \sum_{\mu=0}^{M-1} \overline{Y}_\mu(e^{j(\Omega - \frac{2\pi}{M}\mu)}) \tag{4.21b}$$

$$= \frac{1}{r} \cdot \sum_{\mu=0}^{M-1} X(e^{j\Omega}) \cdot H(e^{j(\Omega - \frac{2\pi}{M}\mu)}) \cdot G(e^{j(\Omega - \frac{2\pi}{M}\mu)}) \tag{4.21c}$$

$$= X(e^{j\Omega}) \cdot \frac{1}{r} \cdot \sum_{\mu=0}^{M-1} H(e^{j(\Omega - \frac{2\pi}{M}\mu)}) \cdot G(e^{j(\Omega - \frac{2\pi}{M}\mu)}) \tag{4.21d}$$

$$= X(e^{j\Omega}) \cdot H_{AS}(e^{j\Omega}). \tag{4.21e}$$

The frequency response $H_{AS}(e^{j\Omega})$ denotes the effective overall response of the short-term analysis–synthesis system.

### 4.1.5 Prototype Filter Design for the Analysis–Synthesis filter bank

From (4.21d) a criterion for perfect reconstruction (neglecting aliasing and interpolation errors but allowing a delay of $k_0$ samples) can be derived,

$$H_{AS}(e^{j\Omega}) = \frac{1}{r} \cdot \sum_{\mu=0}^{M-1} H(e^{j(\Omega - \frac{2\pi}{M}\mu)}) \cdot G(e^{j(\Omega - \frac{2\pi}{M}\mu)}) \tag{4.22a}$$

$$\stackrel{!}{=} e^{-jk_0 \cdot \Omega}, \tag{4.22b}$$

which corresponds to

$$h_{AS} \stackrel{!}{=} \delta(k - k_0) \tag{4.23}$$

in the time domain.

$$q(k) = h(k) * g(k)$$

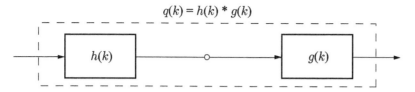

**Figure 4.11** Effective prototype impulse response of the analysis–synthesis system.

Equation (4.22a) can be interpreted as the superposition of $M$ frequency responses

$$Q_\mu(e^{j\Omega}) = H(e^{j(\Omega - \frac{2\pi}{M}\mu)}) \cdot G(e^{j(\Omega - \frac{2\pi}{M}\mu)}) \overset{!}{=} Q(e^{j(\Omega - \frac{2\pi}{M}\mu)}), \tag{4.24}$$

which are weighted by the constant factor $1/r$. The corresponding time domain responses

$$q_\mu(k) = q(k) \cdot e^{j\frac{2\pi}{M}\mu k} \tag{4.25}$$

are modulated versions of the resulting overall prototype impulse response $q(k)$, which is the convolution of $h(k)$ and $g(k)$ as indicated in Figure 4.11. Finally, the overall impulse response $h_{AS}(k)$ of the analysis–synthesis system should be a delay of $k_0$ samples. In analogy to (4.17a) we get

$$h_{AS}(k) = \frac{1}{r} \cdot \sum_{\mu=0}^{M-1} q_\mu(k) \tag{4.26a}$$

$$= q(k) \cdot \frac{1}{r} \cdot \sum_{\mu=0}^{M-1} e^{j\frac{2\pi}{M}\mu k} \tag{4.26b}$$

$$= q(k) \cdot \frac{1}{r} \cdot M \cdot p^{(M)}(k) \tag{4.26c}$$

$$\overset{!}{=} \delta(k - k_0). \tag{4.26d}$$

According to (4.26c), the effective overall impulse response $h_{AS}(k)$ has non zero samples only at times $k = \lambda M$. Therefore, an ideal overall response can be obtained if $q(k)$ with $k_0 = \lambda_0 \cdot M$ satisfies the following condition:

$$q(\lambda M) = \begin{cases} r/M & \lambda = \lambda_0 \\ 0 & \lambda \neq \lambda_0. \end{cases} \tag{4.27}$$

The convolutional product $q(k) = h(k) * g(k)$ should have equidistant zeros and a non zero sample at time instant $k_0$, i.e., $q(k)$ should be the impulse response of an $M$-th band filter. The samples $q(k)$ with $k \neq \lambda \cdot M$ have no direct influence on the effective overall frequency response. They can be chosen to optimize the frequency selectivity $\Delta\Omega$ of the filter bank.

The impulse response $q(k)$ can easily be designed, e.g., by the "modified Fourier approximation" method as described in Section 4.1.1. However, the decomposition of $q(k)$ into $h(k)$ and $g(k)$ is not trivial, especially if identical linear-phase prototypes $h(k) = g(k)$ are used. In this case, the product transfer function $Q(z)$ should have double zeros only. Different exact and approximate solutions have been proposed in [Wackersreuther 1985], [Wackersreuther

1986], [Wackersreuther 1987], [Nguyen 1994], and [Kliewer 1996]; see also [Vaidyanathan 1993b].

An alternative method is to split $Q(z)$ into a minimum-phase response $h(k)$ and a maximum-phase component $g(k)$ (or vice versa) by cepstral domain techniques as proposed in [Boite, Leich 1981].

### 4.1.6   Filter Bank Interpretation of the DFT

The discrete Fourier transform (DFT) may be interpreted as a special case of the short-term spectral analyzer of Figure 4.1. We consider the calculation of the DFT of a block of $M$ samples from an infinite sequence $x(k)$. $M$ samples are extracted from $x(k)$ by applying a sliding window function $w(\lambda)$, $\lambda = 0, 1, \ldots, M - 1$, of finite length so that the most recent sample $x(k)$ is weighted by $w(M - 1)$ and the oldest sample $x(k - M + 1)$ by $w(0)$.

The input sequence to the DFT at time $k$ is defined as

$$x(k - M + 1), \; x(k - M + 2), \; \ldots, x(k - M + 1 + \lambda), \; \ldots, x(k)$$

and the $\mu$-th DFT coefficient $X_\mu(k)$ at time $k$ is given by

$$X_\mu(k) = \sum_{\lambda=0}^{M-1} x(k - M + 1 + \lambda)\, w(\lambda)\, e^{-j\frac{2\pi}{M}\mu\lambda}. \tag{4.28a}$$

For the sake of compatibility with the reference configuration of Figure 4.6, we select a window function, which is the time-reversed version of a prototype lowpass impulse response $h(k)$ of length $M$:

$$w(\lambda) = h(M - 1 - \lambda); \quad \lambda = 0, 1, \ldots, M - 1 \tag{4.28b}$$

and get

$$X_\mu(k) = \sum_{\lambda=0}^{M-1} x(k - M + 1 + \lambda)\, h(M - 1 - \lambda)\, e^{-j\frac{2\pi}{M}\mu\lambda}. \tag{4.28c}$$

With the substitution

$$k - M + 1 + \lambda = \kappa, \quad \text{i.e.,} \quad M - 1 - \lambda = k - \kappa \quad \text{and} \quad \lambda = \kappa - k + M - 1$$

we obtain

$$X_\mu(k) = \sum_{\kappa=k-M+1}^{k} x(\kappa)\, h(k - \kappa)\, e^{-j\frac{2\pi}{M}\mu(\kappa-k+M-1)} \tag{4.28d}$$

$$= e^{j\frac{2\pi}{M}\mu(k+1)} \cdot \sum_{\kappa=-\infty}^{\infty} x(\kappa)\, h(k - \kappa)\, e^{-j\frac{2\pi}{M}\mu\kappa} \tag{4.28e}$$

$$= e^{j\frac{2\pi}{M}\mu(k+1)} \cdot \overline{x}_\mu(k). \tag{4.28f}$$

In (4.28d) we can replace the explicit summation limits by $\pm\infty$ as the window $h(k)$ implicitly performs the necessary limitation. Thus, the final expression for $X_\mu(k)$ is – except for the modulation factor (whose magnitude is 1) – identical to the intermediate signal $\overline{x}_\mu(k)$ of the BP or lowpass short-time analyzer of Figure 4.1 for

$$\Omega_\mu = \frac{2\pi}{M} \cdot \mu \quad \text{and} \quad h_{LP}(k) = h(k)$$

(see also (4.2a) and (4.2c)). Thus, the result

$$X_\mu(k) = \text{DFT}\{x(\kappa)\,h(k-\kappa)\}\,; \qquad \mu = 0, 1, \ldots, M-1$$

of the DFT of any windowed signal segment may be interpreted in terms of modulation, filtering, and decimation. At time instant $k$ we obtain a set of (modulated) short-term spectral values

$$X_\mu(k) = e^{j\frac{2\pi}{M}\mu(k+1)} \cdot \overline{x}_\mu(k), \tag{4.29}$$

where $k$ determines the position of the sliding window (see also Figure 4.2). If the position of the window is shifted between successive DFT calculations by $r = M$, successive DFT calculations do not overlap in time. The "DFT-observation instances" are

$$k = k' \cdot M - 1, \qquad k' \in \mathbb{N}.$$

In this case, which corresponds to *critical decimation*, the modulation factor vanishes

$$X_\mu(k = k' \cdot M - 1) = \overline{x}_\mu(k = k' \cdot M - 1) = x_\mu(k') \tag{4.30}$$

and the DFT delivers – with respect to magnitude and phase – exactly the same samples as the short-term analyzers of Figure 4.1 (also see STFT in Section 3.7).

Note that in (4.30) the decimation grid has a time shift of $\Delta k = -1$, i.e., $k = -1$ and $k = k' \cdot M - 1$ are on the decimation grid, but neither $k = 0$ nor $k = k' \cdot M$.

If the DFT calculation overlaps, e.g., by half a block length (*half-critical decimation*, $r = M/2$), the output samples are calculated at the time instances

$$k = k' \cdot M/2 - 1, \qquad k' \in \mathbb{Z}.$$

Then the modulation factor has to be taken into consideration as follows:

$$X_\mu(k = k' \cdot M/2 - 1) = e^{j\pi\mu k'} \cdot \overline{x}_\mu(k = k' \cdot M/2 - 1)$$
$$= (-1)^{\mu k'} \cdot \overline{x}_\mu(k = k' \cdot M/2 - 1),$$

if we are interested in the samples $x_\mu(k')$ according to Figure 4.1. In both cases, $r = M$ and $r = M/2$, the samples are produced by the DFT on a decimation grid $k' \cdot r - 1$, which is displaced from the origin ($k = 0$) by one sample interval. A block diagram of the sliding-window DFT is given in Figure 4.12. The samples $x(k)$ are fed into a delay chain (delay operator $z^{-1}$) and the delayed samples $x(k - \lambda)$ are multiplied by the window coefficients $w(M - 1 - \lambda) = h(\lambda)$. It is obvious that the delay chain has to be operated at the original sampling rate $f_s$, whereas the window multiplications and the DFT calculations can be carried out at the reduced sampling rate $f_s' = f_s/r$ if decimation by $r$ is applied to the output.

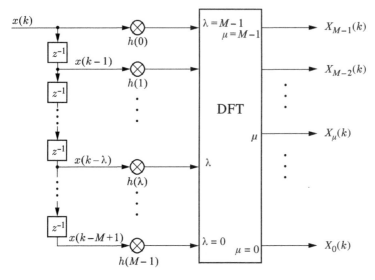

**Figure 4.12** Implementation of the sliding window DFT.

## 4.2 Polyphase Network Filter Banks

The *polyphase network filter bank* (PPN filter bank) is a very efficient implementation of the short-term spectral analyzer and the short-term spectral synthesizer of Sections 4.1.1 and 4.1.3, respectively. The computational complexity of the two systems given in Figures 4.6 and 4.8 can be reduced significantly.

The key points are:

1. The output samples of the analyzer (Figure 4.6) are calculated only at the reduced sampling rate $f_s' = f_s/r$.
2. At each (decimated) time instant $k'$ the complete set of output samples $x_\mu(k')$, $\mu = 0, 1 \ldots, M - 1$, can be obtained by a single FFT of length $M$.
3. The output of the synthesizer (Figure 4.8) can be calculated using one FFT of length $M$ per $r$ samples, which performs interpolation and spectral translation in combination with the impulse response $g(k)$.

If FIR prototype impulse responses $h(k)$ and $g(k)$ are used, there are two equivalent implementations:

A: overlapping of windowed segments
B: polyphase filtering.

Concept B is the more general one, as it can be applied to FIR as well as to IIR prototype impulse responses. The term polyphase network (PPN) is widely used in the literature (e.g., [Bellanger et al. 1976], [Vary 1979], [Vary, Heute 1980], [Vary, Heute 1981], [Vaidyanathan 1990], [Vaidyanathan 1993b]) because different phase characteristics of decimated subsequences and partial impulse responses can be identified, which play an important role.

### 4.2.1 PPN Analysis Filter Bank

In this section, the two equivalent approaches A and B will be derived from the reference short-term spectral analyzer of Figure 4.1 using an FIR prototype impulse response $h_{LP}(k) = h(k)$ of length $L$ ($k = 0, 1, \ldots, L-1$), where $L$ may be larger than $M$, the number of channels. For reasons of simplicity, we define

$$L = N \cdot M; \quad N \in \mathbb{N} \tag{4.31}$$

and append zero samples if the length of $h(k)$ is not an integer multiple of $M$.

**Approach A: Overlapping of Windowed Segments**

The lowpass signal $\bar{x}_\mu(k)$ of the $\mu$-th channel before sampling rate decimation can be formulated equivalently either by BP filtering and complex post-modulation (Figure 4.1a, Eq. (4.2a)) or by complex pre-modulation and lowpass filtering (Figure 4.1b, Eq. (4.2c)) as

$$\bar{x}_\mu(k) = e^{-j\frac{2\pi}{M}\mu k} \cdot \sum_{\kappa=-\infty}^{\infty} x(k-\kappa) \cdot e^{j\frac{2\pi}{M}\mu\kappa} \cdot h(\kappa) \tag{4.32a}$$

$$= \sum_{\kappa=-\infty}^{\infty} x(\kappa) \cdot e^{-j\frac{2\pi}{M}\mu\kappa} \cdot h(k-\kappa). \tag{4.32b}$$

Due to the length $L = N \cdot M$ of the FIR impulse response $h(k)$, the summation index $\kappa$ in (4.32a) is limited to the range

$$\kappa = 0, 1, \ldots, N \cdot M - 1.$$

The key to the derivation of the overlap structure A consists in the index substitution

$$\kappa = v \cdot M + \lambda; \quad \lambda = 0, 1, \ldots, M-1$$
$$v = 0, 1, \ldots, N-1.$$

Thus, (4.32a) can be rearranged as follows:

$$\bar{x}_\mu(k) = e^{-j\frac{2\pi}{M}\mu k} \cdot \sum_{v=0}^{N-1}\sum_{\lambda=0}^{M-1} x(k-vM-\lambda) \cdot h(vM+\lambda) \cdot e^{j\frac{2\pi}{M}\mu\lambda} \tag{4.32c}$$

$$= e^{-j\frac{2\pi}{M}\mu k} \cdot \sum_{\lambda=0}^{M-1}\underbrace{\left(\sum_{v=0}^{N-1} x(k-vM-\lambda) \cdot h(vM+\lambda)\right)}_{u_\lambda(k)} e^{j\frac{2\pi}{M}\mu\lambda} \tag{4.32d}$$

$$= e^{-j\frac{2\pi}{M}\mu k} \cdot \sum_{\lambda=0}^{M-1} u_\lambda(k) \cdot e^{j\frac{2\pi}{M}\mu\lambda} \tag{4.32e}$$

$$= e^{-j\frac{2\pi}{M}\mu k} \cdot \left[\sum_{\lambda=0}^{M-1} u_\lambda(k) \cdot e^{-j\frac{2\pi}{M}\mu\lambda}\right]^* \tag{4.32f}$$

$$= e^{-j\frac{2\pi}{M}\mu k} \cdot \left[\mathrm{DFT}\{u_\lambda(k)\}\right]^* \tag{4.32g}$$

$$= e^{-j\frac{2\pi}{M}\mu k} \cdot \left[U_\mu(k)\right]^* \tag{4.32h}$$

$$= W_M^{\mu k} \cdot \left[U_\mu(k)\right]^* \tag{4.32i}$$

where $[..]^*$ denotes the complex conjugate operation,

$$W_M^{\mu k} = e^{-j\frac{2\pi}{M}\mu k} \tag{4.33}$$

are the complex post-modulation terms, and

$$u_\lambda(k) = \sum_{v=0}^{N-1} x(k - vM - \lambda) \cdot h(vM + \lambda); \quad \lambda = 0, 1, \dots, M-1 \tag{4.34}$$

is the real-valued input sequence to the DFT.

The resulting expression (4.32g) looks like the DFT of the intermediate sequence $u_\lambda(k)$, $\lambda = 0, \dots, M-1$, of length $M$. As the sequence $u_\lambda(k)$ is real-valued, we may use either the DFT or the inverse DFT. In the latter case, a scaling factor of $1/M$ has to be taken into account. Here, we prefer the DFT for the analysis part of the filter bank.

In conclusion, the complete set of samples $\bar{x}_\mu(k)$, $\mu = 0, 1, \dots, M-1$, can be calculated very efficiently for any FIR prototype impulse response $h(\kappa)$ of length $L$ by applying the FFT algorithm to the sequence $u_\lambda(k)$, $\lambda = 0, 1, \dots, M-1$, at the fixed but arbitrary time instant $k$. If we are interested not only in the magnitude samples $|\bar{X}_\mu(k)|$, but also in the magnitude and phase or real and imaginary parts of $\bar{X}_\mu(k)$, respectively, then the post-modulation by $W_M^{\mu k}$ has to be carried out according to (4.32i).

In a pre-processing step, the intermediate sequence $u_\lambda$ according to (4.34) has to be determined, as illustrated in Figure 4.13 by overlapping $N$ windowed segments, each

of length $M$ ($v = 0, 1, \dots, N-1$). The complete pre-processing step requires $M \cdot N$ multiplications and additions which are not more than the number of operations needed to calculate one output sample of a single lowpass filter of length $L = N \cdot M$.

The corresponding block diagram is shown in Figure 4.14 for the special case $L = N \cdot M = 2M$.

So far we have not considered the fact that the narrowband signals $\bar{x}_\mu(k)$ may be represented at a reduced sampling rate. In the algorithm of Figure 4.14, the pre-processing step and the DFT are carried out at each time instant $k$ and the sample rate is decimated afterwards by $r$. However, the output samples can be calculated immediately at the reduced

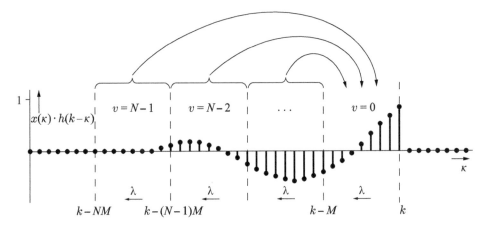

**Figure 4.13** Overlapping of windowed segments; example $M = 8, N = 4$.

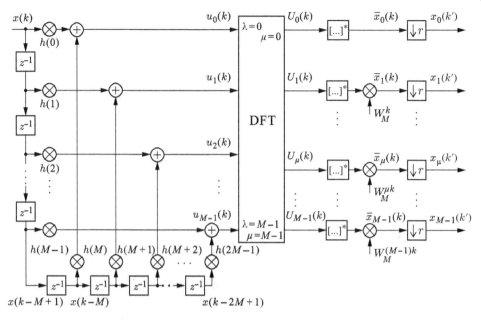

**Figure 4.14** Polyphase network (PPN) filter bank according to approach A; example $L = N \cdot M = 2M$; [...]* denotes complex conjugation.

sampling rate, i.e., the pre-processing step, the DFT, the complex conjugation, and the post-modulation has to be calculated only at the decimated time instances, e.g., at

$$k = k' \cdot r, \tag{4.35}$$

while the delay chain at the input has to be operated at the original sampling rate.

The two decimation cases of special interest are critical decimation ($r = M$) and half-critical decimation ($r = \frac{M}{2}$), with the post-modulation factors

$$W_M^{\mu \cdot k' \cdot r} = e^{-j\frac{2\pi}{M}\mu k' r} = \begin{cases} 1 & r = M \\ (-1)^{\mu k'} & r = M/2. \end{cases} \tag{4.36}$$

Note that in Figure 4.14, the order of the input index $\lambda$ is reversed in comparison to Figure 4.12. This difference is due to the fact that the decimation grid with $k = k' \cdot r$ is not displaced with respect to the origin and the simple post-modulation factors of (4.36) are desired here. If we considered the expression (4.32i) for the decimated instances $k = k' \cdot r - 1$ a post-modulation term

$$e^{-j\frac{2\pi}{M}\mu(k'r-1)} = W_M^{\mu k' \cdot r} \cdot e^{+j\frac{2\pi}{M}\mu}$$

would result. The combined effect of the second factor $e^{+j\frac{2\pi}{M}\mu}$ with the "[...]*-complex conjugation" is equivalent to a cyclic shift and an inversion of the order of the DFT input sequence $u_\lambda(k)$. If this is taken into consideration, it can easily be shown that for $L = M$ the two block diagrams of Figures 4.14 and 4.12 are equivalent as given by (4.29).

In any case, the output samples $x_\mu(k')$ are exactly the same as those of the reference structure of Figure 4.1.

**Approach B: Polyphase Filtering**

The second approach is an alternative interpretation of (4.32) in terms of convolution instead of overlapping weighted segments. This allows us to use an FIR or even an IIR prototype lowpass filter. For an FIR prototype filter, the actual difference lies in the organization of the two nested loops ($v$ and $\lambda$) within the pre-processing step.

The alternative implementation B consists of reorganizing the block diagram of Figure 4.14 and the key equation (4.32) by taking the decimation process into account.

We introduce $M$ sub-sequences

$$\tilde{x}_\lambda(k) = \begin{cases} x(k) & k = v \cdot M + \lambda \\ 0 & \text{else} \end{cases} \quad \lambda = 0, 1, \ldots, M - 1, v = 0, 1, 2, \ldots, \tag{4.37}$$

decimated and up-sampled by $M$ and furthermore $M$ partial impulse responses

$$\tilde{h}_\lambda(k) = \begin{cases} h(k) & k = v \cdot r - \lambda \\ 0 & \text{else} \end{cases} \quad \lambda = 0, 1, \ldots, M - 1, v = 0, 1, 2, \ldots, \tag{4.38}$$

decimated and upsampled by $r$ which are defined on decimation grids with different displacements in time by $\pm\lambda$. There are $M$ partial impulse responses; however, for $r = M/2$ only $M/2$ responses are different ($\tilde{h}_{M/2+i} = \tilde{h}_i$, $i = 0, 1, \ldots, M/2$).

Each intermediate sequence $\tilde{u}_\lambda(k)$ can be interpreted for any fixed index $\lambda$ as a time sequence, i.e., as convolution of the subsequence $\tilde{x}_\lambda(k)$ with the partial impulse response $\tilde{h}_\lambda(k)$

$$\tilde{u}_\lambda(k) = \tilde{x}_\lambda(k) * \tilde{h}_\lambda(k) \begin{cases} \neq 0 & k = k' \cdot r \\ = 0 & \text{else.} \end{cases} \tag{4.39}$$

For the special case $r = M$, Figure 4.15 illustrates exemplarily that the intermediate signals $\tilde{u}_\lambda(k)$ take non zero values only at the decimation instances

$$k = k' \cdot M; \quad k' \in \mathbb{Z}. \tag{4.40}$$

Assuming a causal signal $x(k)$ and a causal impulse response $h(k)$, the first non zero value of $\tilde{x}_\lambda(k)$ is at $k = \lambda$ and the first non zero sample of $\tilde{h}_\lambda(k)$ is at $k = M - \lambda$ ($\lambda = 1, \ldots, M - 1$) or $k = 0$ ($\lambda = 0$). As a result all partial convolutions produce non zero output signals $\tilde{u}_\lambda(k)$ on the same decimation time grid

$$k = k' \cdot M. \tag{4.41}$$

Thus, the partial filters perform different phase shifts, this is why the term *polyphase filtering* was introduced [Bellanger et al. 1976].

For the special case $r = M$ and $L = 2M$, the block diagram of approach B is shown in Figure 4.16. In comparison to Figure 4.14, only the pre-processing has been reorganized, taking the decimation by $r = M$ as well as the post-modulation terms $W_M^{\mu \cdot k' \cdot M} = 1$ (see also (4.36)) into account.

Because of (4.37) and (4.38) the zero samples of $\tilde{h}_\lambda(k)$, $\tilde{x}_\lambda(k)$, and $\tilde{u}_\lambda(k)$ need not be processed. This is illustrated in Figure 4.17. The decimation by $r = M$ takes place at the taps of the delay line with unit delays $T = 1/f_s$, whereas the partial filters and the DFT run at the reduced sample rate $f_s' = f_s/M$.

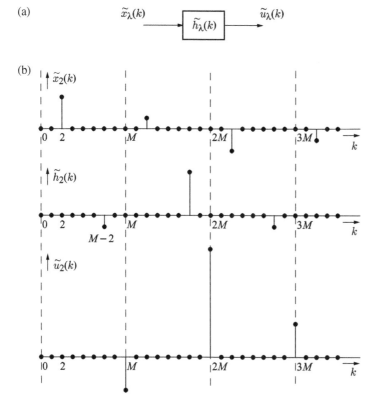

**Figure 4.15** Polyphase filtering: convolution of sub-sequences with partial impulse responses. (a) Block diagram. (b) Example $M = 8$, $\lambda = 2$, $r = M$.

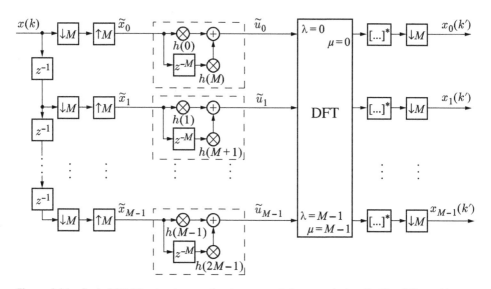

**Figure 4.16** Basic PPN filter bank according to approach B; example $L = N \cdot M = 2M$, $r = M$.

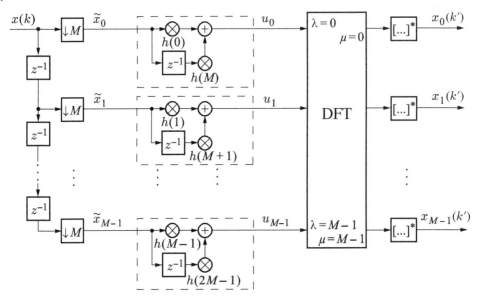

**Figure 4.17** Efficient implementation of the PPN filter bank with partial filters running at the reduced sampling rate $f'_s = f_s/M$; example $L = N \cdot M = 2M$, $r = M$.

In the general case with decimation by $r$, there are $r - 1$ intermediate zero coefficients between two non zero coefficients of $\tilde{h}_\lambda(k)$ and $M - 1$ intermediate zeros between the samples of the subsequences $\tilde{x}_\lambda(k)$. As the input samples to the partial filters, running at the reduced sample rate, are needed at the time instances $k = k' \cdot r$, we first have to decimate the delayed versions of the input signal by $r = M$ and then to upsample them by $m = M/r$.

The solution for the general case producing output samples $x_\mu(k')$ at the reduced sampling rate $f'_s = f_s/r$, with $r = \frac{M}{m}$ and any prototype lowpass filter with impulse response $h(k)$, is shown in Figure 4.18. The decimated partial impulse responses are defined by

$$h_\lambda(k') = h(k' \cdot r - \lambda); \qquad \lambda = 0, 1, \ldots, M - 1. \tag{4.42}$$

Usually half-critical decimation $r = M/2$ is chosen to avoid spectral aliasing in the sub-band signals $x_\mu(k')$. The decimated partial filters $h_\lambda(k')$ and the DFT have to be computed at the reduced sampling rate $f_s/r$ only. For $r = M/2$ the signals $\tilde{x}_\lambda(k' \cdot r)$ are obtained by decimation of the delayed versions of $x(k)$ by $M$ and subsequent upsampling by 2 to achieve the reduced sampling rate. Therefore, we have $M/r - 1 = 1$ zero sample in between two decimated samples.

The PPN approach of Figure 4.18 may be interpreted as a generalization of the sliding-window DFT (Figure 4.12) so that the window multiplication in Figure 4.12 is replaced by a convolution (polyphase filter) with the decimated partial impulse responses $h_\lambda(k')$.

If the prototype lowpass impulse response is of length $L = M$, both systems are identical according to (4.29).

The main advantage of the PPN concept is that the spectral selectivity, i.e., the frequency response $H(e^{j\Omega})$, can be designed independently of the number $M$ of channels and that even IIR prototype filters are possible (e.g., [Vary 1979]). The discussion of the latter possibility is beyond the scope of this book.

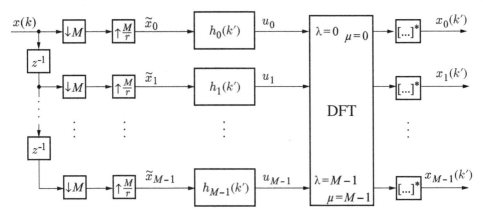

**Figure 4.18** General solution: PPN filter bank with partial filters $h_\lambda(k')$ running at the reduced sampling rate $f_s/r$; $r = \frac{M}{m}$.

Finally, it should be noted that instead of the DFT the *discrete cosine transform* (DCT) or some generalized versions (GDFT, GDCT) (e.g., [Crochiere, Rabiner 1983], [Vary et al. 1998]) can be used.

With the *generalized* DCT the individual BP impulse responses can be formulated as

$$h_\mu^{BP}(k) = h(k) \cos\left(\frac{\pi}{M}(\mu + \mu_0)(k + k_0)\right);$$

$$\mu = 0, 1, \ldots M - 1; \ \mu_0, k_0 \in \{0, 1/2\}. \tag{4.43}$$

### 4.2.2 PPN Synthesis Filter Bank

The short-term spectral synthesizer of Figure 4.8 can be implemented very efficiently using the inverse discrete Fourier transform (IDFT) and a PPN, as shown in Fig. 4.19. The PPN consists of partial impulse responses $g_\lambda(k')$, which are obtained by subsampling the impulse response $g(k)$ of the interpolation filter of the reference structure given in Figure 4.8. The impulse response $g(k)$ can also be regarded as the prototype impulse response of the synthesis filterbank.

The structure of this efficient PPN filter bank can be derived straightforwardly from the reference synthesizer of Figure 4.8. The output signal $y(k)$ is obtained as the superposition of the interpolated and modulated sub-band signals:

$$y(k) = \sum_{\mu=0}^{M-1} y_\mu(k) \tag{4.44a}$$

$$= \sum_{\mu=0}^{M-1} \bar{y}_\mu(k) \cdot e^{j\frac{2\pi}{M}\mu k} \tag{4.44b}$$

$$= \sum_{\mu=0}^{M-1} \sum_{\kappa=0}^{k} \tilde{x}_\mu(\kappa) \cdot g(k - \kappa) \cdot e^{j\frac{2\pi}{M}\mu k}, \tag{4.44c}$$

with the causal FIR $g(k)$ of length $L$, where $L$ might be larger than $M$.

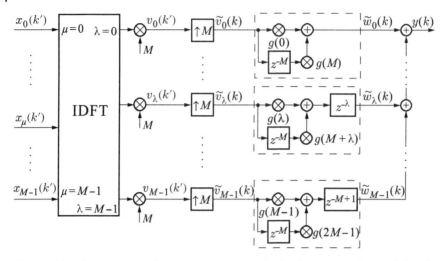

**Figure 4.19** Basic structure of the polyphase synthesis filterbank: example $r = M$, $L = N \cdot M = 2M$, convolution of upsampled subsequences $\tilde{v}_\lambda(k) \neq 0$ only for $k = k' \cdot M$ with partial impulse responses of length $N = 2$.

The key points in the derivation of the structure of the PPN synthesis filter bank of Figure 4.19 from the reference structure of Figure 4.8 are

1. substitution $k = i \cdot M + \lambda$; $\lambda \in \{0, 1, \dots, M-1\}$; $i \in \mathbb{Z}$
2. exchange of the order of the two summations in (4.44c).

Due to the periodicity of the complex exponential function with

$$e^{+j\frac{2\pi}{M}\mu(i \cdot M + \lambda)} = e^{+j\frac{2\pi}{M}\mu\lambda}$$

we obtain

$$y(i \cdot M + \lambda) = \sum_{\kappa=0}^{i \cdot M + \lambda} \underbrace{\left( \sum_{\mu=0}^{M-1} \tilde{x}_\mu(\kappa) \cdot e^{+j\frac{2\pi}{M}\mu\lambda} \right)}_{\tilde{v}_\lambda(\kappa)} \cdot g(i \cdot M + \lambda - \kappa). \qquad (4.44\text{d})$$

For each fixed but arbitrary value of $\kappa$, the expression inside the brackets is the IDFT of the complex samples $\tilde{x}_\mu(\kappa)$, $\mu = 0, 1, \dots, M-1$. If the IDFT routine includes a scaling factor of $1/M$, this can be corrected by subsequent multiplication with $M$ according to

$$\tilde{v}_\lambda(\kappa) = \sum_{\mu=0}^{M-1} \tilde{x}_\mu(\kappa) \cdot e^{+j\frac{2\pi}{M}\mu\lambda}$$

$$= M \cdot \text{IDFT}\left\{ \tilde{x}_\mu(\kappa) \right\}; \quad \kappa = \text{fixed}.$$

It should be noted that $\tilde{x}_\mu(\kappa)$ is an upsampled sequence with (in the general case) $r - 1$ intermediate zero samples between each pair of the decimated samples according to (4.8). The sequence $\tilde{v}_\lambda(\kappa)$ has the same temporal structure

$$\tilde{v}_\lambda(\kappa) \begin{cases} \neq 0 & \kappa = \kappa' \cdot r \\ = 0 & \text{else.} \end{cases} \qquad (4.44\text{e})$$

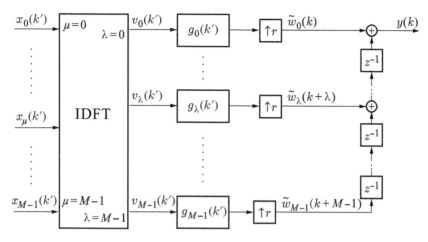

**Figure 4.20** General solution of the efficient polyphase synthesis filterbank: DFT and partial filters are running at the reduced sampling rate $f_s/r$.

Finally, we get for each fixed index $\lambda = 0, 1, \ldots, M - 1$ and variable time or frame index $i$

$$y(i \cdot M + \lambda) = \sum_{\kappa=0}^{i \cdot M + \lambda} \tilde{v}_\lambda(\kappa) \cdot g(i \cdot M + \lambda - \kappa) \qquad (4.44f)$$

$$= \tilde{v}_\lambda(\kappa) * \tilde{g}_\lambda(\kappa). \qquad (4.44g)$$

At the time instances $k = i \cdot M + \lambda$ the output sample $y(k)$ is determined solely by the sequence $\tilde{v}_\lambda(\kappa)$ filtered with the partial impulse response

$$\tilde{g}_\lambda(\kappa) = \begin{cases} g(\kappa) & \kappa = \kappa' \cdot r + \lambda \\ 0 & \text{else} \end{cases} \qquad \lambda = 0, 1, \ldots, M - 1, \qquad (4.45)$$

i.e., a decimated and upsampled version of the interpolator impulse response $g(k)$.

Therefore, we only need to deal with the decimated sequences $x_\mu(\kappa')$ and $v_\lambda(\kappa')$ and to carry out the IDFT every $r$-th sampling interval.

The corresponding block diagram is shown for the special case $r = M$ and $L = 2M$ in Figure 4.19. For the sake of compatibility with (4.44f), the input sequences $\tilde{v}_\lambda(k)$ to the partial filters are obtained here from the sequences $v_\lambda(k')$ by upsampling. However, as $\tilde{v}_\lambda(k)$ as well as the partial impulse $\tilde{g}_\lambda(k)$ responses have $r - 1$ zero samples between each pair of non zero samples, only every $r$-th output sample of each of the partial filters can take non zero values according to

$$\tilde{w}_\lambda(\kappa) \begin{cases} \neq 0 & \kappa = \kappa' \cdot r + \lambda \\ = 0 & \text{else.} \end{cases} \qquad (4.46)$$

The superposition of the filter output samples $\tilde{w}_\lambda(k)$ of Figure 4.19 delivers for $r = M$

$$y(k) = \begin{cases} \sum_{\lambda=0}^{M-1} \tilde{w}_\lambda(k) \; ; & k \neq i \cdot M + \lambda \\ \tilde{w}_\lambda(k) & ; \; k = i \cdot M + \lambda. \end{cases} \qquad (4.47)$$

Furthermore, the polyphase filters can also be run at the decimated sampling rate and upsampling of $v_\lambda(k')$ is not required. This leads to the final and efficient solution as given in Figure 4.20, with

$$g_\lambda(k') = M \cdot g(k' \cdot r + \lambda) \tag{4.48}$$

$$G_\lambda(z) = M \cdot \sum_{k'=0}^{\infty} g_\lambda(k') \cdot z^{-k'}. \tag{4.49}$$

Note that in comparison to Figure 4.19, the scale factor $M$ is now applied by scaling the partial filters for reasons of simplicity. The delays $z^{-\lambda}$ of the individual filter branches are provided by the output delay chain.

The chain of delay elements and adders at the output performs the superposition of the samples $\tilde{w}_\lambda(k)$ according to (4.47).

The conventional IDFT synthesis turns out to be a special case of the structure of Figure 4.20 if the interpolator has rectangular impulse response

$$g(k) = \begin{cases} 1 & 0 \le k < L - 1 \\ 0 & \text{else.} \end{cases} \tag{4.50}$$

Note that in contrast to the conventional analysis and synthesis by DFT and IDFT, the polyphase concept allows to improve significantly the spectral selectivity and the interpolation task with little additional complexity. In the analysis stage, each of the $M$ window multiplications actually has to be replaced by a short convolution with only $N = L/M$ coefficients $h_\lambda(k')$. The typical parameter selection is $N = 2, \dots, 4$. In the synthesis stage, we need $L/r$ multiplications for each of the partial filters $g_\lambda(k')$. The design criteria and procedures for the prototype lowpass $h(k)$ and the interpolator $g(k)$ as described in Sections 4.1.1 and 4.1.4 apply to the PPN analysis and PPN synthesis filter bank without any modification.

## 4.3 Quadrature Mirror Filter Banks

The objective of the quadrature mirror filter (QMF) approach is the spectral decomposition of the signal $x(k)$ into $M = 2^K$ *real-valued* sub-band signals with maximum sampling rate decimation as well as the reconstruction (synthesis) of the signal from the sub-band signals. The analysis and the synthesis filter banks are based on special half-band filters which are called *(QMFs)* (e.g., [Esteban, Galand 1977], [Vaidyanathan 1993b]). These filters are used in a tree structure with decimation/interpolation by $r = 2$ in each stage of the tree.

### 4.3.1 Analysis–Synthesis Filter Bank

We first consider the special case with $M = 2$, which is the basic block of the tree structure.

The input signal is split by a lowpass filter with impulse response $h_{LP}(k)$ and a complementary highpass filter with impulse response $h_{HP}(k)$ into a lowpass signal $\bar{x}_0(k)$ and a highpass signal $\bar{x}_1(k)$. As the lowpass and highpass filters are half-band filters, the sample rate can be decimated by a factor $r = 2$ (see Figure 4.21):

$$x_0(k') = \bar{x}_0(2 \cdot k') \quad ; \quad x_1(k') = \bar{x}_1(2 \cdot k' + \rho) \quad ; \quad \rho \in \{0, 1\}. \tag{4.51}$$

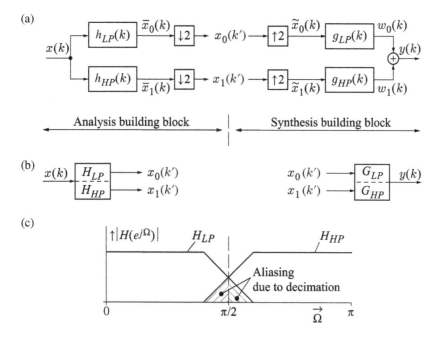

**Figure 4.21** QMF analysis–synthesis bank. (a) Block diagram of basic building block $M = 2$. (b) Simplified block diagram. (c) Schematic of frequency responses.

It is assumed that in the lowpass channel the even-numbered samples are selected and in the highpass channel either the even samples ($\rho = 0$) or the odd samples ($\rho = 1$).

In the synthesis building block the sub-band signals are upsampled by a factor 2 (insertion of zero samples) and interpolated by lowpass and highpass filters with impulse responses $g_{LP}(k)$ and $g_{HP}(k)$. The interpolated signals $w_0(k)$ and $w_1(k)$ are added. As the filters are non ideal, aliasing cannot be avoided (see Figure 4.21c). However, the disturbing aliasing components can be eliminated within the synthesis process, as shown below.

With the basic blocks of Figure 4.21b a tree-structured analysis–synthesis filter bank can be constructed, as illustrated in Figure 4.22 for $M = 2^K = 8$. The input signal $x(k)$ is

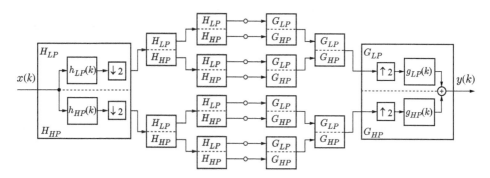

**Figure 4.22** Tree-structured QMF analysis–synthesis bank, $N = 2^K = 8$ channels.

decomposed into $M = 8$ sub-band signals by $K = 3$ stages. In each stage, the sampling rate is decimated by $r = 2$.

## 4.3.2 Compensation of Aliasing and Signal Reconstruction

In what follows the analysis–synthesis blocks of Figure 4.21a will be discussed in detail.

We assume a lowpass filter with impulse response $h_{LP}(k) = h(k)$ and transfer function $H_{LP}(z) = H(z)$. Furthermore, the highpass filter is obtained from the lowpass filter by modulation

$$h_{HP}(k) = e^{\pm j\pi k} \cdot h_{LP}(k) = (-1)^k \cdot h(k) \tag{4.52a}$$

$$H_{HP}(e^{j\Omega}) = H(e^{+j(\Omega \mp \pi)}) \tag{4.52b}$$

$$H_{HP}(z) = H(-z). \tag{4.52c}$$

The two sub-band signals $\overline{x}_0(k)$ and $\overline{x}_1(k)$ are described in the $z$-domain according to Figure 4.21a as

$$\overline{X}_0(z) = X(z) \cdot H(z) \tag{4.53}$$

$$\overline{X}_1(z) = X(z) \cdot H(-z). \tag{4.54}$$

The sampling rate decimation with subsequent upsampling by $r = 2$ can be formulated analytically by multiplication of $\overline{x}_0(k)$ and $\overline{x}_1(k)$ with a sampling function in the time domain:

$$\tilde{x}_0(k) = \overline{x}_0(k) \cdot p^{(2)}(k - \rho) \quad ; \quad \rho \in \{0, 1\} \tag{4.55a}$$

$$\tilde{x}_1(k) = \overline{x}_1(k) \cdot p^{(2)}(k - \rho). \tag{4.55b}$$

The decimation in the two channels can be in phase ($\rho = 0$ or $\rho = 1$ in both channels) or out of phase ($\rho = 0$ in channel 0 and $\rho = 1$ in channel 1, or vice versa) with

$$p^{(2)}(k - \rho) = \frac{1}{2} \left[ 1 + (-1)^\rho \cdot (-1)^k \right] ; \quad \rho \in \{0, 1\}. \tag{4.56}$$

Because $(-1)^k = e^{+j\pi k}$ the $z$-domain representation of (4.55a) and (4.55b) can easily be obtained by using the modulation theorem. For reasons of simplicity, we assume that the lowpass signal $\overline{x}_0(k)$ is decimated by $p^{(2)}(k)$ and the highpass signal $\overline{x}_1(k)$ either by applying $p^{(2)}(k)$ or $p^{(2)}(k - 1)$:

$$\tilde{X}_0(z) = \frac{1}{2} [\overline{X}_0(z) + \overline{X}_0(-z)] \tag{4.57a}$$

$$\tilde{X}_1(z) = \frac{1}{2} [\overline{X}_1(z) + (-1)^\rho \cdot \overline{X}_1(-z)]. \tag{4.57b}$$

The schematics of the frequency characteristics are given in Figure 4.23. Note that after decimation the frequency axis has been normalized to $\Omega' = r \cdot \Omega = 2 \cdot \Omega$ and that the highpass signal occurs in a mirrored version in the baseband $0 \leq \Omega' \leq \pi$.

Within the synthesis building block the two signals $\tilde{x}_0(k)$ and $\tilde{x}_1(k)$ are interpolated using the filters with the $z$-transforms $G_{LP}(z)$ and $G_{HP}(z)$ to obtain the intermediate signals $w_0(k)$ and $w_1(k)$ at the original sampling rate.

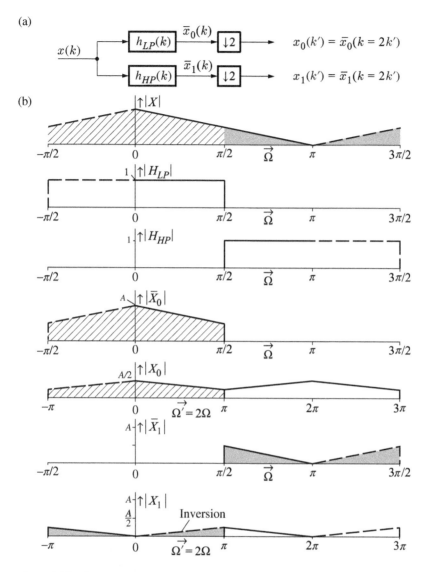

**Figure 4.23** Spectral relations in the QMF analysis block. (a) Block diagram. (b) Schematics of spectra ($\Omega' =$ normalized frequency after decimation).

The reconstructed output signal $y(k)$ is finally given in the $z$-domain by

$$Y(z) = W_0(z) + W_1(z) \tag{4.58a}$$

$$= G_{LP}(z) \cdot \tilde{X}_0(z) + G_{HP}(z) \cdot \tilde{X}_1(z)$$

$$= G_{LP}(z) \cdot \frac{1}{2}[X(z)H(z) + X(-z)H(-z)] \tag{4.58b}$$

$$+ G_{HP}(z) \cdot \frac{1}{2}[X(z)H(-z) + (-1)^p \cdot X(-z)H(z)]$$

$$= \frac{1}{2}X(z) \cdot [G_{LP}(z)H(z) + G_{HP}(z)H(-z)] \tag{4.58c}$$

$$+ \frac{1}{2}X(-z) \cdot [G_{LP}(z)H(-z) + (-1)^p \cdot G_{HP}(z)H(z)].$$

The first part of (4.58d) constitutes the desired signal and the second part the disturbing aliasing component $(X(-z) \hat{=} X(e^{j(\Omega-\pi)}))$. The aliasing component can be compensated as follows:

**(a) Aliasing compensation for $\rho = 0$**

The requirement

$$G_{LP}(z) \cdot H(-z) + G_{HP}(z) \cdot H(z) \stackrel{!}{=} 0 \qquad (4.59)$$

can be fulfilled if

$$G_{LP}(z) = H(z) \text{ and } G_{HP}(z) = -H(-z) = -G_{LP}(-z). \qquad (4.60a)$$

**(b) Aliasing compensation for $\rho = 1$**

The requirement

$$G_{LP}(z) \cdot H(-z) - G_{HP}(z) \cdot H(z) \stackrel{!}{=} 0$$

can be met if

$$G_{LP}(z) = H(z) \text{ and } G_{HP}(z) = H(-z). \qquad (4.60b)$$

The general solution for (a) and (b) is

$$G_{LP}(z) = H(z)$$
$$G_{HP}(z) = -(-1)^{\rho} \cdot H(-z). \qquad (4.60c)$$

Hence, only the lowpass filter with transfer function $H(z)$ has to be designed. Thus, the effective transfer function of the analysis–synthesis system is given by

$$H_{AS}(z) = \frac{1}{2} \left[ H^2(z) - (-1)^{\rho} \cdot H^2(-z) \right]. \qquad (4.60d)$$

Perfect reconstruction of the signal is achieved if the overall frequency response is a delay by $k_0$ samples, i.e.,

$$H_{AS}(z) \stackrel{!}{=} z^{-k_0} \qquad (4.61)$$

according to $y(k) = x(k - k_0)$.

The desired behavior (4.61) can be approximated with an accuracy which is sufficient for practical applications by appropriate design of a linear-phase FIR $h(k)$ of length $L = n + 1$ with

$$h(k) = h(n - k); \quad k = 0, 1, \ldots, n. \qquad (4.62)$$

In the case of $\rho = 0$, it can be shown that an impulse response of even length and in the case of $\rho = 1$ a filter with odd length are needed to avoid $|H_{AS}(e^{j\frac{\pi}{2}})| = 0$.

The desired compensation of the aliasing components does not necessarily require the selection $G_{LP}(z) = H(z)$. A more general condition can be derived from the second part of (4.58d) with

$$\frac{G_{LP}(z)}{G_{HP}(z)} \stackrel{!}{=} -(-1)^{\rho} \frac{H(z)}{H(-z)}. \qquad (4.63)$$

This condition provides new degrees of freedom for the optimization. Various solutions can be found in the literature (e.g., [Jain, Crochiere 1983], [Jain, Crochiere 1984], [Johnston 1980], [Wackersreuther 1987], [Smith, Barnwell 1986]).

### 4.3.3 Efficient Implementation

The QMF analysis and synthesis filter bank with $M = 2$ can be considered as a special case of the PPN filter bank with DFT length $M = 2$. Therefore, the computationally efficient approaches according to Figures 4.17 and 4.20 can be applied here.

For reasons of simplicity, we will consider only the case $\rho = 0$ and the filter design according to (4.60a) with even length $L = n + 1$. Taking into account the sampling rate decimation and the fact that the relation between the lowpass response $h_{LP}(k) = h(k)$ and the highpass response is given in the time domain by (see (4.52a)

$$h_{HP}(k) = (-1)^k \cdot h(k), \tag{4.64}$$

we get (see also Figure 4.21)

$$\bar{x}_0(k = 2k') = \sum_{\kappa=0}^{n} h(\kappa)x(2k' - \kappa)$$

$$= \sum_{\kappa'=0}^{\frac{n-1}{2}} h(2\kappa')x(2k' - 2\kappa') + \sum_{\kappa'=0}^{\frac{n-1}{2}} h(2\kappa' + 1)x(2k' - 2\kappa' - 1)$$

$$= a(k') + b(k') \tag{4.65}$$

$$\bar{x}_1(k = 2k') = \sum_{\kappa=0}^{n} h(\kappa) \cdot (-1)^\kappa \cdot x(2k' - \kappa)$$

$$= a(k') - b(k'). \tag{4.66}$$

Both decimated samples, $\bar{x}_0(k = 2k')$ and $\bar{x}_1(k = 2k')$, can be calculated from the quantities $a(k')$ and $b(k')$. $a(k')$ is the result of the convolution of the even samples of $x(k)$ with the even samples of $h(k)$, and $b(k')$ results from the convolution of the respective odd samples. If the original impulse response is decomposed into its two polyphase components, the overall computational complexity for calculating $\bar{x}_0(2k')$ and $\bar{x}_1(2k')$ is only slightly higher than the complexity of a single convolution of length $n + 1$. The corresponding block diagram for the QMF analysis bank is illustrated in Figure 4.24a.

The derivation of the efficient structure of the synthesis block is slightly more complicated. Again, we consider the case $\rho = 0$ and the filter selection

$$g_{LP}(k) = h(k) \quad ; \quad g_{HP}(k) = -(-1)^k \cdot h(k) \tag{4.67}$$

with even length $L = n + 1$. Furthermore, we have to take into consideration that the interpolator input signals $\tilde{x}_0(k)$ and $\tilde{x}_1(k)$ in Figure 4.21a have non zero values only at even time instances, i.e.,

$$\tilde{x}_\mu(k) = \begin{cases} x_\mu(k') & k = 2k' ; \quad \mu \in \{0, 1\} \\ 0 & k = 2k' + 1. \end{cases} \tag{4.68}$$

(a)

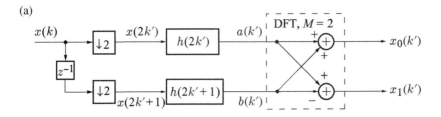

(b)

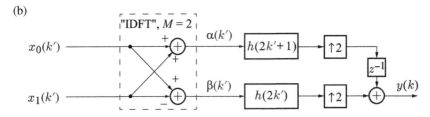

**Figure 4.24** Efficient implementation of QMF bank. (a) Analysis filter bank. (b) Synthesis filterbank.

We get

$$y(k) = \sum_{\kappa=0}^{n} h(\kappa) \cdot \tilde{x}_0(k - \kappa) + \sum_{\kappa=0}^{n} -(-1)^{\kappa} \cdot h(\kappa) \cdot \tilde{x}_1(k - \kappa) \tag{4.69a}$$

$$= \sum_{\kappa'=0}^{\frac{n-1}{2}} h(2\kappa') \cdot [\tilde{x}_0(k - 2\kappa') - \tilde{x}_1(k - 2\kappa')] \tag{4.69b}$$

$$+ \sum_{\kappa'=0}^{\frac{n-1}{2}} h(2\kappa' + 1) \cdot [\tilde{x}_0(k - 2\kappa' - 1) + \tilde{x}_1(k - 2\kappa' - 1)]$$

$$= \begin{cases} \displaystyle\sum_{\kappa'=0}^{\frac{n-1}{2}} h(2\kappa') \cdot \beta\left(k' - \kappa'\right) & k = 2k' \\[4mm] \displaystyle\sum_{\kappa'=0}^{\frac{n-1}{2}} h(2\kappa' + 1) \cdot \alpha\left(k' - \kappa'\right) & k = 2k' + 1. \end{cases} \tag{4.69c}$$

Due to (4.68), the first summation in (4.69c) contributes to the even time instances $k = 2k'$ and the second summation to the odd instances $k = 2k' + 1$ only.

The block diagram of the efficient structure is given in Figure 4.24b. The multiplexing, i.e., interlacing of the even and the odd samples, can be described by upsampling by factor the $r = 2$ in combination with the delay and sum operation.

An example application is the QMF bank of the G.722 wideband speech codec (see Section 8.7.4) with $M = 2$ channels and a prototype lowpass filter with $n = 24$. The frequency responses of the lowpass and highpass filters, and the overall frequency response of the analysis–synthesis filter bank, are given in Figure 4.25.

The QMF concept can easily be modified to achieve non-uniform frequency resolution by leaving out some of the filters of the tree structure. For a QMF bank with $M = 8$, the

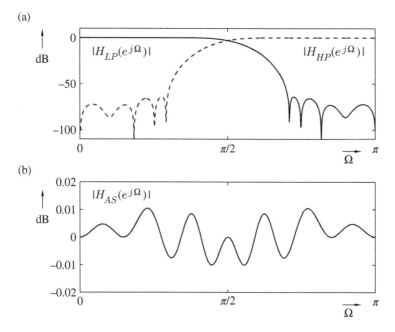

**Figure 4.25** Example: QMF bank of the G.722 wideband speech codec with $M = 2$ channels. (a) Magnitude response of the half-band filters. (b) Resulting overall magnitude response.

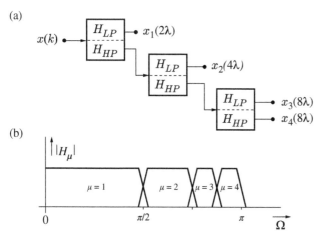

**Figure 4.26** QMF tree structure for non uniform frequency resolution. (a) Block diagram. (b) Schematic of the effective frequency resolution.

structure of the analysis filter bank and a schematic of the frequency response are illustrated in Figure 4.26. It should be noted that this filter bank structure is closely related to the wavelet transform, e.g., [Burrus et al. 1998], [Vetterli, Kovačević 1995], and [Vaidyanathan 1993b].

## 4.4 Filter Bank Equalizer

In this section, a digital filter structure is described for fixed or adaptive spectral modification of speech and audio signals. It is derived from a reference filter bank and is called filter bank equalizer (FBE). Main applications are audio equalization and noise reduction. The signal is filtered in the time domain, while the filter coefficients are calculated in the (short-term) frequency domain. In comparison to the usual short-term spectral-domain processing as discussed in Section 12.3, the proposed structure exhibits distinct advantages with respect to signal delay, prototype-filter design, and complexity. The algorithm allows for uniform, as well as for non-uniform spectral resolution [Vary 2006].

### 4.4.1 The Reference Filter Bank

The task of the FBE can be described by the reference filter bank, as shown in Figure 4.27. The input signal $x(k)$ with discrete time index $k$ is decomposed into $K + 1$ real-valued sub-band signals $y_\mu(k)$ by using $K + 1$ BP filters with impulse responses $h_\mu(k)$ of length $L$:

$$y_\mu(k) = \sum_{\kappa=0}^{L-1} x(k - \kappa) \cdot h_\mu(\kappa); \quad \mu = 0, 1, \dots K. \tag{4.70}$$

The spectral modification of $x(k)$, e.g., for the purpose of audio equalization or adaptive noise reduction, is achieved by applying an individual (fixed or time-variant) gain $g_\mu$ to each sub-band signal $y_\mu(k)$. Finally, the output signal is obtained by superposition of the weighted sub-band signals

$$y(k) = \sum_{\mu=0}^{K} g_\mu \cdot y_\mu(k). \tag{4.71}$$

For the purpose of noise reduction, the time-variant gain factors $g_\mu$ are calculated adaptively according to some spectral gain rule (often denoted as "spectral subtraction"

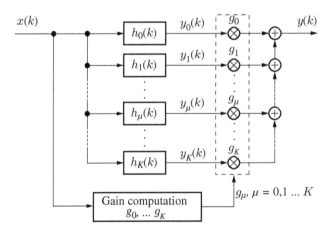

**Figure 4.27** The reference filter bank.

[Boll 1979]), see also Chapter 12. In this case, the gain factors are real-valued and in the range

$$0 \le g_\mu \le 1 \; ; \; \mu = 0, 1, \dots K. \tag{4.72}$$

These gains depend on the instantaneous signal-to-noise ratios of the sub-band signals $y_\mu(k)$. In contrast to conventional "spectral-subtraction," as described in Chapter 12, there is no DFT-based analysis–synthesis filter bank in the signal-path, as the noisy signal is processed without sample-rate decimation in the time domain. The gain multipliers might be calculated either from the sub-band signals $y_\mu(k)$ or, as indicated in Figure 4.27, from the input signal $x(k)$. In the latter case, the same frequency domain procedures known from DFT-based "spectral-subtraction" may be used inside the block *Gain Computation*. A drawback of the reference system of Figure 4.27 is the high computational complexity of $K + 1$ separate filters. In the following section, a much more efficient but equivalent time-domain filter structure, *the FBE*, is derived.

### 4.4.2 Uniform Frequency Resolution

It is assumed that, the $K + 1$ BP filters in Figure 4.27 are derived from an FIR prototype-lowpass $h(k)$ of length $L$ by cosine-modulation:

$$h_\mu(k) = \begin{cases} h(k) & ; \; \mu = 0 \\ h(k) \cdot 2 \cdot \cos(\frac{\pi}{K}\mu(k - k_0)) & ; \; 1 \le \mu \le K - 1 \\ h(k) \cdot (-1)^{(k-k_0)} & ; \; \mu = K, \end{cases} \tag{4.73}$$

where $k_0$ denotes a constant delay or phase parameter, which is required to achieve perfect reconstruction and/or a linear-phase frequency response.

The corresponding frequency responses $H_\mu(e^{j\Omega})$ can be represented with the BP center frequencies $\Omega_\mu = \frac{\pi}{K}\mu$ as follows:

$$H_\mu(e^{j\Omega}) = \begin{cases} H(e^{j\Omega}) & ; \; \mu = 0 \\ H(e^{j(\Omega-\Omega_\mu)}) \cdot e^{-j\Omega_\mu k_0} + H(e^{j(\Omega+\Omega_\mu)}) \cdot e^{+j\Omega_\mu k_0} & ; \; 1 \le \mu \le K - 1 \\ H(e^{j(\Omega-\pi)}) \cdot e^{j\pi k_0} & ; \; \mu = K. \end{cases} \tag{4.74}$$

A schematic representation of the magnitude responses is given in Figure 4.28.

If the gain multipliers $g_\mu$ are assumed to be constant, at least for limited time periods, the filter bank of Figure 4.27 is to be considered a linear time-invariant system. Thus, it can be represented by its overall impulse response $h_s(k)$, which is the sum of the individually weighted impulse responses $g_\mu \cdot h_\mu(k)$

$$h_S(k) = \sum_{\mu=0}^{K} g_\mu \cdot h_\mu(k) \tag{4.75a}$$

$$= g_0 \, h(k) + 2 \sum_{\mu=1}^{K-1} g_\mu \, h(k) \cos\left(\frac{\pi}{K}\mu(k - k_0)\right) + g_K \, h(k)(-1)^{(k-k_0)}$$

$$= h(k) \cdot w(k) \tag{4.75b}$$

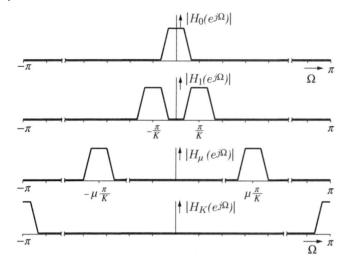

**Figure 4.28** Schematic representation of the magnitude frequency responses of the cosine-modulated BP filters.

with

$$w(k) = g_0 + 2 \sum_{\mu=1}^{K-1} g_\mu \cdot \cos\left(\frac{\pi}{K}\mu(k - k_0)\right) + g_K \cdot (-1)^{(k-k_0)} \tag{4.76}$$

for integer $k_0$. The relation (4.76) might be interpreted as Cosine transformation of the sub-band weights $g_\mu$.

Furthermore, (4.75b) suggests that the complete reference filter-bank can be replaced by a *single* equivalent filter, the *FBE*. The overall impulse response $h_S(k)$ is the product of the prototype lowpass impulse response $h(k)$ with the effective time-domain *weighting function* $w(k)$. The corresponding implementation is shown in Figure 4.29.

Due to properties of the cosine functions, $w(k)$ is symmetric and periodic

$$w(k + \ell \cdot 2K) = w(k) ; \ \ell \in \{1, 2, 3, \dots\}. \tag{4.77}$$

A time limitation is introduced implicitly by the multiplication with $h(k)$.

The reference filter bank of Figure 4.27 can alternatively be described using $M = 2K$ *complex* BP impulse responses $\tilde{h}_\mu(k)$ instead of the $K + 1$ real-valued impulse responses $h_\mu(k)$ (4.73):

$$\tilde{h}_\mu(k) = h(k) \cdot e^{+j\frac{2\pi}{2K}\mu(k-k_0)}; \ \mu = 0, 1, \dots 2K - 1; \ k = 0, 1, \dots L - 1. \tag{4.78}$$

Furthermore, the number of $K + 1$ gain multipliers $g_\mu$, $\mu = 0, 1, \dots K$ is increased to $M = 2K$, $\mu = 0, 1, \dots M - 1 = 2K - 1$ by mirrored repetition

$$g_{2K-\mu} \stackrel{!}{=} g_\mu ; \ \mu = 1, 2, \dots K. \tag{4.79}$$

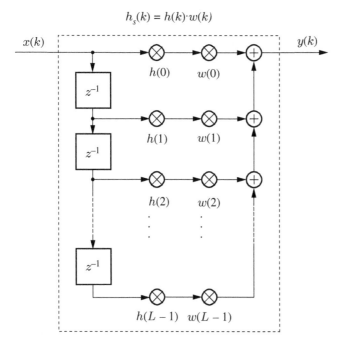

**Figure 4.29** Efficient implementation of the filter bank equalizer (FBE).

Thus, the overall real-valued impulse response of the reference filter bank can also be formulated as

$$h_S(k) = \sum_{\mu=0}^{2K-1} g_\mu \cdot \tilde{h}_\mu(k) \tag{4.80a}$$

$$= h(k) \cdot \sum_{\mu=0}^{2K-1} g_\mu \cdot e^{+j\frac{2\pi}{2K}\mu(k-k_0)} \tag{4.80b}$$

$$= h(k) \cdot w(k). \tag{4.80c}$$

As result, the weighting function $w(k)$ in (4.80c) can be calculated by Inverse Discrete Fourier Transform (IDFT)[1] of length $M = 2K$

$$w(k) = \sum_{\mu=0}^{2K-1} g_\mu \cdot e^{+j\frac{2\pi}{2K}\mu(k-k_0)} \tag{4.81a}$$

$$= \text{IDFT}\left\{ g_\mu \cdot e^{-j\frac{\pi}{K}\mu k_0} \right\}. \tag{4.81b}$$

Equation (4.81b) finally leads to the computationally efficient structure of Figure 4.30.

It can easily be shown, that because of the symmetry (4.79) of the real-valued gain multipliers $g_\mu$, the time-domain weighting functions $w(k)$ in (4.80a) to (4.80c) and (4.75a) to (4.75b) are exactly the same. Interestingly, the complex impulse responses (4.78) can also

---

1 The IDFT as defined in (3.26) would strictly include a scaling factor $1/M = 1/2K$, which is omitted here. This scaling factor can be merged with the prototype impulse response $h(k)$, as shown below.

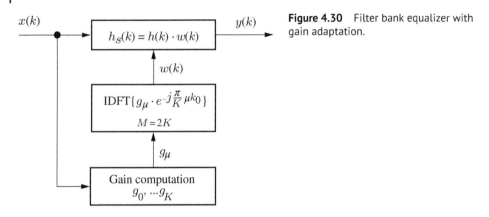

**Figure 4.30** Filter bank equalizer with gain adaptation.

be defined with a negative sign in the exponent. In this case, the time-domain weighting function is obtained [Löllmann, Vary 2005a], [Vary 2006] as

$$w(k) = \text{DFT}\left\{ g_\mu \cdot e^{+j\frac{\pi}{K}\mu k_0} \right\}. \tag{4.82}$$

In the special case of (4.81b) with all $g_\mu = 1$, the filter $h(k) \cdot w(k)$ in Figure 4.30 should deliver a perfect reconstruction of the input (except for a constant delay $k_0$)

$$y(k) \overset{!}{=} x(k - k_0). \tag{4.83}$$

This can be achieved by means of a suitably designed prototype impulse response $h(k)$ of arbitrary length $L$ and a delay parameter

$$k_0 = p \cdot K , \; p \in \{0, 1, 2, \ldots\}. \tag{4.84}$$

Then, the phase term in (4.81b) can be simplified:

$$e^{-j\frac{\pi}{K}\mu k_0} = e^{-j\pi\mu p} = (-1)^{\mu \cdot p} = \begin{cases} (-1)^\mu & ; \; p \text{ odd} \\ 1 & ; \; p \text{ even.} \end{cases} \tag{4.85}$$

Perfect reconstruction of $x(k)$ should be accomplished with unity gains

$$g_\mu = 1, \; \mu = 0, 1 \ldots 2K - 1. \tag{4.86}$$

Then, the window coefficients $w(k)$ (4.81b) with $k_0$ according to (4.84) are given by

$$w(k) = \sum_{\mu=0}^{2K-1} 1 \cdot e^{-j\pi \cdot \mu \cdot p} \cdot e^{+j\frac{2\pi}{2K}\mu \cdot k}. \tag{4.87}$$

The substitution $k = m \cdot K + v$ with $m = 0, 1, \ldots$ and $v = 0, 1, \ldots K - 1$ yields

$$w(k = m \cdot K + v) = \sum_{\mu=0}^{2K-1} 1 \cdot e^{j\pi \cdot \mu \cdot (m-p)} \cdot e^{+j\frac{2\pi}{2K}\mu \cdot v}$$

$$= \begin{cases} 2K \cdot (-1)^{m-p} & ; \; v = 0 \\ 0 & ; \; v \neq 0. \end{cases} \tag{4.88}$$

Therefore, perfect reconstruction can be achieved, if the prototype impulse response $h(k)$ is designed such that it fulfills the following requirements

$$h(k) = \begin{cases} 1/M & ; \ k = k_0 = p \cdot K = p \cdot M/2 \ \text{(i.e., } m = p; \ v = 0) \\ 0 & ; \ k = k_0 \pm \lambda M, \ \lambda \in \{1, 2, 3, \ldots\} \\ \text{arbitrary} & ; \ \text{else.} \end{cases} \qquad (4.89)$$

The prototype impulse response $h(k)$ should take the value $1/M = 1/2K$ at time instant $k = k_0$, and equidistant zeros with minimum distance $2K$. The remaining coefficients of $h(k)$ can be chosen to meet amplitude scaling and spectral resolution requirements.[2] The overall signal delay of this *perfect reconstruction* system ($g_\mu = 1$) is $k_0$, even when $h(k)$ does not have a linear-phase response. In case of a linear-phase prototype impulse response $h(k)$ with length $L = N \cdot 2K$, and a selection of $p = N$, i.e., $k_0 = N \cdot K$, the overall system has a linear phase as well, even for *arbitrary* but symmetric real-valued gain coefficients $g_\mu$.

### 4.4.3 Adaptive Filter Bank Equalizer: Gain Computation

The FBE can be used for adaptive filtering, such as dynamic equalization, noise reduction (see Chapter 12), or near-end listening enhancement (NELE, see Chapter 11). As a representative example, noise reduction will be discussed here.

#### 4.4.3.1 Conventional Spectral Subtraction

The conventional "spectral subtraction" approach to noise reduction (e.g., [Boll 1979], [Ephraim, Malah 1984], [Vary 1985], [Martin 2002], [Hänsler, Schmidt 2004], [Lotter, Vary 2005]) is based on short-term spectral analysis and synthesis using the DFT, i.e., spectral-domain filtering. The concepts are explained in detail in Chapter 12. Spectral subtraction actually means that the DFT coefficients $X_\mu(k)$ of the noisy signal $x(k)$ are multiplied by real-valued gain factors $0 \leq g_\mu \leq 1$

$$Y_\mu(k) = g_\mu \cdot X_\mu(k) \ ; \ \mu = 0, 1, \ldots M - 1. \qquad (4.90)$$

The gain factors $g_\mu(k)$ are derived by some spectral subtraction rule from the magnitude squared DFT coefficients $|X_\mu(k)|^2$ or periodogram (5.111), (5.114), calculated frame by frame, using a sliding time-domain window. The sliding-window DFT can be interpreted, as shown in Section 4.1.6, as a uniform filter bank, where the window is to be understood as the prototype lowpass impulse response $h(k)$ of the filter bank. The magnitude DFT coefficients of the input signal are given for $L \leq M = 2K$ by

$$\left| X_\mu(k) \right| = \left| \sum_{\kappa=0}^{L-1} x(k - \kappa) \cdot h(\kappa) \cdot e^{-j\frac{2\pi}{2K}\mu(\kappa)} \right|. \qquad (4.91)$$

Finally, the modified spectrum $Y_\mu(k)$ is converted back to the time domain by IDFT. The DFT/IDFT resolutions in frequency and time (transform length $M$, window function, and transform overlap) have to be selected such that both the requirements for *signal reconstruction* without aliasing and appropriate *gain computation* are fulfilled.

---

2 Note: The scaling factor $1/M$ in (4.89) can be realized as part of the IDFT.

### 4.4.3.2 Filter Bank Equalizer

The same gain factors $g_\mu$ can be applied to the sub-band-signals $y_\mu(k)$ of the reference filter-bank of Figure 4.27 as well as to the FBE of Figure 4.30. For deriving a relation between these two systems, the reference filter bank with $M = 2K$ *complex* BP impulse responses $\tilde{h}_\mu(k)$ (4.78) of length $L \le M$ is considered again.

These complex BP filters would produce complex sub-band signals

$$\tilde{y}_\mu(k) = \sum_{\kappa=0}^{L-1} x(k-\kappa) \cdot \tilde{h}_\mu(\kappa) \tag{4.92a}$$

$$= \sum_{\kappa=0}^{L-1} x(k-\kappa) \cdot h(\kappa) \cdot e^{+j\frac{2\pi}{2K}\mu(\kappa-k_0)} \tag{4.92b}$$

$$= e^{-j\frac{2\pi}{2K}\mu k_0} \cdot \sum_{\kappa=0}^{L-1} x(k-\kappa) \cdot h(\kappa) \cdot e^{+j\frac{2\pi}{2K}\mu\kappa}. \tag{4.92c}$$

The comparison of (4.92c) with (4.91) reveals the following relation

$$\left|\tilde{y}_\mu(k)\right| = \left|\sum_{\kappa=0}^{L-1} x(k-\kappa) \cdot h(\kappa) \cdot e^{+j\frac{2\pi}{2K}\mu(\kappa-k_0)}\right| \tag{4.93a}$$

$$= \left|e^{-j\frac{2\pi}{2K}\mu k_0}\right| \cdot \left|\sum_{\kappa=0}^{L-1} x(k-\kappa) \cdot h(\kappa) \cdot e^{+j\frac{2\pi}{2K}\mu\kappa}\right| \tag{4.93b}$$

$$= \left|X_\mu^*(k)\right|. \tag{4.93c}$$

As the product $x(k-\kappa) \cdot h(\kappa)$ is real-valued, we may also write

$$\left|\tilde{y}_\mu(k)\right| = |\mathrm{DFT}\{x(k-\kappa) \cdot h(\kappa)\}| \tag{4.94a}$$

$$= \left|X_\mu(k)\right|. \tag{4.94b}$$

For each sampling instant $k$, all samples $|\tilde{y}_\mu(k)|^2$ for $\mu = 0, 1, \ldots M - 1$ could be calculated all at once by a single DFT. The superposition of the $M$ weighted signals $g_\mu \cdot \tilde{y}_\mu(k)$ would result in same real-valued output signal $y(k)$ as produced by the reference filter bank of Figure 4.27. However, the really relevant conclusion from (4.94a) is that the squared magnitudes of the sub-band signals are the same as the squared magnitudes of the DFT coefficients. Thus, the DFT-based spectral subtraction rules for adapting the gain factors $g_\mu$ can be used for the FBE as well.

This concept is illustrated in Figure 4.31. The DFT input samples are the weighted samples $x(k-\kappa) \cdot h(\kappa)$. The linear phase correction of the gains $g_\mu$ by the complex multipliers $e^{-j\frac{2\pi}{2K}\mu k_0}$ according to (4.81a) and (4.81b) can alternatively be introduced in the time domain by a cyclic shift of the weights $w(\mu)$ by $k_0$ samples.

According to the sampling theorem, the samples $\tilde{y}_\mu(k)$ could be calculated at decimated time instances $k = k' = \lambda \cdot r$, $\lambda = 0, 1, \ldots$, because of the reduced bandwidth. In the conventional DFT/IDFT approach of spectral subtraction, the decimation factor $r$ is constrained

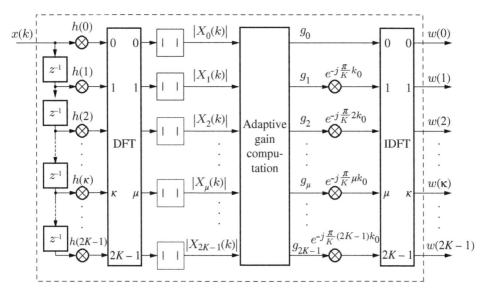

**Figure 4.31** Adaptive gain $g_\mu$ and weight $w(\kappa)$ computation: $L = 2K$; the DFT/IDFT are calculated at the decimated time instances $k = k'$ only.

first of all by the need to re-synthesize the output signal $y(k)$ at the original sampling rate without aliasing effects as far as possible.

In contrast, the FBE approach is not constrained by the signal-synthesis. The filter $h_s(k) = h(k) \cdot w(k)$ in Figure 4.30 is operated at the input sample rate, while for the gain adaptation the sample rate decimation factor $r$ can be chosen to meet only the required time-resolution for the adaptation of the gains $g_\mu$. The DFT of the sequence $x(k' - \kappa) \cdot h(\kappa)$, $\kappa = 0, 1, 2, ...M - 1 = 2K - 1$, can be calculated with a larger decimation factor $r$. While the conventional spectral subtraction with re-synthesis of the output signal $y(k)$, requires typically a decimation factor of $r \leq M/2$, the FBE adaptation could be run even without frame overlap in time, i.e., $r \geq M$. For the energy calculations of the sub-band envelope signals $|\tilde{y}_\mu(k)|^2 = |X_\mu(k)|^2$ a much lower sampling frequency is sufficient. For the gain adaptation, reliable statistical estimates of (short-term) power spectra are required, which could be obtained by using a statistically sufficient number of samples taken even at irregular time instances.

This conclusion is obvious for a prototype impulse response $h(k)$ with length $L \leq M$. If the length $L$ of the impulse response $h(k)$ is shorter than the DFT length $M = 2K$, it can be extended by zero padding.

However, if the prototype lowpass has more than $M = 2K$ coefficients, e.g.,

$$L = N \cdot 2K = N \cdot M$$

the "windowed" signal segment of length $L$ can be converted into a modified sequence $u_\lambda(k')$ with DFT length $M$ by *polyphase processing* of Section 4.2 using the approach A of overlapping of windowed segments as described in Section 4.2.1, equation (4.32e) and Figure 4.13 (e.g., [Vary, Heute 1980], [Vaidyanathan 1993a]).

### 4.4.4  Non-uniform Frequency Resolution

A well-known technique for (non-uniform) frequency warping of digital filters consists in replacing the delay elements $z^{-1}$ by first-order allpass filters $H_A(z)$. This approach has originally been used for the design and implementation of variable digital filters, e.g., for deriving a lowpass filter with cutoff frequency $\Omega_c$ from a prototype lowpass with a normalized cutoff frequency $\Omega_0$ (e.g., [Schüßler 1980], [Schüßler, Kolb 1982]). This warping technique can also be used to achieve non-uniform resolution by DFT (e.g., [Braccini, Oppenheim 1974], [Makur, Mitra 2001]) or by filter banks (e.g. [Vary 1980], [Vary 2006], [Doblinger 1991], [Löllmann, Vary 2005a], [Kates, Arehart 2005]).

This transformation technique is applied here first to the reference filter bank of Figure 4.27. In the $z$-transforms

$$H_\mu(z) = \sum_{k=0}^{L-1} h_\mu(k) \cdot z^{-k} \quad ; \quad \mu = 0, \dots K-1 \tag{4.95}$$

of the BP filter impulse responses, the delay elements $z^{-1}$ are substituted by allpass transfer functions $H_A(z)$

$$H_\mu^t(z) = \sum_{k=0}^{L-1} h_\mu(k) \cdot H_A^k(z) \tag{4.96}$$

with

$$H_A(z) = \frac{z^{-1} - a}{1 - a \cdot z^{-1}} \quad ; -1 < a < +1. \tag{4.97}$$

As the magnitude frequency response of each allpass filter is constant with

$$H_A(e^{j\Omega}) = 1 \cdot e^{-j\varphi_A(\Omega)}, \tag{4.98}$$

the frequency axis $\Omega$ is mapped (warped) to $\Omega^t$ according to the phase characteristic of the allpass element

$$\Omega^t \doteq \varphi_A(\Omega) = 2 \cdot \arctan\left[\frac{1+a}{1-a} \tan\left(\frac{\Omega}{2}\right)\right]. \tag{4.99}$$

Typical frequency warping characteristics are illustrated in Figure 4.32.

Equidistant spacing of the warped frequency axis $\Omega^t$ corresponds to non equidistant spacing of the original frequency axis $\Omega$. For parameter values $a > 0$, a higher frequency resolution is achieved at low frequencies at the expense of a lower resolution at high frequencies. For values $a < 0$, the opposite is true. The original frequency scale is obtained for $a = 0$, as the allpass (4.97) becomes a pure delay element $H_A(z) = z^{-1}$ then.

If this allpass transformation is applied to the reference filter bank of Figure 4.27, each BP filter is transformed individually:

$$H_\mu^t(e^{j\Omega}) = \sum_{\kappa=0}^{L-1} h_\mu(\kappa) \cdot [e^{-j\varphi_A(\Omega)}]^\kappa = h_\mu(e^{j\Omega^t}). \tag{4.100}$$

It should be noted that the original finite impulse response $h_S(k)$ of the overall system is converted into an infinite one.

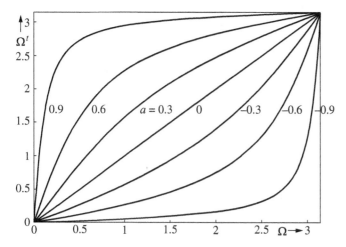

**Figure 4.32**  Non-uniform frequency warping $\Omega \to \Omega^t$ by allpass transformation for parameter values $a$.

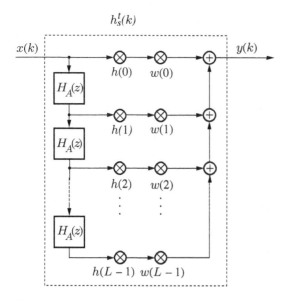

**Figure 4.33**  Frequency warping of the filter bank equalizer by allpass transformation.

With this frequency transformation a very good approximation to the so-called Bark frequency scale (see also Table 2.1) can be achieved for $a = 0.565$ and $K + 1 = 21$ ($M = 2K = 40$) transformed BP filters in the frequency range from 0 to 8 kHz (sampling rate $f_s = 16$ kHz).

This (allpass) transformation technique can also be applied to the FBE $h_s(k)$ of Figures 4.29 and 4.30, as shown in Figure 4.33. Thus, the frequency response is transformed according to the warping characteristics of Figure 4.32, with a correspondingly warped overall impulse response $h_s^t(k)$.

The warped frequency response $H_s^t(e^{j\Omega})$ is the same as if the allpass transformation (4.97) would be applied to each of the BP filters of the reference system of Figure 4.27. Due to the allpass transformation the overall phase response becomes non-linear. If (near) perfect reconstruction with respect to magnitude and phase is required, a phase equalization has to be applied. A solution for the FBE is proposed in [Löllmann, Vary 2005a].

For the DFT-based gain computation, as illustrated in Figure 4.31, allpass filters according to (4.97) have to be used instead of the delay elements $z^{-1}$. A similar structure, which is to be considered as a special case of Figure 4.33, was proposed in [Kates, Arehart 2005] for the use in hearing aids.

### 4.4.5 Design Aspects & Implementation

The main characteristics of the FBE are:

1. Structure:
   The signal $x(k)$ is filtered in the time domain, while the spectral / sub-band gains $g_\mu$ can be calculated in the frequency domain. Thus, the filtering operation and the adaptation are decoupled. This opens new design and implementation aspects. The input signal $x(k)$ is convolved with the effective time-domain impulse responses $h_s(k)$ (uniform spectral resolution, Figure 4.29) or $h_s^t(k)$ (non-uniform spectral resolution, Fig. 4.33) at the original sample rate.
2. Lower design constraints:
   In designing the prototype frequency response, the perfect reconstruction condition can be met easily. Aliasing is not an issue, as no signal re-synthesis is required. Thus, a shorter prototype lowpass / DFT window may be used due less demanding spectral selectivity requirements in terms of transition steepness and stopband attenuation of the frequency response.
3. Reduced algorithmic delay
   As a consequence of decoupling of signal filtering and gain adaptation, the signal delay is not determined by the DFT block length $M = 2K$ or by the decimation factor $r$, but only by the latency of the prototype lowpass. Additional latency reduction can be achieved, by replacing the overall impulse response $h_s(k)$ by a minimum phase filter approximation (FIR or even IIR) [Löllmann, Vary 2007].
4. Choice of the decimation factor $r$:
   With respect to the calculation of the spectral gain factors $g_\mu$, the decimation factor $r$ can be chosen almost freely. Aliasing in the sub-band signals does not play a role, if the spectral subtraction rule is based on the measurement of short-term signal powers in sub-bands.
5. Uniform and non-uniform frequency resolution:
   The uniform spectral resolution of FBE can be converted to a non-uniform resolution, by replacing the delay elements of the time domain filter $h_s(k)$ by allpass filters of first order. This allows to achieve a frequency resolution close to the Bark frequency scale. However, it should be noted that the allpass chain has to be operated without decimation at the original sampling rate.

6. Applications

   The FBE is particular suitable for applications that require low latency. This is the case, for example, in hearing aids. The structures of Figures 4.30 and 4.33 have been implemented for the purpose of noise reduction. The same amount of subjective noise suppression was achieved as with the conventional analysis–synthesis filter bank approach, but with lower latency and complexity.

   Further details can be found in [Löllmann, Vary 2005b], [Löllmann, Vary 2007], [Löllmann, Vary 2008], [Vary 2006], and [Löllmann 2011].

## References

Bellanger, M.; Bonnerot, G.; Coudreuse, M. G. (1976). Digital Filtering by Polyphase Network: Application to Sampling-rate Alteration and Filter Banks, *IEEE Transactions on Acoustics, Speech and Signal Processing*, vol. 24, pp. 109–114.

Boite, R.; Leich, H. (1981). A New Procedure for the Design of High-Order Minimum-Phase FIR Digital or CCD Filters, *EURASIP Signal Processing*, vol. 3, pp. 101–108.

Boll, S. F. (1979). Suppression of Acoustic Noise in Speech Using Spectral Subtraction, *IEEE Transactions on Acoustics, Speech and Signal Processing*, vol. 27, no. 2, pp. 113–120.

Braccini, C.; Oppenheim, A. V. (1974). Unequal Bandwidth Spectral Analysis Using Digital Frequency Warping, *IEEE Transactions on Acoustics, Speech and Signal Processing*, vol. 22, no. 4, pp. 236–244.

Burrus, C. S.; Gopinath, R. A.; Guo, H. (1998). *Introduction to Wavelets and Wavelet Transforms: A Primer*, Prentice Hall, Upper Saddle River, New Jersey.

Crochiere, R. E.; Rabiner, L. R. (1983). *Multirate Digital Signal Processing*, Prentice Hall, Englewood Cliffs, New Jersey.

Dehner, G. (1979). Program for the Design of Recursive Digital Filters, *in Programs for Digital Signal Processing*, IEEE Press, New York.

Doblinger, G. (1991). An Efficient Algorithm for Uniform and Nonuniform Digital Filter Banks, *1991., IEEE International Sympoisum on Circuits and Systems*, Singapore, vol. 1, pp. 646–649.

Ephraim, Y.; Malah, D. (1984). Speech Enhancement Using a Minimum Mean-Square Error Short-Time Spectral Amplitude Estimator, *IEEE Transactions on Acoustics, Speech and Signal Processing*, vol. 32, no. 6, pp. 1109–1121.

Esteban, D.; Galand, C. (1977). Application of Quadrature Mirror Filters to Split-Band Voice Coding Schemes, *Proceedings of the IEEE International Conference on Acoustics, Speech, and Signal Processing (ICASSP)*, pp. 191–195.

Hänsler, H.; Schmidt, G. (2004). *Acoustic Echo and Noise Control - A Practical Approach*, Wiley - Interscience.

Jain, V. K.; Crochiere, R. E. (1983). A Novel Approach to the Design of Analysis-Synthesis Filter Banks, *Proceedings of the IEEE International Conference on Acoustics, Speech, and Signal Processing (ICASSP)*, pp. 228–231.

Jain, V. K.; Crochiere, R. E. (1984). Quadrature-Mirror Filter Design in the Time Domain, *IEEE Transactions on Acoustics, Speech and Signal Processing*, vol. 32, pp. 353–361.

Johnston, J. D. (1980). A Filter Family Designed for Use in Quadrature-Mirror Filter Banks, *Proceedings of the IEEE International Conference on Acoustics, Speech, and Signal Processing (ICASSP)*, Denver, Colorado, USA, pp. 291–294.

Kates, J. M.; Arehart, K. H. (2005). Multi-Channel Dynamic Range Compression Using Digital Frequency Warping, *EURASIP Journal on Applied Signal Processing*, vol. 18, pp. 3003–3014.

Kliewer, J. (1996). Simplified Design of Linear-Phase Prototype Filters for Modulated Filter Banks, *in* G. Ramponi; G. L. Sicuranza; S. Carrato; S. Marsi (eds.), *Signal Processing VIII: Theories and Applications*, Elsevier, Amsterdam, pp. 1191–1194.

Löllmann, H. W. (2011). *Allpass-Based Analysis-Synthesis Filter-Banks: Design and Application*, Dissertation. Aachener Beiträge zu Digitalen Nachrichtensystemen (ABDN), P. Vary (ed.), IND, RWTH Aachen.

Löllmann, H. W.; Vary, P. (2005a). Efficient Non-Uniform Filter-Bank Equalizer, *EUSIPCO*, Antalya, Turkey.

Löllmann, H. W.; Vary, P. (2005b). Generalized Filter-Bank Equalizer for Noise Reduction with Reduced Signal Delay, *Interspeech*, Lisboa, pp. 2105–2108.

Löllmann, H. W.; Vary, P. (2007). Uniform and Warped Low Delay Filter-Banks for Speech Enhancement, *Speech Communication*, vol. 49, no. 7–8, pp. 574–587. Special Issue on Speech Enhancement.

Löllmann, H. W.; Vary, P. (2008). Low Delay Filter-Banks for Speech and Audio Processing, *in* E. Hänsler; G. Schmidt (eds.), *Speech and Audio Processing in Adverse Environments*, Springer-Verlag, Berlin, Heidelberg, Chapter 2, pp. 13–61.

Lotter, T.; Vary, P. (2005). Speech Enhancement by MAP Spectral Amplitude Estimation Using a Super-Gaussian Speech Model, *EURASIP Journal on Applied Signal Processing*, vol. 2005, p. 354850.

Makur, A.; Mitra, S. K. (2001). Warped Discrete Fourier Transform: Theory and Application, *IEEE Transactions on Circuits and Systems I: Fundamental Theory and Applications*, vol. 48, no. 9, pp. 1086–1093.

Martin, R. (2002). Speech Enhancement Using MMSE Short Time Spectral Estimation with Gamma Distributed Speech Priors, *Proceedings of the IEEE International Conference on Acoustics, Speech, and Signal Processing (ICASSP)*, Orlando, Florida, USA, vol. 6, pp. 253–256.

Mitra, K. M. (1998). *Digital Signal Processing - A Computer Based Approach*, McGraw-Hill, New York.

Nguyen, T. (1994). Near-Perfect-Reconstruction Pseudo-QMF Banks, *IEEE Transactions on Signal Processing*, vol. 42, pp. 64–75.

Parks, T. W.; Burrus, C. S. (1987). *Digital Filter Design*, John Wiley & Sons, Ltd., Chichester.

Parks, T. W.; McClellan, J. H.; Rabiner, L. R. (1979). FIR Linear-Phase Filter-Design Program, *in Programs for Digital Signal Processing*, IEEE Press, New York.

Schüßler, H. W. (1980). Implementation of Variable Digital Filters, *Proceedings of the European Signal Processing Conference (EUSIPCO)*, Lausanne, Switzerland, pp. 123–129.

Schüssler, H. W. (1994). *Digitale Signalverarbeitung I*, Springer-Verlag, Berlin (in German).

Schüßler, H. W.; Kolb, H. J. (1982). Variable Digital Filters, *Archiv fur Elektronik und Ubertragungstechnik*, vol. 36, pp. 229–237.

Smith, M. J. T.; Barnwell, T. P. (1986). Exact Reconstruction Techniques for Tree-structured Subband Coders, *IEEE Transactions on Acoustics, Speech and Signal Processing*, vol. 34, pp. 434–441.

Vaidyanathan, P. P. (1990). Multirate Digital Filters, Filter Banks, Polyphase Networks, and Applications: A Tutorial, *Proceedings of the IEEE*, vol. 78, pp. 56–93.

Vaidyanathan, P. P. (1993a). *Multirate Systems and Filter Banks*, Prentice Hall.

Vaidyanathan, P. P. (1993b). *Multirate Systems and Filterbanks*, Prentice Hall, Englewood Cliffs, New Jersey.

Vary, P. (1979). On the Design of Digital Filter Banks Based on a Modified Principle of Polyphase, *International Journal of Electronics and Communications (AEÜ, Archiv für Elektronik und Übertragungstechnik)*, vol. 33, pp. 293–300.

Vary, P. (1980). Digital Filter Banks with Unequal Resolution, *EUSIPCO Short Communication Digest*, Lausanne, Switzerland, pp. 41–42.

Vary, P. (1985). Noise Suppression by Spectral Magnitude Estimation - Mechanism and Theoretical Limits, *Signal Processing (Elsevier)*, vol. 8, no. 4, pp. 378–400.

Vary, P. (2006). An Adaptive Filterbank Equalizer for Speech Enhancement, *Signal Processing*, vol. 86, pp. 1206–1214.

Vary, P.; Heute, U. (1980). A Short-time Spectrum Analyzer with Polyphase-Network DFT, *Signal Processing*, vol. 2, pp. 55–65.

Vary, P.; Heute, U. (1981). A Digital Filter Bank with Polyphase Network and FFT Hardware: Measurements and Applications, *Signal Processing*, vol. 3, pp. 307–319.

Vary, P.; Wackersreuther, G. (1983). A Unified Approach to Digital Polyphase Filter Banks, *International Journal of Electronics and Communications (AEÜ, Archiv für Elektronik und Übertragungstechnik)*, vol. 37, pp. 29–34.

Vary, P.; Heute, U.; Hess, W. (1998). *Digitale Sprachsignalverarbeitung*, B. G. Teubner, Stuttgart (in German).

Vetterli, M.; Kovačević, J. (1995). *Wavelets and Subband Coding*, Prentice Hall, Engelwood Cliffs, New Jersey.

Wackersreuther, G. (1985). On the Design of Filters for Ideal QMF and Polyphase Filter Banks, *International Journal of Electronics and Communications (AEÜ, Archiv für Elektronik und Übertragungstechnik)*, vol. 39, pp. 123–130.

Wackersreuther, G. (1986). Some New Aspects of Filters for Filter Banks, *IEEE Transactions on Acoustics, Speech and Signal Processing*, vol. 34, pp. 1182–1200.

Wackersreuther, G. (1987). *Ein Beitrag zum Entwurf digitaler Filterbänke*, PhD thesis. Ausgewählte Arbeiten über Nachrichtensysteme, vol. 64, H. W. Schüßler (ed.), Universität Erlangen (in German).

# 5

# Stochastic Signals and Estimation

In this chapter, we will review the basic concepts and tools which are required to deal with stochastic signals such as speech signals. Among these are random variables and stochastic processes as well as power spectra and fundamentals of estimation theory. The objective is to provide a compilation of useful concepts and theorems. More extensive discussions of these subjects can be found in many excellent textbooks, for instance, [Papoulis, Unnikrishna Pillai 2001] and [Bishop 2006].

## 5.1  Basic Concepts

### 5.1.1  Random Events and Probability

Modern theory of probability [Kolmogorov 1933] defines the probability $P(A_i)$ of an event $A_i$ on the basis of set-theoretic concepts and axioms, not on the basis of *observed* random phenomena. It thus facilitates the treatment of random processes as it provides a clear conceptual separation between *observed* random phenomena and theoretical *models* of such phenomena.

Given a set $\mathfrak{Q}$ of elementary events $\{\xi_1, \xi_2, \ldots\}$ and a set $\mathfrak{F}$ of subsets $\{A_1, A_2, \ldots\}$ of $\mathfrak{Q}$, we call the subsets $A_i$ in $\mathfrak{F}$ random events. We assume that $\mathfrak{F}$ includes $\mathfrak{Q}$, the empty set $\phi$, any union $\cup A_i$ of subsets, and the complement $\overline{A}_i$ of any subset $A_i$.[1] The complement $\overline{A}_i$ of a subset $A_i$ is defined as the set $\mathfrak{Q}$ without the elementary events in $A_i$. We may then assign a probability measure $P(A_i)$ to a random event $A_i$ in such a way that

- the probability is a non negative real number: $P(A_i) \geq 0$;
- the probability of the certain event is one: $P(\mathfrak{Q}) = 1$;
- the probability that either $A_i$ or $A_j$ occurs is $P(A_i \cup A_j) = P(A_i) + P(A_j)$, provided that the two events $A_i$ and $A_j$ are mutually exclusive.

The last condition is illustrated in Figure 5.1a. When two random events $A_i$ and $A_j$ are disjoint, the probability of $A_i$ or $A_j$ is the sum of the individual probabilities. If these events are not mutually exclusive (Figure 5.1b), the sum of the probabilities of the

---

1 These properties define a $\sigma$-algebra [Kolmogorov 1933].

*Digital Speech Transmission and Enhancement*, Second Edition. Peter Vary and Rainer Martin.
© 2024 John Wiley & Sons Ltd. Published 2024 by John Wiley & Sons Ltd.

(a)                             (b)

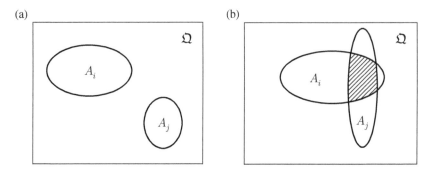

**Figure 5.1** Two random events $A_i$ and $A_j$ as subsets of the set of elementary events $\Omega$ (a) events are mutually exclusive and (b) events are not mutually exclusive.

individual events is not equal to the probability of the event that $A_i$ or $A_j$ occurs. The triple $(\Omega, \mathfrak{F}, P)$ is called a probability space.

The joint probability of two events $A_i$ and $A_j$ is the probability that $A_i$ and $A_j$ occur and is denoted by $P(A_i, A_j) = P(A_j, A_i)$. With respect to the set representation, the joint probability is a measure assigned to the intersection of events $A_i$ and $A_j$ in a space, which contains all possible joint events (product space).

In the general case of $N$ events $A_i$, $i = 1 \ldots N$, we denote the probability of the simultaneous occurrence of these events as $P(A_1, A_2, \ldots, A_N)$.

### 5.1.2 Conditional Probabilities

Conditional probabilities capture the notion of *a priori* information. The conditional probability of an event $B$, given an event $A$, with $P(A) \neq 0$, is defined as the joint probability $P(B, A)$ normalized on the probability of the given event $A$

$$P(B \mid A) = \frac{P(B, A)}{P(A)}. \tag{5.1}$$

When $N$ mutually exclusive events $A_1, A_2, \ldots, A_N$ partition the set of elementary events $\Omega$ in such a way that $P(A_1) + P(A_2) + \cdots + P(A_N) = 1$, we may write the *total probability* of an arbitrary event $B$ as

$$P(B) = P(B \mid A_1)P(A_1) + P(B \mid A_2)P(A_2) + \cdots + P(B \mid A_N)P(A_N).$$

Furthermore, for any of these events $A_i$, we may write

$$
\begin{aligned}
P(A_i \mid B) &= \frac{P(B, A_i)}{P(B)} \\
&= \frac{P(B \mid A_i)P(A_i)}{P(B)} \\
&= \frac{P(B \mid A_i)P(A_i)}{P(B \mid A_1)P(A_1) + P(B \mid A_2)P(A_2) + \cdots + P(B \mid A_N)P(A_N)}.
\end{aligned}
$$

This is also known as *Bayes' theorem*.

### 5.1.3 Random Variables

A random variable $x$ maps elementary events $\xi_i$ onto real numbers and, thus, is the basic vehicle for the mathematical analysis of random phenomena. In our context, random variables are used to represent samples of signals, parameters, and other quantities. A random variable may be continuous or discrete valued. While the former can attain any range of values on the real line, the latter is confined to countable and possibly finite sets of numbers.

Random variables may be grouped into vectors. Vectors of random variables may have two elements (bivariate) or a larger number of elements (multivariate). We may also define complex-valued random variables by mapping the real and the imaginary parts of a complex variable onto the elements of a bivariate vector of real-valued random variables.

### 5.1.4 Probability Distributions and Probability Density Functions

The *(cumulative) distribution function* $P_x$ of a random variable $x$ is defined as the probability that $x$ is smaller than or equal to a threshold value $u$,

$$P_x(u) = P(x \le u). \tag{5.2}$$

We obtain the whole distribution function by allowing $u$ to range from $-\infty$ to $\infty$. Obviously, $P_x$ is non negative and non decreasing with $P_x(-\infty) = 0$. $P_x(\infty)$ corresponds to the certain event, therefore $P_x(\infty) = 1$.

The *probability density function* (PDF) is defined as the derivative (when it exists) of the distribution function with respect to the threshold $u$

$$p_x(u) = \frac{dP_x(u)}{du}. \tag{5.3}$$

A PDF always satisfies

$$\int_{-\infty}^{\infty} p_x(u)\, du = 1. \tag{5.4}$$

Using Dirac impulses, a PDF may be defined for probability distribution functions $P_x(u)$ with discontinuities.

The joint PDF of $N$ random variables $x_1, x_2, \ldots, x_N$ is defined as

$$p_{x_1 \cdots x_N}(u_1, \ldots, u_N) = \frac{d^N P_{x_1 \cdots x_N}(u_1, \ldots, u_N)}{du_1 \ldots du_N}, \tag{5.5}$$

where $P_{x_1 \cdots x_N}(u_1, \ldots, u_N)$ is given by

$$P_{x_1 \cdots x_N}(u_1, \ldots, u_N) = P(x_1 \le u_1, \ldots, x_N \le u_N). \tag{5.6}$$

Given a joint PDF, we compute marginal densities by integrating over one or more variables. For example, given $p_{x_1 \cdots x_N}(u_1, \ldots, u_N)$ we compute the marginal density $p_{x_1 \cdots x_{N-1}}(u_1, \ldots, u_{N-1})$ as

$$p_{x_1 \cdots x_{N-1}}(u_1, \ldots, u_{N-1}) = \int_{-\infty}^{\infty} p_{x_1 \cdots x_N}(u_1, \ldots, u_N)\, du_N. \tag{5.7}$$

### 5.1.5 Conditional PDFs

We define the conditional PDF $p_{x|A}(u \mid A)$ of a random variable $x$ given an event $A$ via the relation

$$p_{x|A}(u \mid A) P(A) = p_{xA}(u, A),$$ (5.8)

where $p_{xA}(u, A)$ is defined as

$$p_{xA}(u, A) = \frac{dP_{xA}(u, A)}{du} = \frac{dP(x \leq u, A)}{du}.$$ (5.9)

The event $A$ may comprise a random variable $y$, which is equal to a value $v$. For a discrete random variable $y$ we may restate the above relation in terms of densities using the simplified notation

$$p_{x|y}(u \mid v) P(y = v) = p_{xy}(u, v).$$ (5.10)

This is called the *mixed form* of the Bayes' theorem, which also holds for continuous random variables, as

$$p_{x|y}(u \mid v) p_y(v) = p_{xy}(u, v).$$ (5.11)

Furthermore, we write the density version of Bayes' theorem as

$$p_{x|y}(u \mid v) p_y(v) = p_{xy}(u, v) = p_{y|x}(v \mid u) p_x(u),$$ (5.12)

or with (5.7) as

$$p_{x|y}(u \mid v) \int_{-\infty}^{\infty} p_{x,y}(u, v) \, du = p_{x|y}(u \mid v) \int_{-\infty}^{\infty} p_{y|x}(v \mid u) p_x(u) \, du.$$ (5.13)

Any joint probability density may be factored into conditional densities as

$$p(x_1, \ldots, x_N) = p(x_1 | x_2, \ldots, x_N) p(x_2, \ldots, x_N)$$
$$= p(x_1 | x_2, \ldots, x_N) p(x_2 | x_3, \ldots, x_N) p(x_3, \ldots, x_N)$$
$$= p(x_1 | x_2, \ldots, x_N) p(x_2 | x_3, \ldots, x_N) \cdots p(x_{N-1} | x_N) p(x_N).$$

For three random variables, $x_1$, $x_2$, and $x_3$, the above *chain rule* simplifies to

$$p(x_1, x_2, x_3) = p(x_1 | x_2, x_3) p(x_2 | x_3) p(x_3).$$ (5.14)

## 5.2 Expectations and Moments

For continuous random variables, the *mean* of a random variable $x$ is given by its *expected value* E $\{x\}$,

$$\mu_x = \text{E}\{x\} = \int_{-\infty}^{\infty} u \, p_x(u) \, du.$$ (5.15)

For discrete random variable $x$, which assumes one of $M$ values $u_i$, $i = 1 \ldots M$, we denote the discrete probability distribution by $p_x(u_i) = P(x = u_i)$ and obtain the mean

$$\mu_x = \sum_{i=1}^{M} u_i \, p_x(u_i). \tag{5.16}$$

The expectation is easily evaluated if the PDF of the random variable is known. In the remainder of this chapter, we will be mostly concerned with continuous random variables. All of these results can easily be adapted to discrete random variables by exchanging the integration for a summation.

More generally, we may write the expectation of any function $f(x)$ of a random variable $x$ as

$$E_x\{f(x)\} = \int_{-\infty}^{\infty} f(u) \, p_x(u) \, du, \tag{5.17a}$$

where, in a more concise notation, we indicate the random variable in the subscript of the expectation operator. Thus, despite the transformation of $x$ into $y = f(x)$, we may still use the PDF $p_x(u)$ to compute the expected value $E_x\{f(x)\}$. More specifically, with

$$f(x) = (x - x_0)^m,$$

we obtain the *m*th *central moment* with respect to $x_0$. For $x_0 = 0$ and $m = 2$, we have the *power* of $x$

$$E_x\{x^2\} = \int_{-\infty}^{\infty} u^2 \, p_x(u) \, du, \tag{5.17b}$$

and for $x_0 = \mu_x$ and $m = 2$ the *variance*

$$\sigma_x^2 = E_x\{(x - \mu_x)^2\} = E_x\{x^2\} - \mu_x^2. \tag{5.17c}$$

### 5.2.1 Conditional Expectations and Moments

The expected value of a random variable $x$ conditioned on an event $y = v$ is defined as

$$E_{x|y}\{x \mid y\} = \int_{-\infty}^{\infty} u \, p_{x|y}(u \mid v) \, du, \tag{5.18}$$

and, when no confusion is possible, also written as $E\{x \mid y\}$. Using the relation

$$E_{x|y}\{f(x) \mid y\} = \int_{-\infty}^{\infty} f(u) \, p_{x|y}(u \mid v) \, du \tag{5.19}$$

and $f(x) = (x - x_0)^m$ we may compute any *conditional* central moment.

### 5.2.2 Examples

In what follows, we will discuss some of the frequently used PDFs and their moments. For an extensive treatment of these and other PDFs, we refer the reader to [Papoulis, Unnikrishna Pillai 2001], [Johnson et al. 1994], and [Kotz et al. 2000].

### 5.2.2.1 The Uniform Distribution

When a random variable $x$ is uniformly distributed in the range $[X_1, X_2]$, its probability density is given by

$$p_x(u) = \begin{cases} \dfrac{1}{X_2 - X_1}, & u \in [X_1, X_2], \\ 0, & u \notin [X_1, X_2]. \end{cases} \tag{5.20}$$

Figure 5.2 illustrates this density. For the uniform density, we find the mean, the power, and the variance as follows:

$$\text{mean:} \qquad \mu_x = \frac{1}{2}\left[X_1 + X_2\right], \tag{5.21}$$

$$\text{power:} \qquad E_x\{x^2\} = \frac{1}{3}\left[X_1^2 + X_1 X_2 + X_2^2\right], \text{ and} \tag{5.22}$$

$$\text{variance:} \qquad \sigma_x^2 = \frac{1}{12}\left(X_2 - X_1\right)^2. \tag{5.23}$$

### 5.2.2.2 The Gaussian Density

The *Gaussian density*, also known as *normal distribution*, is defined as

$$p_x(u) = \frac{1}{\sqrt{2\pi}\sigma_x} \exp\left(-\frac{(u - \mu_x)^2}{2\sigma_x^2}\right), \tag{5.24}$$

and parameterized by its mean $\mu_x$ and variance $\sigma_x^2 > 0$. The Gaussian density is plotted for $\mu_x = 0$ and three different variances in Figure 5.3a. The power of a Gaussian distributed random variable is given by $E\{x^2\} = \sigma_x^2 + \mu_x^2$.

By definition, a complex Gaussian random variable is a pair of real-valued random variables, which are jointly Gaussian distributed. This definition includes the case of two independent real-valued Gaussian random variables which represent the real and the imaginary parts of the complex variable, see (5.131).

### 5.2.2.3 The Exponential Density

For $\sigma_x > 0$, the *(one-sided) exponential density* (see Figure 5.3b) is given by

$$p_x(u) = \begin{cases} \dfrac{1}{\sigma_x} \exp\left(-\dfrac{u}{\sigma_x}\right), & u \geq 0, \\ 0, & u < 0, \end{cases} \tag{5.25}$$

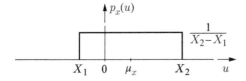

**Figure 5.2** Uniform probability density.

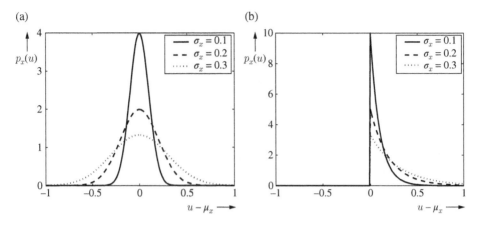

**Figure 5.3** (a) Gaussian probability density functions. (b) Exponential probability density functions.

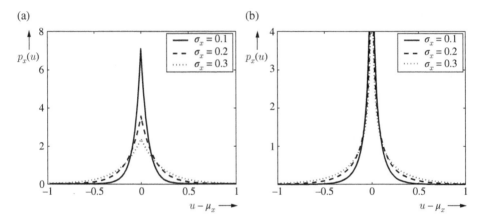

**Figure 5.4** (a) Laplace probability density functions. (b) Gamma probability density functions.

where $\sigma_x^2$ is again the variance of the density function. The mean of an exponentially distributed random variable is given by $\sigma_x$, and hence its power is $2\sigma_x^2$.

### 5.2.2.4 The Laplace Density

For $\sigma_x > 0$, the *two-sided exponential density* (also known as the *Laplace density*) is defined as

$$p_x(u) = \frac{1}{\sqrt{2}\sigma_x} \exp\left(-\sqrt{2}\,\frac{|u - \mu_x|}{\sigma_x}\right), \tag{5.26}$$

where $\mu_x$ is the mean and $\sigma_x^2$ is the variance. The Laplace density function is plotted in Figure 5.4a.

### 5.2.2.5 The Gamma Density

For $\sigma_x > 0$, the *gamma density*[2] (see Figure 5.4b) is given by

$$p_x(u) = \frac{\sqrt[4]{3}}{2\sqrt{2\pi}\sigma_x}\frac{1}{\sqrt{|u - \mu_x|}}\exp\left(-\frac{\sqrt{3}}{2}\frac{|u - \mu_x|}{\sigma_x}\right),\tag{5.27}$$

where $\mu_x$ is the mean and $\sigma_x^2$ is the variance.

### 5.2.2.6 $\chi^2$-Distribution

When $N$ independent and identically distributed zero mean Gaussian random variables with variance $\sigma^2$ are added, the resulting random variable

$$\chi^2 = x_1^2 + x_2^2 + \cdots + x_N^2\tag{5.28}$$

is $\chi^2$-distributed with $N$ degrees of freedom. The $\chi^2$-density is given by

$$p_{\chi^2}(u) = \begin{cases} \dfrac{u^{N/2-1}\exp\left(-\dfrac{u}{2\sigma^2}\right)}{\left(\sqrt{2\sigma^2}\right)^N\Gamma(N/2)}, & u > 0, \\ 0, & u \leq 0, \end{cases}\tag{5.29}$$

where $\Gamma(\cdot)$ is the complete gamma function. With [Gradshteyn, Ryzhik, 2000, Theorem 3.381.4]

$$\int_0^\infty x^n e^{-ax}\,dx = \frac{\Gamma(n+1)}{a^{n+1}}, \quad \mathrm{Re}\{a\} > 0, \quad \mathrm{Re}\{n\} > 0,\tag{5.30}$$

we find the mean of a $\chi^2$-distributed random variable as $\mu_{\chi^2} = N\sigma^2$ and the variance as $\sigma_{\chi^2}^2 = 2N\sigma^4$.

The normalized sum of squares

$$\tilde{\chi}^2 = \frac{1}{N}(x_1^2 + x_2^2 + \cdots + x_N^2)\tag{5.31}$$

is also $\chi^2$-distributed with

$$p_{\tilde{\chi}^2}(u) = \begin{cases} \dfrac{u^{N/2-1}\exp\left(-\dfrac{uN}{2\sigma^2}\right)}{\left(\sqrt{2\sigma^2/N}\right)^N\Gamma(N/2)}, & u > 0, \\ 0, & u \leq 0, \end{cases}\tag{5.32}$$

with mean $\sigma^2$ and variance $\frac{2}{N}\sigma^4$. The same density arises when we divide the sum of $N$ squared independent Gaussians, each of which has the variance $\sigma^2/2$, by $\frac{N}{2}$. This is the case, for example, when we average $K = N/2$ independent magnitude-squared, complex-Gaussian random variables, and when the mean of the squared magnitude of each complex Gaussian is equal to $\sigma^2$, see Section 5.10.1.

---

2 Our definition is a special case of the more general gamma density function as defined in [Johnson et al. 1994].

### 5.2.3 Transformation of a Random Variable

Frequently, we will consider the PDF of a random variable $y = f(x)$, which is a deterministic function of another random variable $x$. To specify the PDF $p_y(u)$, we must first solve $u = f(x)$ for the real roots $x^{\langle k \rangle}$, $k = 1, \ldots, N(u)$, where $N(u)$ depends on the value $u$. When for a specific value of $u$ no real roots exist, we set $N(u) = 0$. The PDF of $y$ is then given by [Papoulis, Unnikrishna Pillai 2001]

$$p_y(u) = \begin{cases} \dfrac{p_x(x^{\langle 1 \rangle})}{|f'(x^{\langle 1 \rangle})|} + \cdots + \dfrac{p_x(x^{\langle N(u) \rangle})}{|f'(x^{\langle N(u) \rangle})|}, & N(u) > 0, \\ 0, & N(u) = 0, \end{cases} \tag{5.33}$$

where $f'(x)$ is the derivative of $f(x)$ with respect to $x$.

To motivate the transformation (5.33), we compute the PDF of the square of a random variable, i.e., $y = f(x) = x^2$ with $f'(x) = 2x$. Since for $u < 0$, the equation $u = x^2$ has no real-valued roots we have $p_y(u) = 0$ for $u < 0$. For $u > 0$, we find two real roots $x^{\langle 1 \rangle} = -\sqrt{u}$ and $x^{\langle 2 \rangle} = +\sqrt{u}$, as shown in Figure 5.5. We can verify the following relation:

$$P(u < y < u + du) \tag{5.34}$$
$$= P(x^{\langle 1 \rangle} - dx_1 < x < x^{\langle 1 \rangle}) + P(x^{\langle 2 \rangle} < x < x^{\langle 2 \rangle} + dx_2)$$
$$= p(x^{\langle 1 \rangle}) \, dx_1 + p(x^{\langle 2 \rangle}) \, dx_2.$$

The summation in (5.34) corresponds to the sum of the gray areas in Figure 5.5.

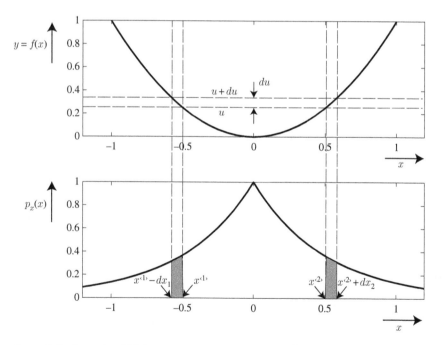

**Figure 5.5** Example of PDF transformation for $y = f(x) = x^2$ with $f'(x) = 2x$ $x^{\langle 1 \rangle} = -\sqrt{u}, x^{\langle 2 \rangle} = +\sqrt{u}$.

Furthermore, we find

$$du = |f'(x^{(i)})|\,dx_i, \quad i = 1, 2, \tag{5.35}$$

i.e.,

$$dx_i = \frac{1}{|f'(x^{(i)})|}\,du,$$

and obtain with

$$p_y(u)\,du = \frac{p_x(x^{(1)})}{|f'(x^{(1)})|}\,du + \frac{p_x(x^{(2)})}{|f'(x^{(2)})|}\,du, \quad u > 0, \tag{5.36}$$

the final result

$$p_y(u) = \begin{cases} \dfrac{p_x(-\sqrt{u})}{2\sqrt{u}} + \dfrac{p_x(+\sqrt{u})}{2\sqrt{u}}, & u > 0, \\ 0, & u \le 0. \end{cases} \tag{5.37}$$

Thus, if $x$ is, e.g., a Gaussian random variable with

$$p_x(u) = \frac{1}{\sqrt{2\pi}\sigma_x}\exp\left(-\frac{u^2}{2\sigma_x^2}\right), \tag{5.38}$$

we obtain the density of $y = x^2$ as a $\chi^2$-distribution with one degree of freedom,

$$p_y(u) = \begin{cases} \dfrac{1}{\sqrt{2\pi u}\sigma_x}\exp\left(-\dfrac{u}{2\sigma_x^2}\right), & u > 0, \\ 0, & u \le 0. \end{cases} \tag{5.39}$$

## 5.2.4 Relative Frequencies and Histograms

In contrast to the abstract concept of probabilities and PDFs, the relative frequencies of events and the histogram are closely linked to experiments and observed random phenomena. When we consider an experiment with $L$ possible outcomes $\{A_1, A_2, \ldots, A_L\}$ and repeat this experiment $N$ times, the relative frequency $\frac{N_i}{N}$ may serve as an estimate of the probability of event $A_i$.

A histogram is a quantized representation of the absolute or relative frequencies of events. Frequently, the events under consideration consist of a random variable $x$ within a given range of values. The histogram divides the range of the variable into discrete bins and displays the combined frequencies of values which fall into one of these bins.

Figure 5.6 depicts two histograms of computer-generated Gaussian noise. While the histogram on the left-hand side displays absolute frequencies, the histogram on the right-hand side uses relative frequencies.

The random variable $x$ and the histogram provide for a numerical representation of events, which makes them accessible for mathematical analysis, such as the computation of average values. We may also fit a PDF to a histogram of relative frequencies. This may then serve as a parametric model for the observed frequencies of the random phenomenon.

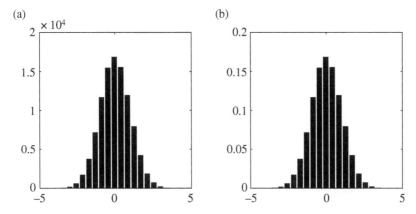

**Figure 5.6** Histogram (21 bins) of 100 000 samples of a computer-generated Gaussian-distributed random variable with zero mean and unit variance. (a) absolute frequencies and (b) relative frequencies.

## 5.3 Bivariate Statistics

As a special case of the multivariate PDF in (5.5) we define *the joint PDF* of two random variables $x$ and $y$ as the derivative of the bivariate probability distribution function $P_{xy}(u, v)$

$$p_{xy}(u, v) = \frac{d^2 P_{xy}(u, v)}{du \, dv},$$  (5.40)

where $P_{xy}$ is given by

$$P_{xy}(u, v) = P(x \le u, y \le v),$$  (5.41)

and $u$ and $v$ are threshold values for $x$ and $y$, respectively.

### 5.3.1 Marginal Densities

We obtain the univariate density of either random variable by integrating $p_{xy}(u, v)$ over the other random variable,

$$p_x(u) = \int_{-\infty}^{\infty} p_{xy}(u, v) \, dv,$$  (5.42)

$$p_y(v) = \int_{-\infty}^{\infty} p_{xy}(u, v) \, du.$$  (5.43)

### 5.3.2 Expectations and Moments

The *cross-correlation* of two random variables $x$ and $y$ is given by

$$\varphi_{xy} = E_{xy}\{x \cdot y\} = \int_{-\infty}^{\infty} \int_{-\infty}^{\infty} u v \, p_{xy}(u, v) \, du \, dv,$$  (5.44)

and the *cross-covariance* by

$$\psi_{xy} = E_{xy}\{(x - \mu_x)(y - \mu_y)\} = \varphi_{xy} - \mu_x\mu_y. \tag{5.45}$$

The normalized cross-covariance $r_{xy}$ (also known as correlation coefficient) is given by

$$r_{xy} = \frac{\psi_{xy}}{\sqrt{\psi_{xx}\psi_{yy}}} = \frac{\psi_{xy}}{\sigma_x\sigma_y}, \tag{5.46}$$

with $|r_{xy}| \leq 1$.

### 5.3.3 Uncorrelatedness and Statistical Independence

Two random variables $x$ and $y$ are referred to as uncorrelated if and only if their cross-covariance is zero, i.e.,

$$E_{xy}\{(x - \mu_x)(y - \mu_y)\} = 0. \tag{5.47}$$

Two random variables $x$ and $y$ are statistically independent if and only if their joint probability density $p_{xy}(u, v)$ factors into the marginal densities $p_x(u)$ and $p_y(v)$,

$$p_{xy}(u, v) = p_x(u)p_y(v). \tag{5.48}$$

Two statistically independent random variables $x$ and $y$ are always uncorrelated since

$$E_{xy}\{(x - \mu_x)(y - \mu_y)\} = \int_{-\infty}^{\infty}\int_{-\infty}^{\infty} (u - \mu_x)(v - \mu_y)p_x(u)p_y(v)\,du\,dv$$

$$= \int_{-\infty}^{\infty}(u - \mu_x)p_x(u)\,du \int_{-\infty}^{\infty}(v - \mu_y)p_y(v)\,dv = 0. \tag{5.49}$$

The converse is in general not true. A notable exception, however, is the bivariate (or multivariate) Gaussian density. Whenever Gaussian random variables are jointly uncorrelated, they are also statistically independent.

When two random variables $x$ and $y$ are statistically independent, the PDF of their sum $z = x + y$ is the convolution of the individual PDFs,

$$p_z(w) = \int_{-\infty}^{\infty} p_x(w - u)p_y(u)\,du. \tag{5.50}$$

As an example, we compute the density function of the sum $z = x_1^2 + x_2^2$ of two squared zero mean Gaussian random variables $x_1$ and $x_2$ which have the same variance $\sigma^2$ and are statistically independent. From (5.39) we recall that, the PDF of $y = x^2$ is given by

$$p_y(u) = \begin{cases} \dfrac{1}{\sqrt{2\pi u}\sigma} \exp\left(-\dfrac{u}{2\sigma^2}\right), & u > 0, \\ 0, & u \leq 0. \end{cases} \tag{5.51}$$

Assuming independence, we compute the density function of $z$ by convolving the density (5.51) with itself. The resulting density is the *exponential density* (5.25), or the more general $\chi^2$-*density* (5.29) in the special case of two degrees of freedom,

$$p_z(u) = \begin{cases} \dfrac{1}{2\sigma^2} \exp\left(-\dfrac{u}{2\sigma^2}\right), & u \geq 0, \\ 0, & u < 0. \end{cases} \tag{5.52}$$

Note that the above density also arises for the squared magnitude of a complex zero mean Gaussian random variable when the real and imaginary parts are independent and of the same variance $\sigma^2/2$.

### 5.3.4 Examples of Bivariate PDFs

Two widely used bivariate density functions are the uniform and the Gaussian density.

#### 5.3.4.1 The Bivariate Uniform Density
Figure 5.7 depicts the joint density of two statistically independent, uniformly distributed random variables $x$ and $y$. If $x$ and $y$ are uniform in $[X_1, X_2]$ and $[Y_1, Y_2]$ respectively, their joint probability density is given by

$$p_{xy}(u, v) = \begin{cases} \dfrac{1}{(X_2 - X_1)(Y_2 - Y_1)}, & u \in [X_1, X_2] \text{ and } v \in [Y_1, Y_2], \\ 0, & \text{elsewhere.} \end{cases} \tag{5.53}$$

#### 5.3.4.2 The Bivariate Gaussian Density
The Gaussian density of two random variables $x$ and $y$ may be parameterized by the means $\mu_x$ and $\mu_y$, the variances $\sigma_x^2$ and $\sigma_y^2$, and the normalized cross-covariance $r_{xy}$. It is given by

$$p_{xy}(u, v) = \frac{1}{2\pi\sigma_x\sigma_y\sqrt{1 - r_{xy}^2}} \tag{5.54}$$

$$\cdot \exp\left(-\frac{(u - \mu_x)^2}{2(1 - r_{xy}^2)\sigma_x^2} - \frac{(v - \mu_y)^2}{2(1 - r_{xy}^2)\sigma_y^2} + r_{xy}\frac{(u - \mu_x)(v - \mu_y)}{(1 - r_{xy}^2)\sigma_x\sigma_y}\right),$$

and depicted in Figure 5.8 for various values of $\sigma_x$, $\sigma_y$, and $r_{xy}$.

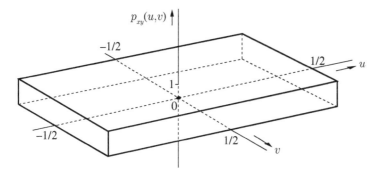

**Figure 5.7** Bivariate uniform probability density of two statistically independent random variables $x$ and $y$ with $X_1 = Y_1 = -0.5$ and $X_2 = Y_2 = 0.5$.

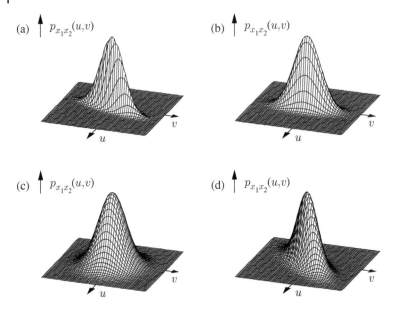

(a) $P_{x_1 x_2}(u,v)$      (b) $P_{x_1 x_2}(u,v)$

(c) $P_{x_1 x_2}(u,v)$      (d) $P_{x_1 x_2}(u,v)$

**Figure 5.8** Bivariate Gaussian densities with $\mu_x = \mu_y = 0$ and (a) $\sigma_x = 0.1$, $\sigma_y = 0.25$, $r_{xy} = -0.75$, (b) $\sigma_x = 0.1$, $\sigma_y = 0.25$, $r_{xy} = 0$, (c) $\sigma_x = \sigma_y = 0.25$, $r_{xy} = 0$, and (d) $\sigma_x = \sigma_y = 0.25$, $r_{xy} = -0.75$.

If $x$ and $y$ are uncorrelated, i.e., $r_{xy} = 0$, the bivariate density factors into two univariate Gaussian densities. Therefore, two uncorrelated Gaussian random variables are also statistically independent. Furthermore, the marginal densities of a bivariate Gaussian density are Gaussian densities.

### 5.3.5 Functions of Two Random Variables

Given two functions $y_1 = f_1(x_1, x_2)$ and $y_2 = f_2(x_1, x_2)$ of two random variables $x_1$ and $x_2$, we find the joint density function of $y_1$ and $y_2$ in terms of the joint density of $x_1$ and $x_2$ as [Papoulis, Unnikrishna Pillai 2001]

$$p_{y_1 y_2}(u, v) = \frac{p_{x_1 x_2}(x_1^{(1)}, x_2^{(1)})}{|\Im(x_1^{(1)}, x_2^{(1)})|} + \cdots + \frac{p_{x_1 x_2}(x_1^{(N)}, x_2^{(N)})}{|\Im(x_1^{(N)}, x_2^{(N)})|}, \tag{5.55}$$

where $x_1^{(k)}$ and $x_2^{(k)}$, $k = 1 \ldots N$, are the real roots of the simultaneous equations $u = f_1(x_1, x_2)$ and $v = f_2(x_1, x_2)$. $|\Im(x_1^{(k)}, x_2^{(k)})|$ is the determinant of the Jacobian matrix

$$\Im(x_1^{(k)}, x_2^{(k)}) = \begin{pmatrix} \dfrac{\partial y_1}{\partial x_1}\Big|_{x_1 = x_1^{(k)}} & \dfrac{\partial y_1}{\partial x_2}\Big|_{x_2 = x_2^{(k)}} \\ \dfrac{\partial y_2}{\partial x_1}\Big|_{x_1 = x_1^{(k)}} & \dfrac{\partial y_2}{\partial x_2}\Big|_{x_2 = x_2^{(k)}} \end{pmatrix}. \tag{5.56}$$

When there is no solution to the simultaneous equations, we obtain $p_{y_1 y_2}(u, v) = 0$.

In the special case of a linear transform with $u = ax_1 + bx_2$, $v = cx_1 + dx_2$, and $|ad - bc| \neq 0$ we find single roots $x_1 = Au + Bv$ and $x_2 = Cu + Dv$ and obtain

$$p_{y_1 y_2}(u, v) = \frac{1}{|ad - bc|} p_{x_1 x_2}(Au + Bv, Cu + Dv). \tag{5.57}$$

As an example, we consider the additive signal model $y = x + n$ where $x$ and $n$ are statistically independent. We are interested in the computation of $p_{xy}(u, v)$ and of $p_{y|x}(v \mid u)$. We define $x_1 = x$ and $x_2 = n$ and have $u = x_1 = x$, $v = x_1 + x_2 = x + n$, $a = c = d = 1$, and $b = 0$. In order to apply (5.55) and (5.57), we introduced the auxiliary function $u = x_1 = x$ which depends on $x_1$ only. The inversion of these equations yields $x_1 = u$ and $x_2 = v - u$. Since $|ad - bc| = 1$, we find

$$p_{xy}(u, v) = p_{xn}(u, v - u) = p_x(u) p_n(v - u) \quad \text{and} \quad p_{y|x}(v \mid u) = p_n(v - u).$$

Note that computing the marginal density $p_y(v)$ leads to the result introduced in (5.50).

## 5.4 Probability and Information

In this section, we briefly summarize some fundamental definitions of information theory. These are treated in more detail in, e.g., [Cover, Thomas 2006].

### 5.4.1 Entropy

The *entropy* $H(x)$ of a discrete random variable $x$ with $x = u_i$, $i = 1 \ldots M$, describes the average uncertainty that one of $M$ values is attained. In the case of a discrete random variable, it is defined with $p_x(u_i) = P(x = u_i)$ as

$$H(x) = -\sum_{i=1}^{M} p_x(u_i) \log(p_x(u_i)), \tag{5.58}$$

and measured in "bits" if the logarithm is computed with respect to base 2, and in "nats" if the logarithm is computed with respect to base $e$. As uncertainty is equivalent to information, $H(x)$ is also a measure of information. The uncertainty of a random variable $x$ with fixed amplitude limits is maximal if the random variable is uniformly distributed within these limits.

For a continuous random variable with a non zero probability density on the interval $[a, b]$ uncertainty is measured in terms of its *differential entropy*

$$H(x) = -\int_{a}^{b} p_x(u) \log(p_x(u)) \, du. \tag{5.59}$$

### 5.4.2 Kullback–Leibler Divergence

The *Kullback–Leibler divergence* $D(p_x \| p_y)$ is a measure of similarity of two discrete distributions $p_x(u)$ and $p_y(u)$,

$$D(p_x \| p_y) = \sum_{i=1}^{M} p_x(u_i) \log\left(\frac{p_x(u_i)}{p_y(u_i)}\right). \tag{5.60}$$

Clearly, the Kullback–Leibler divergence is not symmetric, i.e.,

$$D(p_x \| p_y) \neq D(p_y \| p_x).$$

The Kullback–Leibler divergence is always non negative, and zero if and only if $p_x = p_y$. Frequently, the symmetric *Kullback–Leibler distance*

$$D_S(p_x \| p_y) = \frac{1}{2}\left(D(p_x \| p_y) + D(p_y \| p_x)\right) \tag{5.61}$$

is used.

### 5.4.3 Cross-Entropy

The cross-entropy of two discrete random variables $x$ and $y$ assuming values $x = u_i$ and $y = u_\ell$, $i, \ell \in \{1, \dots, M\}$, is given by

$$H(x, y) = -\sum_{i=1}^{M} p_x(u_i) \log\left(p_y(u_i)\right). \tag{5.62}$$

The cross-entropy is frequently used as an optimization criterion in multi-class classification tasks. Then, $p_x = p_C$ and $p_y = p_{\hat{C}}$ describe the distributions of the ground-truth class labels $C$ and the estimated class labels $\hat{C}$, respectively. We note that

$$H(x, y) = H(x) + D(p_y \| p_x). \tag{5.63}$$

Thus, the cross-entropy is lower bounded by the entropy of the class distribution.

### 5.4.4 Mutual Information

The *mutual information* $I(x, y)$ measures the information about a random variable $x$ conveyed by another random variable $y$. It gives an indication of the dependence of two random variables and may be expressed as the Kullback–Leibler divergence between the joint probability density and the marginal densities of the two random variables, i.e.,

$$I(x, y) = D(p_{xy} \| p_x \cdot p_y) = \sum_{i=1}^{M} \sum_{\ell=1}^{N} p_{xy}(u_i, v_\ell) \log\left(\frac{p_{xy}(u_i, v_\ell)}{p_x(u_i) p_y(v_\ell)}\right). \tag{5.64}$$

The mutual information measure is symmetric in $p_x$ and $p_y$ and non negative. It is equal to zero if the two random variables are statistically independent.

## 5.5 Multivariate Statistics

For the treatment of multivariate statistics, it is convenient to introduce vector notation. We now consider vectors of random variables

$$\mathbf{x} = \begin{pmatrix} x_1 \\ x_2 \\ \vdots \\ x_N \end{pmatrix}, \tag{5.65}$$

and define the first- and second-order statistics in terms of the vector components. The *mean vector* is given by

$$\boldsymbol{\mu}_x = \mathrm{E}\{\mathbf{x}\} = \begin{pmatrix} \mathrm{E}\{x_1\} \\ \mathrm{E}\{x_2\} \\ \vdots \\ \mathrm{E}\{x_N\} \end{pmatrix} = \begin{pmatrix} \mu_{x_1} \\ \mu_{x_2} \\ \vdots \\ \mu_{x_N} \end{pmatrix},$$ (5.66)

and the *covariance matrix* by

$$\begin{aligned} \mathbf{C}_{xx} &= \mathrm{E}\left\{(\mathbf{x} - \boldsymbol{\mu}_x)(\mathbf{x} - \boldsymbol{\mu}_x)^T\right\} \\ &= \begin{pmatrix} \psi_{x_1 x_1} & \psi_{x_1 x_2} & \cdots & \psi_{x_1 x_N} \\ \psi_{x_2 x_1} & \psi_{x_2 x_2} & \cdots & \psi_{x_2 x_N} \\ \vdots & \vdots & \ddots & \vdots \\ \psi_{x_N x_1} & \psi_{x_N x_2} & \cdots & \psi_{x_N x_N} \end{pmatrix}. \end{aligned}$$ (5.67)

The *correlation matrix* $\mathbf{R}_{xx}$ is then computed as

$$\mathbf{R}_{xx} = \mathrm{E}\{\mathbf{x}\mathbf{x}^T\} = \mathbf{C}_{xx} + \boldsymbol{\mu}_x \boldsymbol{\mu}_x^T.$$ (5.68)

### 5.5.1 Multivariate Gaussian Distribution

The multivariate Gaussian probability density of a $N$-dimensional vector-valued random variable $\mathbf{x}$ is defined as

$$\mathfrak{N}(\mathbf{x}; \boldsymbol{\mu}_x, \mathbf{C}_{xx}) = \frac{1}{\sqrt{(2\pi)^N |\mathbf{C}_{xx}|}} \exp\left(-\frac{1}{2}(\mathbf{x} - \boldsymbol{\mu}_x)^T \mathbf{C}_{xx}^{-1}(\mathbf{x} - \boldsymbol{\mu}_x)\right),$$ (5.69)

where $\boldsymbol{\mu}_x$ and $\mathbf{C}_{xx}$ denote the mean vector and the covariance matrix, respectively. $|\mathbf{C}_{xx}|$ is the determinant of the covariance matrix.

For two Gaussian vector-valued variables $\mathbf{x}$ and $\mathbf{y}$ we use the stacked notation

$$\boldsymbol{\mu} = \begin{pmatrix} \boldsymbol{\mu}_x \\ \boldsymbol{\mu}_y \end{pmatrix} \quad \text{and} \quad \mathbf{C} = \begin{pmatrix} \mathbf{C}_{xx} & \mathbf{C}_{xy} \\ \mathbf{C}_{yx} & \mathbf{C}_{yy} \end{pmatrix},$$ (5.70)

and write their joint density

$$p(\mathbf{x}, \mathbf{y}) = p(\mathbf{x}|\mathbf{y})p(\mathbf{y}) = \mathfrak{N}(\mathbf{x}; \boldsymbol{\mu}_{x|y}, \mathbf{C}_{x|y})\mathfrak{N}(\mathbf{y}; \boldsymbol{\mu}_y, \mathbf{C}_{yy})$$ (5.71)

with

$$\begin{aligned} \boldsymbol{\mu}_{x|y} &= \boldsymbol{\mu}_x + \mathbf{C}_{xy}\mathbf{C}_{yy}^{-1}\left(\mathbf{y} - \boldsymbol{\mu}_y\right) \\ \mathbf{C}_{x|y} &= \mathbf{C}_{xx} - \mathbf{C}_{xy}\mathbf{C}_{yy}^{-1}\mathbf{C}_{yx}. \end{aligned}$$

Given a set of $L$ independent, identically distributed observations $\mathbf{x}_\ell$, $\ell = 0, \ldots, L-1$, maximum likelihood (ML) estimates (see Section 5.11) of the parameters of the multivariate Gaussian distribution may be obtained via

$$\hat{\boldsymbol{\mu}}_x = \frac{1}{L}\sum_{\ell=0}^{L-1}\mathbf{x}_\ell \qquad \hat{\mathbf{C}}xx = \frac{1}{L}\sum_{\ell=0}^{L-1}(\mathbf{x}_\ell - \hat{\boldsymbol{\mu}}_x)(\mathbf{x}_\ell - \hat{\boldsymbol{\mu}}_x)^T.$$

### 5.5.2 Gaussian Mixture Models

In many cases, the probability distribution of speech signals or related parameters is not well modeled with a single (multivariate) distribution. Therefore, it is of interest to employ more general parametric models, such as mixture models. A widely used variant is the Gaussian mixture model (GMM)

$$p(\mathbf{x}) = \sum_{i=1}^{M} \alpha_i \, \mathfrak{N}(\mathbf{x}; \boldsymbol{\mu}_x^{\langle i \rangle}, \mathbf{C}_{xx}^{\langle i \rangle}), \tag{5.72}$$

where each $N$-dimensional mixture component is given by

$$\mathfrak{N}(\mathbf{x}; \boldsymbol{\mu}_x^{\langle i \rangle}, \mathbf{C}_{xx}^{\langle i \rangle}) = \frac{1}{\sqrt{(2\pi)^N \mid \mathbf{C}_{xx}^{\langle i \rangle} \mid}} \exp\left(-\frac{1}{2}\left(\mathbf{x} - \boldsymbol{\mu}_x^{\langle i \rangle}\right)^T \left(\mathbf{C}_{xx}^{\langle i \rangle}\right)^{-1} \left(\mathbf{x} - \boldsymbol{\mu}_x^{\langle i \rangle}\right)\right)$$

with

- the *a priori* probability (relative weight) $\alpha_i$ of mixture component $i$,
- the $N \times 1$ mean vector $\boldsymbol{\mu}_x^{\langle i \rangle}$ of component $i$, and
- the $N \times N$ covariance matrix $\mathbf{C}_{xx}^{\langle i \rangle}$ of component $i$.

Unlike the parameter estimation for the single Gaussian distribution, there is no closed-form solution for the maximum likelihood estimation of GMM parameters. GMM parameters are estimated via the iterative *expectation–maximization* (EM) algorithm which provides a general methodology for the maximum likelihood estimation of missing data or parameters [Dempster et al. 1977]. Here, the missing information can be described by an index vector that assigns each data sample $\mathbf{x}_\ell$ to one of the mixture components. Thus, the GMM can also be used in clustering tasks where samples $\mathbf{x}_\ell$ are assigned to classes $C_i$, each of which is represented by one of the GMM components.

The EM algorithm makes use of the concept of *sufficient statistics*, which implies the existence of a function of data samples $\mathbf{x}_\ell$, such that for the given problem knowing the sufficient statistics is just as good as knowing $\mathbf{x}_\ell$. In speech processing, GMMs have become popular through their use as observation densities in hidden Markov models (HMM) and in speaker identification tasks [Reynolds, Rose 1995].

For a given set of $L$ data samples $\mathbf{x}_\ell$, the EM algorithm for the estimation of GMM parameters may be summarized as follows:

- Define number of GMM components $M$, initialize iteration index $m = 0$.
- Initialize the estimated means $\hat{\boldsymbol{\mu}}_x^{\langle i \rangle}(m)$, covariances $\hat{\mathbf{C}}_{xx}^{\langle i \rangle}(m)$, and mixture weights $\hat{\alpha}_i(m)$.
- Compute E-step of the $m$th iteration: for all $i$ and $\ell$,
  - compute probability $\gamma_{i,\ell}$ that observation $\mathbf{x}_\ell$ originates from mixture component $i$ (also known as *responsibility*),

$$\gamma_{i,\ell} = \frac{\hat{\alpha}_i(m) \, \mathfrak{N}(\mathbf{x}_\ell; \hat{\boldsymbol{\mu}}_x^{\langle i \rangle}(m), \hat{\mathbf{C}}_{xx}^{\langle i \rangle}(m))}{\sum\limits_{i=1}^{M} \hat{\alpha}_i(m) \, \mathfrak{N}(\mathbf{x}_\ell; \hat{\boldsymbol{\mu}}_x^{\langle i \rangle}(m), \hat{\mathbf{C}}_{xx}^{\langle i \rangle}(m))},$$

  - and component-wise mean responsibilities,

$$L_i = \sum_{\ell=0}^{L-1} \gamma_{i,\ell}.$$

- Compute $M$ step: update model parameters for iteration $m + 1$

$$\hat{\boldsymbol{\mu}}_x^{(i)}(m+1) = \frac{1}{L_i} \sum_{\ell=0}^{L-1} \gamma_{i,\ell} \mathbf{x}_\ell,$$

$$\hat{\mathbf{C}}_{xx}^{(i)}(m+1) = \frac{1}{L_i} \sum_{\ell=0}^{L-1} \gamma_{i,\ell} (\mathbf{x}_\ell - \hat{\boldsymbol{\mu}}_x^{(i)}(m+1))(\mathbf{x}_\ell - \hat{\boldsymbol{\mu}}_x^{(i)}(m+1))^T,$$

$$\hat{a}_i(m+1) = \frac{L_i}{L}.$$

- Increment iteration index, and

$$m \leftarrow m + 1.$$

- Compute log-likelihood and return to E step if not converged

$$\ln\left( p\left( \mathbf{x}_0, \dots, \mathbf{x}_{L-1} \mid \{\hat{\boldsymbol{\mu}}_x^{(i)}(m), \hat{\mathbf{C}}_{xx}^{(i)}(m), \hat{a}_i(m) \mid \forall i\} \right) \right)$$

$$= \sum_{\ell=0}^{L-1} \ln\left( \sum_{i=1}^{M} \hat{a}_i(m) \, \mathfrak{N}(\mathbf{x}_\ell; \hat{\boldsymbol{\mu}}_x^{(i)}(m), \hat{\mathbf{C}}_{xx}^{(i)}(m)) \right).$$

Finally, we note that GMMs with diagonal covariance matrices $\mathbf{C}_{xx}^{(i)}$ also yield good approximations to a given data distribution. Then, a larger number of GMM components $M$ is required but the overall number of parameters is often significantly reduced. As a consequence, the EM algorithm converges more rapidly.

## 5.6 Stochastic Processes

An indexed sequence of random variables $\{\dots, x(k-1), x(k), x(k+1), \dots\}$, $k \in \mathbb{Z}$, is called a stochastic process. For any index $k$, a random experiment determines the value of $x(k)$. However, the outcome of an experiment at $k = k_1$ may also depend on the outcomes for $k \neq k_1$. Then, the variables are statistically dependent. In our context, stochastic processes are used to model sampled stochastic signals, such as speech signals. An observed speech sample is then interpreted as a specific instantiation of the underlying random process. Stochastic processes are characterized in terms of their univariate distribution function or their moments for each $k$, as well as their multivariate statistical properties. In general, all of these quantities are functions of the sampling time $k$.

### 5.6.1 Stationary Processes

A stochastic process is called *strict sense stationary* if all its statistical properties (such as moments) are invariant with respect to a variation of $k$.

A stochastic process is *wide sense stationary* if its first- and second-order moments are invariant to a variation of the independent variable $k$, i.e.,

$$E\{x(k)\} = \mu_x \quad \forall k \quad \text{and} \tag{5.73}$$

$$E\{x(k)x(k+\lambda)\} = \varphi_{xx}(\lambda) \quad \forall k. \tag{5.74}$$

Clearly, speech signals are neither strict nor wide sense stationary. However, within sufficiently short observation intervals, the first- and second-order statistical properties of speech signals may show only little variations. Speech signals can therefore be considered to be *short-time wide sense stationary.*

### 5.6.2 Auto-Correlation and Auto-Covariance Functions

The *auto-correlation function* quantifies the amount of correlation between the variables of a stochastic process. It is defined as

$$\varphi_{xx}(k_1, k_2) = \mathrm{E}\left\{x(k_1)\, x(k_2)\right\} = \int\limits_{-\infty}^{\infty}\int\limits_{-\infty}^{\infty} u\, v\, p_{x(k_1)x(k_2)}(u, v)\, du\, dv, \tag{5.75}$$

and is related to the *auto-covariance function*

$$\psi_{xx}(k_1, k_2) = \mathrm{E}\left\{(x(k_1) - \mu_x(k_1))\,(x(k_2) - \mu_x(k_2))\right\} \tag{5.76}$$

by

$$\varphi_{xx}(k_1, k_2) = \psi_{xx}(k_1, k_2) + \mu_x(k_1)\mu_x(k_2). \tag{5.77}$$

For a wide sense stationary process, the auto-correlation function depends only on the difference $\lambda = k_2 - k_1$ of indices $k_1$ and $k_2$ and not on their absolute values. Thus, for a wide sense stationary process we may define the auto-correlation function for any $k_1$ as

$$\varphi_{xx}(\lambda) = \mathrm{E}\left\{x(k_1)\, x(k_1 + \lambda)\right\}$$
$$= \int\limits_{-\infty}^{\infty}\int\limits_{-\infty}^{\infty} u\, v\, p_{x(k_1)x(k_1+\lambda)}(u, v)\, du\, dv = \psi_{xx}(\lambda) + \mu_x^2, \tag{5.78}$$

with

$$\psi_{xx}(\lambda) = \mathrm{E}\left\{(x(k_1) - \mu_x)\,(x(k_1 + \lambda) - \mu_x)\right\}. \tag{5.79}$$

The auto-correlation function, as well as the auto-covariance function, are symmetric, i.e., $\varphi_{xx}(\lambda) = \varphi_{xx}(-\lambda)$. The auto-correlation function attains its maximum and its maximum absolute value for $\lambda = 0$, thus $|\varphi_{xx}(\lambda)| \le \varphi_{xx}(0)$.

For $\lambda = 0$, we have

$$\varphi_{xx}(0) = \psi_{xx}(0) + \mu_x^2 = \sigma_x^2 + \mu_x^2. \tag{5.80}$$

In general, we find the moments of a function $g$ of $x(k_1)$ and $x(k_1 + \lambda)$ as

$$\mathrm{E}\left\{g\left(x(k_1), x(k_1 + \lambda)\right)\right\} = \int\limits_{-\infty}^{\infty}\int\limits_{-\infty}^{\infty} g(u, v)\, p_{x(k_1)x(k_1+\lambda)}(u, v)\, du\, dv. \tag{5.81}$$

### 5.6.3 Cross-Correlation and Cross-Covariance Functions

In analogy to the above definitions, the *cross-correlation* and the *cross-covariance functions* may be used to characterize the second-order statistics of two (different) stochastic processes $x(k)$ and $y(k)$ and may be written as

$$\varphi_{xy}(k_1, k_2) = E\left\{x(k_1)\,y(k_2)\right\}$$

$$= \int_{-\infty}^{\infty}\int_{-\infty}^{\infty} u\,v\,p_{x(k_1)y(k_2)}(u, v)\,du\,dv \tag{5.82}$$

$$= \psi_{xy}(k_1, k_2) + \mu_x(k_1)\mu_y(k_2).$$

As before, we may simplify our notation for stationary processes $x(k)$ and $y(k)$

$$\varphi_{xy}(\lambda) = E\left\{x(k_1)y(k_1 + \lambda)\right\} \tag{5.83}$$

and

$$\psi_{xy}(\lambda) = E\left\{\left(x(k_1) - \mu_x\right)\left(y(k_1 + \lambda) - \mu_y\right)\right\}$$
$$= \varphi_{xy}(\lambda) - \mu_x\mu_y. \tag{5.84}$$

### 5.6.4 Markov Processes

We consider again a sequence of random variables $\{x(k), x(k-1), \ldots, x(0)\}$, $k \in \mathbb{Z}$, the description of which entails a high-dimensional statistical model in general. A significant simplification is achieved, if we assume that the time sequence is governed by a (first-order) Markov model. For a first-order Markov process, the random variable $x(k)$ depends only on the preceding variable $x(k-1)$. Therefore, we may write the conditional PDF as

$$p\left(x(k) \mid x(k-1), x(k-2), \ldots, x(0)\right) = p\left(x(k) \mid x(k-1)\right), \tag{5.85}$$

or, equivalently, the joint PDF as

$$p\left(x(k), x(k-1), x(k-2), \ldots, x(0)\right)$$
$$= p\left(x(k) \mid x(k-1)\right) p\left(x(k-1) \mid x(k-2)\right) \cdots p\left(x(1) \mid x(0)\right) p\left(x(0)\right). \tag{5.86}$$

Therefore, when the Markov process is stationary, it is entirely determined by the transition probabilities $p\left(x(k) \mid x(k-1)\right)$ and the prior distribution $p\left(x(0)\right)$.

When the order of the random variables $x(k)$ of a Markov process is reversed, the resulting process is again a Markov process. To show this, we use the Markov property

$$p\left(x(k - K_R + 1), \ldots, x(k) \mid x(k - K_R), x(k - K_R - 1)\right)$$
$$= p\left(x(k - K_R + 1), \ldots, x(k) \mid x(k - K_R)\right), \tag{5.87}$$

and observe with (5.12) for a given constant index offset $K_R$ that

$$p\left(x(k - K_R) \mid x(k - K_R - 1)\right) p\left(x(k - K_R - 1)\right)$$
$$= p\left(x(k - K_R - 1) \mid x(k - K_R)\right) p\left(x(k - K_R)\right). \tag{5.88}$$

Then, we have

$$p\left(x(k - K_R - 1) \mid x(k - K_R), x(k - K_R + 1), \ldots, x(k)\right)$$

$$= \frac{p\left(x(k - K_R - 1), x(k - K_R), x(k - K_R + 1), \ldots, x(k)\right)}{p\left(x(k - K_R), x(k - K_R + 1), \ldots, x(k)\right)} \tag{5.89a}$$

$$= \frac{p\left(x(k - K_R + 1), \ldots, x(k) \mid x(k - K_R), x(k - K_R - 1)\right)}{p\left(x(k - K_R + 1), \ldots, x(k) \mid x(k - K_R)\right)} \tag{5.89b}$$

$$\cdot \frac{p\left(x(k - K_R - 1), x(k - K_R)\right)}{p\left(x(k - K_R)\right)} \tag{5.89c}$$

$$= \frac{p\left(x(k - K_R) \mid x(k - K_R - 1)\right) p\left(x(k - K_R - 1)\right)}{p\left(x(k - K_R)\right)} \tag{5.89d}$$

$$= \frac{p\left(x(k - K_R - 1) \mid x(k - K_R)\right) p\left(x(k - K_R)\right)}{p\left(x(k - K_R)\right)} \tag{5.89e}$$

$$= p\left(x(k - K_R - 1) \mid x(k - K_R)\right),$$

where we applied (5.87) to (5.89b) and (5.88) to (5.89d).

### 5.6.5 Multivariate Stochastic Processes

Similarly to the definition of vectors of random variables, we may define vector-valued (multivariate) stochastic processes as

$$\mathbf{x}(k) = \begin{pmatrix} x_1(k) \\ x_2(k) \\ \vdots \\ x_N(k) \end{pmatrix}. \tag{5.90}$$

The first- and second-order statistics are then also dependent on index $k$. For example, the mean is given by

$$\boldsymbol{\mu}_x(k) = \mathrm{E}\{\mathbf{x}(k)\} = \begin{pmatrix} \mathrm{E}\{x_1(k)\} \\ \mathrm{E}\{x_2(k)\} \\ \vdots \\ \mathrm{E}\{x_N(k)\} \end{pmatrix}, \tag{5.91}$$

and the correlation matrix $\mathbf{R}_{xx}(k_1, k_2)$ of the process $\mathbf{x}(k)$ at indices $k_1$ and $k_2$ by

$$\mathbf{R}_{xx}(k_1, k_2) = \mathrm{E}\left\{\mathbf{x}(k_1)\mathbf{x}^T(k_2)\right\} \tag{5.92}$$

$$= \begin{pmatrix} \mathrm{E}\{x_1(k_1)x_1(k_2)\} & \mathrm{E}\{x_1(k_1)x_2(k_2)\} & \cdots & \mathrm{E}\{x_1(k_1)x_N(k_2)\} \\ \mathrm{E}\{x_2(k_1)x_1(k_2)\} & \mathrm{E}\{x_2(k_1)x_2(k_2)\} & \cdots & \mathrm{E}\{x_2(k_1)x_N(k_2)\} \\ \vdots & \vdots & \ddots & \vdots \\ \mathrm{E}\{x_N(k_1)x_1(k_2)\} & \mathrm{E}\{x_N(k_1)x_2(k_2)\} & \cdots & \mathrm{E}\{x_N(k_1)x_N(k_2)\} \end{pmatrix},$$

which is in general not a symmetric matrix. In analogy to the univariate case, we may also define a covariance matrix, i.e.,

$$\mathbf{C}_{xx}(k_1, k_2) = \mathrm{E}\left\{(\mathbf{x}(k_1) - \boldsymbol{\mu}_x(k_1))(\mathbf{x}(k_2) - \boldsymbol{\mu}_x(k_2))^T\right\}. \tag{5.93}$$

When $\mathbf{x}(k)$ is a stationary process, the above quantities will not depend on the absolute values of the time indices $k_1$ and $k_2$ but on their difference $\lambda = k_2 - k_1$ only.

An interesting special case arises when the elements of the random vector $\mathbf{x}(k)$ are successive samples of one and the same univariate process $x(k)$,

$$\mathbf{x}(k) = \begin{pmatrix} x(k) \\ x(k-1) \\ \vdots \\ x(k-N+1) \end{pmatrix}. \tag{5.94}$$

In this case, the correlation matrix is given by

$$\mathbf{R}_{xx}(k_1, k_2) = \mathrm{E}\left\{\mathbf{x}(k_1)\mathbf{x}^T(k_2)\right\} \tag{5.95}$$

$$= \begin{pmatrix} \varphi_{xx}(k_1, k_2) & \varphi_{xx}(k_1, k_2 - 1) & \cdots & \varphi_{xx}(k_1, k_2 - N + 1) \\ \varphi_{xx}(k_1 - 1, k_2) & \cdots & \cdots & \varphi_{xx}(k_1 - 1, k_2 - N + 1) \\ \vdots & \vdots & \ddots & \vdots \\ \varphi_{xx}(k_1 - N + 1, k_2) & \cdots & \cdots & \varphi_{xx}(k_1 - N + 1, k_2 - N + 1) \end{pmatrix},$$

which is a symmetric matrix for $k_1 = k_2 = k$, i.e., $\mathbf{R}_{xx}(k, k) = \mathbf{R}_{xx}^T(k, k)$.

When the univariate process $x(k)$ is wide sense stationary, the elements of the correlation matrix are independent of the absolute time indices and depend only on the index difference $\lambda = k_2 - k_1$. Then,

$$\mathbf{R}_{xx}(\lambda) = \mathrm{E}\left\{\mathbf{x}(k_1)\mathbf{x}^T(k_1 + \lambda)\right\} \tag{5.96}$$

$$= \begin{pmatrix} \varphi_{xx}(\lambda) & \varphi_{xx}(\lambda - 1) & \cdots & \varphi_{xx}(\lambda - N + 1) \\ \varphi_{xx}(\lambda + 1) & \varphi_{xx}(\lambda) & \cdots & \varphi_{xx}(\lambda - N + 2) \\ \vdots & \vdots & \ddots & \vdots \\ \varphi_{xx}(\lambda + N - 1) & \varphi_{xx}(\lambda + N - 2) & \cdots & \varphi_{xx}(\lambda) \end{pmatrix}.$$

All matrix elements $r_{ij}$ which have the same difference $i - j$ of their row and columns indices are identical. A matrix of this structure is called a *Toeplitz* matrix. For $k_1 = k_2 = k$, i.e., $\lambda = 0$, we obtain a correlation matrix

$$\mathbf{R}_{xx} = \mathrm{E}\left\{\mathbf{x}(k)\mathbf{x}^T(k)\right\} \tag{5.97}$$

$$= \begin{pmatrix} \varphi_{xx}(0) & \varphi_{xx}(-1) & \cdots & \varphi_{xx}(-N + 1) \\ \varphi_{xx}(1) & \varphi_{xx}(0) & \cdots & \varphi_{xx}(-N + 2) \\ \vdots & \vdots & \ddots & \vdots \\ \varphi_{xx}(N - 1) & \varphi_{xx}(N - 2) & \cdots & \varphi_{xx}(0) \end{pmatrix},$$

which is a symmetric Toeplitz matrix. This matrix is completely specified by its first row or column. For a discrete time stochastic process, the correlation matrix is always non negative definite and almost always positive definite [Haykin 1996].

In analogy to the above definitions, the cross-correlation matrix of two vector-valued processes $\mathbf{x}(k)$ and $\mathbf{y}(k)$ is given by

$$\mathbf{R}_{xy}(k_1, k_2) = \mathrm{E}\left\{\mathbf{x}(k_1)\mathbf{y}(k_2)^T\right\} \tag{5.98}$$

with the above special cases defined accordingly.

## 5.7 Estimation of Statistical Quantities by Time Averages

### 5.7.1 Ergodic Processes

Quite frequently we cannot observe more than a single instantiation of a stochastic process. Then, it is not possible to estimate its statistics by averaging over an ensemble of observations. However, if the process is stationary, we might replace ensemble averages by time averages. Whenever the statistics of a stationary random process may be obtained with probability one from time averages over a single observation, the random process is called (strict sense) *ergodic*. This definition implies that specific instances of the random process may not be suited to obtain time averages. These instances, however, occur with probability zero.

Ergodicity is an indispensable prerequisite for many practical applications of statistical signal processing, yet in general it is difficult to prove.

### 5.7.2 Short-Time Stationary Processes

Strictly speaking, the estimation of statistical quantities via time averaging is only admissible when the signal is stationary and ergodic. As speech signals and many noise signals are not stationary and hence not ergodic, we must confine time averages to short segments of the signal where stationarity is not grossly violated. To apply time averaging we must require that the signal is at least *short-time stationary*.

We define the *short-time mean* of $M$ successive samples of a stochastic process $x(k)$ as

$$\bar{x}(k) = \frac{1}{M} \sum_{\kappa=k-M+1}^{k} x(\kappa), \tag{5.99}$$

and the *short-time variance*

$$\hat{\sigma}^2(k) = \frac{1}{M-1} \sum_{\kappa=k-M+1}^{k} (x(\kappa) - \bar{x}(k))^2. \tag{5.100}$$

When $x(k)$ is a wide sense stationary and uncorrelated random process, i.e., $\mathrm{E}\left\{(x(k_i) - \mu_{x(k_i)})(x(k_j) - \mu_{x(k_j)})\right\} = 0$ for all $i \neq j$, it can be shown that these estimates are unbiased, i.e.,

$$\mathrm{E}\left\{\bar{x}(k)\right\} = \mathrm{E}\left\{x(k)\right\} \quad \text{and} \quad \mathrm{E}\left\{\hat{\sigma}^2(k)\right\} = \mathrm{E}\left\{(x(k) - \mu_x)^2\right\}.$$

It is not at all trivial to identify short-time stationary segments. On the one hand one is tempted to make the segment of assumed stationarity as long as possible in order to reduce the error variance of the estimate. On the other hand, the intrinsic non-stationary nature of speech and many noise signals does not allow the use of long averaging segments without introducing a significant bias. Thus, we have to strike a balance between bias and error variance.

Similar averaging procedures may be employed to estimate the cross- and auto-correlation functions on a short-time basis, i.e.,

$$\hat{\varphi}_{xy}(\lambda, k) = \frac{1}{M} \sum_{\kappa=k-M+1}^{k} x(\kappa) y(\kappa + \lambda), \tag{5.101}$$

and

$$\hat{\varphi}_{xx}(\lambda, k) = \frac{1}{M} \sum_{\kappa=k-M+1}^{k} x(\kappa)x(\kappa + \lambda). \tag{5.102}$$

The latter estimate is not symmetric in $\lambda$. A symmetric estimate might be obtained by first extracting a "windowed" signal segment of length $M$

$$\tilde{x}(\kappa) = \begin{cases} x(\kappa), & \text{if } k - M \leq \kappa \leq k, \\ 0, & \text{else,} \end{cases} \tag{5.103}$$

and then computing the sum of products

$$\begin{aligned}
\hat{\varphi}_{xx}(\lambda, k) &= \frac{1}{M - |\lambda|} \sum_{\kappa=k-M+1}^{k} \tilde{x}(\kappa)\tilde{x}(\kappa + \lambda) \\
&= \frac{1}{M - |\lambda|} \sum_{\kappa=0}^{M-1} \tilde{x}(k + \kappa - M + 1)\tilde{x}(k + \kappa + \lambda - M + 1),
\end{aligned} \tag{5.104}$$

for $0 \leq \lambda < M$ on the assumption that the signal $\tilde{x}(k)$ is zero outside this segment. It can be shown that this estimate is unbiased. For a stationary random process and $M \to \infty$ it approaches $\varphi_{xx}(\lambda)$ with probability one.

## 5.8 Power Spectrum and its Estimation

We define the *auto-power spectrum* of a wide sense stationary stochastic process $x(k)$ as the discrete-time Fourier transform of its auto-correlation function,

$$\Phi_{xx}(e^{j\Omega}) = \sum_{\lambda=-\infty}^{\infty} \varphi_{xx}(\lambda) e^{-j\Omega\lambda}, \tag{5.105}$$

and the *cross-power spectrum* of two processes $x(k)$ and $y(k)$ as the Fourier transform of their cross-correlation function

$$\Phi_{xy}(e^{j\Omega}) = \sum_{\lambda=-\infty}^{\infty} \varphi_{xy}(\lambda) e^{-j\Omega\lambda}, \tag{5.106}$$

whenever these transforms exist. The auto-correlation function may be obtained vice versa from an inverse discrete-time Fourier transform of the power spectrum,

$$\varphi_{xx}(\lambda) = \frac{1}{2\pi} \int_{-\pi}^{\pi} \Phi_{xx}(e^{j\Omega}) e^{j\Omega\lambda} \, d\Omega, \tag{5.107}$$

and the signal power from

$$\varphi_{xx}(0) = \frac{1}{2\pi} \int_{-\pi}^{\pi} \Phi_{xx}(e^{j\Omega}) \, d\Omega. \tag{5.108}$$

Note that correlation functions and the corresponding power spectra are considered to be deterministic quantities.

### 5.8.1 White Noise

For an uncorrelated and stationary noise signal $n(k)$, we obtain the auto-correlation $\varphi_{nn}(\lambda) = \mathrm{E}\{n(k)n(k+\lambda)\} = \delta(\lambda)\,\varphi_{nn}(0)$. Thus, we may write

$$\Phi_{nn}(e^{j\Omega}) = \sum_{\lambda=-\infty}^{\infty} \delta(\lambda)\,\varphi_{nn}(0)\,e^{-j\Omega\lambda} = \varphi_{nn}(0). \tag{5.109}$$

As the power spectrum is constant over the full range of frequencies, uncorrelated noise is also called *white noise*.

### 5.8.2 The Periodogram

An estimate of the power spectrum may be obtained from the estimate $\hat{\varphi}_{xx}(\lambda, k)$ of the auto-correlation function as in (5.104) and a subsequent Fourier transform [Oppenheim et al. 1999]. Although this auto-correlation estimate is unbiased, its Fourier transform is not an unbiased estimate of the power spectrum. This is a result of the finite segment length involved in the computation of the auto-correlation.

An estimate of the power spectrum of a wide sense stationary random process may be obtained by computing the Fourier transform of a finite signal segment $x(k)$,

$$X(e^{j\Omega}, k) = \sum_{\ell=0}^{M-1} x(k + \ell - M + 1)\,e^{-j\Omega\ell}, \tag{5.110}$$

where we now include the dependency on the time index $k$ in our notation. The magnitude squared Fourier transform normalized on the transform length is known as the *periodogram* and denoted by

$$I(e^{j\Omega}, k) = \frac{1}{M}|X(e^{j\Omega}, k)|^2. \tag{5.111}$$

The periodogram $I(e^{j\Omega}, k)$ is identical to the normalized magnitude squared discrete Fourier transform (DFT) coefficients of this signal segment at discrete equispaced frequencies $\Omega_\mu = 2\pi\mu/M$. For a real-valued signal $x(k)$, the periodogram may be written as

$$I(e^{j\Omega}, k) = \frac{1}{M}\left|\sum_{\kappa=0}^{M-1} x(k + \kappa - M + 1)\,e^{-j\Omega\kappa}\right|^2 \tag{5.112a}$$

$$= \frac{1}{M}\sum_{\kappa=-M+1}^{M-1}\sum_{\ell=0}^{M-1} \tilde{x}(k + \ell - M + 1)\tilde{x}(k + \ell + \kappa - M + 1)\,e^{-j\Omega\kappa}, \tag{5.112b}$$

where $\tilde{x}(\kappa)$ is defined as in (5.103).

The inner sum in (5.112b) is recognized as an estimate of the auto-correlation function (5.104). Thus, the periodogram corresponds to the Fourier transform of the estimated auto-correlation function multiplied with a triangular window,

$$I(e^{j\Omega}, k) = \sum_{\lambda=-M+1}^{M-1}\left(\frac{M - |\lambda|}{M}\right)\hat{\varphi}_{xx}(\lambda, k)e^{-j\Omega\lambda}. \tag{5.113}$$

Using a tapered analysis window $w(k)$ on the signal segment and a normalization on the window energy, we obtain a *modified periodogram*, which is defined as

$$I_M(e^{j\Omega}, k) = \frac{1}{\sum_{\ell=0}^{M-1} w^2(\ell)} \left| \sum_{\ell=k-M+1}^{k} w(\ell - k + M - 1)x(\ell)e^{-j\Omega\ell} \right|^2.$$

(5.114)

Finally, we note that

- because of the finite limits of summation the periodogram is a biased estimator of the power spectrum,
- the periodogram is asymptotically $(M \to \infty)$ unbiased, and,
- since its variance does not approach zero for $M \to \infty$ the periodogram is not *consistent* [Oppenheim et al. 1999]. In fact, the variance of the periodogram does not depend on the transform length. In the computation of the periodogram, the additional data associated with a larger transform length increases the frequency resolution but does not reduce the variance of the estimator.

### 5.8.3 Smoothed Periodograms

To reduce the variance of the power spectrum estimate, some form of smoothing is required. A variance reduction might be obtained by smoothing successive (modified) periodograms $I_M(e^{j\Omega}, k)$ over time or by smoothing a single periodogram over frequency. In what follows, we briefly outline methods for smoothing along the temporal and the frequency dimensions.

#### 5.8.3.1 Non Recursive Smoothing in Time

The *non recursively smoothed periodogram* is defined as

$$P_{MA}(e^{j\Omega}, k) = \frac{1}{K_{MA}} \sum_{\kappa=0}^{K_{MA}-1} I_M(e^{j\Omega}, k + (\kappa - K_{MA} + 1)r).$$

(5.115)

It averages the periodograms of $K_{MA}$ signal segments of length $M$ obtained via a segment shift of $r$. For a stationary white noise input sequence, we find the variance of the non recursively smoothed modified periodogram as [Martin 2006]

$$\text{var}\left\{P_{MA}(e^{j\Omega})\right\} = \frac{1}{K_{MA}}\text{var}\left\{I_M(e^{j\Omega})\right\}\left(1 + 2\sum_{\ell=1}^{K_{MA}-1} \frac{K_{MA} - \ell}{K_{MA}}\rho(\ell)\right),$$

(5.116)

where $\rho(\ell)$ denotes the squared correlation coefficients of the analysis window

$$\rho(\ell) = \frac{\left(\sum_{k=0}^{M-1} w(k)\,w(k + \ell\,r)\right)^2}{\left(\sum_{k=0}^{M-1} w^2(k)\right)^2}.$$

(5.117)

Thus, for uncorrelated signal segments and $K_{MA} \to \infty$, the variance approaches zero. For a fixed total number of signal samples, the variance of this estimate may be reduced to some extent by overlapping successive signal segments [Welch 1967], such that $K_{MA}$ becomes larger. However, overlap will also increase the correlation coefficient between adjacent segments such that the benefit of large overlap becomes relatively small. For example, instead of using 10 adjacent but non-overlapping segments we may use an overlap of 75% (or alternatively 50%) and thus accommodate 37 (or 19) segments on the same data sequence. When using a Hamming window, we then reduce the variance by a factor of 1.784 (or 1.723 for 50% overlap), while increasing the computational complexity by a factor of almost four (or two).

### 5.8.3.2 Recursive Smoothing in Time

To compute the Welch periodogram, $K_{MA}$ periodograms must be stored. The *recursively smoothed periodogram*,

$$P_{AR}(e^{j\Omega}, k + r) = \alpha\, P_{AR}(e^{j\Omega}, k) + (1 - \alpha)\, I_M(e^{j\Omega}, k + r) \tag{5.118}$$

is more memory efficient. It can be interpreted as an infinite sum of periodograms weighted over time by an exponential window. This corresponds to a first-order low-pass filter and for stability reasons we have $0 < \alpha < 1$. For a stationary white noise input sequence, we find the variance of the recursively smoothed periodogram as [Martin 2006]

$$\mathrm{var}\left\{P_{AR}(e^{j\Omega})\right\} = \mathrm{var}\left\{I_M(e^{j\Omega})\right\} \frac{1 - \alpha}{1 + \alpha}\left(1 + 2\sum_{\ell=1}^{\infty} \alpha^{\ell} \rho(\ell)\right), \tag{5.119}$$

where $\rho(\ell)$ is defined in (5.117).

For $\alpha \approx 0$, little smoothing is applied and hence the variance of the estimate is close to the variance of the modified periodogram. For $\alpha \approx 1$ and stationary signals, the variance of the estimate is small. The above procedure may also be used for smoothing short-time stationary periodograms and for tracking the mean value of the periodograms over larger periods of time. Then, we must strike a balance between smoothing for variance reduction and tracking of non-stationary signal characteristics. Furthermore, comparing (5.119) and (5.116) we may find a relation between $K_{MA}$ and $\alpha$ such that both smoothing methods achieve the same variance. Optimal recursive smoothing in the presence of interference has been explored in more detail in [Taghia et al. 2022].

### 5.8.3.3 Log-Mel Filter Bank Features

It is also common to smooth the modified periodogram not in time but across frequency bins, often aiming at a smaller number of (smoothed) frequency samples and a non-uniform frequency resolution. To this end, the smoothing parameters may be chosen to mimic auditory filters on the mel, ERB, or Bark scale. Typically, the number of aggregated bins $B$ is much smaller than the DFT length $M$. In order to mimic the function of the human auditory system, a logarithmic compression is applied in addition. The resulting features are often used as raw input features in conjunction with data-driven classification or regression approaches. The computation of *log-mel filter bank* features, for instance, comprises the following steps:

1) Define the number of DFT bins $M$ and mel-scale filter bins $B_{mel}$.
2) Compute center frequencies with uniform spacing on the mel scale and transform these back to the linear frequency scale.

3) Define lower and upper edge frequencies $\Omega_{L(b)}$ and $\Omega_{U(b)}$ for the $b$-th mel filter on the linear frequency scale. Here, $L(b)$ and $U(b)$ denote the DFT bins of the lower and the upper band edges.

4) Compute (triangular) mel-filter coefficients $w_{\text{mel}}(\mu, b)$.

5) For the signal frame at time $k$, use the DFT with window $w(k)$ to compute the modified periodogram $I_M(e^{j\Omega_\mu}, k)$ for center frequencies $\Omega_\mu = \frac{2\pi\mu}{M}$.

6) For each mel filter $b$ compute

$$P_{\text{log-mel}}(b) = \log\left( \sum_{\mu=L(i)}^{U(i)} w_{\text{mel}}(\mu, b) I_M(e^{j\Omega_\mu}, k) \right). \tag{5.120}$$

An example of (triangular) mel-filter coefficients $w_{\text{mel}}(\mu, b)$ is shown in Figure 5.9 with coefficients for each filter normalized on the bandwidth of this filter. When these mel-filter coefficients are applied to a succession of DFT frames we obtain a log-mel energy spectrogram, an example of which is shown in Figure 5.10. Clearly, we observe a high spectral resolution at low frequencies such that a number of harmonics related to the fundamental frequency are resolved. At higher frequencies only coarse spectral structures, such as formants are visible.

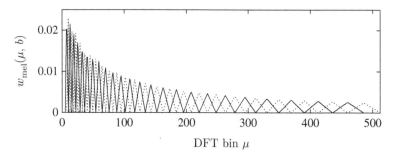

**Figure 5.9** Triangular weights for the computation of 48 mel energy features using 512 magnitude-squared DFT coefficients. Parameters: $f_s = 16\,\text{kHz}$, Hann window of length 1024, DFT length $M = 1024$. Line styles alternate for improved clarity.

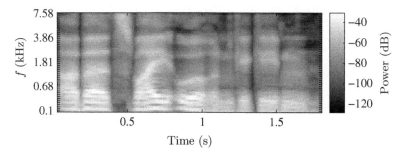

**Figure 5.10** Log-mel spectrogram of a male speech sample with 48 mel bands in the interval between 62.5 Hz and 8 kHz. Parameters: $f_s = 16\,\text{kHz}$, Hann window of length 1024 and overlap 992, DFT length $M = 1024$.

### 5.8.4 Power Spectra and Linear Shift-Invariant Systems

For *linear shift-invariant* systems (see Section 3.3), the output signal $y(k)$ is computed via the convolution (denoted by $*$) of the input signal $x(k)$ and the impulse response $h(k)$,

$$y(k) = x(k) * h(k),\tag{5.121}$$

and also the second-order statistics of the output signal is easily computed. For the auto-correlation of a single linear shift invariant (LSI) system (see Figure 5.11a) we find

$$\varphi_{yy}(\lambda) = \varphi_{xx}(\lambda) * \varphi_{hh}(\lambda),\tag{5.122}$$

and for the cross-correlation of the output signals $y_1(k)$ and $y_2(k)$ of two parallel systems (see Figure 5.11b)

$$\varphi_{y_1 y_2}(\lambda) = \varphi_{x_1 x_2}(\lambda) * \varphi_{h_1 h_2}(\lambda).\tag{5.123}$$

Then, the Fourier transform of the auto-correlations yields the corresponding power spectra

$$\Phi_{yy}(e^{j\Omega}) = \Phi_{xx}(e^{j\Omega})\left|H(e^{j\Omega})\right|^2\tag{5.124}$$

and

$$\Phi_{y_1 y_2}(e^{j\Omega}) = \Phi_{x_1 x_2}(e^{j\Omega})H_1^*(e^{j\Omega})H_2(e^{j\Omega}),\tag{5.125}$$

respectively. For the latter relation, we substitute $u = \ell + \lambda$ and note that

$$\sum_{\lambda=-\infty}^{\infty} \varphi_{h_1 h_2}(\lambda)e^{-j\Omega\lambda} = \sum_{\lambda=-\infty}^{\infty}\sum_{\ell=-\infty}^{\infty} h_1(\ell)h_2(\ell+\lambda)e^{-j\Omega\lambda}$$

$$= \sum_{u=-\infty}^{\infty} h_2(u)e^{-j\Omega u}\sum_{\ell=-\infty}^{\infty} h_1(\ell)e^{j\Omega\ell}$$

$$= H_1^*(e^{j\Omega})H_2(e^{j\Omega}).\tag{5.126}$$

When we add the two output signals such that $y_1(k) + y_2(k) = z(k)$, we have

$$\Phi_{zz}(e^{j\Omega}) = \Phi_{x_1 x_1}(e^{j\Omega})\left|H_1(e^{j\Omega})\right|^2 + \Phi_{x_2 x_2}(e^{j\Omega})\left|H_2(e^{j\Omega})\right|^2\tag{5.127}$$
$$+ \Phi_{x_1 x_2}(e^{j\Omega})H_1^*(e^{j\Omega})H_2(e^{j\Omega}) + \Phi_{x_2 x_1}(e^{j\Omega})H_1(e^{j\Omega})H_2^*(e^{j\Omega}).$$

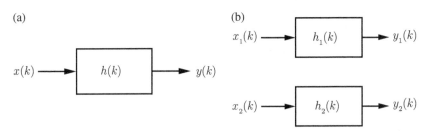

**Figure 5.11** Linear shift-invariant systems: (a) single-input, single-output (SISO) system and (b) two parallel SISO systems.

## 5.9 Statistical Properties of Speech Signals

The statistical properties of speech signals have been thoroughly investigated and are well documented, for example, [Jayant, Noll 1984] and [Brehm, Stammler 1987]. They are of interest, for instance, in the design of optimal quantizers and speech enhancement algorithms. In general, speech signals can be modeled as non-stationary stochastic processes. The power, the correlation properties, and the higher-order statistics vary from one speech sound to the next. For most practical purposes statistical parameters may be estimated on short, quasi-stationary segments of the speech signal.

It is also instructive to consider the long-term histogram of speech signals, as shown in Figure 5.12. Small amplitudes are much more frequent than large amplitudes and the tails of the density do not decay as fast as for a Gaussian signal. Thus, the PDF of speech signals in the time domain is well modeled by *supergaussian* distributions such as the Laplace, gamma, or $K_0$ PDFs [Brehm, Stammler 1987].

## 5.10 Statistical Properties of DFT Coefficients

Frequently, we analyze and process speech signals in the spectral domain using short-term spectral analysis and the DFT. It is therefore of interest to investigate the statistical properties of DFT coefficients. We assume that a signal $x(k)$ is transformed into the frequency domain by applying a window $w(k)$ to a frame of $M$ consecutive samples of $x(k)$ and by computing the DFT of size $M$ on the windowed data. This sliding-window DFT analysis results in a set of frequency domain signals $X_\mu(k)$ (see Chapter 4), which can be written as

$$X_\mu(k) = \sum_{\kappa=0}^{M-1} w(\kappa)x(k - M + 1 + \kappa)e^{-j\frac{2\pi\mu\kappa}{M}}, \qquad (5.128)$$

where $\mu$ is the frequency bin index, $\mu \in \{0, 1, \dots, M - 1\}$. The index $\mu$ is related to the normalized center frequency $\Omega_\mu$ of each frequency bin by $\Omega_\mu = 2\pi\mu/M$. Thus, the complex Fourier coefficients $X_\mu(k)$ constitute a random process.

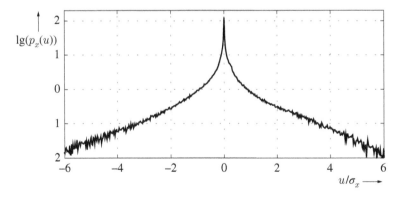

**Figure 5.12** Logarithmic plot of relative histogram of normalized speech amplitudes ($n = 4890872$ speech samples, 1024 histogram bins).

Using the above definitions, individual variables may be split into their real and imaginary parts

$$X_\mu(k) = \text{Re}\{X_\mu(k)\} + j\,\text{Im}\{X_\mu(k)\}, \tag{5.129}$$

or into their magnitude and phase

$$X_\mu(k) = R_\mu(k)\, e^{j\theta_\mu(k)}. \tag{5.130}$$

To simplify notations, we will drop the dependency of all the above quantities on the time index $k$ whenever this is possible.

### 5.10.1 Asymptotic Statistical Properties

For the discussion of asymptotic properties of the DFT coefficients, we assume

- that the transform length $M$ approaches infinity, $M \to \infty$ and
- that the transform length $M$ is much larger than the span of correlation of signal $x(k)$.

The latter condition excludes, for instance, periodic signals from our discussion. If the input signal is sufficiently random, we may conclude from the *central limit theorem* that for $\mu \notin \{0, M/2\}$, the real and imaginary parts of the DFT coefficients $X_\mu$ can be modeled as mutually independent, zero-mean Gaussian random variables [Brillinger 1981] with variance $0.5\,\sigma_{X_\mu}^2 = 0.5\,\text{E}\{|X_\mu|^2\}$, i.e.,

$$p_{\text{Re}\{X_\mu\}}(u) = \frac{1}{\sqrt{\pi}\sigma_{X_\mu}} \exp\left(-\frac{u^2}{\sigma_{X_\mu}^2}\right),$$

$$p_{\text{Im}\{X_\mu\}}(v) = \frac{1}{\sqrt{\pi}\sigma_{X_\mu}} \exp\left(-\frac{v^2}{\sigma_{X_\mu}^2}\right). \tag{5.131}$$

For a real-valued input signal $x(k)$ and $\mu \in \{0, M/2\}$ the imaginary part of $X_\mu$ is zero and the real part is also Gaussian distributed with variance $\sigma_{X_\mu}^2 = \text{E}\{|X_\mu|^2\}$.

For $\mu \notin \{0, M/2\}$, the joint distribution of the real and imaginary parts is given by

$$p_{\text{Re}\{X_\mu\},\,\text{Im}\{X_\mu\}}(u, v) = \frac{1}{\pi\sigma_{X_\mu}^2} \exp\left(-\frac{u^2 + v^2}{\sigma_{X_\mu}^2}\right), \tag{5.132}$$

or with $z = u + jv$ by

$$p_{X_\mu}(z) = \frac{1}{\pi\sigma_{X_\mu}^2} \exp\left(-\frac{|z|^2}{\sigma_{X_\mu}^2}\right). \tag{5.133}$$

Conversion to polar coordinates, $X_\mu = R_\mu e^{j\theta_\mu}$, yields a *Rayleigh density* for the magnitude $|X_\mu| = R_\mu$,

$$p_{R_\mu}(u) = \begin{cases} \dfrac{2u}{\sigma_{X_\mu}^2} \exp\left(-\dfrac{u^2}{\sigma_{X_\mu}^2}\right), & u \geq 0, \\[2mm] 0, & u < 0, \end{cases} \tag{5.134}$$

and a uniform distribution for the principal value of the phase $\theta_\mu$, $0 \le \theta_\mu \le 2\pi$,

$$p_{\theta_\mu}(u) = \begin{cases} \dfrac{1}{2\pi}, & 0 \le u \le 2\pi, \\ 0, & \text{elsewhere.} \end{cases} \tag{5.135}$$

Since for the Gaussian model, the magnitude and the phase are statistically independent, the joint density is the product of the component densities [Papoulis, Unnikrishna Pillai 2001],

$$p_{R_\mu, \theta_\mu}(u, v) = p_{R_\mu}(u) p_{\theta_\mu}(v)$$

$$= \begin{cases} \dfrac{u}{\pi \sigma_{X_\mu}^2} \exp\left(-\dfrac{u^2}{\sigma_{X_\mu}^2}\right), & u \ge 0 \text{ and } 0 \le v \le 2\pi, \\ 0, & \text{elsewhere.} \end{cases} \tag{5.136}$$

Furthermore, each magnitude squared frequency bin $|X_\mu|^2 = R_\mu^2$ is an exponentially distributed random variable with PDF

$$p_{R_\mu^2}(u) = \begin{cases} \dfrac{1}{\sigma_{X_\mu}^2} \exp\left(-\dfrac{u}{\sigma_{X_\mu}^2}\right), & u \ge 0, \\ 0, & u < 0. \end{cases} \tag{5.137}$$

### 5.10.2 Signal-Plus-Noise Model

In applications, such as speech enhancement, we consider an observed signal $x(k) = s(k) + n(k)$, which is the sum of a desired signal $s(k)$ and a noise signal $n(k)$, where $n(k)$ is statistically independent of the desired signal $s(k)$. Obviously, this also leads to an additive noise model in the Fourier or in the DFT domain, $X_\mu(k) = S_\mu(k) + N_\mu(k)$. We now compute the conditional density for the observed DFT coefficients $X_\mu$ given the desired coefficients $S_\mu = A_\mu e^{j\alpha_\mu} = A_\mu \left(\cos(\alpha_\mu) + j\sin(\alpha_\mu)\right)$ using the Gaussian assumption. Since, the desired signal and the noise are additive and statistically independent, the conditional densities for the real and imaginary parts are given by

$$p_{\text{Re}\{X_\mu\}|\text{Re}\{S_\mu\}}(u \mid \text{Re}\{S_\mu\}) = \frac{1}{\sigma_{N_\mu}\sqrt{\pi}} \exp\left(-\frac{(u - A_\mu \cos(\alpha_\mu))^2}{\sigma_{N_\mu}^2}\right)$$

$$p_{\text{Im}\{X_\mu\}|\text{Im}\{S_\mu\}}(v \mid \text{Im}\{S_\mu\}) = \frac{1}{\sigma_{N_\mu}\sqrt{\pi}} \exp\left(-\frac{(v - A_\mu \sin(\alpha_\mu))^2}{\sigma_{N_\mu}^2}\right), \tag{5.138}$$

where we conditioned on the real and the imaginary parts of the desired coefficients. With $z = u + jv$ the conditional joint density is given by

$$p_{\text{Re}\{X_\mu\},\text{Im}\{X_\mu\}|S_\mu}(u, v \mid S_\mu)$$

$$= \frac{1}{\pi \sigma_{N_\mu}^2} \exp\left(-\frac{|z - A_\mu \exp(j\alpha_\mu)|^2}{\sigma_{N_\mu}^2}\right)$$

$$= \frac{1}{\pi \sigma_{N_\mu}^2} \exp\left(-\frac{|z|^2 + A_\mu^2 - 2A_\mu \text{Re}\{\exp(-j\alpha_\mu)z\}}{\sigma_{N_\mu}^2}\right). \tag{5.139}$$

Since a rotation in the complex plane does not change the magnitude

$$|z|^2 = |z\exp(-j\alpha_\mu)|^2 = \text{Re}\{z\exp(-j\alpha_\mu)\}^2 + \text{Im}\{z\exp(-j\alpha_\mu)\}^2, \tag{5.140}$$

the conditional joint density can be also written as

$$p_{\text{Re}\{X_\mu\},\text{Im}\{X_\mu\}|S_\mu}(u,v \mid S_\mu)$$
$$= \frac{1}{\pi\sigma_{N_\mu}^2}\exp\left(-\frac{\left(\text{Re}\{\exp(-j\alpha_\mu)z\} - A_\mu\right)^2 + \text{Im}\{\exp(-j\alpha_\mu)z\}^2}{\sigma_{N_\mu}^2}\right), \tag{5.141}$$

which leads to a *Rician* PDF for the conditional magnitude [Papoulis, Unnikrishna Pillai 2001], [McAulay, Malpass 1980]

$$p_{R_\mu|S_\mu}(u \mid S_\mu) = \begin{cases} \dfrac{2u}{\sigma_{N_\mu}^2}\exp\left(-\dfrac{u^2 + A_\mu^2}{\sigma_{N_\mu}^2}\right) I_0\left(\dfrac{2A_\mu u}{\sigma_{N_\mu}^2}\right), & u \geq 0, \\ 0, & u < 0, \end{cases} \tag{5.142}$$

where $I_0(\cdot)$ denotes the modified Bessel function of the first kind. When no speech is present, the magnitude obeys a Rayleigh distribution as before.

### 5.10.3 Statistics of DFT Coefficients for Finite Frame Lengths

Unlike in Sections 5.10.1 and 5.10.2, we now consider a short transform length $M$ as it is used in mobile communications and other applications. There, the asymptotic assumptions are not well fulfilled, especially for voiced speech sounds which exhibit a high degree of correlation [Martin 2002], [Martin 2005b] and [Martin, Breithaupt 2003]. Clearly, the central limit theorem does not hold when the span of correlation in the signal is larger than the frame length. For short frame lengths, DFT coefficients of speech signals are not Gaussian-distributed and frequency bins will not be independent in general. Hence, in the short-term Fourier domain (frame size < 100ms) the Laplace and gamma densities are much better models for the PDF of the real and imaginary parts of the DFT coefficients than the commonly used Gaussian density. In this section, we will briefly review these densities and provide examples of experimental data.

Let $\text{Re}\{S_\mu\} = S_R$ and $\text{Im}\{S_\mu\} = S_I$ denote the real and the imaginary part of a clean speech DFT coefficient, respectively. To enhance the readability, we will drop both the frame index $k$ and the frequency index $\mu$ and consider a single DFT coefficient at a given time instant. $\sigma_s^2/2$ denotes the variance of the real and imaginary parts of the DFT coefficient. Then, the Laplacian and the gamma densities (real and imaginary parts) are given by

$$p_{S_R}(u) = \frac{1}{\sigma_s}\exp\left(-\frac{2|u|}{\sigma_s}\right)$$
$$\tag{5.143}$$
$$p_{S_I}(v) = \frac{1}{\sigma_s}\exp\left(-\frac{2|v|}{\sigma_s}\right)$$

and

$$p_{S_R}(u) = \frac{\sqrt[4]{3}}{2\sqrt{\pi\sigma_s}\sqrt[4]{2}}|u|^{-\frac{1}{2}}\exp\left(-\frac{\sqrt{3}|u|}{\sqrt{2}\sigma_s}\right)$$

$$p_{S_I}(v) = \frac{\sqrt[4]{3}}{2\sqrt{\pi\sigma_s}\sqrt[4]{2}}|v|^{-\frac{1}{2}}\exp\left(-\frac{\sqrt{3}|v|}{\sqrt{2}\sigma_s}\right),$$

(5.144)

respectively. The gamma density diverges when the argument approaches zero but provides otherwise a good fit to the observed data.

Figures 5.13 and 5.14 plot the histogram of the real part of the DFT coefficients ($M = 256$, $f_s = 8000$ Hz) of clean speech averaged over three male and three female speakers. Since speech has a time-varying power, the coefficients represented in the histogram are selected, such as to range in a narrow power interval. For the depicted histogram, coefficients are selected in a 2 dB wide interval across all frequency bins excluding for the lowest and the highest bins. Thus, the histogram and the corresponding model densities are conditioned on the measured signal-to-noise ratio (SNR) of the spectral coefficients. The full histogram in Figure 5.13 as well as the enlarged section in Figure 5.14 show that indeed the Laplace and gamma densities provide a much better fit to the DFT data than the Gaussian distribution. This result is also reflected by the estimated Kullback–Leibler divergence $D(p_x \| p_y)$ [Kullback 1997] between the histogram data $p_H(x)$ and one of the above PDF candidates $p_{S_R}(u)$. We find that the Kullback–Leibler distance is about 3 times smaller for the gamma density and about 6 times smaller for the Laplace density than for the Gaussian density [Martin 2005]. An even better match of the measured PDF can be obtained with a linear combination of the Laplacian and the gamma model, where the weights 0.7 and 0.3 are

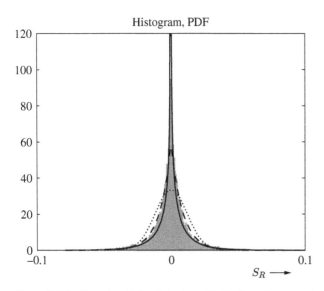

**Figure 5.13** Gaussian (dotted), Laplace (dashed), and gamma (solid) density fitted to a histogram of the real part of clean speech DFT coefficients (shaded, $M = 256$, $f_s = 8000$ Hz). Source: [Martin 2002]; © 2002 IEEE.

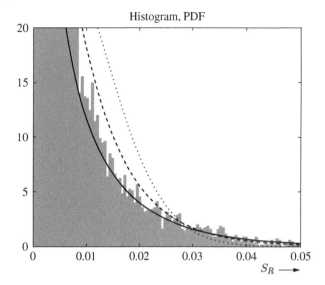

**Figure 5.14** Gaussian (dotted), Laplace (dashed), and gamma (solid) density fitted to a histogram of the real part of clean speech DFT coefficients (shaded, $M = 256, f_s = 8000\,\text{Hz}$). Source: [Martin 2002]; © 2002 IEEE.

assigned to the Laplacian and the gamma model respectively, or with a generalized gamma density [Shin et al. 2005].

Similar approximations may be developed for the magnitude $A = |S|$ of the DFT coefficients. A good fit to the observed data is obtained for the generalized gamma density [Lotter, Vary 2003], [Dat et al. 2005]

$$p_A(u) = \frac{\gamma^{\nu+1}}{\Gamma(\nu+1)} \frac{u^\nu}{\sigma_S^{\nu+1}} \exp\left(-\gamma \frac{u}{\sigma_S}\right),\tag{5.145}$$

where for $\mathrm{E}\left\{A^2\right\} = \sigma_S^2$ the constraint $\gamma = \sqrt{(\nu+1)(\nu+2)}$ applies. The parameters $\nu$, $\mu$ determine the shape of the PDF and thus allow to adapt the underlying PDF of the conditional estimator to the observed distribution.

## 5.11 Optimal Estimation

Quite often we are faced with the task of estimating the value of a random variable $x$ when observations of another random vector $\mathbf{y} = (y_1, \dots, y_N)^T$ are given. The resulting estimator $\hat{x} = f(\mathbf{y})$ maps the observed random vector $\mathbf{y}$ onto the estimated variable $\hat{x}$, which is then in turn also a random variable.

In the context of speech processing, various optimization criteria are used [Vary 2004], which we briefly summarize for continuous random variables as follows:

- The *maximum likelihood* (ML) estimator selects a value in the range of $x$ such that the conditional probability density of the observed variables is maximized, i.e.,

$$\hat{x} = \arg\max_x p_{\mathbf{y}|x}\left(\mathbf{y} \mid x\right).\tag{5.146}$$

- The *maximum a posteriori* (MAP) estimator is defined by

$$\hat{x} = \arg\max_x p_{x|y}(x \mid \mathbf{y}) = \arg\max_x \frac{p_{y|x}(\mathbf{y} \mid x) p_x(x)}{p_y(\mathbf{y})}, \tag{5.147}$$

where now also the *a priori* distribution $p_x(x)$ of the unknown variable $x$ is employed. When the *a priori* density $p_x(x)$ is uniformly distributed across the range of admissible values, the MAP estimator results in the same estimate as the ML estimator since the MAP estimate does not depend on $p_y(\mathbf{y})$.

- In a more general setting, we might strive to determine the multivariate function $\hat{x} = f(\mathbf{y})$ such that the statistical expectation

$$E_{xy}\{C(x,\hat{x})\} = \int_{-\infty}^{\infty} \cdots \int_{-\infty}^{\infty} C(u, f(\mathbf{y})) p_{xy}(u, y_1, \ldots, y_N) \, du \, dy_1 \ldots dy_N \tag{5.148}$$

of a *cost function* $C(x,\hat{x})$ is minimized over the probability spaces of $x$ and $\mathbf{y}$. The most prominent of these estimators is the *minimum mean square error* (MMSE) estimator where $C(x,\hat{x}) = (x - \hat{x})^2$. MMSE solutions to the estimation problem will be discussed in more detail below.

### 5.11.1 MMSE Estimation

We show that the optimal solution $\hat{x}$ in the MMSE sense is given by the expectation of $x$ conditioned on the vector of observations $\mathbf{y} = (y_1, \ldots, y_N)^T$, i.e.,

$$\hat{x} = E_x\{x \mid \mathbf{y}\} = \int_{-\infty}^{\infty} u \, p_{x|y}(u \mid y_1, \ldots, y_N) \, du. \tag{5.149}$$

The optimal solution $\hat{x} = f(\mathbf{y})$ is a function of the joint statistics of the observed variables and $x$. The computation of the mean square error $E_{xy}\{(x - \hat{x})^2\}$ requires averaging over the probability space of $x$ as well as of $\mathbf{y}$. Thus, we might expand the mean square error into

$$E_{xy}\{(x - \hat{x})^2\} = \int_{-\infty}^{\infty} \cdots \int_{-\infty}^{\infty} (u - \hat{x})^2 p_{xy}(u, y_1, \ldots, y_N) \, du \, dy_1 \ldots dy_N \tag{5.150}$$

$$= \int_{-\infty}^{\infty} \cdots \int_{-\infty}^{\infty} (u - \hat{x})^2 p_{x|y}(u \mid y_1, \ldots, y_N) p_y(y_1, \ldots, y_N) \, du \, dy_1 \ldots dy_N.$$

Since PDFs are non negative, it is sufficient to minimize the inner integral

$$\int_{-\infty}^{\infty} (u - \hat{x})^2 p_{x|y}(u \mid y_1, \ldots, y_N) \, du, \tag{5.151}$$

for any given vector $\mathbf{y}$ of observations. Setting the first derivative with respect to $\hat{x}$ to zero then yields the desired result

$$\hat{x} = \int_{-\infty}^{\infty} u \, p_{x|y}(u \mid y_1, \ldots, y_N) \, du = E_x\{x \mid \mathbf{y}\}. \tag{5.152}$$

In general, $\hat{x}$ is a non-linear function of the observed values $\mathbf{y}$.

### 5.11.2 Estimation of Discrete Random Variables

In analogy to the above estimation methods, we may develop MAP and MMSE estimators also for discrete-valued random variables. To this end, we enumerate all possible values $x_\kappa$, $\kappa \in \{0, \ldots, \mathcal{K}\}$ and observations $\mathbf{y}_\ell$, $\ell \in \{0, \ldots, \mathcal{L}\}$. Then, given the observation $\mathbf{y}_\ell$ we may write the *a posteriori* probability as

$$
P\left(x_\kappa \mid \mathbf{y}_\ell\right) = \frac{P\left(\mathbf{y}_\ell \mid x_\kappa\right) P\left(x_\kappa\right)}{P(\mathbf{y}_\ell)}
$$

$$
= \frac{P\left(\mathbf{y}_\ell \mid x_\kappa\right) P\left(x_\kappa\right)}{\sum_{\kappa=0}^{\mathcal{K}} P\left(\mathbf{y}_\ell \mid x_\kappa\right) P\left(x_\kappa\right)}, \tag{5.153}
$$

and consequently the MAP estimator as

$$
\hat{x} = \arg\max_\kappa P\left(x_\kappa \mid \mathbf{y}_\ell\right). \tag{5.154}
$$

Similarly, we obtain for the MMSE estimator with (5.153)

$$
\hat{x} = \mathrm{E}_x\{x \mid \mathbf{y}_\ell\} = \sum_{\kappa=0}^{\mathcal{K}} x_\kappa \, P(x_\kappa \mid \mathbf{y}_\ell)
$$

$$
= \frac{\sum_{\kappa=0}^{\mathcal{K}} x_\kappa \, P(\mathbf{y}_\ell \mid x_\kappa) P(x_\kappa)}{\sum_{\kappa=0}^{\mathcal{K}} P(\mathbf{y}_\ell \mid x_\kappa) P(x_\kappa)}. \tag{5.155}
$$

### 5.11.3 Optimal Linear Estimator

We now simplify the estimation procedure by constraining the estimate $\hat{x}$ to be a linear combination of the observed data, i.e.,

$$
\hat{x} = \mathbf{h}^T \mathbf{y} = \mathbf{y}^T \mathbf{h}, \tag{5.156}
$$

where $\mathbf{h}^T = (h_1, \ldots, h_N)$ is a vector of constant weights. The expansion of $\mathrm{E}\left\{(x - \hat{x})^2\right\}$ leads to

$$
\mathrm{E}\left\{(x - \hat{x})^2\right\} = \mathrm{E}\left\{x^2\right\} - 2\mathrm{E}\left\{x\mathbf{y}^T\right\}\mathbf{h} + \mathbf{h}^T \mathrm{E}\left\{\mathbf{y}\mathbf{y}^T\right\} \mathbf{h}, \tag{5.157}
$$

and the minimization of the mean square error for an invertible auto-correlation matrix $\mathbf{R}_{yy}$ to

$$
\mathbf{h} = \mathbf{R}_{yy}^{-1} \mathbf{r}_{xy}, \tag{5.158}
$$

where the cross-correlation vector $\mathbf{r}_{xy}$ is defined as

$$
\mathbf{r}_{xy} = \left(\mathrm{E}\left\{xy_1\right\}, \ldots, \mathrm{E}\left\{xy_N\right\}\right)^T. \tag{5.159}
$$

Thus, for a given vector of observations $\mathbf{y}$, we compute the estimated value as

$$
\hat{x} = \mathbf{y}^T \mathbf{R}_{yy}^{-1} \mathbf{r}_{xy}. \tag{5.160}
$$

In contrast to the general non-linear solution, the probability densities of the signal and the noise are not involved in the computation of the linearly constrained solution. The weight vector $\mathbf{h} = \mathbf{R}_{yy}^{-1} \mathbf{r}_{xy}$ is a function of second-order statistics but not a function of the observed vector $\mathbf{y}$ itself.

The linear estimator may be extended to the more general non-homogeneous case

$$\hat{x} = \mathbf{h}^T \mathbf{y} + a. \tag{5.161}$$

We obtain

$$\mathbf{h} = \mathbf{C}_{yy}^{-1} \mathbf{c}_{xy} = (\mathbf{R}_{yy}^{-1} - \boldsymbol{\mu}_y \boldsymbol{\mu}_y^T)^{-1} (\mathbf{r}_{xy} - \boldsymbol{\mu}_y \boldsymbol{\mu}_x) \tag{5.162}$$

and

$$a = \mu_x - \boldsymbol{\mu}_y^T \mathbf{C}_{yy}^{-1} \mathbf{c}_{xy}. \tag{5.163}$$

The estimate is therefore given by

$$\hat{x} = (\mathbf{y}^T - \boldsymbol{\mu}_y) \mathbf{C}_{yy}^{-1} \mathbf{c}_{xy} + \mu_x. \tag{5.164}$$

If we assume an additive signal plus noise model, $y = x + n$, where the signal $x$ and the noise $n$ are mutually uncorrelated signals with $E\{x\} = \mu_x$, $E\{n\} = 0$, and $E\{xn\} = 0$, we may further simplify the above result. It is especially instructive to interpret the solution in the case of a single observation $y$. Then, we have

$$\hat{x} = (y - \mu_x) \frac{\sigma_x^2}{\sigma_x^2 + \sigma_n^2} + \mu_x = (y - \mu_x) \frac{\xi}{1 + \xi} + \mu_x, \tag{5.165}$$

where $\xi = \sigma_x^2 / \sigma_n^2$ is called the *a priori* SNR. For $\xi \gg 1$, we have $\hat{x} \approx y$. The estimate is approximately equal to the observed variable. For a low *a priori* SNR, $\xi \gtrsim 0$, we have $\hat{x} \approx \mu_x$. In this case, the best estimate is the unconditional mean of the desired variable $x$.

### 5.11.4 The Gaussian Case

We will show that for the additive noise model, a scalar observation $y$, and jointly Gaussian signals the non-linear MMSE estimator is identical to the linearly constrained estimator. Since $x$ and $n$ are uncorrelated and have zero mean, the PDF of $y = x + n$ is given by

$$p_y(v) = \frac{1}{\sqrt{2\pi(\sigma_x^2 + \sigma_n^2)}} \exp\left(-\frac{v^2}{2(\sigma_x^2 + \sigma_n^2)}\right)$$

because the convolution of two Gaussians is a Gaussian. The conditional density $p_{y|x}(v \mid u) = p_n(v - u)$ and the density of the undisturbed signal $x$ may be written as

$$p_n(v - u) = \frac{1}{\sqrt{2\pi}\sigma_n} \exp\left(-\frac{(v - u)^2}{2\sigma_n^2}\right) \tag{5.166}$$

and

$$p_x(u) = \frac{1}{\sqrt{2\pi}\sigma_x} \exp\left(-\frac{u^2}{2\sigma_x^2}\right), \tag{5.167}$$

respectively. The product $p_{y|x}(v \mid u) p_x(u)$ may now be written as

$$p_{y|x}(v \mid u) p_x(u) \tag{5.168}$$

$$= \frac{1}{2\pi \sqrt{\sigma_x^2 \sigma_n^2}} \exp\left(-\frac{\sigma_x^2 + \sigma_n^2}{2\sigma_x^2 \sigma_n^2}\left(u - v\frac{\sigma_x^2}{\sigma_x^2 + \sigma_n^2}\right)^2\right) \exp\left(\frac{v^2}{2\sigma_n^2}\left(\frac{\sigma_x^2}{\sigma_x^2 + \sigma_n^2} - 1\right)\right),$$

where the first exponential is recognized as a Gaussian density and the second exponential does not depend on $u$ anymore. Then, the non-linear MMSE estimate is given for any specific value $v$ of the observed variable $y$ by

$$\hat{x}(v) = \int_{-\infty}^{\infty} u\, p_{x|y}(u \mid v)\, du$$

$$= \frac{1}{p_y(v)} \int_{-\infty}^{\infty} u\, p_{y|x}(v \mid u)\, p_x(u)\, du$$

$$= \frac{\sigma_x^2}{\sigma_x^2 + \sigma_n^2}\, v = \frac{\xi}{1 + \xi}\, v \tag{5.169}$$

with $\xi$ as defined above. This result can be extended to multiple observations and multivariate estimates as well.

### 5.11.5 Joint Detection and Estimation

In a practical application, the desired signal $x$ may not always be present in a observed noisy signal $y = x + n$. The optimal estimator must be adapted to this uncertainty about the presence of the desired signal and must deliver an optimal estimate regardless of whether the signal is present or not. To consider a more general setting, we now assume that there are two versions $x_0$ and $x_1$ of the desired signal, which are present in the observed signal with *a priori* probabilities $P(H^{(0)})$ and $P(H^{(1)}) = 1 - P(H^{(0)})$ respectively, where we denote the presence of $x_0$ and the presence of $x_1$ as the two hypotheses

$$H^{(0)}: \quad x = x_0 \quad \text{and} \tag{5.170}$$

$$H^{(1)}: \quad x = x_1. \tag{5.171}$$

The two versions of the signal $x$ are treated as random variables with possibly different PDFs. With these assumptions, the PDF of $x$ may be written as

$$p_x(u) = p_{x|H^{(0)}}(u \mid H^{(0)})P(H^{(0)}) + p_{x|H^{(1)}}(u \mid H^{(1)})P(H^{(1)}). \tag{5.172}$$

Again, we use a quadratic cost function $C(x, \hat{x}) = (\hat{x} - x)^2$ where $\hat{x}$ is in general a function of the observed variable $y = x + n$. We minimize the total (Bayes) cost

$$\mathfrak{J} = \int_{-\infty}^{\infty}\int_{-\infty}^{\infty} C(u, \hat{x}(v))p_{xy}(u, v)\, du\, dv$$

$$= \int_{-\infty}^{\infty}\int_{-\infty}^{\infty} (\hat{x}(v) - u)^2 \left( p_{xy|H^{(0)}}(u, v \mid H^{(0)})P(H^{(0)}) \right.$$

$$\left. + p_{xy|H^{(1)}}(u, v \mid H^{(1)})P(H^{(1)}) \right)\, du\, dv. \tag{5.173}$$

Setting the first derivative of the inner integral with respect to $\hat{x}$ to zero, we obtain

$$\int_{-\infty}^{\infty} (\hat{x}(v) - u)\left( p_{xy|H^{(0)}}(u, v|H^{(0)})P(H^{(0)}) + p_{xy|H^{(1)}}(u, v|H^{(1)})P(H^{(1)}) \right) du = 0,$$

and substituting

$$p_{xy|H^{(0)}}(u,v \mid H^{(0)}) \, P(H^{(0)}) = p_{y|H^{(0)}}(v \mid H^{(0)}) \, p_{x|y,H^{(0)}}(u \mid v, H^{(0)}) \, P(H^{(0)})$$

$$p_{xy|H^{(1)}}(u,v \mid H^{(1)}) \, P(H^{(1)}) = p_{y|H^{(1)}}(v \mid H^{(1)}) \, p_{x|y,H^{(1)}}(u \mid v, H^{(1)}) \, P(H^{(1)})$$

yields

$$\hat{x}(v) \left[ p_{y|H^{(0)}}(v \mid H^{(0)}) \, P(H^{(0)}) + p_{y|H^{(1)}}(v \mid H^{(1)}) \, P(H^{(1)}) \right]$$

$$= p_{y|H^{(0)}}(v \mid H^{(0)}) \, P(H^{(0)}) \int_{-\infty}^{\infty} u \, p_{x|y,H^{(0)}}(u \mid v, H^{(0)}) \, du$$

$$+ p_{y|H^{(1)}}(v \mid H^{(1)}) \, P(H^{(1)}) \int_{-\infty}^{\infty} u \, p_{x|y,H^{(1)}}(u \mid v, H^{(1)}) \, du. \tag{5.174}$$

We introduce the generalized likelihood ratio

$$\Lambda(v) = \frac{p_{y|H^{(1)}}(v \mid H^{(1)}) \, P(H^{(1)})}{p_{y|H^{(0)}}(v \mid H^{(0)}) \, P(H^{(0)})}, \tag{5.175}$$

and obtain the solution [Middleton, Esposito 1968]

$$\hat{x}(v) = \mathrm{E}_x\{x \mid v, H^{(0)}\} \, \frac{1}{1+\Lambda(v)} + \mathrm{E}_x\{x \mid v, H^{(1)}\} \, \frac{\Lambda(v)}{1+\Lambda(v)}. \tag{5.176}$$

The joint MMSE detection and estimation problem leads to a linear combination of the MMSE estimators for the two hypotheses $H^{(0)}$ and $H^{(1)}$. The weights of the two estimators are in the range $[0, 1]$ and are determined as a function of the generalized likelihood ratio $\Lambda(v)$. Since, the likelihood ratio is in general a continuous function of the observed data $v$, (5.176) comprises a *soft decision* weighting of the two conditional estimators.


## 5.12 Non-Linear Estimation with Deep Neural Networks

In recent years, the use of statistical models in speech-signal processing has shifted from low-dimensional analytic models to more general data-driven machine-learning approaches. These are mostly based on deep neural networks (DNNs), which are able to model data distributions with a high degree of detail. While artificial neural networks have been known for quite some time, their recent success is enabled by several important developments, such as

- the feasibility of gradient back-propagation in deeply nested networks with millions of network parameters,
- the availability of huge amounts of data recorded via ubiquitous microphones and cameras in smartphones and other devices, and corresponding resources in social media and the world wide web,
- the availability of massively parallel computers utilizing graphics processing units (GPUs) for parameter optimization, and
- the emergence of software frameworks that relieve the user from hand-crafted gradient computations and provide appropriate parameter optimization procedures.

With DNNs, high-dimensional and intrinsically non-linear statistical models are available. Most notably, these can be used to model the joint distribution $p(\mathbf{x}) = p(x_1, x_2, \ldots, x_N)$ of a vector-valued random variable $\mathbf{x} = (x_1, x_2, \ldots, x_N)^T$. While many of these approaches have been initially developed for classification tasks, also estimation ("regression") tasks have been successfully tackled. In a classification application with $M$ classes $C_i$, the DNN may be trained to model the posterior probability

$$p(C_i \mid \mathbf{x}) = \frac{p(\mathbf{x} \mid C_i) p(C_i)}{p(\mathbf{x})} = \frac{p(\mathbf{x}, C_i)}{p(\mathbf{x})}, \quad i = 1, \ldots, M, \tag{5.177}$$

and then be used to determine the class membership for given data samples. This entails training a model for the joint distribution of input samples and class labels and constitutes a *generative* model as the joint distribution may be sampled to generate new input samples for any class [Bishop 2006]. In other applications, it might suffice to simply train the network to map any input sample $\mathbf{x}_\ell$ directly to class labels $C_i$, which is then known as a *discriminative* model.

In what follows, we provide a short overview of basic network components and DNN structures.

### 5.12.1 Basic Network Components

#### 5.12.1.1 The Perceptron

The basic element of many DNNs is the artificial neuron, which is inspired by the mechanism of neural transduction in a single biological neuron, as shown in Figure 5.15. In a biological neural cell, dendrites collect the incoming signals and the cell generates an electric activation, which is passed on to neighboring cells via biochemical transmitter substances at the synaptic terminals. A single layer of such neurons is also known as the *perceptron* [Rosenblatt 1957]. When each neuron is connected to any element of the input vector $\mathbf{x}(k)$, this is also called a *dense* layer.

A single artificial neuron is defined by a non-linear input–output mapping

$$y(k) = f(\mathbf{w}^T \mathbf{x}(k) + b), \tag{5.178}$$

where $\mathbf{x}^T(k) = (x_1(k), \ldots, x_N(k))$ denotes a vector of input data, $\mathbf{w}^T = (w_1, \ldots, w_N)$ is a vector of adjustable weights, and $b$ is an adjustable additive bias. $f$ is known as the *activation function* which, in its original formulation, is a threshold function with threshold $T$

$$f_T(z) = \begin{cases} 1, & z \geq T, \\ 0, & z < T. \end{cases} \tag{5.179}$$

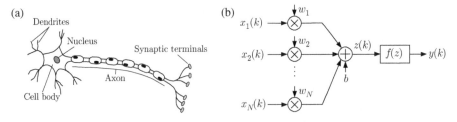

**Figure 5.15** A computational model of neural transduction. (a) Simplified sketch of neural cell and (b) block diagram of artificial neuron.

However, this activation function is less suitable for numeric parameter optimization via gradient back-propagation and gradient descent as it is not differentiable in $z = T$ and leads to vanishing gradients elsewhere. Therefore, smooth activation functions are preferred, a selection of which and their corresponding derivatives are summarized below:

- the logistic (sigmoid) function

$$f_\sigma(z) = \frac{1}{1 + \exp(-z)}, \qquad \frac{df_\sigma(z)}{dz} = f_\sigma(z)(1 - f_\sigma(z)), \qquad (5.180)$$

- the hyperbolic tangens function

$$f_{tanh}(z) = \tanh(z), \qquad \frac{df_{tanh}(z)}{dz} = 1 - f_{tanh}^2(z), \qquad (5.181)$$

- the rectified linear unit (ReLU)

$$f_{ReLU}(z) = \max(0, z), \qquad \frac{df_{ReLU}(z)}{dz} = \begin{cases} 1, & z > 0, \\ 0, & z < 0, \\ \text{undefined}, & z = 0, \end{cases} \qquad (5.182)$$

- the modified rectified linear unit (MReLU)

$$f_{MReLU}(z) = \min(\max(0, z), a), \qquad \frac{df_{MReLU}(z)}{dz} = \begin{cases} 0, & z < 0 \,|\, z > a, \\ 1, & 0 < z < a, \\ \text{undef.}, & z = 0 \,|\, z = a. \end{cases} \qquad (5.183)$$

We note that any continuous function $h : \mathbb{R}^N \to \mathbb{R}$ may be approximated by a linear combination of these basic elements, yielding a function $\hat{h}$,

$$\hat{h}(\mathbf{x}) = \sum_{i=1}^{I} \alpha_i f_b(\mathbf{w}_i^T \mathbf{x} + b_i), \qquad (5.184)$$

where $\mathbf{x}, \mathbf{w}_i \in \mathbb{R}^N$, $\alpha_i, b_i \in \mathbb{R}$, and $f_b$ refers to any bounded activation function satisfying certain smoothness constraints, for example, the sigmoid function or the modified rectified linear unit [Cybenko 1989], [Qi et al. 2019]. The absolute approximation error depends on the number $I$ of neurons and is in the order of

$$\left| \hat{h}(\mathbf{x}) - h(\mathbf{x}) \right| = \mathcal{O}\left( \frac{1}{\sqrt{I}} \right). \qquad (5.185)$$

Thus, networks of basic elements may approximate arbitrary functions with any precision. They act as universal function approximators and may in theory fit any type of data. However, in practice a large number of neurons also requires large amounts of training data. Since in some application domains training data can be scarce, large networks may be severely over-parametrized. Then, the trained regression $\hat{h}$ will follow random variations of the training data and will not generalize well to unseen data, an effect which is known as "over-fitting." Also the choice of activation functions and the application of suitable normalization procedures are critical to ensure the stability of gradient computations and the convergence of training. Therefore, a good balance between the number of parameters and the amount of training data needs to be preserved and further regularization is necessary to improve the performance on previously unseen data samples. As described in Section 5.12.1.2, the use of a relatively small convolutional kernel constitutes a successful strategy for the reduction of parameters.

#### 5.12.1.2 Convolutional Neural Network

Convolutional neural networks (CNNs) have been devised to capture structures in multi-dimensional data with a low number of trainable network coefficients. They have been introduced for feature extraction in 2D image data [Fukushima 1980], [LeCun et al. 1998] but have also widely been used in the context of speech and audio signals. Here, the input data ("receptive field") most often consist of spectrogram patches or other forms of time-frequency representations. The computation of a CNN entails shifting one or multiple *kernels* comprising network weights across an array of data samples. This is visualized in Figure 5.16 which depicts the $\ell$th filter kernel with coefficients $w_\ell(i,j)$ acting on a 2D array of data points. In this example, the kernel consists of $4 \times 2 + 1 = 9$ trainable weights (including a scalar bias) and it slides across the input data to produce an array of output samples $y_\ell(m, n)$. Thus, the $\ell$th output channel of a CNN is computed as

$$y_\ell(m, n) = f\left( \sum_{j=0}^{J-1} \sum_{i=0}^{I-1} w_\ell(i,j) x(m - i, n - j) + b_\ell \right), \tag{5.186}$$

where additional considerations such as zero-padding are necessary at the boundaries of the data patch. Quite frequently, several kernels $w_\ell(i,j)$ are arranged in parallel such that different signal features may be extracted in one CNN layer resulting in several output channels. In conjunction with additional measures for data compression, for instance *sub-sampling* or *pooling* operations, CNNs provide a highly flexible and scalable tool for data analysis. We furthermore note that appropriate *inverse* layers (transposed convolutional layers) for kernel-based data interpolation are available as well.

### 5.12.2 Basic DNN Structures

The basic elements of neural networks are typically arranged in a layered fashion resulting in a deep network. Thus, these networks are composed of an input layer, at least one *hidden*

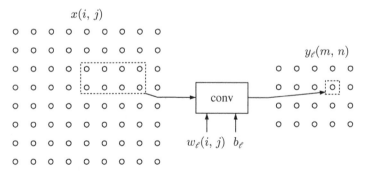

**Figure 5.16** Single output channel of a convolutional neural network layer with decimation. Data samples are indicated by circles and are arranged in a rectangular pattern. "conv" denotes the convolution operation, as shown in (5.186).

layer, and an output layer. Over the past years, a variety of network structures have been devised, the most prominent of which are summarized below (see also, e.g., [Goodfellow et al. 2016], [Aggarwal 2018]).

### 5.12.2.1 Fully-Connected Feed-Forward Network

The multi-layer perceptron (MLP) is composed of multiple concatenated instances of the perceptron. When each neuron in one layer connects to any other neuron in the following layer, the multi-layered structure forms a *fully connected* network. While the fully connected network shown in Figure 5.17 features the same number of neurons in each layer, the number of neurons may vary from layer to layer in general. The overall structure of an MLP is simple but also results in a relatively large number of network coefficients. The processing of a single layer can be expressed as a matrix multiplication.

### 5.12.2.2 Autoencoder Networks

A widely used structure for joint data compression and generation is the autoencoder (AE) [Goodfellow et al. 2016] with input vector $\mathbf{x} = (x_1, \ldots, x_N)^T$ and output vector $\mathbf{y} = (y_1, \ldots, y_N)^T$, as shown in Figure 5.18. The basic idea of an AE network is to reconstruct the input data at the output with minimal distortion. The distortion is measured by a loss function, such as the mean square error between input and output. Since no additional labeled data are needed, it constitutes a framework for self-supervised learning. The AE is composed of encoding and decoding networks and implements an information bottleneck, frequently referred to as the *latent representation* or *code*, in between these stages.

The minimization of, for instance, the mean square error $E\left\{||\mathbf{x} - \mathbf{y}||^2\right\}$, forces the neural network to represent the most salient signal features in the bottleneck layer. This leads to a compressed latent data representation, which may be viewed as a non-linear variant of a principal component analysis (PCA). The information bottleneck method is not only a powerful compression and feature extraction method but can also be used as a generative model. The variational autoencoder (VAE) [Kingma, Welling 2014], most notably, uses the encoder network to estimate the parameters of a given distribution (e.g., multivariate

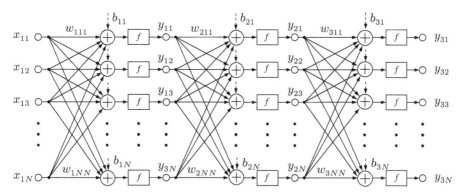

**Figure 5.17** Fully connected deep neural network with three layers, also known as multi-layered perceptron. $f$ denotes the activation function which may vary from layer to layer.

Gaussian) and then regenerates the input via sampling from this distribution and the decoder network.

In a practical implementation, the AE is often augmented with *skip connections*, which improve the information transfer from encoding to decoding layers, as indicated in Figure 5.18. Skip connections help to avoid vanishing gradients in back-propagation computations. Skip connections may be implemented either by enlarging the input dimension of layers in the decoding stage, or, by using a weighted addition of the output of one encoding layer to the input of the corresponding decoding layer. The latter will not increase the dimension of the decoding layer and is thus computationally more efficient.

### 5.12.2.3 Recurrent Neural Networks

As in traditional signal processing systems, such as auto-regressive filters, efficient storage and recursive use of information is also of great importance in DNNs. This is especially true for speech and audio signals where much information resides in the extended sequences of signal samples and the underlying symbolic information.

For learning from sequential data, a variety of recurrent neural networks (RNN) have been devised. A simple RNN structure may be defined as [Aggarwal 2018]

$$\mathbf{h}(k) = f_h \left( \mathbf{W}_x \mathbf{x}(k) + \mathbf{W}_h \mathbf{h}(k-1) + \mathbf{b}_h \right) \tag{5.187}$$

$$\mathbf{y}(k) = f_y \left( \mathbf{W}_y \mathbf{h}(k) + \mathbf{b}_y \right) \tag{5.188}$$

with $N$-dimensional input vectors $\mathbf{x}(k) \in \mathbb{R}^N$, $L$-dimensional output vectors $\mathbf{y}(k) \in \mathbb{R}^L$, and $M$-dimensional state vectors $\mathbf{h}(k) \in \mathbb{R}^M$, as shown in Figure 5.19. $\mathbf{W}_x$, $\mathbf{W}_h$, and $\mathbf{W}_y$ are weight matrices of appropriate dimensions, $\mathbf{b}_h$ and $\mathbf{b}_y$ are bias vectors (not explicitly shown in Figure 5.19), and $f_h$ and $f_y$ denote activation functions as introduced in Section 5.12.1.1. These activation functions are applied to each element of their respective input vector. Note that this simple RNN unit can be considered as a non-linear extension of a linear state-space model. In DNNs, multiple instances of these units may be concatenated or combined with feed-forward or convolutional layers.

Slightly different definitions of the simple RNN model have also been proposed, see, e.g., [Elman 1990]. As with all recursive systems, there are stability issues that need to be controlled through a proper selection of network weights. As a consequence, the basic RNN and

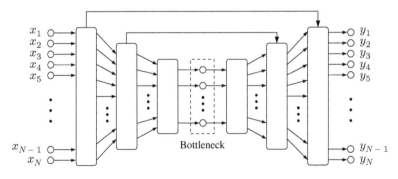

**Figure 5.18** Autoencoder network with bottleneck and skip connections.

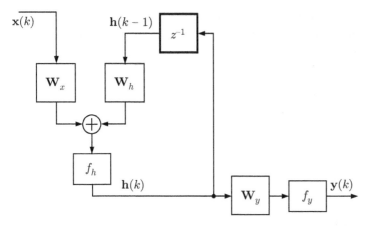

**Figure 5.19** A basic recurrent neural network unit with input vector $\mathbf{x}(k)$, output vector $\mathbf{y}(k)$, and state vector $\mathbf{h}(k)$. The $z^{-1}$ block denotes a vector-valued memory cell. Bias vectors are not shown.

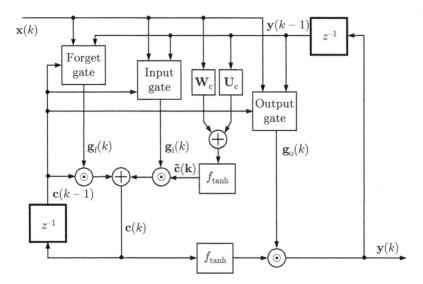

**Figure 5.20** Long short-term memory (LSTM) unit. The $z^{-1}$ blocks denote vector-valued memory cells.

deep variations thereof are difficult to train via back-propagation through time as gradients may have a tendency to either diverge or to vanish [Aggarwal 2018].

Two widely used networks for processing sequential data that mitigate the vanishing-gradient problem are the *long short-term memory* (LSTM) unit [Hochreiter, Schmidhuber 1997] and the *gated recursive* unit (GRU) [Cho et al. 2014]. Both control the consumption of new data and the internal circulation of past data via multiplicative gating signals.

In contrast to the simple RNN unit, the LSTM network (see Figure 5.20) uses the internal *cell state* $\mathbf{c}(k)$ to capture long-term information, and uses three gates, the input, the output, and the forget gate, to control the information flow where the latter had been added to the LSTM network in [Gers et al. 1999].

In the general case [Gers et al. 2002], [Graves 2013], [Chung et al. 2014], a gating signal $\mathbf{g}_z(k)$ at time $k$ is computed as a linear combination of the input, past output, and state data followed by an activation function (here, we chose the sigmoid function and omit optional bias vectors),

$$\mathbf{g}_z(k) = f_\sigma\left(\mathbf{W}_z\mathbf{x}(k) + \mathbf{U}_z\mathbf{y}(k-1) + \mathbf{V}_z\mathbf{c}(k)\right), \tag{5.189}$$

where $\mathbf{x}(k)$ is the input and $\mathbf{y}(k-1)$ the past output of the unit. The identifier z represents either the input (i), output (o), or forget (f) gate and is also applied to the corresponding weight matrices. Typically, $\mathbf{V}_z$ is a diagonal matrix. When $\odot$ denotes element-wise multiplication, the output (also known as activation) of an LSTM unit with cell memory $\mathbf{c}(k)$ at time $k$ is given by

$$\mathbf{y}(k) = \mathbf{g}_o(k) \odot f_{\tanh}(\mathbf{c}(k)), \tag{5.190}$$

where $\mathbf{g}_o(k)$ is an output gate that modulates the visibility of the internal state. $\mathbf{c}(k)$ is updated via a forget gate $\mathbf{g}_f(k)$ and an input gate $\mathbf{g}_i(k)$,

$$\mathbf{c}(k) = \mathbf{g}_f(k) \odot \mathbf{c}(k-1) + \mathbf{g}_i(k) \odot \widetilde{\mathbf{c}}(k), \tag{5.191}$$

where the update $\widetilde{\mathbf{c}}(k)$ is derived from the input and the past output

$$\widetilde{\mathbf{c}}(k) = f_{\tanh}(\mathbf{W}_c\mathbf{x}(k) + \mathbf{U}_c\mathbf{y}(k-1)). \tag{5.192}$$

As a result and unlike a linear first-order recursive system or the simple RNN, the LSTM unit may learn to balance newer and older information such that salient long-term features of the signal are not immediately overwritten by newly incoming signal information and long-term dependencies are preserved. This mechanism allows LSTMs to capture patterns in signals that are not apparent in shorter sequences. Different variants have been devised. For instance, to reduce the overall number of trainable parameters in the gating functions, the cell state vector may be organized into blocks each of which is then controlled by a single gating value.

A somewhat simpler recursive network that may perform just as well as the LSTM on certain tasks is known as gated recursive unit (GRU) [Cho et al. 2014], [Chung et al. 2014]. As shown in Figure 5.21, the internal memory state is represented by the output (activation), which is a simple linear combination of the previous output $\mathbf{y}(k-1)$ and a new contribution $\mathbf{h}(k)$

$$\mathbf{y}(k) = (1 - \mathbf{g}_u(k)) \odot \mathbf{y}(k-1) + \mathbf{g}_u(k) \odot \mathbf{h}(k), \tag{5.193}$$

where $\odot$ denotes element-wise multiplication and the use of the update gate $\mathbf{g}_u(k) = f_\sigma(\mathbf{W}_u\mathbf{x}(k) + \mathbf{U}_u\mathbf{h}(k-1))$ is reminiscent of the parameter of a first-order recursive filter. The update is composed of input and output activation data

$$\mathbf{h}(k) = f_{\tanh}\left(\mathbf{W}_h\mathbf{x}(k) + \mathbf{U}_h\left(\mathbf{r}(k) \odot \mathbf{y}(k-1)\right)\right) \tag{5.194}$$

where $\mathbf{r}(k) = f_\sigma\left(\mathbf{W}_r\mathbf{x}(k) + \mathbf{U}_r\mathbf{y}(k-1)\right)$ is a vector of reset gates.

The GRU is often preferred because it has fewer parameters and requires less computational resources than LSTMs and is thus easier to train, especially on smaller data sets.

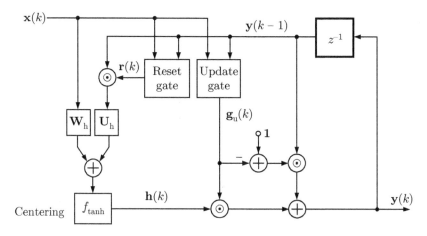

**Figure 5.21** Gated recursive unit. The $z^{-1}$ block denotes a vector-valued memory cell and $\odot$ an element-wise multiplication of two vectors.

### 5.12.2.4 Time Delay, Wavenet, and Transformer Networks

Finally, we note that modeling of long temporal dependencies has been also successfully accomplished via hierarchical network structures. A successful solution is the time-delay neural network (TDNN) [Waibel et al. 1989] which evaluates a succession of time frames using the sliding-window technique whereas the data within the window is combined via trained network weights. While the sliding window at any specific position encompasses only local context, the succession of all windows provides information across the full extent of the data sequence. Then, the second layer may now combine the output of a certain number of windows at different positions as to increase the temporal context. This procedure may be followed in several steps until the desired degree of aggregation is achieved. Similar to CNNs, the sliding window of the TDNN forces the network to learn shift-invariant features of the input sequence. The computational and the memory consumption of the process may be further optimized by introducing decimation in time [Peddinti et al. 2015], i.e., skipping adjacent windows of a lower layer in the aggregation of the next higher layer. TDNNs have been successfully employed, e.g., in speaker verification tasks [Synder et al. 2018], [Desplanques et al. 2020].

The general idea of TDNNs, i.e., extending the receptive field across a wider range of input samples, is also implemented in the *Wavenet* [van den Oort 2016]. The Wavenet consumes the input data in a hierarchical fashion via *dilated convolutions*. These convolutions employ in each layer an increasingly decimated version of the input sequence. Unlike TDNNs, they produce output samples at the same rate as the input signal.

More recently, also non-recursive attention-based models such as the *Transformer* have become popular for sequence-to-sequence modeling tasks [Vaswani et al. 2017]. They provide an attention mechanism which learns to focus on salient parts of the input sequence. Transformer models have also been successfully used in speaker separation tasks [Subakan et al. 2020].

### 5.12.2.5 Training of Neural Networks

Training procedures of network weights use a large number of examples of input–output mappings derived from real-world observations or from data generated via simulations (also known as *data augmentation*). The training process requires the definition of a cost function that measures the deviation of network output from given desired output **z** (also known as *training target* or *labels*). The adaptation of network weights is governed by the derivative of the cost function with respect to the adjustable weights and biases. The networks are designed such that the derivatives can be computed across all layers using the generalized chain rule of the *back-propagation* procedure [Goodfellow et al. 2016], [Aggarwal 2018]. Networks may consist of millions of neurons and are typically trained on a parallel computer using GPUs and a software framework that relieves the user from manual gradient computations.

The objective of the weight update is to minimize an objective function $\mathfrak{L}(\theta)$ where $\theta$ denotes the set of all network parameters and $\mathfrak{L}(\theta)$ is differentiable with respect to $\theta$. Typically, the objective function and hence the gradients derived via back-propagation are noisy as they are evaluated on sampled data (mini-batches) or depend on unknown additional factors. The purpose of the learning rule is to optimize the network parameters in a fast and stable way. We briefly outline two widely used methods.

### 5.12.2.6 Stochastic Gradient Descent (SGD)

The stochastic gradient descent adapts network parameters in the direction of the negative instantaneous gradient of the loss function

$$\theta(m) = \theta(m-1) - \eta(m) \, \nabla_\theta \mathfrak{L} \, (\mathbf{z}, \theta(m-1)), \tag{5.195}$$

where $m$ is the iteration index and $\eta(m)$ the learning rate. In practice, the gradient $\mathbf{g}(m) = \nabla_\theta \mathfrak{L} \, (\mathbf{z}, \theta(m-1))$ is based on a batch of training samples and is thus a stochastic quantity.

The learning rate $\eta(m)$ is frequently adapted in line with training progress. To achieve fast convergence it is beneficial to start the learning process with a large learning rate. When $\theta(m)$ is already close to the optimal solution the learning rate must be reduced to small values in order to zero in on the optimum solution.

### 5.12.2.7 Adaptive Moment Estimation Method (ADAM)

The adaptive moment estimation method (ADAM) [Kingma, Ba 2014] estimates the first and second-order moments of the stochastic gradient $\mathbf{g}(m)$ and uses a scaled and smoothed gradient in its updates. The moment estimation is based on linear first-order recursive systems

$$\mu_g(m+1) = \beta_1 \mu_g(m) + \left(1 - \beta_1\right) \mathbf{g}(m), \tag{5.196}$$

$$\mathbf{p}_g(m+1) = \beta_2 \mathbf{p}_g(m) + \left(1 - \beta_2\right) \mathbf{g}^2(m), \tag{5.197}$$

and an additional compensation to account for an all-zeros initialization. The ADAM adaptive learning rule is then given by

$$\theta(m) = \theta(m-1) - \eta(m)\frac{\mu_g(m)}{\sqrt{\mathbf{p}_g(m) + \epsilon}} \tag{5.198}$$

with $\eta(m) = \alpha \, \sqrt{1 - \beta_2^m}/(1 - \beta_1^m)$ and $\epsilon$ a small constant. $\alpha$ is a stepsize parameter that can also be varied over iterations to improve convergence.

Unlike the stepsize control of linear adaptive filters, the determination of the best performing learning rate for training a given DNN and its adjustment according to learning progress (known as learning rate schedule) requires "trial and error" experiments. This is mainly due to the fact that network parameters may have a diverse amplitude range, which is not explicitly considered in the update rule.

## References

Aggarwal, C. (2018). *Neural Networks and Deep Learning*, Springer Nature, Cham, Switzerland.

Bishop, C. M. (2006). *Pattern Recognition and Machine Learning*, Information Science and Statistics, Springer-Verlag, New York.

Brehm, H.; Stammler, W. (1987). Description and Generation of Spherically Invariant Speech-Model Signals, *Signal Processing*, vol. 12, pp. 119–141.

Brillinger, D. R. (1981). *Time Series: Data Analysis and Theory*, Holden-Day, New York.

Cho, K.; van Merrienboer, B.; Bahdanau, D.; Bengio, Y. (2014). On the Properties of Neural Machine Translation: Encoder-Decoder Approaches, arXiv, https://arxiv.org/abs/1409.1259.

Chung, J.; Gulcehre, C.; Cho, K.; Bengio, Y. (2014). Empirical Evaluation of Gated Recurrent Neural Networks on Sequence Modeling, arXiv, https://arxiv.org/abs/1412.3555.

Cover, T. M.; Thomas, J. A. (2006). *Elements of Information Theory*, 2nd edn, John Wiley & Sons, Inc., Hoboken, New Jersey.

Cybenko, G. (1989). Approximation by Superpositions of a Sigmoidal Function, *Mathematics of Control, Signals, and Systems*, vol. 2, no. 4, pp. 303–314.

Dat, T. H.; Takeda, K.; Itakura, F. (2005). Generalized Gamma Modeling of Speech and its Online Estimation for Speech Enhancement, *Proceedings of the IEEE International Conference on Acoustics, Speech, and Signal Processing (ICASSP)*, vol. IV, pp. 181–184.

Dempster, A.; Laird, N.; Rubin, D. (1977). Maximum Likelihood from Incomplete Data via the EM Algorithm, *Journal of the Royal Statistical Society: Series B (Methodological)*, vol. 39, no. 1, pp. 1–38.

Desplanques, B.; Thienpondt, J.; Demuynck, K. (2020). ECAPA-TDNN: Emphasized Channel Attention Propagation and Aggregation in TDNN Based Speaker Verification, Interspeech 2020, 21st Annual Conference of the International Speech Communication Association (ISCA), *Virtual Event, Shanghai*, China, 25–29 October 2020, pp. 3830–3834.

Elman, J. (1990). Finding Structure in Time, *Cognitive Science*, vol. 14, no. 2, pp. 179–211.

Fukushima, K. (1980). Neocognitron: A Self-organizing Neural Network Model for a Mechanism of Pattern Recognition Unaffected by Shift in Position, *Biological Cybernetics*, vol. 36, pp. 193–202.

Gers, F.; Schmidhuber, J.; Cummins, F. (1999). Learning to Forget: Continual Prediction with LSTM, *1999 Ninth International Conference on Artificial Neural Networks ICANN 99. (Conf. Publ. No. 470)*, vol. 2, pp. 850–855.

Gers, F.; Schraudolph, N.; Schmidhuber, J. (2002). Learning Precise Timing with LSTM Recurrent Networks, *Journal of Machine Learning Research*, vol. 3, 01, pp. 115–143.

Goodfellow, I.; Bengio, Y.; Courville, A. (2016). *Deep Learning*, The MIT Press, Cambridge, Massachusetts.

Gradshteyn, I. S.; Ryzhik, I. M. (2000). *Table of Integrals, Series, and Products*, 6th edn, Academic Press, San Diego, California.

Graves, A. (2013). Generating Sequences With Recurrent Neural Networks.

Haykin, S. (1996). *Adaptive Filter Theory*, 3rd edn, Prentice Hall, Englewood Cliffs, New Jersey.

Hochreiter, S.; Schmidhuber, J. (1997). Long Short-Term Memory, *Neural Computation*, vol. 9, no. 8, pp. 1735–1780.

Jayant, N. S.; Noll, P. (1984). *Digital Coding of Waveforms*, Prentice Hall, Englewood Cliffs, New Jersey.

Johnson, N. L.; Kotz, S.; Balakrishnan, N. (1994). *Continuous Univariate Distributions*, John Wiley & Sons, Ltd., Chichester.

Kingma, D. P.; Ba, J. (2014). Adam: A Method for Stochastic Optimization, arXiv, https://arxiv.org/abs/1412.6980.

Kingma, D.; Welling, M. (2014). Auto-Encoding Variational Bayes, *2nd International Conference on Learning Representations, ICLR 2014, Banff, AB, Canada, April 14–16, 2014, Conference Track Proceedings*.

Kolmogorov, A. (1933). *Grundbegriffe der Wahrscheinlichkeitsrechnung*, Ergebnisse der Mathematik und ihrer Grenzgebiete, Springer-Verlag, Berlin (in German).

Kotz, S.; Balakrishnan, N.; Johnson, N. L. (2000). *Continuous Multivariate Distributions*, 2nd edn, vol. 1, John Wiley & Sons, Ltd., Chichester.

Kullback, S. (1997). *Information Theory and Statistics*, Dover, New York. Republication of the 1968 edition.

LeCun, Y.; Bottou, L.; Bengio, Y.; Haffner, P. (1998). Gradient-Based Learning Applied to Document Recognition, *Proceedings of the IEEE*, vol. 86, no. 11, pp. 2278–2324.

Lotter, T.; Vary, P. (2003). Noise Reduction by Maximum A Posteriori Spectral Amplitude Estimation with Supergaussian Speech Modeling, *Proceedings of the International Workshop on Acoustic Echo and Noise Control (IWAENC)*.

Martin, R. (2002). Speech Enhancement Using MMSE Short Time Spectral Estimation with Gamma Distributed Speech Priors, *Proceedings of the IEEE International Conference on Acoustics, Speech, and Signal Processing (ICASSP)*, Orlando, Florida, USA, pp. 253–256.

Martin, R. (2005). Speech Enhancement Based on Minimum Mean Square Error Estimation and Supergaussian Priors, *IEEE Transactions on Speech and Audio Processing*, vol. 13, no. 5, pp. 845–856.

Martin, R. (2006). Bias Compensation Methods for Minimum Statistics Noise Power Spectral Density Estimation, *Signal Processing*, vol. 86, no. 6, pp. 1215–1229.

Martin, R.; Breithaupt, C. (2003). Speech Enhancement in the DFT Domain Using Laplacian Speech Priors, *Proceedings of the International Workshop on Acoustic Echo and Noise Control (IWAENC)*.

McAulay, R. J.; Malpass, M. L. (1980). Speech Enhancement Using a Soft-Decision Noise Suppression Filter, *IEEE Transactions on Acoustics, Speech and Signal Processing*, vol. 28, no. 2, pp. 137–145.

Middleton, D.; Esposito, R. (1968). Simultaneous Optimum Detection and Estimation of Signals in Noise, *IEEE Transactions on Information Theory*, vol. 14, no. 3, pp. 434–444.

Oppenheim, A.; Schafer, R. W.; Buck, J. R. (1999). *Discrete-Time Signal Processing*, 2nd edn, Prentice Hall, Englewood Cliffs, New Jersey.

Papoulis, A.; Unnikrishna Pillai, S. (2001). *Probability, Random Variables, and Stochastic Processes*, 4th edn, McGraw-Hill, New York.

Peddinti, V.; Povey, D.; Khudanpur, S. (2015). A Time Delay Neural Network Architecture for Efficient Modeling of Long Temporal Contexts, *Proceedings Interspeech 2015*, pp. 3214–3218.

Qi, J.; Du, J.; Siniscalchi, S. M.; Lee, C.-H. (2019). A Theory on Deep Neural Network Based Vector-to-Vector Regression With an Illustration of Its Expressive Power in Speech Enhancement, *IEEE/ACM Transactions on Audio, Speech, and Language Processing*, vol. 27, no. 12, pp. 1932–1943.

Reynolds, D.; Rose, R. (1995). Robust Text-Independent Speaker Identification Using Gaussian Mixture Speaker Models, *IEEE Transactions on Speech and Audio Processing*, vol. 3, no. 1, pp. 72–83.

Rosenblatt, F. (1957). The Perceptron – A Perceiving and Recognizing Automaton, *Technical Report 85-460-1*, Cornell Aeronautical Laboratory.

Shin, J. W.; Chang, J.; Kim, N. S. (2005). Statistical Modeling of Speech Signals Based on Generalized Gamma Distribution, *IEEE Signal Processing Letters*, vol. 12, no. 3, pp. 258–261.

Subakan, C.; Ravanelli, M.; Cornell, S.; Bronzi, M.; Zhong, J. (2020). Attention is All You Need in Speech Separation, arXiv, https://arxiv.org/abs/2010.13154.

Snyder, D.; Garcia-Romero, D.; Sell, G.; Povey, D.; Khudanpur, S. (2018). X-vectors: Robust DNN Embeddings for Speaker Recognition, *Proceedings of the IEEE International Conference on Acoustics, Speech, and Signal Processing (ICASSP)*, Calgary, Alberta, Canada, pp. 5329–5333.

Taghia, J.; Neudek, D.; Rosenkranz, T.; Puder, H.; Martin, R. (2022). First-Order Recursive Smoothing of Short-Time Power Spectra in the Presence of Interference, *IEEE Signal Processing Letters*, vol. 29, pp. 239–243.

van den Oort, A.; Dieleman, S.; Zen, H.; Simonyan, K.; Vinyals, O.; Graves, A.; Kalchbrenner, N.; Senior, A.; Kavukcuoglu, K. (2016). WaveNet: A generative model for raw audio, https://arXiv.org/abs/1609.03499.

Vary, P. (2004). Advanced Signal Processing in Speech Communication, *Proceedings of the European Signal Processing Conference (EUSIPCO)*, Vienna, Austria, pp. 1449–1456.

Vaswani, A.; Shazeer, N.; Parmar, N.; Uszkoreit, J.; Jones, L.; Gomez, A.; Kaiser, L.; Polosukhin, I. (2017). Attention is All You Need, *31st Conference on Neural Information Processing Systems (NIPS 2017)*.

Waibel, A.; Hanazawa, T.; Hinton, G.; Shikano, K.; Lang, K. (1989). Phoneme Recognition Using Time-Delay Neural Networks, *IEEE Transactions on Acoustics, Speech, and Signal Processing*, vol. 37, no. 3, pp. 328–339.

Welch, P. D. (1967). The Use of Fast Fourier Transform for the Estimation of Power Spectra: A Method Based on Time Averaging over Short, Modified Periodograms, *IEEE Transactions on Audio and Electroacoustics*, vol. 15, no. 2, pp. 70–73.

# 6

# Linear Prediction

This chapter is concerned with the estimation of the spectral envelope of speech signals and its parametric representation. By far the most successful technique, known as *linear predictive* analysis, is based on autoregressive (AR) modeling. Linear prediction (LP) enables us to estimate the coefficients of AR filters and is thus closely related to the model of speech production. It is in fact a key component of all speech compression algorithms, e.g., [Jayant, Noll 1984] and [Vary et al. 1998]. AR modeling, in conjunction with linear prediction, not only is used successfully in speech coding, but has also found numerous applications in spectral analysis, in speech recognition, and in the enhancement of noisy signals. The application of LP techniques in speech coding is quite natural, as (simplified) models of the vocal tract correspond to AR filters. The underlying algorithmic task of LP modeling is to solve a set of linear equations. Fast and efficient algorithms, such as the Levinson–Durbin algorithm [Durbin 1960], [Makhoul 1975], are available and will be explained in detail. Speech signals are stationary only for relatively short periods of 20 to 400 ms. To account for the variations of the vocal tract, the prediction filter must be adapted on a short-term basis. We discuss block-oriented and sequential methods.

## 6.1 Vocal Tract Models and Short-Term Prediction

At the outset of our considerations, we recall that speech may be modeled as the output of a linear, time-varying filter excited either by periodic pulses or by random noise, as shown in Figure 6.1.

The filter with impulse response $h(k)$, which is assumed time invariant at first, is excited with the signal

$$v(k) = g \cdot u(k), \tag{6.1}$$

where the gain factor $g$ controls the amplitude, and thus, the power of the excitation signal $v(k)$.

To synthesize *voiced* segments, a periodic impulse sequence

$$u(k) = \sum_{i=-\infty}^{+\infty} \delta(k - iN_0) \tag{6.2}$$

*Digital Speech Transmission and Enhancement*, Second Edition. Peter Vary and Rainer Martin.
© 2024 John Wiley & Sons Ltd. Published 2024 by John Wiley & Sons Ltd.

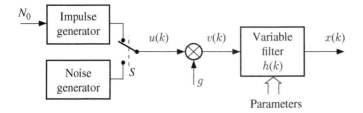

**Figure 6.1** Discrete time model of speech production. $N_0$: pitch period, $S$: voiced/unvoiced decision, $g$: gain, $h(k)$: impulse response, $x(k)$: speech signal, $v(k)$: excitation signal.

with the period $N_0$ is used; to synthesize *unvoiced* segments, a white noise signal $u(k)$ with variance $\sigma_u^2 = 1$ is applied.

In general, the relation between the excitation signal $v(k)$ and the output signal $x(k)$ is described in the time domain by the difference equation

$$x(k) = \sum_{i=0}^{m'} b_i\, v(k-i) - \sum_{i=1}^{m} c_i\, x(k-i). \tag{6.3}$$

The output signal $x(k)$ is generated by a linear combination of $m'+1$ excitation samples $v(k-i)\,(i=0,1,2,\dots,m')$, and $m$ past output values $x(k-i)\,(i=1,2,\dots,m)$. In the $z$-domain, this leads with $c_0 = 1$ to the transfer function

$$\frac{X(z)}{V(z)} = H(z) = \frac{\displaystyle\sum_{i=0}^{m'} b_i\, z^{-i}}{\displaystyle\sum_{i=0}^{m} c_i\, z^{-i}} = \frac{z^m \displaystyle\sum_{i=0}^{m'} b_{m'-i}\, z^i}{z^{m'} \displaystyle\sum_{i=0}^{m} c_{m-i}\, z^i}. \tag{6.4}$$

This function has the form of a general, recursive digital filter with infinite impulse response (IIR). Depending on the choice of the coefficients, the following signal models can be distinguished:

### 6.1.1 All-Zero Model

For $c_0 = 1$ and $c_i \equiv 0\,(i = 1, 2, \dots, m)$, the filter is purely non recursive, i.e., in the time domain we obtain

$$x(k) = \sum_{i=0}^{m'} b_i\, v(k-i), \tag{6.5a}$$

and in the frequency domain

$$H(z) = \frac{\displaystyle\sum_{i=0}^{m'} b_{m'-i}\, z^i}{z^{m'}} = \frac{1}{z^{m'}} \cdot \prod_{i=1}^{m'} (z - z_{0i}). \tag{6.5b}$$

The transfer function has an $m'$-th order pole at $z = 0$ and it is determined by its zeros $z_{0i}$ alone. This model is known as the *all-zero* model or *moving-average* model (MA model).

## 6.1.2 All-Pole Model

With $b_0 \neq 0$ and $b_i \equiv 0$ $(i = 1, 2, \ldots, m')$ a recursive filter with

$$x(k) = b_0\, v(k) - \sum_{i=1}^{m} c_i\, x(k-i) \tag{6.6a}$$

and

$$H(z) = b_0\, \frac{z^m}{\displaystyle\sum_{i=0}^{m} c_{m-i}\, z^i} = b_0\, \frac{z^m}{\displaystyle\prod_{i=1}^{m}(z - z_{\infty i})} \tag{6.6b}$$

results.

Except for the $m$-th order zero at $z = 0$, this filter has only poles $z_{\infty i}$. Therefore, it is called an *all-pole* model. In signal theory, (6.6a) is associated with an *autoregressive process* or *AR process*. This process corresponds to the model of speech production introduced in Section 2.4, neglecting the nasal tract and the glottal and labial filters. The corresponding discrete-time filter is the equivalent model of the lossless acoustical tube, as illustrated in Figure 6.2.

In this case, the effective transfer function of (6.4) for $b_0 = 1$ is

$$H(z) = \frac{1}{1 - C(z)} \tag{6.7a}$$

with

$$C(z) = -\sum_{i=1}^{m} c_i \cdot z^{-i}. \tag{6.7b}$$

## 6.1.3 Pole-Zero Model

The general case is described by (6.3) and (6.4) and represents a mixed *pole–zero* model. In statistics, this is called an *autoregressive moving-average* model (ARMA model).

This would be the adequate choice, if we include the pharynx (C), the nasal cavity (D), and the mouth cavity (E) separately in our model, as indicated in Figure 6.3a. As the radiations of the nostrils and the lips superimpose each other, we have a parallel connection of the tubes (D) and (E) and a serial connection of the two with tube (C).

(a)     (b)

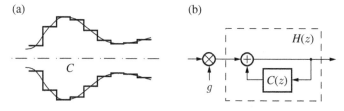

**Figure 6.2** All-pole filter as model of the lossless tube. (a) Acoustical tube model and (b) corresponding digital AR filter.

(a)

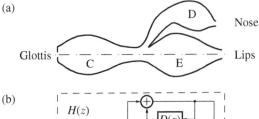

Glottis

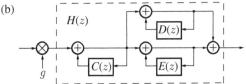

D

Nose

Lips

**Figure 6.3** Pole–zero filter as model of connected lossless tube sections.
(a) Acoustical tube model and
(b) corresponding digital ARMA filter.

(b)

$H(z)$

$D(z)$

$C(z)$

$E(z)$

$g$

The three tubes can be modeled by three individual AR filters with the transfer functions

$$H_C(z) = \frac{1}{1 - C(z)}, \quad H_D(z) = \frac{1}{1 - D(z)}, \quad H_E(z) = \frac{1}{1 - E(z)}, \tag{6.8}$$

where $D(z)$ and $E(z)$ are defined analog to (6.7b).

Neglecting acoustic interactions between the tubes, we obtain the overall transfer function as shown in Figure 6.3b,

$$H(z) = \frac{1}{1 - C(z)} \cdot \left( \frac{1}{1 - D(z)} + \frac{1}{1 - E(z)} \right) \tag{6.9a}$$

$$= \frac{2 - D(z) - E(z)}{(1 - C(z))\,(1 - D(z))\,(1 - E(z))} \tag{6.9b}$$

$$= \frac{\prod\limits_{i=1}^{m'} (z - z_{0i})}{\prod\limits_{i=1}^{m} (z - z_{\infty i})}, \tag{6.9c}$$

having poles and zeros.

However, in connection with coding and synthesis of speech signals, the *all-pole* model is most frequently used in practice. This would mean that, the nasal cavity is neglected and sections (C) and (E) of Figure 6.3a can jointly be modeled by one single all-pole model of corresponding degree.

The predominance of the all-pole model in practical applications will be further justified by the following considerations (e.g., [Deller Jr. et al. 2000]). First, the general pole–zero filter according to (6.4), which is assumed to be causal and stable, will be examined. In the z-plane, the poles are located inside the unit circle, while the zeros may also lie outside the unit circle. Such a situation is illustrated in Figure 6.4a for two pole–zero pairs.

The transfer function $H(z)$ can now be split into a *minimum-phase* system $H_{\min}(z)$ and an allpass transfer function $H_{Ap}(z)$ according to

$$H(z) = H_{\min}(z) \cdot H_{Ap}(z). \tag{6.10}$$

In order to do so, the zeros $z_0$ and $z_0^*$ outside the unit circle are first reflected into the circle to the positions $1/z_0^*$ and $1/z_0$. Then, they are compensated by poles at the same (reflected) position, as depicted in Figure 6.4b. At $z = w$ and $z = w^*$, we now have a zero, as well as a

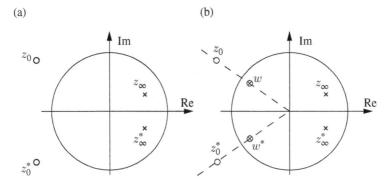

**Figure 6.4** Pole–zero locations. (a) Original filter $H(z)$ and (b) decomposition into minimum-phase and allpass filter. o : location of a zero, × : location of a pole.

pole. The two locations of the zeros and the poles at

$$w = \frac{1}{z_0^*} \qquad \text{and} \qquad w^* = \frac{1}{z_0} \tag{6.11}$$

are assigned to a minimum-phase system having zeros inside the unit circle

$$H_{\min}(z) = \frac{(z - w)(z - w^*)}{(z - z_\infty)(z - z_\infty^*)} . \tag{6.12}$$

The two zeros outside the unit circle and the poles at $w$ and $w^*$ form an allpass system

$$H_{\text{Ap}}(z) = \frac{(z - z_0)(z - z_0^*)}{(z - w)(z - w^*)} , \tag{6.13}$$

with a constant magnitude

$$\left| H_{\text{Ap}}(z = e^{j\Omega}) \right| = |z_0|^2 . \tag{6.14}$$

For the synthesis of speech, it is sufficient to realize only the minimum-phase system, since speech perception is relatively insensitive to phase changes caused by allpass filtering. Therefore, a minimum-phase pole–zero filter can be used to model the speech synthesis process according to Figure 6.1. This leads to three important consequences:

1. Since the poles and zeros are located within the unit circle, a stable inverse filter exists with

$$H_{\min}^{-1}(z) = \frac{1}{H_{\min}(z)} . \tag{6.15}$$

   The vocal tract model filter $1/H_{\min}(z)$ can, thus, be inverted to recover an excitation signal.

2. Every minimum-phase pole–zero filter can be described exactly by an all-pole filter of infinite order, which in turn can be approximated by an $m$-th order filter (e.g., [Marple Jr. 1987]). This justifies the use of an all-pole filter for the synthesis of speech in practical applications.

3. The coefficients of an all-pole filter can be derived from the speech samples $x(k)$ by solving a set of linear equations. Fast algorithms are available to fulfill this task (see Section 6.3).

In accordance with these considerations, the model of speech production is usually based on an all-pole filter. The coefficients of this filter can be determined, as will be shown below, using *linear prediction* techniques. The prediction implies the above mentioned inverse filtering of the speech signal $x(k)$, so that, apart from the filter parameters, an excitation signal $v(k)$ according to the model in Figure 6.1 can be obtained for speech synthesis or coding.

The all-pole model, which is defined by difference equation (6.6a), shows that successive samples $x(k)$ are correlated. With given coefficients $c_i$, each sample $x(k)$ is determined by the preceding samples $x(k-1)$, $x(k-2)$, ... , $x(k-m)$, and $v(k)$, which is also called *innovation*. For reasons of simplicity, we assume that $b_0 = g = 1$. Therefore, it must be possible to estimate or predict the present sample $x(k)$ despite the contribution of the innovation $v(k)$ by a weighted linear combination of previous samples:

$$\hat{x}(k) = \sum_{i=1}^{n} a_i\, x(k-i).$$
(6.16)

This operation is called linear prediction (LP). The predicted signal $\hat{x}(k)$ is a function of the unknown coefficients $a_i$ and the previous samples $x(k-i)$, $i = 1, 2, \dots, n$. The difference

$$d(k) \doteq x(k) - \hat{x}(k)$$
(6.17)

is termed *prediction error* signal.

The prediction task and the generation of the error signal can be described, as shown in Figure 6.5, by non-recursive filtering of signal $x(k)$. The overall filter with input $x(k)$ and output $d(k)$ is called the *LP analysis* filter. As the number $n$ of predictor coefficients $a_i$, $i = 1, 2, \dots, n$, is relatively small, the memory of the predictor is quite short according to

$$n \cdot T = n \cdot \frac{1}{f_s}.$$
(6.18)

For $n = 8$ and $f_s = 8\,\text{kHz}$ a typical value is $n \cdot T = 1\,\text{ms}$. Therefore, the predictor of Figure 6.5 is usually called the *short-term predictor*.

Referring to (6.6a) and (6.16) for $n = m$ we obtain

$$d(k) = x(k) - \hat{x}(k)$$
(6.19a)

$$= v(k) - \sum_{i=1}^{m} c_i\, x(k-i) - \sum_{i=1}^{m} a_i\, x(k-i)$$
(6.19b)

$$= v(k) - \sum_{i=1}^{m} \left(c_i + a_i\right) x(k-i).$$
(6.19c)

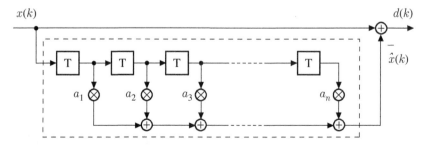

**Figure 6.5** Linear prediction with a non-recursive filter of order *n*.

If we could adjust the predictor coefficients $a_i$ in such a way that

$$a_i = -c_i,$$
(6.20a)

we obtain

$$d(k) = v(k).$$
(6.20b)

In this case, the LP analysis filter would compensate the vocal tract filter. Therefore, the coefficients $a_i$ implicitly describe the (instantaneous) spectral envelope of the speech signal. In Section 6.2, it will be shown how to calculate optimal coefficients $a_i$ by performing a system identification according to (6.20a).

## 6.2 Optimal Prediction Coefficients for Stationary Signals

In this section, it will be shown that prediction coefficients $a_i$, which minimize the mean square error

$$E\{d^2(k)\} = E\{(x(k) - \hat{x}(k))^2\},$$
(6.21)

also obey (6.20a). Furthermore, an algorithm for calculating the coefficients $a_i$ from the speech signal $x(k)$ is derived.

For the sake of simplicity, it is first presumed that the unknown coefficients $c_i$ of the model filter, or the impulse response $h(k)$ of the vocal tract respectively, are time invariant and that the order $m = n$ is known. Furthermore, a real-valued, stationary, uncorrelated, and zero-mean excitation signal $v(k)$ (white noise, unvoiced stationary speech) is assumed. Then, the case of periodic pulse excitation $v(k)$ (voiced stationary speech) will be considered as well.

### 6.2.1 Optimum Prediction

Within the optimization procedure, the auto-correlation functions of the sequences $v(k)$ and $x(k)$ and their cross-correlation are of interest:

$$\varphi_{vv}(\lambda) = E\{v(k)\, v(k \pm \lambda)\} = \begin{cases} \sigma_v^2 & \text{for } \lambda = 0 \\ 0 & \text{for } \lambda = \pm 1, \pm 2, \ldots \end{cases}$$
(6.22)

$$\varphi_{xx}(\lambda) = E\{x(k)\, x(k \pm \lambda)\} = \sigma_v^2 \sum_{i=0}^{\infty} h(i)\, h(i \pm \lambda)$$
(6.23)

$$\varphi_{vx}(\lambda) = E\{v(k)\, x(k + \lambda)\} = E\left\{ \sum_{i=0}^{\infty} h(i)\, v(k + \lambda - i)\, v(k) \right\}$$
(6.24a)

$$= \sum_{i=0}^{\infty} h(i)\, \varphi_{vv}(\lambda - i) = h(\lambda)\, \sigma_v^2.$$
(6.24b)

The predictor coefficients $a_i$, $i = 1, 2, \ldots, n$ are chosen to minimize the mean square prediction error according to (6.21).

Taking (6.16) and (6.17) into account, the partial derivative of (6.21) with respect to the coefficient $a_\lambda$ is set to 0

$$\frac{\partial E\{d^2(k)\}}{\partial a_\lambda} = E\left\{2\,d(k)\,\frac{\partial d(k)}{\partial a_\lambda}\right\} \tag{6.25a}$$

$$= E\{-2\,d(k)\,x(k-\lambda)\} \overset{!}{=} 0\,;\quad \lambda = 1, 2, \ldots, n \tag{6.25b}$$

and the second derivation results in

$$\frac{\partial^2 E\{d^2(k)\}}{\partial a_\lambda^2} = E\left\{2x^2(k-\lambda)\right\} \geq 0, \tag{6.25c}$$

proving that the solution is a minimum. Equation (6.25b) has a single solution

$$a_i = -c_i, \tag{6.26}$$

as with (6.19), (6.20b), (6.24b), (6.25b), and the assumed causality $(h(-\lambda) = 0$; $\lambda = 1, 2, \ldots, n)$ the following holds

$$E\{-2\,d(k)\,x(k-\lambda)\} = -2\,E\{v(k)\,x(k-\lambda)\} \tag{6.27a}$$

$$= -2\,\varphi_{vx}(-\lambda) = 0. \tag{6.27b}$$

The prediction process compensates the model filter, as was shown with (6.20b). The complete filter of Figure 6.5 with the input signal $x(k)$ and the output signal $d(k)$ can be interpreted as the inverse of the vocal tract model.

If, in case of voiced segments, the excitation signal is not a white noise but a periodic pulse train

$$v(k) = g \cdot \sum_{i=-\infty}^{\infty} \delta\left(k - i \cdot N_0\right) \tag{6.28}$$

with $N_0 > n$ then (6.22) is valid for $\lambda = 0, \pm1, \pm2, \ldots, \pm(N_0 - 1)$ with

$$\sigma_v^2 = \frac{g^2}{N_0}. \tag{6.29}$$

Then (6.27b) is fulfilled for the interesting range of $\lambda = 1, 2, \ldots, n < N_0$ too.

With the above assumptions, the power minimization of the prediction error implies in the unvoiced as well as in the voiced case a system identification process, as the coefficients $a_i$ of the optimal non-recursive predictor match the coefficients $c_i$ of the recursive model filter.

Equation (6.25b) can be transformed into an explicit rule to calculate the predictor coefficients $a_j$. For this purpose, $d(k)$ and $x(k)$ according to (6.17) and (6.16) are inserted into (6.25b)

$$-2\,E\{d(k)\,x(k-\lambda)\} = -2\,E\left\{\left(x(k) - \sum_{i=1}^{n} a_i\,x(k-i)\right) x(k-\lambda)\right\} \tag{6.30a}$$

$$= -2\,\varphi_{xx}(\lambda) + 2\sum_{i=1}^{n} a_i\,\varphi_{xx}(\lambda - i) \overset{!}{=} 0. \tag{6.30b}$$

For $\lambda = 1, 2, \ldots, n$, this results in the so-called *normal equations* in vector-matrix notation

$$
\begin{pmatrix} \varphi_{xx}(1) \\ \varphi_{xx}(2) \\ \vdots \\ \varphi_{xx}(n) \end{pmatrix} = \begin{pmatrix} \varphi_{xx}(0) & \varphi_{xx}(-1) & \cdots & \varphi_{xx}(1-n) \\ \varphi_{xx}(1) & \varphi_{xx}(0) & \cdots & \varphi_{xx}(2-n) \\ \vdots & \vdots & \ddots & \vdots \\ \varphi_{xx}(n-1) & \varphi_{xx}(n-2) & \cdots & \varphi_{xx}(0) \end{pmatrix} \begin{pmatrix} a_1 \\ a_2 \\ \vdots \\ a_n \end{pmatrix}
\tag{6.30c}
$$

or

$$
\boldsymbol{\varphi}_{xx} = \mathbf{R}_{xx}\, \mathbf{a},
\tag{6.30d}
$$

respectively, where $\boldsymbol{\varphi}_{xx}$ denotes the correlation vector and $\mathbf{R}_{xx}$ the correlation matrix. $\mathbf{R}_{xx}$ is a real-valued, positive-definite Toeplitz matrix.

The solution of the normal equations provides the optimal coefficient vector

$$
\mathbf{a}_{\text{opt}} = \mathbf{R}_{xx}^{-1}\, \boldsymbol{\varphi}_{xx}.
\tag{6.31}
$$

The power of the prediction error in the case of optimal predictor coefficients can be calculated explicitly:

$$
\sigma_d^2 = E\{(x(k) - \hat{x}(k))^2\}
\tag{6.32a}
$$

$$
= E\{x^2(k) - 2x(k)\hat{x}(k) + \hat{x}^2(k)\}.
\tag{6.32b}
$$

In vector notation

$$
\hat{x}(k) = \mathbf{a}^T \mathbf{x}(k-1)
$$

with $\mathbf{a} \doteq (a_1, a_2, \ldots, a_n)^T$ and $\mathbf{x}(k-1) \doteq (x(k-1), x(k-2), \ldots, x(k-n))^T$, we get

$$
\sigma_d^2 = \sigma_x^2 - 2\mathbf{a}^T \boldsymbol{\varphi}_{xx} + \mathbf{a}^T \mathbf{R}_{xx}\, \mathbf{a}.
\tag{6.32c}
$$

Inserting (6.30d) results in

$$
\sigma_d^2 = \sigma_x^2 - \mathbf{a}^T \boldsymbol{\varphi}_{xx}
\tag{6.32d}
$$

$$
= \sigma_x^2 - \sum_{\lambda=1}^{n} a_\lambda\, \varphi_{xx}(\lambda).
\tag{6.32e}
$$

Alternatively, (6.32d) can be reformulated with (6.31) into

$$
\sigma_d^2 = \sigma_x^2 - \boldsymbol{\varphi}_{xx}^T \mathbf{R}_{xx}^{-1} \boldsymbol{\varphi}_{xx},
\tag{6.32f}
$$

where the optimal coefficients $a_i$ are not explicitly needed to calculate $\sigma_d^2$.

The power ratio

$$
G_p = \frac{\sigma_x^2}{\sigma_d^2}
\tag{6.33}
$$

is called the *prediction gain*. This gain is a measure for the bit rate reduction that can be achieved through predictive coding techniques (cf. Section 8.3.3). Figure 6.6 depicts the prediction gain for two speech signals with a duration of 30 seconds each as a function of the predictor order $n$. For each predictor order $n = 1, 2, \ldots, 30$, a set of coefficients which is optimal for the respective complete signal was calculated according to (6.31).

It becomes apparent that for predictor orders $n \geq 2$ the prediction gain increases rather slowly. The achievable prediction gain depends to a certain extent on the speaker.

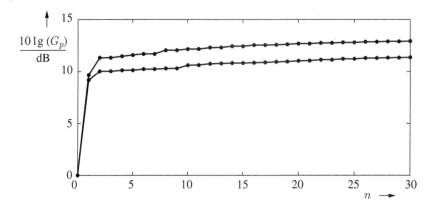

**Figure 6.6** Logarithmic prediction gain $10 \lg \left(\sigma_x^2/\sigma_d^2\right)$ of a time-invariant prediction filter for two different speakers (signal length $30\,\text{s}, f_s = 8\,\text{kHz}$).

## 6.2.2 Spectral Flatness Measure

The achievable prediction gain is related to the *spectral flatness* of the signal $x(k)$. No prediction gain can be achieved for white noise which has a constant power spectral density

$$\Phi_{xx}(e^{j\Omega}) = \text{const.} = \sigma_x^2. \tag{6.34}$$

This power spectral density is completely flat; the signal is not correlated. With increasing correlation the prediction gain increases while the spectral flatness decreases.

A first measure of spectral flatness is the average of the logarithmic normalized power spectral density, e.g., [Markel, Gray 1976] and [Sluijter 2005]

$$\text{sf}_x = \frac{1}{2\pi} \int\limits_{-\pi}^{\pi} \ln\left[\frac{\Phi_{xx}\left(e^{j\Omega}\right)}{\sigma_x^2}\right] d\Omega \tag{6.35}$$

with

$$\sigma_x^2 = \frac{1}{2\pi} \int\limits_{-\pi}^{\pi} \Phi_{xx}\left(e^{j\Omega}\right) d\Omega. \tag{6.36}$$

The measure (6.35) is bounded according to

$$-\infty < \text{sf}_x \le 0$$

where the maximum value $\text{sf}_x = 0$ applies in case of white noise.

A more convenient spectral flatness measure $\gamma_x^2$ with

$$0 \le \gamma_x^2 \le 1 \tag{6.37}$$

is obtained from (6.35), e.g., [Makhoul, Wolf 1972] and [Markel, Gray 1976] by

$$\gamma_x^2 = \exp\left(\text{sf}_x\right) \tag{6.38}$$

$$= \exp\left[\frac{1}{2\pi} \int\limits_{-\pi}^{\pi} \ln\left(\Phi_{xx}(e^{j\Omega})\right) d\Omega - \ln\left(\sigma_x^2\right)\right]. \tag{6.39}$$

Taking (6.36) into account, the widely used form of the spectral flatness measure results:

$$\gamma_x^2 = \frac{\exp\left(\frac{1}{2\pi}\int_{-\pi}^{\pi} \ln\left(\Phi_{xx}(e^{j\Omega})\right)\,d\Omega\right)}{\frac{1}{2\pi}\int_{-\pi}^{\pi} \Phi_{xx}(e^{j\Omega})\,d\Omega}.$$

(6.40)

In the case of $\Phi_{xx}(e^{j\Omega}) = \text{const.} = \sigma_x^2$ we obtain $\gamma_x^2 = 1$.

The expression (6.40) is the ratio of the geometric and the arithmetic mean of the power spectral density $\Phi_{xx}(e^{j\Omega})$. This becomes more obvious when the integrals in (6.40) are approximated by summations and when the power spectral density is replaced by the periodogram, i.e., the squared magnitude of the discrete Fourier transform of a segment of $x(k)$ of length $M$:

$$\gamma_x^2 \simeq \frac{\exp\left(\frac{1}{M}\sum_{\mu=0}^{M-1} \ln |X(e^{j\Omega_\mu})|^2\right)}{\frac{1}{M}\sum_{\mu=0}^{M-1} |X(e^{j\Omega_\mu})|^2}$$

(6.41a)

$$= \frac{\left[\prod_{\mu=0}^{M-1} |X(e^{j\Omega_\mu})|^2\right]^{\frac{1}{M}}}{\frac{1}{M}\sum_{\mu=0}^{M-1} |X(e^{j\Omega_\mu})|^2}, \quad \Omega_\mu = \frac{2\pi}{M}\mu, \; \mu = 0, 1, \ldots, M-1.$$

(6.41b)

The spectral flatness measure is related to the achievable prediction gain as shown below.

We consider the LP analysis filter with input $x(k)$ and output $d(k)$ as shown in Figure 6.7. If $x(k)$ is a random process with power spectral density $\Phi_{xx}(e^{j\Omega})$, the following relations are valid:

$$\Phi_{dd}\left(e^{j\Omega}\right) = |A_0\left(e^{j\Omega}\right)|^2 \cdot \Phi_{xx}\left(e^{j\Omega}\right)$$

(6.42)

$$\frac{1}{2\pi}\int_{-\pi}^{\pi} \ln\left(\Phi_{dd}\left(e^{j\Omega}\right)\right)\,d\Omega = \frac{1}{2\pi}\int_{-\pi}^{\pi} \ln |A_0\left(e^{j\Omega}\right)|^2\,d\Omega$$

(6.43)

$$+ \frac{1}{2\pi}\int_{-\pi}^{\pi} \ln\left(\Phi_{xx}\left(e^{j\Omega}\right)\right)\,d\Omega.$$

(6.44)

It can be shown that the average value of the logarithm of $|A_0(z)|$ is zero if $A_0(z)$ has all its zeros within the unit circle, e.g., [Markel, Gray 1976] and [Itakura, Saito 1970]

$$\frac{1}{2\pi}\int_{-\pi}^{\pi} \ln |A_0\left(e^{j\Omega}\right)|^2\,d\Omega = 0.$$

(6.45)

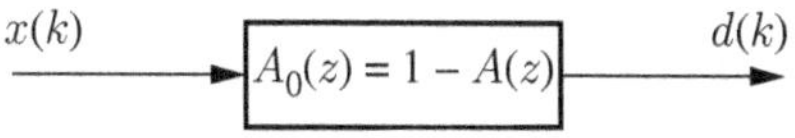

$x(k)$      $\boxed{A_0(z) = 1 - A(z)}$      $d(k)$

**Figure 6.7** LP analysis filter with $A_0(z) = 1 - A(z) = 1 - \sum_{i=1}^{n} a_i \cdot z^{-i} = \frac{1}{z^n}\prod_{i=1}^{n}(z - z_i)$.

Therefore, the ratio of the spectral flatness measures of $x(k)$ and $d(k)$ can be written as

$$\frac{\gamma_x^2}{\gamma_d^2} = \frac{\exp\left(\frac{1}{2\pi}\int\limits_{-\pi}^{\pi} \ln\left(\Phi_{xx}\left(e^{j\Omega}\right)\right) \, d\Omega\right)}{\sigma_x^2} \cdot \frac{\sigma_d^2}{\exp\left(\frac{1}{2\pi}\int\limits_{-\pi}^{\pi} \ln\left(\Phi_{dd}\left(e^{j\Omega}\right)\right) \, d\Omega\right)}$$

$$= \frac{\sigma_d^2}{\sigma_x^2} = \frac{1}{G_p} \, .$$

For perfect prediction (requiring infinite predictor degree) the residual signal $d(k)$ has a perfectly flat spectrum with $\gamma_d^2 = 1$. Thus the following relation is valid

$$\lim_{n\to\infty}\left\{\frac{1}{G_p}\right\} = \gamma_x^2 \, . \qquad (6.46)$$

The spectral flatness measure $\gamma_x^2$ is identical to the inverse of the theoretically maximum prediction gain.

## 6.3 Predictor Adaptation

Speech signals can be considered as stationary only for relatively short time intervals between 20 and 400 ms; thus, the coefficients of the model filter change quickly. Therefore, it is advisable to optimize the predictor coefficients frequently. Here, we distinguish between block-oriented and sequential methods.

### 6.3.1 Block-Oriented Adaptation

The optimization of the coefficients $a_i$ is performed for short signal segments (also called blocks or frames) which consist of $N$ samples each. Considering the vocal tract variations, the block length is usually set to $T_B = 10 - 30$ ms. With a sampling frequency of $f_s = 8$ kHz, this corresponds to a block size of $N = 80 - 240$ signal samples.

Basically, the solution (6.31) for stationary signals can be used for the block-oriented adaptation. At time instant $k_0$ the coefficient vector

$$\mathbf{a}(k_0) = (a_1(k_0), a_2(k_0), \dots, a_n(k_0))^T$$

is determined. As $x(k)$ is not limited to $N$ samples, we can, for the optimization, limit either the input signal $x(k)$ or the prediction error signal $d(k)$ to a finite time interval. These two cases are often referred to in the literature as the *auto-correlation method* and *covariance method* [Makhoul 1975]. The terms *stationary approach* and *non-stationary approach* or *auto-* and *cross-correlation method*, however, are more appropriate.

The distinction between the two methods is illustrated in Figure 6.8. For the basic version of the *auto-correlation method* (Figure 6.8a) a frame $\tilde{x}(k)$ which includes the last $N$ samples up to the time instant $k_0$ is extracted from the sequence $x(k)$ using a window $w(k)$ of length $N$

$$\tilde{x}(k) = x(k)\,w(k_0 - k), \qquad (6.47)$$

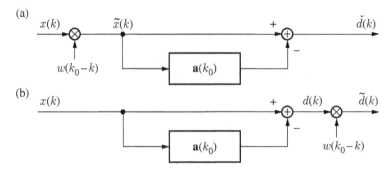

**Figure 6.8** Block-oriented adaptation. (a) Auto-correlation method (stationary method) and (b) covariance method (non-stationary method).

while for the *covariance method* (Figure 6.8b) the finite segment is extracted from the prediction error signal

$$\tilde{d}(k) = d(k)\,w(k_0 - k)\,.\tag{6.48}$$

These two different definitions of the time-limited segment lead to the two above-mentioned alternative approaches, which will be discussed in the next two sections 6.3.1.1 and 6.3.1.2.

### 6.3.1.1 Auto-Correlation Method

The signal $\tilde{x}(k)$ attains non-zero values only within the time interval

$$k_1 = k_0 - N + 1 \le k \le k_0\,.\tag{6.49}$$

Thus, $\tilde{x}(k)$ has finite energy. Due to the finite order $n$ of the non-recursive predictor, the prediction error signal $\check{d}(k)$ is limited to the finite interval

$$k_1 = k_0 - N + 1 \le k \le k_0 + n = k_2\tag{6.50}$$

and also has finite energy. Therefore, the optimization procedure can be simplified by minimizing the energy of the signal $\check{d}(k)$

$$\sum_{k=k_1}^{k_2} \check{d}^2(k) \overset{!}{=} \min\,.\tag{6.51}$$

In analogy to the derivation for stationary signals, the solution (6.31) is valid by replacing the auto-correlation function

$$\varphi_{xx}(\lambda) = \mathrm{E}\{x(k)\,x(k+\lambda)\}$$

by the short-term auto-correlation which may be defined for signals with finite energy as

$$r_\lambda = \sum_{k=k_1}^{k_0} \tilde{x}(k)\tilde{x}(k+\lambda)\tag{6.52a}$$

$$= \sum_{k=k_1}^{k_0-\lambda} x(k)x(k+\lambda) \quad \forall\,\lambda = 0, 1, \dots, n\,,\tag{6.52b}$$

Due to the symmetry

$$r_\lambda = r_{-\lambda} \tag{6.53}$$

the normal equations (6.30c) can be rewritten as

$$
\begin{pmatrix} r_1 \\ r_2 \\ r_3 \\ r_4 \\ \vdots \\ r_n \end{pmatrix}
=
\begin{pmatrix}
r_0 & r_1 & r_2 & r_3 & \cdots & r_{n-1} \\
r_1 & r_0 & r_1 & r_2 & \cdots & r_{n-2} \\
r_2 & r_1 & r_0 & r_1 & \cdots & r_{n-3} \\
r_3 & r_2 & r_1 & r_0 & \cdots & r_{n-4} \\
\vdots & \vdots & \vdots & \vdots & \ddots & \vdots \\
r_{n-1} & r_{n-2} & r_{n-3} & r_{n-4} & \cdots & r_0
\end{pmatrix}
\begin{pmatrix} a_1 \\ a_2 \\ a_3 \\ a_4 \\ \vdots \\ a_n \end{pmatrix}
\tag{6.54a}
$$

or according to (6.30d)

$$\mathbf{r} = \mathbf{R}\,\mathbf{a} \tag{6.54b}$$

respectively, where $\varphi_{xx}$ is replaced by $\mathbf{r}$ and $\mathbf{R}_{xx}$ by $\mathbf{R}$. Like the correlation matrix $\mathbf{R}_{xx}$, the short-term correlation matrix $\mathbf{R}$ has a symmetric Toeplitz structure. This facilitates the use of very efficient algorithms such as the Levinson–Durbin algorithm (see Section 6.3.1.3) for the solution of the set of equations.

Because of the formal analogy with the solution (6.31) for stationary signals, the term *stationary approach* is used. For the calculation of the short-term auto-correlation values $r_\lambda$, often non-rectangular windows $w(k)$ are used, e.g., the Hamming window or asymmetric windows, e.g., [Barnwell III 1981], which give more weight to the most recent samples than to the older ones.

### 6.3.1.2 Covariance Method
For the so-called *covariance method*, the energy of the error signal $\tilde{d}(k)$ in Figure 6.8b is minimized over the finite interval of length $N$

$$\sum_{k=k_1}^{k_0} \tilde{d}^2(k) = \sum_{k=k_1}^{k_0} d^2(k) \stackrel{!}{=} \min . \tag{6.55}$$

Compared to (6.51), the upper limit of the sum changes. Note that, in addition to this, the signal $x(k)$ is now applied to the filter which is not restricted to the time interval $k_1 \le k \le k_0$.
With

$$d(k) = x(k) - \sum_{i=1}^{n} a_i\, x(k - i), \tag{6.56}$$

the partial derivative of (6.55) with respect to a specific coefficient $a_\lambda$ yields

$$\frac{\partial}{\partial a_\lambda} \sum_{k=k_1}^{k_0} d^2(k) = 2 \sum_{k=k_1}^{k_0} d(k) \frac{\partial d(k)}{\partial a_\lambda}; \qquad \lambda = 1, 2, \ldots, n \tag{6.57a}$$

$$= -2 \sum_{k=k_1}^{k_0} d(k) x(k - \lambda) \tag{6.57b}$$

$$= -2 \sum_{k=k_1}^{k_0} \left( x(k) - \sum_{i=1}^{n} a_i x(k - i) \right) x(k - \lambda) \stackrel{!}{=} 0. \tag{6.57c}$$

For $\lambda = 1, 2, \ldots, n$ this results in

$$\sum_{k=k_1}^{k_0} x(k) x(k - \lambda) = \sum_{i=1}^{n} a_i \sum_{k=k_1}^{k_0} x(k - i) x(k - \lambda). \tag{6.58}$$

With the abbreviation

$$\hat{r}_{i,\lambda} = \sum_{k=k_1}^{k_0} x(k - i) x(k - \lambda) \tag{6.59}$$

and the symmetry

$$\hat{r}_{i,\lambda} = \hat{r}_{\lambda,i}, \tag{6.60}$$

we obtain the set of equations

$$\begin{pmatrix} \hat{r}_{0,1} \\ \hat{r}_{0,2} \\ \hat{r}_{0,3} \\ \vdots \\ \hat{r}_{0,n} \end{pmatrix} = \begin{pmatrix} \hat{r}_{1,1} & \hat{r}_{1,2} & \hat{r}_{1,3} & \cdots & \hat{r}_{1,n} \\ \hat{r}_{1,2} & \hat{r}_{2,2} & \hat{r}_{2,3} & \cdots & \hat{r}_{2,n} \\ \hat{r}_{1,3} & \hat{r}_{2,3} & \hat{r}_{3,3} & \cdots & \hat{r}_{3,n} \\ \vdots & \vdots & \vdots & \ddots & \vdots \\ \hat{r}_{1,n} & \hat{r}_{2,n} & \hat{r}_{3,n} & \cdots & \hat{r}_{n,n} \end{pmatrix} \begin{pmatrix} a_1 \\ a_2 \\ a_3 \\ \vdots \\ a_n \end{pmatrix}, \tag{6.61a}$$

or, in compact notation,

$$\hat{\mathbf{r}} = \hat{\mathbf{R}} \mathbf{a}. \tag{6.61b}$$

In contrast to the stationary approach (6.54a), the calculation of the short-term correlation values $\hat{r}_{i,\lambda}$ is not shift invariant

$$\hat{r}_{i,\lambda} \neq \hat{r}_{i+i_0, \lambda+i_0}. \tag{6.62}$$

Therefore, this method of calculating the predictor coefficients is also called the *non-stationary approach*.

While for the auto-correlation method the number of product terms $x(i) x(i + \lambda)$ used in (6.52b) decreases for increasing $\lambda$, the computation of $\hat{r}_{\lambda,i}$ is always based on $N$ product terms. Consequently, in addition to the $N$ samples from $k = k_1$ to $k = k_0$, $n$ preceding values $x(k)$ are needed for this approach. The covariance method thus provides more precise estimates of the short-term correlation than the auto-correlation method. The resulting matrix $\hat{\mathbf{R}}$ is still symmetric, but the Toeplitz form of $\mathbf{R}$ is lost. Thus, the matrix inversion requires more complex methods, such as the Cholesky decomposition. It should be noted that stability of the resulting synthesis filter cannot be guaranteed. However simple stabilizing methods exist, such as adding a small constant to the diagonal of $\hat{\mathbf{R}}$.

Various recursion algorithms are available to solve the set of equations for both the auto-correlation method and the covariance method. Here, the Levinson–Durbin algorithm for the auto-correlation method will be developed as one possible solution. Comparable results are provided by similar algorithms, such as those of Schur, Burg, Le Roux, and Gueguen [Kay 1988], [Marple Jr. 1987].

### 6.3.1.3 Levinson–Durbin Algorithm

In this section we will derive the *Levinson–Durbin* algorithm for the auto-correlation method [Levinson 1947], [Durbin 1960]. This algorithm starts from a known solution $\mathbf{a}^{(p-1)}$ of the set of equations (6.54a) for the predictor order $(p - 1)$ leading to the solution $\mathbf{a}^{(p)}$ for the order $p$. Beginning with the (trivial) solution for $p = 0$ we can find the solution for the actually desired predictor order $n$ iteratively with low computational effort. As an example, we will now examine the step from $p - 1 = 2$ to $p = 3$.

In order to simplify the notation, we replace the predictor coefficients by

$$\alpha_i^{(p)} \doteq -a_i^{(p)} \; ; \; i \in \{1, 2, \dots, p\} . \tag{6.63}$$

For $p = 2$, the normal equations

$$\begin{pmatrix} r_1 \\ r_2 \end{pmatrix} + \begin{pmatrix} r_0 & r_1 \\ r_1 & r_0 \end{pmatrix} \cdot \begin{pmatrix} \alpha_1^{(2)} \\ \alpha_2^{(2)} \end{pmatrix} = \begin{pmatrix} 0 \\ 0 \end{pmatrix} \tag{6.64a}$$

can be written equivalently as

$$\begin{pmatrix} r_1 & r_0 & r_1 \\ r_2 & r_1 & r_0 \end{pmatrix} \cdot \begin{pmatrix} 1 \\ \alpha_1^{(2)} \\ \alpha_2^{(2)} \end{pmatrix} = \begin{pmatrix} 0 \\ 0 \end{pmatrix} . \tag{6.64b}$$

In analogy to (6.32e), the following expression holds for the short-term energy of the prediction error:

$$E^{(2)} = r_0 + \sum_{i=1}^{2} \alpha_i^{(2)} r_i . \tag{6.65}$$

Thus, (6.64b) can be extended to

$$\begin{pmatrix} r_0 & r_1 & r_2 \\ r_1 & r_0 & r_1 \\ r_2 & r_1 & r_0 \end{pmatrix} \cdot \begin{pmatrix} 1 \\ \alpha_1^{(2)} \\ \alpha_2^{(2)} \end{pmatrix} = \begin{pmatrix} E^{(2)} \\ 0 \\ 0 \end{pmatrix} . \tag{6.66a}$$

or in vector-matrix notation

$$\mathbf{R}^{(3)} \cdot \boldsymbol{\alpha}^{(2)} = \mathbf{e}^{(2)} . \tag{6.66b}$$

Because of the symmetry property of the correlation matrix of dimension $(p+1) \times (p+1)$, the following representation is valid, too:

$$
\begin{pmatrix} r_0 & r_1 & r_2 \\ r_1 & r_0 & r_1 \\ r_2 & r_1 & r_0 \end{pmatrix} \cdot \begin{pmatrix} \alpha_2^{(2)} \\ \alpha_1^{(2)} \\ 1 \end{pmatrix} = \begin{pmatrix} 0 \\ 0 \\ E^{(2)} \end{pmatrix}.
$$
(6.67)

For $p = 3$, we now try the following solution approach:

$$
\alpha^{(3)} \doteq \begin{pmatrix} 1 \\ \alpha_1^{(3)} \\ \alpha_2^{(3)} \\ \alpha_3^{(3)} \end{pmatrix} = \begin{pmatrix} 1 \\ \alpha_1^{(2)} \\ \alpha_2^{(2)} \\ 0 \end{pmatrix} + k_3 \begin{pmatrix} 0 \\ \alpha_2^{(2)} \\ \alpha_1^{(2)} \\ 1 \end{pmatrix}
$$
(6.68)

with the unknown constant $k_3$. Obviously, the coefficients $\alpha_1^{(3)}$, $\alpha_2^{(3)}$, and $\alpha_3^{(3)}$ are only a function of the respective coefficients for $p = 2$ and $k_3$.

Extending (6.66b) from $p = 2$ to 3 results in

$$
\mathbf{R}^{(4)} \, \alpha^{(3)} = \mathbf{e}^{(3)}
$$
(6.69a)

and with (6.68) we get

$$
\begin{pmatrix} r_0 & r_1 & r_2 & r_3 \\ r_1 & r_0 & r_1 & r_2 \\ r_2 & r_1 & r_0 & r_1 \\ r_3 & r_2 & r_1 & r_0 \end{pmatrix} \cdot \left[ \begin{pmatrix} 1 \\ \alpha_1^{(2)} \\ \alpha_2^{(2)} \\ 0 \end{pmatrix} + k_3 \begin{pmatrix} 0 \\ \alpha_2^{(2)} \\ \alpha_1^{(2)} \\ 1 \end{pmatrix} \right]
$$

$$
= \begin{pmatrix} E^{(2)} \\ 0 \\ 0 \\ q \end{pmatrix} + k_3 \begin{pmatrix} q \\ 0 \\ 0 \\ E^{(2)} \end{pmatrix} \stackrel{!}{=} \begin{pmatrix} E^{(3)} \\ 0 \\ 0 \\ 0 \end{pmatrix}
$$
(6.69b)

with the known quantity $q = r_3 + r_2 \, \alpha_1^{(2)} + r_1 \, \alpha_2^{(2)}$.

For the determination of $k_3$ and $E^{(3)}$ we exploit (6.69b) and find the conditions

$$
E^{(2)} + k_3 \, q = E^{(3)}
$$
$$
q + k_3 \, E^{(2)} = 0
$$

or

$$
k_3 = -\frac{q}{E^{(2)}}
$$
(6.70a)

$$
E^{(3)} = E^{(2)} \left( 1 + k_3 \frac{q}{E^{(2)}} \right)
$$

$$
= E^{(2)} \left( 1 - k_3^2 \right)
$$
(6.70b)

respectively. Using $k_3$ as expressed in (6.70a), $\alpha^{(3)}$ can be calculated by (6.68).

In the next step, the solution for $p = 4$ can be evaluated. The coefficient $k_4$ and the energy of the prediction error $E^{(4)}$ are determined in analogy to (6.68)–(6.70b).

The complete algorithm is summarized in general form. Starting from the solution for $p = 0$, i.e., no prediction, the solution for $p = n$ is computed in $n$ steps.

1. Computation of $n + 1$ values $r_i$ of the short-term auto-correlation.
2. $p = 0$, i.e., no prediction or

$$d(k) = x(k)$$
$$E^{(0)} = r_0$$
$$\alpha_0^{(0)} \doteq 1.$$

3. For $p \geq 1$, computation of

(a) $\quad q = \sum_{i=0}^{p-1} \alpha_i^{(p-1)} r_{p-i}$

$\quad k_p = -\dfrac{q}{E^{(p-1)}}$

(b) $\alpha_0^{(p)} = 1$

$\quad \alpha_i^{(p)} = \alpha_i^{(p-1)} + k_p \, \alpha_{p-i}^{(p-1)} \quad \forall \, 1 \leq i \leq p - 1$

$\quad \alpha_p^{(p)} = k_p$

(c) $E^{(p)} = E^{(p-1)} (1 - k_p^2)$

(d) $\quad p = p + 1.$

4. Repetition of step 3, if $p \leq n$.
5. $a_i = -\alpha_i^{(n)} \quad \forall \, 1 \leq i \leq n$.

The parameters $k_p$ are called reflection coefficients. Besides the sign, they are identical to those reflection coefficients which characterize the tube model of the vocal tract developed in Chapter 2 ($k_i = -r_i$). The first reflection coefficient of the tube model corresponds to $k_0 = 1$.

Reflection coefficients have a number of interesting properties:

- They are limited by one in magnitude, i.e., $|k_p| \leq 1$. This is a consequence of using proper auto-correlation sequences as input to the Levinson–Durbin algorithm. Therefore, it can be concluded that the short-term energy $E^{(p)}$ is reduced from iteration to iteration, i.e.,

$$E^{(p)} = E^{(p-1)} (1 - k_p^2) \leq E^{(p-1)}. \tag{6.71}$$

- The transfer function

$$1 - A(z) = 1 - \sum_{i=1}^{n} a_i z^{-i}$$

of the LP analysis filter has all of its zeros inside the unit circle, if and only if $|k_p| < 1 \; \forall \, p$ [Hayes 1996]. The reflection coefficients are therefore key for checking the stability of the inverse LP filter and for the development of inherently stable filter structures. In fact, the well-known Schur–Cohn stability test [Haykin 1996] employs the Levinson–Durbin recursion.

- Given the reflection coefficients $k_p$ ($p = 1 \ldots n$), the AR model coefficients $a_i = -\alpha_i^{(n)}$ ($i = 1 \ldots n$, $a_0 = 1$) can be calculated by the Levinson–Durbin algorithm.
- Given the AR model coefficients $a_i = -\alpha_i^{(n)}$, the reflection coefficients $k_p$ can be computed by the following recursion:
  1. Initialization with $\alpha_i^{(n)} = -a_i$ for $1 \leq i \leq n$.
  2. For $p = n, n-1, \ldots, 1$, computation of

  (a) $\quad k_p = \alpha_p^{(p)}$

  (b) $\quad \alpha_i^{(p-1)} = \dfrac{\alpha_i^{(p)} - k_p \, \alpha_{p-i}^{(p)}}{1 - k_p^2}; \quad 1 \leq i \leq p-1.$

Since the Levinson–Durbin algorithm provides the predictor coefficients, as well as the reflection coefficients, the predictor can be realized alternatively in direct form (Figure 6.9a) or, e.g., in the so-called lattice structure (Figure 6.9b).

The effect of the prediction is illustrated for one example in Figure 6.10. It depicts the speech signal $x(k)$ and the prediction error signal $d(k)$ in the time and frequency domain, as well as, the corresponding magnitude responses of the LP analysis and synthesis filters. In this example, the $n = 8$ predictor coefficients were calculated every 20 ms using the auto-correlation method and a rectangular window of length $N = 160$. The LP analysis filter performs a reduction of the dynamic range (Figure 6.10b) in the time domain and spectral flattening (Figure 6.10d, *whitening effect*) in the frequency domain. The corresponding frequency responses of the analysis and synthesis filter are shown in Figure 6.10e and 6.10f.

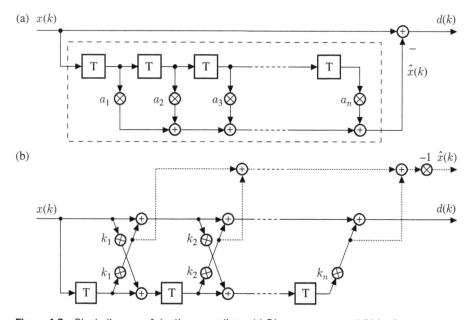

**Figure 6.9** Block diagram of the linear predictor. (a) Direct structure and (b) lattice structure.

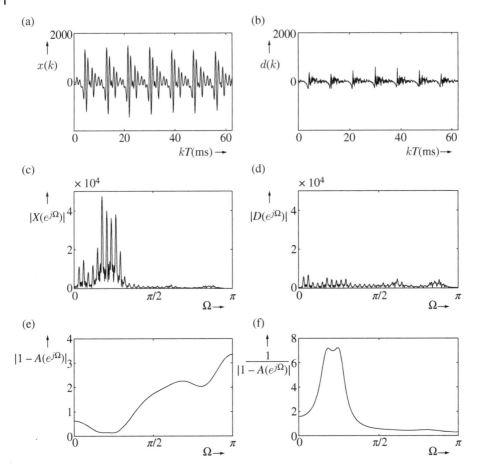

**Figure 6.10** Example of the effect of linear prediction with block adaptation. (a) Speech signal $x(k)$, (b) prediction error signal $d(k)$, (c) short-term spectral analysis of the speech signal, (d) short-term spectral analysis of the prediction error signal, (e) magnitude response of the LP analysis filter, (f) magnitude response of the LP synthesis filter.

Obviously, the LP synthesis filter describes the spectral envelope of $x(k)$, i.e., the frequency response of the vocal tract filter. The LP analysis filter produces an error signal $d(k)$ which still has a quasi periodic structure. Furthermore, the spectral envelope of the error signal is almost flat (see Figure 6.10d).

Figure 6.11 shows the achievable prediction gain as a function of the predictor order $n$ for two speech signals. The predictors were adapted by the auto-correlation method. In comparison to the results obtained for a time-invariant predictor as in Figure 6.6, a distinctly higher prediction gain results due to the block adaptation. Furthermore, it can be observed that with an adaptive adjustment, the prediction gain saturates at a filter order of $n = 8$–$10$. An additional increase of the prediction order provides no appreciable further gain. This confirms the vocal tract model of Section 2.4.

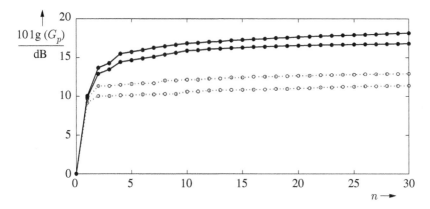

**Figure 6.11** Logarithmic prediction gain for block adaptation with auto-correlation-based predictors for two speakers (solid) and for a time-invariant prediction filter (dashed, Figure 6.6); (signal length 30s, $N = 160$, $f_s = 8$ kHz).

### 6.3.2 Sequential Adaptation

With the block-oriented adaptation the predictor coefficients $a_i$ ($i = 1, 2, \ldots, n$) are recalculated for blocks of $N$ samples. In this section an alternative method, the *least-mean-square* (LMS) algorithm, will be derived, in which the coefficients are sequentially adapted in each sample interval.

In a first step, we consider a predictor with a single fixed coefficient $a$. The power of the prediction error signal $d(k) = x(k) - a x(k-1)$ can be expressed as

$$\sigma_d^2 = \sigma_x^2 - 2a\, \varphi_{xx}(1) + a^2\, \sigma_x^2$$

according to (6.32c). The power $\sigma_d^2$ is a second-order function of the coefficient $a$, which is depicted in Figure 6.12.

With (6.31), the minimum mean square error is reached in point C for

$$a_{\text{opt}} = \frac{\varphi_{xx}(1)}{\varphi_{xx}(0)} = \frac{\varphi_{xx}(1)}{\sigma_x^2}.$$

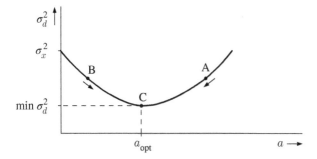

**Figure 6.12** Error function of the adaptive filter.

Starting from points A or B, the minimum (point C) can be approached iteratively by taking the gradient

$$\nabla = \frac{\partial \sigma_d^2}{\partial a} = -2\,\varphi_{xx}(1) + 2\,a\,\varphi_{xx}(0)$$

into consideration. After inserting $a_{opt}$ in the above equation, we obtain

$$\nabla = 2\,\varphi_{xx}(0) \cdot (a - a_{opt}).$$

The gradient is proportional to the difference of the instantaneous, i.e., time-variant coefficient $a$ from the optimum $a_{opt}$. To reduce the mean square error $\sigma_d^2$, the instantaneous coefficient $a(k)$ must thus be corrected in the direction of the negative gradient according to

$$a(k+1) = a(k) - \vartheta \cdot \nabla. \tag{6.72}$$

Here, the constant $\vartheta$ denotes the stepsize, which controls the size of the incremental correction.

This procedure, which is known in the literature as the *steepest descent algorithm*, can be generalized and applied to the $n$-th order prediction as follows. With the signal vector

$$\mathbf{x}(k-1) = (x(k-1), x(k-2), \ldots, x(k-n))^T \tag{6.73}$$

and an arbitrary but fixed coefficient vector

$$\mathbf{a}(k) = \mathbf{a} = (a_1, a_2, \ldots, a_n)^T, \tag{6.74}$$

the prediction $\hat{x}(k)$ can be described as

$$\hat{x}(k) = \sum_{i=1}^{n} x(k-i)a_i \tag{6.75a}$$

$$= \mathbf{a}^T \mathbf{x}(k-1), \tag{6.75b}$$

and the following expression results for the power of the prediction error:

$$\sigma_d^2 = E\left\{ \left( x(k) - \mathbf{a}^T \mathbf{x}(k-1) \right)^2 \right\}. \tag{6.76}$$

The gradient with respect to the coefficient vector $\mathbf{a}$ becomes

$$\nabla = -2\,E\left\{ \left( x(k) - \mathbf{a}^T \mathbf{x}(k-1) \right) \mathbf{x}(k-1) \right\} \tag{6.77a}$$

$$= -2\,\boldsymbol{\varphi}_{xx} + 2\,\mathbf{R}_{xx}\,\mathbf{a}. \tag{6.77b}$$

The gradient indicates the direction of the steepest ascent of the mean square error. For an iterative minimization of the mean square error, the instantaneous coefficient vector $\mathbf{a}(k)$ must consequently be corrected in the direction of the negative gradient. In analogy to (6.72), this results in

$$\mathbf{a}(k+1) = \mathbf{a}(k) + 2\,\vartheta\left( \boldsymbol{\varphi}_{xx} - \mathbf{R}_{xx}\,\mathbf{a}(k) \right). \tag{6.78}$$

The steepest descent algorithm requires knowledge of the auto-correlation values $\varphi_{xx}(\lambda)$ for $\lambda = 0, 1, \ldots, n$, in the form of the correlation vector $\boldsymbol{\varphi}_{xx}$ and the auto-correlation matrix $\mathbf{R}_{xx}$.

One member of the family of *stochastic gradient algorithms* is the so-called *LMS* algorithm. This algorithm is particularly interesting for practical applications, as the auto-correlation values are not explicitly required.

For the LMS algorithm, a simple estimator for the mean square error $\sigma_d^2$ is used, i.e., the *instantaneous squared error*

$$\hat{\sigma}_d^2(k) = d^2(k) \tag{6.79a}$$

$$= \left(x(k) - \mathbf{a}^T(k)\mathbf{x}(k-1)\right)^2. \tag{6.79b}$$

In analogy to (6.77b), the *instantaneous gradient* results in

$$\hat{\nabla} = -2\underbrace{\left(x(k) - \mathbf{a}^T(k)\mathbf{x}(k-1)\right)}_{d(k)}\mathbf{x}(k-1) \tag{6.80a}$$

$$= -2\,d(k)\mathbf{x}(k-1), \tag{6.80b}$$

where instead of $\mathbf{R}_{xx}$ and $\boldsymbol{\varphi}_{xx}$ the prediction error $d(k)$ and the state variables $\mathbf{x}(k-1)$ are needed (Figure 6.13).

Consequently, the LMS algorithm for the coefficient vector reads

$$\mathbf{a}(k+1) = \mathbf{a}(k) + 2\,\vartheta\,d(k)\,\mathbf{x}(k-1) \tag{6.81a}$$

or for the individual coefficient

$$a_i(k+1) = a_i(k) + 2\,\vartheta\,d(k)\,x(k-i) \quad \forall\, 1 \le i \le n \tag{6.81b}$$

respectively, with the effective stepsize parameter $2\,\vartheta$.

For stability reasons the stepsize parameter must be limited to the range

$$0 < \vartheta < \frac{1}{\|\mathbf{x}(k-1)\|^2}$$

(e.g., [Haykin 1996]).

Assuming a stationary AR process $x(k)$, the coefficient vector $\mathbf{a}$ converges with a sufficiently small stepsize toward the optimal solution according to (6.31)

$$\mathbf{a} \;\to\; \mathbf{a}_{\text{opt}} = \mathbf{R}_{xx}^{-1}\,\boldsymbol{\varphi}_{xx}.$$

Due to its low complexity, the LMS algorithm is of great practical significance (see also Section 8.3.4 and Chapter 14). A time-variant stepsize is often used to improve the convergence behavior.

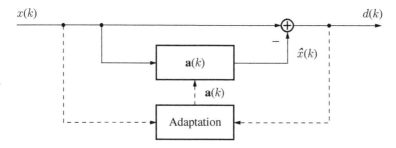

**Figure 6.13** Predictor with sequential adaptation.

Numerous further adaptation algorithms are discussed in the literature, which differ regarding their convergence characteristics and their complexity. Here, the *recursive least-square* (RLS) algorithm is mentioned as an example. It can be derived from the steepest descent algorithm as well. In this case, instead of the auto-correlation values $\varphi_{xx}(\lambda)$, estimated values $\hat{\varphi}_{xx}(\lambda)$ are used, which are determined by recursive computation with exponential windowing. This method is characterized by a high convergence speed. The improvement in performance, however, is achieved at the expense of a large increase in computational complexity, i.e., the number of operations per iteration grows with the square of the filter order $n$. In comparison, for the LMS algorithm, there is only a linear increase in complexity (e.g., Chapter 14, [Haykin 1996]).

## 6.4 Long-Term Prediction

As shown in Section 6.3 by example of Figure 6.10, we can extract the spectral envelope of $x(k)$ by short-term prediction using $n = 8$–$10$ coefficients. According to the speech production model (Figure 6.1) the resulting LP synthesis filter represents the vocal tract, and the prediction error signal $d(k)$ the excitation. Thus, the remaining quasi periodic spectral fine structure as in Figure 6.10d is determined by the excitation generator of the speech production model. This periodic structure is associated with the long-term correlation of the speech signal $x(k)$ or of the prediction error signal $d(k)$ respectively, as illustrated in Figure 6.14.

The short-term prediction discussed in Section 6.3 exploits the short-term correlation $\varphi_{xx}(\kappa)$ ($\kappa = 0 \ldots n$, with $n = 8$–$10$). Obviously, the prediction error signal $d(k)$ (Figure 6.10b) still exhibits long-term correlation, which is due to the pitch period $T_0 = 1/f_0$ of voiced segments. With fundamental frequencies in the range of $50\,\mathrm{Hz} \leq f_0 \leq 250\,\mathrm{Hz}$ and a sampling rate of $f_s = 8\,\mathrm{kHz}$ the periods $T_0$ have lengths of $N_0 = 32$–$160$ samples. As $T_0$ is large in comparison to $n \cdot T$ (memory of the short-term predictor), the prediction over the time span $T_0$ is called *long-term prediction* (LTP).

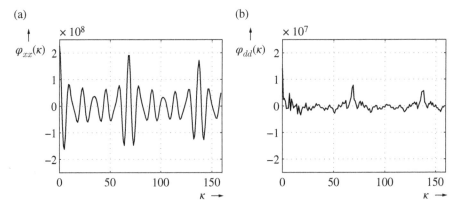

**Figure 6.14** Auto-correlation function of the (a) speech signal $x(k)$ and (b) error signal $d(k)$, for the example signals depicted in Figure 6.10.

The high correlation of subsequent signal periods can be used for a further improvement of the prediction gain by estimating the most recent signal period from the preceding one. If the instantaneous period length $N_0$ is known, the long-term prediction error signal can be calculated as follows:

$$d'(k) = d(k) - b \cdot d(k - N_0) = d(k) - \hat{d}(k) \tag{6.82}$$

with a weighting factor $b$.

Figure 6.15a depicts the block diagram of the corresponding LTP-analysis filter. The frequency response can be computed as follows:

$$1 - P(e^{j\Omega}) = 1 - b \cdot e^{-jN_0\Omega} \tag{6.83a}$$

$$= \sqrt{\left(1 - b \cdot \cos\left(N_0\Omega\right)\right)^2 + b^2 \cdot \sin^2\left(N_0\Omega\right)} \cdot e^{-j\varphi_P(\Omega)} \tag{6.83b}$$

(a)
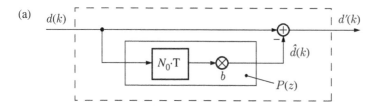

(b)
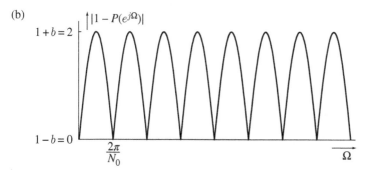

(c)
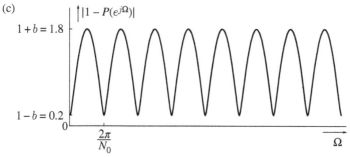

**Figure 6.15** Long-term prediction (LTP). (a) Block diagram of the LTP-analysis filter, (b) magnitude response $|1 - P(e^{j\Omega})|$ for $b = 1$, and (c) magnitude response $|1 - P(e^{j\Omega})|$ for $b = 0.8$.

with

$$\varphi_P = -\arctan\left(\frac{b \cdot \sin\left(N_0\Omega\right)}{1 - b \cdot \cos\left(N_0\Omega\right)}\right). \tag{6.83c}$$

The magnitude response is of particular interest. In the special case $b = 1$, with

$$|1 - P(e^{j\Omega})| = 2 \cdot \left|\sin\left(\frac{N_0\Omega}{2}\right)\right| \tag{6.83d}$$

the plot outlined in Figure 6.15b results in equidistant zeros at

$$\Omega_i = \frac{2\pi}{N_0} i, \quad i \in \mathbb{Z}. \tag{6.84}$$

Accordingly, the LTP-analysis filter is a comb filter with equally spaced notches at $\Omega_i$. As the length of the instantaneous pitch period $T_0 = 1/f_0$ is, in general, not an exact integer multiple of the sampling interval $T$, the notches are not necessarily exactly at the normalized fundamental frequency $\Omega_0 = 2\pi/N_0$ and its harmonics $\Omega_i$.

The two parameters $N_0$ and $b$ are chosen in such a way as to minimize the energy

$$\sum_k d'^2(k) = \sum_k \left(d(k) - b \cdot d(k - N_0)\right)^2 \tag{6.85}$$

of the error signal $d'(k)$ for short blocks.

In analogy to short-term prediction, the summation limits in (6.85) can be chosen according to either the auto-correlation method or the covariance method (see Section 6.3.1).

In what follows, the covariance method is applied. The energy of the error signal $d'(k)$ is minimized over an interval of length $L$. Most frequently, the length of the interval is $L \cdot T = 5$ ms or $L = 40$ for $f_s = 8$ kHz, respectively.

For each fixed but arbitrary value $N_0$, the optimal coefficient $b$ can be determined by minimizing the energy of the error signal

$$\frac{\partial}{\partial b} \sum_{k=k_0-L+1}^{k_0} d'^2(k) = \sum_{k=k_0-L+1}^{k_0} - 2d(k - N_0)\left(d(k) - b \cdot d\left(k - N_0\right)\right) \overset{!}{=} 0. \tag{6.86a}$$

With the abbreviation $k_1 = k_0 - L + 1$ we finally get

$$b = \frac{\sum\limits_{k=k_1}^{k_0} d(k)\, d(k - N_0)}{\sum\limits_{k=k_1}^{k_0} d^2(k - N_0)} = \frac{R(N_0)}{S(N_0)}, \tag{6.86b}$$

where $R(N_0)$ is the short-term auto-correlation function for $\lambda = N_0$ according to the covariance method (cf. Section 6.3.1.2) and $S(N_0)$ is the energy of the current frame of the error signal $d(k)$.

Inserting the optimal coefficient $b$ into (6.85), the resulting error energy can be computed as a function of the parameter $N_0$

$$\sum_{k=k_0-L+1}^{k_0} d'^2(k) = S(0) - \frac{R^2(N_0)}{S(N_0)}. \tag{6.87}$$

This expression can be utilized to determine the optimal delay $N_0$. In (6.87), only the second term depends on $N_0$. Thus, in a first step this term is maximized through variation of $N_0$ in the relevant range of, for example, $32 \leq N_0 < 160$. In this range $N_0$ can take, e.g., $2^7 = 128$ different values, which allows coding with 7 bits only. Subsequently, the weighting coefficient $b$ can be determined for the delay $N_0$ according to (6.86b). This coefficient can be quantized quite coarsely (see also Section 7.7).

A signal example is depicted in Figure 6.16, showing the input signal $x(k)$, the first error signal $d(k)$ after short-term prediction, and the second error signal $d'(k)$ after long-term prediction. By the use of the second predictor, a further significant dynamic reduction is achieved. Chapter 8 will reveal how this additional prediction gain can be exploited for bit rate reductions in the sense of model-based and psychoacoustically motivated source coding. The effect of the two-stage prediction is illustrated once more in the frequency domain by Figure 6.17.

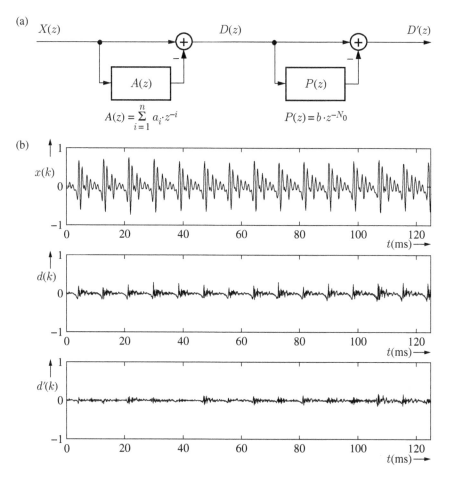

**Figure 6.16** Example of short-term and long-term prediction. (a) Block diagram and (b) time signals.

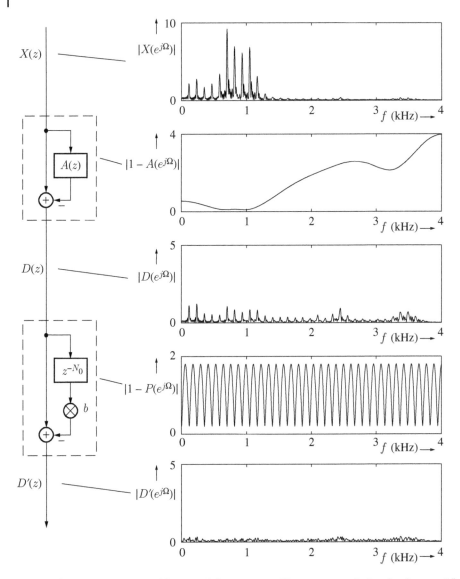

**Figure 6.17** Example: spectral impact of short-term and long-term prediction for the vowel "a."

Essentially, the formant structure is removed by means of block adaptive short-term prediction or through filtering with the transfer function $1 - A(z)$, leaving a spectrally flattened
version of the input signal with an almost periodic respectively harmonic structure over a
wide range of frequencies. The subsequent processing with the LTP-analysis filter with the
transfer function $1 - P(z)$ causes a further power reduction and an almost complete elimination of the harmonic structure, finally resulting in a spectrally flat noise-like residual
signal. In this example, the short-term predictor of order $n = 8$ was adapted every 20 ms
($N = 160$ samples), while the parameters of the long-term predictor were computed every
5 ms ($L = 40$ samples).

Recall that the delay $N_0$ represents in principle the pitch period, which is approximated by an integer multiple of the sampling interval. For this reason, general approaches for pitch detection may be used [Hess 1983].

In fact, however, the minimization according to (6.87) is based on a criterion which does not aim at approximating the true pitch period, but at minimizing the energy of the prediction error.

As a consequence, the lowest error energy might in some circumstances be achieved with a delay $N_0$, which does not correspond to the true, but to, for instance, half of the actual pitch period (or to twice the fundamental frequency).

Just as for the short-term prediction, the prediction gain can be improved by increasing the filter order. Often a long-term predictor with three coefficients according to

$$P(z) = b_{-1} \cdot z^{-(N_0-1)} + b_0 \cdot z^{-N_0} + b_{+1} \cdot z^{-(N_0+1)} \tag{6.88}$$

is used. With its interpolating effect, this predictor generally provides an improved prediction, as the true pitch period in most cases is not an integer multiple of the sampling interval or the fundamental frequency $f_0$ is not an integer fraction of the sampling frequency, respectively. In this case, however, three coefficients must be transmitted, which requires a correspondingly higher bit rate.

A similar improvement of the prediction gain can also be achieved with a predictor with one coefficient as in Figure 6.15, if the sampling frequency of the first residual signal $d(k)$ is increased by interpolation. In this way, the time resolution is improved accordingly. In the literature, this technique is called *high-resolution LTP analysis* (see [Marques et al. 1990] and [Marques et al. 1989]). Interpolating by a factor of 4, the word length for the representation of the parameter $N_0$ grows only by 2 bits. Regarding the bit rate, the high-resolution LTP analysis must therefore be preferred to the multiple-tap long-term prediction according to (6.88).

In analogy to the short-term predictor structures, which will be discussed in Section 8.3, the long-term prediction can alternatively be implemented as a forward predictor (*open-loop*) or backward predictor (*closed-loop*). When quantizing the error signal, these two alternatives differ regarding the quantization error on the receiver side (cf. Section 8.3.3).

# References

Barnwell III, T. P. (1981). Recursive Windowing for Generating Autocorrelation Coefficients for LPC Analysis, *IEEE Transactions on Acoustics, Speech and Signal Processing*, vol. 29, no. 5, pp. 1062–1066.

Deller Jr., J. R.; Proakis, J. G.; Hansen, J. H. L. (2000). *Discrete-Time Processing of Speech Signals*, 2nd edn, IEEE Press, New York.

Durbin, J. (1960). The Fitting of Time-Series Models, *Revue de l'Institut International de Statistique*, vol. 28, no. 3, pp. 233–243.

Hayes, M. H. (1996). *Advanced Digital Signal Processing*, John Wiley & Sons, Ltd., Chichester.

Haykin, S. (1996). *Adaptive Filter Theory*, Prentice Hall, Englewood Cliffs, New Jersey.

Hess, W. J. (1983). *Pitch Determination of Speech Signals*, Springer-Verlag, Berlin.

Itakura, F.; Saito, S. (1970). A Statistical Method for Estimation of Speech Spectral Density and Formant Frequencies, *Electronics and Communications in Japan*, vol. 52-A, pp. 36–43.

Jayant, N. S.; Noll, P. (1984). *Digital Coding of Waveforms*, Prentice Hall, Englewood Cliffs, New Jersey.

Kay, S. (1988). *Modern Spectral Estimation*, Prentice Hall, Englewood Cliffs, New Jersey.

Levinson, N. (1947). The Wiener RMS (Root Mean Square) Error Criterion in Filter Design and Prediction, *Journal of Mathematical Physics*, vol. 25, no. 4, pp. 261–278.

Makhoul, J. (1975). Linear Prediction: A Tutorial Review, *IEEE Proceedings*, vol. 63, pp. 561–580.

Makhoul, J.; Wolf, J. (1972). Linear Prediction and the Spectral Analysis of Speech, *Technical Report 2304*, Bolt, Beranek, and Newman, Inc., Boston, Massachusetts.

Markel, J. D.; Gray, A. H. (1976). *Linear Prediction of Speech*, Springer-Verlag, Berlin, Heidelberg, New York.

Marple Jr., S. L. (1987). *Digital Spectral Analysis: With Applications*, Prentice Hall, Englewood Cliffs, New Jersey.

Marques, J. S.; Tribolet, J. M.; Trancoso, I. M.; Almeida, L. B. (1989). Pitch Prediction with Fractional Delays in CELP Coding, *Proceedings of European Conference on Speech Technology*, Paris, France, pp. 509–512.

Marques, J. S.; Trancoso, I. M.; Tribolet, J. M.; Almeida, L. B. (1990). Improved Pitch Prediction with Fractional Delays in CELP Coding, *Proceedings of the IEEE International Conference on Acoustics, Speech, and Signal Processing (ICASSP)*, Albuquerque, New Mexico, USA, pp. 665–668.

Sluijter, R. J. (2005). *The Development of Speech Coding and the First Standard Coder for Public Mobile Telephony*, PhD thesis, Technical University Eindhoven.

Vary, P.; Heute, U.; Hess, W. (1998). *Digitale Sprachsignalverarbeitung*, B. G. Teubner, Stuttgart (in German).

# 7

# Quantization

## 7.1 Analog Samples and Digital Representation

In the context of quantization, we first of all have to deal with a sequence of samples $s_a(t = kT)$ of an analog signal $s_a(t)$, which is a function of the continuous-time variable $t$. The samples $s_a(kT)$ at the discrete time instants $kT = k/f_s$ are continuous quantities, where $T$ is the *sampling period*, i.e., the time between successive samples and $f_s = 1/T$ is the *sampling frequency*. By quantization we convert each sample $s_a(kT)$ into a quantized version $s(k)$ which can take only one out of $K_0 = 2^{w_0}$ different discrete values.

The sequence of samples $s(k)$ is now discrete with respect to time and amplitude.

The complete process of analog-to-digital conversion (A/D conversion), which is shown in Figure 7.1a, consists of the three steps:

- lowpass filtering according to the sampling theorem with cutoff frequency $f_c < 0.5 f_s$,
- sampling at frequency $f_s$, and
- quantization with word length $w_0$, i.e., $K_0 = 2^{w_0}$ different quantization levels $s_i$, with $i = 0, \dots, K_0 - 1$.

Quantization is not only needed for A/D conversion, but also whenever any parameter $x(k')$ which has been extracted from the signal samples $s(k)$ has to be represented at a reduced word length $w < w_0$, e.g., for the purpose of source encoding (compression). It is assumed that, the parameter $x(k')$ is calculated by processing blocks of samples $s(k)$, where $k'$ is denoting the block time index (see Figure 7.1b).

Various quantization techniques are available whose basic principles will be discussed in this chapter. To simplify the representation, no distinction will be made between signal samples $s(k)$ and parameters $x(k')$. We will consider either values $x$ which are applied to a scalar quantizer with $K = 2^w$ quantizer reproduction levels or vectors $\mathbf{x} = (x_1, x_2, \dots, x_L)^T$ consisting of $L$ values $x_\lambda$, $\lambda = 1, 2, \dots, L$ which are applied to a vector quantizer with $K = 2^w$ quantizer reproduction vectors. The quantity $x$ might represent, for instance, a signal sample which is provided by an A/D converter with $w_0 = 13$ bit resolution and must be represented with a word length of $w = 8$ bits. The parameter $x$ might be, for example, a predictor coefficient, which has been calculated with a digital signal processor using 16 bit fixed-point arithmetic. For the purpose of transmission, this parameter has to be represented, e.g., with a word length of $w = 5$.

*Digital Speech Transmission and Enhancement*, Second Edition. Peter Vary and Rainer Martin.
© 2024 John Wiley & Sons Ltd. Published 2024 by John Wiley & Sons Ltd.

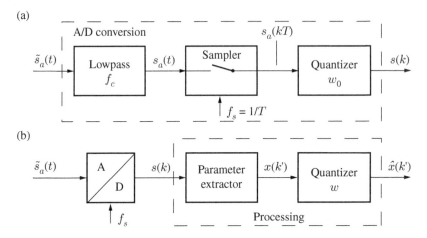

**Figure 7.1** A/D conversion and parameter extraction. (a) A/D conversion of an analog signal: lowpass filtering, sampling at $f_s$, and quantization with $K_0 = 2^{w_0}$ levels. (b) Parameter extraction: block processing at time instances $k'$ and quantization with $K = 2^w$ levels.

## 7.2 Uniform Quantization

First, we consider a symmetric quantizer which maps the input range

$$x_{\min} \le x \le +x_{\max} \qquad \text{with } x_{\min} = -x_{\max}, \quad x_{\max} > 0, \tag{7.1}$$

to the output range

$$\hat{x}_{\min} \le \hat{x} \le +\hat{x}_{\max} \qquad \text{with } \hat{x}_{\min} = -\hat{x}_{\max}, \quad \hat{x}_{\max} > 0, \tag{7.2}$$

where the quantized value, denoted by $\hat{x}$, can take one out of $K = 2^w$ quantizer reproduction levels (in short *quantization levels*) $\hat{x}_i$. The amplitude range is subdivided into $K = 2^w$ uniform intervals of width

$$\Delta x = \frac{2 x_{\max}}{K} = \frac{x_{\max}}{2^{w-1}}, \tag{7.3}$$

where $\Delta x$ is called the *quantizer stepsize*. The quantization operation can be described, as depicted in Figure 7.2a, by a staircase function

$$\hat{x} = f(x) \in \left\{ \hat{x}_i = \pm(2i+1) \cdot \frac{\Delta x}{2}; \ i = 0, 1, 2, \ldots \right\} \tag{7.4}$$

representing the *quantizer characteristic*. The quantized value $\hat{x}$ differs from $x$ by the *quantization error* or *quantization noise*

$$e = \hat{x} - x = f(x) - x, \tag{7.5}$$

according to

$$\hat{x} = e + x. \tag{7.6}$$

The quantization operation can be modeled by the quantization noise $e(k)$, which is added to $x(k)$ (see Figure 7.2b). In Figure 7.2c, $x(k)$, $\hat{x}(k)$, and $e(k)$ are plotted for a sinusoidal signal. In this example, the sampling rate is much higher than the frequency of the

(a)

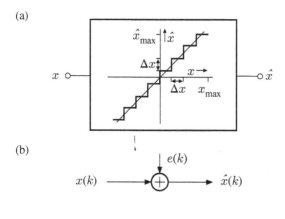

(b)

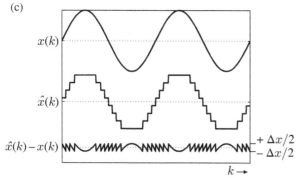

(c)

**Figure 7.2** Description of a uniform quantization operation. (a) Quantizer characteristic for $w = 3$, i.e., $K = 8$. (b) Quantizer model with additive quantization noise. (c) Example: sinusoidal signal.

sinusoidal signal, which results in a smooth shape of $x(k)$. Due to the coarse quantization with $K = 8$ levels and the relatively high sampling rate, each step of the staircase-shaped signal $\hat{x}(k)$ consists of several identical quantizer reproduction levels.

The quantizer of Figure 7.2 is not able to represent $x = 0$ exactly, as the smallest magnitude of $|\hat{x}|$ is $\frac{\Delta x}{2}$. With a slight modification of $f(x)$, the value $\hat{x} = 0$ can be allowed. However, in this case, the symmetry is lost if the number $K$ of quantization intervals is even.

Three different quantizer characteristics are depicted in Figure 7.3 for $w = 4$ or $K = 16$, respectively. Figure 7.3a represents the symmetric case of a *midrise* quantizer with eight quantization levels in the positive and negative range. In contrast to this, the *midtread* quantizer of Figure 7.3b allows the accurate representation of $x = 0$, but takes seven levels in the positive and eight levels in the negative range. The magnitude truncation characteristic of Figure 7.3c is symmetric and allows the accurate representation of $x = 0$.

For the graphs in Figure 7.3a,b, the quantization is performed by a rounding operation, whereas the quantizer characteristic in Figure 7.3c corresponds to magnitude truncation. The parameters of these quantization operations are compiled in Figure 7.3.

For both characteristics of Figure 7.3a,b, the quantization error is limited to

$$|e| \leq \frac{\Delta x}{2}. \tag{7.7}$$

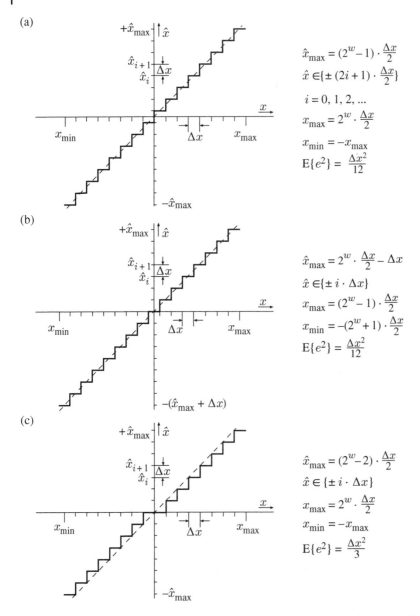

(a)

$$\hat{x}_{\max} = (2^w - 1) \cdot \frac{\Delta x}{2}$$

$$\hat{x} \in \{\pm (2i + 1) \cdot \frac{\Delta x}{2}\}$$

$$i = 0, 1, 2, \ldots$$

$$x_{\max} = 2^w \cdot \frac{\Delta x}{2}$$

$$x_{\min} = -x_{\max}$$

$$E\{e^2\} = \frac{\Delta x^2}{12}$$

(b)

$$\hat{x}_{\max} = 2^w \cdot \frac{\Delta x}{2} - \Delta x$$

$$\hat{x} \in \{\pm i \cdot \Delta x\}$$

$$x_{\max} = (2^w - 1) \cdot \frac{\Delta x}{2}$$

$$x_{\min} = -(2^w + 1) \cdot \frac{\Delta x}{2}$$

$$E\{e^2\} = \frac{\Delta x^2}{12}$$

(c)

$$\hat{x}_{\max} = (2^w - 2) \cdot \frac{\Delta x}{2}$$

$$\hat{x} \in \{\pm i \cdot \Delta x\}$$

$$x_{\max} = 2^w \cdot \frac{\Delta x}{2}$$

$$x_{\min} = -x_{\max}$$

$$E\{e^2\} = \frac{\Delta x^2}{3}$$

**Figure 7.3** Uniform quantizer characteristics with $w = 4$. (a) Uniform *midrise* quantizer: $\hat{x}_i = \pm (2i + 1)\frac{\Delta x}{2}$. (b) Uniform *midtread* quantizer: $\hat{x}_i = \pm i \Delta x$. (c) Uniform quantizer with magnitude truncation: $\hat{x}_i = \pm i \Delta x$.

In contrast, for the characteristic in Figure 7.3c, the maximum quantization error is twice as large, i.e.,

$$\max |e| = \Delta x. \tag{7.8}$$

The quantization characteristics depicted in Figure 7.3 can be represented in the range $x_{\min} \leq x \leq x_{\max}$ analytically as follows:

$$\hat{x} = f_a(x) = \text{sign}(x) \cdot \Delta x \cdot \left[ \text{int} \left( \frac{|x|}{\Delta x} \right) + 0.5 \right], \tag{7.9a}$$

$$\hat{x} = f_b(x) = \text{sign}(x) \cdot \Delta x \cdot \text{int} \left( \frac{|x|}{\Delta x} + 0.5 \right), \tag{7.9b}$$

$$\hat{x} = f_c(x) = \text{sign}(x) \cdot \Delta x \cdot \text{int} \left( \frac{|x|}{\Delta x} \right), \tag{7.9c}$$

where int(x) the integer part of x.

Although there is a deterministic relation between the actual sample $x(k)$ and the resulting quantization error $e(k)$, the quantization error is usually modeled as a statistical quantity. We assume that the mean value of the sequence $x(k)$ is zero and that $x(k)$ has a symmetric probability density function (PDF)

$$p_x(+u) = p_x(-u). \tag{7.10}$$

We are interested in quantifying the performance of the quantizer in terms of the resulting signal-to-noise ratio (SNR).

With the PDF $p_x(u)$ of signal $x(k)$, we obtain the power $S$ of the signal as

$$S = \text{E}\left\{ x^2 \right\} = \int_{-\infty}^{+\infty} u^2 \, p_x(u) du \tag{7.11}$$

and due to the quantization characteristic $\hat{x} = f(x)$ for the quantization noise power $N$

$$N = \text{E}\left\{ e^2(x) \right\} = \int_{-\infty}^{+\infty} (f(u) - u)^2 \, p_x(u) du. \tag{7.12}$$

We assume a symmetric quantization characteristic as shown, for example, in Figure 7.3a. The overload amplitude for this type of quantizer characteristic is

$$\pm x_{\max} = \pm \left( \hat{x}_{\max} + \frac{\Delta x}{2} \right). \tag{7.13}$$

Exploiting the symmetry, the following equation is valid for the quantization noise power:

$$N = 2 \int_0^{x_{\max}} (f(u) - u)^2 p_x(u) du + 2 \int_{x_{\max}}^{\infty} (\hat{x}_{\max} - u)^2 p_x(u) du \tag{7.14a}$$

$$= N_Q + N_O. \tag{7.14b}$$

As $x(k)$ is not necessarily limited to the dynamic range $\pm x_{\max}$, the noise power can be divided into two components $N_Q$ and $N_O$, which are caused by quantization $(Q)$ or by overload and clipping $(O)$, respectively. In the following, we will assume that no overload errors occur or that this effect can be neglected, i.e., $N_O = 0$ applies.

In the positive range, the given uniform midrise quantizer has $\frac{K}{2} = 2^{w-1}$ quantization levels

$$\hat{x}_i = f(x) = i \cdot \Delta x - \frac{\Delta x}{2}, \qquad i = 1, 2, \ldots, \frac{K}{2}, \tag{7.15}$$

which are assigned to the *i*-th interval

$$(i-1) \cdot \Delta x \leq x < i \cdot \Delta x \tag{7.16a}$$

or,

$$\hat{x}_i - \frac{\Delta x}{2} \leq x < \hat{x}_i + \frac{\Delta x}{2}. \tag{7.16b}$$

The contributions of the individual quantization intervals to the total quantization noise power $N_Q$ according to (7.14a) result from integration over each interval. We obtain

$$N_Q = 2 \sum_{i=1}^{\frac{K}{2}} \int_{(i-1)\cdot\Delta x}^{i\cdot\Delta x} (\hat{x}_i - u)^2 \, p_x(u) du. \tag{7.17}$$

With the substitution $z = u - \hat{x}_i$ this expression can be simplified to

$$N_Q = 2 \sum_{i=1}^{\frac{K}{2}} \int_{-\frac{\Delta x}{2}}^{+\frac{\Delta x}{2}} z^2 \, p_x(z + \hat{x}_i) dz. \tag{7.18}$$

The quantization noise power depends on the PDF $p_x(u)$ of the sequence $x(k)$. For the special case of uniform distribution, it can be easily computed. With

$$\max |x| = x_{\text{max}}, \tag{7.19}$$

$$p_x(u) = \frac{1}{2\,x_{\text{max}}}, \quad -x_{\text{max}} \leq u \leq +x_{\text{max}}, \tag{7.20}$$

$$\Delta x = \frac{2\,x_{\text{max}}}{K}, \tag{7.21}$$

the total quantization noise power results in

$$N_Q = 2 \sum_{i=1}^{\frac{K}{2}} 2 \int_0^{+\frac{\Delta x}{2}} z^2 \, \frac{1}{2\,x_{\text{max}}} dz \tag{7.22a}$$

$$= K \, \frac{1}{x_{\text{max}}} \, \frac{1}{3} z^3 \Big|_0^{\frac{\Delta x}{2}} \tag{7.22b}$$

$$= \frac{\Delta x^2}{12}, \tag{7.22c}$$

and the signal power is

$$S = \int_{-x_{\text{max}}}^{+x_{\text{max}}} x^2 \, \frac{1}{2\,x_{\text{max}}} dx \tag{7.22d}$$

$$= \frac{1}{3} x_{\text{max}}^2. \tag{7.22e}$$

With $K = 2^w$ and (7.21), we obtain the SNR

$$\frac{\text{SNR}}{\text{dB}} = 10 \lg \left( \frac{S}{N_Q} \right) \tag{7.23a}$$

$$= w \, 20 \lg (2) \tag{7.23b}$$

$$\approx 6 \, w. \tag{7.23c}$$

This is the so-called *6-dB-per-bit rule* which, however, is only accurate for this special case.

For the general case, an approximation can be provided. If the quantizer stepsize is sufficiently small ($\Delta x \ll x_{\text{max}}$) and if $p_x(u)$ is sufficiently smooth, the PDF can be approximated by its value in the middle of the interval as in

$$p_x(z + \hat{x}_i) \approx p_x(\hat{x}_i) \quad \text{for} \quad -\frac{\Delta x}{2} \leq z < +\frac{\Delta x}{2}. \tag{7.24}$$

Then, we find that

$$N_Q \approx \frac{\Delta x^2}{12} \tag{7.25}$$

approximately holds too, independently of the PDF $p_x(u)$. This approximation corresponds to the assumption that within the quantization intervals the signal is uniformly distributed. Thus, the quantization noise $e(k)$ arises with uniform distribution $p_e(u)$ in the interval

$$-\frac{\Delta x}{2} < e \leq +\frac{\Delta x}{2}, \tag{7.26}$$

and its power

$$N_Q = \text{E} \left\{ e^2 \right\} = \int_{-\infty}^{+\infty} u^2 \, p_e(u) du \tag{7.27}$$

is also given by (7.25). In conclusion, (7.25) can be applied even in the case of a non-uniform PDF $p_x(u)$, if the quantization is sufficiently small, i.e., the word length $w$ is sufficiently large, $p_x(u)$ is sufficiently smooth, and overload effects are negligibly small.

If, furthermore, the signal power $S$ is normalized to the squared overload amplitude $x_{\text{max}}^2$ of the quantizer input by the use of the *form factor F*

$$S = F \, x_{\text{max}}^2, \tag{7.28}$$

the SNR is approximately described by

$$\frac{\text{SNR}}{\text{dB}} = 10 \lg \left( \frac{S}{N_Q} \right) \approx w \, 20 \lg (2) + 10 \lg (3F) \tag{7.29a}$$

$$= w \, 6.02 + 10 \lg(3) + 10 \lg(F). \tag{7.29b}$$

The form factor $F$, which can be considered as normalized power, depends on the shape of the PDF.

Figure 7.4 depicts the general behavior of the SNR as a function of the form factor $F$ (normalized power). The SNR is a linear function of the signal level, i.e., of the logarithmic signal power

$$10 \lg \left( \frac{S}{x_{\text{max}}^2} \right) = 10 \lg (F). \tag{7.30}$$

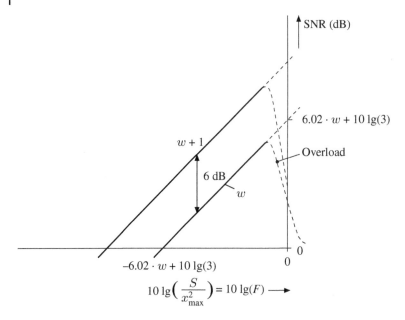

**Figure 7.4** Signal-to-noise ratio for uniform quantization as a function of form factor *F*.

In case of overload, the SNR drops rapidly with increasing signal power as indicated in Figure 7.4. By scaling the amplitude of $x(k)$ or the overload amplitude $\pm x_{max}$ of the quantizer, the overload effects can be kept small. In principle, the quantizer may be derived in such a way that in (7.14a) the contributions caused by quantization and overload are balanced

$$N_Q = N_O. \tag{7.31}$$

This case will not be discussed further here; for more details see, e.g., [Jayant, Noll 1984].

The impact of form factor *F* is shown in Table 7.1 for different full range signals (no quantizer overload, (a), (b), (c)), as well as for signals with small overload probability ($P = 0.001$, (d), (e)).

**Table 7.1** Influence of the form factor *F* on the SNR for uniform quantization (see also (7.29a)).

| Probability distribution $p_u(x)$ | F | 10 lg (3F) |
|---|---|---|
| (a) Uniform distribution | 1/3 | 0 |
| (b) Distribution of a sinusoidal signal | 1/2 | +1.76 |
| (c) Triangular probability density function | 1/6 | −3.01 |
| (d) Laplace probability density function | ≈1/24 | ≈−9 |
| (e) Measured distribution of speech signals | 1/300 to 1/20 | −20 to −8 |

Overload of the quantizer with probability $P = 0.001$.

Finally, it should be noted that for uniform quantization the *6-dB-per-bit rule* (7.23c) is generally a good approximation. However, a constant penalty in the SNR can be experienced, due to the shape of the PDF. If small signal values occur much more often than large ones, or if the signal amplitude is so low that only a part of the dynamic range of the quantizer is exploited, the penalty can be substantial.

## 7.3 Non-uniform Quantization

For uniform quantization, the SNR according to (7.29a) is proportional to the signal level, hence it becomes smaller with decreasing signal power. However, especially in speech signals, small sample values are particularly frequent, corresponding to a PDF, which can be approximated, for instance, by a Laplacian PDF, a gamma PDF, or by spherically invariant models, e.g., [Brehm, Stammler 1987] and [Jayant, Noll 1984] (see Section 5.9). In this case, the resulting low SNRs can be improved by using a quantizer with a non-uniform amplitude resolution, which reduces the width of the quantization intervals in the low-amplitude region and allows larger intervals otherwise. A corresponding approach using *signal compression* is depicted in Figure 7.5.

Prior to the actual quantization, the values $x(k)$ are non-linearly transformed with a compressor characteristic

$$y = g(x),\qquad(7.32)$$

in such a way that for an unchanged signal range ($\pm y_{max} = \pm x_{max}$) small amplitudes are amplified more than large ones. Subsequently, uniform quantization with $K = 2^w$ quantization levels $\hat{y}_i$, $i = 1, 2, \ldots, K$, is applied as in Section 7.2, which is modeled by adding a uniformly distributed, spectrally white noise signal $e(k)$ of power

$$N_Q = \frac{\Delta y^2}{12}; \qquad \Delta y = \frac{2\,y_{max}}{K},\qquad(7.33)$$

to the compressed signal $y$. Without loss of generality, we will in the following assume a normalized signal representation with

$$x_{max} = y_{max} = 1.\qquad(7.34)$$

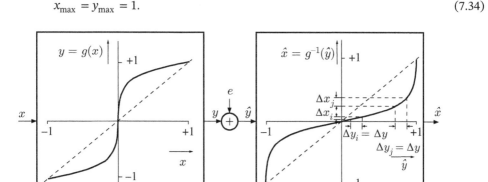

**Figure 7.5** Principle of quantization with companding ($x_{max} = y_{max} = 1$).

The non-linear distortion caused by the compressor characteristic should be removed by the inverse characteristic of the expander

$$\hat{x} = g^{-1}(\hat{y}),\tag{7.35}$$

before the quantized values $\hat{y}(k)$ are applied to any signal processing algorithm, such as digital filtering.

The compressor characteristics with its instantaneous amplification is reversed as can be seen in Figure 7.5. Thus, the lower quantization levels $\hat{y}_i$ are attenuated more than the higher ones. This yields a non-uniform amplitude resolution of the values $\hat{x}(k)$, where the effective quantization levels are given by

$$\hat{x}_i = g^{-1}(\hat{y}_i).\tag{7.36}$$

As a result the magnitude of the effective quantization noise $e(k)$ depends strongly on the amplitude of the input signal $x(k)$. The combination of compressor and expander is commonly called *compander*.

The relation between the different signals is illustrated in Figure 7.6. The values $y(k)$ are quantized uniformly using a midrise quantizer

$$\hat{y} = f(y),\tag{7.37}$$

according to Figure 7.3b with word length of $w = 5$ and $w = 8$. The relatively fine amplitude resolution near the origin and the coarser resolution for larger signal values can clearly be seen.

For the quantized values $\hat{y}$ and $\hat{x}$, we have

$$\hat{y} = \hat{y}_i \quad \forall \quad \hat{y}_i - \frac{\Delta y}{2} \leq y < \hat{y}_i + \frac{\Delta y}{2}\tag{7.38a}$$

or, respectively,

$$\hat{x} = g^{-1}(\hat{y}_i) = \hat{x}_i \quad \forall \quad g^{-1}\left(\hat{y}_i - \frac{\Delta y}{2}\right) \leq x < g^{-1}\left(\hat{y}_i + \frac{\Delta y}{2}\right).\tag{7.38b}$$

The quantization stepsizes $\Delta x_i$ of the effective quantization intervals now depend on the magnitude of $x$. By simple geometrical considerations, it can be shown that the different quantization stepsizes $\Delta x_i$ are approximately determined by the gradient

$$\frac{d\left(g^{-1}(\hat{y})\right)}{d\hat{y}} = \left(g^{-1}(\hat{y})\right)'\tag{7.39}$$

of the inverse characteristic $g^{-1}(\hat{y})$ or, respectively, by the reciprocal $1/g'(\hat{x})$ of the gradient $g'(\hat{x})$ and the constant quantizer stepsize $\Delta y$:

$$\Delta x_i \approx \frac{\Delta y}{g'(\hat{x}_i)}.\tag{7.40}$$

A suitable criterion for the development of a compressor characteristic is the requirement for a constant relative quantization error. The quantization stepsize should therefore be proportional to the signal magnitude,

$$\Delta x(x) \approx \frac{\Delta y}{g'(x)} \sim |x|,\tag{7.41}$$

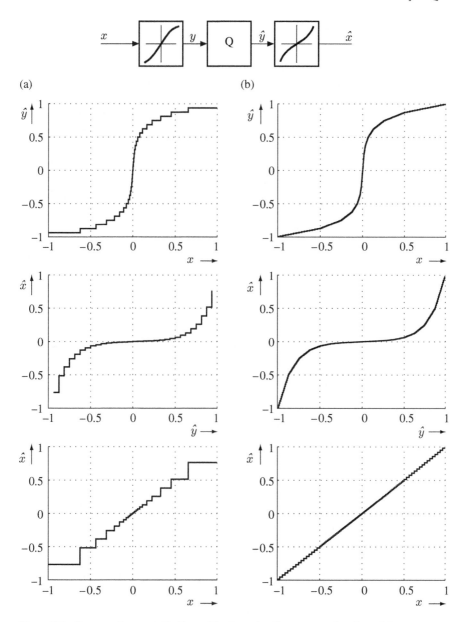

**Figure 7.6** Companding quantization with piecewise linear approximation of the compressor and the expander characteristic (A-law characteristic, see also Figure 7.9). (a) $w = 5$ and (b) $w = 8$.

as far as this is achievable with $K$ quantization intervals. For simplicity, we first consider only positive values of $x$ and extend the resulting characteristic $g(x)$ later on to the negative range, such that a symmetric characteristic is obtained, as illustrated in Figure 7.5.

According to (7.41) the gradient $g'(x)$ should obey

$$\frac{1}{g'(x)} \overset{!}{=} c\,x; \qquad c = \text{const}; \qquad x > 0. \tag{7.42a}$$

Thus, we get

$$g(x) = \int \frac{1}{c\,x}\,dx = c_1 + c_2 \ln(x) \tag{7.42b}$$

with appropriate constants $c_1$ and $c_2$. The desired compressor characteristic $g(x)$ is a logarithmic function, the expander characteristic an exponential function.

However, the function $\ln(x)$ is only defined for positive values and diverges for $x \to 0$. Consequently, the purely logarithmic compressor characteristic (7.42b) is not practical. For this reason, the characteristic defined by (7.42b) needs some pragmatic modifications. Two approximations to logarithmic quantization have found wide use as international standards providing almost a constant relative quantization error.

For the fixed (wire-line) digital telephone networks in Europe, the so-called *A-law characteristic* was defined as

$$g_A(x) = \begin{cases} \text{sign}(x) \cdot \dfrac{1 + \ln(A\,|x|)}{1 + \ln(A)}, & \dfrac{1}{A} < |x| \leq +1, \\[3mm] \dfrac{A\,x}{1 + \ln(A)}, & -\dfrac{1}{A} \leq x \leq +\dfrac{1}{A}. \end{cases} \tag{7.43}$$

Near the origin, in the range

$$-\frac{1}{A} \leq x \leq +\frac{1}{A}, \tag{7.44}$$

the A-law compressor characteristic is based on a linear characteristic and is logarithmic beyond that. At $x = 1/A$, both characteristics meet smoothly without discontinuity. For the negative range, the characteristic is mirrored.

For the linear part of the quantization characteristic a gradient of

$$g_A'(0) = 16 \tag{7.45}$$

is chosen, resulting in a parameter value $A = 87.56$. In accordance with (7.40), the effective quantization stepsizes $\Delta x_i$ in the linear region

$$-\frac{1}{A} \leq x \leq +\frac{1}{A}, \tag{7.46}$$

i.e., for *small signals*, are thus reduced by the factor $2^{-4}$. This corresponds to an increase of the SNR in the linear region by

$$\Delta\text{SNR} = 20 \lg(2^4) = 24.082\,\text{dB}. \tag{7.47}$$

According to the *6-dB-per-bit rule*, a uniform quantizer would need $\Delta w = 4$ additional bits for the same SNR for *small signals*.

In the digital telephone systems of North America and Japan, the so-called *$\mu$-law characteristic* is utilized to approximate (7.42b) in a slightly different way. This compressor characteristic is described by a single continuous function as

$$g_\mu(x) = \text{sign}(x)\,\frac{\ln(1 + \mu\,|x|)}{\ln(1 + \mu)} \quad \text{with} \quad \mu = 255. \tag{7.48}$$

For relatively small signals $x(k)$ or low values $\mu$, the compressor characteristic evolves as a linear function according to $\ln(1 + \mu|x|) \approx \mu|x|$. For $\mu|x| \gg 1$, it is logarithmic since $\ln(1 + \mu|x|) \approx \ln(\mu|x|)$. The gradient of the compressor characteristic at the origin is

$$g_\mu'(0) = \frac{\mu}{\ln(1 + \mu)} = 45.99. \tag{7.49}$$

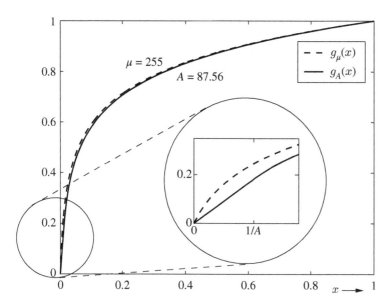

**Figure 7.7** Comparison of A-law and $\mu$-law compressor characteristics.

For the relatively small signals, the effective SNR increases by

$$\Delta \text{SNR} = 20 \lg(45.99) = 33.25\,\text{dB}. \tag{7.50}$$

The $A$-law characteristic and the $\mu$-law characteristic are very similar. They are illustrated in Figure 7.7.

The logarithmic companding results in a SNR which is to a large extent independent of the signal level or the power of the sequence $x(k)$, respectively.

We now calculate the achievable SNR. Presuming a uniform distribution of $x$ within each quantization interval, the individual quantization noise power is $\Delta x_i^2/12$. The contribution of the $i$-th quantization interval to the total power of the quantization noise is

$$N_{Qi} = \frac{\Delta x_i^2}{12} P_i, \tag{7.51}$$

where $x(k)$ takes a value in the $i$-th quantization interval with a probability $P_i$. Under the assumption of a symmetric PDF $p_x(+u) = p_x(-u)$, the total quantization noise power $N_Q$ can be calculated with (7.40) and (7.42a):

$$N_Q = 2 \sum_{i=1}^{\frac{K}{2}} N_{Qi} \tag{7.52a}$$

$$= 2 \sum_{i=1}^{\frac{K}{2}} \frac{\Delta y^2}{12} c^2 \hat{x}_i^2 P_i \tag{7.52b}$$

$$\approx \frac{\Delta y^2}{12} c^2 S. \tag{7.52c}$$

Hence, the noise power is proportional to the signal power. With

$$\Delta y = \frac{2 y_{\max}}{2^w} = 2^{-(w-1)}, \quad y_{\max} = 1, \tag{7.53}$$

the SNR

$$\frac{\text{SNR}}{\text{dB}} = 10 \lg \left( \frac{S}{N_Q} \right) \approx w \, 20 \lg(2) + 10 \lg(3) - 20 \lg(c) \tag{7.54a}$$

$$= w \, 6.02 + 10 \lg(3) - 20 \lg(c) \tag{7.54b}$$

becomes independent of the signal power. It now contains the constant $c$ instead of the form factor $F$ in (7.29a). Applying (7.42a), constant $c$ amounts to

$$c_A = (1 + \ln A) \approx 5.47, \tag{7.55}$$

for the $A$-law characteristic in the range $1/A \le |x| \le 1$, and to

$$c_\mu \approx \ln(1 + \mu) \approx 5.55, \tag{7.56}$$

for the $\mu$-law characteristic for $\mu x \gg 1$. With both characteristics, a similar SNR is thus obtained using (7.54a):

$$\text{SNR}_A \approx 6 \cdot w - 9.99 \, \text{dB}, \tag{7.57a}$$

$$\text{SNR}_\mu \approx 6 \cdot w - 10.11 \, \text{dB}. \tag{7.57b}$$

The SNR again satisfies a 6-dB-per-bit rule, which is almost independent of the signal level. This independence "costs" approximately 10 dB compared to (7.23c) obtained for uniform quantization, i.e., for signals with uniform distribution and matching the peak-to-peak range of the quantizer. However, for uniform quantization the SNR depends on the signal level. With uniform quantization substantial reductions must be expected in practice, according to level variations corresponding to the form factor $F$. With the same peak-to-peak range, the logarithmic compander, therefore, proves indeed to be superior to the uniform quantizer over a wide range of amplitudes (see Figure 7.8).

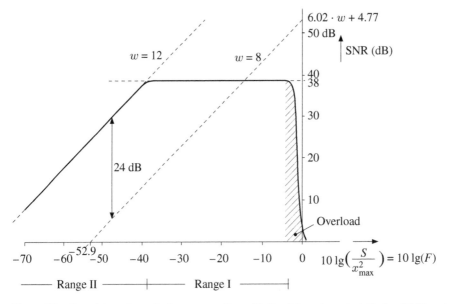

**Figure 7.8** Signal-to-noise ratio for companding with the $A$-law characteristic $A = 87.56$ and a word length of $w = 8$.

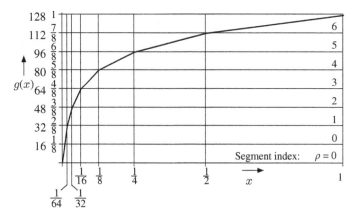

**Figure 7.9**   The 13-segment $A$-law characteristic.

The approximations (7.57a,b) do not apply to very small values of $|x|$. Close to the origin, we have a uniform quantization with an effective quantization stepsize of

$$\Delta x = \frac{\Delta y}{g'(x = 0)},\tag{7.58}$$

and with the gradients given in (7.45) and (7.49). For the $A$-law characteristic, Figure 7.8 shows the SNR as a function of the signal power $S$ for a word length of $w = 8$, which is common in the digital telecommunication network. In range I, the SNR amounts to approximately 38 dB as derived in (7.57a). In range II with $|x| \leq 1/A \approx 0.011$, i.e., for a range of approximately 1% of the maximum amplitude, the following expression applies in analogy to (7.29a) with consideration of (7.47):

$$10 \lg \left( \frac{S}{N_Q} \right) \approx w \; 20 \; \lg(2) + 10 \; \lg(3F) + 24 \; \text{dB}.\tag{7.59}$$

In this range, the SNR is comparable to that of a uniform quantizer, which, however, has a word length of $w = 12$. The corresponding improvement by 24 dB is also termed *compander gain*.

In practice, the $A$-law characteristic as well as the $\mu$-law characteristic are each realized by a piecewise linear approximation which is depicted for the $A$-law characteristic in Figure 7.9.

The range $-1 \leq g \leq +1$ is divided into 16 equally spaced intervals, in which the $A$-law characteristic is approximated by straight lines. In the four innermost intervals $(0 < |x| < 1/64)$, the $A$-law characteristic is almost linear (7.43). Thus in these four intervals the compressor characteristic can be approximated by one line. As a result, the overall characteristic is approximated by $16 - 3 = 13$ segments of distinct slope. Due to the increasing length by a factor of 2 for each successive segment, the slopes of adjacent segments differ by a factor of 2.

For quantization with a word length of $w = 8$ bits, the first bit denotes the algebraic sign, the next 3 bits encode the respective segment, and the last 4 bits indicate the quantization level within the segment.

The effective quantization stepsize in the lowest segment amounts to

$$\Delta x_{min} = \frac{1/128}{2^{-4}} = 2^{-11},$$ (7.60)

and in the highest segment to

$$\Delta x_{max} = \frac{1/2}{2^{-4}} = 2^{-5}.$$ (7.61)

The resulting non-uniform quantization characteristic has already been depicted in Figure 7.6b.

Quantization according to the 13-segment characteristic can also be achieved by uniform pre-quantization with $w_0 \geq w + 4$ and subsequent code conversion to the word length $w$. This coding law is summarized in Table 7.2 for $w_0 = 12$ and $w = 8$.

The 13-segment $A$-law coding rule can be derived on a bit level if we start with a 12 bit sign–magnitude representation

$$x = \text{sign}\{x\} \cdot |x|.$$ (7.62)

We denote by $0 \leq a \leq 7$, the number of leading zeros of the binary representation of $|x|$ and by $c$ the next 4 bits as indicated in Table 7.2. The remaining $d$ bits are neglected. Finally, the binary representation of $\hat{y}$ for $w = 8$ is obtained, as shown in Table 7.3.

**Table 7.2** Coding law of the 13-segment $A$-law characteristic – $a$: number of leading zeros following the sign bit; $b$: binary code of $7 - a$; $c$: the last four digits, if $a = 7$, or, the first four digits behind the leading 1, if a < 7; $d$: neglected digits.

| | | Binary representation | |
|---|---|---|---|
| Segment index | Range | $\lvert x \rvert$ | $\lvert \hat{y} \rvert$ |
| $\rho$ | | $\overbrace{a \quad\quad c}$ | $\overbrace{b \quad c}$ |
| 0 | $0 \leq \lvert x \rvert < 2^{-7}$ | 0000000.... | 000.... |
| 0 | $2^{-7} \leq \lvert x \rvert < 2^{-6}$ | 0000001.... | 001.... |
| 1 | $2^{-6} \leq \lvert x \rvert < 2^{-5}$ | 000001....- | 010.... |
| 2 | $2^{-5} \leq \lvert x \rvert < 2^{-4}$ | 00001.....-- | 011.... |
| 3 | $2^{-4} \leq \lvert x \rvert < 2^{-3}$ | 0001.....--- | 100.... |
| 4 | $2^{-3} \leq \lvert x \rvert < 2^{-2}$ | 001.....----- | 101.... |
| 5 | $2^{-2} \leq \lvert x \rvert < 2^{-1}$ | 01.....------ | 110.... |
| 6 | $2^{-1} \leq \lvert x \rvert < 2^{-0}$ | 1.....------ | 111.... |
| | | $\underbrace{a \quad c} \quad d$ | $\underbrace{b \quad c}$ |

**Table 7.3** Bit allocation of 13-segment $A$-law quantization.

| | 1 bit | 3 bits | 4 bits |
|---|---|---|---|
| $\hat{y}$ : | $\text{sign}\{x\}$ | $7 - a$ | $c$ |

This form of signal quantization which fulfills the international standard ITU G.711 [ITU-T Rec. G.711 1972] is the basis for digital speech transmission in the European digitized telecommunication networks with a bit rate of $B = w \cdot f_s = 64$ kbit/s per voice channel (Integrated Services Digital Network [ISDN]).

## 7.4 Optimal Quantization

The quantizers discussed so far work with uniform or non-uniform stepsizes. The quantizer levels $\hat{x}_i$ are in the middle or at the edge of the intervals (see Figure 7.3). In principle, the *interval limits* or *decision levels* $x_i$ and the *quantizer representation levels* $\hat{x}_i$ for $i \in \{1, 2, \ldots, 2^w\}$ can be chosen arbitrarily as indicated in Figure 7.10. In particular, they can be determined such that for a given signal PDF $p_x(u)$ the maximal SNR is obtained. In other words, the (scalar) optimal quantizer minimizes the power $N_Q$ of the quantization error.

For non-uniformly distributed signals, a non-uniform resolution of the amplitude is to be expected; for signals like speech finer quantization of small amplitudes and coarser quantization of large amplitudes is desired. The characteristic should thus indeed show similarities to that of logarithmic companding. However, it might differ, as it is generated from a different minimization criterion and does not aim for PDF-independent SNR. The underlying optimization problem was solved in [Lloyd 1982] and [Max 1960]. This solution is called the *Lloyd–Max quantizer*.

In analogy to (7.17), the power of the quantization noise amounts to

$$N = \sum_{i=1}^{2^w} \int_{x_{i-1}}^{x_i} (\hat{x}_i - u)^2 \, p_x(u) du. \tag{7.63}$$

Necessary conditions for determining the interval limits $x_i$ and the representation levels $\hat{x}_i$ ($i = 1, \ldots, 2^w$) are provided by partial derivatives. Taking into account that the outer limits $x_0$ and $x_K$ ($K = 2^w$) have to be treated separately, we obtain for $k = 1, 2, \ldots, 2^w - 1$

$$\frac{\partial N}{\partial x_k} = (\hat{x}_k - x_k)^2 \, p_x(x_k) - (\hat{x}_{k+1} - x_k)^2 \, p_x(x_k) \overset{!}{=} 0. \tag{7.64a}$$

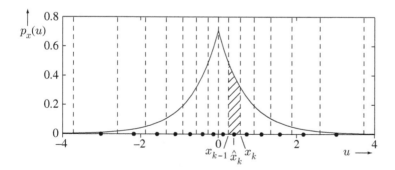

**Figure 7.10** Relation between decision levels $x_k$ and quantizer representation levels $\hat{x}_k$.

This yields

$$x_k = \frac{\hat{x}_k + \hat{x}_{k+1}}{2}, \tag{7.64b}$$

and secondly for $k = 1, 2, \ldots, 2^w - 1$,

$$\frac{\partial N}{\partial \hat{x}_k} = 2 \int_{x_{k-1}}^{x_k} (\hat{x}_k - u)\, p_x(u)\mathrm{d}u \overset{!}{=} 0 \tag{7.64c}$$

results in

$$\hat{x}_k = \frac{\int_{x_{k-1}}^{x_k} u\, p_x(u)\mathrm{d}u}{\int_{x_{k-1}}^{x_k} p_x(u)\mathrm{d}u}. \tag{7.64d}$$

Hence, the optimal interval representatives $\hat{x}_i$ correspond to the centers of gravity of the quantization intervals. The optimal interval limits $x_k$ are located midway between two adjacent representation levels with the two outer interval limits $x_0$ and $x_K$ being exceptions. The latter are given by the range of $x$, for example, $x_0 = -\infty$ and $x_K = +\infty$. The conditions (7.64b) and (7.64d) can be numerically solved for arbitrary PDFs $p_x(u)$. The achievable improvement of the SNR (see also examples from Table 7.5) depends on the shape of the PDF $p_x(u)$ (e.g., [Jayant, Noll 1984]).

## 7.5 Adaptive Quantization

An alternative for reducing the dependency of the SNR on the (instantaneous) quantizer load is to use a uniform quantizer with $K$ quantizer representation levels but with dynamical adaptation of the quantizer stepsize $\Delta x$.

Two basic solutions exist, which are designated in Figure 7.11 as quantization with forward adaptation (*adaptive quantization forward* (AQF)) or with backward adaptation (*adaptive quantization backward* (AQB)) [Jayant, Noll 1984]. In both cases

$$\hat{x}(k) = \mathrm{sign}\,(x(k))\; Z(k)\, \frac{\Delta x(k)}{2}, \qquad Z(k) \in \{1, 3, 5, \ldots\} \tag{7.65}$$

applies to the quantized values, if a symmetric quantization characteristic according to Figure 7.3a is assumed.

With the AQF method, $\Delta x(k)$ is adjusted blockwise and transmitted (or, respectively, stored) as additional side information. Because of the extra required bit rate a relatively large block length of, for instance, $N = 128$ at $f_s = 8\,\mathrm{kHz}$ is chosen. With the AQB method, there is no side information, as the quantizer stepsize is derived from $Z(k-1)$, which for undisturbed transmission is also available at the receiving side.

With both methods, the instantaneous power of $x(k)$ or $\hat{x}(k)$ is estimated and the stepsize is adjusted proportionally to the estimated standard deviation $\hat{\sigma}_x(k)$:

$$\Delta x(k) = c\, \hat{\sigma}_x(k), \qquad c = \mathrm{const.} \tag{7.66}$$

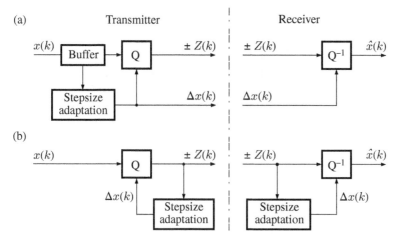

**Figure 7.11** Adaptive quantization. (a) Adaptive quantization with forward estimation (AQF). (b) Adaptive quantization with backward estimation (AQB).

With the AQF method, the variance estimation is performed once on blocks of $N$ samples according to

$$\hat{\sigma}_x^2(k) = \frac{1}{N} \sum_{i=0}^{N-1} x^2(k_0 + i), \quad k = k_0, k_0 + 1, \dots, k_0 + N - 1. \tag{7.67}$$

In the AQB method, however, $\sigma_x^2$ is estimated recursively using the already available quantized value $\hat{x}(k-1)$

$$\hat{\sigma}_x^2(k) = \alpha \, \hat{\sigma}_x^2(k-1) + (1 - \alpha) \, \hat{x}^2(k-1), \quad 0 < \alpha < 1. \tag{7.68}$$

The stepsize $\Delta x$ can be adjusted more frequently than with the AQF method. This is illustrated for a speech signal in Figure 7.12.

For the AQB method, an algorithm has been proposed in [Jayant 1973], which can be realized very efficiently. Due to (7.66), the following expression holds:

$$\frac{\Delta x(k)}{\Delta x(k-1)} = \frac{\hat{\sigma}_x(k)}{\hat{\sigma}_x(k-1)} \doteq M(k-1). \tag{7.69}$$

The term $M(k-1)$ will be called the *stepsize multiplier* from now on. Equation (7.69) in combination with (7.68), (7.66), and (7.65) yields

$$M^2(k-1) = \frac{\hat{\sigma}_x^2(k)}{\hat{\sigma}_x^2(k-1)} \tag{7.70a}$$

$$= \alpha + (1 - \alpha) \, Z^2(k-1) \, \frac{c^2}{4} \tag{7.70b}$$

or,

$$M(k-1) = \sqrt{\alpha + (1 - \alpha) \, Z^2(k-1) \, \frac{c^2}{4}}. \tag{7.70c}$$

Hence, the stepsize multiplier depends only on $Z$. As $Z$ takes only $2^{w-1}$ different values, $M(k-1)$ can be computed according to (7.70c) in advance and stored in a table.

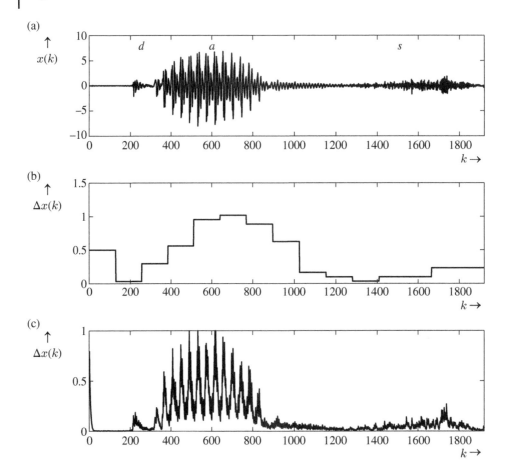

**Figure 7.12** Example of the adaptation of the stepsize $\Delta x(k)$. (a) Speech signal "das" (German). (b) Stepsize with forward adaptation (AQF). (c) Stepsize with backward adaptation (AQB).

The computational steps required at time instant $k$ at the transmitting and receiving sides are summarized below:

1. Computation of the new stepsize

$$\Delta x(k) = M(k-1) \cdot \Delta x(k-1). \tag{7.71a}$$

2. Quantization of $x(k)$ or, respectively, determination of $Z$ according to

$$\hat{x}(k) = \text{sign}(x(k)) \cdot Z(k) \cdot \frac{\Delta x(k)}{2}, \qquad Z(k) \in \{1, 3, 5, \ldots\} \tag{7.71b}$$

with

$$Z(k) = 2 \cdot \text{int}\left(\frac{x(k)}{\Delta x(k)}\right) + 1 \tag{7.71c}$$

(see also (7.9a)).

3. Determination of the stepsize multiplier for $k + 1$ by selecting the corresponding value

$$M(k) = f(Z(k)) \tag{7.71d}$$

from a table.

**Table 7.4** Stepsize multipliers $M = f(Z(k))$ for adaptive quantization of speech signals [Jayant 1973], [Jayant, Noll 1984]. PCM: Pulse Code Modulation (A-law or $\mu$-law quantization). DPCM: Differential PCM.

| | | $Z =$ | 1 | 3 | 5 | 7 | 9 | 11 | 13 | 15 |
|---|---|---|---|---|---|---|---|---|---|---|
| | $w = 2$ | $M =$ | 0.60 | 2.20 | | | | | | |
| PCM | 3 | | 0.85 | 1.00 | 1.00 | 1.50 | | | | |
| | 4 | | 0.80 | 0.80 | 0.80 | 0.80 | 1.20 | 1.60 | 2.00 | 2.40 |
| | 2 | | 0.80 | 1.60 | | | | | | |
| DPCM | 3 | | 0.90 | 0.90 | 1.25 | 1.75 | | | | |
| | 4 | | 0.90 | 0.90 | 0.90 | 0.90 | 1.20 | 1.60 | 2.00 | 2.40 |

**Table 7.5** SNRs for quantization with $w = 4$ for the example given in Figure 7.13.

| Quantization | SNR (dB) | SNRseg (dB) |
|---|---|---|
| Uniform | 11.34 | 2.42 |
| A-law characteristic | 13.52 | 11.53 |
| $\mu$-law characteristic | 13.34 | 12.02 |
| Optimal quantizer | 18.34 | 7.64 |
| AQF | 19.49 | 18.26 |
| AQB | 20.15 | 19.41 |

Stepsize multipliers optimized for the adaptive quantization of speech signals are listed in Table 7.4 for speech (PCM). It should be noted that the optimization for prediction error signals in differential PCM (DPCM) gives somewhat different values [Jayant 1973].

In conclusion, the different quantization methods are represented comparatively for a short speech signal in Table 7.5 which shows the mean SNR and the segmental SNR[1] values obtained for the example from Figure 7.13. The form factor according to (7.28) is $F = 0.0177$ in this case (see also Table 7.1).

With SNR values of 13.5 or 13.3 dB, respectively, the quantizer for companding with the A-law or $\mu$-law characteristic provides distinctively better results in comparison to uniform quantization. Compared to the quantization process with companding, the fixed optimal quantizer adjusted to the PDF of this signal segment achieves a mean SNR, which is further

---

1 The segmental SNR SNRseg is defined as the average of the short-term SNR

$$\text{SNR}(k') = 10\lg\left(\frac{\hat{\sigma}_x^2(k')}{\hat{\sigma}_e^2(k')}\right),$$

where $\hat{\sigma}_x^2(k')$ and $\hat{\sigma}_e^2(k')$ are determined as short-term powers of the signal $x(k)$ and the quantization noise $e(k)$ for blocks of length $N$, while $k'$ denotes the block index.

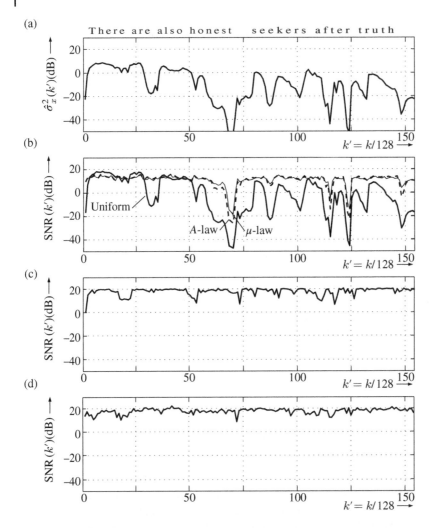

**Figure 7.13** Short-term power and SNR for various quantizers with $w = 4$. (a) Short-term power of the signal (Eq. (7.67), $N = 128$, $|x(k)|_{max} = 8$). (b) Blockwise computed SNR for uniform quantization and for companding with the $A$-law or $\mu$-law characteristic. (c) Blockwise computed SNR for AQF. (d) Blockwise computed SNR for AQB.

improved by approximately 5 dB. The best result is obtained by AQB with an SNR value of approximately 20 dB.

## 7.6 Vector Quantization

### 7.6.1 Principle

So far we have discussed *scalar quantization*. For the individual distributions of signal or parameter values $x$, the suitable quantization intervals and quantizer reproduction levels $\hat{x}_i$

were identified. This procedure can be generalized: $L$ values combined to an $L$-dimensional vector,

$$\mathbf{x} = (x_1, x_2, \ldots, x_L)^T,$$ (7.72)

are allocated to one of $K$ possible $L$-dimensional *quantization cells*, and are replaced by a corresponding *quantizer representation vector*

$$\hat{\mathbf{x}}_i = (\hat{x}_{i,1}, \hat{x}_{i,2}, \ldots, \hat{x}_{i,L})^T.$$ (7.73)

This procedure is called *vector quantization* (VQ) [Gersho, Gray 1992]. With $L = 1$ the scalar quantization is included as a special case. The allocation of $\mathbf{x}$ to a suitable quantization cell, the *Voronoi cell*, is addressed by the cell index $i$. The corresponding representation vector is indexed by $i$ in the *code book* consisting of $K$ *code vectors* $\hat{\mathbf{x}}_i$ (quantizer reproduction vector, $i = 1, 2, \ldots, K$).

In analogy to scalar quantization, vector quantization can be realized in the $L$-dimensional vector space with uniform as well as with non-uniform resolution. For the two-dimensional case, two vector quantizers with uniform and non-uniform resolution and $K = 25$ are depicted in Figure 7.14.

With a given code book, the vector quantization task is to replace an input vector $\mathbf{x}$ by the *most similar* vector $\hat{\mathbf{x}} = \hat{\mathbf{x}}_{i_{opt}}$. The choice is based on a distance or error measure $d(\mathbf{x}, \hat{\mathbf{x}})$, such that the condition

$$d(\mathbf{x}, \hat{\mathbf{x}}_{i_{opt}}) = \min_i d(\mathbf{x}, \hat{\mathbf{x}}_i)$$ (7.74)

is fulfilled. Thus, the boundaries of the Voronoi cells are implicitly determined.

Since, the code book is known at the receiver, only the code book index $i_{opt}$, not the quantized vector $\hat{\mathbf{x}}_{i_{opt}}$, is transmitted. The basic procedure is illustrated in Figure 7.15.

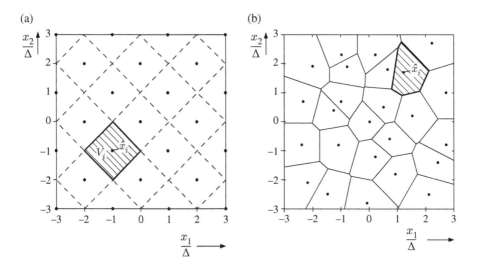

**Figure 7.14** Vector quantization: example with $K = 25$ vectors of dimension $L = 2$. (a) Uniform resolution ($D_2$-lattice). (b) Non uniform resolution.

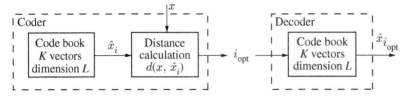

**Figure 7.15** Principle of vector quantization.

If the code book includes $K = 2^w$ vectors $\hat{\mathbf{x}}_i$ of dimension $L$, the selected index $i_{opt}$ and hence, indirectly, the chosen vector can be coded with

$$w = \mathrm{ld}\,(K) \text{ bits.} \tag{7.75}$$

With respect to a single component $x_\lambda$ of vector $\mathbf{x}$, an effective word length of

$$\overline{w} = \frac{\mathrm{ld}\,(K)}{L} \text{ [bits per component}\,x_\lambda] \tag{7.76}$$

results. With $K = 2^{10} = 1024$ and $L = 40$, which are typical dimensions in prediction error signal coding (see Section 8.5.3), only $1/4$ bit per value $x_\lambda$ has to be transmitted.

Regarding the choice of the distance measure $d(\mathbf{x}, \hat{\mathbf{x}}_i)$, different possibilities exist. Note that, the vector $\mathbf{x}$ may contain, e.g., either speech samples or some model-based codec parameters. If the speech signal is reconstructed from the quantized vectors $\hat{\mathbf{x}}$, the error vectors

$$\mathbf{e} = \hat{\mathbf{x}} - \mathbf{x} \tag{7.77}$$

affect the subjective speech quality in both cases differently. Therefore, different distance measures should be applied which should preferably correspond to the psychoacoustic perception.

For quantizing *signal* vectors, frequently the *squared error distortion measure* (Euclidian norm)

$$d(\mathbf{x}, \hat{\mathbf{x}}_i) = \frac{1}{L}(\mathbf{x} - \hat{\mathbf{x}}_i)^T(\mathbf{x} - \hat{\mathbf{x}}_i) \tag{7.78a}$$

$$= \frac{1}{L}\sum_{\mu=1}^{L}(x_\mu - \hat{x}_{i\mu})^2, \qquad i = 1, 2, \dots, K \tag{7.78b}$$

is minimized. This corresponds to selecting the nearest neighbor $\hat{\mathbf{x}}_i$ of $\mathbf{x}$ in the $L$-dimensional vector space.

Alternatively, the *weighted mean square error*

$$d(\mathbf{x}, \hat{\mathbf{x}}_i) = \frac{1}{L}(\mathbf{x} - \hat{\mathbf{x}}_i)^T \cdot \mathbf{W} \cdot (\mathbf{x} - \hat{\mathbf{x}}_i) \tag{7.79}$$

is applied, with $\mathbf{W}$ representing a symmetric, positive-definite matrix of dimension $L \times L$. The errors of the individual vector components can, for instance, be weighted differently by a diagonal matrix $\mathbf{W}$.

For quantizing coefficient sets of linear predictors, usually different distance measures are utilized, e.g., the Itakura–Saito distance [Itakura, Saito 1968], which is defined as

$$d(\mathbf{x}, \hat{\mathbf{x}}_i) = \frac{(\mathbf{x} - \hat{\mathbf{x}}_i)^T\,\mathbf{R}^{(n+1)}\,(\mathbf{x} - \hat{\mathbf{x}}_i)}{\mathbf{x}^T\,\mathbf{R}^{(n+1)}\,\mathbf{x}}. \tag{7.80}$$

Here, vector $\mathbf{x}$ of dimension $L = n + 1$ includes the non-quantized predictor coefficients $a_\lambda$ and vector $\hat{\mathbf{x}}_i$ the quantized representatives $\hat{a}_{i\lambda}$ according to

$$\mathbf{x} = (1, -a_1, -a_2, \ldots, -a_n)^T \tag{7.81a}$$

$$\hat{\mathbf{x}}_i = (1, -\hat{a}_{i1}, -\hat{a}_{i2}, \ldots, -\hat{a}_{in})^T. \tag{7.81b}$$

$\mathbf{R}^{(n+1)}$ denotes the squared auto-correlation matrix of dimension $(n + 1) \times (n + 1)$ of the signal segment for which the optimal predictor coefficients $a_\lambda$ have been computed.

### 7.6.2 The Complexity Problem

Vector quantization might be computationally very intensive, as the input vector $\mathbf{x}$ must be compared to all $K$ code vectors $\hat{\mathbf{x}}_i$ in order to minimize a distance measure according to (7.74). This is called *full search*.

The required computational effort shall be estimated for the squared error distortion measure. In the search of the nearest neighbor $\hat{\mathbf{x}}_i$ to input vector $\mathbf{x}$ in (7.78a) the division by $L$ can be omitted since it is constant for every code book entry. Per distance computation, $L$ differences, $L$ squares, and $(L - 1)$ additions must then be computed. This results in $3L - 1$ operations per vector $\hat{\mathbf{x}}_i$, i.e., in total

$$Op = (3L - 1)K \tag{7.82a}$$

$$= (3L - 1)2^{L\bar{w}}. \tag{7.82b}$$

The computational complexity increases exponentially with the effective word length $\bar{w}$ according to (7.76). For real-time implementations of VQ encoders with the sampling rate $f_s = \frac{1}{T}$, these operations must be performed in the time period

$$\tau = LT = \frac{L}{f_s}. \tag{7.83}$$

This leads to the computational complexity

$$CC = \frac{Op}{\tau} = \frac{3L - 1}{L} K f_s \approx 3K f_s. \tag{7.84}$$

For the typical dimensions $K = 1024$ and $f_s = 8\,\text{kHz}$, this results in

$$CC \approx 24.6\,\text{MOPS (Mega Operations Per Second)}. \tag{7.85}$$

Taking the computational capacity of state-of-the-art signal processors into account, this value is already a substantial load. This restricts the application of vector quantizers with full search. This example with $K = 1024$ and $L = 40$ (i.e., $\tau = 5\,\text{ms}$ at $f_s = 8\,\text{kHz}$) yields $\bar{w} = 0.25$ bits per sample $x(k)$. The SNR which can be achieved with such low bit rates depends, as will be shown below, on the statistical properties of the sequence $x(k)$.

Because of the complexity problem, the implementation of larger vector code books requires modifications which allow a reduction of the computational complexity. For this, fast search algorithms, e.g., with tree topology, structured code books, or cascaded vector quantizers, have been proposed (e.g., [Gray 1984], [Makhoul et al. 1985]), which, however, in general only provide suboptimal results. Theoretical bounds of hierarchical, i.e., cascaded, vector quantizers are discussed in [Erdmann 2004] and [Erdmann, Vary 2004].

### 7.6.3 Lattice Quantization

With respect to complexity, *lattice quantizers* – a special class of structured vector quantization encoders – are of particular interest. Their code vectors are given by regular grid or lattice points in the $L$-dimensional vector space.

The simple example of the $D_2$-*lattice* has already been depicted in Figure 7.14a. The positions of the code vectors can be analytically described so that there is no need to store a code book. Furthermore, fast algorithms which render a full search superfluous can be developed. As an example, the $D_L$-lattice will be discussed here. The points of this type of lattice fulfill the two conditions (see also Figure 7.14a) that all vector components are integer multiples of a smallest unit $\Delta$ and that additionally the sum of the components is an even multiple of $\Delta$:

$$\hat{x}_\mu = i\,\Delta; \qquad i \in \mathbb{Z} \tag{7.86a}$$

$$\sum_{\mu=1}^{L} \hat{x}_\mu = 2\,m\,\Delta; \qquad m \in \mathbb{Z}. \tag{7.86b}$$

Because of these conditions, the vector quantization can be reduced to simple-component, scalar rounding operations.

First, all components of the signal vector $\mathbf{x}$ are mathematically rounded to integer multiples of $\Delta$. If the resulting component sum is even, the quantized vector $\hat{\mathbf{x}}$ has already been found. If the component sum is odd, the component which shows the biggest rounding error is rounded in the "wrong direction." Thus, the condition of the even component sum is fulfilled.

The advantage of easy realization, however, is offset by the disadvantage that lattice quantizers are only optimal for uniformly distributed vectors $\mathbf{x}$. Lattice quantizers can be considered as vector generalizations of uniform scalar quantizers and can analogously be combined with companding.

A detailed presentation of the theory of lattice vector quantization can be found, for example, in [Conway, Sloane 1988].

### 7.6.4 Design of Optimal Vector Code Books

Optimal vector quantization is the $L$-dimensional generalization of the Lloyd–Max quantizer discussed in Section 7.4.

The scalar series $x(k)$ is replaced by the vector series $\mathbf{x}(k)$, which is described by the $L$-dimensional joint probability density $p_\mathbf{x}(\mathbf{u}) = p_\mathbf{x}(u_1, u_2, \dots, u_L)$. In analogy to (7.63), the $K$ code vectors $\hat{\mathbf{x}}_i$ must be chosen such that the expected error value

$$\mathrm{E}\,\{d(\mathbf{x}, \hat{\mathbf{x}})\} = \sum_{i=1}^{K} \int_{V_i} d(\mathbf{u}, \hat{\mathbf{x}}_i)\,p_\mathbf{x}(\mathbf{u})\mathrm{d}\mathbf{u} \tag{7.87}$$

becomes minimal.

The partial differentiation with respect to the representation vector $\hat{\mathbf{x}}_k$ provides a necessary condition in analogy to Section 7.4. When choosing the squared error distortion measure according to (7.78a), the optimal representation vectors $\hat{\mathbf{x}}_k$ correspond to the centers of gravity (*centroids*) of the Voronoi cells.

In general, a mathematical relation between the $K$ code vectors $\hat{\mathbf{x}}_i$ and the $L$-dimensional PDF cannot be formulated. But there is an elegant iterative design procedure, the *Linde–Buzo–Gray (LBG) algorithm* [Linde et al. 1980], which exists in two alternative versions (A) and (B).

### Algorithm (A)

Code book optimization (A) begins with a random start code book and improves this iteratively by means of *training vectors* $\mathbf{x}$, until the decrease of the average distortion is below a certain limit or has reached a minimum. The algorithm consists of the following steps:

Step 0: (a) Choose a start code book consisting of $K$ random vectors $\hat{\mathbf{x}}_i$ (or "uniform" lattice vectors) of dimension $L$.

(b) Set $m = 1$.

Step 1: (a) Quantize the training sequence $\mathbf{x}(k)$, $k = 1, 2, \ldots, K_T$, with $K_T \gg K$ and compute the average distortion

$$D_m = \frac{1}{K_T} \sum_{k=1}^{K_T} d\left(\mathbf{x}(k), \hat{\mathbf{x}}_{i_{\text{opt}}}\right). \tag{7.88}$$

(b) Terminate the iteration if the relative difference between $D_m$ and its previous value $D_{m-1}$ is sufficiently small

$$\frac{|D_{m-1} - D_m|}{D_m} < \epsilon. \tag{7.89}$$

Step 2: (a) Replace the old code vectors $\hat{\mathbf{x}}_i$ by the centroids of those training vectors $\mathbf{x}(k)$ which have been allocated to the old vectors $\hat{\mathbf{x}}_i$ (in generalization of (7.64d)).

(b) $m = m + 1$. Go to step 1.

The algorithm (A) will generally not deliver the best code book but find a local minimum of the quantization noise power. The choice of the start code book determines which local minimum will be achieved. This is why alternative procedures have been proposed.

### Algorithm (B)

The second design algorithm (B) starts with a single representation vector $\hat{\mathbf{x}}_1$, which is the centroid of $K_T$ training vectors $\mathbf{x}$. Within each iteration the code book is split and applied to algorithm (A) as start code book. The aim of the special "splitting" procedure is to obtain a better start code book for algorithm (A) in the *last* iteration. However, it will still not guarantee that the global minimum of the quantization noise power is achieved.

Step 1: Set $\kappa = 1$ and determine the center of gravity of all $K_T$ training vectors $\mathbf{x}$.

$\rightarrow \{\hat{\mathbf{x}}_i \mid i = 1\}$

Step 2: Split $\{\hat{\mathbf{x}}_i \mid i = 1, \ldots, \kappa\}$ using a small difference vector $\boldsymbol{\Delta}$, which is chosen arbitrarily.

$\rightarrow \{\hat{\mathbf{x}}_i - \boldsymbol{\Delta}; \ \hat{\mathbf{x}}_i + \boldsymbol{\Delta} \mid i = 1, \ldots, \kappa\}$

Step 3: (a) Run algorithm (A) with $\{\hat{\mathbf{x}}_i - \boldsymbol{\Delta}; \hat{\mathbf{x}}_i + \boldsymbol{\Delta} \mid i = 1, \ldots, \kappa\}$ as start code book resulting in $2\kappa$ optimized representation vectors.

$$\rightarrow \{\hat{\mathbf{x}}_i \mid i = 1, \ldots, 2\kappa\}$$

(b) Set $\kappa = 2\kappa$.

Step 4: (a) If $\kappa < K$ return to step 2,

(b) otherwise the optimized vector code book is obtained.

$$\rightarrow \{\hat{\mathbf{x}}_i \mid i = 1, \ldots, \kappa\}$$

The splitting algorithm (B) has a distinct advantage if the code book indices have to be transmitted over a channel with bit errors. For the transmission of any selected index a bit pattern of length $w$ has to be assigned ($K = 2^w$). A single bit error on the transmission link might produce the index of a different code book entry $\hat{\mathbf{x}}_j$, which might have a very large distance to the desired entry $\hat{\mathbf{x}}_i$. Robust index assignment can be achieved if we combine the index assignment with the splitting procedure. As soon as we split one vector we add another address bit. Thus, we can guarantee that a single bit error will produce an index corresponding to the neighborhood relations. In [Goertz 1999], Chapter 5, it is shown that this index assignment is nearly optimal for transmission with bit errors.

Examples of code book optimization with the LBG algorithm are presented in Figure 7.16. For a vector length of $L = 2$, code books with $K = 256$ vectors, respectively Voronoi cells, were designed. The first case (Figure 7.16a) is based on a non-correlated sequence $x(k)$ with Gaussian distribution, the second on a first-order Gauss–Markov source, i.e., an auto-regressive (AR) process (see also Section 6.1) with a first-order recursive filter. The feedback coefficient was $a = 0.9$. The code books were trained with 100 000 vectors each, with the termination criterion set to a value of $10^{-6}$ for the difference from one iteration to the next according to step 1 (b).

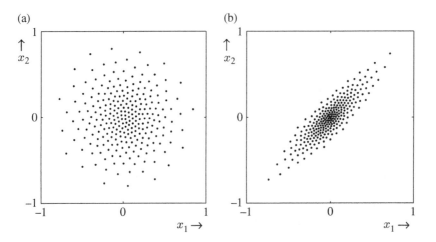

**Figure 7.16** Examples of code book designs with the LBG algorithm ($L = 2$, $K = 256$ or, respectively, $\overline{w} = 4$). (a) Gaussian source. (b) First-order Gauss–Markov source ($a = 0.9$).

**Table 7.6** SNRs for vector quantization with LBG code books, uncorrelated and first-order correlated Gaussian source ($a = 0.9$), 100 000 training vectors, $w = L\overline{w} = 10$.

| Vector dimension $L$ | 2 | 5 | 10 | 20 | 40 |
|---|---|---|---|---|---|
| Effective word length $\overline{w}$ | 5 | 2 | 1 | 0.5 | 0.25 |
| Gaussian source (dB) | 25.93 | 10.18 | 4.94 | 2.41 | 1.17 |
| Gauss–Markov source (dB) | 29.56 | 15.99 | 11.54 | 8.68 | 6.34 |

It can clearly be seen in Figure 7.16a that the cell sizes are adjusted to the PDF and that in Figure 7.16b the correlation leeds to a higher cell density in the neighborhood of the diagonal of the $(x_1, x_2)$-plane. This is also underlined by the SNR, which results in 20.87 dB for the Gaussian source and in 24.05 dB for the correlated Gauss–Markov source. Evidently, the vector quantizer is able to exploit the correlation for an improvement of the SNR.

For $L\overline{w} = 10$, Table 7.6 shows the SNRs obtained with LBG code books, with variations of the effective word length in the range of

$$0.25 \leq \overline{w} \leq 5. \tag{7.90}$$

Again, the uncorrelated Gaussian source and the correlated first-order Gauss–Markov source ($a = 0.9$) are compared.

The table shows that the vector quantizer exploits the correlation for an improvement of the SNR. Overall, however, for the given dimensions, relatively low SNR values are obtained.

For this reason, vector quantizers are generally not used for the direct signal quantization. Rather, they are utilized to quantize prediction error signals (see Section 8.5.1) or other sets of parameters within model-based codec algorithms.

### 7.6.5 Gain–Shape Vector Quantization

For the vector quantization discussed so far, we assumed that the code book consists of quantizer representation vectors which are representative signal waveforms or representative sets of parameters. In signal quantization, it may happen that the same signal shapes can occur with different amplitudes, e.g., if the volume of the speech signal is changed. Hence, the code book would have to contain vectors $\hat{x}_i$ with the same *shape* and a different *gain*. This might possibly lead to a considerable increase in the size and complexity of the code book.

One possible solution consists of normalizing each input vector by means of a scaling factor derived from $\mathbf{x}$ (e.g., the biggest vector component $x_\mu$). Per input vector, one additional scaling factor must then be transmitted.

Better results can be obtained if each code book vector $\hat{x}_i$ is adjusted to each respective input vector $\mathbf{x}$ with an individually optimized *gain factor* $g_i$ derived from $\mathbf{x}$ and $\hat{x}_i$. This method is termed *gain–shape* vector quantization.

Utilizing the squared error distortion measure, the gain factor $g_i$ for each arbitrary but fixed vector $\hat{\mathbf{x}}_i$ can be computed by minimizing the mean square error

$$d_i = d(\mathbf{x}, g_i\hat{\mathbf{x}}_i) = \frac{1}{L}\|\mathbf{x} - g_i\hat{\mathbf{x}}_i\|^2 \tag{7.91a}$$

$$= \frac{1}{L}\sum_{\mu=1}^{L}(x_\mu - g_i\hat{x}_{i\mu})^2 \overset{!}{=} \min, \qquad i = 1, 2, \dots, K. \tag{7.91b}$$

After setting the partial derivative with respect to the unknown $g_i$ to zero

$$\frac{\partial d_i}{\partial g_i} = -\frac{2}{L}\sum_{\mu=1}^{L}(x_\mu - g_i\hat{x}_{i\mu})\hat{x}_{i\mu} \overset{!}{=} 0, \tag{7.92}$$

the solution

$$g_{i,\text{opt}} = \frac{\sum_{\mu=1}^{L}x_\mu\hat{x}_{i\mu}}{\sum_{\mu=1}^{L}\hat{x}_{i\mu}^2} = \frac{\mathbf{x}^T\hat{\mathbf{x}}_i}{\|\hat{\mathbf{x}}_i\|^2} \tag{7.93}$$

results. If we insert the optimum gain factor (7.93) in (7.91a), the mean square error can be calculated explicitly for each code book vector $\hat{\mathbf{x}}_i$:

$$d_i = \frac{1}{L}\|\mathbf{x} - g_{i,\text{opt}}\hat{\mathbf{x}}_i\|^2 \tag{7.94a}$$

$$= \frac{1}{L}\left\|\mathbf{x} - \frac{\mathbf{x}^T\hat{\mathbf{x}}_i}{\|\hat{\mathbf{x}}_i\|^2}\hat{\mathbf{x}}_i\right\|^2 \tag{7.94b}$$

$$= \frac{1}{L}\left[\|\mathbf{x}\|^2 - \frac{(\mathbf{x}^T\hat{\mathbf{x}}_i)^2}{\|\hat{\mathbf{x}}_i\|^2}\right]. \tag{7.94c}$$

Hence, we could in a first step evaluate (7.94c) for each index $i$ and identify the best code book vector $\hat{\mathbf{x}}_i$, which minimizes this expression. As the squared norm of $\mathbf{x}$ is constant, (7.94c) can be minimized by maximizing the second term:

$$\frac{(\mathbf{x}^T\hat{\mathbf{x}}_i)^2}{\|\hat{\mathbf{x}}_i\|^2} \overset{!}{=} \max. \tag{7.95}$$

Then, in a second step, the corresponding optimum gain factor $g_{i,\text{opt}}$ has to be calculated according to (7.93) only for the selected best code book vector. Prior to transmission, this gain factor must be quantized.

## 7.7 Quantization of the Predictor Coefficients

The speech quality of vocoders using linear predictive coding (LPC) also depends on how precisely the spectral envelope of the speech signal is matched by the frequency response of the synthesis filter, especially in the neighborhood of the formant frequencies.

This accuracy is determined by three factors: the LP analysis algorithm (Chapter 6), the filter order, and finally, the quantization of the coefficients. Generally, the predictor coefficients are computed with relatively high precision, e.g., in fixed-point arithmetic with a word length of 16 bits. If the coefficients were directly transmitted as 16 bit numbers with a filter order of $n = 10$ and a block length of 20 ms, a bit rate of 160 bits/20 ms $=$ 8 kbit/s would be needed. Such a high accuracy is not required for the representation of the LP coefficients. They can be quantized at a significantly reduced bit rate. The effect of the quantization error on the frequency response and possibly on the stability of the synthesis filter strongly depends on the filter structure which is used. We can use various equivalent types, such as the direct structure, the lattice structure, the ladder structure, or the cascade of second-order filters. For the quantization of the filter coefficients, the scalar and vector methods discussed in this chapter can in principle be used. As regards the bit rate reduction, a variety of specific solutions can be found in the literature, which differ in terms of complexity and consider the statistical properties of the coefficients of the underlying filter type in different ways.

A detailed overview, from which the following numerical examples are taken, can be found in [Kleijn, Paliwal 1995].

In order to evaluate the quality of an LPC quantizer, an objective measure or a distance measure, which is preferably independent of the chosen filter structure, is required. A common measure is the *mean spectral distortion* of the logarithmic frequency response of the synthesis filter. If the frequency responses of the synthesis filter for non quantized and quantized coefficients are termed $H(e^{j\Omega})$ and $\hat{H}(e^{j\Omega})$, a spectral distance measure $SD$ can be determined for each single speech frame as follows:

$$SD = \sqrt{\frac{1}{2\pi} \int_{-\pi}^{\pi} \left[ 10 \lg |H(e^{j\Omega})|^2 - 10 \lg |\hat{H}(e^{j\Omega})|^2 \right]^2 \, d\Omega}. \tag{7.96}$$

Evaluating the spectral distortion $SD$ for all frames in the test data and computing its average value over many frames gives the mean spectral distance $\overline{SD}$. Transparency is given for $\overline{SD} \leq 1 \, \text{dB}$ (e.g., [Sugamura, Farvardin 1988], [Atal et al. 1989], [Kleijn, Paliwal 1995]).

The quantization methods to be discussed are of interest not only for the LPC vocoder, but also especially for the hybrid coding methods of Section 8.5.

### 7.7.1 Scalar Quantization of the LPC Coefficients

The coefficients $a_i$ of a predictor in direct form must be represented very precisely in order to guarantee the stability of the synthesis filter. Besides this, each coefficient has to be quantized with the same precision, because each coefficient shows a similar impact on the frequency response. Using individual scalar optimal quantizers (see Section 7.4) with 6 bits/coefficient $a_i$, i.e., 60 bits/frame corresponding to a bit rate of $B = 3$ kbit/s, a mean spectral distortion $\overline{SD}$ of 1.83 dB has been determined in [Kleijn, Paliwal 1995] for a representative speech database. For 25% of the frames, an unstable synthesis filter resulted. Consequently, this type of quantization is not used in practice.

### 7.7.2 Scalar Quantization of the Reflection Coefficients

The reflection coefficients $k_i$ of the acoustic tube model of speech production or of the corresponding digital lattice and ladder filters result from solving the normal equations (6.54-a), e.g., by means of the Levinson–Durbin algorithm (Section 6.4.1.3). An advantage of these structures is a guaranteed stability if the quantized parameters $\tilde{k}_i$ fulfill the condition

$$-1 < \tilde{k}_i < 1.$$

In addition, not all the coefficients have to be represented with the same precision. A non-uniform bit allocation is allowed, where, for instance, the first reflection coefficient $k_1$ is quantized with 6 bits and the last coefficient $k_{10}$ with only 2 bits. If individual optimal quantizers are adjusted to the PDF of the respective reflection coefficient, the contribution of each single quantization interval to the mean square quantization error is the same. The effects of the quantization errors on the spectral distortion according to (7.96), however, strongly depend on the actual value of the reflection coefficient. This effect can be described by a U-shaped spectral sensitivity curve with its maximal values at $k_i = \pm 1$ (e.g., [Makhoul, Viswanathan 1975]). For this reason, large values of $|k_i|$ should be quantized more accurately. This aim is achieved by means of a non-linear transformation of each coefficient.

Two appropriate transformations are the inverse sine

$$S_i = \frac{2}{\pi} \arcsin(k_i), \tag{7.97}$$

and the inverse hyperbolic tangent

$$L_i = \operatorname{arctanh}(k_i) \tag{7.98a}$$

$$= \frac{1}{2} \ln\left(\frac{1+k_i}{1-k_i}\right); \quad |k_i| < 1. \tag{7.98b}$$

Both transformation characteristics are depicted in Figure 7.17.

There is a direct relation between the reflection coefficient $k_i$ and the cross-sectional areas $A_i$ and $A_{i+1}$ of two successive tube segments as discussed in Section 2.4 (approach A):

$$k_i = \frac{A_i - A_{i+1}}{A_i + A_{i+1}}. \tag{7.98c}$$

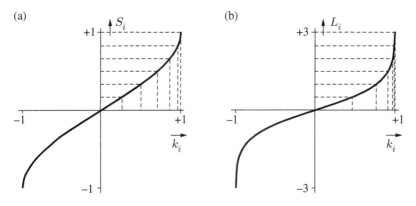

**Figure 7.17** Non-linear transformation of the reflection coefficients. (a) $S_i = \frac{2}{\pi} \arcsin(k_i)$. (b) $L_i = \frac{1}{2} \ln\left(\frac{1+k_i}{1-k_i}\right)$.

**Table 7.7** Scalar quantization of the reflection coefficients (according to [Kleijn, Paliwal 1995]).

| Coefficient | Mean spectral distortion [dB] $\overline{SD}$ | Bits/frame (10 coefficients) |
|---|---|---|
| $k_i$ | 1.02 | 34 |
| $S_i$ | 1.04 | 32 |
| $L_i$ | 1.04 | 32 |

Inserting (7.98c) in (7.98b) yields

$$L_i = \frac{1}{2} \ln \left( \frac{A_i}{A_{i+1}} \right). \tag{7.98d}$$

The transformed coefficient $L_i$ is proportional to the logarithm of the quotient of adjacent cross-sectional areas. Therefore, the coefficients $L_i$ are named *log area ratios* (LARs). On the other hand, the inverse-sine transformation offers with $S_i$ the *arcsine reflection coefficients* (ASRCs).

By means of uniform quantization of the transformed values $S_i$ or $L_i$, the desired better resolution for large values of $k_i$ is obtained. This represents a near-optimal but low-complexity alternative to optimal quantization. The second solution can be found in the *full-rate codec* of the Global System for mobile communication system (GSM, see also Section 8.7.1). Here, the logarithm is approximated by linear segments. The $n = 8$ coefficients $L_i$ are uniformly quantized with different word lengths of $w = 3-6$ (e.g., [Vary et al. 1988]). In total, the set of coefficients is coded with 36 bits per 20 ms resulting in a bit rate of 1.8 kbit/s.

Table 7.7 shows the number of bits which is required to achieve a mean spectral distortion of $\overline{SD} \approx 1$ dB with optimal quantization of $n = 10$ reflection coefficients or transformed coefficients, respectively. Obviously, the use of the LAR or ASRC representation provides a gain of 2 bit/frame in comparison to the reflection coefficients.

### 7.7.3 Scalar Quantization of the LSF Coefficients

The *line spectral frequencies* (LSFs) [Itakura 1975], [Soong, Juang 1984], [Sugamura, Itakura 1986], [Kabal, Ramachandran 1986] represent another method to ensure stability of the all-pole synthesis filter after LPC quantization. Furthermore, the LSF coefficients have some favorable properties, which can be exploited for quantization at low bit rates.

The basis of the LSF representation is a stability theorem for recursive digital filters and the decomposition of the polynomial

$$G(z) = 1 - \sum_{i=1}^{n} a_i z^{-i} \tag{7.99a}$$

$$= \sum_{i=0}^{n} \alpha_i z^{-i}, \tag{7.99b}$$

with

$$\alpha_0 = 1,$$ (7.100a)

$$\alpha_i = -a_i; \quad i = 1, 2, \ldots, n,$$ (7.100b)

into a mirror polynomial

$$P(z) = G(z) + z^{-(n+1)} G(z^{-1}),$$ (7.101a)

and an anti-mirror polynomial

$$Q(z) = G(z) - z^{-(n+1)} G(z^{-1}).$$ (7.101b)

The mirror and the anti-mirror properties are characterized by

$$P(z) = z^{-(n+1)} P(z^{-1}) \quad \text{and}$$ (7.102)

$$Q(z) = -z^{-(n+1)} Q(z^{-1}),$$ (7.103)

respectively. The polynomial $G(z)$ can be reconstructed as follows:

$$G(z) = \frac{1}{2}[P(z) + Q(z)].$$ (7.104)

Thus, $G(z)$ can also be described by the zeros of $P(z)$ and $Q(z)$. It can be shown that $G(z)$ is minimum phase and thus, the stability of the synthesis filter

$$H(z) = \frac{1}{G(z)}$$ (7.105)

is guaranteed [Itakura 1975], [Schüssler 1994], if the following hold:

- the zeros $z_{p_i}$ and $z_{q_i}$ of the polynomials $P(z)$ and $Q(z)$ are located on the unit circle of the $z$-plane

$$z_{p_i} = e^{j\omega_{p_i}} \quad z_{q_i} = e^{j\omega_{q_i}}; \quad i \in \{0, 1, \ldots, n\};$$

- the zero positions $\omega_{p_i}$ and $\omega_{q_i}$ of $P(z)$ and $Q(z)$, respectively, are interleaved on the unit circle, as shown in Figure 7.18 (see also (7.107)).

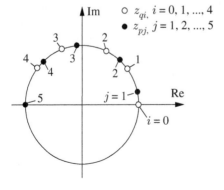

$\circ \; z_{qi}, \; i = 0, 1, \ldots, 4$
$\bullet \; z_{pj}, \; j = 1, 2, \ldots, 5$

**Figure 7.18** LSF parameters of an eighth order predictor.

These two properties can be exploited to represent the zero positions $\omega_{p_i}$ and $\omega_{q_i}$ ($i \in \{0, 1, \ldots, n\}$) of the LSF parameters with a relatively low bit rate.

First, we assume an even filter order $n$. $P(z)$ has a zero at $z = e^{j\pi} = -1$ and $Q(z)$ one at $z = e^{j0} = +1$, which are denoted by $\omega_{p_{\frac{n}{2}+1}} = \pi$ and $\omega_{q_0} = 0$. The locations of the remaining zeros depend on $G(z)$. These zeros occur in $\frac{n}{2}$ interleaved, complex conjugate pairs. The $2n + 2$ zeros of $P(z)$ and $Q(z)$ fulfill the following condition:

$$P(z): \quad 0 < \omega_{p_1} < \omega_{p_2} < \cdots < \omega_{p_{\frac{n}{2}}} < \omega_{p_{\frac{n}{2}+1}} = \pi \tag{7.106a}$$

$$Q(z): \quad 0 = \omega_{q_0} < \omega_{q_1} < \cdots < \omega_{q_{\frac{n}{2}}} < \pi. \tag{7.106b}$$

The complex conjugate zeros in the lower $z$-plane are not considered here, as they can be reconstructed from the others at the receiver. According to the relation

$$0 = \omega_{q_0} < \omega_{p_1} < \omega_{q_1} < \omega_{p_2} < \omega_{q_2} < \cdots < \omega_{q_{\frac{n}{2}}} < \omega_{p_{\frac{n}{2}+1}} = \pi, \tag{7.107}$$

the following notation is introduced:

$$0 = \omega_0 < \omega_1 < \omega_2 < \cdots < \omega_n < \omega_{n+1} = \pi. \tag{7.108}$$

Since the first value $\omega_{q_0} = \omega_0 = 0$ and the last value $\omega_{p_{\frac{n}{2}+1}} = \omega_{n+1} = \pi$ are fixed, $n$ LSF parameters $\omega_1, \ldots, \omega_n$ must be quantized. Instead of a straightforward scalar quantization of the LSF parameters, their sequential ordering can be exploited by scalar quantization of the differences of successive LSFs.

A more advanced vector quantization scheme which achieves a spectral distortion of only 1.04 dB for $n = 10$ and 28 bit/frame was proposed in [Xie, Adoul 1995]. The vector consisting of 10 LSF parameters $\omega_1 \ldots \omega_{10}$ is split into four groups:

$$\boldsymbol{\omega}_A = (\omega_3, \quad \omega_7)$$

$$\boldsymbol{\omega}_B = \left( \frac{\omega_1}{\hat{\omega}_3}, \quad \frac{\hat{\omega}_3 - \omega_2}{\hat{\omega}_3} \right)$$

$$\boldsymbol{\omega}_C = \left( \frac{\omega_4 - \hat{\omega}_3}{\hat{\omega}_7 - \hat{\omega}_3}, \quad \frac{\omega_5 - \hat{\omega}_4}{\hat{\omega}_7 - \hat{\omega}_3}, \quad \frac{\hat{\omega}_7 - \omega_6}{\hat{\omega}_7 - \hat{\omega}_3} \right)$$

$$\boldsymbol{\omega}_D = \left( \frac{\omega_8 - \hat{\omega}_7}{\pi - \hat{\omega}_7}, \quad \frac{\omega_9 - \hat{\omega}_8}{\pi - \hat{\omega}_7}, \quad \frac{\pi - \omega_{10}}{\pi - \hat{\omega}_7} \right).$$

Four different quantizers are used. For the reference vector $\boldsymbol{\omega}_A$, an LBG-trained vector quantizer with a code book of size 64 (6 bits) is used, which delivers $\hat{\omega}_3$ and $\hat{\omega}_7$. The normalized vectors $\boldsymbol{\omega}_B$, $\boldsymbol{\omega}_C$, and $\boldsymbol{\omega}_D$ are quantized with dedicated lattice quantizers with 5, 9, and 8 bits respectively, where the quantities $\hat{\omega}_4$ and $\hat{\omega}_8$ are obtained through a constraint of the lattice quantizer. For the details, the reader is referred to [Xie, Adoul 1995]. This quantizer is of practical significance as this concept has been adopted in various speech coding standards.

A further reduction of the bit rate for the LSF coefficients can be achieved by exploiting the interframe correlation of sets of coefficients. With a spectral distortion of approx. 1.2 dB, the approach in [Kataoka et al. 1996] needs 18 bit/frame, i.e., an average of only 1.8 bit/coefficient.

# References

Atal, B. S.; Cox, R.; Kroon, P. (1989). Spectral Quantization and Interpolation for CELP Coders, *Proceedings of the IEEE International Conference on Acoustics, Speech, and Signal Processing (ICASSP)*, Glasgow, Scotland, pp. 69–72.

Brehm, H.; Stammler, W. (1987). Description and Generation of Spherically Invariant Speech-Model Signals, *Signal Processing*, vol. 12, pp. 119–141.

Conway, J. H.; Sloane, N. J. A. (1988). *Sphere Packings, Lattices and Groups*, Springer-Verlag, New York.

Erdmann, C. (2004). *Hierarchical Vector Quantization: Theory and Application to Speech Coding*, PhD thesis. *Aachener Beiträge zu digitalen Nachrichtensystemen*, vol. 19, P. Vary (ed.), RWTH Aachen University.

Erdmann, C.; Vary, P. (2004). Performance of Multistage Vector Quantization in Hierarchical Coding, *European Transactions on Telecommunications*, vol. 15, no. 4, pp. 363–372.

Gersho, A.; Gray, R. M. (1992). *Vector Quantization and Signal Compression*, Kluwer Academic, Boston, Massachusetts, Dordrecht, London.

Goertz, N. (1999). *Aufwandsarme Qualitätsverbesserungen bei der gestörten Übertragung codierter Sprachsignale*, PhD thesis, University of Kiel (in German).

Gray, R. M. (1984). Vector Quantization, *IEEE ASSP Magazine*, vol. 1, no. 2, pp. 4–29.

Itakura, F. (1975). Line Spectral Representation of Linear Prediction Coefficients of Speech Signals, *Journal of the Acoustical Society of America*, vol. 57, no. 1, p. S35.

Itakura, F.; Saito, S. (1968). Analysis Synthesis Telephony Based on the Maximum Likelihood Method, *Proceedings of the 6th International Congress of Acoustics*, Tokyo, Japan.

ITU-T Rec. G.711 (1972). Pulse Code Modulation (PCM) of Voice Frequencies, International Telecommunication Union (ITU).

Jayant, N. S. (1973). Adaptive Quantization with a One Word Memory, *Bell System Technical Journal*, vol. 52, no. 7, pp. 1119–1144.

Jayant, N. S.; Noll, P. (1984). *Digital Coding of Waveforms*, Prentice Hall, Englewood Cliffs, New Jersey.

Kabal, P.; Ramachandran, R. P. (1986). The Computation of Line Spectral Frequencies Using Chebyshev Polynomials, *IEEE Transactions on Acoustics, Speech and Signal Processing*, vol. 34, no. 6, pp. 1419–1426.

Kataoka, A.; Moriya, T.; Hayashi, S. (1996). An 8 kb/s Conjugate Structure CELP (CS-CELP) Speech Coder, *IEEE Transactions on Speech and Audio Processing*, vol. 4(6), pp. 401–411.

Kleijn, W. B.; Paliwal, K. K. (eds.) (1995). *Speech Coding and Synthesis*, Elsevier, Amsterdam.

Linde, Y.; Buzo, A.; Gray, R. M. (1980). An Algorithm for Vector Quantizer Design, *IEEE Transactions on Communications*, vol. 28, no. 1, pp. 84–95.

Lloyd, S. P. (1982). Least Squares Quantization in PCM, *IEEE Transactions on Information Theory*, vol. 28, pp. 129–136.

Makhoul, J.; Viswanathan, R. (1975). Quantization Properties of Transmission Parameters in Linear Predictive Systems, *IEEE Transactions on Acoustics, Speech and Signal Processing*, vol. 23, no. 3, pp. 309–321.

Makhoul, J.; Roucos, S.; Gish, H. (1985). Vector Quantization in Speech Coding, *IEEE Proceedings*, vol. 73, no. 11, pp. 1551–1588.

Max, J. (1960). Quantizing for Minimum Distortion, *IRE Transactions on Information Theory*, vol. 6, pp. 7–12.

Schüssler, H. W. (1994). *Digitale Signalverarbeitung I*, Springer-Verlag, Berlin (in German).

Soong, F.; Juang, B. (1984). Line Spectrum Pair (LSP) and Speech Data Compression, *Proceedings of the IEEE International Conference on Acoustics, Speech, and Signal Processing (ICASSP)*, San Diego, California, USA, pp. 1.10.1–1.10.4.

Sugamura, N.; Farvardin, N. (1988). Quantizer Design in LSP Speech Analysis and Synthesis, *Proceedings of the IEEE International Conference on Acoustics, Speech, and Signal Processing (ICASSP)*, New York, USA, pp. 398–401.

Sugamura, N.; Itakura, F. (1986). Speech Analysis and Synthesis Methods Developed at ECL in NTT – From LPC to LSP, *Speech Communication*, vol. 5, pp. 199–215.

Vary, P.; Hellwig, K.; Hofmann, R.; Sluyter, R. J.; Galand, C.; Rosso, M. (1988). Speech Codec for the European Mobile Radio System, *Proceedings of the IEEE International Conference on Acoustics, Speech, and Signal Processing (ICASSP)*, New York, USA, pp. 227–230.

Xie, M.; Adoul, J. (1995). Fast and Low Complexity LSF Quantization Using Algebraic Vector Quantizer, *Proceedings of the IEEE International Conference on Acoustics, Speech, and Signal Processing (ICASSP)*, Detroit, Michigan, USA, pp. 716–719.

# 8

# Speech Coding

The main application of speech coding is telephony. In the traditional fixed and wireless telephone networks, speech signals are limited to the so-called *telephone bandwidth*, which causes the typical telephone sound of *narrowband (NB) telephony*. However, due to technological advances, more powerful coding techniques are meanwhile available, which allow significant improvements in terms of voice quality and speech intelligibility.

In this chapter, we will develop a comprehensive and unified description of the most important speech coding techniques for traditional NB telephony as well as for high quality telephony. The basic concepts, such as LPC *vocoder*, *differential pulse code modulation* (DPCM), and *code-excited linear prediction* (CELP), as well as advanced coding standards, will be discussed.

## 8.1 Speech-Coding Categories

The minimum demands on the telephone bandwidth were specified in the CCITT Red Book from 1961: at the lower and the upper cutoff frequencies of $f_L = 300\,\text{Hz}$ and $f_U = 3.4\,\text{kHz}$, the analog filters in the transmission path may attenuate the signal by no more than $10\,\text{dB}$ with regard to the level at the frequency of $800\,\text{Hz}$ [ITU-T Rec. G.132 1988], [ITU-T Rec. G.151 1988] as shown in Figure 8.1a.

This specification of the *telephone bandwidth* dates back to the analog age when long range transmission was based on frequency division multiplexing (FDM) with a frequency grid of $4\,\text{kHz}$.

By using single side band modulation, e.g., $10\,800$ telephone channels could be transmitted on a coaxial cable or a high-frequency radio-relay link. The frequency spacing of $4\,\text{kHz}$ was specified to realize as many as possible telephone channels within the given overall transmission bandwidth of the FDM system, while maintaining a reasonable speech quality and intelligibility. The specification of the lower limit of $300\,\text{Hz}$ was due to the need of channel separation by bandpass filtering and due to sub-audio (below $300\,\text{Hz}$) signalling tones. The selection of the cutoff frequencies of the *telephone bandpass* was based on subjective listening tests. According to this specification, the intelligibility of (meaningless) syllables is about 91% and the comprehension of sentences is about 99% [Schmidt, Brosze 1967], [Brosze et al. 1992], [Terhardt 1998]. However, listening experiments have shown that a certain increase of the acoustic bandwidth significantly improves not only the

*Digital Speech Transmission and Enhancement*, Second Edition. Peter Vary and Rainer Martin.
© 2024 John Wiley & Sons Ltd. Published 2024 by John Wiley & Sons Ltd.

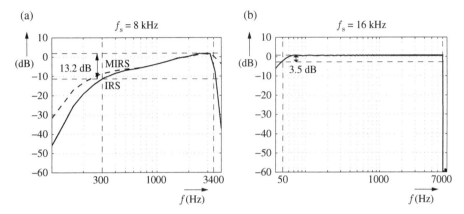

**Figure 8.1** (a) Magnitude response of the *intermediate reference systems* (IRS), and the *modified intermediate reference systems* (MIRS) for $f_s = 8$ kHz [ITU-T Rec. P.48 1993], [ITU-T Rec. P.830 1996] and (b) Magnitude response for wideband handsfree telephony terminals for $f_s = 16$ kHz [ITU-T Rec. P.341 1998].

perceived speech quality [Krebber 1995], [Voran 1997] but also the intelligibility of, for example, unvoiced and fricative speech sounds.

Since then, the public telephone networks have been converted almost completely to digital transmission techniques. For compatibility reasons, the speech signals are sampled at a sampling rate of $f_s = 8$ kHz and the samples are quantized using non-linear companding according to the A-law or $\mu$-law characteristic (see Section 7.3) with 8 bits per sample, resulting in the bit rate of 64 kbit/s. This has been specified in the international standards [ITU-T Rec. G.711 1972] and [ITU-T Rec. G.712 1988].

For the early cellular phone systems, a further limitation of the frequency range was specified in order to reduce the amount of disturbing low-frequency background noise. In the GSM system, for instance, the sensitivity of both the sending and the receiving terminal should provide an attenuation of at least 12 dB below 100 Hz [ETSI Rec. GSM 03.50 1998].

As we know from everyday life, the intelligibility of telephone speech seems to be sufficient, at least for a normal conversation, although the fundamental frequency of speech is not transmitted in most cases. We become aware of the limited intelligibility of syllables if we are forced to understand unknown words or names. In these cases we often need to spell a word, especially to distinguish between certain unvoiced or plosive phones, such as /f/ and /s/ or /p/ and /t/.

A second restriction of the subjectively perceived speech quality is that some speaker-specific features are not transmitted over the telephone. Speaker transparency is frequently limited.

These restrictions can be overcome with the introduction of *wideband (WB) telephony*, which is characterized by the extended frequency band from 50 Hz to 7.0 kHz. Then, the sampling frequency has to be increased from $f_s = 8$ kHz to $f_s' = 16$ kHz, and the required resolution for A/D conversion has to be improved from $w = 8$ bits (non-uniform A-law or $\mu$-law) to $w' = 14$ bits (uniform) per sample. The network operators' marketing term for WB speech is HD voice. In what follows, conventional telephone speech will be called *NB speech* (as opposed to *WB speech*).

**Table 8.1** Speech coding categories.

| Category | | Bandwidth | Sample rate |
|---|---|---|---|
| NB: | Narrowband | 0.20 – 3.4 kHz | $f_s = 8$ kHz |
| WB: | Wideband | 0.05 – 7.0 kHz | $f_s = 16$ kHz |
| SWB: | Super wideband | 0.05 – 14.0 kHz | $f_s = 32$ kHz |
| FB: | fullband | 0.02 – 20.0 kHz | $f_s = 48$ kHz |

While in fixed digital networks, NB speech signals are predominantly sampled at a rate of $f_s = 8$ kHz and quantized non-uniformly at 64 kbit/s ($A$-law or $\mu$-law characteristic, $w = 8$ bit/sample, see Section 7.3), this bit rate is not available in mobile networks.

Moreover, there is an increasing demand for a wider frequency band to improve the syllable intelligibility and the listening comfort in voice and video calls.

For speech coding, the bandwidth and quality categories have been specified, as shown in Table 8.1.

For any kind of signal processing, such as coding, the analog signal has to be frequency-limited by some pre-filter as part of analog-to-digital (ADC) conversion and some post-filter as part of digital-to-analog (DAC) conversion. Figure 8.1 shows the corresponding frequency characteristics for NB and WB coding ([ITU-T Rec. P.48 1993], [ITU-T Rec. P.830 1996]). A minimum stopband attenuation of at least 25 dB has to be achieved above 4.6 kHz. At the low frequencies, there are no hard restrictions, allowing to reduce the lower frequency limit to, e.g., 200 Hz.

The typical target bit rates for coding of NB and WB speech in mobile networks are in the range of $B = \overline{w} \cdot f_s = 4$–32 kbit/s, i.e., on the average $0.5 \leq \overline{w} \leq 2$ bits per speech sample.

The majority of speech-coding algorithms is based on *linear prediction* with an underlying *model of speech production* (see Chapter 2) and implicitly on *masking properties* of the auditory system.

In contrast, coding schemes for music signals are explicitly based on models of the human auditory system, since appropriate source models do not exist. High audio quality with a bandwidth of about 16 kHz (i.e., $f_s = 32$ kHz) can be obtained with effectively $\overline{w} = 2$–4 bit/sample or $B = \overline{w} \cdot f_s = 64$–192 kbit/s.

Speech-coding algorithms have to be optimized with regard to the following requirements:

- high speech quality
- low bit rate
- low complexity
- limited signal delay.

These criteria that partly contradict each other must be balanced according to application requirements. They are highly interrelated, as, for example, a constant speech quality may be obtained at a reduced data rate but at the expense of increased complexity and/or increased signal delay.

Speech-coding algorithms can be subdivided into three main categories:

(a) Waveform coding: reproduction of signal samples
(b) Vocoders: model based coding
(c) Hybrid coding: Mixture of (a) and (b).

In *waveform coding* [Jayant, Noll 1984], bit rate reduction is achieved through (fixed or adaptive) quantization of time domain samples of the speech signal or some intermediate signal(s), such as a prediction error signal or subband signals. In comparison to plain quantization of speech samples, better results are achieved by applying a (fixed or adaptive) linear predictive filter (LP filter), which is adjusted according to the correlation properties of the signal. The LP filter can be considered to be a whitening filter (see Section 6.3). The resulting reduction of the signal dynamics in the time and the frequency domain can be quantified by the prediction gain. In time domain coding algorithms, quantization and predictive filtering are generally adaptive processes. At the receiver end, the signal is reconstructed by applying the (quantized) residual signal to the synthesis filter. Both the synthesis filter as well as the inverse quantizer can be adjusted by backward or closed-loop adaptation. No side information about the quantizer and the synthesis filter needs to be transmitted. Only the quantized residual signal is transmitted. The effective quantization error can be spectrally shaped within certain limits in order to maximize the subjective speech quality, exploiting the psychoacoustic masking effect (see Section 8.3.3).

For $f_s = 8$ kHz and a target bit rate of $B = 32$ kbit/s, i.e., 4 bit per sample, the predominant waveform coding scheme is *adaptive differential pulse code modulation* (ADPCM), which allows a reconstruction of the signal waveform with a signal-to-noise ratio of SNR $= 30$–$35$ dB. ADPCM (see Section 8.3.4) is used in digital cordless phones and in circuit multiplication equipment (two voice channels within 64 kbit/s).

In contrast to waveform coding, *parametric vocoders (vocoders)* do not encode the waveform, but a set of model parameters. This implies the realization of a speech production model as discussed in Section 6.1. The time-variant synthesis filter in the receiver can be interpreted as a model of the vocal tract. Its excitation signal (glottis signal) is delivered from a controlled generator. On the transmission side, the parameters of this model, i.e., the coefficients of the filter and the control parameters of the generator, are extracted from the speech samples by analysis procedures and transmitted in quantized form. At a typical bit rate of $B = 2.4$ kbit/s, parametric coders produce a clearly intelligible but somewhat synthetic speech.

An interesting compromise between these two concepts is the *hybrid coding* approach. Like in the vocoder, the parameters of a time-variant synthesis filter are transmitted as side information, whereas the excitation signal is generated similarly to waveform coding by the quantized prediction error signal (residual signal). In consideration of the auditory system, it is possible to quantize the residual signal quite coarsely with respect to amplitude and time resolution. However, large quantization errors do not permit a derivation of the synthesis filter coefficients from the quantized residual signal at the receiver. Subjectively, telephone quality can almost be achieved with 0.75–1.5 bit/sample, whereas the measurable signal-to-noise ratio might only be 10 dB. The hybrid approach is widely used in digital mobile radio systems, e.g., [Goldberg, Riek 2000], [Hanzo et al. 2001], [Chu 2003], [Kondoz 2004], and [Sluijter 2005] (see Section 8.5).

A common feature of the three coding schemes is the time-varying synthesis filter, which more or less approximates the vocal tract transfer function. The necessary processing at the transmitter is based on linear prediction (LP). Therefore, the generic term *linear predictive coding* (LPC) is widely used for any codec of the three classes. The attribute LP is often used to characterize special variants of the hybrid approach, such as CELP codecs (*code excited linear prediction*, see Section 8.5.3).

An alternative classification distinguishes between time-domain algorithms using linear prediction and frequency-domain algorithms based on short-term spectral analysis. The frequency domain algorithms, which rely more on auditory models, require at least 2 bit/sample and are especially suitable for music signals (e.g., [Rao, Hwang 1996], [Brandenburg, Bosi 1997], [MPEG-2 1997], [Vary et al. 1998]). Due to the underlying model of speech production, the predictive time-domain algorithms are especially suitable for speech coding with effective bit rates of less than 2 bit/sample.

## 8.2 Model-Based Predictive Coding

For speech coding with medium to low bit rates, i.e., with effective word lengths of

$$\overline{w} = \frac{B}{f_s} \le 2\,\text{bit/sample}\,,$$

model-based time-domain methods are widely used [Atal 1982]. The international codec standards for telecommunications almost exclusively rely on the simplifying model of speech production derived in Section 2.4. Some properties of the auditory system are exploited as well, especially spectral masking of quantization errors (see Section 2.5).

According to the classification given in Section 8.1, a common feature is an LP-analysis filter at the transmitter and a synthesis filter at the receiver end. For telephone applications, the most interesting classes are waveform coding and hybrid coding. Both concepts are based on the model of speech production, as illustrated in Figure 8.2.

The transmitter processes speech samples $x(k)$, which are considered to be produced by an autoregressive process (AR process). It is assumed that these samples have been generated from an excitation sequence $v(k)$ by a purely recursive time-variant filtering process (vocal tract filter).

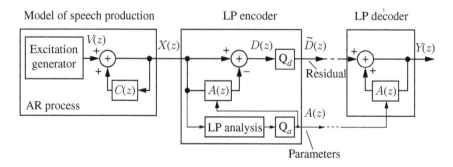

**Figure 8.2** Basic principle of model-based predictive coding.

In Chapter 6, it was shown that, if the model is strictly valid, optimal linear prediction in terms of the minimum mean square prediction error implicitly performs a system identification of the vocal tract filter. Hence, according to Figure 8.2, the $z$-transform $D(z)$ of the residual signal $d(k)$ is described by

$$D(z) = \frac{1 - A(z)}{1 - C(z)} V(z). \tag{8.1}$$

For perfect system identification, i.e., for $A(z) = C(z)$, the residual signal $d(k)$ is identical to the excitation signal $v(k)$ of the model filter. The analysis filter on the transmission side with the transfer function

$$G(z) = 1 - A(z) \tag{8.2}$$

is the inverse of the vocal tract filter.

The residual signal $d(k)$ is transmitted to the receiver in quantized form and is then used as the excitation signal for the synthesis filter

$$H(z) = \frac{1}{1 - A(z)}, \tag{8.3}$$

which corresponds to the vocal tract filter. Thus, the speech signal is resynthesized at the receiving side (decoding) according to the speech production model.

However, we must assume that

(1) the utilized model of speech production is only approximately valid;
(2) the estimated filter parameters of $A(z)$ do not perfectly match those of $C(z)$.

Nevertheless, with this approach the bit rate can be efficiently reduced. Even for inaccurate estimates of the filter parameters the frequency responses of the filters on the transmitting and receiving sides are exactly inverse to each other:

$$G(z) \cdot H(z) = 1. \tag{8.4}$$

If we do not quantize the residual signal $d(k)$, the output signal and the input signal are identical:

$$y(k) = x(k). \tag{8.5}$$

There is no signal delay. The key to bit rate reduction is that rather coarse quantization can be applied to the residual signal $d(k)$. If the quantized residual signal

$$\tilde{d}(k) = d(k) + \Delta(k) \tag{8.6}$$

is used to re-synthesize the speech signal, the output signal $y(k)$ consists of the original $x(k)$ and a filtered version of the quantization noise:

$$y(k) = x(k) + r(k). \tag{8.7}$$

The effective quantization noise $r(k)$ has (almost) the same spectral shape as the signal $x(k)$. Therefore, subjective perception is significantly improved by the psychoacoustic effect of masking.

The approach illustrated in Figure 8.2 covers most of the predictive time-domain concepts for speech coding. However, essential differences exist regarding the quantization of the residual signal (fixed or adaptive, scalar or vector quantization, error criterion) and the prediction type (sequential or block adaptation, with or without long-term prediction).

## 8.3 Linear Predictive Waveform Coding

These codecs use at the transmitter a linear prediction *analysis filter* (LP analysis filter) $G(z) = 1 - A(z)$ and at the receiver end a linear prediction *synthesis filter* (LP synthesis filter) $H(z) = \frac{1}{1-A(z)}$. For historical reasons, often the technical term *DPCM* (Differential Pulse Code Modulation) is used in contrast to PCM.

### 8.3.1 First-Order DPCM

According to the classification given in Section 8.1, the adaptive version of DPCM, $ADPCM$, can be attributed to the class of waveform coders.

The objective of linear prediction is bit rate reduction. The analog signal $x_a(t)$ will be digitized with the sampling rate $f_s$ and a sufficient word length $w_0 \geq 12$. Thus, the initial bit rate is $B_0 = w_0 \cdot f_s$. With linear prediction, we generate the residual signal $d(k)$. The subsequent quantization of $d(k)$ with shorter word length $w < w_0$ will lead to the reduced bit rate of $B = w \cdot f_s$.

The simplest version is a *DPCM system* with a first-order predictor as depicted in Figure 8.3. The current signal sample $x(k)$ is predicted by weighting the preceding $x(k-1)$ with coefficient $a$:

$$\hat{x}(k) = a \cdot x(k-1). \tag{8.8}$$

The predictor coefficient $a$ can be optimized as described in Chapter 6 by means of block adaptation (Section 6.3.1) or sequential adaptation using, for example, the LMS algorithm (Section 6.3.2). Here, we will first look at the block adaptive approach, which is also referred to as *adaptive predictive coding* (APC). The coefficient $a$, which is optimal in terms of the minimum mean square prediction error (see Eq. (6.31)), is

$$a_{opt} = \frac{\varphi_{xx}(1)}{\varphi_{xx}(0)}. \tag{8.9}$$

In practice, the auto-correlation values $\varphi_{xx}(i)$ are replaced by short-term estimates. The effect of a DPCM system with a first-order predictor which is optimal for a voiced signal segment of block length $N = 160$ ($N \cdot T = 20$ ms) is depicted in Figure 8.4. Compared to the input signal $x(k)$, the dynamic range of the residual signal $d(k)$ is distinctively reduced. The corresponding prediction gain (see Section 6.2)

$$G_p = \frac{\varphi_{xx}(0)}{\varphi_{dd}(0)} = \frac{\varphi_{xx}^2(0)}{\varphi_{xx}^2(0) - \varphi_{xx}^2(1)} \geq 1 \tag{8.10}$$

can be exploited for shortening the word length to $w < w_0$ under certain circumstances as shown in Section 8.3.3.

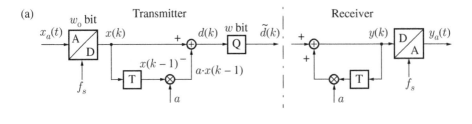

(a)

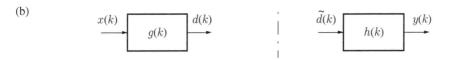

(b)

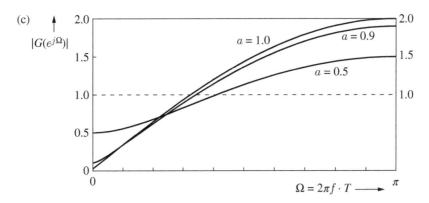

(c)

**Figure 8.3** Differential pulse code modulation (DPCM) of first order. (a) Block diagram, (b) Equivalent filters, (c) Magnitude response of the analysis filter ($f_s = 1/T$).

First, we will analyze the system performance in the frequency domain for the special case when no additional quantization is applied, i.e., $w = w_0$.

For a constant coefficient $a$ (constant for the duration of the signal block), the impulse responses can be determined for the transmitter and the receiver as suggested in Figure 8.3:

$$g(k) = \begin{cases} 1 & k = 0 \\ -a & k = 1 \\ 0 & \text{otherwise} \end{cases} \tag{8.11a}$$

$$h(k) = \begin{cases} a^k & k \geq 0 \\ 0 & k < 0. \end{cases} \tag{8.11b}$$

Through $z$-transformation of these impulse responses, we obtain the frequency responses

$$G(e^{j\Omega}) = 1 - a \cdot e^{-j\Omega} \tag{8.12a}$$

or

$$|G(e^{j\Omega})| = \sqrt{1 + a^2 - 2a \cos \Omega}, \tag{8.12b}$$

and for $|a| < 1$

$$H(e^{j\Omega}) = \sum_{k=0}^{\infty} a^k \cdot e^{-jk\Omega} \tag{8.13a}$$

$$= \frac{1}{1 - a \cdot e^{-j\Omega}} = \frac{1}{G(e^{j\Omega})} . \tag{8.13b}$$

Since, the filter at the receiving side is inverse to the filter on the transmitting side, the output signal $y(k)$ and the input signal $x(k)$ are identical if $d(k)$ is not quantized, i.e., if $\tilde{d}(k) = d(k)$.

The magnitude response of the transmission filter for $a = 1$ is of special interest:

$$|G(e^{j\Omega})| = \sqrt{2 \cdot (1 - \cos \Omega)} \tag{8.14a}$$

$$= 2 \cdot \left| \sin \left( \frac{\Omega}{2} \right) \right| . \tag{8.14b}$$

The magnitude response outlined in Figure 8.3c for $a = 1$ is almost linear at low frequencies. This behavior approximates the magnitude response of the differentiator. The extreme values of the magnitude response are

$$|G(e^{j \cdot 0})| = |1 - a| \quad \text{and} \quad |G(e^{j \cdot \pi})| = |1 + a|.$$

The logarithmic prediction gain of the example in Figure 8.4 is

$$10 \lg(G_p) = -10 \lg(1 - a_{opt}^2) \approx 10.5 \, \text{dB}.$$

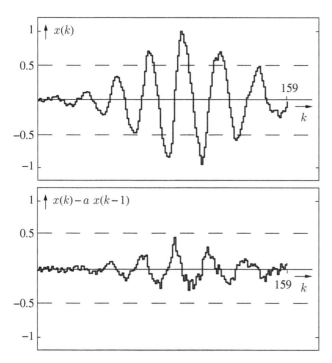

**Figure 8.4** Example of block adaptive DPCM ($N = 160, f_s = 8$ kHz, $a = 0.9545$).

If the achievable bit rate reduction increases with the prediction gain according to the 6-dB-per-bit rule, about 1.5 bit/sample could be saved for a gain of 10.5 dB compared to PCM. This interdependency will be studied in Section 8.3.3.

### 8.3.2 Open-Loop and Closed-Loop Prediction

So far we have discussed the simple DPCM system with a single coefficient $a$. In Chapter 6, it was shown that the prediction gain can be improved by increasing the filter order $n$ and by adaptation of the predictor coefficients $a_i(k)$ $(i = 1, \ldots, n)$ to the time-varying characteristics of speech.

The structure of the DPCM system with an $n$-th order prediction filter is shown in Figure 8.5. As the prediction signal

$$\hat{x}(k) = \sum_{i=1}^{n} a_i(k) \cdot x(k - i) \tag{8.15}$$

is produced from preceding input samples $x(k - i)$, this structure is called *forward-prediction* or *open-loop prediction*.

The time-variant coefficients of the transmitter must be known at the receiver end. However, the transmission of the prediction coefficients requires a considerable part of the bit rate saved by quantization of $d(k)$.

Therefore, it would be attractive if the vector $\mathbf{a}(k) = \left(a_1(k), a_2(k), \ldots, a_n(k)\right)^T$ of the predictor coefficients could be re-calculated at the receiver without transmitting any *side information*. This can be achieved with the modified predictor structure of Figure 8.6

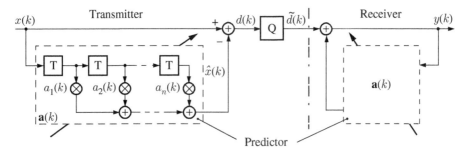

**Figure 8.5** DPCM system with *open-loop prediction* of *n*-th order.

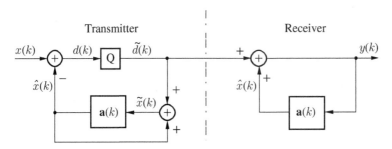

**Figure 8.6** DPCM system with adaptive backward predictor (*closed-loop prediction*).

(see Section 8.3.4), if the residual signal $d(k)$ is quantized with sufficient accuracy. The modification compared to Figure 8.5 is that the estimated signal $\hat{x}(k)$ at the transmitter is derived in the same way as at the receiver. This variation is called *backward-prediction* or *closed-loop prediction*, as the predictor and the quantizers are located within a loop.

If the residual signal $d(k)$ is not quantized, then the open- and closed-loop structures produce exactly the same estimated signal $\hat{x}(k)$. In this case, we get

$$\tilde{d}(k) = d(k) = x(k) - \hat{x}(k), \tag{8.16a}$$

i.e.,

$$x(k) = d(k) + \hat{x}(k) \tag{8.16b}$$

and

$$\tilde{x}(k) = \tilde{d}(k) + \hat{x}(k) = x(k). \tag{8.16c}$$

Hence, the estimated signal for closed-loop prediction

$$\hat{x}(k) = \sum_{i=1}^{n} a_i(k) \cdot \tilde{x}(k-i) = \sum_{i=1}^{n} a_i(k) \cdot x(k-i) \tag{8.16d}$$

is identical to the result produced by open-loop prediction (8.15) (see Figure 8.5). Furthermore, with closed-loop prediction the decoder output signal $y(k)$ is identical to the intermediate signal $\tilde{x}(k)$ even if the residual signal $\tilde{d}(k)$ is quantized. However, if the residual signal $d(k)$ is quantized, open- and closed-loop prediction produce different signals $\hat{x}(k)$. Thus, in both cases quantization affects the output signal $y(k)$ differently. This will be discussed in the next Section 8.3.3.

### 8.3.3 Quantization of the Residual Signal

In this section, we will show how the quantization of the residual signal $d(k)$ affects the output signal $y(k)$ and how the prediction gain can be exploited for bit rate reduction.

#### 8.3.3.1 Quantization with Open-Loop Prediction

The quantization of the residual signal $d(k)$ will be described by additive quantization noise $\Delta(k)$. Assuming uniform quantization with mathematical rounding, the quantization error can be expressed (see Section 7.2) as uniformly distributed noise in the range

$$-\frac{\Delta d}{2} < \Delta(k) \le +\frac{\Delta d}{2}, \tag{8.17a}$$

with the power

$$\sigma_\Delta^2 = \frac{(\Delta d)^2}{12}, \tag{8.17b}$$

and a constant noise power spectral density (*white noise*)

$$\Phi_{\Delta\Delta}(e^{j\Omega}) = \frac{(\Delta d)^2}{12}. \tag{8.17c}$$

Here, $\Delta d$ denotes the quantizer's stepsize.

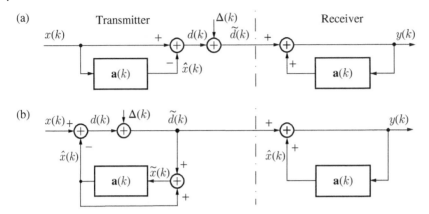

**Figure 8.7** DPCM with quantization of the residual signal. (a) Open-loop prediction (forward prediction) and (b) Closed-loop prediction (backward prediction).

According to Figure 8.7a, the quantized residual signal $\tilde{d}(k)$ consists of the two components $d(k)$ and $\Delta(k)$. Due to the linearity of the receiving filter, the output signal $y(k)$ consists of the sum of the filtered versions of these two components. Since the transmitting and the receiving filters are inverse to each other, we get the original signal $x(k)$ and a filtered version $r(k)$ of the noise $\Delta(k)$ according to

$$y(k) = x(k) + r(k). \tag{8.18}$$

The *reconstruction error* $r(k)$ is thus a spectrally shaped version of the white quantization noise $\Delta(k)$. As shown in Section 6.1, the optimization of the analysis filter implies a system identification of the vocal tract. With adaptive linear prediction, the frequency response of the synthesis filter, therefore, approximates the instantaneous frequency response of the vocal tract filter. For this reason, the spectrum of the reconstruction error follows the spectral envelope of the speech signal. From a psychoacoustic point of view, this is advantageous as the quantization error is masked to a certain extent by the speech signal itself.

In this context, the significance of the prediction gain for the achievable signal-to-noise ratio and the required bit rate is of particular interest. In order to clarify this issue, we will look at the system in the frequency domain. We assume that the prediction is perfect in such a way that a spectrally flat residual signal $d(k)$ with constant power spectral density

$$\Phi_{dd}(e^{j\Omega}) = \text{const.} = \varphi_{dd}(0) \tag{8.19}$$

results. This assumption holds, as shown in Section 6.1, for unvoiced segments. With respect to the spectral envelope, and especially when utilizing a long-term predictor (see Section 6.5, Figure 6.17), this assumption applies to voiced segments as well.

Furthermore, the relation between the power spectral densities of the residual signal and the input signal is given by

$$\Phi_{dd}(e^{j\Omega}) = \Phi_{xx}(e^{j\Omega}) \cdot |1 - A(e^{j\Omega})|^2. \tag{8.20a}$$

With (8.19) we get

$$\frac{\varphi_{dd}(0)}{|1 - A(e^{j\Omega})|^2} = \Phi_{xx}(e^{j\Omega}). \tag{8.20b}$$

The integration over the frequency interval $-\pi \leq \Omega \leq +\pi$ with

$$\varphi_{dd}(0)\frac{1}{2\pi}\int_{-\pi}^{\pi}\frac{1}{|1-A(e^{j\Omega})|^2}\,d\Omega = \frac{1}{2\pi}\int_{-\pi}^{\pi}\Phi_{xx}(e^{j\Omega})\,d\Omega = \varphi_{xx}(0) \tag{8.20c}$$

yields the power $\varphi_{xx}(0)$ of signal $x(k)$. Dividing by $\varphi_{dd}(0)$ leads to the prediction gain

$$\frac{\varphi_{xx}(0)}{\varphi_{dd}(0)} = \frac{1}{2\pi}\int_{-\pi}^{\pi}\frac{1}{|1-A(e^{j\Omega})|^2}\,d\Omega = G_p. \tag{8.21}$$

Equation (8.21) integrates the squared magnitude response

$$|H(e^{j\Omega})|^2 = \frac{1}{|1-A(e^{j\Omega})|^2} \tag{8.22}$$

of the synthesis filter.

The signal-to-noise ratio of the uniform quantizer is given by

$$\left(\frac{S}{N}\right)_d = \frac{\varphi_{dd}(0)}{\frac{(\Delta d)^2}{12}}. \tag{8.23}$$

The noise power $\varphi_{rr}(0)$ at the receiver output results from integration over the respective noise power spectral density using (8.17c), (8.21), and (8.22)

$$\varphi_{rr}(0) = \frac{1}{2\pi}\int_{-\pi}^{\pi}\Phi_{\Delta\Delta}(e^{j\Omega}) \cdot |H(e^{j\Omega})|^2\,d\Omega = \frac{(\Delta d)^2}{12} \cdot G_p. \tag{8.24}$$

Hence, with (8.21) and (8.23), the signal-to-noise ratio $\left(\frac{S}{N}\right)_y$ of the output signal is given by

$$\left(\frac{S}{N}\right)_y = \frac{\varphi_{xx}(0)}{\varphi_{rr}(0)} \tag{8.25a}$$

$$= \frac{\varphi_{xx}(0)}{\frac{(\Delta d)^2}{12} \cdot G_p} = \frac{G_p \cdot \varphi_{dd}(0)}{\frac{(\Delta d)^2}{12} \cdot G_p} \tag{8.25b}$$

$$= \frac{\varphi_{dd}(0)}{\frac{(\Delta d)^2}{12}} = \left(\frac{S}{N}\right)_d. \tag{8.25c}$$

As a result, the signal-to-noise ratio is not improved by open-loop prediction. The prediction gain *cannot* be used to improve the *objective* signal-to-noise ratio. Due to the psychoacoustic masking effect, however, the *subjective* speech quality is significantly improved by spectral shaping of the quantization error, which can also be exploited for bit rate reduction.

### 8.3.3.2 Quantization with Closed-Loop Prediction

The analysis of the structure in Figure 8.7b is carried out in the time domain. Assuming an error-free transmission, we get

$$y(k) = \tilde{x}(k) = \hat{x}(k) + \tilde{d}(k) \tag{8.26a}$$

$$= \hat{x}(k) + d(k) + \Delta(k) \tag{8.26b}$$

$$= x(k) + \Delta(k). \tag{8.26c}$$

In contrast to open-loop prediction, the reconstruction error $r(k)$ equals the spectrally white quantization noise $\Delta(k)$. Consequently, $r(k)$ appears as a spectrally white noise signal at the output of the receiver.

In this case, the prediction gain can be used to improve the signal-to-noise ratio

$$\left(\frac{S}{N}\right)_y = \frac{\varphi_{xx}(0)}{\varphi_{\Delta\Delta}(0)} \tag{8.27a}$$

$$= \frac{\varphi_{xx}(0)}{\varphi_{dd}(0)} \cdot \frac{\varphi_{dd}(0)}{\varphi_{\Delta\Delta}(0)} \tag{8.27b}$$

$$= G_p \cdot \left(\frac{S}{N}\right)_d. \tag{8.27c}$$

Compared to the quantizer output, the signal-to-noise ratio $\text{SNR}_y$ of the receiver output is increased by the prediction gain $G_p$ (see (8.21))

$$\text{SNR}_y = 10\lg\left(\frac{S}{N}\right)_y = 10\lg\left(\frac{S}{N}\right)_d + 10\lg G_p. \tag{8.27d}$$

This fact can be exploited for objective quality improvement, but also for bit rate reduction. In view of the 6-dB-per-bit rule (see Section 7.2), this implies that closed-loop prediction provides a $10\lg G_p/6$ bit advantage over PCM. Two views of the word length gain are equivalent:

1. SNR improvement for $w = w_0$:
   If we compare plain PCM and closed-loop DPCM both using the same word length $w$, the quantizer of the DPCM system has a lower peak-to-peak load since $\sigma_d < \sigma_x$. Therefore, we can adapt the quantization stepsize ($\Delta d < \Delta x$) and achieve a quantization noise power, which is reduced according to the prediction gain.
2. Constant SNR but $w < w_0$:
   In practice, a certain target performance of $\text{SNR}_y$ is required. Because of the relation

$$\text{SNR}_y = \text{SNR}_{\bar{d}} + 10\lg G_p \tag{8.27e}$$

we design the quantizer for a target signal-to-noise ratio $\text{SNR}_{\bar{d}}$, which can be reduced by $\Delta\text{SNR} = 10\lg G_p$ compared to open-loop prediction. In accordance with (7.23-b) this implies a possible word length reduction by

$$\Delta w_p = \frac{\Delta\text{SNR}}{20\lg 2} = \frac{10\lg G_p}{20\lg 2} = \frac{1}{2}\operatorname{ld} G_p. \tag{8.27f}$$

Finally, it should be noted that white quantization noise is in principle less pleasant than the spectrally shaped noise. However, as the power of the white noise in closed-loop prediction may be significantly lower than that of the colored noise in open-loop prediction, a comparable subjective performance is achieved.

### 8.3.3.3 Spectral Shaping of the Quantization Error

As for the psychoacoustic aspects, open-loop and closed-loop prediction represent two extreme cases with respect to the reconstruction error. In the first case, the noise spectrum follows the spectral envelope of the speech signal, while the signal-to-noise ratio is not improved. The logarithmic distance between the power spectral densities of the speech signal and the reconstruction error is independent of the frequency. In the second case, a white quantization noise remains; however, the signal-to-noise ratio is increased by the prediction gain. In order to exploit to some extent both the psychoacoustic masking effect and the prediction gain, a compromise between these two extremes is desirable. The signal-to-noise ratio should be improved to a certain extent. At the same time, the noise spectrum should be matched to the spectrum of the desired signal in such a way that, compared to the open-loop prediction, an increased distance results in the range of the formant frequencies, whereas a smaller distance is permissible in the "spectral valleys." This objective can be achieved by the technique of *noise shaping* [Schroeder et al. 1979].

The starting point for the derivation of such a structure is the closed-loop DPCM system according to Figure 8.8a. To highlight the functional relations between the input, the output and the quantization noise of the DPCM system, we prefer to use $z$-transform representations instead of power spectral densities. The $z$-transform of a finite segment of the quantization noise $\Delta(k)$ is denoted by $\Delta(z)$.

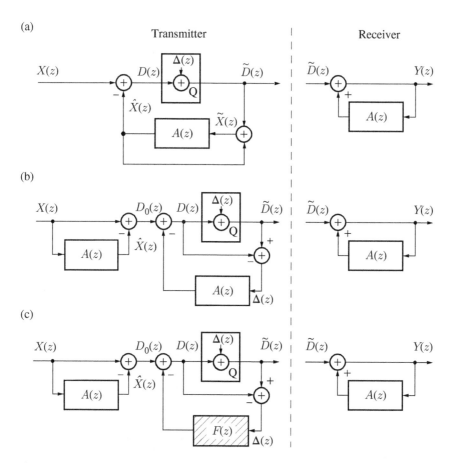

**Figure 8.8** Spectral shaping of the quantization noise (*noise shaping*). (a) Closed-loop DPCM with quantization in the loop, (b) Alternative but equivalent structure, (c) Generalization of (b).

For the $z$-transform $\tilde{D}(z)$, according to Figure 8.8a we obtain

$$\tilde{D}(z) = X(z)(1 - A(z)) + \Delta(z)(1 - A(z)) . \tag{8.28a}$$

Figure 8.8b shows an alternative but equivalent structure, which is also described by (8.28a). The two noise components $\Delta(z)$ and $-\Delta(z) \cdot A(z)$ are added to the signal component

$$D_0(z) = X(z)(1 - A(z)) .$$

The first part results from quantizing $d(k)$, while the second one is obtained by filtering the difference between the input and the output of the quantizer. The behavior of closed-loop prediction (Figure 8.8a) can therefore be achieved by open-loop prediction and feedback of the filtered quantization error (Figure 8.8b). However, this alternative structure has the advantage that the reconstruction error can be explicitly influenced to a certain extent. As shown in Figure 8.8c, a filter function $F(z)$ can be utilized for the noise feedback:

$$\tilde{D}(z) = X(z)(1 - A(z)) + \Delta(z)(1 - F(z)) . \tag{8.28b}$$

Finally, we get for the receiver output

$$Y(z) = X(z) + \Delta(z)\frac{1 - F(z)}{1 - A(z)} . \tag{8.29}$$

For the choice of $F(z)$, it has to be considered that delay-less loops cannot be implemented. A suitable function $F(z)$ can be derived from $A(z)$. According to [Schroeder et al. 1979], we choose

$$F(z) = A(z/\gamma) \quad \text{with} \quad 0 \leq \gamma \leq 1. \tag{8.30}$$

With the parameter $\gamma$, the adaptation of the noise spectrum to the spectrum of the speech signal can be performed with respect to the desired compromise. The effect of this choice can be explained by the zeros $z_{0i}$ of the LP-analysis filter. The product form

$$1 - A(z) = \frac{1}{z^n} \prod_{i=1}^{n}(z - z_{0i}) \tag{8.31a}$$

leads to

$$1 - F(z) = 1 - A(z/\gamma) = \frac{1}{z^n} \prod_{i=1}^{n}(z - \gamma \cdot z_{0i}). \tag{8.31b}$$

If a positive factor $\gamma < 1$ is chosen, the magnitude $|z_{0i}|$ of each zero is reduced, while the angles $\varphi_{0i}$ are maintained:

$$\tilde{z}_{0i} = \gamma \cdot z_{0i} = \gamma \cdot |z_{0i}| \cdot e^{j\varphi_{0i}} . \tag{8.32}$$

The extrema of the resulting frequency response are less distinct, since the zeros that are inside the unit circle (see Chapter 6) are moved toward the origin of the $z$-plane. The special case of closed-loop prediction is covered with $\gamma = 1$, i.e., $F(z) = A(z)$, as shown in (8.31b). The second extreme case of open-loop prediction is obtained for the choice of $\gamma = 0$, i.e., $1 - F(z) = \frac{1}{z^n} \cdot z^n = 1$.

The effective noise shaping is illustrated by an example in Figure 8.9. Represented are the squared magnitude responses of the synthesis filter for a finite signal segment,

$$|H(e^{j\Omega})|^2 = \frac{1}{|1 - A(e^{j\Omega})|^2}, \tag{8.33}$$

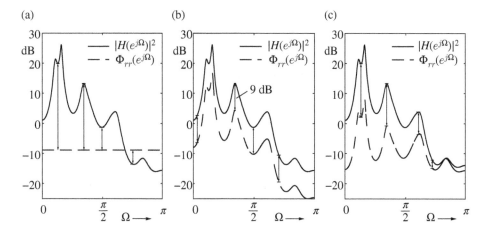

**Figure 8.9** Noise shaping for the example of a vowel. (a) Closed-loop prediction, $\gamma = 1$, (b) Open-loop prediction, $\gamma = 0$, (c) Noise shaping, $\gamma = 0.7$.

and the power spectral density of the reconstruction error,

$$\Phi_{rr}(e^{j\Omega}) = \Phi_{\Delta\Delta}(e^{j\Omega}) \cdot \left| \frac{1 - A\left(\frac{1}{\gamma} e^{j\Omega}\right)}{1 - A(e^{j\Omega})} \right|^2, \tag{8.34}$$

for three values of $\gamma$. According to (8.17c) with

$$\Phi_{\Delta\Delta}(e^{j\Omega}) = \frac{(\Delta d)^2}{12}, \tag{8.35}$$

we assumed a constant noise power spectral density due to uniform quantization, and for the purpose of illustration we exemplarily assume a quantizer with a signal-to-noise ratio of the quantized residual signal $\tilde{d}(k)$ of

$$\text{SNR}_{\tilde{d}} = 10\lg\left( \frac{\varphi_{dd}(0)}{\frac{(\Delta d)^2}{12}} \right) = 9\,\text{dB and } \varphi_{dd}(0) = 1. \tag{8.36}$$

For $\gamma = 0$ (open-loop prediction, Figure 8.9b), we obtain $\text{SNR}_y = \text{SNR}_{\tilde{d}}$ and the noise spectrum follows the magnitude response with a constant distance given by the uniform quantization (in this case 9 dB).

For $\gamma = 1$ (closed-loop prediction, Figure 8.9a), and the same quantizer stepsize $\Delta d$, a constant white noise spectrum results. Compared to the open-loop prediction the signal-to-noise ratio is increased by the logarithmic prediction gain $10\lg G_p$ (see (8.21)).

The DPCM system with noise shaping (Figure 8.9c) shows that in the lower formant frequency range the spectral signal-to-noise ratio has markedly increased in comparison to open-loop prediction. Despite the SNR decrease at higher frequencies, noise shaping provides the best *subjective* speech quality. This is due to the psychoacoustic masking effect. Compared to closed-loop prediction, however, the objectively measured signal-to-noise ratio deteriorates.

The effect on the speech signal is depicted for a second example in Figure 8.10. A fixed, $n = 8$-th order predictor was utilized in the structure according to Figure 8.8c. The quantizer

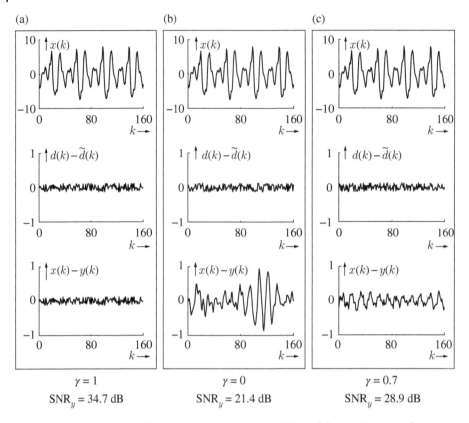

(a)                          (b)                          (c)

$\gamma = 1$                    $\gamma = 0$                    $\gamma = 0.7$

$SNR_y = 34.7$ dB           $SNR_y = 21.4$ dB           $SNR_y = 28.9$ dB

**Figure 8.10** Example of differential coding according to Figure 8.8c (predictor: $n = 8$, quantizer: $w = 5$, voiced speech). (a) Closed-loop prediction, $\gamma = 1$, (b) Open-loop prediction, $\gamma = 0$, (c) Noise shaping, $\gamma = 0.7$.

with a word length of $w = 5$ was adjusted to the reduced dynamic range of the residual signal $d(k)$.

From Figure 8.10, we can infer that the best objective match between the original signal $x(k)$ and output signal $y(k)$ is achieved with closed-loop prediction. The difference $x(k) - y(k)$ is the same as $d(k) - \tilde{d}(k)$ (see also (8.26c)). The largest reconstruction error energy $\sum_k (x(k) - y(k))^2$, i.e., the lowest $SNR_y$, results from open-loop prediction. However, the best auditory impression is obtained by quantization with noise shaping. Here, the energy of the reconstruction error is higher than with closed-loop prediction but smaller than with open-loop prediction. The parameter $\gamma$ controls the noise shaping intensity and ranges typically in $0.6 \leq \gamma \leq 0.9$.

### 8.3.4 ADPCM with Sequential Adaptation

Discussions in this section will refer to a system called *ADPCM* with adaptive prediction and adaptive quantization. When applying the *least-mean-square* (LMS) algorithm (see Section 6.3.2) and adaptive quantization backwards (AQB, see Section 7.5), the adaptation of the predictor and the quantizer can be carried out recursively. Apart from the quantized residual signal $\tilde{d}(k)$, no additional information must be transmitted to the receiver. Furthermore, better results in terms of bit rate reduction and speech quality compared to DPCM are obtained.

We will look at the ADPCM structure according to Figure 8.11 with closed-loop prediction and quantization within the loop.

For the derivation of the LMS algorithm, we will in a first step assume that $d(k)$ is not quantized:

$$\tilde{d}(k) = d(k). \tag{8.37}$$

We consider the *instantaneous power*

$$\hat{\sigma}_d^2(k) = d^2(k) \tag{8.38a}$$

$$= \left(x(k) - \mathbf{a}^T(k)\tilde{\mathbf{x}}(k-1)\right)^2 \tag{8.38b}$$

and calculate the instantaneous gradient $\hat{\nabla}(k)$ with respect to the coefficient vector $\mathbf{a}(k)$ in analogy to Section 6.3.2, (6.79-b), and (6.80-b):

$$\hat{\nabla} = \frac{\partial d^2(k)}{\partial \mathbf{a}(k)} = -2\underbrace{\left(x(k) - \mathbf{a}^T(k)\tilde{\mathbf{x}}(k-1)\right)}_{d(k)}\tilde{\mathbf{x}}(k-1) \tag{8.39a}$$

$$= -2\, d(k)\,\tilde{\mathbf{x}}(k-1), \tag{8.39b}$$

with the vector notation

$$\tilde{\mathbf{x}}(k-1) = (\tilde{x}(k-1), \tilde{x}(k-2), \dots, \tilde{x}(k-n))^T \tag{8.39c}$$

$$\mathbf{a}(k) = \left(a_1(k), a_2(k), \dots, a_n(k)\right)^T. \tag{8.39d}$$

The new set of prediction coefficients $\mathbf{a}(k+1)$ is computed by updating the present vector $\mathbf{a}(k)$ in a direction opposite to that of the instantaneous gradient $\hat{\nabla}$. The corresponding simple recursive algorithm

$$\mathbf{a}(k+1) = \mathbf{a}(k) + 2\,\vartheta\, d(k)\,\tilde{\mathbf{x}}(k-1) \tag{8.40a}$$

or, accordingly for the individual coefficient with index $i = 1, \dots, n$,

$$a_i(k+1) = a_i(k) + 2\,\vartheta\, d(k)\,\tilde{x}(k-i) \tag{8.40b}$$

represents the LMS algorithm (see Section 6.3.2). The effective stepsize $2\,\vartheta$ controls the rate of adaptation and the stability (see (6.81-b)).

In the case of error-free transmission, the required information for the adaptation is also available at the receiver. Due to the structural correspondence,

$$\tilde{x}(k) = y(k) \tag{8.41}$$

is also valid if the residual signal $d(k)$ is quantized, i.e., we replace $d(k)$ by $\tilde{d}(k)$ at the transmitting side in (8.40a). The predictor at the receiving side can be adjusted synchronously to the filter on the transmitting side, if in (8.40a) we replace $d(k)$ by $\tilde{d}(k)$ and $\tilde{x}(k)$ by $y(k)$.

ADPCM with a word length of $w = 4$, provides almost the same speech quality as scalar logarithmic quantization of $x(k)$ with $w = 8$ (A-law PCM).

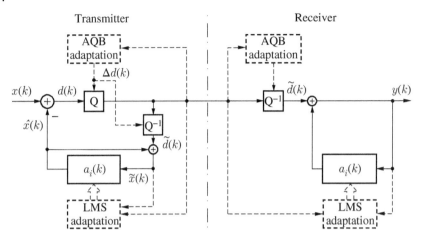

**Figure 8.11** ADPCM with sequential adaptation.

The speech codec used in digital cordless phones according to the *DECT standard* (ITU-T/G.726) is based on the ADPCM system illustrated in Figure 8.11. In this concept, a prediction filter with poles and zeros is applied. The adaptation of the filter is carried out using the so-called *algebraic sign–LMS* algorithm for implementational simplicity. Each adjustment is proportional to the negative of an estimate of the gradient according to

$$a_i(k+1) = a_i(k) + 2\,\vartheta\,\text{sign}\{d(k)\}\,\text{sign}\{\tilde{x}(k-i)\}\,. \tag{8.42}$$

The adaptation does not require multiplications, which is advantageous for implementation in *application-specific integrated circuits* (ASICs).

## 8.4 Parametric Coding

### 8.4.1 Vocoder Structures

The second class of speech-coding algorithms comprises the parametric coders, generally called *vocoders* (*voice coders*). Vocoders provide the greatest reduction of bit rate: effectively 0.1–0.5 bit/sample can be achieved. Although speech sounds mostly synthetic, a sufficient intelligibility is obtained. "Speaker transparency" is only given to a limited degree. In contrast to waveform coding, an exact signal reproduction is not the main objective; in particular, phase information is not considered. Therefore, the perceived quality of the synthesized speech signal cannot be quantified by objective distortion measures, such as the signal-to-noise ratio.

A common characteristic of the different time-domain vocoders is a signal analysis procedure for extracting perceptually significant parameters from the speech signal. Using these parameters, which contain information about

- the instantaneous frequency response of the vocal tract filter with impulse response $h_0(k)$ and
- the excitation signal $v(k)$,

an output signal is synthesized in the decoder. In the synthesis procedure, the excitation signal is built frame by frame, either by a noise signal or by a periodic impulse sequence,

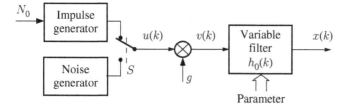

**Figure 8.12** Parametric decoding of a vocoder. $S$ : voiced/unvoiced switch, $N_0$ : pitch period, $g$ : gain.

depending on whether the speech frame is classified as voiced or unvoiced. Some more advanced schemes provide also mixed voiced/unvoiced excitation modes. The fundamental differences of the vocoder variants lie in the structure of the synthesis filter and in the analysis of the filter parameters. Figure 8.12 shows a generic description of the synthesis procedure of a vocoder, i.e., the decoder at the receiving side.

The basic structure corresponds to the discrete time model of speech production according to Section 6.1. The time-variant synthesis filter is excited by $v(k)$ with either a noise-like or a periodic structure. The development of the different vocoders is closely linked to the technology available at the respective time. In what follows, we will first briefly discuss two early variants of the vocoder approaches. More detailed descriptions can be found in, for example, [Rabiner, Schafer 1978] and [Sluijter 2005].

The oldest vocoder is the so-called *channel vocoder*, which was originally realized in analog technology with $n = 10$ channels [Dudley 1939], in order to transmit speech signals in analog form with a markedly reduced bandwidth. The basic principle is depicted in Figure 8.13.

The synthesis filter consists of a parallel arrangement of bandpass filters (Figure 8.13b). The corresponding output signals are added up to form signal $\hat{x}(k)$. All bandpass filters are excited with the same signal $u(k)$, which is individually scaled by quantized time-variant factors $\tilde{g}_i$, $i = 1, \ldots, n$. The gain factors $g_i$ are determined on the transmitting side by measuring the envelope of the short-term power in each frequency band (Figure 8.13a), while using the same bandpass filters on the transmitting and receiving sides. Due to the relatively slow changes of the envelope, the scaling factors are transmitted with a highly reduced sampling frequency (subsampling by the factor $r$). If, at the receiving side, the quantized gain factors $\tilde{g}_i$ are combined with the bandpass filters, the overall filter can be interpreted as an approximation of the vocal tract filter. Consequently, the desired frequency response is approximated by a parallel arrangement of $n$ bandpass filters with fixed center frequencies and bandwidths but different and variable gain factors $\tilde{g}_i$.

Another vocoder variant is the *formant vocoder* (e.g., [Rosenberg et al. 1971], [Rabiner, Schafer 1978]). In contrast to the channel vocoder, the spectral envelope does not result from measuring the energy in fixed frequency bands, but from an explicit determination of the formant frequencies $F_i$ and formant bandwidths $B_i$. The synthesis filter can be realized as a cascade or, as depicted in Figure 8.14, as a parallel connection of second-order filters.

With this concept, bit rates of $B < 1$ kbit/s can be realized. However, this entails fundamental limitations on the naturalness of speech. With a correct analysis of the formant frequencies, a better speech quality is achieved compared to the channel vocoder. The formant analysis is especially problematic if two formants lie close together. Linear prediction or cepstral analysis is suitable to determine the formants.

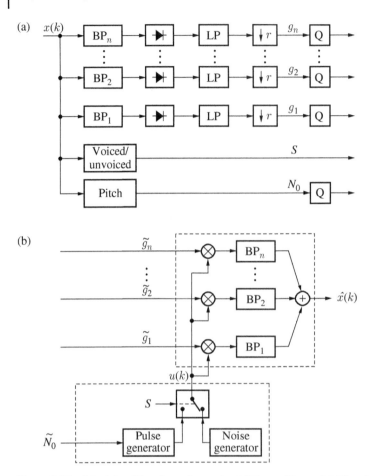

**Figure 8.13** Principle of the channel vocoder. (a) Transmitter and (b) Receiver.

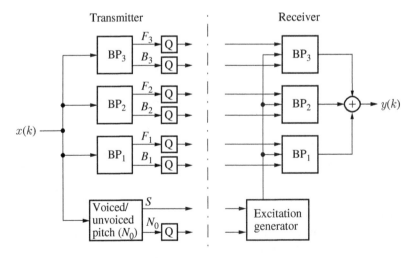

**Figure 8.14** Principle of the formant vocoder.

Apart from these time-domain vocoders, related approaches exist, which represent frequency-domain methods, such as the *phase vocoder* [Flanagan, Golden 1966] and the *cepstrum vocoder* (e.g., [Rabiner, Schafer 1978], [O'Shaughnessy 2000]), which will not be discussed here.

### 8.4.2 LPC Vocoder

*Linear predictive coding* is based on the all-pole model of speech production. The simplest variant is the *LPC vocoder*, which corresponds to the model of speech production derived in Section 2.4 and simplified in Section 6.1 (Figure 6.1). The essential properties of this model are:

- the nasal tract is not considered, and the vocal tract is approximated by lossless tube segments or by a minimum-phase all-pole filter accordingly;
- based on a voiced/unvoiced classification of short signal frames of 20–30 ms duration, the excitation or the glottis signal is approximated by a periodic impulse sequence or a noise signal.

In contrast to the waveform coding of Section 8.3, the coding strategy depends on a dynamic signal classification. Often, a predictor or a synthesis filter of reduced filter order is utilized in unvoiced frames so that the mean bit rate can be reduced. The basic structure of the LPC vocoder is illustrated in Figure 8.15.

A typical dimensioning will be explained by means of the so-called *LPC-10 algorithm* according to [Campbell et al. 1989] and [Tremain 1982].

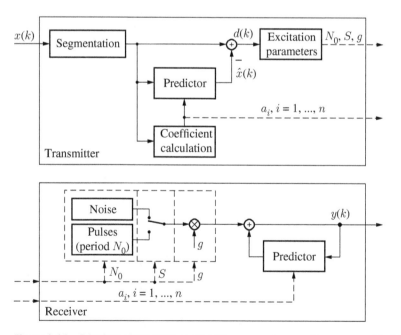

**Figure 8.15** Principle of the LPC vocoder (The parameter quantizers are omitted here for simplicity).

The speech signal is sampled with a frequency of $f_s = 8\,\mathrm{kHz}$ and segmented into frames of 22.5 ms duration ($N = 180$ samples). Using the covariance method (see Section 6.4), 10 reflection coefficients in the case of voiced frames and 4 reflection coefficients for unvoiced frames are computed.

The coefficients are quantized in the form of the *log area ratios* (see Section 7.7), where the first two coefficients are non uniformly quantized with 5 bits each, while the remaining coefficients are represented with uniform resolution and word lengths of $w = 2, \dots, 5$. The complete set of coefficients is coded with 41 bits or 20 bits in the case of voiced or unvoiced frames, respectively. For the pitch period and the voiced/unvoiced decision 7 bits are used, 5 bits for the logarithmic quantization of the gain factor and 1 bit for the synchronization. Thus, the bit rates for both voicing modes result in

$$\frac{54\,\mathrm{bits}}{22.5\,\mathrm{ms}} = 2.4\,\mathrm{kbit/s}\ (\text{voiced}) \quad \text{and} \quad \frac{33\,\mathrm{bits}}{22.5\,\mathrm{ms}} = 1.47\,\mathrm{kbit/s}\ (\text{unvoiced}).$$

The basic delay of this codec amounts to approximately 90 ms. Despite the reduced naturalness of the resynthesized speech, a relatively high intelligibility is achieved. The LPC-10 algorithm was primarily developed for encrypted transmission using modems in non-public analog telephone networks.

## 8.5 Hybrid Coding

### 8.5.1 Basic Codec Concepts

The main application areas for source coding of speech signals with bit rates below the 32 kbit/s of standard ADPCM are digital mobile radio communication, speech storage, and "line multiplication," i.e., channel sharing by speech signal compression. Of particular interest are coders with 0.5–1.5 bits per sample, i.e., for telephone speech with a bit rate of $B = 4$–$12\,\mathrm{kbit/s}$.

In these applications, hybrid speech codecs are used almost exclusively. These codecs take a position between waveform coding and parametric coding. As a common feature, the coefficients of a synthesis filter are transmitted as *side information*, and the residual signal is approximated quite coarsely with respect to the amplitude and/or time resolution. The literature shows a wide range of concepts, and a vast variety of different codec variants.

In this section, we will develop a uniform conceptual description of these approaches. The following sections 8.5.1.1 and 8.5.1.2 will deal with typical variants, while some selected codec standards are outlined in Section 8.7.

In hybrid codecs, short-term as well as long-term prediction are common; the decoder structure is shown in Figure 8.16.

The decoder consists of a cascade of a long-term prediction (LTP) and a linear prediction (LP) synthesis filter, which is excited by the quantized residual signal $\tilde{d}'(k)$. The predictors $A(z)$ and $P(z)$ are time variant and, as described in Chapter 6, adjusted blockwise. The coefficients are quantized as explained in Section 7.7. Since the filters are generally held constant for the length of a signal frame or subframe, respectively, their time variation will not be considered here for simplicity of the presentation. The typical frame duration is 20 ms. Frames may be divided into subframes of, e.g., 5 ms each.

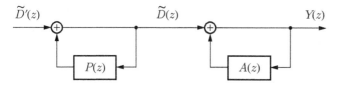

$\tilde{D}'(z)$    $\tilde{D}(z)$    $Y(z)$

**Figure 8.16**  Structure of the hybrid decoder. $A(z)$: Short-term predictor (LP), $P(z)$: Long-term predictor (LTP).

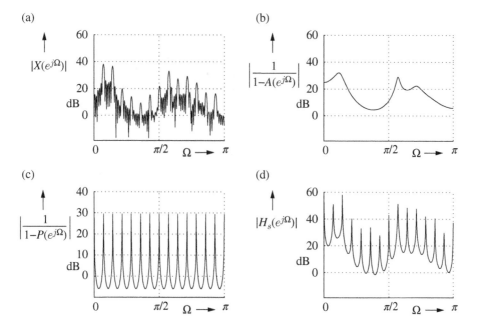

**Figure 8.17**  Functionality of the hybrid decoder. (a) Short-term spectral analysis $|X(e^{j\Omega})|$ for a voiced segment (20 ms), (b) Magnitude response of the LP-synthesis filter $|1/(1 - A(e^{j\Omega}))|$, (c) Magnitude response of the LTP-synthesis filter $|1/(1 - P(e^{j\Omega}))|$, (d) Magnitude response of the cascade of both filters $|H_s(e^{j\Omega})| = |1/(1 - P(e^{j\Omega}))| \cdot |1/(1 - A(e^{j\Omega}))|$.

The basic functionality of the decoder will be illustrated by the following example: Figure 8.17 shows the short-term spectral analysis of a (synthetic) voiced speech segment (20 ms), the magnitude responses of the LTP-synthesis filter, the LP-synthesis filter, and the overall filter. The first filter stage (LTP), which is adapted every subframe, shows a distinctive comb filter characteristic in voiced segments. Starting from a flat spectrum $\tilde{D}'(z)$ of the excitation signal, all spectral components located at multiples of the estimated fundamental frequency, i.e., at

$$\Omega_i = \frac{2\pi}{N_0} i; \quad i = (0), 1, 2, \dots, \tag{8.43}$$

are amplified, and the components in between are attenuated. The result is signal $\tilde{d}(k)$ with an approximately spectrally flat envelope and a discrete, harmonic structure. Finally, with the second filter stage (LP) the spectral envelope of the speech segment is formed on $\tilde{d}(k)$ by the LP-synthesis filter.

The codec concepts to be discussed here basically differ in the predictor structures at the transmitter and in the representation of the residual signal by scalar or vector quantization.

### 8.5.1.1 Scalar Quantization of the Residual Signal

Basic schemes using scalar quantization are shown in Figure 8.18. The effect of the quantizer will be described in the time domain by additive quantization noise $\Delta(k)$ according to

$$\tilde{d}'(k) = d'(k) + \Delta(k). \tag{8.44}$$

Scalar quantization is applied to the residual signal $d'(k)$ either after two-stage open-loop prediction (Figure 8.18a) or inside a closed prediction loop (Figure 8.18b–d). In all the cases, the quantized residual signal is termed $\tilde{d}'(k)$ and the same decoder of Figure 8.16 is utilized.

If the quantizer is switched off, i.e., $\tilde{d}'(k) = d'(k)$, the three structures in Figure 8.18a–c provide the same residual signal. Moreover, for identical LP and LTP coefficients on the transmitting and receiving side, the output signal $y(k)$ equals the input signal $x(k)$, since the filter operations are inverse to each other.

In contrast to this, with quantization of the residual signal $d'(k)$, marked differences in the reconstruction error occur due to the different positions of the quantizer.

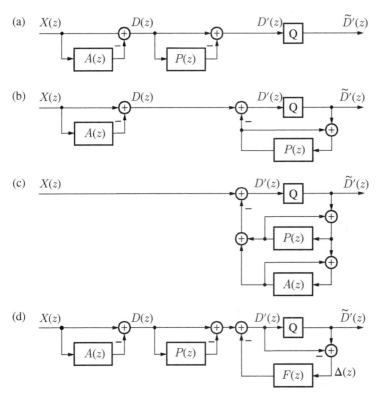

**Figure 8.18** Hybrid coding with scalar quantization of the residual signal. (a) LP and LTP open loop, (b) LP open loop and LTP closed loop, (c) LP and LTP closed loop, (d) LP and LTP with noise shaping.

As the predictors are adapted to signal frames of finite length, the following analysis can be performed for a signal $x(k)$ of limited duration, for which the $z$-transform $X(z)$ exists.

Without committing to a certain type of quantizer, the quantization error $\Delta(k)$ can thus be described by

$$\Delta(z) = \tilde{D}'(z) - D'(z) \tag{8.45}$$

in the $z$-domain. In analogy to the concept of noise shaping derived in Section 8.3.3, it can be shown that the performance of the block diagrams of Figure 8.18a–c can be described exactly by the structure of Figure 8.18d with an adequate choice of the *noise shaping* filter $F(z)$. Of particular interest is the resulting reconstruction error in the output signal of the decoder according to Figure 8.16.

**Open-Loop Short-Term and Long-Term Prediction**   This case corresponds to the choice

$$F(z) = 0$$

resulting in

$$\tilde{D}'(z) = X(z)\,(1 - A(z))\,(1 - P(z)) + \Delta(z)$$

$$Y(z) = X(z) + \frac{\Delta(z)}{(1 - P(z))\,(1 - A(z))}\,.$$

The spectrum of the quantization error is weighted with the frequency response of the cascaded synthesis filters. Applying psychoacoustic criteria, this noise shaping is advantageous, especially when the spectral envelope of the quantization error is flat. However, the prediction gain cannot be exploited to improve the signal-to-noise ratio, since both predictors are applied in the open-loop mode (see Section 8.3.3.1).

**Open-Loop Short-Term and Closed-Loop Long-Term Prediction**   This case corresponds to the choice

$$F(z) = P(z)$$

resulting in

$$\tilde{D}'(z) = X(z)\,(1 - A(z))\,(1 - P(z)) + \Delta(z)\,(1 - P(z))$$

$$Y(z) = X(z) + \frac{\Delta(z)}{(1 - A(z))}\,.$$

The spectrum of the quantization error is weighted with the spectral envelope of the input signal. Since the long-term predictor is inside a closed loop, only its corresponding part of the prediction gain can be used to improve the signal-to-noise ratio.

**Closed-Loop Short-Term and Long-Term Prediction**   This case corresponds to the choice

$$F(z) = A(z) + P(z) - A(z)\,P(z)$$

resulting in

$$\tilde{D}'(z) = X(z)\,(1 - A(z))\,(1 - P(z))$$
$$\qquad + \Delta(z)\,(1 - A(z) - P(z) + A(z)\,P(z))$$
$$\qquad = X(z)\,(1 - A(z))\,(1 - P(z)) + \Delta(z)\,(1 - A(z))\,(1 - P(z))$$
$$Y(z) = X(z) + \Delta(z)\,.$$

Reconstruction and quantization error are identical. Due to the two-step closed-loop prediction, the total prediction gain can be used entirely to improve the signal-to-noise ratio (see Section 8.3.3.2).

***Short-Term and Long-Term Prediction with Noise Shaping*** This is the most general case with

$$\tilde{D}'(z) = X(z)\,(1 - A(z))\,(1 - P(z)) + \Delta(z)\,(1 - F(z))$$

$$Y(z) = X(z) + \Delta(z)\,\frac{1 - F(z)}{(1 - P(z))\,(1 - A(z))}\,.$$

In correspondence with spectral noise shaping according to Section 8.3.3.3, an auditory improvement can be achieved exploiting the masking effect of the human ear (see Section 2.6) by only partly using the prediction gain for the objective improvement. Different options for $F(z)$ and its effect on the reconstruction error are shown in Table 8.2.

With respect to psychoacoustic aspects, the choice of $\gamma_1 < 1$ and $\gamma_2 = 1$ is favorable. In this case, the spectral weighting corresponds to the conventional noise shaping (see Section 8.3.3.3), while additionally the prediction gain of the LTP loop is used for an objective improvement of the signal-to-noise ratio.

### 8.5.1.2 Vector Quantization of the Residual Signal

A target bit rate of effectively only 0.5–1.5 bits per sample suggests vector quantization of the residual signal. Note that due to prediction, the residual signal is more or less decorrelated. For complexity reasons, only *gain–shape* vector quantization (Section 7.6.5) is applicable. Basically, code books for normalized residual signal vectors must be designed. The statistical analysis of speech material shows that the normalized residual signal vectors follow a multivariate Gaussian distribution to a good approximation. This almost does not depend on the speaker. Due to the non-uniform distribution, we can thus obtain better results by vector quantization than by scalar quantization despite decorrelation, as shown in Section 7.6.4. With the available low bit rates, however, only relatively low signal-to-noise ratios can be achieved. Therefore, the error criterion or the effective distance measure, respectively, is of great importance.

The introduction of noise shaping techniques (see Section 8.5.1.1) offers a considerable quality improvement with respect to psychoacoustic criteria. Here, the masking effect in speech perception is implicitly exploited. This approach can also be applied to vector quantization, resulting in what is called *analysis-by-synthesis* coding in the literature. It is the

**Table 8.2** Effect of the noise shaping filter $F(z)$; $0 \le (\gamma_1, \gamma_2) \le 1$.

| $F(z)$ | Reconstruction error |
|---|---|
| $A(z/\gamma_1)$ | $\Delta(z)\,\dfrac{1}{1 - P(z)}\,\dfrac{1 - A(z/\gamma_1)}{1 - A(z)}$ |
| $P(z/\gamma_2)$ | $\Delta(z)\,\dfrac{1 - P(z/\gamma_2)}{1 - P(z)}\,\dfrac{1}{1 - A(z)}$ |
| $A(z/\gamma_1) + P(z/\gamma_2) - A(z/\gamma_1)\,P(z/\gamma_2)$ | $\Delta(z)\,\dfrac{1 - P(z/\gamma_2)}{1 - P(z)}\,\dfrac{1 - A(z/\gamma_1)}{1 - A(z)}$ |

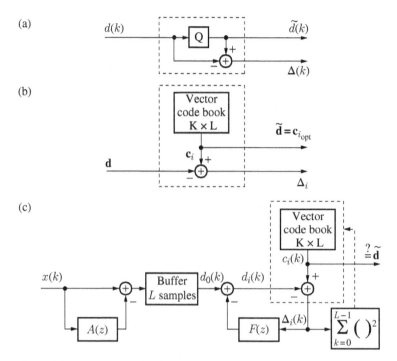

**Figure 8.19** Vector quantization with spectral noise shaping. (a) Scalar quantization of samples $d(k)$, (b) Vector quantization of vector **d**, (c) Vector quantization with noise shaping. (Note: For reasons of simplicity, without long-term prediction).

basis of most codec standards (see also Section 8.7). In what follows, we will develop the approach, starting from scalar quantization with noise shaping according to Figure 8.18d.

In order to simplify the representation, we will disregard the long-term prediction in a first step and use plain vector quantization instead of gain–shape vector quantization. The starting point is Figure 8.19a. It shows the relation between the non-quantized scalar residual signal $d(k)$, the quantized value $\tilde{d}(k)$, and the resulting error

$$\Delta(k) = \tilde{d}(k) - d(k). \tag{8.46}$$

For the vector approach, $L$ samples of $d(k)$ are combined to a vector **d**, which can be vector-quantized as described in Figure 8.19b. In simplified notation, we start with a segment of the sequence $x(k)$, which for $k = 0, 1, 2, \ldots, L - 1$ corresponds to the respective vector elements and append outside this interval only zero values. Due to the linearity of the filters, the performance of the complete structure can be described correctly by segmenting the input sequence $x(k)$, separately filtering these segments, and superposing the partial reactions. The superposition is achieved if the filter states are continued when going from one segment to the next.

The code book contains $K$ vectors $\mathbf{c}_i$ ($i = 1, 2, \ldots, K$) of dimension $L$. The best vector with respect to the smallest mean square error will be called

$$\tilde{\mathbf{d}} = \mathbf{c}_{i_{\text{opt}}}. \tag{8.47}$$

For each code book vector $\mathbf{c}_i$ an individual error vector

$$\Delta_i = \mathbf{c}_i - \mathbf{d} \tag{8.48}$$

can be calculated. This procedure applies to the case of open-loop prediction without noise shaping.

If we want to use noise shaping in combination with vector quantization of the residual, we have to apply the vector quantizer in analogy to Figure 8.8 within the loop as shown in Figure 8.19c.

However, the vector $\mathbf{d}$ is not completely available in advance. Each element $d(k)$ of vector $\mathbf{d}$ depends on the preceding elements

$$\Delta_i(k - \kappa) = c_i(k - \kappa) - d(k - \kappa), \quad \kappa = 1, 2, \ldots \tag{8.49}$$

of the error vector $\Delta_i$. For this reason, the vector $\mathbf{d}$ develops sample by sample for each $\mathbf{c}_i$ in the process of vector quantization. For each code book vector $\mathbf{c}_i$ a different vector $\mathbf{d}_i$ might ensue. For this reason, the $L$ input samples $d_0(k)$ must be buffered so that the filter loop can be processed for each code book vector $\mathbf{c}_i$, where the filter states of $F(z)$ have to be set to the original start values. Within the search procedure, the optimal excitation vector $\mathbf{c}_{i_{\mathrm{opt}}}$ can be found among $K$ entries by minimizing the mean square error. Obviously, the same mechanism is effective for the spectral shaping of the quantization error as for scalar quantization.

However, the practical realization of this principle is based on a different structure, which can be derived from Figure 8.19c by means of the following considerations.

It is sufficient to look at only one segment to find the relations between $x(k)$, $d_0(k)$, $c_i(k)$, and $\Delta_i(k)$. According to Figure 8.19c with the $z$-transforms of the respective sequences, we have

$$\Delta_i(z) = -X(z) \frac{1 - A(z)}{1 - F(z)} + C_i(z) \frac{1}{1 - F(z)} \tag{8.50a}$$

$$= -X(z) \frac{1 - A(z)}{1 - F(z)} + C_i(z) \frac{1}{1 - A(z)} \frac{1 - A(z)}{1 - F(z)} \tag{8.50b}$$

$$= \left[ Y_i(z) - X(z) \right] \frac{1 - A(z)}{1 - F(z)} \tag{8.50c}$$

with

$$Y_i(z) = C_i(z) \frac{1}{1 - A(z)}. \tag{8.51}$$

The result of (8.50c) is depicted in Figure 8.20.

With respect to a perceptually weighted spectral error criterion, the optimal vector quantization of the residual signal can be found within a search process. As a result, we obtain an "optimal" replica $y_{i_{\mathrm{opt}}}(k)$, which matches the speech signal $x(k)$ with minimum error. Note

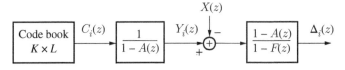

**Figure 8.20** Principle of an analysis-by-synthesis coder.

that the $z$-transform of the effective weighting filter $W(z) = (1 - A(z))/(1 - F(z))$ refers to the inverse of the corresponding noise shaping filter.

This procedure, which is equivalent to Figure 8.19c, is defined in the literature by the generic term *analysis-by-synthesis* coding, and in this special form as CELP coding (*code-excited linear prediction coding*, see also Section 8.5.3) [Schroeder, Atal 1985].

This coding approach is computationally intensive since $K$ different replicas $y_i(k)$ must be synthesized for each signal segment $x(k)$. In order to encode the residual signal with effectively 0.5 bits per sample corresponding to a bit rate of $B = 0.5$ bits $\cdot$ 8 kHz $= 4$ kbit/s, a code book with

$$\frac{\mathrm{ld}(K)}{L} = 0.5 \tag{8.52}$$

is required. For example, for $L = 20\,(\hat{=}\,2.5\,\mathrm{ms})$, the size of the code book already amounts to $K = 2^{10} = 1024$.

Based on these concepts of scalar and vector quantization of the residual signal, the next Sections 8.5.2, and 8.5.3 will introduce different codec structures that provide comparable speech quality, and which differ considerably in terms of the bit rate and complexity. The concrete variants of the important codec standards are outlined in Section 8.7.

## 8.5.2 Residual Signal Coding: RELP

The first concept to be dealt with follows the *APC* scheme with block adaptation (see Section 8.3.1). The residual signal produced by short-term and long-term prediction is applied to a scalar quantizer (see Figure 8.18a). The core problem of this approach becomes evident in the following consideration. If the target bit rate is, for instance, $B = 15$ kbit/s, only a bit rate of about 12 kbit/s is available for the coding of the residual signal, since the quantization of the LP and LTP filter parameters requires a bit rate of about 3 kbit/s (see Section 7.7 and 6.4). Thus, with a sampling frequency of $f_s = 8$ kHz only 1.5 bits per sample can be utilized. However, the power of the resulting quantization error $\Delta(k)$ would be so high that, even with noise shaping, no acceptable speech quality could be obtained.

A comparison with the LPC vocoder (see Section 8.4.2) reveals that it is not necessary to exactly reconstruct the residual signal at the receiver, as perception is insensitive to certain changes in this signal. Extracting and coding perceptually relevant aspects of the residual signal, such as

- the correct temporal volume contour,
- the correct (quasi-)periodicity in voiced segments, and
- the noise-like character in unvoiced segments

provides communications-quality speech coders for the 2.4–8 kbit/s range.

Assuming that the lowest speech frequencies carry the perceptually most important information, the relevant characteristics can mostly be reconstructed from a baseband of the residual signal extracted by a lowpass filter as shown in Figure 8.21. This coding scheme is called the *baseband–RELP codec* [Un, Magill 1975]. For reasons of simplicity, the long-term predictor is not considered here.

The lowpass signal $d_{LP}(k)$ is decimated and transmitted in quantized form. In the receiver, the missing high frequencies are reconstructed by means of spectrally shifted versions of the baseband.

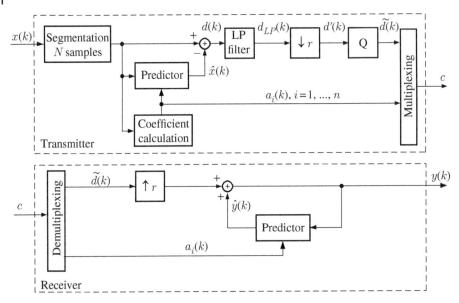

**Figure 8.21** Principle of the baseband–RELP codec (RELP: *Residual Excited Linear Prediction*).

The prediction coefficients $a_i(k)$ $(i = 1, 2, \ldots, n)$ depend on time index $k$ and are adapted blockwise every $N$ samples, e.g., $N = 160$ ($\hat{=} 20$ ms) (see Section 6.3.1). The segmentation into frames is needed to buffer the speech samples for LP coefficient calculation and subsequent filtering. As a result, the codec introduces an *algorithmic delay* of at least one frame, i.e., $N$ sampling intervals. This delay is increased by the time needed for the computation of the coefficients, the filter operations, and the signal delay caused by the lowpass. In a practical realization, such a codec will cause a total delay of $1.25 N$ to $2 N$ cycles (here 25–40 ms). To simplify the representation, we will assume that the coefficients $a_i(k)$ already exist in quantized form (see Section 7.7), i.e., the allocation of the quantized values to the bit patterns and the corresponding decoding at the receiver end are not explicitly described here.

A generic feature of the baseband–RELP codec is that the prediction error signal $d(k)$ is applied to a lowpass filter with subsequent decimation by a factor $r$. The lowpass has a cutoff frequency of $\Omega_c = \pi/r$. Thus, the sampling frequency at the output can be reduced by factor $r$ without spectral aliasing. The factor $r$ is generally set to $r = 3$ or $r = 4$. At $f_s = 8$ kHz, the baseband signal $d'(k)$ has a bandwidth of $4/3$ kHz or 1 kHz, respectively, and a sampling frequency of $f'_s = 8/3$ kHz or $f'_s = 2$ kHz, respectively. If, for example, a bit rate of 12 kbit/s is available for coding the residual signal, each sample $d'(k)$ can be encoded quite accurately with $12/8 \cdot r = 12/8 \cdot 3 = 4.5$ bits or $12/2 = 6$ bits, respectively. The operation of lowpass filtering, downsampling and scalar quantization can be interpreted as joint time *and* amplitude quantization.

At the receiving side, the sampling frequency is increased back to the original rate by replacing the missing samples with zeros. Thus, the baseband spectrum is mirrored $(r - 1)$ times and overlapped. Disregarding the quantization (i.e., for $\tilde{d}(k) = d'(k)$) this process can be described exactly by introducing the decimation sequence $p(k)$ as follows:

$$\tilde{d}(k) = d'(k) = d_{LP}(k) \cdot p(k) = \begin{cases} d_{LP}(k) & k = \lambda \cdot r \\ 0 & k \neq \lambda \cdot r \end{cases} \quad \lambda \in \mathbb{N}_0 \tag{8.53}$$

with

$$p(k) = \frac{1}{r} \sum_{i=0}^{r-1} e^{+j\frac{2\pi}{r}ik} = \begin{cases} 1 & k = \lambda \cdot r \\ 0 & k \neq \lambda \cdot r. \end{cases} \tag{8.54}$$

It is obviously not necessary to transmit the zero samples of $\tilde{d}(k)$. Because of the periodicity of the complex exponential function, $p(k)$ can be written as the superposition of $r$ complex exponentials with the frequencies $\Omega_i = \frac{2\pi}{r} i$ $(i = 0, 1, \dots, r-1)$. According to the modulation theorem, the spectrum of $d'(k)$ results as a superposition of spectrally shifted versions of the baseband spectrum $D_{LP}(e^{j\Omega})$

$$D'(e^{j\Omega}) = \frac{1}{r} \sum_{i=0}^{r-1} D_{LP}(e^{j(\Omega - \frac{2\pi}{r}i)}). \tag{8.55}$$

The spectral interrelation is shown schematically in Figure 8.22.

The result is a spectrally more or less flat broadband excitation signal $\tilde{d}(k)$, which is applied to the synthesis filter. This excitation signal essentially shows the perceptually relevant properties obtained if the predictor produces a spectrally flat residual signal $d(k)$. For unvoiced sounds, a broadband, noise-like residual signal emerges, while for periodic segments $\tilde{d}(k)$ shows a discrete line spectrum. Furthermore, in the baseband, i.e., for

$$0 \leq \Omega \leq \frac{\pi}{r}, \tag{8.56}$$

the codec provides transparent transmission except for the quantization noise $\Delta(k) = \tilde{d}(k) - d'(k)$.

Compared to the LPC vocoder, the RELP concept improves the speech quality considerably. The perceptually relevant aspects, which are essential for the naturalness of the speech signal, are, at least in the baseband, transmitted transparently. For the transition between different types of sounds, a mixed voiced/unvoiced excitation is also possible. However, especially for high-pitched voices of women and children, a disturbing metallic or vocoder-like sound becomes noticeable in voiced segments due to the lack of preserving the harmonic structure. In voiced segments, discrete line spectra are generated by the

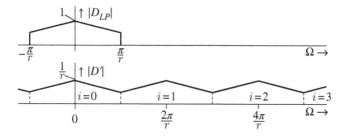

**Figure 8.22** Generation of the excitation signal in the baseband–RELP decoder.

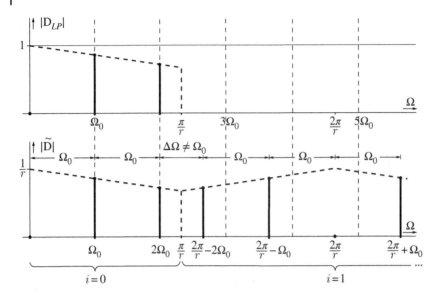

**Figure 8.23** Spectral mirroring in the RELP concept.

spectral mirroring, as shown in Figure 8.23. As a result, the spectral components outside the baseband generally do not occur at multiples of the fundamental frequency $\Omega_0$. This effect is less interfering with male voices, since most of the harmonics that hold the biggest part of the energy fall into the baseband. Due to the lowpass filtering, this codec principle is not suited for voiceband data and music signals.

The speech quality can be considerably improved if a long-term predictor is added. During the signal synthesis in voiced segments, the comb filter characteristic of the LTP-synthesis filter amplifies spectral components at multiples of the fundamental frequency and attenuates components lying in between.

### 8.5.3 Analysis by Synthesis: CELP

*CELP* is the most important concept for medium to low bit rate speech codecs. Based on the original proposal [Schroeder, Atal 1985], many variations exist, e.g., [Atal et al. 1991], [Furui, Sondhi 1992], [Goldberg, Riek 2000], [Hanzo et al. 2001], [Chu 2003], and [Kondoz 2004], which can be found in most of the speech codec standards used today in telecommunications, e.g., the GSM half-rate codec (GSM-HR), the GSM enhanced full-rate codec (GSM-EFR), the GSM/UMTS/LTE adaptive multi-rate codec (AMR), or the 5G enhanced voice services codec (EVS).

#### 8.5.3.1 Principle
The principle of analysis-by-synthesis coding was derived in Section 8.5.1 from the concept of predictive coding with noise shaping by replacing the scalar quantization of the residual signal $d(k)$ with vector quantization.

From these considerations, the structure outlined in Figure 8.20 resulted, which is depicted once more in detail in Figure 8.24a. For reasons of simplicity, we will again neglect the long-term prediction in the beginning.

(a)

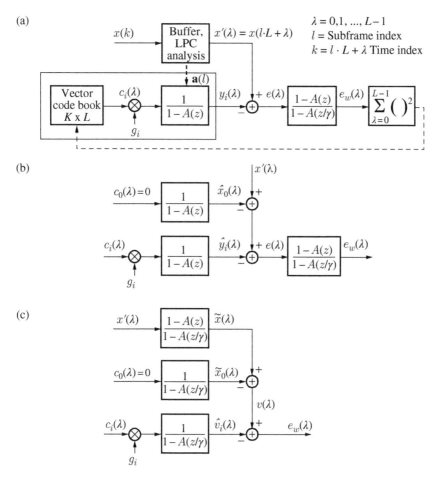

(b)

(c)

**Figure 8.24** Principle of the CELP codec. (a) Basic structure of the CELP concept, (b) Blockwise processing ($\lambda = 0, 1, \ldots, L - 1$) with ringing of the synthesis filter, (c) Structure equivalent to (b) with reduced complexity.

The encoder on the transmission side contains a complete decoder. For each of the $K$ code book vectors

$$\mathbf{c}_i = \left( c_i(0), c_i(1), \ldots, c_i(L-1) \right)^T; \quad i = 1, 2, \ldots, K \tag{8.57}$$

$L$ values $y_i(\lambda)$, $\lambda = 0, 1, \ldots, L - 1$, are provisionally synthesized. The reconstruction error

$$e(\lambda) = x'(\lambda) - y_i(\lambda); \quad \lambda = 0, 1, \ldots, L - 1 \tag{8.58}$$

is spectrally weighted so that the masking effect of human hearing is implicitly exploited (see also Section 8.5.1). The computationally intensive search of the best vector $\mathbf{c}_{i_{\mathrm{opt}}}$ is based on minimizing the mean square spectrally weighted reconstruction error $e_w(\lambda)$.

### 8.5.3.2 Fixed Code Book

Since, the vector code book according to Figure 8.24 was originally obtained from a stochastic sequence with Gaussian distribution [Schroeder, Atal 1985], the terms *stochastic* code book or *fixed code book* are commonly used. The following explanations show how the computational complexity of the stochastic code book search and the respective gain factor $g_i$ can be considerably reduced by modifying the structure of Figure 8.24a.

In CELP coders with the stochastic code book, short sequences of the residual waveform, i.e., blocks of $L$ samples, are coded. At the output of the (time-variant) recursive synthesis filter with the transfer function

$$H(z) = \frac{1}{1 - A(z)}, \tag{8.59}$$

each selected optimal excitation vector $\mathbf{c}_{i_{opt}}$ contributes to the reconstructed signal $y_i$. Due to the ringing of the synthesis filter, this contribution is not limited to the time interval of length $L$. Therefore, when coding the $l$-th signal segment

$$x'(\lambda) = x(l \cdot L + \lambda), \quad \lambda = 0, 1, \ldots, L - 1, \tag{8.60}$$

the ringing of the preceding block must be taken into account. In Figure 8.24b, this contribution is denoted $\hat{x}_0(\lambda)$. It is generated by means of a second synthesis filter, which is excited by a zero sequence $c_0(\lambda)$. At the beginning of the $l$-th block, its state variables are set to those which the synthesis filter achieved at the end of the $(l-1)$-th block for the best excitation vector. For the $l$-th block, the ringing contribution $\hat{x}_0(\lambda)$ from the synthesis filter of the $(l-1)$-th block can be precalculated. Then, the modified sequence $x'(\lambda) - \hat{x}_0(\lambda)$ is the target for the search of the new excitation vector $\mathbf{c}_i$.

Due to the linearity of the filter operations, the contribution of each excitation sequence $c_i(\lambda)$ to the weighted error sequence $e_w(\lambda)$ can be described by filtering with a cascade of two filters, i.e., the synthesis and the weighting filter. This yields the effective transmission function

$$\tilde{H}(z) = \frac{1}{1 - A(z)} \cdot \frac{1 - A(z)}{1 - A(z/\gamma)} \tag{8.61a}$$

$$= \frac{1}{1 - A(z/\gamma)} = \frac{1}{1 - F(z)}. \tag{8.61b}$$

The result is a marked reduction in computational complexity since now each excitation vector only needs to be filtered once.

With the same linearity argument, the contributions of the signal samples $x'(\lambda)$ and the ringing contribution $\hat{x}_0(\lambda)$ included in $e_w(\lambda)$ can be determined. The result is depicted in Figure 8.24c. In principle, the weighting filter is shifted to the three signal branches via the two summation points.

The new target signal $v(\lambda)$, which is only computed once for each interval, must be approximated by the sequence $\hat{v}_i(\lambda)$ ($\lambda = 0, 1, \ldots, L - 1$) with respect to the minimal mean square error by choosing the best excitation vector $\mathbf{c}_i$ and the optimal scaling factor $g_i$.

Part of the problem can be solved in closed form. If according to (8.61b) the impulse response of the weighted synthesis filter $1/(1 - A(z/\gamma))$ is denoted by $\tilde{h}(k)$, we have

$$\hat{v}_i(\lambda) = \sum_{\kappa=0}^{L-1} \tilde{h}(\lambda - \kappa) \cdot g_i \cdot c_i(\kappa); \quad \lambda = 0, 1, \ldots, L - 1. \tag{8.62}$$

For simplification, the following vector/matrix notation is introduced:

$$\hat{\mathbf{v}}_i = (\hat{v}_i(0), \hat{v}_i(1), \ldots, \hat{v}_i(L - 1))^T, \tag{8.63a}$$

$$\mathbf{v} = (v(0), v(1), \ldots, v(L - 1))^T, \tag{8.63b}$$

$$\mathbf{H} = \begin{pmatrix} \tilde{h}(0) & 0 & 0 & \cdots & 0 \\ \tilde{h}(1) & \tilde{h}(0) & 0 & \cdots & 0 \\ \tilde{h}(2) & \tilde{h}(1) & \tilde{h}(0) & \cdots & 0 \\ \vdots & \vdots & \vdots & \ddots & \vdots \\ \tilde{h}(L-1) & \tilde{h}(L-2) & \tilde{h}(L-3) & \cdots & \tilde{h}(0) \end{pmatrix}. \tag{8.63c}$$

Now the optimization criterion can be formulated as the squared norm of the reconstruction error vector:

$$\min_{i=1,\ldots,K} \sum_{\lambda=0}^{L-1} \left( v(\lambda) - \hat{v}_i(\lambda) \right)^2 = \min_{i=1,\ldots,K} \| \mathbf{v} - \hat{\mathbf{v}}_i \|^2. \tag{8.64a}$$

Hence, the term

$$E_i = \| \mathbf{v} - \hat{\mathbf{v}}_i \|^2 \tag{8.64b}$$

$$= \| \mathbf{v} - g_i \mathbf{H} \mathbf{c}_i \|^2 \tag{8.64c}$$

must be minimized.

By partial differentiation of $E_i$ with respect to $g_i$, the optimal gain factor is determined for each excitation vector $\mathbf{c}_i$:

$$g_i = \frac{\mathbf{v}^T \mathbf{H} \mathbf{c}_i}{\| \mathbf{H} \mathbf{c}_i \|^2}. \tag{8.65}$$

Inserting the optimal gain factor $g_i$ in (8.64c) leads to an expression for the minimal error energy that can be achieved with each excitation vector $\mathbf{c}_i$

$$E_i = \| \mathbf{v} \|^2 - \frac{(\mathbf{v}^T \mathbf{H} \mathbf{c}_i)^2}{\| \mathbf{H} \mathbf{c}_i \|^2}. \tag{8.66}$$

In view of the optimization, the term $\| \mathbf{v} \|^2$ is constant so that the criterion reduces to the maximization of the fraction

$$\frac{(\mathbf{v}^T \mathbf{H} \mathbf{c}_i)^2}{\| \mathbf{H} \mathbf{c}_i \|^2} \overset{!}{=} \max. \tag{8.67}$$

The search for the maximum through variation of $i$ leads to the best excitation vector $\mathbf{c}_{i_{opt}}$. Then, the respective optimal gain factor $g_{i_{opt}}$ is explicitly calculated according to (8.65).

The computational complexity of the code book search can easily be estimated if, in view of a realization with programmable signal processors, a multiplication with subsequent addition is counted as one operation per instruction cycle (multiply–accumulate operation, MAC). For the division, we assume a complexity of 16 instruction cycles.

Listed in Table 8.3 are the different numbers of arithmetic operations for the code book search according to (8.65) and (8.67).

The term $\tilde{\mathbf{y}}_i = \mathbf{v}^T \mathbf{H}$ is calculated only once. Taking into account the triangular shape of $\mathbf{H}$, see (8.63c), the product $\mathbf{v}^T \mathbf{H}$ does not require $L^2$ but only $L(L+1)/2$ operations. The numerator and denominator of (8.65), which occur as intermediate results when utilizing (8.67), are not recomputed.

Under these assumptions the required mean computational effort results in

$$\mathrm{CE} = \frac{K(L(L+5)/2 + 17) + 16}{L} f_s \tag{8.68}$$

**Table 8.3** Processing steps and number of arithmetic operations for the code book search $i = 1, 2, \ldots K$ and subframe vector length $L$.

| (1) | $\tilde{\mathbf{y}}_i = \mathbf{H}\mathbf{c}_i$ in (8.67) | $K \cdot L(L+1)/2$ |
|---|---|---|
| (2) | $(\mathbf{v}^T \tilde{\mathbf{y}}_i)^2$ in (8.67) | $K \cdot (L+1)$ |
| (3) | $\|\tilde{\mathbf{y}}_i\|^2 = \|\mathbf{H}\mathbf{c}_i\|^2$ in (8.67) | $K \cdot L$ |
| (4) | Division (8.67) | $K \cdot 16$ |
| (5) | Division (8.65) | $1 \cdot 16$ |
| | Sum: | $K(L(L+5)/2 + 17) + 16$ |

operations per second. Hence, a typical dimensioning with $f_s = 8$ kHz, $K = 256$ and $L = 40$ requires a high computational effort of approximately $46.1 \cdot 10^6$ operations per second.

Numerous approaches to reduce this computational complexity have been proposed in the literature. One successful approach is the choice of code vectors with only a few non-zero components. If, for example, each vector $\mathbf{c}_i$ contains only four non-zero pulses, the number of arithmetic operations for the denominator in (8.67) (see Table 8.3b)) decreases from $K \cdot L(L+1)/2$ to $K \cdot 4 \cdot L$. Thus, for $L = 40$ the dominating part of the computational complexity for the code book search is reduced by a factor of 5 (e.g., [Adoul et al. 1987]).

A widely used approach is the choice of a structured code book, which is called the *algebraic code book* (ACELP) [Laflamme et al. 1990]. A typical version can be found in the ITU codec G.729 [ITU-T Rec. G.729 1996], [Salami et al. 1997a]. The excitation sequence $c_i(\lambda)$ of length $L = 40$ contains only four non-zero pulses in the positions $\lambda_0, \lambda_1, \lambda_2$, and $\lambda_3$

$$c_i(\lambda) \in \{+1, 0, -1\}; \quad \lambda = 0, 1, \ldots, 39. \tag{8.69}$$

The 40 possible positions for non-zero pulses are divided into four tracks with 8 or 16 positions as indicated in Table 8.4.

With the signs $s_0, \ldots, s_3$ of the four selected pulses we construct the excitation sequence

$$c_i(\lambda) = s_0 \,\delta(\lambda - \lambda_0) + s_1 \,\delta(\lambda - \lambda_1) + s_2 \,\delta(\lambda - \lambda_2) + s_3 \,\delta(\lambda - \lambda_3) \tag{8.70}$$

$$\delta(\lambda) = \begin{cases} 1 & \lambda = 0 \\ 0 & \text{else.} \end{cases}$$

**Table 8.4** ACELP code book: positions for non zero pulses.

| Pulse positions | Sign | Tracks | Number of bits |
|---|---|---|---|
| $\lambda_0$ | $s_0$ | 0, 5, 10, 15, 20, 25, 30, 35 | 3+1 |
| $\lambda_1$ | $s_1$ | 1, 6, 11, 16, 21, 26, 31, 36 | 3+1 |
| $\lambda_2$ | $s_2$ | 2, 7, 12, 17, 22, 27, 32, 37 | 3+1 |
| $\lambda_3$ | $s_3$ | 3, 8, 13, 18, 23, 28, 33, 38<br>4, 9, 14, 19, 24, 29, 34, 39 | 4+1 |

This sparse ternary code book allows an efficient search procedure with significantly reduced complexity.

### 8.5.3.3 Long-Term Prediction, Adaptive Code Book

So far, the principle of CELP coding has been discussed without consideration of long-term prediction. To exploit the highly periodic nature of speech signals, especially in voiced speech segments, a long-term predictor is essential. Figure 8.25a shows the CELP codec of Figure 8.24a extended by an LTP-synthesis filter

$$H_{LTP}(z) = \frac{1}{1 - P(z)} = \frac{1}{1 - b_j \cdot z^{-N_0}} \tag{8.71}$$

as described in Section 8.5.1 and Figure 8.16.

Thus, the excitation sequence $u'(\lambda)$ of the LP-synthesis filter

$$H_{LP}(z) = \frac{1}{1 - A(z)} \tag{8.72}$$

is given by

$$u'(\lambda) = u(l \cdot L + \lambda) = g_i \cdot c_i(\lambda) + b_j \cdot u'(\lambda - N_0), \quad j = N_0. \tag{8.73a}$$

(a)

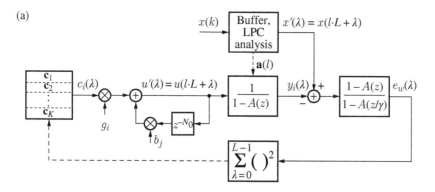

(b)

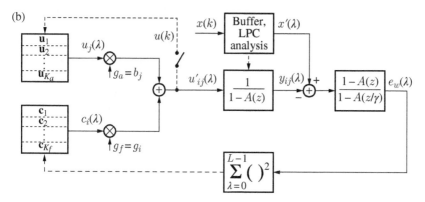

**Figure 8.25** CELP codec with short-term and long-term prediction. (a) Conventional realization of the LTP-synthesis filter, (b) Realization of the LTP loop by means of an adaptive code book.

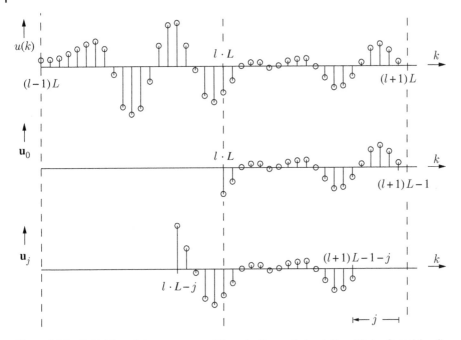

**Figure 8.26** Definition of the vectors $\mathbf{u}_j$ of the adaptive code book ($L = 20, j = 0$, and $j = 4$).

As shown in Figure 8.26, we combine $L$ consecutive samples of $u(k)$ according to

$$u'(\lambda - j) = u(l \cdot L + \lambda - j) \doteq u_j(\lambda); \; j = \text{const.}, \; \lambda = 0, 1, \dots, L - 1 \tag{8.73b}$$

to a vector

$$\mathbf{u}_j = (u_j(0), u_j(1), \dots, u_j(L - 1))^T \tag{8.73c}$$

so that (8.73a) yields

$$\mathbf{u}' \equiv \mathbf{u}_0 = g_i \, \mathbf{c}_i + b_j \, \mathbf{u}_{N_0} . \tag{8.73d}$$

This vector superposition is illustrated in Figure 8.25b. A generalization that still needs to be justified is made by substituting the pitch parameter $N_0$ with an index $j$ ($j = 1, 2, \dots, K_a$). In this concept, the LTP-synthesis filter is replaced by a pseudo code book containing vectors $\mathbf{u}_j$ of length $L$, which are overlapping segments of the recent past of the LP excitation signal. In order to distinguish the two partial contributions $g_i \, \mathbf{c}_i$ and $b \, \mathbf{u}_j$ according to (8.73d), double indexing is introduced according to $u'_{ij}$, indicating the two contributions $c_i$ and $u_j$. Furthermore, the weighting factors are replaced by $g_a = b_j$ and $g_f = g_i$. With the vector components $u_0(\lambda)$, $c_i(\lambda)$, and $u_j(\lambda)$ from (8.73d), the excitation sequence $u'_{ij}(\lambda)$ of Figure 8.25 is generated:

$$u'_{ij}(\lambda) \equiv u_0(\lambda) = g_i \, c_i(\lambda) + b_j \, u_j(\lambda) \tag{8.73e}$$

$$= g_f \, c_i(\lambda) + g_a \, u_j(\lambda) \tag{8.73f}$$

with $\lambda = 0, 1, \dots, L - 1$.

For $j = N_0$, the terms (8.73a) and (8.73e) obviously coincide. The advantage of the modified structure is that the contribution of the LTP loop can be treated in the same way as that of the fixed code book. In particular, in an analysis-by-synthesis procedure, the LTP parameters $b_j$ and $N_0$ or $g_a$ and $j_{opt}$, respectively, can be determined just like the parameters $g_f = g_i$ and $i_{opt}$. This procedure is called *closed-loop* LTP. Since, the content of the second code book changes depending on the signal, it is called the *adaptive code book*. Due to the *closed-loop* search of the LTP parameters $g_a$ and $j_{opt}$, considerably better speech quality and a higher SNR of the synthesized speech signal $y(k)$ by 2–5 dB are achieved [Singhal, Atal 1984].

Finally, we note:

- In order to reach the optimal result, a complete search of all combinations of the two excitation vectors $\mathbf{u}_j$ and $\mathbf{c}_i$ with the respective optimal weighting factors $g_a$ and $g_f$ would have to be performed. Since, this procedure has an extremely high computational complexity for the common code book dimensions, a suboptimal solution is chosen by a sequential search. First, in analogy to the criterion given in (8.67) and (8.65), the optimal adaptive code book entry and gain are determined. For the search, we can make use of the fact that two successive vectors $\mathbf{u}_j$ and $\mathbf{u}_{j+1}$ consist of the same elements except for the first and last element of $\mathbf{u}_j$ and $\mathbf{u}_{j+1}$, respectively. The selected code book entry is scaled and filtered by the LP-synthesis filter and subtracted from the speech signal. The resulting modified target vector is used for the stochastic code book search by applying (8.67) and (8.65).
- While searching for the best contribution $\mathbf{u}_j$, the adaptive code book is not altered for the current subframe of length $L$. Only after completing the search for the best vector $\mathbf{c}_i$ from the fixed code book is the adaptive codebook updated.
- Strictly speaking, the equivalence between Figure 8.25a,b is only given for $j_{opt} = N_0 \geq L$. Due to the block processing with subframe length $L$ and the sequential search in the adaptive and fixed code book, depending on the constellation of $j$ and $\lambda$ no samples $u(k)$ are available for $N_0 < L$. When searching the adaptive code book, the contribution from the fixed code book is not yet known. In this case, the missing entries of the adaptive code book are generated, for example, from the periodic repetition of the last $N_0$ samples of the excitation sequence $u(k)$ of the preceding frame (subframe index $l - 1$).

## 8.6 Adaptive Postfiltering

Predictive speech coding with effectively 0.5 to 2 bits per sample is the key element of many commercial applications, such as mobile communication. This can be largely attributed to the technique of noise shaping or spectral error weighting, respectively, where at the speech encoder the reconstruction error is shaped. This spectral masking improves the perceptual speech quality significantly. However, admitting larger reconstruction errors in the region of the formants leads to an increased error power. The signal-to-noise ratio over frequency takes its highest values in the intervals with spectral peaks, while it is significantly worse in the spectral valleys (see also Figure 8.9). Typically, the noise around the spectral peaks is below the masking threshold, while in valley regions it is not. Consequently, at low bit rates spectral error weighting alone is not sufficient to completely mask the noise so that an audible speech-dependent noise remains.

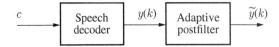

**Figure 8.27** Adaptive postfiltering of the decoded speech signal.

At the output of the speech decoder the effect of the non-masked quantization noise can be reduced to a certain extent through an adaptive postfilter (see Figure 8.27). Since most of the perceived noise components come from spectral valleys, the postfilter attenuates the frequency components between pitch harmonics, as well as the components between formants. In speech perception, the formants and local spectral peaks are much more important than spectral valley regions. Therefore, by attenuating the components in spectral valleys, the postfilter only introduces minimal distortion in the speech signal, while a substantial noise reduction can be achieved.

In the literature, various postfilters that exploit this effect have been proposed. A unifying presentation, on which the following considerations are based, can be found in [Chen, Gersho 1995]. Different variants of this general approach are contained in various codec standards (see Section 8.7).

The desired characteristics of the instantaneous frequency response of an adaptive postfilter are depicted for a voiced signal segment in Figure 8.28.

The frequency response of the postfilter follows the spectral envelope and the spectral fine structure in such a way that the formants and the local maxima are preferably not altered, whereas the spectral valleys between the formants and pitch harmonic peaks are attenuated. The filter parameters can generally be derived from the coefficients of the speech decoder's LTP- and LP-synthesis filters, or it can be determined by signal $y(k)$.

The required attenuation of the spectral valleys can be performed separately for the spectral envelope and for the spectral fine structure. Consequently, a short-term and a long-term postfilter with the transfer functions $H_A(z)$ and $H_P(z)$ are cascaded to achieve the desired characteristics:

$$H_{PF}(z) = H_A(z) \cdot H_P(z). \tag{8.74}$$

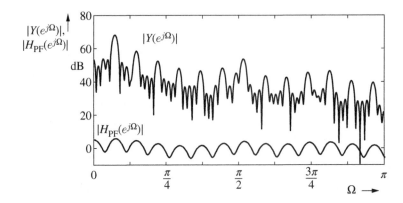

**Figure 8.28** Signal spectrum $|Y(e^{j\Omega})|$ and magnitude response $|H_{PF}(e^{j\Omega})|$ of an adaptive postfilter.

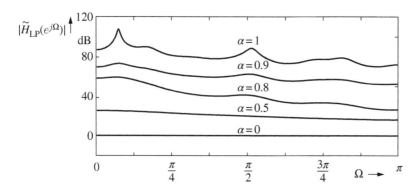

**Figure 8.29** Magnitude response of $\tilde{H}_{LP}(e^{j\Omega})$ for different $\alpha$; adjacent curves are raised by 20 dB each.

For the construction of $H_A(z)$, the pole radii of the LP-synthesis filter are scaled with a factor $\alpha$ according to

$$\tilde{H}_{LP}(z) = \frac{1}{1 - A(z/\alpha)}; \qquad 0 \le \alpha \le 1. \tag{8.75}$$

As a result, the poles are moved radially toward the origin of the $z$-plane. The effect on the magnitude response is shown in Figure 8.29.

With decreasing $\alpha$ the resonances become less pronounced as well. However, the magnitude responses also show a tilt toward higher frequencies, i.e., the relative intensity of the formants will change due to the postfilter, which is not desired.

This tilt can be largely avoided by means of the following transfer function:

$$H_A(z) = g_A \cdot \frac{1 - A(z/\beta)}{1 - A(z/\alpha)} \cdot (1 - \gamma \cdot z^{-1}); \qquad 0 \le \alpha, \beta, \gamma \le 1. \tag{8.76}$$

Here $g_A$ denotes a (time-variant) scaling factor that ensures that the power of the signal remains unchanged by the filtering process.

On a logarithmic scale, the relation between the terms $(1 - A(z/\beta))$ and $(1 - A(z/\alpha))$ appears as the difference between two curves of Figure 8.29. Thus, the spectral tilt is already partly compensated. With a fixed (or adaptive) coefficient $\gamma$, the last product term in (8.76), i.e., $(1 - \gamma \cdot z^{-1})$, performs the remaining tilt compensation.

Figure 8.30a,b exemplarily depicts the speech spectrum and the resulting magnitude response $|H_A(e^{j\Omega})|$ of the short-term postfilter.

The long-term postfilter $H_P(z)$ is either derived from the long-term predictor $P(z)$ or obtained by means of a renewed LTP analysis of the decoded signal $y(k)$. The latter approach is preferable if the LTP parameters (gain factor $b$ or $g_a$, respectively, and delay $N_0$) on the transmission side are determined by an analysis-by-synthesis procedure. Due to the error criterion, the delay $N_0$ does not necessarily match the instantaneous pitch period of the signal in this case.

The periodic fine structure of Figure 8.28 can be realized using the following approach:

$$H_P(z) = g_P \cdot \frac{1 + \varepsilon \cdot z^{-N_0}}{1 - \eta \cdot z^{-N_0}}; \qquad 0 \le \varepsilon, \eta \le 1, \tag{8.77}$$

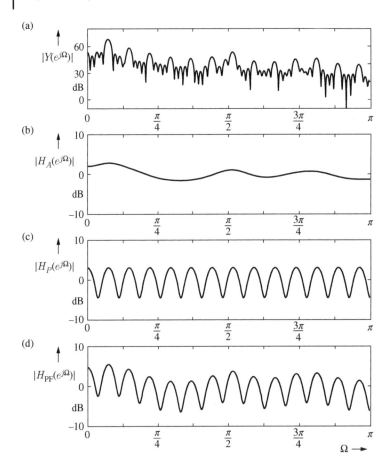

**Figure 8.30**  (a) Magnitude spectrum of speech $|Y(e^{j\Omega})|$, (b) Magnitude response of the short-term postfilter $|H_A(e^{j\Omega})|$ (c) Magnitude response of the long-term postfilter $|H_P(e^{j\Omega})|$ (d) Magnitude response of the complete postfilter $|H_{PF}(e^{j\Omega})|$ ($\alpha = 0.8, \beta = 0.5, \gamma = 0.2, \varepsilon = 0.4, \eta = 0.05$).

with the scaling factor $g_P$. The transfer function $H_P(z)$ of the pole–zero postfilter has its poles at

$$z_{\infty i} = \rho \cdot e^{j\Omega_{\infty i}} \text{ with } \rho = {}^{N_0}\!\sqrt{\eta}; \ \Omega_{\infty i} = \frac{2\pi}{N_0} i \tag{8.78}$$

and its zeros at

$$z_{0i} = \zeta \cdot e^{j\Omega_{0i}} \text{ with } \zeta = {}^{N_0}\!\sqrt{\varepsilon}; \ \Omega_{0i} = \frac{\pi}{N_0}(2i+1) \tag{8.79}$$

for $i = 0, 1, \ldots, N_0 - 1$.

[Chen, Gersho 1995] propose to adjust the coefficients $\varepsilon$ and $\eta$ depending on the LTP parameter $b$ (see, e.g., (6.83-a)) as follows:

$$\varepsilon = c_1 \cdot f(b), \eta = c_2 \cdot f(b); \ 0 \le c_1, c_2 < 1; \ c_1 + c_2 = 0.5 \tag{8.80}$$

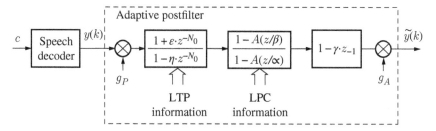

**Figure 8.31** Speech decoder with adaptive postfilter.

with

$$f(b) = \begin{cases} 0 & b < c_3 \\ b & c_3 \le b \le 1 \\ 1 & 1 < b \end{cases} \quad 0 < c_3 < 1. \tag{8.81}$$

In unvoiced segments, in which $b$ or $g_a$ takes very small values, the long-term postfilter is turned off. The parameters $\alpha, \beta, \gamma, \varepsilon, \eta, c_1, c_2$, and $c_3$ must be experimentally adjusted to the respective codec. In general, $c_2$ is set to very small values or even zero so that the recursive part in (8.77) is canceled. A typical choice is: $\alpha = 0.8, \beta = 0.5, c_1 = 0.5, c_2 = 0.0$, and $c_3 = 0.6$.

The scaling factor $g_P$ in (8.77) is adjusted in such a way that the short-term power of $y(k)$ is approximately not altered by the filtering with $H_P(z)$. The factor $g_P$ is determined within relatively small time intervals, e.g., according to [Chen, Gersho 1995],

$$g_P = \frac{1 - \eta/b}{1 + \varepsilon/b}. \tag{8.82}$$

The second scaling factor $g_A$ (see also Figure 8.31) ensures that the powers of $y(k)$ and $\tilde{y}(k)$ match with a time resolution that lies in the range of the duration of a syllable.

All in all, this concept of adaptive postfiltering provides a further, in some circumstances significant, reduction of the audible quantization distortions. However, this postfiltering can cause considerable signal distortions if the signal runs through a chain of two or more codecs (tandem operation). In this case, it is better to switch off the postfilter.

## 8.7 Speech Codec Standards: Selected Examples

For different application requirements, there are codec standards that differ with respect to speech quality, bit rate $B$, complexity, and signal delay. Table 8.5 provides an overview of the most common codecs.

Two popular NB codecs are the ADPCM codec ITU-T / G.726 (1985), mainly used in cordless DECT telephone sets, and the GSM Full Rate codec (1988) for cellphones. The latter one, which is based on the RELP principle, has been the mandatory codec for GSM cell phones.

In modern smart phones, more powerful NB CELP codecs are used, i.e., the Enhanced Full Rate codec (EFR,1996), and the Adaptive Multi-Rate codec (AMR, 1998).

**Table 8.5** Selected standards for speech communication.

| Codec | Standard | B/(kbit/s) |
|---|---|---|
| **ADPCM:** | ITU-T/G.726 | 32 |
| Adaptive Differential Pulse Code Modulation | | (16, 24, 40) |
| **LD-CELP:** Low-Delay CELP Speech Coder | ITU-T/G.728 | 16 |
| **CS-ACELP:** | ITU-T/G.729 | 8 |
| Conjugate-Structure Algebraic CELP Codec | | |
| **Split band ADPCM:** | ITU-T G.722 | 64 (48, 56) |
| 7 kHz Audio Coding within 64 kbit/s | | |
| **GSM-FR:** Full Rate Speech Transcoding | ETSI 06.10 | 13 |
| **GSM-HR:** Half Rate Speech Transcoding | ETSI 06.20 | 5.6 |
| **GSM-EFR:** | ETSI 06.60 | 12.2 + 0.8 |
| Enhanced Full Rate Speech Transcoding | | |
| **GSM-AMR:** | ETSI 06.90 | 4.75–12.2 |
| Adaptive Multi-Rate Speech Transcoding | | |
| **AMR-WB:** Adaptive Multi-Rate | ETSI/3GPP TS 126.190 | 6.6–23.85 |
| Wideband Speech Transcoding | ITU-T/G.722.2 | |
| **AMR-WB⁺:** Extended | ETSI/3GPP TS 126.290 | 12.8–38.4 |
| Adaptive Multi-Rate Wideband Codec | | |
| **EVS:** Codec for Enhanced Voice Services | ETSI/3GPP TS 126.445 | 5.9–128.0 |
| **IMBE:** | INMARSAT (IMBE) | 4.15 |
| Improved Multi-Band Excitation Codec | | |
| **OPUS:** | IETF RFC 6716 | 6.0 – 510 |

ITU: International Telecommunication Union; ETSI: European Telecommunications Standards Institute;
TIA: Telecommunications Industry Association; IETF: Internet Engineering Task Force;
3GPP: 3rd Generation Partnership Project; TS: Technical Specification

In 1985, a first WB speech codec for digital telephony and teleconferencing with bit rates of 64, 56, and 48 kbit/s was specified by CCITT [ITU-T Rec. G.722 1988]. For the sake of compatibility with the existing terminals, this standard comprises a dedicated signaling procedure for the automatic detection of the capabilities of the far-end terminal and a fall-back mode to NB telephony ($A$-law, $\mu$-law).

This codec is used, e.g., for internet telephony with cordless telephones that are based on the CAT-iq standard (Cordless Advanced Technology – internet and quality).

In 1999, a second wideband codec [ITU-T Rec. G.722.1 1999] was introduced that produces almost comparable speech quality at reduced bit rates of 32 and 24 kbit/s.

A breakthrough was the standardized *adaptive multi-rate wideband* (AMR-WB, 2001) speech codec, specified by the 3GPP initiative (3rd Generation Partnership Project) and by ETSI for cellular networks. This 7 kHz codec has eight different modes with bit rates from 6.6 up to 23.85 kbit/s with increasing quality [Salami et al. 1997a], [3GPP TS 26.190 2001]. This codec has also been adopted by ITU [ITU-T Rec. G.722.2 2002]. A further extension has been specified by 3GPP with the AMR-WB+ codec (2005), which is an extension of the AMR-WB codec in terms of the frequency range up to 16 kHz and bit rates up to 32 kbit/s. The aforementioned AMR-WB codec is part of the AMR-WB+ standard [3GPP TS 26.290 2005].

The latest codec standard is the EVS codec (Enhanced Voice Services, 2014), a versatile multi-rate audio codec for mobile and multi-media systems that has been optimized for a wide range of bit rates, bandwidths and qualities [3GPP TS 126.445 2014], [Dietz et al. 2015]. Due to its flexibility, the EVS codec is appropriate for speech signals, music signals, and mixed speech/music signals.

### 8.7.1 GSM Full-Rate Codec

The concept of residual excited linear prediction (RELP) has gained great significance for digital mobile radio communications [Sluijter 2005]. The *full-rate codec* of the European GSM (*Global System for Mobile communications*) standard [Vary et al. 1988], [Rec. GSM 06.10 1992] is based on this principle as shown in Figure 8.32.

The basic structure corresponds to the concept given in Figure 8.18b. Short-term prediction is implemented as an open-loop filter with order $n = 8$, while the long-term predictor (LTP) works in a closed loop. The baseband lowpass filter, the decimator, and the scalar quantizer are located within the LTP loop.

Thus, the prediction gain of the long-term prediction contributes to reducing the effective quantization noise power. Furthermore, the following details must be considered.

The second residual signal designated by $e(k)$ is processed blockwise. The lowpass filter is a linear-phase FIR filter with $m = 11$ coefficients. The filtering process is implemented as

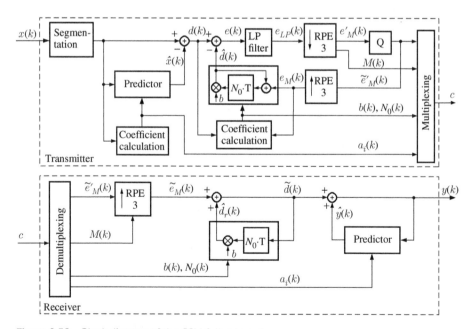

**Figure 8.32** Block diagram of the GSM full-rate codec.

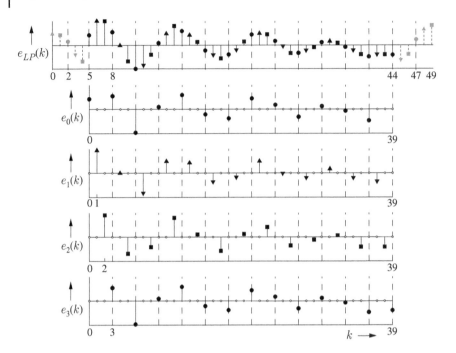

**Figure 8.33** Adaptive decimation according to the RPE principle.

a *block filter*, where each $L = 40$ samples $e(k)$ are supplemented by $m - 1$ zeros and the filtered version $e_{LP}(k)$, consisting of $L + m - 1 = 50$ values, is calculated. In the block named RPE (*Regular Pulse Excitation*), an adaptive downsampling by a factor $r = 3$ is applied.[1] As depicted in Figure 8.33, the downsampling is adaptive in the sense that four different possibilities exist in principle, which vary in the decimation grid starting at $k = 0$, $k = 1$, $k = 2$, or $k = 3$. The sequence $e_{LP}(k)$ is divided into four subsequences $e_i(k)$, with the sequences $e_0(k)$ and $e_3(k)$ differing only in the first and in the last sample.

From the $L = 40$ central samples of $e_{LP}(k)$, $(k = 5, \dots, 44)$, 13 samples are selected with regard to the best subsequence according to the energy criterion

$$E_M = \max_i \sum_{\lambda=0}^{12} e_{LP}^2(i + \lambda \cdot 3 + 5); \quad i = 0, 1, 2, 3. \tag{8.83}$$

Besides the subsequence $e_M = e_{i_{opt}}$, the grid position $M = i_{opt}$ is determined every 5 ms as well. This quasi-random variation helps to avoid the tonal–metallic sound of the baseband–RELP codec. However, especially for high-pitched voices, a slight roughness of the reconstructed speech signal is produced [Sluyter et al. 1988].

The 13 selected samples are quantized with AQF (see Section 7.5) by normalizing and uniformly quantizing them with eight levels (3 bits). For the block maximum, a logarithmic quantizer with $2^6$ levels is used, while the grid position is represented by 2 bits. For the

---

1 This particular variation of the RELP structure with block filtering and adaptive downsampling results as a special case of the so-called RPE method [Kroon et al. 1986]. It is an analysis-by-synthesis coding process, where pulse-shaped excitation sequences are optimized on a regular grid.

residual signal, this leads to a bit rate of

$$B' = (13 \cdot 3 + 6 + 2)\,\text{bit}/5\,\text{ms} \tag{8.84a}$$

$$= 9.4\,\text{kbit/s}. \tag{8.84b}$$

The LP coefficients are coded as LARs with 36 bits/20 ms = 1.8 kbit/s (see Section 7.7) and the LTP parameters $N_0$ and $b$ with $(7 + 2)$ bits/5 ms = 1.8 kbit/s. This results in a total bit rate of $B = 13.0$ kbit/s.

The example of Figure 8.34 provides an insight into the mechanism of the codec. For the different intermediate signals of Figure 8.32, a short-term spectral analysis was performed. A polyphase filter bank with a 3 dB channel bandwidth of 4 kHz/128 = 31.25 Hz and a 40 dB bandwidth of 62.5 Hz was utilized. The same linear scale was used for all magnitude responses.

Starting from the spectrum of the input signal $x$, we can clearly observe in the residual signal $d$ the spectral whitening effect of the LP-analysis filter in Figure 8.34b. Regarding the spectrum of the LTP residual signal $e$ (Figure 8.34c), the closed-loop structure containing a lowpass filter with a cutoff frequency of 4/3 kHz must be considered. Hence, a further prediction gain is noticeable only in the range up to $\approx 1.33$ kHz. The LTP excitation signal $e_M$ essentially contains the baseband of the residual signal $e$, which is transmitted to the receiver and broadened by time-varying spectral folding (Figure 8.34d).

The harmonic structure is regained in the residual $\tilde{d}$ by means of the LTP-synthesis filter (Figure 8.34e), and the spectral envelope is reconstructed through the LP-synthesis filtering (Figure 8.34f).

The comparison of Figure 8.34a–f shows a transparent transmission in the range up to approx. 1.33 kHz, whereas the spectral amplitude characteristic of the higher-frequency components is only approximately reproduced. Due to the properties of human hearing, a relatively high perceptual quality can be obtained, while the computational complexity is relatively small in comparison to the class of CELP codecs, which is the subject of Sections 8.7.2 to 8.7.6.

### 8.7.2 EFR Codec

Most of the recent standards in speech coding are based on the concept of CELP. Two representative examples are the GSM-enhanced full-rate codec (EFR) and the GSM-adaptive multi-rate codec (AMR). In comparison to the first GSM full-rate codec as described in Section 8.5.2, the EFR codec [ETSI Rec. GSM 06.60 1996], [Järvinen et al. 1997] gives a substantially better quality, which is almost equivalent to the speech quality of ADPCM at 32 kbit/s [ITU-T Rec. G.726 1990].

A simplified block diagram of the EFR decoder is given in Figure 8.35. The decoder structure contains the basic CELP elements of a fixed and an adaptive code book with individual gain factors $g_f$ and $g_a$. In addition, the decoded output signal $\hat{s}(k)$ is processed by an adaptive postfilter to improve the subjectively perceived speech quality (see Section 8.6). The frame length is 20 ms, and twice per frame 10 LP coefficients are calculated, which are transformed for quantization into the LSF representation. A dedicated vector matrix quantization scheme is applied for each set of 20 LSF coefficients [ETSI Rec. GSM 06.60 1996].

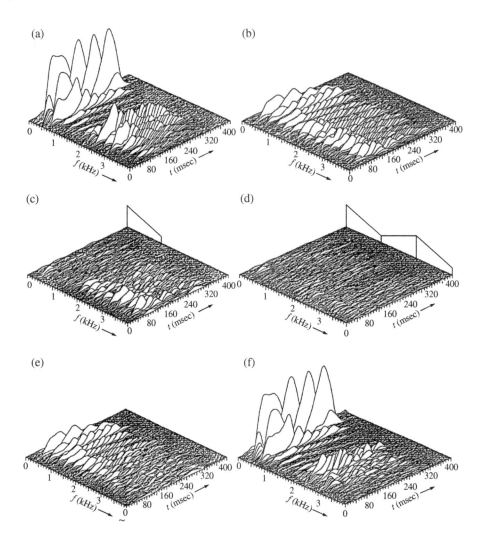

(a)

(b)

(c)

(d)

(e)

(f)

**Figure 8.34** Example of the GSM full-rate codec; short-term spectral analysis for the syllable "De" (female voice). (a) Encoder input $x$, (b) LPC residual $d$, (c) LTP residual $e$, (d) LTP excitation $e_M$, (e) LTP excitation $d$, and (f) decoder output $y$. Source: [Vary, Hofmann 1988].

The EFR codec is based on a fixed ACELP [Salami et al. 1997b] with a subframe length of $L = 40$ consisting of effectively $K = 2^{35}$ vectors of dimension 40 with 10 non zero values $s_\mu \in \{+1, -1\}$ each:

$$c_i(\lambda) = \sum_{\mu=0}^{9} s_\mu \cdot \delta(\lambda - \lambda_\mu); \qquad s_\mu \in \{+1, -1\}. \tag{8.85}$$

The code vectors are constructed, as shown in Table 8.6, from five tracks of eight interleaved positions, with two pulses per track. The signs of the two pulses are encoded with

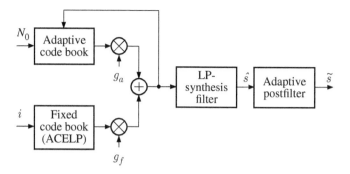

**Figure 8.35** Simplified block diagram of the enhanced full-rate decoder (GSM-EFR).

**Table 8.6** ACELP code book of the EFR codec.

| Pulse position | Sign | Positions | Bits |
|---|---|---|---|
| $\lambda_0, \lambda_5$ | $s_0, s_5$ | 0, 5, 10, 15, 20, 25, 30, 35 | $2 \times 3 + 1$ |
| $\lambda_1, \lambda_6$ | $s_1, s_6$ | 1, 6, 11, 16, 21, 26, 31, 36 | $2 \times 3 + 1$ |
| $\lambda_2, \lambda_7$ | $s_2, s_7$ | 2, 7, 12, 17, 22, 27, 32, 37 | $2 \times 3 + 1$ |
| $\lambda_3, \lambda_8$ | $s_3, s_8$ | 3, 8, 13, 18, 23, 28, 33, 38 | $2 \times 3 + 1$ |
| $\lambda_4, \lambda_9$ | $s_4, s_9$ | 4, 9, 14, 19, 24, 29, 34, 39 | $2 \times 3 + 1$ |

only one bit. This bit indicates the sign of the first pulse, whereas the sign of the second pulse depends on its position relative to the first pulse.

The bit allocation is summarized in Table 8.7.

### 8.7.3 Adaptive Multi-Rate Narrowband Codec (AMR-NB)

The second CELP example is the adaptive multi-rate narrowband codec (AMR-NB) [ETSI Rec. GSM 06.90 1998], [Bruhn et al. 1999], [Ekudden et al. 1999], [Järvinen 2000], which was originally designed for GSM and the Universal Mobile Telecommunications System (UMTS) and is also used in LTE (Long-term Evolution) networks.

The analog NB speech signal (200 Hz ... 3.4 kHz) is sampled at the rate $f_s = 8$ kHz. The codec can be considered as an extension of the EFR codec. The AMR codec has eight differ-ent bit rates from 4.75 kbit/s to 12.2 kbit/s. The mode of the highest bit rate is identical to the EFR codec. The overall structure is very similar to the block diagram of Figure 8.35. The dif-ferent bit rates are primarily achieved by using different ACELP code books. The objective of the AMR codec is to increase error robustness for better speech quality in adverse channel conditions. In GSM, the codec can be used in the half-rate and the full-rate channel. In the latter case, the gross bit rate including channel coding is 22.8 kbit/s. The bit rate allocation between speech coding and channel coding is controlled dynamically by the network. The codec mode can be switched every 20 ms. If the channel gets worse, the bit rate for chan-nel coding (error protection) is increased at the expense of the bit rate for speech coding. In the GSM full-rate channel, the AMR codec extends the lower radio channel C/I limit

**Table 8.7** Bit allocation of the EFR codec.

| Parameter | Subframes 1 and 2 (5 ms each) | Subframes 3 and 4 (5 ms each) | Bits per frame 20 ms |
|---|---|---|---|
| 2 × 10 LPC coefficients | | | 38 |
| (*line spectral frequencies*) | | | |
| Adaptive code book | | | |
| • Delay $N_0$ | 9 | 6 | 30 |
| • Gain $g_a$ | 4 | 4 | 16 |
| Fixed code book (ACELP) | | | |
| • Pulse positions and signs | 35 | 35 | 140 |
| • Gain $g_s$ | 5 | 5 | 20 |
| Bits every 20 ms | | | 244 |

(Carrier-to-Interference limit) of the EFR codec of C/I = 9 dB down to about C/I ≥ 4–7 dB. The introduction of this codec into existing GSM networks improved significantly the coverage, especially in buildings.

A simulation example is shown in Figure 8.36.

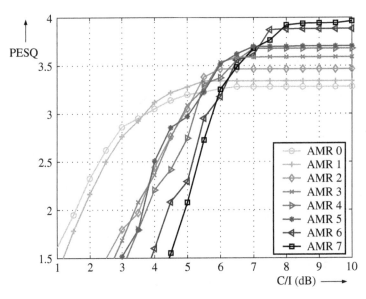

**Figure 8.36** Speech quality improvement in adverse radio channel conditions by using the AMR codec; GSM channel TU50, average PESQ values for 70 seconds of speech. PESQ: *Perceptual Evaluation of Speech Quality*, C/I: Carrier-to-Interference ratio.

The AMR codec was adopted by the 3GPP (3rd Generation Partnership Project) as the default speech codec for 3G wideband CDMA systems (UMTS, CDMA2000, LTE) [3GPP TS 26.090 2001].

### 8.7.4 ITU-T/G.722: 7 kHz Audio Coding within 64 kbit/s

This is the very first ITU-T standard G.722 for WB coding with an increased bandwidth of $0.05\,\mathrm{kHz} \leq f \leq 7.0\,\mathrm{kHz}$ [Taka, Maitre 1986], [ITU-T Rec. G.722 1988]. The codec was designed for audio- and video-conferencing over Integrated Services Digital Networks (ISDN).

The analog WB signal is sampled at $f_s = 16\,\mathrm{kHz}$ and split into two subbands using a QMF filterbank with a prototype lowpass filter degree of $n = 24$ (see Section 4.3) and sampling rate decimation by a factor of 2. Both subband signals are encoded separately by ADPCM techniques (see Section 8.3.4) with $w = 4, 5$, or 6 bits (or 32, 40, 48 kbit/s) in the lower subband and $w = 2$ (or 16 kbit/s) in the upper subband. The algorithmic delay is only 1.5 ms.

A block diagram is shown in Figure 8.37.

### 8.7.5 Adaptive Multi-Rate Wideband Codec (AMR-WB)

A further development of the AMR concept is the Adaptive Multirate Wideband Codec (AMR-WB) standard [3GPP TS 26.190 2001], [Bessette et al. 2002], which was designed for mobile radio systems, such as GSM or LTE. The analog wideband speech signal with audio frequencies 50 Hz ... 7.0 kHz is sampled at $f_s = 16\,\mathrm{kHz}$. Figure 8.38 shows a high-level block diagram of the encoder. The digital input signal $x(k)$ is split into a *lowband* signal $s_{lb}(k)$ (0 ... 6.4 kHz) and a *highband* signal $s_{hb}(k)$ (6.4 ... 7.0 kHz).

The lowband signal $s_{lb}(k)$ is decimated from $f_s = 16.0\,\mathrm{kHz}$ to $f'_s = 12.8\,\mathrm{kHz}$ and encoded with an ACELP encoder.

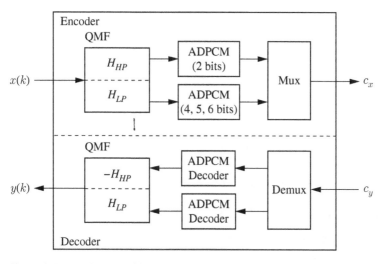

**Figure 8.37**  Split band ADPCM wideband codec ITU-T G.722.

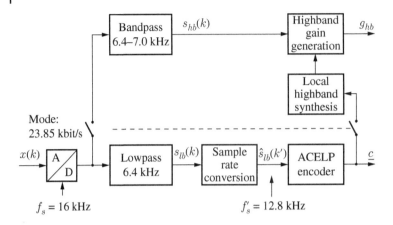

**Figure 8.38** AMR-WB encoder.

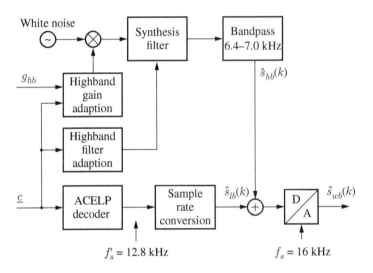

**Figure 8.39** AMR-WB decoder.

The highband signal $s_{hb}(k)$ is substituted in the decoder by artificial *bandwidth extension* (BWE) as proposed in [Paulus 1996] (see also Chapter 10) as shown in Figure 8.39. The BWE module consists of a white noise generator, a gain multiplier, a highband synthesis filter and a 6.4...7.0 kHz bandpass filter.

The AMR-WB codec offers nine different bitrates from 6.6 kbit/s to 22.85 kbit/s. The BWE module is supported by gain side information only in the highest bitrate of 22.85 kbit/s. The highband synthesis filter is derived from the lowband LP-filter. The bit allocation per 20 ms frame is summarized in Table 8.8.

The main algorithmic features of the AMR-WB codec can be summarized as follows:

- ACELP encoding of the lowband signal $s_{lb}(k)$ similar as in the AMR-NB codec.
- Algebraic code book with four tracks with 16 positions
- 16 LP filter coefficients per 20 ms frame.

**Table 8.8** AMR-WB bit allocation per frame of 20 ms.

| Parameter | Codec mode (kbit/s) | | | | | | | | |
|---|---|---|---|---|---|---|---|---|---|
| | 6.60 | 8.85 | 12.65 | 14.25 | 15.85 | 18.25 | 19.85 | 23.05 | 23.85 |
| VAD flag | 1 | 1 | 1 | 1 | 1 | 1 | 1 | 1 | 1 |
| LTP filtering flag | 0 | 0 | 4 | 4 | 4 | 4 | 4 | 4 | 4 |
| ISP | 36 | 46 | 46 | 46 | 46 | 46 | 46 | 46 | 46 |
| Pitch delay | 23 | 36 | 30 | 30 | 30 | 30 | 30 | 30 | 30 |
| Algebraic code | 48 | 80 | 144 | 176 | 208 | 256 | 288 | 352 | 352 |
| Gains | 24 | 24 | 28 | 28 | 28 | 28 | 28 | 28 | 28 |
| High-band energy | 0 | 0 | 0 | 0 | 0 | 0 | 0 | 0 | 16 |
| Total per frame | 132 | 177 | 253 | 285 | 317 | 365 | 397 | 461 | 477 |

- BWE with or without side information in the range 6.4 – 7.0 kHz.
- BWE side information (gain) only for the highest bitrate.
- derivation of the highband synthesis filter from the lowband synthesis filter.
- asymmetric analysis window of length 30 ms.
- conversion of LP coefficients to ISP representation for quantization and interpolation (immittance spectral frequencies, instead of LSP).
- subframe size of 64 samples.
- special error weighting filter.
- joint vector quantization of adaptive and fixed vector code book gains.

For further details, the reader is referred to [3GPP TS 26.190 2001], [Bessette et al. 2002]. In the case of clean speech without background noise and without transmission errors, the six highest modes (23.85–14.25kbit/s) offer a speech quality which is equal to or better than that of the wire-line wideband codec G.722 ([ITU-T Rec. G.722 1988], split band ADPCM, 48, 56, and 64 kbit/s). The speech quality of the 12.65 kbit/s mode is at least equal to G.722 at 56 kbit/s, while the 8.85kbit/s mode is still comparable to G.722 at 48 kbit/s. The two lowest modes (8.85kbit/s and 6.6 kbit/s) are used only for adverse channel conditions or during network congestion. The AMR-WB codec has also been adopted as an ITU-T standard for multimedia applications [ITU-T Rec. G.722.2 2002].

### 8.7.6 Codec for Enhanced Voice Services (EVS)

The Enhanced Voice Services (EVS) codec is a versatile multi-rate audio codec for mobile and multi media systems that has been optimized for a wide range of bit rates, bandwidths, and qualities [3GPP TS 126.445 2014], [Bruhn et al. 2015], [Dietz et al. 2015]. Due to its flexibility, the EVS codec is appropriate for speech signals, music signals, and mixed speech/music signals. The codec supports the four bandwidths narrowband (NB), wideband (WB), super wideband (SWB), fullband (FB), and the corresponding sampling rates (see also Table 8.1).

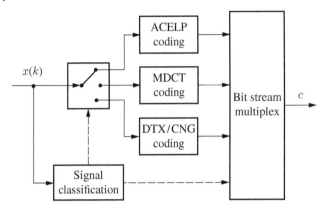

**Figure 8.40** Simplified block diagram of the EVS codec.

Besides operation modes with fixed bit rates, the standard comprises a source controlled rate scheme, a voice/sound activity detector, a comfort noise generation system, an error concealment mechanism to combat the effects of transmission errors and lost packets, e.g., [3GPP TS 126.447 2014], [Lecomte et al. 2015] and a jitter buffer management for Voice over IP (VoIP) transmission.

The codec combines speech-model-based ACELP methods and waveform coding techniques using the Modified Discrete Cosine Transform (MDCT) as shown in Figure 8.40. By signal frame classification, a decision is taken to use either the ACELP or the MDCT coding method. ACELP coding is mainly used for speech and background noise with low latency and low-to-medium bit rates, while MDCT coding is preferably applied to music signals with higher bit rates and less latency restriction.

The MDCT coding mode is selected whenever a signal is not speech or the bit rate for encoding is high enough for the transform coder. The EVS codec supports two MDCT coding modes, MDCT of Transform Coded Excitation (MDCT-TCX) and High Quality MDCT (MDCT-HQ). Both operate with different configurations depending on the available bit rate for encoding.

The MDCT-TCX mode uses algebraic vector quantization in the transform domain. For shaping of the quantization noise, the DCT coefficients $X_\mu(k)$ are multiplied before vector quantization by the inverse frequency response of the weighted LP-synthesis filter.

The MDCT-HQ coding variant is mainly applied to WB, SWB, and FB music signals at bit rate between 32.0 kbit/s and 128.0 kbit/s.

For speech coding, the ACELP codec is selected that offers six specific coding modes: *voiced, unvoiced, transition, inactive, audio, generic*. The coding modes differ in the number and combination of codebooks for encoding the LP excitation signal. For each frame, the actual coding mode is determined by signal classification.

The coding mode can be switched for every signal frame of 20 ms in terms of the coding technology (ACELP or MDCT) and bit rates. Besides the *primary* EVS coding modes, the codec comprises modes which are compatible with the 9 bit rates of the AMR-WB codec. These modes are called the AMR-WB interoperable modes (AMR-WB-IO). The bit rates of the EVS primary mode are summarized in Table 8.9.

The EVS codec can be operated in the discontinuous transmission mode (DTX). The objective of DTX is to reduce the bit rate during inactive signal periods, which contain

**Table 8.9** EVS primary modes.

| Bit rate kbit/s | Bandwidths |
| --- | --- |
| 5.9 (VBR) | NB, WB |
| 7.2 | NB, WB |
| 8.0 | NB, WB |
| 9.6 | NB, WB, SWB |
| 13.2 | NB, WB, SWB |
| 16.4 | NB, WB, SWB, FB |
| 24.4 | NB, WB, SWB, FB |
| 32.0 | NB, WB, SWB, FB |
| 48.0 | NB, WB, SWB, FB |
| 64.0 | NB, WB, SWB, FB |
| 96.0 | NB, WB, SWB, FB |
| 128 | NB, WB, SWB, FB |

only background noise. The default parameter setting is to transmit only every 8th frame a so-called SID frame (Silence Insertion Descriptor) which carries the information for producing in the decoder artificial background noise using a Comfort Noise Generator (CNG). Comfort noise is generated by applying a random excitation signal to an LP synthesis filter, which models the spectral envelope of the background noise.

The ACELP codebook configurations as summarized in Table 8.10 give an impression of the great variety of the EVS CELP coding modes, especially as the dimensions of the the different codebooks are not only depending on signal classification, but also on the available bit rate. It should be noted that there are in addition a few special cases, which are not covered by this table.

**Table 8.10** ACELP codebooks of EVS primary mode.

| Signal class | Coding mode | Codebook combination |
| --- | --- | --- |
| Voiced | VC | ACB + FCB |
| Unvoiced | UC | 2 Gaussian CBs |
| Transition | TC | glottal CB + FCB |
| Generic | GC | ACB + FCB |
| | | ACB + TDCB + FCB (32/64 kbit/s) |
| Audio | AC | ACB + FCB + TDCB |
| Inactive | IC | ACB + Gaussian CB (9.6 kbit/s) |
| | | ACB + FCB |
| | | ACB + FCB + TDCB (32/64 kbit/s) |

ACB: Adaptive Code Book (CB); FCB: Fixed algebraic CB; TDCB: Transform Domain CB

The audio coding mode (AC) is intended for *generic audio signal coding* (GSC) at low bit rates without introducing more delay than the ACELP structure requires. The GSC mode is used only at 12.8 kHz internal sampling rate, and the excitation may be encoded with 4 subframes, 2 subframes, or 1 subframe per frame depending on the bit rate or the signal type. GSC is a hybrid time-domain/frequency-domain CELP coding scheme. The excitation to the LP synthesis filter comprises two contributions in the time-domain, i.e., from the adaptive codebook and from the fixed algebraic codebook. A third contribution is obtained by vector quantization in the frequency domain. The difference between the LP-residual signal and the two time-domain contributions is transformed by a cosine transformation to the frequency domain. Then, the quantized spectral coefficients are re-transformed to the time domain and added to the two other time-domain contributions.

Further details can be found in [3GPP TS 126.445 2014].

### 8.7.7 Opus Codec IETF RFC 6716

The Opus codec was published in 2012 by the Internet Engineering Task Force (IETF) as Internet standard RFC 6716. The specification document [Valin, Terriberry 2012] characterizes the codec as follows:

"OPUS is designed to handle a wide range of interactive audio applications, including Voice over IP, videoconferencing, in-game chat, and even live, distributed music performances. It scales from low bitrate narrowband speech at 6 kbit/s to very high quality stereo music at 510 kbit/s. OPUS uses both Linear Prediction (LP) and the Modified Discrete Cosine Transform (MDCT) to achieve good compression of both speech and music."

The main features are:

- Totally open and royalty-free standard.
- Highly versatile and flexible speech-audio codec.
- Composed of a SILK codec [Vos et al. 2009] and a CELT codec [Valin et al. 2010].
- Bit rates from 6 to 510 kb/s, frame sizes from 2.5 to 60 ms.
- Sampling rates from 8 kHz (narrowband) to 48 kHz (fullband).
- Support for both constant bit-rate (CBR) and variable bit-rate (VBR).
- Audio bandwidth from narrowband to full-band.
- Support for speech and music and for mono and stereo.
- Support for up to 255 channels (multistream frames).
- Dynamically adjustable bitrate, audio bandwidth, and frame size.
- Good loss robustness and packet loss concealment (PLC).
- Floating point and fixed-point implementations.

The SILK layer of the OPUS codec performs hybrid coding. The speech/audio signal is reconstructed by applying a coded excitation signal to a long-term prediction filter and a short-term prediction filter. The SILK decoder is similar to a CELP decoder but uses, e.g., special measures for increasing robustness against frame erasures.

The CELT layer (Constrained Energy Lapped Transform) of the OPUS codec performs waveform coding and is based on the MDCT with partially overlapping windows of 5–22.5 ms. The main principle behind CELT is that the MDCT spectrum is divided into

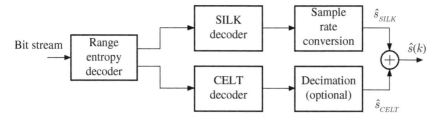

**Figure 8.41** High level block diagram of the OPUS decoder.

bands that (roughly) follow the Bark scale, i.e., the scale of the critical bands of the auditory system [Fastl, Zwicker 2007].

Both codec layers are combined as shown in Figure 8.41. The OPUS decoder consists of two main blocks: the SILK decoder and the CELT decoder. At any given time, one or both of the SILK and CELT decoders may be active. The output of the OPUS decoder is the sum of the outputs from the SILK and CELT decoders with proper sample rate conversion and delay compensation on the SILK layer, and optional decimation (when decoding to sample rates less than 48 kHz) on the CELT layer [Valin, Terriberry 2012].

## References

3GPP TS 126.445 (2014). Codec for Enhanced Voice Services (EVS); Detailed algorithmic description, 3GPP.

3GPP TS 126.447 (2014). Codec for Enhanced Voice Services (EVS; Error concealment of lost packets), 3GPP.

3GPP TS 26.090 (2001). Adaptive Multi-Rate (AMR) Speech Codec; Transcoding Functions, 3GPP.

3GPP TS 26.190 (2001). AMR Wideband Speech Codec; Transcoding Functions, 3GPP.

3GPP TS 26.290 (2005). Extended Adaptive Multi-Rate Wideband (AMR-WB+) Codec, 3GPP.

Adoul, J. P.; Mabilleau, P.; Delprat, M.; Morisette, S. (1987). Fast CELP Coding Based on Algebraic Codes, *Proceedings of the IEEE International Conference on Acoustics, Speech, and Signal Processing (ICASSP)*, Dallas, Texas, USA, pp. 1957–1960.

Atal, B. S. (1982). Predictive Coding of Speech at Low Bit Rates, *IEEE Transactions on Communications*, vol. 30, pp. 600–614.

Atal, B. S.; Cuperman, V.; Gersho, A. (eds.) (1991). *Advances in Speech Coding*, Kluwer Academic, Boston, Massachusetts.

Bessette, B.; Salami, R.; Lefebvre, R.; Jelinek, M.; Rotola-Pukkila, J.; Vainio, J.; Mikkola, H.; Järvinen, K. (2002). The Adaptive Multirate Wideband Speech Codec (AMR-WB), *IEEE Transactions on Speech and Audio Processing*, vol. 10, no. 8, pp. 620–636.

Brandenburg, K.; Bosi, M. (1997). Overview of MPEG Audio: Current and Future Standards for Low Bit-rate Audio Coding, *Journal of the Audio Engineering Society*, vol. 45, pp. 4–21.

Brosze, O.; Schmidt, K. O.; Schmoldt, A. (1992). Der Gewinn an Verständlichkeit beim "Fernsprechen", *Nachrichtentechnische Zeitschrift (NTZ)*, vol. 15, no. 7, pp. 349–352 (in German).

Bruhn, S.; Blöcher, P.; Hellwig, K.; Sjöberg, J. (1999). Concepts and Solutions for Link Adaptation and Inband Signaling for the GSM AMR Speech Coding Standard, *IEEE Vehicular Technology Conference*, Houston, Texas, USA, vol. 3, pp. 2451–2455.

Bruhn, S.; Pobloth, H.; Schnell, M.; Grill, B.; Gibbs, J.; Miao, L.; Järvinen, K.; Laaksonen, L.; Harada, N.; Naka, N.; Ragot, S.; Proust, S.; Sanda, T.; Varga, I.; Greer, C.; Jelínek, M.; Xie, M.; Usai, P. (2015). Standardization of the new 3GPP EVS codec, *2015 IEEE International Conference on Acoustics, Speech and Signal Processing (ICASSP)*, pp. 5703–5707.

Campbell, J. P.; Welch, V. C.; Tremain, T. E. (1989). The New 4800 bps Voice Coding Standard, *Proceedings Military Speech Technology*, pp. 64–70.

Chen, J. H.; Gersho, A. (1995). Adaptive Postfiltering for Quality Enhancement of Coded Speech, *IEEE Transactions on Speech and Audio Processing*, vol. 3, no. 1, pp. 59–71.

Chu, W. C. (2003). *Speech Coding Algorithms*, Wiley-Interscience, New York.

Dietz, M.; Multrus, M.; Eksler, V.; Malenovsky, V.; Norvell, E.; Pobloth, H.; Miao, L.; Wang, Z.; Laaksonen, L.; Vasilache, A.; Kamamoto, Y.; Kikuiri, K.; Ragot, S.; Faure, J.; Ehara, H.; Rajendran, V.; Atti, V.; Sung, H.; Oh, E.; Yuan, H.; Zhu, C. (2015). Overview of the EVS Codec Architecture, *2015 IEEE International Conference on Acoustics, Speech and Signal Processing (ICASSP)*, pp. 5698–5702.

Dudley, H. (1939). The Vocoder, *Bell Labs Record*, vol. 17, pp. 122–126.

Ekudden, E.; Hagen, R.; Johansson, I.; Svedberg, J. (1999). The AMR Speech Coder, *Proceedings of the IEEE Workshop on Speech Coding*, Porvoo, Finland, pp. 117–119.

ETSI Rec. GSM 03.50 (1998). *Digital Cellular Telecommunications System (Phase 2+); Transmission Planning Aspects of the Speech Service in the GSM Public Land Mobile (PLMN) System, Version 8.1.0*, European Telecommunications Standards Institute.

ETSI Rec. GSM 06.60 (1996). *Digital Cellular Telecommunications System (Phase 2+); Enhanced Full Rate (EFR) Speech Transcoding*, European Telecommunications Standards Institute.

ETSI Rec. GSM 06.90 (1998). *Digital Cellular Telecommunications System (Phase 2+); Adaptive Multi-Rate (AMR) speech transcoding*, European Telecommunications Standards Institute.

Fastl, H.; Zwicker, E. (2007). *Psychoacoustics*, 2nd edn, Springer-Verlag, Berlin.

Flanagan, J. L.; Golden, R. (1966). Phase Vocoder, *Bell System Technical Journal*, vol. 45, pp. 1493–1509.

Furui, S.; Sondhi, M. M. (1992). *Advances in Speech Signal Processing*, Marcel Dekker, New York.

Goldberg, R.; Riek, L. (2000). *A Practical Handbook of Speech Coders*, CRC Press, New York.

Hanzo, L.; Somerville, F. C. A.; Woodard, J. P. (2001). *Voice Compression and Communications*, Wiley-Interscience, New York.

ITU-T Rec. G.132 (1988). ITU-T Recommendation G.132, Attenuation Performance, in Blue Book, vol. Fascicle III.1, General Characteristics of International Telephone Connections and Circuits, International Telecommunication Union (ITU).

ITU-T Rec. G.151 (1988). General Performance Objectives Applicable to all Modern International Circuits and National Extension Circuits, in Blue Book, vol. Fascicle III.1, General Characteristics of International Telephone Connections and Circuits, International Telecommunication Union (ITU).

ITU-T Rec. G.711 (1972). Pulse Code Modulation (PCM) of Voice Frequencies, International Telecommunication Union (ITU).

ITU-T Rec. G.712 (1988). Performance Characteristics of PCM Channels Between 4-wire Interfaces at Voice Frequencies, International Telecommunication Union (ITU).

ITU-T Rec. G.722 (1988). 7 kHz Audio Coding within 64 kbit/s, *Recommendation G.722*, International Telecommunication Union (ITU), pp. 269–341.

ITU-T Rec. G.722.1 (1999). Coding at 24 and 32 kbit/s for Hands-free Operation in Systems with Low Frame Loss, International Telecommunication Union (ITU).

ITU-T Rec. G.722.2 (2002). Wideband Coding of Speech at Around 16 kbits/s Using Adaptive Multi-Rate Wideband (AMR-WB), International Telecommunication Union (ITU).

ITU-T Rec. G.726 (1990). 40, 32, 24, 16 kbit/s Adaptive Differential Pulse Code Modulation (ADPCM), *Recommendation G.726*, International Telecommunication Union (ITU).

ITU-T Rec. G.729 (1996). Coding of Speech at 8 kbit/s Using Conjugate-Structure Algebraic-Code-Excited Linear-Prediction (CS-ACELP), *Recommendation G.729*, International Telecommunication Union (ITU).

ITU-T Rec. P.341 (1998). Transmission characteristics for wideband (150-7000,Hz) digital hands-free telephony terrminals, *Recommendation P.341*, International Telecommunication Union (ITU).

ITU-T Rec. P.48 (1993). Specification for an Intermediate Reference System, *Recommendation P.48*, International Telecommunication Union (ITU).

ITU-T Rec. P.830 (1996). Subjective Performance Assessment of Telephone-Band and Wideband Digital Codecs. Annex D – Modified IRS Send and Receive Characteristics, *Recommendation P.830*, International Telecommunication Union (ITU).

Järvinen, K. (2000). Standardization of the Adaptive Multi-Rate Codec, *Proceedings of the European Signal Processing Conference (EUSIPCO)*, Tampere, Finland.

Järvinen, K.; Vainio, J.; Kapanen, P.; Salami, R.; Laflamme, C.; Adoul, J. P. (1997). GSM Enhanced Full Rate Speech Codec, *Proceedings of the IEEE International Conference on Acoustics, Speech, and Signal Processing (ICASSP)*, Munich, Germany, pp. 771–774.

Jayant, N. S.; Noll, P. (1984). *Digital Coding of Waveforms*, Prentice Hall, Englewood Cliffs, New Jersey.

Kondoz, A. M. (2004). *Digital Speech*, John Wiley & Sons, Inc., New York.

Krebber, W. (1995). *Sprachübertragunsqualität von Fernsprech-Handapparaten*, PhD thesis, RWTH Aachen University (in German).

Kroon, P.; Deprettere, E. F.; Sluyter, R. (1986). Regular-Pulse Excitation - A Novel Approach to Effective and Efficient Multipulse Coding of Speech, *IEEE Transactions on Acoustics, Speech and Signal Processing*, vol. 34, no. 5, pp. 1054–1063.

Laflamme, C.; Adoul, J.-P.; Su, H.; Morisette, S. (1990). On Reducing Computational Complexity of Codebook Search in CELP Coder Through the Use of Algebraic Codes, *Proceedings of the IEEE International Conference on Acoustics, Speech, and Signal Processing (ICASSP)*, Albuquerque, New Mexico, USA.

Lecomte, J.; Vaillancourt, T.; Bruhn, S.; Sung, H.; Peng, K.; Kikuiri, K.; Wang, B.; Subasingha, S.; Faure, J. (2015). Packet-loss concealment technology advances in EVS, *2015 IEEE International Conference on Acoustics, Speech and Signal Processing (ICASSP)*, pp. 5708–5712.

MPEG-2 (1997). Advanced Audio Coding, AAC International Standard IS 13818-7, ISO/IEC JTC1/SC29 WG11, MPEG.

O'Shaughnessy, D. (2000). *Speech Communications*, IEEE Press, New York.

Paulus, J. (1996). *Codierung breitbandiger Sprachsignale bei niedriger Datenrate*, PhD thesis. Aachener Beiträge zu digitalen Nachrichtensystemen, vol. 6, P. Vary (ed.), RWTH Aachen University (in German).

Rabiner, L. R.; Schafer, R. W. (1978). *Digital Processing of Speech Signals*, Prentice Hall, Englewood Cliffs, New Jersey.

Rao, K. R.; Hwang, J. J. (1996). *Techniques & Standards for Image, Video & Audio Coding*, Prentice Hall, Upper Saddle River, New Jersey.

Rec. GSM 06.10 (1992). Recommendation GSM 06.10 GSM Full Rate Speech Transcoding, ETSI TC-SMG.

Rosenberg, A. E.; Schafer, R. W.; Rabiner, L. R. (1971). Effects of Smoothing and Quantizing the Parameters of Formant-Coded Voiced Speech, *Journal of the Acoustical Society of America*, vol. 50, no. 6, pp. 1532–1538.

Salami, R.; Laflamme, C.; Bessette, B.; Adoul, J. P. (1997a). Description of ITU-T Rec. G.729 Annex A: Reduced Complexity 8 kbit/s CS-ACELP Coding, *Proceedings of the IEEE International Conference on Acoustics, Speech, and Signal Processing (ICASSP)*, Munich, Germany.

Salami, R.; Laflamme, C.; Bessette, B.; Adoul, J. P. (1997b). ITU-T G.729 Annex A: Reduced Complexity 8 kb/s CS-ACELP Codec for Digital Simultaneous Voice and Data, *IEEE Communications Magazine*, vol. 35, no. 9, pp. 56–63.

Schmidt, K. O.; Brosze, O. (1967). *Fernsprech-Übertragung*, Fachverlag Schiele und Schön, Berlin (in German).

Schroeder, M. R.; Atal, B. S. (1985). Code-Excited Linear Prediction (CELP): High-Quality Speech at Very Low Bit Rates, *Proceedings of the IEEE International Conference on Acoustics, Speech, and Signal Processing (ICASSP)*, Tampa, Florida, USA, pp. 937–940.

Schroeder, M. R.; Atal, B. S.; Hall, J. (1979). Optimizing Digital Speech Coders by Exploiting Masking Properties of the Human Ear, *Journal of the Acoustical Society of America*, vol. 66, pp. 1647–1652.

Singhal, S.; Atal, B. S. (1984). Improving Performance of Multi-pulse LPC Coders at Low Bit Rates, *Proceedings of the IEEE International Conference on Acoustics, Speech, and Signal Processing (ICASSP)*, San Diego, California, USA, pp. 9–12.

Sluijter, R. J. (2005). *The Development of Speech Coding and the First Standard Coder for Public Mobile Telephony*, PhD thesis, Technical University Eindhoven.

Sluyter, R. J.; Vary, P.; Hofmann, R.; Hellwig, K. (1988). A Regular-pulse Excited Linear Predictive Codec, *Speech Communication*, vol. 7, no. 2, pp. 209–215.

Taka, M.; Maitre, X. (1986). CCITT Standardizing Activities on Speech Coding, *IEEE International Conference on Acoustics, Speech, and Signal Processing*, vol. 11, pp. 817–820.

Terhardt, E. (1998). *Akustische Kommunikation: Grundlagen mit Hörbeispielen*, Springer-Verlag, Berlin.

Tremain, T. E. (1982). The Government Standard Linear Predictive Coding Algorithm: LPC-10, *Speech Technology*, vol. 1, pp. 40–49.

Un, C. K.; Magill, D. T. (1975). The Residual-Excited Linear Prediction Vocoder with Transmission Rate Below 9.6 kbit/s, *IEEE Transactions on Communications*, vol. 23, pp. 1466–1474.

Valin, J.; Terriberry, T. (2012). Definition of the OPUS Audio Codec IETF 6716, *Technical report*, Internet Engineering Task Force (IETF).

Valin, J.; Terriberry, T.; Maxwell, G.; Montgomery, C. (2010). Constrained-Energy Lapped Transform (CELT) Codec, *Technical report*. IETF draft http://www.celt-codec.org/.

Vary, P.; Hofmann, R. (1988). Sprachcodec für das Europäische Funkfernsprechnetz, *Frequenz*, vol. 42, pp. 85–93 (in German).

Vary, P.; Hellwig, K.; Hofmann, R.; Sluyter, R. J.; Galand, C.; Rosso, M. (1988). Speech Codec for the European Mobile Radio System, *Proceedings of the IEEE International Conference on Acoustics, Speech, and Signal Processing (ICASSP)*, New York, USA, pp. 227–230.

Vary, P.; Heute, U.; Hess, W. (1998). *Digitale Sprachsignalverarbeitung*, B. G. Teubner, Stuttgart (in German).

Voran, S. (1997). Listener Ratings of Speech Passbands, *Proceedings of the IEEE Workshop on Speech Coding*, Pocono Manor, Pennsylvania, USA, pp. 81–82.

Vos, K.; Jensen, S.; Soerensen, K. (2009). Silk Speech Codec, https://tools.ietf.org/html/draft-vos-silk-00, *Technical report*.

# 9

# Concealment of Erroneous or Lost Frames

Digital speech, audio, and video communication over noisy channels usually comprises source and channel coding. The objectives of source coding are to achieve a certain quality of the decoded signal under given constraints with respect to bit rate, latency, and complexity. The task of channel encoding is to correct bit errors caused by noise and interference on the transmission channel. Channel encoders are designed and optimized for a typically expected minimum quality of the transmission channel. However, in practice, more adverse channel conditions occur, and the error correction scheme is overloaded such that residual bit errors remain after channel decoding. This may result in severe degradations of the signal and subjectively annoying effects, which can be reduced or even eliminated by means of *error concealment* (e.g., [Feldes 1993], [Feldes 1994], [Gerlach 1993], [Skoglund, Hedelin 1994], [Gerlach 1996], [Fingscheidt et al. 1998], [Fingscheidt 1998], [Fingscheidt, Vary 2001]).

A similar situation occurs in packet voice transmission via the fixed or the mobile Internet ("Voice over IP"). Within the real-time constraints, frames, or packets of bits may arrive too late or may have too many bit errors and have thus to be *declared as lost or erased* (e.g., [Jayant, Christensen 1981], [Goodman et al. 1986], [Valenzuela, Animalu 1989], [Sereno 1991], [Erdöl et al. 1993], [Stenger et al. 1996], [Clüver 1996], [Martin et al. 2001]). For the compensation of variable arrival times of packets, a so-called *jitter buffer* is used at the receiving end. As shown in Figure 9.1, the jitter buffer is organized as a ring buffer or a "carrousel buffer," which is written at variable packet arrival times and read at constant speed as required by the decoder. The jitter buffer causes some delay, which might be exploited for the concealment of lost packets by interpolation between the parameters before and after the packet loss.

For speech-audio source encoding with bit rates between 4 kbit/s and 64 kbit/s, the signals are processed frame by frame. Narrowband telephone speech is sampled at a rate of $f_s = 8$ kHz. The frame covers typically a duration of 20 ms and consists of $N = 160$ samples $s(k)$. For wideband speech coding, a sample rate of $f_s = 16$ kHz is used with $N = 320$ samples per 20 ms.

The encoders deliver per frame of speech-audio samples quantized excitation or subband samples, quantized filter coefficients, and gain factors, which are transmitted as bit packets or frames of bits. The reasoning behind error concealment is that practical source codecs for speech, audio, and video are not ideal, due to delay and complexity constraints. As Shannon

*Digital Speech Transmission and Enhancement*, Second Edition. Peter Vary and Rainer Martin.
© 2024 John Wiley & Sons Ltd. Published 2024 by John Wiley & Sons Ltd.

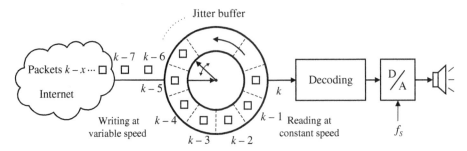

**Figure 9.1** Jitter buffer.

pointed out already, this source coding suboptimality should be exploited at the receiver "to combat noise" [Shannon 1948]. Residual correlation between codec parameters can be exploited to improve the subjective quality in case of erased frames or frames with residual bit errors.

In this chapter, we will first discuss basic ideas for error concealment. The term *error concealment* will be used both for the substitution of single erroneous parameters in a frame and for the substitution of a complete frame that is erroneous, erased, or lost.

Then, a generic approach will be presented, which is called *soft decision source decoding* [Fingscheidt 1998], [Fingscheidt, Vary 2001], [Martin, Heute, Antweiler (eds) 2008] (Chapter 10, by T. Fingscheidt). It can be applied to any parametric source decoder as it relies on residual source redundancy, reliability information from the channel (or channel decoder), and optimum parameter estimation (e.g., [Feldes 1993], [Feldes 1994], [Gerlach 1993], [Gerlach 1996]), [Martin, Heute, Antweiler (eds) 2008] (Chapter 11, by S. Heinen, M. Adrat) or joint source-channel decoding [Martin, Heute, Antweiler (eds) 2008] (Chapter 12, by S. Heinen, T. Hindelang, Chapter 13, by M. Adrat, T. Clevorn, and L. Schmalen).

Finally, a few standardized error concealment examples will be described that are applied in the mobile radio systems GSM (*Global System for Mobile Communications*), UMTS (*universal mobile telecommunications system*), and LTE (*long-term evolution system*).

## 9.1 Concepts for Error Concealment

If a received frame of bits contains residual bit errors that have not been corrected by channel decoding or if a complete frame is marked as erased due to too many bit errors or too late arrival, the subjective speech quality can significantly be improved by concealment techniques. A block diagram of a transmission system with error concealment as part of the decoding processing is shown in Figure 9.2.

The speech source encoder typically converts 20 ms frames consisting of samples $s$ into frames of bits $x$. Scalar parameters and vectors of parameters of the codec, such as gain factors and vectors of filter coefficients, respectively, are quantized and encoded by small groups of bits. It turns out that single-bit errors degrade the quality of the decoded speech signal differently, depending on the subjective importance of the disturbed parameter and on the importance of the disturbed bit within the bit group. A most significant bit of a parameter is rated as more sensitive than the least significant bit.

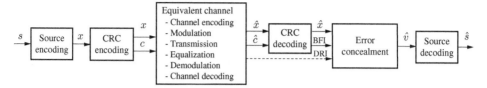

**Figure 9.2** Transmission with error or erasure concealment. $s$: speech sample, $x$: information bit, $c$: parity bit, $\hat{x}$: received information bit after channel decoding, $\hat{c}$: received parity bit after channel decoding, $\hat{v}$: concealed parameter, $\hat{s}$: decoded speech sample, CRC: Cyclic redundancy check, BFI: Bad frame indication, DRI: Decoder reliability information.

Thus, often *unequal error protection* is applied. The bits of the frame are subdivided into 2 or 3 classes of different importance, which get different levels of error protection by channel encoding. In addition, a cyclic redundancy check (CRC) is applied to a *subgroup of the most important bits* by adding a few parity bits $c$ before channel encoding. At the receiver, residual errors in the most important subgroup can be detected by applying (after channel decoding) the same CRC to the corrected bits $\hat{x}$ and by comparing the result $\tilde{c}$ with the received parity bits $\hat{c}$. In the basic version, the outcome of this comparison is finally used to set a *bad frame indication* flag BFI, a binary flag, for controlling the error concealment process of individual parameters or the erasure concealment of the complete frame.

It should be noted that there are various variations and more sophisticated solutions with binary or real-valued quality indicators. In mobile radio systems such as GSM, UMTS, and LTE, the channel decoder delivers for each bit $\hat{x}$, or each group of bits or for the total frame of bits *decoder reliability information* (DRI) which is either a binary flag or a soft value for concealment of individual parameters or complete frames.

In voice over IP systems, so-called jitter buffers, retransmissions of an erased frame, or incremental redundancy for better channel decoding may be used instead of error concealment. This alternative to error concealment will not be discussed here as it introduces additional delay, which is often not allowed.

### 9.1.1 Error Concealment by Hard Decision Decoding

The signal $s(\kappa)$ is segmented into frames of $N$ samples $s_k(\lambda) = s(\kappa = k \cdot N + \lambda)$, $\lambda = 0, 1, \ldots, N - 1$, where $k$ is denoting here the frame index. For each frame of samples $s_k(\lambda)$, a set of *codec parameters*, such as predictor coefficients, gain factors and pitch lags, is calculated by means of *parameter analysis*. For simplicity, we will consider a representative scalar parameter (e.g., a gain factor or a filter coefficient) $\tilde{v}_k \in \mathbb{R}$ from this set, as indicated in Figure 9.3.

We will consider scalar parameters and scalar quantizers. All the concepts to be discussed below can easily be extended to vector quantization.

The parameter $\tilde{v}_k$ is applied to the quantizer Q with $2^w$ quantizer reproduction levels. The $2^w$ levels can be addressed by $w$ bits. Therefore, the quantized version $v_k \in \{v^{(i)} \in \mathbb{R}, i = 0, 1, \ldots, 2^w - 1\}$ of each sample of the quantized parameter $\tilde{v}_k$ is mapped to a corresponding bit pattern $\mathbf{x}_k \in \{\mathbf{x}^{(i)}, i = 0, 1, \ldots, 2^w - 1\}$ of length $w$ (BM: Bit Mapping).

The transmission of any quantized codec parameter $v_k$, respectively, the corresponding bit pattern $\mathbf{x}_k \in \{\mathbf{x}^{(i)}, i = 0, 1, \ldots, 2^w - 1\}$ of length $w$ over the noisy channel will be

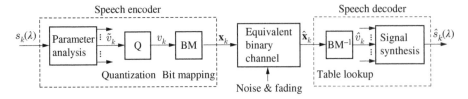

**Figure 9.3** Hard decision speech decoding. $k$: frame index (time), $s_k(\lambda)$: speech sample $s(k \cdot N + \lambda)$, $\tilde{v}_k$: real-valued source codec parameter in frame $k$, $v_k$: quantized parameter $v_k \in \{v^{(i)}, i = 0, 1, \ldots 2^w - 1\}$, $\mathbf{x}_k$: bit pattern (vector) of length $w$, $\hat{\mathbf{x}}_k$: received bit vector of length $w$, Q: quantization, BM: bit mapping.

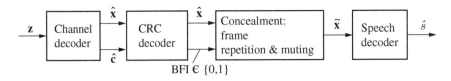

**Figure 9.4** Error concealment with hard decision source decoding.

described here by the equivalent binary symmetric channel with input vectors $\mathbf{x}_k$ and output vectors $\hat{\mathbf{x}}_k$. The $w$ bits $x_k(\kappa)$ (bit index $\kappa = 1, \ldots, w$) of each vector $\mathbf{x}_k$ are transmitted sequentially.

The equivalent channel might consist of any combination of the noisy analog channel with channel (de)coding, (de)modulation, and equalization. Due to the channel noise, the received bit combination $\hat{\mathbf{x}}_k$ is possibly not identical to the transmitted one. In the hard decision decoding scheme of Figure 9.3 the received bit combination $\hat{\mathbf{x}}_k$ is applied to table-lookup decoding (*inverse bit mapping* (BM$^{-1}$) scheme). Thus, in the case of residual bit error(s) a wrong table entry is selected. The decoded parameter $\hat{v}_k$ is finally used within the synthesis algorithm to reconstruct signal samples $\hat{s}_k(\lambda)$. Basic error concealment for hard decision decoding is controlled by a binary BFI flag as indicated in Figure 9.4. If the BFI flag is set, the concealment procedure uses two simple measures

- repetition of the last good frame
- step-wise muting of the decoded output signal in case of consecutive bad frames.

Substitution of a single lost frame is mostly not noticed by the user. The purpose of muting the output in the case of several lost frames is to indicate to the user the breakdown of the channel by graceful degradation. This approach is used, e.g., for the GSM full rate codec as described in more detail in Section 9.2.1

### 9.1.2 Error Concealment by Soft Decision Decoding

In this section, we will derive a concept for exploiting soft information within the source decoding process. The *soft decision* source decoding (e.g., [Fingscheidt, Vary 1997], [Fingscheidt, Vary 2001]), [Martin, Heute, Antweiler (eds) 2008], (Chapter 10, by T. Fingscheidt), should be compatible with existing transmission systems so that we do not have to modify the transmitter. In the literature, there are various proposals for

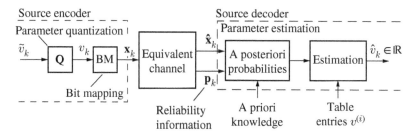

**Figure 9.5** Error concealment with soft decision source decoding. $k$: frame index (time), $\tilde{v}_k$: real-valued source codec parameter at time $k$, $v_k$: quantized parameter $v_k \in \{v^{(i)}, i = 0, 1, \ldots, 2^w - 1\}$, $\mathbf{x}_k$: bit pattern, $\mathbf{x}_k \in \{\mathbf{x}^{(i)}, i = 0, 1, \ldots, 2^w - 1\}$, vector of length $w$, $\hat{\mathbf{x}}_k$: received bit vector of length $w$, $\mathbf{p}_k$: vector of instantaneous bit error rates at time $k$, Q: quantization, BM: bit mapping.

error concealment exploiting reliability and/or a priori information (e.g., [Gerlach 1993], [Gerlach 1996], [Feldes 1993], [Feldes 1994], [Wong et al. 1984], [Sundberg 1978]). The presentation given here follows [Fingscheidt, Vary 2001]. As a reference, we consider the basic solution of Figure 9.3 with hard decision source decoding at the receiver by table lookup.

In the soft decision approach, we replace the table lookup module by a parameter estimator as illustrated in Figure 9.5. The codec parameter $\tilde{v}_k$ at time instant $k$ is quantized according to $v_k = \mathbf{Q}[\tilde{v}_k]$ with $v_k \in \{v^{(i)}, i = 0, 1, \ldots, 2^w - 1\}$ (Quantization Table) and can be represented by the quantization table index $i$. At the time instant $k$, a bit combination

$$\mathbf{x}_k = (x_k(1), x_k(2), \ldots, x_k(\kappa), \ldots, x_k(w)) \tag{9.1}$$

consisting of $w$ bits is assigned via BM to each quantized parameter $v_k$ (or quantization table index $i$). The bits $x_k(\kappa) \in \{-1, +1\}$ are assumed to be bipolar. Due to the channel noise, the received bit combination $\hat{\mathbf{x}}_k$ is possibly not identical to the transmitted one. In the conventional hard decision decoding scheme of Figure 9.3, the received bit combination $\hat{\mathbf{x}}_k$ is decoded by table lookup (inverse bit mapping (BM$^{-1}$) scheme). Thereafter, the decoded parameter $\hat{v}_k$ is used within the specific source decoder algorithm to reconstruct samples $\hat{s}(k \cdot N + \lambda) = \hat{s}_k(\lambda), \lambda = 0, 1, 2 \ldots N - 1$ of the speech signal.

The soft decision estimation process takes the following information into account

- $\hat{\mathbf{x}}_k$: received bits (after channel decoding)
- $\mathbf{p}_k = (p_k(1), \ldots, p_k(w))$: vector of instantaneous reliabilities on bit level
- $P(v^{(i)})$: the probabilities of the quantizer levels $v^{(i)}$ (*1D a priori knowledge*)
- $P(v^{(i)}|v^{(j)})$: parameter dependencies in time (*2D a priori knowledge*).

The core of the soft decision decoding algorithm consists of

- *step 1:* calculation of $2^w$ a posteriori probabilities

    (a) with 1D a priori knowledge (distribution)
    $P(\mathbf{x}^{(i)} \mid \hat{\mathbf{x}}_k)$ with $i \in \{0, 1, \ldots, 2^w - 1\}$
    (b) with 2D a priori knowledge (correlation)
    $P(\mathbf{x}^{(i)} \mid \hat{\mathbf{x}}_k, \hat{\mathbf{x}}_{k-1})$ with $i \in \{0, 1, \ldots, 2^w - 1\}$

- *step 2*: estimation of a real-valued parameter $\hat{v}_k$

  (a) maximum a posteriori estimation (MAP)
  (b) minimum mean square estimation (MS).

  These steps will be described below. We start with the description of the second step.

### 9.1.3 Parameter Estimation

The a posteriori terms $P(\mathbf{x}^{(i)} \mid \hat{\mathbf{x}}_k)$ or $P(\mathbf{x}^{(i)} \mid \hat{\mathbf{x}}_k, \hat{\mathbf{x}}_{k-1})$ can be computed from $\mathbf{p}_k$, the vector of the (instantaneous) bit error rates and the 1D or 2D parameter a priori knowledge by using Bayes theorem. This will be explained in Section 9.1.4.

For any received bit pattern $\hat{\mathbf{x}}_k$, the probabilities $P(\mathbf{x}^{(i)} \mid \hat{\mathbf{x}}_k)$, $i = 0, 1, \ldots 2^w - 1$ quantify the reliabilities that the pattern $\mathbf{x}^{(i)}$ would have been transmitted while the bit pattern $\hat{\mathbf{x}}_k$ has been received. The probabilities $P(\mathbf{x}^{(i)} \mid \hat{\mathbf{x}}_k, \hat{\mathbf{x}}_{k-1})$ consider two successively received bit patterns $\hat{\mathbf{x}}_k$ and $\hat{\mathbf{x}}_{k-1}$.

Once all a posteriori terms have been computed for the received bit pattern $\hat{\mathbf{x}}_k$, the transmitted parameter value can be estimated. For different parameters, different estimation error criteria can be used, reflecting the impact of a parameter error on the subjective quality of the decoded signal.

For most of the speech, audio, and video codec parameters the minimum mean square (MS) error criterion is appropriate. In the case of speech, these parameters may be signal samples, spectral coefficients, filter coefficients, gain factors, etc. However, for some parameters, such as the pitch period, the maximum a posteriori (MAP) estimator should be applied.

#### 9.1.3.1 MAP Estimation

The MAP estimator follows the criterion

$$\hat{v}_k = v^{(v)} \quad \text{with} \quad v = \arg\max_i P(\mathbf{x}^{(i)} \mid \hat{\mathbf{x}}_k) \qquad \text{(1D knowledge)} \qquad (9.2a)$$

$$\hat{v}_k = v^{(v)} \quad \text{with} \quad v = \arg\max_i P(\mathbf{x}^{(i)} \mid \hat{\mathbf{x}}_k, \hat{\mathbf{x}}_{k-1}) \qquad \text{(2D knowledge)}. \qquad (9.2b)$$

MAP estimation minimizes the probability of an erroneously decoded parameter [Melsa, Cohn 1978]. The decoded parameter $\hat{v}_k$ equals one of the code book/quantization table entries. In the case of error-free transmission, only one of the $2^w$ a priori probabilities takes the value 1, all the others are zero. In this situation, the MAP decoder selects the same table entry as the hard decision decoder.

#### 9.1.3.2 MS Estimation

The optimum decoded parameter in a minimum mean square error sense (see also Chapter 5) equals

$$\hat{v}_k = \sum_{i=0}^{2^w - 1} v^{(i)} \cdot P(\mathbf{x}_k^{(i)} \mid \hat{\mathbf{x}}_k) \qquad \text{(1D knowledge)} \qquad (9.3a)$$

$$\hat{v}_k = \sum_{i=0}^{2^w - 1} v^{(i)} \cdot P(\mathbf{x}_k^{(i)} \mid \hat{\mathbf{x}}_k, \hat{\mathbf{x}}_{k-1}) \qquad \text{(2D knowledge)}. \qquad (9.3b)$$

According to the orthogonality principle of linear MS estimation (see, e.g., [Melsa, Cohn 1978]), the variance of the estimation error $e_0 = \hat{v}_k - v_k$ is $\sigma_e^2 = \sigma_v^2 - \sigma_{\hat{v}}^2$ with $\sigma_v^2$ being the variance of the undisturbed parameter $v_k$ and $\sigma_{\hat{v}}^2$ denoting the variance of the estimated parameter $\hat{v}_k$. Because $\sigma_e^2 \geq 0$ we can state that the variance of the estimated parameter is smaller than or equals the variance of the error-free parameter.

For the worst-case channel with $p_0 = 0.5$, the a posteriori probabilities simplify to $P(\mathbf{x}^{(i)} \mid \hat{\mathbf{x}}_k, \ldots) = P(\mathbf{x}^{(i)})$. If in this case the unquantized parameter $\tilde{v}_k$ as well as the quantization table entries $v_k^{(i)}$ are distributed symmetrically around zero, the MS estimated parameter according to Eq. (9.3) is attenuated to zero (by weighted averaging, i.e., conditional expectation of $v$). Such symmetries are often found for gain factors in speech and audio encoders. Thus, the MS estimation of gain factors results in an inherent muting mechanism providing a *graceful degradation* of the signal quality with decreasing channel quality. This is one of the main advantages of soft decision source decoding.

On the other hand, if the channel is free of errors ($p_e = 0$) and $\mathbf{x}^{(\kappa)}$ has been transmitted, then all the parameter transition probabilities are zero except $P(\hat{\mathbf{x}}_k \mid \mathbf{x}^{(\kappa)}) = 1$. This yields $P(\mathbf{x}^{(\kappa)} \mid \hat{\mathbf{x}}_k, \ldots) = 1$ while all other a posteriori probabilities become zero. As a consequence, the MS estimator also yields the correct parameter value $\hat{v}_k = v_k$. This is equivalent to bit exactness in clear channel situations.

Finally, it should be mentioned that all of the algorithms discussed above can be used in the case of vector quantization. The only difference to scalar quantization lies in the estimation step. Let us consider a $P$-tuple of codec parameters $\tilde{\mathbf{v}}_k \in \mathbb{R}^P$, encoded by $w$ bits. The quantized parameter vector is then $Q[\tilde{\mathbf{v}}_k] = \mathbf{v}_k$ with $\mathbf{v}_k \in$ CB (code book). MAP estimation simply yields the most probable parameter vector $\hat{\mathbf{v}}_k \in \{\mathbf{v}^{(i)}, \; i = 0, 1, \ldots, 2^w - 1\}$ instead of a scalar, whereas MS estimation can be formulated as

$$\hat{\mathbf{v}}_k = \sum_{i=0}^{2^w - 1} \mathbf{v}^{(i)} \cdot P(\mathbf{x}^{(i)} \mid \hat{\mathbf{x}}_k, \ldots). \tag{9.4}$$

The MS estimator is to be considered as a weighted sum of all code book entries.

### 9.1.4 The A Posteriori Probabilities

The a posteriori probabilities can be extended as shown below to take into consideration $K$ previous and/or $L$ future bit patterns $\hat{\mathbf{x}}_n$ with $n = k, k - 1, \ldots, k - K$ and/or with $n = k + 1$, $\ldots, k + L$, which are received later. The latter solution requires the introduction of some delay into the decoding process. The corresponding general a posteriori probability reads

$$P(\mathbf{x}_k = \mathbf{x}^{(i)} \mid \ldots, \hat{\mathbf{x}}_k, \ldots) =$$
$$P(\mathbf{x}_k = \mathbf{x}^{(i)} \mid \hat{\mathbf{x}}_{k+L}, \hat{\mathbf{x}}_{k+L-1}, \ldots, \hat{\mathbf{x}}_{k+1}, \hat{\mathbf{x}}_k, \hat{\mathbf{x}}_{k-1}, \hat{\mathbf{x}}_{k-2}, \ldots, \hat{\mathbf{x}}_{k-K}). \tag{9.5}$$

Due to its computational complexity and the achievable quality improvements, the most relevant cases are $K = 0$ and $K = 1$, both with $L = 0$.

For simplicity, we will consider here the most recent bit vector $\hat{\mathbf{x}}_k$ only, i.e., $L = 0$ and $K = 0$ in (9.5). The extension to the 2D case with $K = 1$ is straightforward. In order to obtain the a posteriori probabilities $P(\mathbf{x}_k = \mathbf{x}^{(i)} \mid \hat{\mathbf{x}}_k)$ for $i \in \{0, 1, \ldots, 2^w - 1\}$, we apply Bayes' theorem:

$$P(\mathbf{x}_k, \hat{\mathbf{x}}_k) = P(\mathbf{x}_k \mid \hat{\mathbf{x}}_k) \cdot P(\hat{\mathbf{x}}_k) = P(\hat{\mathbf{x}}_k \mid \mathbf{x}_k) \cdot P(\mathbf{x}_k). \tag{9.6}$$

The probabilities $P(\hat{\mathbf{x}}_k)$ are not known. They depend on the a priori probabilities $P(\mathbf{x}_k = \mathbf{x}^{(\kappa)})$ and on the transition probabilities $P(\hat{\mathbf{x}}_k \mid \mathbf{x}_k = \mathbf{x}^{(\kappa)})$. Thus, these probabilities can be computed via the marginal distribution as follows:

$$P(\hat{\mathbf{x}}_k) = \sum_{\kappa=0}^{2^w-1} P(\hat{\mathbf{x}}_k \mid \mathbf{x}_k = \mathbf{x}^{(\kappa)}) \cdot P(\mathbf{x}_k = \mathbf{x}^{(\kappa)}). \tag{9.7}$$

Finally, we find the solution for the a posteriori probabilities

$$P(\mathbf{x}_k = \mathbf{x}^{(i)} \mid \hat{\mathbf{x}}_k) = \frac{P(\hat{\mathbf{x}}_k \mid \mathbf{x}_k = \mathbf{x}^{(i)}) \cdot P(\mathbf{x}_k = \mathbf{x}^{(i)})}{\displaystyle\sum_{\kappa=0}^{2^w-1} P(\hat{\mathbf{x}}_k \mid \mathbf{x}_k = \mathbf{x}^{(\kappa)}) \cdot P(\mathbf{x}_k = \mathbf{x}^{(\kappa)})}. \tag{9.8}$$

For the calculation of (9.7) and (9.8) we need the following probabilities:

- a priori probabilities $P(\mathbf{x}_k = \mathbf{x}^{(\kappa)})$
- transition probabilities $P(\hat{\mathbf{x}}_k \mid \mathbf{x}_k = \mathbf{x}^{(\kappa)})$, $\kappa = 0, 1, \ldots, 2^w - 1$

which will be discussed below.

### 9.1.4.1 The A Priori Knowledge

For the computation of (9.8), we need a priori knowledge about the quantized parameter in terms of the $2^w$ probabilities $P(\mathbf{x}^{(i)}) = P(v^{(i)})$, $i = 0, 1, \ldots, 2^w - 1$, i.e., the histogram of the quantized parameter.

### 9.1.4.2 The Parameter Distortion Probabilities

Knowing the (instantaneous) bit error probabilities $p_k(\kappa)$, $\kappa = 1, \ldots, w$, of the known received bit $\hat{x}_k(\kappa)$, we get the *bit* transition probabilities for any transmitted bit $x_k(\kappa)$ as

$$P(\hat{x}_k(\kappa) \mid x_k(\kappa) = x^{(i)}(\kappa)) = \begin{cases} 1 - p_k(\kappa) & \text{if } \hat{x}_k(\kappa) = x^{(i)}(\kappa) \\ p_k(\kappa) & \text{if } \hat{x}_k(\kappa) \neq x^{(i)}(\kappa). \end{cases} \tag{9.9}$$

If we consider an equivalent channel with independent bit errors (memoryless), the *parameter distortion probability* reads

$$P(\hat{\mathbf{x}}_k \mid \mathbf{x}_k = \mathbf{x}^{(i)}) = \prod_{\kappa=0}^{w-1} P(\hat{x}_k(\kappa) \mid x^{(i)}(\kappa)). \tag{9.10}$$

This term includes the channel characteristics and provides the probability of a transition from any possibly transmitted bit combination $\mathbf{x}^{(i)}$, $i \in \{0, 1, \ldots, 2^w - 1\}$, at time $k$, to the known received bit combination $\hat{\mathbf{x}}_k$.

In real-life applications, the assumption of a memoryless equivalent channel can be a coarse approximation, even if an interleaving scheme is employed within inner channel coding. However, the achievable error concealment based on (9.10) might still be very effective.

For the clarification of (9.10), we consider the calculation of the parameter distortion probability by way of example.

---

**Example:**

$w = 3$,
$\mathbf{p}_k = (p_0, p_0, p_0)$,
$\hat{\mathbf{x}}_k = (+1, +1, -1)$ and
$i = 2$, i.e., $\mathbf{x}_k = \mathbf{x}^{(2)} = (+1, -1, +1)$
$P(\hat{\mathbf{x}}_k \mid \mathbf{x}_k = \mathbf{x}^{(2)}) = (1 - p_0) \cdot p_0^2$.

---

The bit error rate is assumed to be constant during the transmission of the 3 bits. We are interested in the probability that at time instant $k$ the bit pattern $\mathbf{x}_k^{(2)} = \mathbf{x}^{(2)} = (+1, -1, +1)$, i.e., $i = 2$, is transmitted and that we receive the bit pattern $\hat{\mathbf{x}}_k = (+1, +1, -1)$. In this case, we would obviously receive one bit without error and two erroneous bits. The corresponding parameter transition probability for the quantization table entry $i = 2$ is therefore given by

$$P(\hat{\mathbf{x}}_k \mid \mathbf{x}_k = \mathbf{x}^{(2)}) = (1 - p_0) \cdot p_0^2. \tag{9.11}$$

### 9.1.5 Example: Hard Decision vs. Soft Decision

In a final example, the alternative concepts of hard decision and soft decision concealment are compared with each other for the GSM full-rate speech decoder [Rec. GSM 06.10 1992], [Fingscheidt, Scheufen 1997a], [Fingscheidt, Scheufen 1997b].

For the soft decision approach, the consideration of either 1D knowledge or 2D knowledge can be supported by entropy calculations of the quantized parameters. If we model a parameter as a zeroth-order Markov process (1D knowledge), $2^w$ probabilities $P(\mathbf{x}_k = \mathbf{x}^{(i)})$ with $i \in \{0, 1, \dots, 2^w - 1\}$ have to be stored in the decoder. With the entropy defined as

$$H(\mathbf{x}_k) = -\sum_{i=0}^{2^w - 1} P(\mathbf{x}^{(i)}) \log_2 (P(\mathbf{x}^{(i)})), \tag{9.12}$$

the redundancy of $\Delta R = w - H(\mathbf{x}_k)$ can be used for error concealment.

If a parameter is modeled as a first-order Markov process (2D knowledge), the $2^{2w}$ probabilities $P(\mathbf{x}_k = \mathbf{x}^{(i)} \mid \mathbf{x}_{k-1} = \mathbf{x}^{(j)})$ with $i, j \in \{0, \dots, 2^w - 1\}$ have to be stored in the decoder. Then, a redundancy of $\Delta R = w - H(\mathbf{x}_k \mid \mathbf{x}_{k-1})$ can be used for error concealment, with the conditional entropy $H(\mathbf{x}_k \mid \mathbf{x}_{k-1})$ (see, e.g., [Cover, Thomas 1991]).

In Table 9.1, some codec parameters and their corresponding entropy values are listed. It is obvious that for most of the parameters first-order a priori knowledge will be helpful. Exceptions are the RPE grid and the RPE pulses (see also Section 8.7.1). For reasons of simplicity, each codec parameter is modeled as a first-order Markov process.

For example, LAR no. 1 (Log Area Ratio, see Chapter 8), which is one of the subjectively most important parameters, provides a redundancy of 1.54 bits. This is more than 25%. Even speech codecs with lower bit rates than the GSM full-rate codec provide a high amount of residual redundancy within the spectral parameters: e.g., [Alajaji et al. 1996] found about 29% of redundancy for the line spectral frequencies (LSFs) of the FS 1016 CELP [FS 1016 CELP 1992] due to non-uniform distribution (zeroth-order a priori knowledge) and due to intraframe correlation (first-order a priori knowledge concerning correlation to LSFs of the same frame).

**Table 9.1** Number of bits $w$, entropy $H(\mathbf{x}_k)$, conditional entropy $H(\mathbf{x}_k \mid \mathbf{x}_{k-1})$ of the parameters of the GSM full-rate codec [Fingscheidt, Scheufen 1997a], [Rec. GSM 06.10 1992], [Fingscheidt, Scheufen 1997b]

| $\left[\dfrac{\text{bit}}{\text{parameter}}\right]$ | LAR no. | | | | | | | |
|---|---|---|---|---|---|---|---|---|
| | 1 | 2 | 3 | 4 | 5 | 6 | 7 | 8 |
| $w$ | 6 | 6 | 5 | 5 | 4 | 4 | 3 | 3 |
| $H(\mathbf{x}_k)$ | 5.43 | 4.88 | 4.75 | 4.53 | 3.73 | 3.76 | 2.84 | 2.88 |
| $H(\mathbf{x}_k \mid \mathbf{x}_{k-1})$ | 4.46 | 4.29 | 4.18 | 4.09 | 3.37 | 3.39 | 2.49 | 2.46 |

| $\left[\dfrac{\text{bit}}{\text{parameter}}\right]$ | LTP | | RPE | | |
|---|---|---|---|---|---|
| | Lag | Gain | Grid | Max. | Pulse |
| $w$ | 7 | 2 | 2 | 6 | 3 |
| $H(\mathbf{x}_k)$ | 6.31 | 1.88 | 1.96 | 5.39 | 2.86 |
| $H(\mathbf{x}_k \mid \mathbf{x}_{k-1})$ | 5.75 | 1.74 | 1.96 | 4.29 | 2.86 |

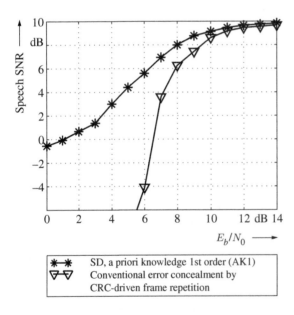

**Figure 9.6** SNR performance of GSM full-rate soft decision speech decoding, TU50 (MS estimation). Source: Adapted from Vary and Fingscheidt [2001].

| | |
|---|---|
| ✳—✳ | SD, a priori knowledge 1st order (AK1) |
| ▽—▽ | Conventional error concealment by CRC-driven frame repetition |

It turns out that the non-integer GSM codec parameters can be well estimated using an MS estimator. In contrast to this, the estimation of a pitch period (LTP lag) or the RPE grid position should be performed by an MAP estimator.

Depicted in Figure 9.6 are the results of a complete GSM simulation using the COSSAP GSM library [COSSAP 1995] with speech and channel (de)coding, (de)interleaving, (de)modulation, a channel model, and equalization. The channel model represents a typical case of an urban area (TU) with six characteristic propagation paths [Rec. GSM 05.05

1992] and a user speed of 50 km/h (TU50). Soft-output channel decoding is carried out by the algorithm of Bahl et al. [Bahl et al. 1974]. The conventional reference GSM decoder performs error concealment by a frame repetition (FR) algorithm as proposed in [Rec. GSM 06.11 1992]. The bad frame indicator (BFI) is simply set by the evaluation of the CRC code.

The SNR is surely not the optimum measure for speech quality. However, also informal listening tests reveal the superiority of the soft decision speech decoder in comparison to the conventional decoding scheme in all situations of vehicle speeds and C/I ratios. Error concealment by soft decision speech decoding provides quite a good subjective speech quality down to C/I = 6 dB, whereas conventional FR produces severe distortions at C/I = 7 dB. Even long error bursts caused by a low vehicle speed can be decoded with sufficient quality. In the soft decision decoding simulation, the annoying clicks of hard decision decoding in the case of CRC failures and the synthetic sounds of the FR disappear completely and turn into slightly noisy or modulated speech. This enhances the perceived speech quality significantly.

A very attractive feature of this soft decision source decoding technique is the fact that it can be applied to any source coding algorithm without modifications at the transmitter side. In clear channel conditions, bit exactness is always preserved.

## 9.2 Examples of Error Concealment Standards

### 9.2.1 Substitution and Muting of Lost Frames

The GSM standards on substitution and muting of lost frames [Rec. GSM 06.11 1992], [Rec. GSM 06.21 1995], [Rec. GSM 06.61 1996] propose simple mechanisms such as parameter repetition and step-by-step muting. They are driven by the binary BFI flag that marks the current received frame as good or bad. The BFI flag can be understood as binary reliability information that may initiate the substitution of a complete frame, even if only a few bits have been disturbed. Conversely, the BFI may declare a frame reliable although some bits are incorrect.

As an example of standardized error concealment, we will briefly discuss the solution [Rec. GSM 06.11 1992] for the GSM full-rate codec (FR, see also Chapter 8 and Section 8.7.1). The FR encoder produces frames of 260 bits $\mathbf{x}$ every 20 ms, resulting in a source bit rate of 13.0 kbit/s. In the GSM cellular system, the source and the channel encoding process are organized at the transmitter as shown in the block diagram of Figure 9.7a.

These bits are grouped, as illustrated in Figure 9.7b, according to their auditive sensitivity with respect to bit errors into $50 + 132 = 182$ class I bits and 78 class II bits. The 50 subjectively most important (class Ia) bits are protected by three CRC bits. The class Ia bits usually belong to the most important bits of various speech codec parameters. A systematic cyclic block code with the generator polynomial

$$G_{CRC}(D) = 1 + D + D^3 \tag{9.13}$$

is used, where $D$ denotes the delay operator. Thus, the CRC and the resulting BFI give a measure of frame reliability rather than parameter or bit reliability. Class II comprises mostly the numerically least significant bits (LSBs).

(a)

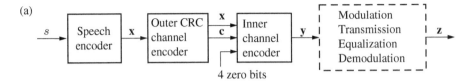

(b)

1.  Speech codec: 260 bits **x** arranged according to bit error sensitivity

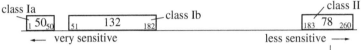

2.  Outer code: 3 CRC bits for error detection: $G(D) = 1 + D + D^3$

3.  Inner code: convolutional encoding of 189 bits (rate 1/2, memory 4), no coding of 78 class II bits

(c)

**Figure 9.7** Standard error concealment of the GSM full-rate speech codec GSM 06.10. (a) Block diagram of the encoder. (b) Bit classes and error protection. (c) Block diagram of the decoder.

The inner encoder performs feed-forward convolutional encoding of the 50-class Ia bits, the three parity check bits, and the remaining 132 class Ib bits; there is no coding of the remaining 78 class II bits, which are not protected at all. The generator polynomials of the convolutional encoder of rate $r = 1/2$ and memory 4 are given by

$$G_1(D) = 1 + D^3 + D^4$$
$$G_2(D) = 1 + D + D^3 + D^4. \tag{9.14}$$

According to the degree of the polynomials (9.14) and to the length of the encoder memory, four 0 bits are appended to the 182 class I bits to drive the encoder into the zero state. In total, 189 bits are applied to the convolutional encoder, which delivers $2 \cdot 189 = 378$ bits. Thus, the inner encoder produces $378 + 78 = 456$ bits every 20 ms. The gross bit rate is 456 bits/20 ms = 22.8 kbit/s.

Following the original standard [Rec. GSM 06.11 1992], error concealment is controlled by the binary BFI, as indicated in Figure 9.7c, while the BFI is produced by the outer channel decoder. The parity check bits are recalculated from the received 50 class Ia bits $\hat{\mathbf{x}}$, which possibly have residual bit errors. Then they are compared to the received parity bits $\hat{\mathbf{c}}$. If these do not coincide, residual bit errors are assumed within the class Ia bits, and the error concealment procedure is activated via the BFI flag. If one single frame is marked as bad, the complete frame of 260 bits $\hat{x}$ is replaced by the previous bit frame. If several consecutive bad frames occur, the last good frame is repeated while the following frames are gradually attenuated and muted after 320 ms (16 frames). In practice, it is not sufficient to derive the BFI flag just from the CRC check, as multiple bit errors might occur which cannot be detected by this simple mechanism. It is up to the manufacturer of the mobile phone to use additional information for generating the BFI information, such as the metric of the inner channel decoder (DRI) and the received field strength. There are some proposals to enhance the reliability information (e.g., [Sereno 1991], [Su, Mermelstein 1992], [Heikkila et al. 1993], [Järvinen, Vainio 1994]). Alternatively, BFIs are not generated explicitly. In [Minde et al. 1993] weighting factors are computed to perform parameter substitution by weighted summation over previous frames.

More sophisticated concealment techniques have been designed for the enhanced full-rate (EFR) codec [Rec. GSM 06.61 1996], and the GSM adaptive multi-rate (AMR) speech codec [ETSI Rec. GSM 06.90 1998], which are also being used in the UMTS and in the LTE cellular system. In both cases, the BFI flags of the present and the previous frame are considered. They control the concealment process, which is based on a state machine with seven states. Depending on the state, certain parameters of the codec are replaced by attenuated counterparts from the previous frame or by averaged values.

All these empirical algorithms improve the perceived speech quality significantly in adverse channel conditions, and they are a major part of the whole error protection concept. If the error concealment were switched off, the speech quality would not be acceptable in many everyday transmission conditions.

This has been the motivation to derive a theoretical framework leading to optimal error concealment techniques in terms of parameter estimation or soft decision source decoding. The optimum estimators should systematically exploit the residual redundancy of the source encoder and all kinds of quality information which can be made available at the receiver.

### 9.2.2 AMR Codec: Substitution and Muting of Lost Frames

The objective of frame substitution is to conceal the effect of lost frames, while the purpose of muting the output in the case of several lost frames is to indicate the breakdown of the channel to the user and to avoid possible annoying sounds as a result of ongoing frame substitution.

The recommendation GSM 06.91 "Substitution and muting of lost frames for Adaptive Multi Rate (AMR) speech traffic channels" [ETSI Rec. GSM 06.91 1998] defines a frame substitution and muting procedure which should be used at each receiving end, i.e., in the mobile station as well as in the radio subsystem (RSS) of the network. The recommendation distinguishes between speech frames and silence descriptor (SID) frames.

SID frames are used in the discontinuous transmission (DTX) mode, which allows the radio transmitter to be switched off most of the time during speech pauses:

- to save power in the mobile station and
- to reduce the overall interference level on the radio channel.

When speech is absent, the signal synthesis in the decoder is different from the case when normal speech frames are received. The synthesis of an artificial noise based on the received non-speech parameters is termed comfort noise generation.

The normal repetition frequency of the 20 ms speech frame is 50 times per second. In contrast, the SID frames, which carry the information for synthesizing the background noise, are updated only every $8^{th}$ frame, i.e., 6.25 times per second or every 160 ms.

Different procedures are used for speech frames and SID frames. The core of the recommended procedure for speech frames will be described below.

The AMR narrowband codec (see also Section 8.7.3) offers eight different bit rates from 4.75 ... 12.2 kbit/s. Each codec mode employs an 8 bit CRC for detecting bad frames.

The bits of a frame are grouped into sensitivity classes A, B, and C. Class A contains the bits which are most sensitive to errors. Any error in class A bits would result in a corrupted speech frame which should not be decoded without applying appropriate error concealment. For the purpose of detecting residual errors after channel decoding in class A, eight CRC parity bits are provided and transmitted as part of the encoded speech frame.

Classes B and C contain bits, which cause, in the case of an increasing bit error rate, a gradually reduced speech quality. However, decoding of a speech frame with a few class B and/or C bit errors is possible without too annoying artifacts. Class B bits are more sensitive to errors than class C bits. This different error sensitivity is taken into account in the design of the channel coding scheme for error correction applied to the combined bit classes A,B, and C, including the eight CRC bits. The subdivision into the bit classes A,B, and C is summarized in Table 9.2.

At the receiving end (see Figure 9.2), the CRC is repeated and compared with the received parity bits. If residual errors in class A are detected, the frame is declared as lost and the BFI is set.

The channel decoder delivers in addition to the corrected bits, a DRI (see Figure 9.2), which is in the case of the AMR codec, the binary PDI flag (*potentially degraded frame indication*). If the BFI flag is not set, the PDI flag might indicate bit errors in classes B and C or even errors in class A which are not detected due to the limited capabilities (Hamming distance) of the CRC check. This situation might occur in adverse channel conditions. The

**Table 9.2** Bit classes A,B,C for each AMR coding mode.

| Frame type | AMR coding mode | Total number of bits | Class A | Class B | Class C |
|---|---|---|---|---|---|
| 0 | 4.75 | 95 | 42 | 53 | 0 |
| 1 | 5.15 | 103 | 49 | 54 | 0 |
| 2 | 5.90 | 118 | 55 | 63 | 0 |
| 3 | 6.70 | 134 | 58 | 76 | 0 |
| 4 | 7.40 | 148 | 61 | 87 | 0 |
| 5 | 7.95 | 159 | 75 | 84 | 0 |
| 6 | 10.2 | 204 | 65 | 99 | 40 |
| 7 | 12.2 | 244 | 81 | 103 | 60 |

Source: 3GPP TS 126.101 2002/European Telecommunications Standards Institute.

error correction scheme is based on recursive systematic convolutional (RSC) coding with puncturing to obtain unequal error protection of the three bit classes at the given eight different bit rates.

The following example solution for substitution and muting is taken from the recommendation GSM 06.91 [ETSI Rec. GSM 06.91 1998].

The AMR error concealment proposal is based on a state machine with seven states as shown in Figure 9.8.

The system starts in state 0. Each time a bad frame is detected, the state counter is incremented by one and is saturated when it reaches 6. Each time a good speech frame is detected, the state counter is reset to zero, except when it is in state 6, where we set the state counter to 5. The state indicates the quality of the channel: the larger the value of the state counter, the worse the channel quality is. In addition to this state machine, the bad frame flag from the previous frame is checked (prevBFI). The processing depends on the value of the state-variable. In states 0 and 5, the processing also depends on previous BFI (PrevBFI).

**Case 1: BFI = 0, PrevBFI = 1, S = 0, or S = 5:**

No error is detected in the received frame, but the previous frame was bad. The adaptive and fixed codebook gains $g_a$ and $g_f$ are limited by their values used for the last received good subframe.

**Figure 9.8** State machine for controlling the bad frame substitution. Source: Adapted from ETSI Rec. GSM 06.91, 1998.

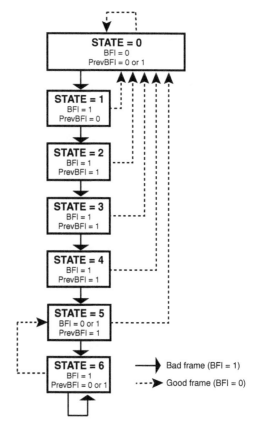

Notation:

$g_a(0)$, $g_f(0)$ = current received adaptive and fixed gains,

$g_a(-1)$, $g_f(-1)$ = last good adaptive and fixed gains.

$$g_a = \begin{cases} g_a(0) & ; \quad g_a(0) \leq g_a(-1) \\ g_a(-1) & ; \quad g_a(0) > g_a(-1) \end{cases}$$

$$g_f = \begin{cases} g_f(0) & ; \quad g_f(0) \leq g_f(-1) \\ g_f(-1) & ; \quad g_f(0) > g_f(-1). \end{cases}$$

### Case 2: BFI = 1, PrevBFI = 0 or 1, S = 1 ... 6:

The CRC has detected residual error(s) in the most important class A bits. The adaptive and fixed code book gains $g_a$ and $g_f$ are replaced by attenuated values from the previous subframes. The memory of the $g_f$ gain predictor is replaced by the average value of the past four values.

Notation: g(0)= current received gain, g(−1) = last good gain $g_a$ or $g_f$.

$$g_a = \begin{cases} C_a(S) \cdot g_a(-1) & ; \quad g_a(-1) \leq \text{median5}(ga) \\ C_a(S) \cdot \text{median5}(ga) & ; \quad g_a(-1) > \text{median5}(ga) \end{cases}$$

$$g_f = \begin{cases} C_f(S) \cdot g_f(-1) & ; \quad g_f(-1) \leq \text{median5}(gf) \\ C_f(S) \cdot \text{median5}(gf) & ; \quad g_f(-1) > \text{median5}(gf). \end{cases}$$

The median5 is calculated by sorting the last five elements in an ascending order and selecting the sample in the middle, i.e., the third largest value. $C_a(S)$ and $C_f(S)$ are fixed weighting coefficients which are dependent on the state $S$.

The higher the state value $S$ is, the more the gains are attenuated.

The LTP-lag values are replaced by the past value from the fourth subframe of the previous frame.

The fixed code book excitation is taken from the erroneous frame.

The past LSFs are used by shifting their values toward their mean

$$\text{LSF1}(i) = \text{LSF2}(i) = \alpha \cdot \text{LSF2}_{\text{past}}(i) + (1 - \alpha) \cdot \text{LSF}_{\text{mean}}(i), \quad i = 1, \ldots 9, \tag{9.15}$$

with $\alpha = 0.95$ and

LSF1($i$):       LSF set of the first half of the frame (10 ms)
LSF2($i$):       LSF set of the second half of the frame
LSF2$_{\text{past}}$($i$):   second LSF set of the past frame
LSF$_{\text{mean}}$:       average LSF set.

It should be noted that this proposal is not mandatory and that more sophisticated solutions can be implemented. In mobile phones, both channel coding and speech coding are implemented on the same processor; thus, soft information from the channel decoder, i.e., real-valued DRI, see Figure 9.2 can be made available for individual bits, groups of bits, or the total frame. It is up to the manufacturer to use more sophisticated concealment methods which exploit soft information from the channel decoder and a priori knowledge about parameter distributions (see also Section 9.1.2).

### 9.2.3 EVS Codec: Concealment of Lost Packets

The enhanced voice services (EVS) codec was designed for packet-switched mobile communication networks, in particular for voice over LTE (VoLTE). Multiple measures have been taken to mitigate the impact of lost packets. Frames may be erroneous due to transmission errors, or frames may be lost or delayed in the packet transport network. A frame erasure is signaled to the decoder by setting the BFI.

The EVS concealment measures depend on the instantaneous frame characteristics and on the bit rate (5.9 ... 128.0 kbit/s). This implies frame classification for frame erasure concealment (FEC) with five categories:

- unvoiced
- unvoiced transition
- onset
- voiced
- voiced transition.

Besides classification for FEC, the EVS codec distinguishes between six distinct *coding modes* optimized for the frame classes: *unvoiced, voiced, transition, audio, inactive, and generic*. The coding mode is signaled to the decoder as part of the frame of coded bits. The FEC classification can partly be derived from the coding mode.

In this section, the key ideas of the EVS error concealment procedures are briefly explained, extracted from the detailed description of the corresponding 3GPP standard [3GPP TS 26.447 2014], [Lecomte et al. 2015].

Error propagation after a frame loss may severely degrade the quality of the reconstructed speech signal. Due to a frame loss, the content of the adaptive code book which is based on past frames, may become different from the content of the adaptive code book of the encoder.

The impact of a lost frame depends on the nature of the speech segment in which the erasure occurred. If the erasure occurs in a speech onset or a transition, the effect of the erasure can propagate through several frames.

The most critical situation is that the erased frame number $k - 1$ is a voiced onset. In this case, the next good voiced frame $k$ cannot use the outdated information from the adaptive code book as the first pitch period is missing. Thus, errors propagate into the following frames, and it may take several frames until the synchronization of the adaptive code book is re-established.

One of the measures to mitigate error propagation and to ensure fast recovery of the content of the adaptive code book of the decoder is to use in the encoder the transition coding mode (TC) for each frame *after a voiced onset*. In the TC mode, the adaptive code book is replaced with a code book of eight different glottal impulse shapes of length $P = 17$. The best pulse shape and the corresponding gain are identified with the same analysis-by-synthesis search method as used for the fixed code book.

Furthermore, precise control of the speech energy is a very important issue in FEC. The importance of the energy control becomes evident when a normal operation is resumed after a *series of erased frames*. To enhance the frame erasure performance, side information consisting of concealment/recovery parameters may be sent in the bitstream to the decoder.

One of several options is to transmit energy information of the last frame $k - 1$ along with the actual frame number $k$. If the decoding is delayed by one frame, an erased frame can be recovered using this side information and parameters can estimated by interpolation between the good frames $k - 2$ and $k$.

As the EVS decoder includes a jitter buffer management (JBM) for compensating variable transmission delay of packets, the frames required for concealment and decoding are available.

The interpolation from frame to frame (and from subframe to subframe) applies not only to energy and gain parameters, but also to the pitch period and to the LP filter parameters which are represented by the LSFs.

At the decoder, the LSF parameters of a lost frame number $k - 1$ can be interpolated from frame $k - 2$ to frame $k$, if a delay by one frame is allowed.

However, if the LSF parameters of frame $k$ are not yet available, the LSF parameters of the concealed frame $k - 1$ have to be extrapolated using the LSF parameters of the last good frame $k - 2$. In this case, the general idea is to fade the last good LSF parameters toward an adaptive mean of the LSF vector. This is comparable with the LSF interpolation used for error concealment in the AMR decoder, as described in (9.15).

In the EVS decoder, a second LSF vector is derived based on the last good LSF vector, but faded to the LSF representation of the comfort noise estimate. In case of a series of consecutive frame losses, this second set of LSFs is used to slowly fade from the last good speech frame to comfort noise. It prevents complete silence observed in standard muting mechanisms in case of a burst of packet losses. This concealment strategy can be summarized as a convergence of the signal energy and the spectral envelope to the estimated parameters of the background noise.

## 9.3 Further Improvements

Further improvements are obtained if we replace in Figure 9.2 the outer parity check code (CRC) by a more powerful block code [Fingscheidt et al. 1999] or even by a non-linear block code, e.g., [Heinen, Vary 2000], [Heinen, Vary 2005], [Hindelang et al. 2000], and [Heinen 2001], which is not intended for explicit channel decoding at the receiver but for increasing the a priori knowledge in support of the soft decision parameter estimation process.

This combination of soft decision source and channel decoding is actually an attractive alternative to conventional error concealment based on parity check coding. Furthermore, if the residual correlation of the codec parameters exceeds a certain value, soft decision source decoding with support by outer channel block coding can even outperform conventional channel coding [Hindelang et al. 2000].

In any case, this concept is also advantageous in combination with conventional inner channel coding, even if the residual parameter correlation is low.

The concept of soft decision source decoding opens up also possibilities for iterative source-channel decoding, e.g., [Farvardin 1990], [Görtz 2000], [Perkert et al. 2001], [Adrat et al. 2002], and [Martin, Heute, Antweiler 2008]. This approach of joint and iterative source-channel decoding is called *turbo error-concealment* [Adrat 2003]. The decoding process is based on the turbo principle [Berrou, Glavieux 1996], as illustrated in Figure 9.9.

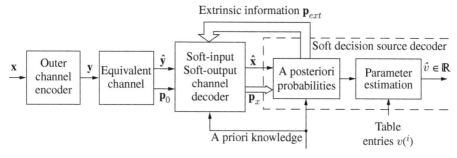

**Figure 9.9** The concept of iterative source-channel decoding.

One of the two component decoders is a channel decoder, the other is a soft decision source decoder. The inner soft-input soft-output (SISO) channel decoder provides *extrinsic information* to the soft decision source decoder which itself extracts extrinsic information on the bit level from the parameter a posteriori probabilities and feeds it back to the channel decoder. After terminating the iterations, the final step consists of estimating the codec parameter as described in Section 9.1.3 using the resulting reliability information (bit error probabilities) on the parameter level.

A more detailed treatment of these ideas can be found in, e.g., the companion book "Advances in Digital Speech Transmission," [Martin, Heute, Antweiler 2008] in the chapters on "Joint Source-Channel Coding":

Chapter 10 "Parameter Models and Estimators in Soft Decision Source Decoding" by T. Fingscheidt

Chapter 11 "Optimal MMSE Estimation for Vector Sources with Spatially and Temporally Correlated Elements" by S. Heinen, M. Adrat

Chapter 12 "Source Optimized Channel Codes & Source Controlled Channel Decoding" by S. Heinen, T. Hindelang

Chapter 13 "Iterative Source-Channel Decoding & Turbo DeCodulation" by A. Adrat, T. Clevorn, L. Schmalen

## References

3GPP TS 126.101 (2002). Adaptive Multi-Rate (AMR) speech codec frame structure, 3GPP.

3GPP TS 26.447 (2014). Codec for Enhanced Voice Services (EVS); Error Concealment of Lost Packets, 3GPP.

Adrat, M. (2003). *Iterative Source-Channel Decoding for Digital Mobile Communications*, PhD thesis. *Aachener Beiträge zu digitalen Nachrichtensystemen*, vol. 16, P. Vary (ed.), RWTH Aachen University.

Adrat, M.; Hänel, R.; Vary, P. (2002). On Joint Source-Channel Decoding for Correlated Sources, *Proceedings of the IEEE International Conference on Acoustics, Speech, and Signal Processing (ICASSP)*, Orlando, Florida, USA.

Alajaji, F. I.; Phamdo, N. C.; Fuja, T. E. (1996). Channel Codes that Exploit the Residual Redundancy in CELP-Encoded Speech, *IEEE Transactions on Speech and Audio Processing*, vol. 4, no. 5, pp. 325–336.

Bahl, L. R.; Cocke, J.; Jelinek, F.; Raviv, J. (1974). Optimal Decoding of Linear Codes for Minimizing Symbol Error Rate, *IEEE Transactions on Information Theory*, vol. 20, pp. 284–287.

Berrou, C.; Glavieux, A. (1996). Near Optimum Error Correcting Coding and Decoding: Turbo-Codes, *IEEE Transactions on Communications*, vol. 44, pp. 1261–1271.

Clüver, K. (1996). An ATM Speech Codec With Improved Reconstruction of Lost Cells, *Signal Processing VIII, Proceedings of the European Signal Processing Conference (EUSIPCO)*, Trieste, Italy, pp. 1641–1643.

COSSAP (1995). *Model Libraries Volume 3*, Synopsys Inc., Mountain View, California.

Cover, T. M.; Thomas, J. A. (1991). *Elements of Information Theory*, John Wiley & Sons, Inc., New York.

Erdöl, N.; Castelluccia, C.; Zilouchian, A. (1993). Recovery of Missing Speech Packets Using the Short-Time Energy and Zero-Crossing Measurements, *IEEE Transactions on Speech and Audio Processing*, vol. 1, no. 3, pp. 295–303.

ETSI Rec. GSM 06.90 (1998). *Digital Cellular Telecommunications System (Phase 2+); Adaptive Multi-Rate (AMR) speech transcoding*, European Telecommunications Standards Institute.

ETSI Rec. GSM 06.91 (1998). *Digital Cellular Telecommunications System (Phase 2+); Substitution and muting of lost frames for Adaptive Multi Rate (AMR) speech traffic channels*, European Telecommunications Standards Institute.

Farvardin, N. (1990). A Study of Vector Quantization for Noisy Channels, *IEEE Transactions on Information Theory*, vol. 36, no. 4, pp. 799–809.

Feldes, S. (1993). Enhancing Robustness of Coded LPC-Spectra to Channel Errors by Use of Residual Redundancy, *Proceedings of EUROSPEECH*, Berlin, Germany, pp. 1147–1150.

Feldes, S. (1994). *Restredundanzbasierte Rekonstruktionsverfahren bei der digitalen Sprachübertragung in Mobilfunksystemen*, PhD thesis, TH Darmstadt, Germany (in German).

Fingscheidt, T. (1998). *Softbit-Sprachdecodierung in digitalen Mobilfunksystemen*, PhD thesis. *Aachener Beiträge zu digitalen Nachrichtensystemen*, vol. 9, P. Vary (ed.), RWTH Aachen University (in German).

Fingscheidt, T.; Scheufen, O. (1997a). Error Concealment in the GSM System by Softbit Speech Decoding, *Proceedings of the 9th Aachen Symposium "Signaltheorie"*, Aachen, Germany, pp. 229–232.

Fingscheidt, T.; Scheufen, O. (1997b). Robust GSM Speech Decoding Using the Channel Decoder's Soft Output, *Proceedings of EUROSPEECH*, Rhodes, Greece, pp. 1315–1318.

Fingscheidt, T.; Vary, P. (1997). Robust Speech Decoding: A Universal Approach to Bit Error Concealment, *Proceedings of the IEEE International Conference on Acoustics, Speech, and Signal Processing (ICASSP)*, Munich, Germany, vol. 3, pp. 1667–1670.

Fingscheidt, T.; Vary, P. (2001). Softbit Speech Decoding: A New Approach to Error Concealment, *IEEE Transactions on Speech and Audio Processing*, vol. 9, no. 3, pp. 240–250.

Fingscheidt, T.; Vary, P.; Andonegui, J. A. (1998). Robust Speech Decoding: Can Error Concealment Be Better Than Error Correction? *Proceedings of the IEEE International Conference on Acoustics, Speech, and Signal Processing (ICASSP)*, Seattle, Washington, USA, vol. 1, pp. 373–376.

Fingscheidt, T.; Heinen, S.; Vary, P. (1999). Joint Speech Codec Parameter and Channel Decoding of Parameter Individual Block Codes (PIBC), *IEEE Workshop on Speech Coding*, Porvoo, Finland, pp. 75–77.

FS 1016 CELP (1992). Details to Assist in Implementation of Federal Standard 1016 CELP, US Department of Commerce – National Technical Information Service.

Gerlach, C. G. (1993). A Probabilistic Framework for Optimum Speech Extrapolation in Digital Mobile Radio, *Proceedings of the IEEE International Conference on Acoustics, Speech, and Signal Processing (ICASSP)*, Minneapolis, Minnesota, vol. 2, pp. 419–422.

Gerlach, C. G. (1996). *Beiträge zur Optimalität in der codierten Sprachübertragung*, PhD thesis. *Aachener Beiträge zu digitalen Nachrichtensystemen*, vol. 5, P. Vary (ed.), RWTH Aachen University (in German).

Goodman, D. J.; Lockhart, G. B.; Wasem, O. J.; Wong, W.-C. (1986). Waveform Substitution Techniques for Recovering Missing Speech Segments in Packet Voice Communications, *IEEE Transactions on Acoustics, Speech and Signal Processing*, vol. 34, no. 6, pp. 1440–1448.

Görtz, N. (2000). Iterative Source-Channel Decoding using Soft-In/Soft-Out Decoders, *Proceedings of the International Symposium on Information Theory (ISIT)*, Sorrento, Italy, p. 173.

Heikkila, I.; Jokinen, H.; Ranta, J. (1993). A Signal Quality Detecting Circuit and Method for Receivers in the GSM System, European Patent # 0648032 A 1.

Heinen, S. (2001). *Quellenoptimierter Fehlerschutz für digitale Übertragungskanäle*, PhD thesis. *Aachener Beiträge zu digitalen Nachrichtensystemen*, vol. 14, P. Vary (ed.), RWTH Aachen University (in German).

Heinen, S.; Vary, P. (2000). Source Optimized Channel Codes (SOCCs) for Parameter Protection, *Proceedings of the International Symposium on Information Theory (ISIT)*, Sorrento, Italy.

Heinen, S.; Vary, P. (2005). Source Optimized Channel Coding for Digital Transmission Channels, *IEEE Transactions on Communications*, vol. 53, no. 4, pp. 592–600.

Hindelang, T.; Heinen, S.; Vary, P.; Hagenauer, J. (2000). Two Approaches to Combined Source-Channel Coding: A Scientific Competition in Estimating Correlated Parameters, *International Journal of Electronics and Communications (AEÜ)*, vol. 54, no. 6, pp. 364–378.

Järvinen, K.; Vainio, J. (1994). Detection of Defective Speech Frames in a Receiver of a Digital Speech Communication System, World Patent # WO 96/09704.

Jayant, N. S.; Christensen, S. W. (1981). Effects of Packet Losses in Waveform Coded Speech and Improvements Due to an Odd-Even Sample-Interpolation Procedure, *IEEE Transactions on Communications*, vol. 29, no. 2, pp. 101–109.

Lecomte, J.; Vaillancourt, T.; Bruhn, S.; Sung, H.; Peng, P.; Kikuiri, K.; Wang, B.; Subasingha, S.; Faure, J. (2015). Packet-Loss Concealment Technology Advances in EVS, *Proceedings of the IEEE International Conference on Acoustics, Speech, and Signal Processing (ICASSP)*, pp. 5708–5712.

Martin, R.; Hoelper, C.; Wittke, I. (2001). Estimation of Missing LSF Parameters Using Gaussian Mixture Models, *Proceedings of the IEEE International Conference on Acoustics, Speech, and Signal Processing (ICASSP)*, Salt Lake City, Utah, USA.

Martin, R.; Heute; U.; Antweiler; C. (eds.) (2008). *Advances in Digital Speech Transmission*, John Wiley & Sons.

Melsa, J. L.; Cohn, D. L. (1978). *Decision and Estimation Theory*, McGraw-Hill Kogakusha, Tokyo.

Minde, T. B.; Mustel, P.; Nilsson, H.; Lagerqvist, T. (1993). Soft Error Correction in a TDMA Radio System, World Patent # WO 95/16315.

Perkert, R.; Kaindl, M.; Hindelang, T. (2001). Iterative Source and Channel Decoding for GSM, *Proceedings of the IEEE International Conference on Acoustics, Speech, and Signal Processing (ICASSP)*, Salt Lake City, Utah, USA, pp. 2649–2652.

Rec. GSM 05.05 (1992). Recommendation GSM 05.05 Radio Transmission and Reception, ETSI TC-SMG.

Rec. GSM 06.10 (1992). Recommendation GSM 06.10 GSM Full Rate Speech Transcoding, ETSI TC-SMG.

Rec. GSM 06.11 (1992). Recommendation GSM 06.11 Substitution and Muting of Lost Frames for Full Rate Speech Traffic Channels, ETSI TC-SMG.

Rec. GSM 06.21 (1995). European Digital Cellular Telecommunications System Half Rate Speech Part 3: Substitution and Muting of Lost Frames for Half Rate Speech Traffic Channels (GSM 06.21), ETSI TM/TM5/TCH-HS.

Rec. GSM 06.61 (1996). Digital Cellular Telecommunications System: Substitution and Muting of Lost Frames for Enhanced Full Rate (EFR) Speech Traffic Channels (GSM 06.61), ETSI TC-SMG.

Sereno, D. (1991). Frame Substitution and Adaptive Post-Filtering in Speech Coding, *Proceedings of EUROSPEECH*, Genoa, Italy, pp. 595–598.

Shannon, C. E. (1948). A Mathematical Theory of Communication, *Bell System Technical Journal*, vol. 27, pp. 379–423.

Skoglund, M.; Hedelin, P. (1994). Vector Quantization Over a Noisy Channel Using Soft Decision Decoding, *Proceedings of the IEEE International Conference on Acoustics, Speech, and Signal Processing (ICASSP)*, Adelaide, Australia, vol. 5, pp. 605–608.

Stenger, A.; Younes, K. B.; Reng, R.; Girod, B. (1996). A New Error Concealment Technique for Audio Transmission with Packet Loss, *Signal Processing VIII, Proceedings of the European Signal Processing Conference (EUSIPCO)*, Trieste, Italy, pp. 1965–1968.

Su, H.-Y.; Mermelstein, P. (1992). Improving the Speech Quality of Cellular Mobile Systems Under Heavy Fading, *Proceedings of the IEEE International Conference on Acoustics, Speech, and Signal Processing (ICASSP)*, San Francisco, California, USA, pp. II 121–124.

Sundberg, C.-E. (1978). Soft Decision Demodulation for PCM Encoded Speech Signals, *IEEE Transactions on Communications*, vol. 26, no. 6, pp. 854–859.

Valenzuela, R. A.; Animalu, C. N. (1989). A New Voice-Packet Reconstruction Technique, *Proceedings of the IEEE International Conference on Acoustics, Speech, and Signal Processing (ICASSP)*, Glasgow, UK, pp. 1334–1336.

Wong, W. C.; Steele, R.; Sundberg, C.-E. (1984). Soft Decision Demodulation to Reduce the Effect of Transmission Errors in Logarithmic PCM Transmitted over Rayleigh Fading Channels, *AT&T Bell Laboratories Technical Journal*, vol. 63, no. 10, pp. 2193–2213.

# 10

# Bandwidth Extension of Speech Signals

The characteristic sound of *telephone speech* with its restricted audio quality goes back to the frequency bandwidth limitation of the old analog *narrowband (NB) telephone*. In the stepwise conversion process of the world-wide telephone networks from analog to digital technology, the limited frequency band of 300 Hz to 3.4 kHz [ITU-T Rec. G.712 1988] has been retained for reasons of compatibility.

Due to the progress of digital signal processing technology, speech codecs for the so-called *wideband (WB) telephony* have been standardized with a bandwidth of $B = 7$ kHz and a sampling rate of $f_s = 16$ kHz. A first WB speech codec for digital telephony and teleconferencing was specified by CCITT already in 1985, [ITU-T Rec. G.722 1988]. For mobile telephony, appropriate WB-codecs with lower bit rates have been developed, such as, e.g., the *adaptive multi-rate WB* (AMR-WB) codec [3GPP TS 26.190 2001]. The latest codec standard for cellphones is the enhanced voice services (EVS) codec, which has been optimized for a wide range of bit rates, bandwidths, and qualities [3GPP TS 126.445 2014] (see also Section 8.7).

Meanwhile, WB codecs have been introduced in some networks. This happens mainly in the context of network conversion from circuit-switched to packet-switched technology, which offers more flexibility w.r.t. to coding and signaling. WB transmission requires new telephone sets with better electro-acoustic front-ends, improved A/D and D/A converters, and new speech codecs. In addition, appropriate signaling procedures are needed for detection and activation of the WB capability.

In cellular radio networks, further modifications are necessary, since speech-codec-specific error protection (channel coding) is required in the mobile phones as well as in the base stations.

This ongoing NB-to-WB conversion process will take a long transition period in which many terminals and/or parts of the networks have not yet been equipped with the WB capability.

If a NB telephone is used at the far-end of a communication link, a WB telephone at the near-end has to be operated in the NB fall-back mode.

The attractiveness of new WB services and devices would be increased, if the received NB speech signal could be improved by means of *bandwidth extension* (BWE). The application of BWE is illustrated in Figure 10.1 for a digital phone at the *far-end* using a NB speech encoder and a WB terminal at the *near-end*. The far-end terminal is operating at the NB sampling rate of $f_s' = 8$ kHz, while the near-end terminal uses the sampling rate of

*Digital Speech Transmission and Enhancement*, Second Edition. Peter Vary and Rainer Martin.
© 2024 John Wiley & Sons Ltd. Published 2024 by John Wiley & Sons Ltd.

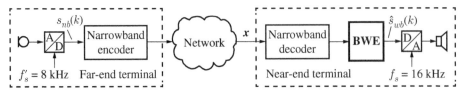

**Figure 10.1** Wideband telephony with bandwidth extension (BWE) at the near-end.

$f_s = 16$ kHz. The BWE module increases the bandwidth of the decoded narrowband signal artificially from $B' = 3.4$ kHz to $B = 7.0$ kHz. The required parameters for BWE are either derived from the decoded NB signal or transmitted explicitly as part of the encoded bit frame **x**.

As a next step in the network evolution, speech/audio codecs for *super wideband* signals (SWB, $B = 14$ kHz, $f_s = 32$ kHz) and *fullband* signals (FB, $B = 20$ kHz, $f_s = 48$ kHz) have been standardized (see also Section 8.7 on coding standards).

The mismatch between NB, WB, and fullband speech signals is illustrated in Figure 10.2 for an unvoiced /s/ speech sample. The figure depicts a spectral comparison of the original

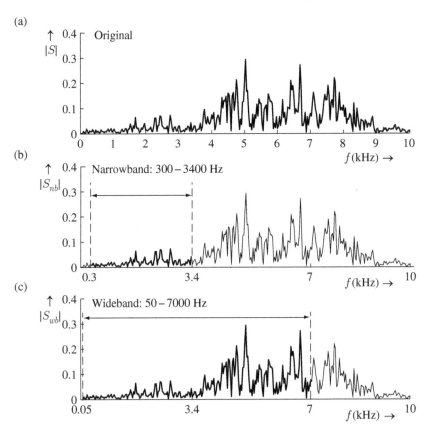

**Figure 10.2** Example of a magnitude spectrum of an unvoiced speech sound. (a) $S$: original speech, (b) $S_{nb}$: narrowband telephone speech, and (c) $S_{wb}$: wideband telephone speech.

(fullband) speech with the corresponding NB and WB versions. A closer look reveals that the NB speech signal may lack significant parts of the spectrum and that even the difference between wideband speech and original speech is still clearly noticeable. The frequency range of wideband telephone voice is comparable to the bandwidth of AM radio broadcast, which offers excellent speech quality and intelligibility while the quality of music signals is still limited. However, the original speech signal covers a frequency range, which is still significantly wider than $B = 7.0$ kHz.

Thus, BWE of WB signals to SWB and the extension of SWB to FB seem also to be promising BWE applications, e.g., [Yamanashi et al. 2011], [Geiser, Vary 2013], [Atti et al. 2015], and [Bachhav et al. 2018].

Furthermore, it should be mentioned that BWE is of special interest for video conferencing with participants connected to the conference bridge with different bandwidths: NB, WB, SWB, or even FB. In the conference bridge, the lower bandwidths could be increased by BWE for those links using a wider bandwidth.

Two basic BWE concepts will be discussed below:

- BWE *without* WB side information:
  The parameters for bandwidth extension are extracted from the received signal with the lower bandwidth (or from the corresponding encoded signal parameters). This approach does neither require modifications of the far-end terminal nor of the network.
- BWE *with* WB side information:
  As part of a modified speech encoder, a small amount of *WB side information* is extracted from the WB input signal and transmitted to a modified decoder, which comprises a BWE module.

Obviously, the first approach without side information is much more challenging than the second approach with WB side information.

In this context, it should be mentioned that bandwidth extension of WB speech signals beyond 7.0 kHz using the source-filter model or using deep learning techniques is significantly less demanding than the extension from NB to WB. In e.g., [Su et al. 2021] it is shown that bandwidth extension from WB to FB by deep learning techniques can produce an audio quality, which is comparable to FB recordings at a sample rate of $f_s = 48$ kHz.

In all these cases, the objective of bandwidth extension is to increase the frequency bandwidth, improve the perceived speech or audio quality, and eventually improve the speech intelligibility by artificially adding some *high-frequency* and *low-frequency* components.

In the following, the two BWE concepts will be discussed on the basis of the source-filter model of speech production.

## 10.1 BWE Concepts

Early BWE proposals used simple (analog or digital) signal processing techniques without taking the model of speech production into account. Probably the first proposal for BWE was made as early as 1933 by [Schmidt 1933], who tried to extend the speech bandwidth by non-linear processing. In 1972, a first application was explored by the BBC [Croll 1972], aiming at the improvement of telephone speech contributions to radio broadcast programs. Two different signal processing techniques were used for the regeneration of the lower frequencies (80–300 Hz) and the higher frequencies (3.4–7.6 kHz). Artificial low-frequency components were produced by rectification of the NB signal and by applying a low-pass filter with a cutoff frequency of $f_c = 300$ Hz. The high-frequency components were inserted as bandpass noise, which was modulated by the spectral speech energy in the range 2.4–3.4 kHz. However, these methods were too simplistic to produce consistently improved speech signals as the level of the low-frequency content turned out to be often too high or too low and the high-frequency extension introduced a noisy disturbance.

A digital signal processing approach was proposed by P. J. Patrick [Patrick 1983]. In his experiments, he made a distinction between unvoiced and voiced sounds. During voiced segments, a harmonic spectrum extension beyond 3.4 kHz was derived. By FFT techniques, an amplitude scaled version of the spectrum between 300 Hz and 3.4 kHz was inserted, taking the fundamental frequency into account. This synthetic extension had spectral components at multiples of the fundamental frequency. Due to mismatch of the synthetic components w.r.t. level, and spectral envelope, these pure signal processing concepts do not provide consistently natural sounding speech.

In the context of audio transmission using coding by *adaptive pulse code modulation* (ADPCM) with a sample rate of 16 kHz and 4 bits per sample, [Dietrich 1984] found that the subjective quality can be improved by allowing the output filter following the D/A conversion to deliberately produce some aliasing via a wide transition region of the output low-pass filter from 8 to 12 kHz.

Advanced implementations based on this idea have also been proposed in the context of audio coding. This idea is also known as *spectral band replication* (SBR) in the context of MPEG audio coding [Larsen et al. 2002], [Larsen, Aarts 2004]. SBR has been used to enhance the coding efficiency of MP3 (MP3Pro) and Advanced Audio Coding (aacPlus), e.g., [Dietz et al. 2002], [Ehret et al. 2003], [Gröschel et al. 2003], and [Homm et al. 2003].

It should be noted that these simple techniques work quite well at frequencies above 8 kHz but that they fail in the interesting frequency range 3.4–7.0 kHz.

A breakthrough was achieved in 1994 by the proposals of H. Carl [Carl 1994], H. Carl and U. Heute [Carl, Heute 1994], Y. M. Cheng, D. O'Shaughnessy, and P. Mermelstein [Cheng et al. 1994], and Y. Yoshida and M. Abe [Yoshida, Abe 1994]. They explicitly took the model of speech production into consideration.

These approaches exploit the redundancy of the speech production mechanism, as well as specific properties of auditory perception. These pioneering proposals should be regarded as the starting point of a series of further investigations, e.g., [Avendano et al. 1995], [Iyengar et al. 1995], [Epps, Holmes 1999], [Enbom, Kleijn 1999], [Jax, Vary 2000], [Park, Kim 2000], [Fuemmeler et al. 2001], [Kornagel 2001], [Jax, Vary 2002,2003], [Jax 2002], [Jax 2004], [Jax, Vary 2006], [Geiser et al. 2007], [Martin, Heute, Antweiler 2008] (Chapter 9, by P. Jax), [Bauer, Fingscheidt 2008,2009], [Abel, Fingscheidt 2016,2018], [Schlien et al. 2018], [Schlien 2019], and [Abel 2020].

## 10.2   BWE using the Model of Speech Production

In this chapter, basic BWE concepts with and without side information for the extension of high frequencies will be explained. These methods may also be used for the extension of low frequencies.

The widely used signal processing architecture, which is based on the source-filter model of speech production is illustrated in Figure 10.3 for the example of BWE from narrowband (NB, 3.4 kHz) to wideband (WB, 7.0 kHz). The processing takes place in two separate frequency bands:

- a lowpass band (lowband, 0.0–4.0 kHz) and
- a highpass band (highband, 4.0–7.0 kHz).

The bitstream $\mathbf{x}$ of the encoded NB signal is applied to the speech decoder to produce the decoded NB signal $s(k')$ at the sampling rate of $f_s' = 8\,\text{kHz}$. In the lowpass band, the decoded NB signal $s(k')$ is upsampled by a factor of 2 to $f_s = 16\,\text{kHz}$ and interpolated by low-pass filtering with a cut-off frequency of $f_c = 4\,\text{kHz}$, resulting in the oversampled *lowband signal* $s_{lb}(k)$.

The decoded NB signal $s(k')$ and/or some parameters extracted from $\mathbf{x}$, possibly containing BWE side information, are passed on to the BWE module which produces the *highband signal* $s_{hb}(k)$ at a sampling rate of $f_s = 16\,\text{kHz}$. The BWE module comprises an excitation source, time-varying gain factors $c$, $g$, and a highband spectral envelope filter $\hat{H}_{hb}(z)$. The excitation signal $e_{hb}(k)$ is scaled with a gain factor $c$, which controls the *frame energy*. The scaled excitation $\tilde{d}_{hb}(k)$ is applied to the highband envelope filter $\hat{H}_{hb}(z) = \dfrac{1}{1 - \hat{A}_{hb}(z)}$ for producing the preliminary highband signal $\tilde{s}_{hb}(k)$. The preliminary signal may need additional scaling by a gain factor $g$ to compensate for amplification/attenuation changes

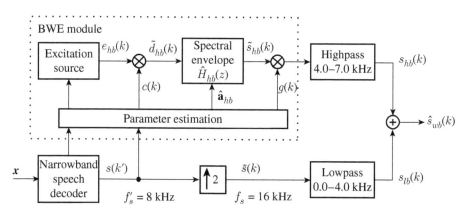

**Figure 10.3**   BWE concept for narrowband-to-wideband extension using the soure-filter-model of speech production. **x**: received bit stream with or without side-information for BWE; $c(k)$: excitation gain (short-term); $g(k)$: temporal envelope gain (medium term); $s(k')$: narrowband signal ($f_s' = 8\text{kHz}$); $e_{hb}(k)$: highband excitation signal ($f_s = 16\text{kHz}$); $s_{hb}(k)$: highband signal; $s_{lb}(k)$: interpolated narrowband signal $s(k)$; $\hat{s}_{wb}(k)$: wideband signal.

caused by the highband spectral envelope filter, and to adjust the *energy across frames*, i.e., the energy time envelope.

The final highband signal $s_{hb}(k)$ is extracted with a highpass filter with a cut-off frequency of $f_c = 4\,\text{kHz}$. Note that this highpass filter would not be required if the excitation signal $e_{hb}(k)$ is already limited to the highband band frequencies.

A further variation of the concept of Figure 10.3 is to produce first a highband signal $\tilde{s}_{hb}(k')$ at the lower sampling rate $f_s'$ and to upsample and interpolate this signal then by a factor of $r = 2$.

Finally, the superposition of the lowband signal $s_{lb}(k)$ and the highband signal $s_{hb}(k)$ delivers the artificially extended wideband signal

$$\hat{s}_{wb}(k) = s_{lb}(k) + s_{hb}(k). \tag{10.1}$$

This BWE concept can be subdivided into *three separate estimation tasks* for the extension band, e.g., [Carl 1994], [Jax, Vary 2003], and [Larsen, Aarts 2004]:

- the excitation signal including a gain $c$
- the spectral envelope estimate $|\hat{H}_{hb}(z)|$
- the energy time envelope $g$ across frames.

It turns out that the estimation of the energy time envelope $g$ is the most critical part, e.g., [Geiser 2012], [Schlien 2019], and [Abel 2020]. The second most important task is the estimation of the spectral envelope $|\hat{H}_{hb}(z)|$, while the estimation of the excitation signal $\tilde{d}_{hb}(k)$ is less critical, e.g., [Carl 1994], [Carl, Heute 1994], [Fuemmeler et al. 2001], [Kornagel 2001], [Jax, Vary 2003], and [Jax 2004].

### 10.2.1  Extension of the Excitation Signal

According to the simplifying linear model of speech production, the excitation signal of the vocal tract filter can be derived from the speech signal as a *residual signal d* by linear prediction (LP). The LP residual signal is more or less spectrally flat. In voiced speech segments, it consists of sinusoids at multiples of the fundamental (pitch) frequency where all harmonics have almost the same amplitude. During unvoiced segments, the excitation is more or less white noise.

Figure 10.4 shows for NB-to-WB extension how to derive a highband excitation signal $e_{hb}(k)$ and the gain scaled highband excitation signal $\tilde{d}_{hb}(k)$ from the NB LP residual signal $d(k')$.

The NB signal $s(k')$, sampled at $f_s' = 8\,\text{kHz}$ is applied to the analysis filter $1 - A_{lb}(z)$ to produce the LP residual signal $d(k')$. The residual signal is interpolated by upsampling to $f_s = 16\,\text{kHz}$ and low-pass filtering to obtain an oversampled version $d_{lb}(k)$ of the lowband LP residual (Figure 10.4a). The residual signal $d_{lb}(k)$ is almost flat within the NB telephone frequency band and almost zero outside as indicated in Figure 10.4b.

Due to these properties, the true highband excitation signal can approximately be replicated by modulation, i.e., frequency shifting by $\Omega_M$ and highpass filtering of the lowband residual signal $d_{lb}(k)$ [Carl 1994], [Fuemmeler et al. 2001], and [Kornagel 2001]. This approach is illustrated in Figure 10.4c, d.

The choice of $\Omega_M$ is discussed below. The real-valued modulation delivers

$$\hat{d}_{wb}(k) = d_{lb}(k)\,2\cos(\Omega_M k) \tag{10.2}$$

(a)

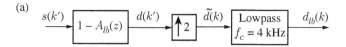

(b)

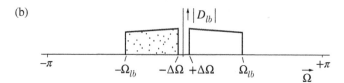

(c)

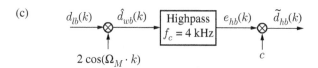

(d)

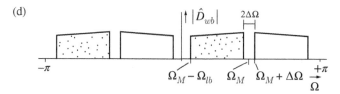

(e)

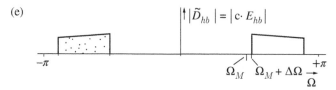

**Figure 10.4** Derivation of a highband excitation signal $\tilde{d}_{hb}(k)$ from the narrowband/lowband LP residual signal $d(k')$. (a) Generation of the oversampled LP residual signal $d_{lb}(k)$. (b) Schematic spectrum of the oversampled narrowband/lowband residual signal $d_{lb}(k)$. (c) Generation of a highband excitation signal $e_{hb}(k)$ by modulation and filtering according to (10.2) and (10.4). (d) Schematic spectrum of the modulated narrowband residual signal $\hat{d}_{wb}(k)$ with modulation frequency $\Omega_M$ in the range $\Omega_{lb} \leq \Omega_M \leq \pi - \Omega_{lb}$. (e) Schematic spectrum of the highband excitation signal $\tilde{d}_{hb}(k)$.

consisting of *two* frequency shifted components

$$\hat{D}_{wb}e^{j\Omega} = D_{lb}(e^{j(\Omega - \Omega_M)}) + D_{lb}(e^{j(\Omega + \Omega_M)}). \tag{10.3}$$

A highband source signal $e_{hb}(k)$ is obtained by filtering of $\hat{d}_{wb}(k)$ with the highpass impulse response $h_{HP}(k)$ (Figure 10.4c, d).

$$e_{hb}(k) = \hat{d}_{wb}(k) * h_{HP}(k). \tag{10.4}$$

The residual energies of a WB signal $s_{wb}(k)$ in the highband and the lowband may be different. A gain factor $c$ has to be applied to obtain an appropriate highband excitation estimate

$$\tilde{d}_{hb}(k) = c \cdot e_{hb}(k) \tag{10.5}$$

as indicated in Figure 10.4c.

For the selection of the modulation frequency $\Omega_M$ several options exist:

- Modulation with the *Nyquist* frequency, i.e., $\Omega_M = \pi$ and $\cos(\Omega_M k) = (-1)^k$. This is the method of *spectral mirroring* as proposed in [Un, Magill 1975] and [Makhoul, Berouti 1979] for LP speech coding (see also Section 8.5.2). The highband excitation spectrum becomes a spectrally reversed version of the lowband residual.

    However, this simple modulation by alternating sign reversal has a major disadvantage. Voiced NB speech signals have a fundamental pitch frequency $\Omega_p$ and several harmonics at $m \cdot \Omega_p$ up to the frequency limit $\Omega_{lb}$. The discrete spectral components of the extended frequency band beyond $\Omega_{lb}$, which are generated by sign switching are in general not harmonics of the fundamental frequency $\Omega_p$ if the modulation frequency $\Omega_M = \pi$ is not a multiple of $\Omega_p$, resulting in non-harmonic distortions.

- A solution is to choose the modulation frequency $\Omega_M$ as an integer multiple of the estimated instantaneous pitch frequency $\tilde{\Omega}_p$ [Fuemmeler et al. 2001], e.g.,[1]

$$\Omega_M = \left\lceil \frac{\Omega_{3.4}}{\tilde{\Omega}_p} \right\rceil \tilde{\Omega}_p . \tag{10.6}$$

This method guarantees that the harmonics in the extended frequency band will always match the harmonic structure of the lowband. However, the pitch-adaptive modulation method is quite sensitive to small estimation errors of the pitch frequency, because these are significantly enlarged by the factor $m$. Therefore, a reliable pitch estimator is needed.

A further problem seems to be the spectral gap of $2\Delta\Omega$ between the maximum frequency $\Omega_{lb}$ of the lowband and the minimum frequency $\Omega_M + \Delta\Omega$ of the highband (see Figure 10.4d). This problem can be solved by minimizing the width of the gap. The modulation frequency has to be chosen again as a multiple of $\Omega_p$ such that the highband excitation signal $\tilde{d}_{hb}(k)$ provides more or less a seamless harmonic continuation of the lowband spectrum.

However, informal listening reveals that the human ear is surprisingly insensitive w.r.t. magnitude and phase distortions at frequencies above 3.4 kHz. A spectral gap in a speech spectrum of up to a width of 1 kHz is almost inaudible. Furthermore, a moderate misalignment of the harmonic structure of speech at frequencies beyond 3.4 kHz does not significantly degrade the subjective quality of the enhanced speech signal. A reasonable compromise between subjective quality and computational complexity is the modulation with the fixed frequency of $\Omega_M = \Omega_{lb}$.

It should be noted that several other proposals for the extension of the excitation signal can be found in the literature. These techniques include, for example, non-linearities [Valin, Lefebvre 2000], [Kornagel 2003] or synthetic multiband excitation (MBE vocoder) [Chan, Hui 1996].

## 10.2.2  Spectral Envelope Estimation

In this section, different versions of a WB spectral envelope estimator and its training procedure will be discussed. For the training, frames of WB speech signals $s_{wb}(k)$ as well

---

1  $\lceil x \rceil$ denotes the smallest integer larger than $x$.

as the corresponding NB signals $s_{nb}(k)$ are available. The signals $s_{nb}(k)$ are NB versions of $s_{wb}(k)$, degraded both by the restricted frequency band and by quantization/coding noise. The spectral envelope of a WB signal frame is described by the filter coefficients of the speech production model in terms of the LP-coefficients $a(i), i = 1, 2, \ldots, n$, the corresponding reflection coefficients $k_i, i = 1, 2, \ldots, n$, the cepstral coefficients $c(i), i = 0, 1, 2, \ldots$ or any other related representation, see also Chapter 2, Section 3.8.1, and Chapter 6.

The spectral envelope estimator is based on two vector codebooks A and B containing LP coefficient vectors $\mathbf{a} = \mathbf{a}_{wb}$ and $\mathbf{b} = \mathbf{a}_{nb}$, which can be derived from WB signals as indicated in the upper part of Figure 10.5.

In an off-line pre-training phase, the WB codebook A with $N$ entries is optimized like a vector quantizer (see also Section 7.6) using a representative sequence of WB LP coefficient vectors $\mathbf{a}$, while the NB codebook B with $M$ entries is designed from the corresponding NB LP coefficients $\mathbf{b}$.

The entries of codebook A are denoted as $\mathbf{a}_v = (a_v(1), a_v(2), \ldots, a_v(n))$ with $v = 1, 2, \ldots, N$ where $n$ is the degree of the WB LP filter, while codebook B comprises $M$ vectors $\mathbf{b}_\mu = (b_\mu(1), b_\mu(2), \ldots, b_\mu(m))$, $\mu = 1, 2, \ldots, M$ where $m$ is the degree of the NB LP filter.

In a second training phase, the a priori probabilities $P(\mathbf{b}_\mu)$ and the posterior probabilities $P(\mathbf{a}_v|\mathbf{b}_\mu)$ are measured using the pre-trained codebooks A and B and the training sequences as shown in the lower part of Figure 10.5. The posterior probabilities $P(\mathbf{a}_v|\mathbf{b}_\mu)$ describe the conditional probability of the WB vector $\mathbf{a}_v$ given the NB vector $\mathbf{b}_\mu$.

At the receiver side, codebook A and the actual NB LP coefficients $\mathbf{b} \in (\mathbf{b}_\mu, \mu = 1, 2, \ldots, M)$ are available. The NB coefficient vector $\mathbf{b}_\mu$ is either transmitted as part of the

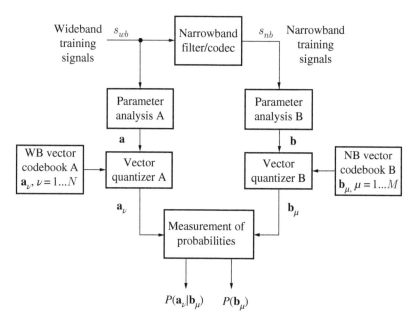

**Figure 10.5** Off-line training of probabilities for spectral envelope estimation.

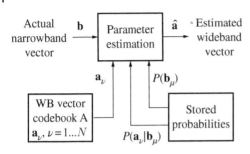

**Figure 10.6** On-line wideband spectral envelope estimation â.

encoded bit frame or can be derived from the received NB signal $s_{nb}(k)$. A WB estimate â of the true WB parameter vector $\mathbf{a}$ is calculated taking the present observation $\mathbf{b} = \mathbf{b}_\mu$ and the statistical knowledge gained in the off-line training process into account as indicated in Figure 10.6.

The joint statistics of $\mathbf{a}_v$ and $\mathbf{b}_\mu$ can be evaluated as

$$P(\mathbf{a}_v, \mathbf{b}_\mu) = P(\mathbf{b}_\mu) \cdot P(\mathbf{a}_v | \mathbf{b}_\mu). \tag{10.7}$$

The estimation process should produce the best possible â in a statistical sense, e.g., by minimizing some error criterion based on a reasonable cost function.

A cost function $C(\mathbf{a}_v, \hat{\mathbf{a}})$ as described in, e.g., [Melsa, Cohn 1978] assigns a cost value to the estimate â with respect to each instance $\mathbf{a}_v$ of the WB parameter vector.

From the analytic expression of the total average cost

$$\rho_0 = E\{C(\mathbf{a}_v, \hat{\mathbf{a}})\} = \sum_{\mu=1}^{M} \sum_{v=1}^{N} C(\mathbf{a}_v, \hat{\mathbf{a}}) \cdot P(\mathbf{a}_v, \mathbf{b}_\mu), \tag{10.8}$$

the optimal estimate $\hat{\mathbf{a}} = f(\mathbf{b})$ can be derived. For the initially unknown estimate â, the individual cost values $C(\mathbf{a}_v, \hat{\mathbf{a}})$ are weighted by their corresponding joint probabilities $P(\mathbf{a}_v, \mathbf{b}_\mu)$ and summed up for every combination of $v$ and $\mu$. From this general expression, the estimation rule $\hat{\mathbf{a}} = f(\mathbf{b}_\mu)$ can be found by minimizing $\rho_0$. With (10.7), we re-write (10.8) as follows:

$$\rho_0 = \sum_{\mu=1}^{M} \left( \sum_{v=1}^{N} C(\mathbf{a}_v, \hat{\mathbf{a}}) \cdot P(\mathbf{a}_v | \mathbf{b}_\mu) \right) P(\mathbf{b}_\mu). \tag{10.9}$$

As the set of probabilities $P(\mathbf{b}_\mu)$ is fixed and as all values are non-negative, the minimum of $\rho_0$ can be found by minimizing the inner summation for every observation $\mathbf{b}_\mu$ [Melsa, Cohn 1978], i.e., the conditional expectation

$$\rho_1 = E\left\{C(\mathbf{a}_v, \hat{\mathbf{a}}) | \mathbf{b}_\mu\right\} = \sum_{v=1}^{N} C(\mathbf{a}_v, \hat{\mathbf{a}}) \cdot P(\mathbf{a}_v | \mathbf{b}_\mu). \tag{10.10}$$

### 10.2.2.1 Minimum Mean Square Error Estimation

Choosing as cost function the squared Euclidean error norm, i.e.,

$$C(\mathbf{a}_v, \hat{\mathbf{a}}) = (\mathbf{a}_v - \hat{\mathbf{a}})^T (\mathbf{a}_v - \hat{\mathbf{a}}), \tag{10.11}$$

the minimization of $\rho_1$ with respect to $\hat{a}$ by partial derivation

$$\nabla_{\hat{a}} \left[ \sum_{v=1}^{N} (\mathbf{a}_v - \hat{\mathbf{a}})^T (\mathbf{a}_v - \hat{\mathbf{a}}) \cdot P(\mathbf{a}_v | \mathbf{b}_\mu) \right] = -\sum_{v=1}^{N} 2(\mathbf{a}_v - \hat{\mathbf{a}}) \cdot P(\mathbf{a}_v | \mathbf{b}_\mu) \overset{!}{=} 0 \qquad (10.12)$$

with $\sum_{v=1}^{N} P(\mathbf{a}_v | \mathbf{b}_\mu) = 1$ leads to the *minimum mean square error* (MMSE) or conditional mean estimator:

$$\hat{\mathbf{a}} = \mathrm{E}\{\mathbf{a}_v | \mathbf{b}_\mu\} = \sum_{v=1}^{N} \mathbf{a}_v \cdot P(\mathbf{a}_v | \mathbf{b}_\mu). \qquad (10.13)$$

It should be noted that the solution (10.13) might become rather complex with increasing codebook sizes $N$ and $M$. However, as the accuracy of the spectral envelope estimation in the highpass band of Figure 10.3 is not too critical, rather small codebooks with just a few dozens of entries are sufficient. Therefore, the codebooks A and B can be designed independently from the LP parameter codebook of the NB (lowband) speech codec. Alternatively, the time domain signals $s_{wb}$ and $s_{nb}$ can be used for training rather small numbers of dedicated codebook vectors $\mathbf{a}_v$ and $\mathbf{b}_\mu$.

### 10.2.2.2 Conditional Maximum *A Posteriori* Estimation

Another possibility for weighting the estimation errors in (10.9) is the uniform cost function

$$C(\mathbf{a}_v, \hat{\mathbf{a}}) = \begin{cases} 0 & ||\mathbf{a}_v - \hat{\mathbf{a}}|| < \epsilon \\ 1 & \text{else}. \end{cases} \qquad (10.14)$$

To minimize the sum in (10.10) according to this cost function, we must select the codebook vector $\hat{\mathbf{a}} = \mathbf{a}_v$ with the maximum conditional probability $P(\mathbf{a}_v | \mathbf{b}_\mu)$ since then its contribution to the sum in (10.10) is zero. Thus, the estimate $\hat{\mathbf{a}}$ is obtained by searching the maximum of the conditional probability

$$\hat{\mathbf{a}} = \arg\max_{\mathbf{a}_v} P(\mathbf{a}_v | \mathbf{b}_\mu). \qquad (10.15)$$

Instead of the LP coefficients, any other representation of the LP filter could used within the estimation process.

### 10.2.2.3 Extensions

In the literature, different proposals for estimating the WB spectral envelope have been made, which may be considered as variations, specializations, or extensions of the described conditional estimation framework.

In [Carl 1994], the wideband LP coefficients are obtained by fixed *code book mapping* between the coefficient set of the NB speech (observation $\mathbf{b}_\mu$) and a WB code book (quantized target vectors $\mathbf{a}_v$). The selection of the most probable entry of the code book, i.e., the most probable WB coefficient, is to be considered as a special case of conditional *maximum a posteriori* (MAP) estimation.

In [Park, Kim 2000], parameter vectors with continuous amplitudes are considered which may be described by power spectra. The statistics of these power spectra are approximated by *Gaussian mixture models* (GMMs, see Section 5.5.2) [Reynolds, Rose 1995], [Vaseghi 1996].

In [Jax, Vary 2000], generalized posteriori probabilities are employed, which take not only the LP parameters of the NB signal into account but a feature vector **b**, which is derived from the upsampled and interpolated NB speech signal. This feature vector comprises the LP parameters in terms of cepstral LP coefficients and a combination of several other features such as frame energy, zero crossing rate, pitch period, spectral flatness, and voicing. This feature vector **b** of dimension $m' > m$ is to be considered as a generalization of the *quantized observation* vector $\mathbf{b}_\mu$, $\mu = 1, 2, \ldots, M$, in (10.13) and (10.16). The statistics of the features are described by GMMs which are included in the a posteriori probabilities.

As the vocal tract changes rather slowly, the LP-coefficients of successive frames exhibit correlation and do not change arbitrarily. Therefore, a *Hidden Markov Model* (HMM) [Rabiner 1989], [Papoulis 1991] is used in [Jax, Vary 2000] for tracing temporal transitions between two adjacent frames. The HMM states are given by the entries of the pre-trained codebook A containing the vectors of cepstral coefficients of wideband speech.

Finally, the conditional estimate is calculated according to the MMSE criterion [Jax 2002, 2004], [Jax, Vary 2003]. In [Bauer, Fingscheidt 2009] a variation is proposed which includes phonetic aspects in the state classification.

In the literature, also deep neural networks (DNN) approaches can be found for the codebook-based estimation framework. In, e.g., [Bauer et al. 2014] parts of the HMM/GMM approach are replaced by DNNs. [Abel, Fingscheidt 2016] and [Abel, Fingscheidt 2018] show how to replace the GMM-based calculation of observation probabilities by a DNN. A next step is to skip observation probabilities and use a DNN for the direct determination of the a posteriori probabilities $P(\mathbf{a}_v | \mathbf{b}_\mu)$ [Abel, Fingscheidt 2018].

Finally, even the source-filter model of BWE is discarded and the extended spectrum is calculated directly by DNN support, e.g., [Li, Lee 2015], [Liu et al. 2015], [Wang et al. 2015], [Abel et al. 2018], and [Abel 2020].

### 10.2.2.4 Simplifications

Alternative simplified methods, which are not based on the conditional estimation framework of Figures 10.5 and 10.6 have been proposed for the application in mobile phones in, e.g., [3GPP TS 26.190 2001] (AMR-WB codec) and [Schlien 2019].

The AMR-WB codec uses for the rather small extension band from 6.6 to 7.0 kHz a spectral envelope filter, which is derived from the baseband LP filter (0.05–6.3 kHz) by spectral stretching. This solution is explained in Section 10.3.2. In [Schlien et al. 2018] and [Schlien 2019], a wideband spectral envelope estimate is derived by interpolation of the acoustic tube model of the lowband LP synthesis filter.

Both of these alternatives are not as accurate as, e.g., the conditional MMSE estimate (10.13). However, in [Schlien 2019] it is shown by objective and subjective evaluations of *complete* BWE systems for NB-to-WB extension, that these envelope inaccuracies hardly affect the quality of the total BWE system.

### 10.2.3 Energy Envelope Estimation

The achievable improvement by bandwidth extension in terms of perceived speech quality and intelligibility depends on the accuracy of the BWE parameters. Thus, some more or less perceivable artifacts are generated if no auxiliary side information or only a small amount of auxiliary side information is available.

The question is, to which extent the inaccuracies of different parameters contribute to these artifacts. The highband excitation source signal $e_{hb}(k)$ is the least sensitive part, as already mentioned above, while the spectral envelope has been rated traditionally as the crucial item, e.g., [Jax, Vary 2000] and [Bauer, Fingscheidt 2008]. More recently, the *energy time envelope* in terms of the gains $c(k)$ and $g(k)$ has been identified as being actually the most critical part, e.g., [Geiser et al. 2007], [Schlien 2019], and [Abel 2020].

In [Schlien 2019], an "oracle experiment" is described: Even when the BWE algorithm uses the true spectral highband filter $\hat{H}_{hb}(e^{j\Omega}) = H_{hb}(e^{j\Omega})$, artifacts appear if an excitation signal $\tilde{d}_{hb}(k)$ (see Figure 10.3) with a suboptimal energy time envelope is employed. In a second oracle experiment in [Schlien 2019], a *spectrally flat envelope filter* with

$$\hat{H}_{hb}(z) = \frac{1}{1 - A_{hb}(z)} = 1.0,$$

i.e., $A(z) = 0$ was used, while the true highband excitation $e_{hb}(k)$ with the *correct energy envelope factors* $c \cdot g$ was applied (see Figure 10.3). In listening tests, this oracle set up outperformed surprisingly the reference BWE solution based on HMM estimation of the spectral envelope.

In [Abel, Fingscheidt 2016] the authors report that the quality enhancement achieved with a DNN approach in comparison to a state-of-the art HMM approach was also mainly due to the superior DNN highband energy estimation, while the fidelity of spectral envelope estimation played a marginal role only. The temporal energy envelope is adjusted by the two gain factors $c(k)$ and $g(k)$. The short-term gain $c$ controls the energy of the highband excitation signal $e_{hb}(k)$ and can be adjusted, e.g., per subframe of 5 ms for compensating amplification/attenuation variations due to the frequency response of the highband spectral envelope filter $\hat{H}_{hb}(z)$.

The second highband gain $g$ is applied, e.g., over one or several frames of typically 20 ms each for adjusting the medium-term energy envelope. It is obvious that both gains could be implemented as a combined single gain. However, the separate implementation of $c$ and $g$ before and after the envelope filters has the advantage that abrupt changes of the excitation gain $c$ are smoothed out by the envelope filter.

There are many possibilities how to estimate appropriate factors $c(k)$ and $g(k)$. All the statistical methods for estimating the spectral envelope as addressed in Section 10.2.2 can be used for estimating $c(k)$ and $g(k)$ as well.

As an example, a solution for the gain factor $c$ will be described here. In analogy to the MMSE estimation of the spectral envelope (10.13), the gain $c$ can be estimated as illustrated in Figures 10.7 and 10.8.

We assume that the energy $D$ of the highband residual $\tilde{d}_{hb}(k)$ can be estimated from the energy $E$ of the lowband residual $d_{lb}(k)$. For the training of the codebooks in Figure 10.8, the true energies can be measured by splitting WB signals into highband signals $s_{hb}(k)$ and lowband signals $s_{lb}(k)$ and by calculating the LP residual signals $d_{hb}(k)$ and $d_{lb}(k)$ using two LP analysis filters $1 - A_{hb}(z)$ and $1 - A_{lb}(z)$ as shown in Figure 10.9.

In Section 10.2.1, it is shown how to derive a highband excitation signal

$$\tilde{d}_{hb}(k) = c \cdot e_{hb}(k) \tag{10.16}$$

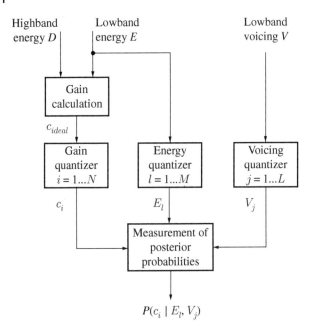

**Figure 10.7** Off-line training of posterior probabilities $P(c_i|E_l, V_j)$.

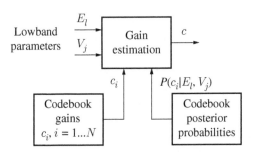

**Figure 10.8** On-line estimation of gain $c$.

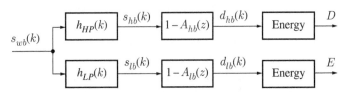

**Figure 10.9** Measurement of highband and lowband residual energies D and E. $s_{wb}(k)$: wideband signal; $s_{lb}(k)$: lowband signal; $s_{hb}(k)$: highband signal; $d_{lb}(k)$: lowband residual; $d_{hb}(k)$: highband residual.

from the spectrally flat lowband residual signal $d_{lb}(k)$ by modulation, filtering, and scaling (see Figure 10.4c). In this case, the following energy relation for a frame or subframe of length $K$ should be valid, as the energy of $e_{hb}(k)$ is the same as the energy of $d_{lb}(k)$.

$$c^2 \cdot \sum_{k=0}^{K-1} e_{hb}^2(k) \overset{!}{=} \sum_{k=0}^{K-1} d_{hb}^2(k). \tag{10.17}$$

For the off-line training of posterior probabilities, an ideal scaling factor $c_{ideal}$ can be determined by statistical measurements as shown in Figure 10.7:

$$c_{ideal} = \sqrt{\frac{\sum_{k=0}^{K-1} d_{hb}^2(k)}{\sum_{k=0}^{K-1} e_{hb}^2(k)}} = \sqrt{\frac{\sum_{k=0}^{K-1} d_{hb}^2(k)}{\sum_{k=0}^{K-1} d_{lb}^2(k)}} = \sqrt{\frac{D}{E}}. \tag{10.18}$$

As the highband-to-lowband energy ratio $D/E$ of unvoiced and voiced frames differ, a voicing measure $V$ with a few (at least two) levels $V_j, j = 1, 2, \ldots, L$ is taken additionally as a classification feature into consideration. This feature is also available in the BWE decoder and can be included in the posterior probabilities.

This concept can easily be adapted for the estimation of the second gain $g$. Furthermore, additional features can be included which improve the estimation accuracy, e.g., [Schlien 2019].

## 10.3 Speech Codecs with Integrated BWE

Artificial BWE is closely related to speech coding. In fact, some very special and effective variants of BWE techniques have been used as integral part of various NB speech codecs for many years.

BWE has also been applied in the context of WB speech coding, e.g., [Paulus, Schnitzler 1996], [Paulus 1996], [Schnitzler 1999], [Taori et al. 2000], [Erdmann et al. 2001], and [3GPP TS 26.190 2001] (AMR-WB codec), [3GPP TS 126.445 2014] (EVS codec). In these approaches, CELP coding (see also Section 8.5.3) is applied to speech components up to about 6 kHz, and BWE is used to synthesize a supplementary signal for the highband, which covers the frequency range from about 6.0 to 7.0 kHz. The extension is partly supported by transmitting different amounts of side information, which controls the spectral envelope and the level of noise excitation in the extension band. Similar approaches have also been introduced in MPEG audio coding as *SBR* [Dietz et al. 2002], [Gröschel et al. 2003]. SBR has also been used to enhance the coding efficiency of MP3 (MP3pro), *advanced audio coding* (AAC, HE-AAC), e.g., [Dietz et al. 2002], [Ehret et al. 2003], [Gröschel et al. 2003], [Homm et al. 2003], and [Herre, Dietz 2008].

### 10.3.1 BWE in the GSM Full-Rate Codec

BWE can be quite successful if some information about the spectral envelope including gain information is transmitted, while the extension of the excitation can be performed at the receiver without additional side information.

This idea has been used for coding *NB telephone speech* for quite a long time to achieve bit rates below 16 kbit/s with moderate computational complexity. A prominent example in this respect is the GSM full-rate codec [Rec. GSM 06.10 1992] (see also Section 8.7.1).

The excitation signal $d(k)$ is transmitted with a bandwidth even smaller than the telephone bandwidth by applying low-pass filtering and sample rate decimation. In this way, more bits can be assigned to each of the residual samples, which are transmitted. The basic concept, which was originally proposed by [Makhoul, Berouti 1979], is called *baseband* residual excited linear prediction (RELP). It is illustrated in Figure 10.10 (see also Section 8.5.2).

The prediction error signal $d(k)$ or *residual* signal is obtained by LP using the NB analysis filter $A(z)$

$$1 - A(z) = 1 - \sum_{v=1}^{n} a(v) \cdot z^{-v}$$

to produce the residual signal

$$d(k) = x(k) - \sum_{v=1}^{n} a(v) \cdot x(k - v).$$

Because of the limited bit rate, the bandwidth is reduced by applying a low-pass filter with a normalized cut-off frequency of $\Omega_c = \pi/r$ to the residual signal $d(k)$. The decimation

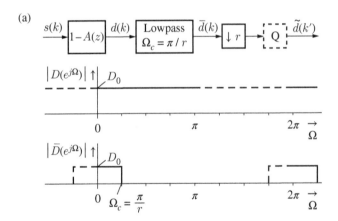

Figure 10.10 Schematic principle of the RELP codec ($r = 4$, $Q$ = quantization). (a) Encoder and (b) decoder.

factor $r \in \{2, 3, \dots\}$ for the sample rate decimation of the low-pass filtered residual $\bar{d}(k)$ is fixed according to

$$\tilde{d}(k') = \bar{d}(k = k' \cdot r).$$

The number of samples, which have to be quantized (operation Q in Figure 10.10) and coded is therefore reduced by $r$ (see also Section 8.5.2) and more bits are available for the quantized representation of the remaining residual samples $\tilde{d}(k')$. At the receiving end, the missing samples are replaced by zeros, i.e., upsampling by a factor $r$

$$\hat{d}(k) = \begin{cases} \tilde{d}(k') & k = k' \cdot r \\ 0 & \text{else} \end{cases}$$

and thus the baseband of the residual signal, which occupies the frequency range $-\pi/r \leq \Omega \leq \pi/r$, is repeated $r$ times and scaled by $1/r$ with center frequencies

$$\Omega_i = \frac{2\pi}{r} \cdot i; \qquad i = 0, 1, 2, \dots, r-1. \tag{10.19}$$

The spectrum of the bandwidth extended excitation signal is given by

$$\hat{D}(e^{j\Omega}) = \frac{1}{r} \cdot \sum_{i=0}^{r-1} D\left(e^{j(\Omega - \Omega_i)}\right). \tag{10.20}$$

This method yields reasonable results due to the insensitivity of the human ear at higher frequencies, especially for unvoiced sounds. In this case, the regenerated excitation signal $\hat{d}(k)$ is a noise signal with a flat spectral envelope. If the input signal is voiced, the spectral repetitions according to (10.20) will deliver a spectrum with discrete components and a flat envelope. However, the discrete components are not necessarily harmonics of the fundamental frequency. Therefore, this type of speech codec produces a slightly metallic sound, especially for female voices.

The extension of the excitation signal by spectral repetition or spectral folding as given by (10.20) is called *high-frequency regeneration* [Makhoul, Berouti 1979], which may be considered as BWE of the low-pass filtered excitation signal $\tilde{d}(k')$ from $\Omega_c = \pi/r$ to $\Omega = \pi$. As the residual signal $d(k)$ has a flat spectrum, a separate gain for scaling the amplitude of the excitation signal in the extension band is not required.

The overall spectral envelope of the speech segment is reconstructed from the flat excitation by applying the synthesis filter $1/(1 - A(z))$, which covers the whole frequency range $0 \leq \Omega \leq \pi$. There is no need for estimating the spectral envelope in the *extension band* $\frac{\pi}{r} \leq \Omega \leq \pi$. The transmission of the coefficients of $a(v)$ with $v = 1, 2, \dots, n$ may be considered as the transmission of *auxiliary information* for the construction of the decoded signal in the extension band.

### 10.3.2 BWE in the AMR Wideband Codec

The standardized AMR-WB codec [3GPP TS 26.190 2001] includes a BWE module as shown for the decoder in Figure 10.11. The highband signal $s_{hb}(k)$ is substituted by a noise signal as proposed in [Paulus 1996]. The bandwidth extension module consists of a white noise generator, a gain multiplier, a highband synthesis filter and a 6.4 − 7.0 kHz bandpass filter.

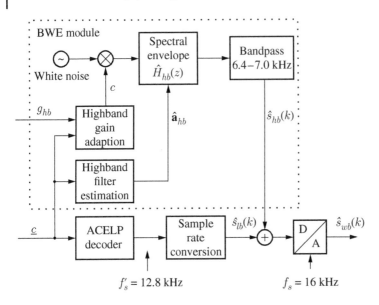

**Figure 10.11** AMR-WB decoder with integrated BWE from 6.4 to 7.0 kHz.

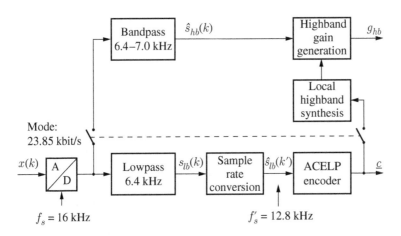

**Figure 10.12** AMR-WB encoder.

The AMR-WB codec offers nine different bitrates from 6.6 to 22.85 kbit/s (see also Section 8.7.5). In the encoder, shown in Figure 10.12, the lowband signal $s_{lb}(k)$ is decimated from $f_s = 16\,\text{kHz}$ to $f'_s = 12.8\,\text{kHz}$ and encoded with an ACELP encoder. The lowband covers the frequency range $0 \leq f \leq 6.4\,\text{kHz}$, while the highband range is $6.4 \leq f \leq 7.0\,\text{kHz}$.

Only in the highest bitrate mode of 23.85 kbit/s, the BWE module is supported by *side information* in terms of a highband gain $g_{hb}$. For the lower bit rates the highband gain is estimated from the lowband. For the highest bit rate, the highband gain is measured in the encoder in relation to the gain of the lowband signal.

The highband LP synthesis filter is derived from the lowband LP synthesis filter

$$H_{lb}(z) = \frac{1}{1 - A_{lb}(z)} \tag{10.21}$$

as a weighted and frequency-scaled version according to:

$$\hat{H}_{hb}(z) = H_{lb}(z/0.8) = \frac{1}{1 - A_{lb}(z/0.8)}. \tag{10.22}$$

Stretching over frequency is achieved by just operating the low-band filter $H_{lb}(z)$ at the increased sampling rate $f_s = 16\,\text{kHz}$. $A_{lb}(z)$ is computed from the decimated lowband signal $\hat{s}_{lb}(k')$ with sampling rate $f'_s = 12.8\,\text{kHz}$ while the highband LP filter $\hat{H}_{hb}(z)$ is operated at $f_s = 16\,\text{kHz}$. Effectively, this means that the frequency response $FR_{16}(f)$ of $\hat{H}_{hb}(z)$ is related to the stretched frequency response $FR_{12.8}(f)$ of $H_{lb}(z)$ by

$$FR_{16}(f) = FR_{12.8}\left(\frac{12.8}{16}f\right) = FR_{12.8}(0.8 \cdot f). \tag{10.23}$$

This means that the range $5.1 \ldots 5.6\,\text{kHz}$ of the lowband in the $f_s = 12.8\,\text{kHz}$ domain is mapped to the $6.4 \ldots 7.0\,\text{kHz}$ range of the highband in the $16\,\text{kHz}$ domain.

Note that the bandwidth of highband signal $\hat{s}_{hb}(k)$ in the AMR-WB codec is rather small. Thus, the correspondingly small portion of the scaled and stretched frequency response of the lowband LP filter $H_{lb}(z)$ can sufficiently shape the white excitation noise. This simplification is allowed due to the psychoacoustic properties of the auditory system, especially the loudness perception in critical bands (see also Section 2.6).

### 10.3.3 BWE in the ITU Codec G.729.1

The ITU-T recommendation G.729.1, e.g., [ITU-T Rec. G.729.1 2006], [Ragot et al. 2007], and [Varga et al. 2009], describes a *hierarchical* scalable speech and audio coding algorithm, which produces an embedded bitstream structured in 12 layers corresponding to 12 bit rates from 8 to 32 kbit/s.

The bitrate can be adjusted on the transmission link or at the decoder without the need for signaling by neglecting higher layers. The higher the bit rate of the layered bit stream, the better the audio quality. Depending on the used bit rate, the audio bandwidth is either *narrowband* (NB, 3.4 kHz) or *wideband* (WB, 7.0 kHz).

The core layer 1 consists of the 8 kbit/s narrowband CELP codec following the ITU-T standard G.729 [ITU-T Rec. G.729 1996] with 160 bits per 20 ms frame. With layers 1+2 (12 kbit/s, 240 bits per frame), the NB core codec is improved by spending 78 bits for a second fixed codebook and 2 bits for error protection. Layers 1+2+3 (14 kbit/s, 280 bits per frame) include BWE information. The additional bitrate of 40 bits/20 ms = 2 kbit/s is used for transmitting side information for bandwidth extension in the decoder from NB to WB (1.65 kbit/s, 33 additional bits per frame). The remaining 0.35 kbit/s (7 bits per frame) are for error protection. The higher layers 4…12 use transform domain coding for further improvement of the wideband quality. The BWE side information of layer 3 provides a coarse description of the temporal and spectral energy envelopes of the highband signal ($4.0 - 7.0\,\text{kHz}$) [ITU-T Rec. G.729.1 2006], [Geiser et al. 2007]. This information is generated by the encoder.

In the encoder, the WB input signal $s_{wb}(k)$, which is sampled at $f_s = 16\,\text{kHz}$, is split into two sub-bands of 4.0 kHz each, using a quadratur mirror filter (QMF) bank with implicit sample rate decimation to $f_s' = 8\,\text{kHz}$ (see also Section 4.3).

The lowband is encoded in frames of 10 ms according to the layers 1 and 2 specifications. The highband signal $s_{hb}(k')$ is analyzed in 20 ms frames consisting of 160 samples. These are subdivided into 16 segments of 10 samples and the time envelope parameters $T(n), n = 0, 1, \ldots, 15$ are calculated as logarithmic segment energies

$$T(n) = \frac{1}{2}\text{ld}\left(\sum_{\kappa=0}^{9} s_{hb}^2(\kappa + n \cdot 10)\right), \quad n = 0, 1, \ldots 15. \tag{10.24}$$

The mean of the logarithmic energies

$$M_T = \frac{1}{16}\sum_{n=0}^{15} T(n), \tag{10.25}$$

is subtracted from the segment energies. Then, the 16 zero-mean time envelope parameters $T(n) - M_T$ are split into two sequences of length 8 for 7 bits vector quantization. The logarithmic mean of the $M_T$ is quantized with 5 bits.

The spectral envelope of the 3.0 kHz wide highband signal is computed in terms of 12 logarithmic sub-band energies $F(i), i = 0, 1, \ldots, 11$ by using a windowed DFT. The $i$-th sub-band spans a bandwidth of three DFT bins. After removing the mean $M_T$ (log. scaling factor), the 12 parameters are split into three vectors of length 4, which are applied to 3 vector quantizers with 5+5+4 bits. The total number of bits for the time envelope and the frequency envelope amounts to 19+14=33. They are transmitted as side information.

The decoder high-level block diagram is shown in Figure 10.13. The algorithm is called *time domain bandwidth extension* (TDBWE) [Geiser et al. 2007], [Martin, Heute, Antweiler 2008], (Chapter 8 by B. Geiser, S. Ragot, H. Taddei).

The lowband signal $s_{lb}(k)$ is obtained by CELP decoding and interpolation of the NB signal $s(k')$ by a factor of 2. The highband signal $\tilde{s}_{hb}(k')$ is first produced at a sample rate of $f_s' = 8$ kHz, using the temporal and the spectral side information $T(n)$ and $F(i)$.

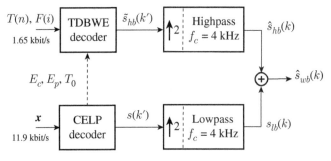

**Figure 10.13** ITU-T G.729.1 decoder with time domain bandwidth extension (TDBWE). **x**: bit frame (20 ms) of the narrowband CELP codec; $T(n)$: time envelope $n = 0, 1, \ldots, 15$; $F(i)$: frequency envelope $i = 0, 1, \ldots, 11$; $E_c$: energy of fixed codebook contribution; $E_p$: energy of adaptive codebook contribution; $T_0$: pitch period; $\hat{s}_{hb}(k)$: highband signal; $s_{lb}(k)$: interpolated narrowband signal $s(k')$; $\hat{s}_{wb}(k)$: wideband signal.

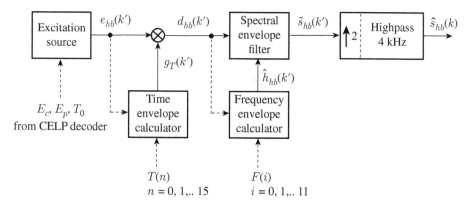

**Figure 10.14** Generation of highband signal $\hat{s}_{hb}(k)$ in the ITU-T G.729.1 decoder. $k'$ = time index at sampling rate $f'_s$ = 8kHz; $k$ = time index at sampling rate $f_s$ = 16kHz.

Furthermore, the energies $E_c$ and $E_p$ of the fixed and the adaptive codebook contributions and the actual pitch period $T_0$, provided by the CELP decoder, are taken into consideration. The final highband signal $\hat{s}_{hb}(k)$ is calculated by interpolation and the extended WB signal $\hat{s}_{wb}(k)$ is obtained by superposition of $\hat{s}_{hb}(k)$ and $s_{lb}(k)$.

The core of the TDBWE algorithm, as shown in Figure 10.14, is based on the *source-filter model* of Figure 10.3. The main difference w.r.t. the basic BWE concept as described in Figure 10.1 is that auxiliary information $(T(n), F(i), E_c, E_p, T_0)$ is available and that the preliminary highband signal $\tilde{s}_{hb}(k')$ is interpolated to $f_s$ = 16 kHz to obtain the final highband signal $\hat{s}_{hb}(k)$.

The highband excitation signal $e_{hb}(k')$ is composed of a mixture of noise and periodic pitch pulses. It is generated taking the instantaneous pitch period $T_0$ and as voicing measure the relative energies $E_p/E_c$ into account for adjusting the voiced–unvoiced mixture of the highband excitation signal $e_{hb}(k')$.

The time envelope calculator delivers the gain factors $g_T(k')$, which are derived from the time envelope parameters $T(n)$.

The highband spectral envelope filter is described by a linear phase impulse response $\hat{h}_{hb}(k')$ of length 33, which is implemented as a filterbank equalizer with 12 channels (see also Section 4.4). The corresponding frequency response is constructed from the 12 spectral envelope magnitudes $F(i)$, considering the spectral energies of the gain scaled excitation signal $d_{hb}(k')$.

The obtained speech quality is rated to be comparable to WB speech codecs used in mobile communication systems [Geiser et al. 2007].

## References

3GPP TS 126.445 (2014). Codec for Enhanced Voice Services (EVS), 3GPP.

3GPP TS 26.190 (2001). AMR Wideband Speech Codec; Transcoding Functions, 3GPP.

Abel, J. (2020). *DNN-Based Artificial Bandwidth Extension -Enhancement and Instrumental Assessment of Speech Quality*, Dissertation, IFN, TU Braunschweig.

Abel, J.; Fingscheidt, T. (2016). Artificial Bandwidth Extension Using Deep Neural Networks for Wideband Spectral Envelope Estimation, *Proceedings International Workshop on Acoustic Signal Enhancement (IWAENC)*, Xi'an, China, pp. 1–5.

Abel, J.; Fingscheidt, T. (2018). Artificial Bandwidth Extension Using Deep Neural Networks for Spectral Envelope Estimation, *IEEE/ACM Transactions on Audio, Speech, and Language Processing*, vol. 26, pp. 71–83.

Abel, J.; Strake, M.; Fingscheidt, T. (2018). A Simple Cepstral Domain DNN Approach to Artificial Speech Bandwidth Extension, *2018 IEEE International Conference on Acoustics, Speech and Signal Processing (ICASSP)*, Calgary, Alberta, Canada, pp. 5469–5473.

Atti, V.; Krishnan, V.; Dewasurendra, D.; Chebiyyam, V.; Subasingha, S.; Sinder, D. J.; Rajendran, V.; Varga, I.; Gibbs, J.; Miao, L.; Grancharov, V.; Pobloth, H. (2015). Super-Wideband Bandwidth Extension for Speech in the 3GPP EVS Codec, *2015 IEEE International Conference on Acoustics, Speech and Signal Processing (ICASSP)*, Brisbane, Australia, pp. 5927–5931.

Avendano, C.; Hermansky, H.; Wan, E. A. (1995). Beyond Nyquist: Towards the Recovery of Broad-Bandwidth Speech from Narrow-Bandwidth Speech, *Proceedings of EUROSPEECH*, Madrid, Spain, vol. 1, pp. 165–168.

Bachhav, P.; Todisco, M.; Evans, N. (2018). Efficient Super-Wide Bandwidth Extension Using Linear Prediction Based Analysis-Synthesis, *2018 IEEE International Conference on Acoustics, Speech and Signal Processing (ICASSP)*, Calgary, Alberta, Canada, pp. 5429–5433.

Bauer, P.; Fingscheidt, T. (2008). An HMM-Based Artificial Bandwidth Extension Evaluated by Cross-Language Training and Test, *Proceedings of the IEEE International Conference on Acoustics, Speech and Signal Processing (ICASSP)*, Las Vegas, Nevada, USA, pp. 4589–4592.

Bauer, P.; Fingscheidt, T. (2009). A Statistical Framework for Artificial Bandwidth Extension Exploiting Speech Waveform and Phonetic Transcription, *Proceedings of the EUSIPCO*, Glasgow, Scotland, pp. 1839–1843.

Bauer, P.; Abel, J.; Fingscheidt, T. (2014). HMM-Based Artificial Bandwidth Extension Supported by Neural Networks, *Proceedings of IWAENC*, Juan le Pins, France, pp. 1–5.

Carl, H. (1994). *Untersuchung verschiedener Methoden der Sprachkodierung und eine Anwendung zur Bandbreitenvergrößerung von Schmalband-Sprachsignalen*, PhD thesis. *Arbeiten über Digitale Signalverarbeitung*, vol. 4, U. Heute (ed.), Ruhr-Universität Bochum (in German).

Carl, H.; Heute, U. (1994). Bandwidth Enhancement of Narrow-Band Speech Signals, *Proceedings of the European Signal Processing Conference (EUSIPCO)*, Edinburgh, Scotland, vol. 2, pp. 1178–1181.

Chan, C.-F.; Hui, W.-K. (1996). Wideband Enhancement of Narrowband Coded Speech Using MBE Re-Synthesis, *Proceedings of the International Conference on Signal Processing (ICSP)*, Beijing, China, vol. 1, pp. 667–670.

Cheng, Y.; O'Shaughnessy, D.; Mermelstein, P. (1994).Statistical Recovery of Wideband Speech from Narrowband Speech, *IEEE Transactions on Speech and Audio Processing*, vol. 2, no. 4, pp. 544–548.

Croll, M. G. (1972). Sound-Quality Improvement of Broadcast Telephone Calls, Technical Report 1972/26, The British Broadcasting Corporation (BBC).

Dietrich, M. (1984). Performance and Implementation of a Robust ADPCM Algorithm for Wideband Speech Coding with 64kbit/s, *Proceedings of the International Zürich Seminar on Digital Communications*, Zurich, Switzerland.

Dietz, M.; Liljeryd, L.; Kjorling, K.; Kunz, O. (2002). Spectral Band Replication: A Novel Approach in Audio Coding, *112th Convention of the Audio Engineering Society (AES)*, Munich, Germany.

Ehret, A.; Dietz, M.; Kjorling, K. (2003). State-of-the-Art Audio Coding for Broadcasting and Mobile Applications, *114th Convention of the Audio Engineering Society (AES)*, Amsterdam, The Netherlands.

Enbom, N.; Kleijn, W. B. (1999). Bandwidth Expansion of Speech Based on Vector Quantization of the Mel Frequency Cepstral Coefficients, *Proceedings of the IEEE Workshop on Speech Coding*, Porvoo, Finland, pp. 171–173.

Epps, J.; Holmes, W. H. (1999). A New Technique for Wideband Enhancement of Coded Narrowband Speech, *Proceedings of the IEEE Workshop on Speech Coding*, Porvoo, Finland, pp. 174–176.

Erdmann, C.; Vary, P.; Fischer, K.; Xu, W.; Marke, M.; Fingscheidt, T.; Varga, I.; Kaindl, M.; Quinquis, C.; Kovesi, B.; Massaloux, D. (2001). A Candidate Proposal for a 3GPP Adaptive Multi-Rate Wideband Speechcodec, *Proceedings of the IEEE International Conference on Acoustics, Speech, and Signal Processing (ICASSP)*, Salt Lake City, Utah, USA, pp. 757–760.

Fuemmeler, J. A.; Hardie, R. C.; Gardner, W. R. (2001). Techniques for the Regeneration of Wideband Speech from Narrowband Speech, *EURASIP Journal on Applied Signal Processing*, vol. 4, pp. 266–274.

Geiser, B. (2012). *High-Definition Telephony over Heterogeneous Networks*, Dissertation. Aachener Beiträge zu Digitalen Nachrichtensystemen (ABDN), P. Vary (ed.), IND, RWTH Aachen.

Geiser, B.; Vary, P. (2013). Artificial Bandwidth Extension of Wideband Speech by Pitch-Scaling of Higher Frequencies, *in* M. Horbach (ed.), *Workshop Audiosignal- und Sprachverarbeitung (WASP)*, Gesellschaft für Informatik e.V. (GI), vol. 220 of *Lecture Notes in Informatics (LNI)*, pp. 2892–2901. Workshop at the 43rd Annual Meeting of the German Society for Computer Science.

Geiser, B.; Jax, P.; Vary, P.; Taddei, H.; Schandl, S.; Gartner, M.; Guillaume, C.; Rago, S. (2007). Bandwidth Extension for Hierarchical Speech and Audio Coding in ITU-T Rec. G.729.1, *IEEE Transactions on Audio, Speech, and Language Processing*, vol. 15, no. 8, pp. 2496–2509.

Gröschel, A.; Schug, M.; Beer, M.; Henn, F. (2003). Enhancing Audio Coding Efficiency of MPEG Layer-2 with Spectral Band Replication for DigitalRadio (DAB) in a Backwards Compatible Way, *114th Convention of the Audio Engineering Society (AES)*, Amsterdam, The Netherlands.

Herre, J.; Dietz, M. (2008). MPEG-4 high-efficiency AAC coding [Standards in a Nutshell], *IEEE Signal Processing Magazine*, vol. 25, no. 3, pp. 137–142.

Homm, D.; Ziegler, T.; Weidner, R.; Bohm, R. (2003). Implementation of a DRM Audio Decoder (aacPlus) on ARM Architecture, *114th Convention of the Audio Engineering Society (AES)*, Amsterdam, The Netherlands.

ITU-T Rec. G.712 (1988). Performance Characteristics of PCM Channels Between 4-wire Interfaces at Voice Frequencies, International Telecommunication Union (ITU).

ITU-T Rec. G.722 (1988). 7 kHz Audio Coding within 64 kbit/s, *Recommendation G.722*, International Telecommunication Union (ITU), pp. 269–341.

ITU-T Rec. G.729 (1996). Coding of Speech at 8 kbit/s Using Conjugate-Structure Algebraic-Code-Excited Linear-Prediction (CS-ACELP), *Recommendation G.729*, International Telecommunication Union (ITU).

ITU-T Rec. G.729.1 (2006). An 8-32 kbit/s scalable wideband coder bitstream interoperable with G.729, *Recommendation G.729.1*, International Telecommunication Union (ITU).

Iyengar, V.; Rabipour, R.; Mermelstein, P.; Shelton, B. R. (1995). Speech Bandwidth Extension Method and Apparatus, US patent no. 5455888.

Jax, P. (2002). *Enhancement of Bandlimited Speech Signals: Algorithms and Theoretical Bounds*, PhD thesis. Aachener Beiträge zu digitalen Nachrichtensystemen, vol. 15, P. Vary (ed.), RWTH Aachen University.

Jax, P. (2004). Bandwidth Extension for Speech, *in* E. Larsen; R. M. Aarts (eds.), *Audio Bandwidth Extension*, John Wiley & Sons, Ltd., Chichester, Chapter 6, pp. 171–236.

Jax, P.; Vary, P. (2000). Wideband Extension of Telephone Speech Using a Hidden Markov Model, *Proceedings of the IEEE Workshop on Speech Coding*, Delavan, Wisconsin, USA, pp. 133–135.

Jax, P.; Vary, P. (2002). An Upper Bound on the Quality of Artificial Bandwidth Extension of Narrowband Speech Signals, *Proceedings of the IEEE International Conference on Acoustics, Speech, and Signal Processing (ICASSP)*, Orlando, Florida, USA, vol. 1, pp. 237–240.

Jax, P.; Vary, P. (2003). On Artificial Bandwidth Extension of Telephone Speech, *Signal Processing*, vol. 83, no. 8, pp. 1707–1719.

Jax, P.; Vary, P. (2006). Bandwidth Extension of Speech Signals: A Catalyst for the Introduction of Wideband Speech Coding?, *IEEE Communications Magazine*, vol. 44, no. 5, pp. 106–111.

Kornagel, U. (2001). Spectral Widening of the Excitation Signal for Telephone-Band Speech Enhancement, *Proceedings of the International Workshop on Acoustic Echo and Noise Control (IWAENC)*, Darmstadt, Germany, pp. 215–218.

Kornagel, U. (2003). Improved Artificial Low-Pass Extension of Telephone Speech, *Proceedings of the International Workshop on Acoustic Echo and Noise Control (IWAENC)*, Kyoto, Japan, pp. 107–110.

Larsen, E. R.; Aarts, R. M. (eds.) (2004). *Audio Bandwidth Extension - Application of Psychoacoustics, Signal Processing and Loudspeaker Design*, John Wiley & Sons, Ltd., Chichester.

Larsen, E. R.; Aarts, R. M.; Danessis, M. (2002). Efficient High-Frequency Bandwidth Extension of Music and Speech, *112th Convention of the Audio Engineering Society (AES)*, Munich, Germany.

Li, K.; Lee, C.-H. (2015). A Deep Neural Network Approach to Speech Bandwidth Extension, *Proceedings of the IEEE International Conference on Acoustics, Speech, and Signal Processing (ICASSP)*, Brisbane, Australia, pp. 4395–4399.

Liu, B.; Tao, J.; Wen, Z.; Li, Y.; Bukhari, D. (2015). A Novel Method for Artificial Bandwidth Extension Using Deep Architectures, *Proceedings of INTERSPEECH*, Dresden, Germany, pp. 2598–2602.

Makhoul, J.; Berouti, M. (1979). High-Frequency Regeneration in Speech Coding Systems, *Proceedings of the IEEE International Conference on Acoustics, Speech, and Signal Processing (ICASSP)*, Washington, District of Columbia, USA, pp. 208–211.

Martin, R.; Heute, U.; Antweiler, C. (eds.) (2008). *Advances in Digital Speech Transmission*, John Wiley & Sons, Inc.

Melsa, J. L.; Cohn, D. L. (1978). *Decision and Estimation Theory*, McGraw-Hill, Kogakusha, Tokyo.

Papoulis, A. (1991). *Probability, Random Variables, and Stochastic Processes*, 3rd edn, McGraw-Hill, New York.

Park, K. Y.; Kim, H. S. (2000). Narrowband to Wideband Conversion of Speech Using GMM-Based Transformation, *Proceedings of the IEEE International Conference on Acoustics, Speech, and Signal Processing (ICASSP)*, Istanbul, Turkey, vol. 3, pp. 1847–1850.

Patrick, P. J. (1983). *Enhancement of Bandlimited Speech Signal*, PhD thesis, Loughborough University of Technology.

Paulus, J. (1996). *Codierung breitbandiger Sprachsignale bei niedriger Datenrate*, PhD thesis. Aachener Beiträge zu digitalen Nachrichtensystemen, vol. 6, P. Vary (ed.), RWTH Aachen University (in German).

Paulus, J.; Schnitzler, J. (1996). 16 kbit/s Wideband Speech Coding Based on Unequal Subbands, *Proceedings of the IEEE International Conference on Acoustics, Speech, and Signal Processing (ICASSP)*, Atlanta, Georgia, USA, pp. 651–654.

Rabiner, L. R. (1989). A Tutorial on Hidden Markov Models and Selected Applications in Speech Recognition, *Proceedings of the IEEE*, vol. 77, no. 2, pp. 257–286.

Ragot, S.; Kovesi, B.; Trilling, R.; Virette, D.; Duc, N.; Massaloux, D.; Proust, S.; Geiser, B.; Gartner, M.; Schandl, S.; Taddei, H.; Gao, Y.; Shlomot, E.; Ehara, H.; Yoshida, K.; Vaillancourt, T.; Salami, R.; Lee, M. S.; Kim, D. Y. (2007). ITU-T G.729.1: An 8-32 kbit/s Scalable Coder Interoperable with G.729 for Wideband Telephony and Voice Over IP, *IEEE International Conference on Acoustics, Speech and Signal Processing, (ICASSP)*, Honolulu, Hawaii, USA, vol. IV, pp. 529–532.

Rec. GSM 06.10 (1992). Recommendation GSM 06.10 GSM Full Rate Speech Transcoding, ETSI TC-SMG.

Reynolds, D. A.; Rose, R. C. (1995). Robust Text-Independent Speaker Identification Using Gaussian Mixture Speaker Models, *IEEE Transactions on Speech and Audio Processing*, vol. 3, no. 1, pp. 72–83.

Schlien, T. (2019). *HD Telephony by Artificial Bandwidth Extension - Quality, Concepts and Complexity*, PhD thesis. Aachener Beiträge zu Digitalen Nachrichtensystemen (ABDN), P. Vary (ed.), RWTH Aachen University, Germany.

Schlien, T.; Jax, P.; Vary, P. (2018). Acoustic Tube Interpolation for Spectral Envelope Estimation in Artificial Bandwidth Extension, *ITG-Fachtagung Sprachkommunikation*.

Schmidt, K. O. (1933). Neubildung von unterdrückten Sprachfrequenzen durch ein nichtlinear verzerrendes Glied, *Telegraphen- und Fernsprech-Technik*, vol. 22, no. 1, pp. 13–22 (in German).

Schnitzler, J. (1999). *Breitbandige Sprachcodierung: Zeitbereichs- und Frequenzbereichskonzepte*, PhD thesis. *Aachener Beiträge zu digitalen Nachrichtensystemen*, vol. 12, P. Vary (ed.), RWTH Aachen University (in German).

Su, J.; Wang, Y.; Finkelstein, A.; Jin, Z. (2021). Bandwidth extension is all you need, *Proceedings of the IEEE International Conference on Acoustics, Speech and Signal Processing (ICASSP)*, pp. 696–700.

Taori, R.; Sluijter, R. J.; Gerrits, A. J. (2000). Hi-BIN: An Alternative Approach to Wideband Speech Coding, *Proceedings of the IEEE International Conference on Acoustics, Speech, and Signal Processing (ICASSP)*, Istanbul, Turkey, vol. 2, pp. 1157–1160.

Un, C. K.; Magill, D. T. (1975). The Residual-Excited Linear Prediction Vocoder with Transmission Rate Below 9.6 kbit/s, *IEEE Transactions on Communications*, vol. 23, pp. 1466–1474.

Valin, J.-M.; Lefebvre, R. (2000). Bandwidth Extension of Narrowband Speech for Low Bit-Rate Wideband Coding, *IEEE Workshop on Speech Coding*, Delavan, Wisconsin, USA, pp. 130–132.

Varga, I.; Proust, S.; Taddei, H. (2009). ITU-T G.729.1 Scalable Codec for New Wideband Services, *IEEE Communications Magazine*, vol. 47, no. 10, pp. 131–137.

Vaseghi, S. V. (1996). *Advanced Signal Processing and Digital Noise Reduction*, John Wiley & Sons, Ltd., Chichester, and B. G. Teubner, Stuttgart.

Wang, Y.; Zhao, S.; Liu, W.; Li, M.; Kuang, J. (2015). Speech bandwidth extension based on deep neural networks, *Proceedings of INTERSPEECH*, Dresden, Germany, pp. 2593–2597.

Yamanashi, T.; Oshikiri, M.; Ehara, H. (2011). Low Bit-Rate High-Quality Audio Encoding and Low Complexity Bandwidth Extension Technologies for ITU-T G.718/G.718-SWB, *2011 IEEE International Conference on Communications (ICC)*, Kyoto, Japan, pp. 1–5.

Yoshida, Y.; Abe, M. (1994). An Algorithm to Reconstruct Wideband Speech from Narrowband Speech Based on Codebook Mapping, *Proceedings of the International Conference on Spoken Language Processing (ICSLP)*, Yokohama, Japan, pp. 1591–1594.

# 11

# NELE: Near-End Listening Enhancement

Mobile phones are often used in noisy environments where a listener at the *near-end* perceives a mixture of the clean far-end speech and the acoustical background noise. In this situation, *speech intelligibility* might be degraded and *listening effort* has to be increased. The same happens while listening to a public address system, e.g., at a train station or in an airplane cabin. The degree to which speech perception is affected depends not only on the absolute sound pressure levels and the signal-to-noise ratio (SNR), but also on the spectral and temporal characteristics of speech and background noise. The degradation of speech perception can be explained and quantified by psychoacoustical masking, e.g., [Fastl, Zwicker 2007], see Section 2.6.3.

People speaking face-to-face, automatically adapt to the background noise, e.g., by raising their voices, speaking slower, and with a higher fundamental frequency, to enhance the audibility of their voice. This is called the *Lombard effect* [Lombard 1911].

In a non-face-to-face situation, the perception of the far-end speech can technically be improved by modifying the loudspeaker signal at the *near-end*. This is called *near-end listening enhancement* (NELE), a technical term, coined in [Sauert, Vary 2006]. NELE processing does not mimic the Lombard effect but tries to improve speech perception in terms of intelligibility and listening effort by spectral modifications of the received far-end speech signal $s(k)$, e.g., [Sauert, Vary 2006], [Sauert, Vary 2010a], [Sauert 2014], [Niermann 2019], and [Niermann, Vary 2021] as indicated in Figure 11.1. The terms *far-end* and *near-end* refer to the locations of the talker and the listener, respectively.

NELE algorithms modify the received far-end speech signal $s(k)$ taking the instantaneous spectral characteristics of the near-end background noise $n(k)$ into account. The noise is captured, e.g., with the microphone of a mobile phone, a communication headset, or an extra microphone attached to the NELE module of a public address system. The signal $s(k)$ is modified to produce the enhanced signal $u(k)$, which is then played out through the loudspeaker of the mobile phone or the public address system.

The speech usually comes from a human speaker at the far-end side of the communication link as shown in the left part of Figure 11.1. In public address systems, the far-end speech may also be played out from a storage.

The most simplistic NELE approach would be to increase the sound volume of the loudspeaker. However, this is constrained by the loudspeaker and by the human ear: High sound

*Digital Speech Transmission and Enhancement*, Second Edition. Peter Vary and Rainer Martin.
© 2024 John Wiley & Sons Ltd. Published 2024 by John Wiley & Sons Ltd.

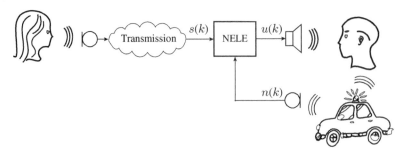

**Figure 11.1** Concept of near-end listening enhancement (NELE).

pressure levels can destroy the loudspeaker, may be harmful to the ear, or may cause at least hearing discomfort. The allowed limited increase of the volume is often insufficient to achieve the desired improvement. The trivial spectrally flat amplification of the loudspeaker signal is thus not considered here. The more challenging questions are:

- How can we improve speech perception by a frequency-dependent allocation of a given limited power budget?
- Can we improve perception even without increasing the total power?

Mostly, NELE is achieved by applying spectral gains to the short-term speech spectrum. The gains may be obtained, e.g., by maximizing an objective intelligibility measure. A popular measure is the *Speech Intelligibility Index* (SII), defined in [ANSI S3.5 1997] and employed for the NELE purpose in, e.g., [Sauert, Vary 2009], [Sauert, Vary 2010a], [Taal et al. 2013], [Sauert et al. 2014], and [Kleijn et al. 2015].

In the literature, there are further approaches. The authors in [Kleijn, Hendriks 2015] consider a speech communication model and determine spectral gains by optimizing the information-theoretical mutual information (see Section 5.4.4). Also dynamic range compressors (DRC) can be used to enhance the speech perception in noise, e.g., in [Zorila et al. 2012] and [Schepker et al. 2015]. In [Petkov, Kleijn 2015], speech is modified such that the processed noisy speech, as perceived by a listener, has the same temporal dynamics as the unprocessed clean speech. In [Aubanel, Cooke 2013] and [Cooke et al. 2019], temporal masking is considered in addition to spectral masking.

The NELE concept, which improves situations where the near-end listener is disturbed by local noise, has been expanded by [Hendriks et al. 2015], [Petkov, Stylianou 2016], and [Grosse, van de Par 2017] to consider also reverberation at the near-end. Multizone speech reinforcement is proposed in [Crespo, Hendriks 2014], where public address systems with several playback zones and crosstalk between the zones are considered. Finally, an overview on frequency domain NELE approaches can be found in [Kleijn et al. 2015].

NELE solutions for public address systems with strong crosstalk between the loudspeaker and the microphone at the near-end and NELE in the context of mobile communication with noise and noise reduction at the *far-end* have been studied in, e.g., [Niermann et al. 2016b], [Niermann et al. 2017] .

## 11.1 Frequency Domain NELE (FD)

Frequency domain NELE improves the listening performance by spectrally redistributing the speech energy, taking the masking threshold into account which is caused by the near-end background noise, e.g., [Sauert, Vary 2006]. This allows to improve intelligibility and to reduce listening effort with limited or even without any increase of the volume. If it is not allowed to increase the overall volume, the energy increase at some frequencies has to be compensated by the attenuation of some other frequency components. However, if an additional energy/power headroom is available, this additional budget should also be used for the redistribution of spectral energy. The basic frequency domain NELE structure is given in Figure 11.2. The received speech signal $s(k)$ and the near-end noise signal $n(k)$ are spectrally analyzed either frame-by-frame using, e.g., the Discrete Fourier Transformation (DFT) with $K_{DFT}$ bins or by using a (non-uniform) analysis filter bank with $K_{FB}$ sub-bands.

The frequency domain/sub-band domain representations are denoted by $S_\mu(\lambda)$ and $N_\mu(\lambda)$ with *frame or decimated time index* $\lambda$ and DFT or filter bank (FB) *frequency index* $\mu = 0,1 \ldots K_{DFT/FB}$.

Here, the NELE strategies are formulated in terms of DFT spectra with uniform frequency resolution with $K_{DFT} = K$ spectral bins. The implementation may be based on groups of combined DFT bins according to the Bark frequency scale (see Table 2.1), as NELE processing is related to the masking effect. Alternatively, the NELE algorithm may be implemented with a non-uniform time-domain FB.

The psychoacoustical masking, caused by the near-end noise $N_\mu(\lambda)$, can be estimated by calculating a masking threshold $M_\mu(\lambda)$. It is assumed that the noise characteristics can be analyzed either during far-end speech pauses or continuously if the crosstalk from the loudspeaker signal $u(k)$ to the microphone signal $n(k)$ is suppressed, e.g., by echo cancellation (see Chapter 14).

Finally, the (re-)allocation of spectral energies is achieved by applying real-valued spectral NELE weights $W_\mu(\lambda)$ to the spectrum $S_\mu(\lambda)$ of the received far-end signal:

$$U_\mu(\lambda) = W_\mu(\lambda) \cdot S_\mu(\lambda). \tag{11.1}$$

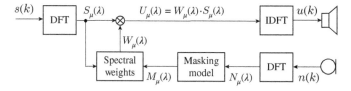

**Figure 11.2** The frequency domain NELE approach $k$ = time index, $\lambda$ = frame index, $\mu$ = DFT frequency index.

The NELE weights are derived from DFT periodograms[1] $|X_\mu(\lambda)|^2$ smoothed over time or from corresponding sub-band energies. In the following, smoothed periodograms/sub-band energies are denoted by

$$\Phi_{X_\mu}(\lambda) = \overline{|X_\mu(\lambda)|^2}, \tag{11.2}$$

where $X$ is a placeholder for $S, N$, or $M$. Smoothing over time/frame index $\lambda$ provides a smooth evolution of the NELE weights. Smoothing over frequency bins $\mu$ for achieving a non-uniform spectral resolution is performed to match the critical bands (see Section 2.6.2) and to avoid too strong spectral changes especially in case of very narrow bandpass noise signals.

The spectral NELE weights $W_\mu(\lambda)$ are determined as a function of smoothed spectral/sub-band speech energies $\Phi_{S_\mu}(\lambda) = \overline{|S_\mu(\lambda)|^2}$ and masking energy thresholds $\Phi_{M_\mu}(\lambda) = \overline{|M_\mu(\lambda)|^2}$,

$$W_\mu(\lambda) = f\left(\Phi_{S_\mu}(\lambda), \Phi_{M_\mu}(\lambda)\right) \geq 0. \tag{11.3}$$

### 11.1.1 Speech Intelligibility Index NELE Optimization

Early proposals, e.g., [Sauert, Vary 2010a], [Sauert, Vary 2010b], [Sauert et al. 2014], and [Hendriks et al. 2015], choose the NELE weights $W_\mu(\lambda)$ such that the speech intelligibility is optimized, subject to a power constraint.

The SII standard as specified in [ANSI S3.5 1997] defines a method for computing a physical measure that is highly correlated with the intelligibility of speech. The SII is calculated from acoustical measurements of speech and noise levels in critical bands. For NELE processing, mainly situations with strong background noise are of interest. Therefore, the following simplifications are assumed, which do not imply any relevant limitations:

- Self-masking of the received speech signal $s(k)$ is neglected, which accounts for the masking of higher speech frequencies by lower speech frequencies.
- The masking spectrum level is higher than the hearing threshold.

The SII calculations are based on the *equivalent spectrum levels*[2], which can be calculated as the sub-band powers of the critical bands, normalized to the reference sound pressure of $p_0 = 20\mu\text{Pa}$ and the corresponding frequency bandwidths.

---

1 The periodogram of a signal $x(k)$, which is considered to be an estimate of the power spectral density, is usually defined as $\frac{1}{K}|X_\mu(\lambda)|^2$ with DFT length $K$, see Section 5.8.2. The scaling factor or normalization factor $\frac{1}{K}$ can be omitted in the derivations of the gains $W_\mu(\lambda)$ as it cancels out in the calculations of spectral SNR values. Thus, no distinction is made here between short-term power and short-term energy. Within NELE processing, the constrained absolute power of the output signal $u(k)$ is controlled by a separate single gain factor which is either included in the calculation of the weighting factors or which is applied to the modified time domain signal.

2 The *equivalent spectrum level* is defined as the spectrum level measured at the point corresponding to the center of the listeners head, with the listener absent. For the NELE application, the spectra of the loudspeaker signal $s(k)$ and the microphone signal $n(k)$ are considered instead.

For the sake of brevity, the normalization and the dependency on the frame index $\lambda$ is omitted in the following:

$$\Phi_{S\mu} = \Phi_{S_\mu}(\lambda),$$
$$\Phi_{M\mu} = \Phi_{M_\mu}(\lambda).$$

The SII for a speech signal $s(k)$ in the presence of a background noise $n(k)$ is then calculated as the sum of contributions from 21 non-uniform critical bands as

$$SII(s) = \sum_{\mu=1}^{21} I_\mu \cdot A_\mu \left(\Phi_{S\mu}, \Phi_{M\mu}\right), \tag{11.4a}$$

where

$$0 \leq A_\mu \left(\Phi_{S\mu}, \Phi_{M\mu}\right) \leq 1, \tag{11.4b}$$

is called the *band audibility function* of the critical band $\mu$. The speech and masking levels are determined with the spectral resolution of the first 21 critical bands from 100 Hz to 9.5 kHz (see Section 2.6.2, Table 2.1). The *band importance function* $I_\mu$ ([ANSI S3.5 1997], Table 1), which satisfies the constraint

$$\sum_{\mu=1}^{21} I_\mu = 1, \tag{11.4c}$$

quantifies the relative importance of sub-band $\mu$ with regard to the total speech intelligibility. Thus, the SII can take values

$$0 \leq SII \leq 1. \tag{11.4d}$$

A value of SII $\geq 0.75$ is considered to be good, while an SII $\leq 0.45$ is considered poor. The band audibility function is decomposed into two factors

$$A_\mu \left(\Phi_{S\mu}, \Phi_{M\mu}\right) = LD_\mu \left(\Phi_{S\mu}\right) \cdot MF_\mu \left(\Phi_{S\mu}, \Phi_{M\mu}\right). \tag{11.5}$$

The *level distortion factor* $LD_\mu = LD_\mu \left(\Phi_{S\mu}\right)$ quantifies the decrease of intelligibility due to overload caused by too high presentation levels of $s(k)$ while the *masking factor* $MF_\mu = MF_\mu \left(\Phi_{S\mu}, \Phi_{M\mu}\right)$ models the *loss of intelligibility* due to masking of speech by noise.

The behavior of the two factors $MF_\mu$ and $LD_\mu$ and the band audibility $A_\mu$ (11.5) are illustrated in Figure 11.3a for a weak masker with $10\log\left(\Phi_{M\mu}\right) = 5$ dB and in Figure 11.3b for strong masking with $10\log\left(\Phi_{M\mu}\right) = 50$ dB as a function of the speech level $10\log\left(\Phi_{S\mu}\right)$.

In case of a *weak masker*, there is no masking, i.e., $MF_\mu = 1.0$ as long as the speech level $\Phi_{S\mu}$ is at least by 15 dB stronger than the masking level $\Phi_{M\mu}$, but lower than the overload level $OL_\mu$ i.e.,

$$10\log\left(\Phi_{S\mu}\right) \geq 10\log\left(\Phi_{M\mu}\right) + 15 \, \text{dB}$$
$$10\log\left(\Phi_{S\mu}\right) \leq 10\log\left(OL_\mu\right) = 45 \, \text{dB}.$$

Partial masking occurs if the speech-to-masker ratio is outside this range, and total masking with $MF_\mu = 0.0$ if the speech level is by at least 15 dB lower than the masker.

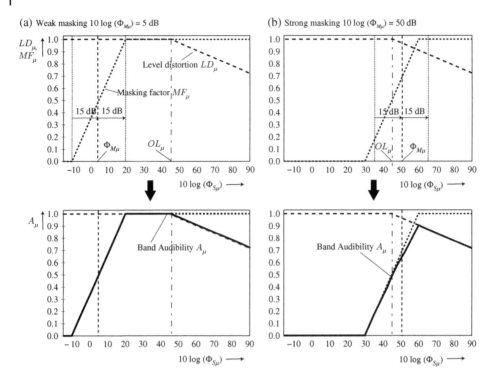

**Figure 11.3** Band audibility $A_\mu = LD_\mu \cdot MF_\mu$ for critical band $\mu = 2$ with overload level $10 \log \left( OL_\mu \right) = 45$ dB and two different masking levels. Source: [Sauert 2014], © 2014 IEEE.

Furthermore, if the speech level increases in this band beyond the overload level of $10 \log \left( OL_\mu \right) = 45$ dB, speech distortions ($LD_\mu < 1$) occur increasingly due to overload effects of the auditory system. In the second critical band this level is about +45 dB.[3]

The corresponding relations for a *strong masker* are illustrated in Figure 11.3b.

For SII optimized NELE processing, individual weights $W_\mu(\lambda)$, $\mu = 1, \ldots 21$ are applied according to Figure 11.2 and Eq. (11.1) to the critical bands:

$$U_\mu(\lambda) = W_\mu(\lambda) \cdot S_\mu(\lambda).$$

The weights must be chosen such that the speech intelligibility SII($U$) of the modified signal $u(k)$ is maximized, taking the background noise into account under fulfillment of a given power constraint $P_{max}$

$$SII(U) = \arg\max_{W_\mu(\lambda)} \sum_{\mu=1}^{21} I_\mu \cdot A_\mu \left( W_\mu^2(\lambda) \cdot \Phi_{S\mu}, \Phi_{M\mu} \right) \tag{11.6a}$$

$$\sum_{\mu=1}^{21} W_\mu^2(\lambda) \cdot \Phi_{S_\mu}(\lambda) \overset{!}{\leq} P_{max}. \tag{11.6b}$$

---

3 As already mentioned, the necessary normalization/calibration of the spectrum levels to the reference sound pressure of 20 $\mu$Pa is omitted here for the sake of brevity.

If the masking noise is limited to a very narrow frequency band, a bounding constraint

$$0 \leq W_{\mu}(\lambda) \leq W_{\max},$$ (11.7)

has to be applied to avoid overly large distortions of the speech spectrum.

The optimization problem (11.6a,b) may be solved, e.g., by numerical iterations [Sauert, Vary 2010a] or by semi-analytical Lagrange calculations, e.g., [Sauert, Vary 2010b].

A more detailed description can be found in [Sauert, Vary 2010a], [Sauert et al. 2014], and [ANSI S3.5 1997].

### 11.1.1.1 SII-Optimized NELE Example

An example is illustrated in Figure 11.4. The SII optimization was carried out for all speech files of the TIMIT database [DARPA 1990], with a total duration of 5.4 hours, disturbed by *factory 1* noise from the NOISEX-92 database [Varga, Steeneken 1993]. The sampling rate was $f_s = 8\,\mathrm{kHz}$.

Prior to processing, the speech database was scaled to an overall sound pressure level of 62.35 dB as specified in [ANSI S3.5 1997] for normal voice effort. The input SNR before NELE processing was adjusted via the sound pressure level of the noise.

Finally, the average SII for each SNR value was calculated over all noise files. The numerical SII optimization of [Sauert, Vary 2010a], [Sauert 2014] improves the average SII considerably *without increasing the average power*.

For the iterative numerical optimization, the band audibility function $A_{\mu}$ is approximated by a strictly concave function to ensure convergence to a global maximum. For this purpose, the limitations to zero of the level distortion factor $LD_{\mu}$ and of the masking factor $MF_{\mu}$ are omitted. The solution of the optimization function took about 3.5 iterations for low SNRs up to 5 dB and about 14 iterations for high SNRs above 15 dB. Due to the lowpass (LP) characteristic of the noise, the spectral speech power is shifted toward higher frequencies, leading to a change in tone color but resulting in improved intelligibility.

In general, the speech spectrum is modified by the time varying spectral weights $W_{\mu}(\lambda)$. This causes more or less audible effects, ranging from tone coloration to severe distortions. The degree of coloring or distortion depends on the SNR before modification, the power

**Figure 11.4** Example SII optimization *without increase of output power* as a function of the SNR prior to NELE processing. Source: Adapted from [Sauert, Vary 2010a] and [Sauert 2014], TIMIT speech data base, factory 1 noise from NOISEX-92.

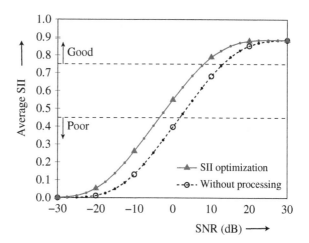

constraint $P_{max}$, the spectral energy distribution $\Phi_{S\mu}$ of the speech signal, and the spectral shape of the masking threshold $\Phi_{M\mu}$ caused by the noise signal $n(k)$.

It should be noted that severe distortions can be avoided by clipping the spectral gains at the expense of less listening enhancement. A more detailed study of these relations can be found in [Sauert 2014].

## 11.1.2  Closed-Form Gain-Shape NELE

As an alternative to SII optimization, NELE processing can also be performed using closed-form mathematical functions for the calculation of the spectral weights $W_\mu(\lambda)$. These formulas use also the masking energy threshold $\Phi_{M_\mu}(\lambda)$ but do not require iterative optimization [Niermann et al. 2016a], [Niermann 2019], [Niermann, Vary 2021]. While the objective is not explicit optimization of the SII, the resulting methods still have a performance similar to SII-optimized NELE.

These NELE functions modify the spectral shape by a frequency-dependent *shape function* $H_\mu(\lambda)$, while the overall energy is adjusted by a frequency-independent *gain factor* $g(\lambda)$. The spectral weights are formulated as

$$W_\mu(\lambda) = g(\lambda) \cdot H_\mu(\lambda), \tag{11.8}$$

and the modified spectrum becomes

$$U_\mu(\lambda) = W_\mu(\lambda) \cdot S_\mu(\lambda) \tag{11.9a}$$
$$= g(\lambda) \cdot H_\mu(\lambda) \cdot S_\mu(\lambda) \tag{11.9b}$$
$$= g(\lambda) \cdot \tilde{U}_\mu(\lambda). \tag{11.9c}$$

The energy can be adjusted by $g(\lambda)$ in the time domain (or in the frequency domain) by scaling the spectrally shaped intermediate signal $\tilde{u}(k)$

$$u(k) = g(\lambda) \cdot \tilde{u}(k), \ \ \lambda \cdot r \leq k < (\lambda + 1) \cdot r, \tag{11.10}$$

where $r$ denotes the DFT frame shift. If an analysis-synthesis filter bank is used instead of DFT – IDFT, $r$ denotes the sample rate decimation factor. The basic concept of gain-shape frequency domain NELE is illustrated in Figure 11.5.

The spectral shaping component $H_\mu(\lambda)$ is specified as a closed-form function of the spectral short-term speech power $\Phi_{S_\mu}(\lambda)$ and the masking threshold $\Phi_{M_\mu}(\lambda)$. The gain factor $g(\lambda)$ controls the energy of frame $\lambda$ consisting of $K$ samples.

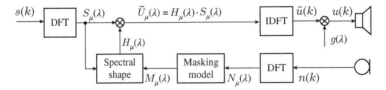

**Figure 11.5**  The gain-shape frequency domain (FD) NELE approach. $\lambda =$ frame index, $\mu =$ frequency index.

Using the DFT notation and Parseval's theorem, the energy of the speech signal $s(k)$ before spectral modification is given either in time domain or in the frequency domain,

$$E_S(\lambda) = \sum_{\kappa=0}^{K-1} s^2(\lambda \cdot r - K + \kappa) \tag{11.11a}$$

$$= \frac{1}{K} \sum_{\mu=0}^{K-1} |S_\mu(\lambda)|^2. \tag{11.11b}$$

The energy of the modified signal $u(k)$ is given by

$$E_U(\lambda) = g^2(\lambda) \cdot E_{\tilde{U}}(\lambda) \tag{11.12a}$$

$$= g^2(\lambda) \cdot \sum_{\kappa=0}^{K-1} \tilde{u}^2(\lambda \cdot r - K + \kappa) \tag{11.12b}$$

$$= g^2(\lambda) \cdot \frac{1}{K} \sum_{\mu=0}^{K-1} |\tilde{U}_\mu(\lambda)|^2. \tag{11.12c}$$

If no or only a limited increase of the output energy is allowed according to

$$E_U(\lambda) \overset{!}{=} (1 + \Delta)^2 \cdot E_S(\lambda); \ 0 \le \Delta \le \Delta_{\max}, \tag{11.13}$$

the gain factor $g(\lambda)$ can be calculated easily as

$$g(\lambda) = (1 + \Delta) \cdot \sqrt{\frac{E_S(\lambda)}{E_{\tilde{U}}(\lambda)}}. \tag{11.14}$$

It should be noted that the calculation of the gain factor $g(\lambda)$ according to (11.14), (11.11a), and (11.12b) is not restricted to the DFT frame length $K$. The short-term energy can be adjusted as a moving average (MA) over any reasonable period of time. Alternatively, recursive sample-by-sample autoregressive (AR) averaging might be applied as

$$\hat{E}_S(k) = (1 - \alpha) \cdot \hat{E}_S(k-1) + \alpha \cdot s^2(k) \tag{11.15a}$$

$$\hat{E}_{\tilde{U}}(k) = (1 - \alpha) \cdot \hat{E}_{\tilde{U}}(k-1) + \alpha \cdot \tilde{u}^2(k), \tag{11.15b}$$

with a smoothing factor $0 < \alpha < 1$, e.g., $\alpha = 0.9$.

Two closed-form NELE-weight functions have been proposed in [Niermann 2019] and [Niermann, Vary 2021] that follow two opposite objectives. They are appropriate for specific relative spectral distributions of speech and noise as well as for specific ranges of the SNR of the signal $s(k)$ in relation to the near-end noise $n(k)$. The available energy is allocated *either* more to the strongly masked *or* more to the weakly masked frequency bins.

1) Spectral shaping function *NoiseProp*:
   Amplification of frequency bins with *strong masking* at the expense of less amplification or even attenuation of frequency bins with *low masking*. The spectral weights are defined as a function of the smoothed power spectra as

$$W_\mu^2(\lambda) = g^2(\lambda) \cdot \left( \frac{\Phi_{M_\mu}(\lambda)}{\Phi_{S_\mu}(\lambda)} \right)^p, \ p \in [0,1]. \tag{11.16}$$

2) Spectral shaping function *NoiseInverse*:
Amplification of frequency bins with *low masking* at the expense of less amplification or even attenuation of frequency bins with *strong masking*

$$W_\mu^2(\lambda) = g^2(\lambda) \cdot \left( \frac{1}{\Phi_{S_\mu}(\lambda) \cdot \Phi_{M_\mu}(\lambda)} \right)^p, \quad p \in [0,1].$$
(11.17)

In both cases, the amount of modification can be controlled by the tuning parameter $p \in [0,1]$. There is no modification (besides scaling with $g(\lambda)$) for $p = 0$ and maximum modification for $p = 1$. The effect of $p < 1$ on the weights $W_\mu(\lambda)$ can also be understood as compression. An exponent $p < 1$ causes an increase or a decrease of $W_\mu^2(\lambda)$ depending on whether the expression in the parentheses is smaller or larger than 1.

The totally allowed energy per frame $\lambda$ of output samples $u(k)$ is adjusted via the factor $g(\lambda)$ as specified in (11.14).

In both cases, weight limitation has to be used to avoid excessive amplification due to division by a very small denominator.

### 11.1.2.1 The NoiseProp Shaping Function

The *NoiseProp* strategy (11.16) is to modify the speech periodogram such that it exceeds the masking energy threshold $\Phi_{M_\mu}(\lambda)$:

$$\Phi_{U_\mu}(\lambda) = W_\mu^2(\lambda) \cdot \Phi_{S_\mu}(\lambda) > \Phi_{M_\mu}(\lambda).$$
(11.18)

The same applies correspondingly to sub-band signals, if a filter bank is used instead of a DFT. The masking threshold is the lowest level at which a test tone, here a frequency component of the processed speech, can be perceived in the noisy background. Thus, the modified speech signal $u(k)$ from the loudspeaker can be perceived at frequency $\mu$ if the *signal-to-mask ratio* (SMR) is larger than one

$$SMR_\mu(\lambda) = \frac{\Phi_{U_\mu}(\lambda)}{\Phi_{M_\mu}(\lambda)} = \frac{W_\mu(\lambda)^2 \cdot \Phi_{S_\mu}(\lambda)}{\Phi_{M_\mu}(\lambda)} \overset{!}{>} 1.$$
(11.19)

However, with regard to intelligibility, the speech level should surpass the masking level not only slightly, but by a sufficient margin. The SII, for instance, predicts maximum speech intelligibility if the speech level is by 15 dB larger than the masking threshold for all frequencies [ANSI S3.5 1997] (see also Figure 11.3). If sufficient power is available, the ideal target would be $10 \log (SMR_\mu(\lambda)) = 15$ dB.

Insertion of (11.16) into (11.19) for the case of maximum modification, i.e., $p = 1.0$ results in

$$SMR_\mu(\lambda) = g^2(\lambda),$$
(11.20)

which shows that strategy (11.16) achieves an SMR that is constant over frequency $\mu$ for each frame index $\lambda$ and which can be controlled by $g(\lambda)$.

In case of $p = 1$, we have

$$\Phi_{U_\mu}(\lambda) = W_\mu^2(\lambda) \cdot \Phi_{S_\mu}(\lambda) \tag{11.21a}$$

$$= g^2(\lambda) \cdot \left(\frac{\Phi_{M_\mu}(\lambda)}{\Phi_{S_\mu}(\lambda)}\right)^p \cdot \Phi_{S_\mu}(\lambda) \tag{11.21b}$$

$$= g^2(\lambda) \cdot \Phi_{M_\mu}(\lambda). \tag{11.21c}$$

Thus, the power spectrum, i.e., the smoothed periodogram of the modified signal $u(k)$, would become strictly proportional to the masking spectrum $\Phi_{M_\mu}(\lambda)$. This is why this NELE strategy is called *NoiseProp*.

However, this reshaping with $p = 1$ might result in an overly strong modification. Choosing a tuning parameter $p < 1.0$, reduces the amplitude dynamics of the spectral weights $W_\mu(\lambda)$. By informal listening it has been verified that the range $p = 0.5 \dots 0.8$ [Niermann 2019] is a good trade-off between natural sounding speech and listening enhancement performance. It keeps most of the speech characteristics because the modified speech periodogram $|U(\lambda, \mu)|^2$ is shaped like the masking spectrum only on the average. Thus, the short-term speech characteristics and the spectral fine structure are preserved when the noise is stationary or slowly changing over time.

The main advantage of *NoiseProp* (11.16) is its capability to completely overcome masking, unless there is a limiting power constraint. Under the condition of complete unmasking *NoiseProp* needs less output power in comparison to a spectrally flat amplification. NELE processing increases intelligibility at the expense of some "spectral coloring" of the speech signal, which is mostly not perceived as unpleasant. However, if the available output power is still too low to reach the target SMR, the gain $g(\lambda)$ has to be decreased and thus speech masking increases correspondingly.

The *NoiseProp* strategy is suitable only, if sufficient output frame energy $E_U(\lambda)$ is available, including eventually some headroom $\Delta$ according to (11.13).

In case of medium SNR, the *NoiseProp* strategy and the NELE approach based on the optimization of the SII (as described in Section 11.1.1) sound very similar.

### 11.1.2.2 The NoiseInverse Strategy

When the available output power is too low for the *NoiseProp* strategy (11.16) and the speech is hardly understandable, the better strategy is to selectively attenuate some strongly masked sub-bands of the speech and to use the released power to raise other sub-bands over the masking threshold. This is the objective of the *NoiseInverse* strategy (11.17). Inserting (11.17) into (11.19) results in

$$SMR_\mu(\lambda) = g^2(\lambda) \cdot \frac{\Phi_{S_\mu}(\lambda)}{\left(\Phi_{S_\mu}(\lambda) \cdot \Phi_{M_\mu}(\lambda)\right)^p}, \quad p \in [0, 1]. \tag{11.22a}$$

In the extreme case of maximum modification ($p = 1$), the *SMR* and thus the power spectrum of $u(k)$ become *inverse* to the masking threshold

$$\Phi_{U_\mu}(\lambda) = W_\mu^2(\lambda) \cdot \Phi_{S_\mu}(\lambda) \tag{11.23a}$$

$$= g^2(\lambda) \cdot \frac{1}{\left(\Phi_{S_\mu}(\lambda) \cdot \Phi_{M_\mu}(\lambda)\right)^p} \cdot \Phi_{S_\mu}(\lambda) \tag{11.23b}$$

$$= g^2(\lambda) \cdot \frac{1}{\Phi_{M_\mu}(\lambda)}. \tag{11.23c}$$

The speech signal $s(k)$ is amplified where the masking threshold and thus the noise is weak and it is attenuated where the noise is strong. Therefore, this is called *NoiseInverse*. By shaping $|U_\mu(\lambda)|^2$ on the average proportional to the inverse of the masking level $\Phi_{M_\mu}(\lambda)$, the speech bins, which are strongly masked get attenuated. In these sub-bands, high effort in terms of power would be necessary to overcome the masking threshold. Consequently, it is reasonable to give them up and save power. Contrarily, in sub-bands with lower relative masking levels, the effort to elevate the speech level above the masking threshold is much lower.

Therefore, the saved power is implicitly redistributed to sub-bands where the masking power is low. The amount of the totally allowed power, i.e., energy per DFT frame is adjusted via the gain $g(\lambda)$. Again, $p = 0.5 \dots 0.8$ is a good trade-off between natural sounding speech and the listening enhancement performance.

### 11.1.2.3 Gain-Shape Frequency Domain NELE Example

An example evaluation of the closed-form gain-shape frequency domain (FD) *NoiseProp* and *NoiseInverse* strategies in terms of the SII is shown in Figure 11.6 for the noise types *white*, *lowpass*, and *highpass* at different SNRs of the received signal $s(k)$ in relation to the *near-end* noise $n(k)$. These experiments were carried out *without increase of the output power*, i.e., solely relying on spectral energy redistribution ($\Delta = 0$ in (11.13)).

The noise signals were generated by filtering white noise signals at a sampling rate of $f_s = 16$ kHz. The lowpass noise is generated with a first-order recursive filter with a coefficient $a = 0.9$. The *highpass* noise signal was generated with an FIR filter having a cut-off frequency of $f_c = 4$ kHz and a stop band attennuation of 20 dB. For the calculation of the weights $W_\mu(\lambda)$, uniform DFT frequency bins ($K_{DFT} = 512$) were grouped according to the Bark scale. More details can be found in [Niermann, Vary 2021] and [Niermann 2019].

For white noise (Figure 11.6a), it can be seen that *NoiseProp* and *NoiseInverse* perform equally well. This is due to the constant masking threshold $\Phi_{M_\mu}(\lambda)$ over frequency.

For lowpass noise (Figure 11.6b), *NoiseInverse* outperforms *NoiseProp* as the spectral envelopes of speech and noise have similar characteristics (spectral tilt). The available power is too low to raise the speech spectrum with the *NoiseProp* strategy noticeably above the masking threshold. In contrast, the *NoiseInverse* strategy achieves a clearly perceivable improvement by amplification of the higher, less masked frequencies. This results in a more intelligible highpass sound.

In the case of highpass noise (Figure 11.6c) and SNRs larger than approx. $-2.5$ dB, *NoiseProp* performs better than *NoiseInverse* since the available output power is sufficient to

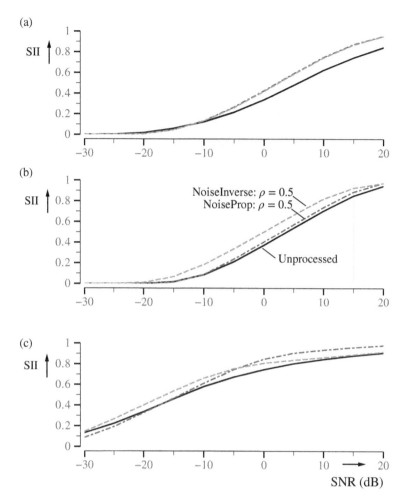

**Figure 11.6** Example of gain-shape frequency domain (FD) NELE: Comparison of the *Speech Intelligibility Index* SII for the strategies *NoiseProp* and *NoiseInverse* and three different near-end noise signals. Note: *No increase of overall speech power*, i.e., $\Delta = 0$ in (11.14) (a) White noise, (b) lowpass noise, $a = 0.9$, and (c) highpass noise, cut-off 4 kHz, stop band attenuation 20 dB. Source: adapted from [Niermann, Vary 2021], © 2021 IEEE.

improve unmasking at higher frequencies. If the SNR is lower, *NoiseInverse* achieves better SII results, which shows that this strategy is the better choice in case of low SNR conditions with strongly restricted output power. Since the noise is analyzed frame-by-frame, the algorithms are able to track non-stationary noise.

In [Niermann, Vary 2021], a NELE *listening test* (23 participants) at the limit of intelligibility is described. By using the FD NoiseInverse approach without increasing the power, the intelligibility of 25 phonetically balanced words was raised on the average from 40% to 63%.

## 11.2 Time Domain NELE (TD)

NELE processing can also be performed partly or completely in the time domain (TD). In [Sauert, Vary 2010a], [Sauert, Vary 2010b], the SII-based NELE weights are calculated by means of two warped filter banks applied to $s(k)$ and $n(k)$, while the modification of $s(k)$ is achieved with a filter bank equalizer (see also Section 4.4). The NELE weights are calculated, as described in Section 11.1.1, according to the SII optimization approach.

Alternatively, the NELE weights may be calculated using the closed-form gain-shape functions of Section 11.1.2 while filtering operation can be performed with a filter bank equalizer, too.

Both NELE variants may be considered as a mixture of FD and TD processing.

### 11.2.1 NELE Processing using Linear Prediction Filters

An alternative pragmatic TD approach is proposed in [Niermann et al. 2016c], [Niermann 2019], and [Niermann, Vary 2021], which neither needs a frequency transformation/filter bank nor the calculation of a masking threshold. The design is motivated by the two gain-shape FD-NELE strategies (11.16) and (11.17) of Section 11.1. The resulting TD algorithm is not a one-to-one translation of the FD gain-shape approach, but the performance is very similar.

Instead of calculating the speech and noise periodograms in the FD, corresponding spectral envelopes are used, which are realized by TD filters. The filter coefficients are determined as linear prediction (LP) coefficients.

The two FD strategies (11.16) and (11.17), which motivate the TD approach, are both comprising a speech and a noise component.

The FD-*NoiseProp* gain rule (11.16) can be expressed for $p = 1$ as

$$W_\mu^2(\lambda) = g^2(\lambda) \cdot \underbrace{\Phi_{S_\mu}(\lambda)^{-1}}_{\text{speech-dependent}} \cdot \underbrace{\Phi_{M_\mu}(\lambda)}_{\text{noise-dependent}} . \tag{11.24}$$

The FD-*NoiseInverse* gain rule (11.17) can be expressed for $p = 1$ as

$$W_\mu^2(\lambda) = g^2(\lambda) \cdot \underbrace{\Phi_{S_\mu}(\lambda)^{-1}}_{\text{speech-dependent}} \cdot \underbrace{\Phi_{M_\mu}(\lambda)^{-1}}_{\text{noise-dependent}} . \tag{11.25}$$

As suggested by (11.24) and (11.25), time domain versions of the proposed NELE strategies can be realized by filtering the received far-end speech signal $s(k)$ with a cascade of two filters which depend on the far-end speech signal and on the near-end background noise $n(k)$. The time-domain NELE concept is illustrated in Figure 11.7. The first filter $A(z)$ realizes a smoothed version of the inverse of the spectral envelope of the far-end speech signal in terms of the LP coefficients $a(m)$ that are adapted with the far-end speech signal $s(k)$. This type of filter is a so-called *LP analysis filter*.

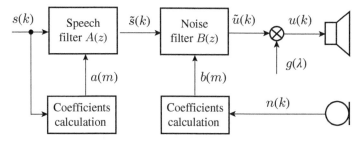

**Figure 11.7** Concept of the time domain (TD) NELE approach. Source: [Niermann 2019], adapted from [Niermann, Vary 2021].

The *time variable* speech-dependent filter is a finite impulse response LP analysis filter with transfer function

$$A(z) = 1 - A_0\left(\frac{z}{\gamma}\right) \tag{11.26a}$$

$$= 1 - \sum_{m=1}^{M_s} a(m)\gamma^m z^{-m}, \tag{11.26b}$$

where $M_s$ denotes the order of the speech predictor $A_0(z)$. The filter can be controlled by the tuning parameter $0 \le \gamma \le 1$ in a way that is known from CELP coding and post-filtering, as described in Section 8.6 [Chen, Gersho 1995]. The tuning parameter $\gamma$ of the speech dependent filter is the counterpart of the control parameter $p$ of the FD gain-shape NELE approach. The amount of modification can be adjusted: we have no modification at all if $\gamma = 0$ and maximum modification if $\gamma = 1$.

It is a known property of LP analysis filters that their magnitude response is inversely proportional to the spectral envelope of the speech frame, as described in Section 6.4. Applied to speech signals, the spectral envelope has mostly a highpass characteristic and reduces the amplitude dynamics of the TD signal and of the spectral components.

The second filter $B(z)$ should either describe the spectral envelope of the masking threshold according to (11.24) or its inverse according to (11.25). As mentioned before, the FD strategies are just taken as motivation for the TD approach. Thus, for simplicity, the spectral envelope (or its inverse) of the *near-end* noise signal $n(k)$ is taken instead of the masking threshold $\Phi_{M_\mu}(\lambda)$.

The spectral envelope of the noise signal can be described by an *LP synthesis* filter while the inverse of the envelope corresponds to an *LP analysis* filter. In both cases, the coefficients $b(m)$ can be adapted with the *near-end* noise signal $n(k)$.

A generic *time variable* noise-dependent filter for both the *NoiseProp* and the *NoiseInverse* strategy can be formulated as

$$B(z) = \frac{1 - B_0\left(\frac{z}{\alpha}\right)}{1 - B_0\left(\frac{z}{\beta}\right)} \tag{11.27a}$$

$$= \frac{1 - \sum_{m=1}^{M_n} b(m)\alpha^m z^{-m}}{1 - \sum_{m=1}^{M_n} b(m)\beta^m z^{-m}} \tag{11.27b}$$

where $M_n$ denotes the order of the noise predictor $B_0(z)$. The control parameters $\alpha$ and $\beta$, can be used to switch between the two NELE strategies:

$$TD - NoiseProp : \quad \alpha = 0, \ \beta = 1$$
$$TD - NoiseInverse : \quad \alpha = 1, \ \beta = 0.$$

Furthermore, the parameters $\alpha, \beta \in [0,1]$ allow to choose an adjustable mixture of the two strategies.

The generic noise-dependent filter $B(z)$ is realized as a pole-zero filter consisting of a finite impulse response filter (FIR, nominator of (11.27b)) and an infinite impulse response filter (IIR, denominator of (11.27b)). The magnitude response of the FIR part amplifies the speech at frequencies where the noise power is low and vice versa (*NoiseInverse*), whereas the IIR part amplifies the speech where the noise power is large and vice versa (*NoiseProp*).

In contrast to the FD NELE strategies of Section 11.1, the adaptation of this filter does neither rely on the calculation of a masking threshold nor on the SII, as all the filter coefficients are derived directly from the TD signals, i.e., the received far-end speech $s(k)$ and the *near-end* noise $n(k)$.

Any technique known from speech coding, such as the Levinson–Durbin algorithm (see also Section 6.4), can be used to calculate the coefficients $a(m)$ and $b(m)$. The required autocorrelation values can be determined either by block processing causing a block latency or by recursive sample-by-sample processing without additional delay.

The computational complexity can be reduced further, if the coefficients are determined by iterative Least Mean Square (LMS) adaptations. In this case, neither the autocorrelation calculation nor the Levinson–Durbin recursion is required.

The example frequency responses of a combined TD NELE filter

$$F(z) = g(k) \cdot A(z) \cdot B(z), \tag{11.28}$$

and its components are illustrated in Figure 11.8. The speech-dependent filter $A(z)$ with $\gamma = 0.5$ takes the role of an adaptive pre-emphasis filter. The two components of the noise-dependent filter $B(z)$ (11.27b) are illustrated separately for different settings of $\alpha$ and $\beta$. The gain factor $g(k)$ adjusts the output power according to a given constraint.

An example SII evaluation of the TD versions of the strategies *NoiseProp* and *NoiseInverse* and their combination is illustrated in Figure 11.9. For an unbiased comparison, these experiments were carried out again *without power amplification*, i.e., $\Delta = 0$.

In case of *LP noise* (Figure 11.9a), the mixture and the *NoiseInverse* strategy clearly outperforms *NoiseProp* as speech and noise have similar spectral characteristics (spectral tilt). For the *NoiseProp* strategy, the available power is too low to raise the speech spectrum noticeably in relation to the masking threshold. In contrast, the *NoiseInverse* strategy achieves a perceivable improvement by amplification of the higher, less masked frequencies. The mixture that gives more emphasis on the *NoiseInverse* component ($\alpha = 1.0$) than on the *NoiseProp* component ($\beta = 0.8$) actually outperforms the *NoiseInverse* strategy.

In the case of *highpass noise* (Figure 11.9b) the *NoiseInverse* strategy and the mixture perform almost identical and perform, up to an SNR of approx. 8-10 dB, better than the *NoiseProp* strategy since the available output power is too low to improve unmasking by *NoiseProp*.

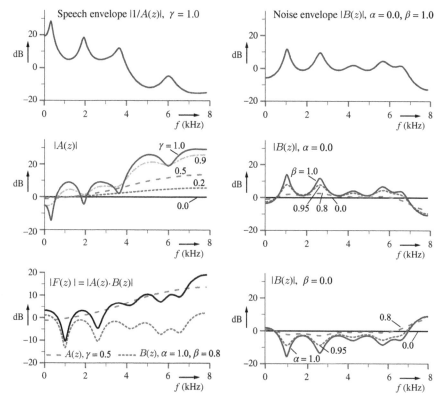

**Figure 11.8** Example frequency responses of the time domain (TD) NELE approach: Speech filter $|A(z)|$, noise filter $|B(z)|$ and composite NELE filter $|F(z)|$ for $f_s = 16$ kHz and various parameter settings. Source: adapted from [Niermann, Vary 2021], © 2021 IEEE.

The comparison between Figures 11.6 and 11.9 exhibits that the TD and the FD approaches show a very similar behavior.

Finally, an experimental comparison of TD-NELE and FD-NELE by objective SII measurements and by informal listening tests should be mentioned [Niermann, Vary 2021]. The evaluation was done for white noise, car noise (midsize car2, 130 km/h, binaural, [ETSI Guide EG 202 396-1 2008]), and babble noise [Varga, Steeneken 1993] at SNR conditions in the range −10 dB … +20 dB. For an unbiased comparison, the output power was not increased.

As FD-NELE reference, the SII optimized FD approach of Section 11.1.1, Figure 11.4 [Sauert, Vary 2010a], [Sauert 2014] was taken. Since the SII evaluation criterion is part of the reference algorithm, it is not surprising that the reference algorithm performed the best. However, the much simpler gain-shape FD-NELE approach *FD NoiseInverse* with $p = 0.5$ was rated in terms of the SII and in terms of the subjective intelligibility only marginally lower. Even the least complex TD-NELE approach *TD mixed* with $\gamma = 0.5$, $\alpha = 1.0$, $\beta = 0.8$), performed almost as good as the reference algorithm.

Compared to the FD approach, the distinctive advantages of the TD solution are the lower complexity and a *lower signal delay* especially in case of sample-by-sample filter adaptation.

(a)

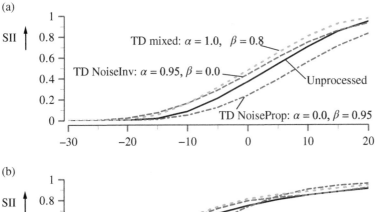

(b)

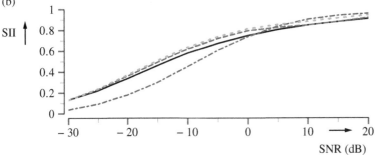

**Figure 11.9** Example of TD NELE: Comparison of the *Speech Intelligibility Index* SII for the strategies *NoiseProp* and *NoiseInverse* and two different near-end noise signals. Note: *No increase of speech power*, i.e., $\Delta = 0$ in (11.14). (a) LP noise, $a = 0.9$ (b) highpass noise, cut-off 4 kHz, stop band attenuation 20 dB. Source: adapted from [Niermann, Vary 2021], © 2021 IEEE.

## References

ANSI S3.5 (1997). Methods for Calculation of the Speech Intelligibility Index.

Aubanel, V.; Cooke, M. (2013). Information-Preserving Temporal Reallocation of Speech in the Presence of Fluctuating Maskers, *Proceedings of INTERSPEECH*, pp. 3592–3596.

Chen, J. H.; Gersho, A. (1995). Adaptive Postfiltering for Quality Enhancement of Coded Speech, *IEEE Transactions on Speech and Audio Processing*, vol. 3, no. 1, pp. 59–71.

Cooke, M.; Aubanel, V.; Lecumberri, M. L. G. (2019). Combining Spectral and Temporal Modification Techniques for Speech Intelligibility Enhancement, *Computer Speech and Language*, vol. 55, pp. 26–39.

Crespo, J. B.; Hendriks, R. C. (2014). Multizone Speech Reinforcement, *IEEE Transactions on Audio, Speech, and Language Processing*, vol. 22, no. 1, pp. 54–66.

DARPA (1990). The DARPA TIMIT Acoustic-Phonetic Continuous Speech Corpus CD-ROM, National Institute of Standards and Technology (NIST).

ETSI Guide EG 202 396-1 (2008). *Speech Processing, Transmission and Quality Aspects (STQ); Speech Quality Performance in the Presence of Background Noise; Part 1: Background Noise Simulation Technique and Background Noise Database*, European Telecommunications Institute.

Fastl, H.; Zwicker, E. (2007). *Psychoacoustics*, 2nd edn, Springer-Verlag, Berlin.

Grosse, J.; van de Par, S. (2017). A Speech Preprocessing Method Based on Overlap-Masking Reduction to Increase Intelligibility in Reverberant Environments, *Journal of the Audio Engineering Society*, vol. 65, no. 1/2, pp. 31–41.

Hendriks, R. C.; Crespo, J. B.; Jensen, J.; Taal, C. H. (2015). Optimal Near-End Speech Intelligibility Improvement Incorporating Additive Noise and Late Reverberation Under an Approximation of the Short-Time SII, *IEEE Transactions on Audio, Speech, and Language Processing*, vol. 23, no. 5, pp. 851–862.

Kleijn, W. B.; Hendriks, R. C. (2015). A Simple Model of Speech Communication and Its Application to Intelligibility Enhancement, *IEEE Signal Processing Letters*, vol. 22, no. 3, pp. 303–307.

Kleijn, W. B.; Crespo, J. B.; Hendriks, R. C.; Petkov, P. N.; Sauert, B.; Vary, P. (2015). Optimizing Speech Intelligibility in a Noisy Environment: A unified view, *IEEE Signal Processing Magazine*, vol. 32, no. 2, pp. 43–54.

Lombard, E. (1911). Le signe de elevation de la voix, *Annales des maladies de l'oreille, du larynx*, vol. 37, pp. 101–119.

Niermann, M. (2019). *Digital Enhancement of Speech Perception in Noisy Environments*, Dissertation. Aachener Beiträge zu Digitalen Nachrichtensystemen (ABDN), P. Vary (ed.), Institute of Communication Systems (IND/IKS), RWTH Aachen University.

Niermann, M.; Vary, P. (2021). Listening Enhancement in Noisy Environments: Solutions in Time and Frequency Domain, *IEEE Transactions on Audio, Speech, and Language Processing*, vol. 29, pp. 699–709.

Niermann, M.; Jax, P.; Vary, P. (2016a). Near-End Listening Enhancement by Noise-Inverse Speech Shaping, *Proceedings of the European Signal Processing Conference (EUSIPCO)*, EURASIP, pp. 2390–2394.

Niermann, M.; Jax, P.; Vary, P. (2016b). Noise Estimation for Speech Reinforcement in the Presence of Strong Echoes, *Proceedings of the IEEE International Conference on Acoustics, Speech, and Signal Processing (ICASSP)*, IEEE, Shanghai, China, pp. 5920–5924.

Niermann, M.; Thierfeld, C.; Jax, P.; Vary, P. (2016c). Time Domain Approach for Listening Enhancement in Noisy Environments, *in ITG Conference Speech Communication*, VDE Verlag GmbH, pp. 282–286.

Niermann, M.; Jax, P.; Vary, P. (2017). Joint Near-End Listening Enhancement and Far-End Noise Reduction, *Proceedings of the IEEE International Conference on Acoustics, Speech, and Signal Processing (ICASSP)*, IEEE, New Orleans, USA, pp. 4970–4974.

Petkov, P. N.; Kleijn, W. B. (2015). Spectral Dynamics Recovery for Enhanced Speech Intelligibility in Noise, *IEEE Transactions on Audio, Speech, and Language Processing*, vol. 23, no. 2, pp. 327–338.

Petkov, P. N.; Stylianou, Y. (2016). Adaptive Gain Control for Enhanced Speech Intelligibility Under Reverberation, *IEEE SPL*, vol. 23, no. 10, pp. 1434–1438.

Sauert, B. (2014). *Near-End Listening Enhancement: Theory and Application*, Dissertation. Aachener Beiträge zu Digitalen Nachrichtensystemen (ABDN), P. Vary (ed.), IND, RWTH Aachen University.

Sauert, B.; Vary, P. (2006). Near End Listening Enhancement: Speech Intelligibility Improvement in Noisy Environments, *Proceedings of IEEE International Conference on Acoustics, Speech, and Signal Processing (ICASSP)*, IEEE, Toulouse, France, pp. 493–496.

Sauert, B.; Vary, P. (2009). Near End Listening Enhancement Optimized with Respect to Speech Intelligibility Index, *Proceedings of the European Signal Processing Conference (EUSIPCO)*, EURASIP, Hindawi Publ., pp. 1844–1848.

Sauert, B.; Vary, P. (2010a). Near End Listening Enhancement Optimized with Respect to Speech Intelligibility Index and Audio Power Limitations, *Proceedings of the European Signal Processing Conference (EUSIPCO)*, EURASIP, pp. 1919–1923.

Sauert, B.; Vary, P. (2010b). Recursive Closed-Form Optimization of Spectral Audio Power Allocation for Near end Listening Enhancement, *in ITG Conference Speech Communication*, VDE Verlag GmbH.

Sauert, B.; Heese, F.; Vary, P. (2014). Real-Time Near-End Listening Enhancement for Mobile Phones, *IEEE International Conference on Acoustics, Speech, and Signal Processing (ICASSP)*, IEEE. Show and Tell Demonstration.

Schepker, H.; Rennies, J.; Doclo, S. (2015). Speech-in-Noise Enhancement Using Amplification and Dynamic Range Compression Controlled by the Speech Intelligibility Index, *Journal of the Acoustical Society of America*, vol. 138, no. 5, pp. 2692–2706.

Taal, C. H.; Jensen, J.; Leijon, A. (2013). On Optimal Linear Filtering of Speech for Near-End Listening Enhancement, *IEEE Signal Processing Letters*, vol. 20, no. 3, pp. 225–228.

Varga, A.; Steeneken, H. J. (1993). Assessment for Automatic Speech Recognition: II. NOISEX-92: A Database and an Experiment to Study the Effect of Additive Noise on Speech Recognition Systems, *Speech Communication*, vol. 12, no. 3, pp. 247–251.

Zorila, T.; Kandia, V.; Stylianou, Y. (2012). Speech-in-Noise Intelligibility Improvement Based on Power Recovery and Dynamic Range Compression, *Proceedings of the European Signal Processing Conference (EUSIPCO)*, pp. 2075–2079.

# 12

# Single-Channel Noise Reduction

This chapter is concerned with algorithms for reducing additive noise when only a single microphone signal is available. More general dual-channel and multi-microphone beam-forming techniques will be discussed in Chapters 13 and 15. The first part of this chapter gives an introduction to the basic principles and implementation aspects. We will introduce the Wiener filter and "spectral subtraction," as well as noise power spectral density estima-tion techniques. The second part is more advanced and dedicated to non-linear optimal estimators. These estimators explicitly use the probability density function of short-time spectral coefficients and are thus better able to utilize the information available a priori. We will present maximum likelihood (ML), maximum a posteriori, and minimum mean square estimators for the estimation of the complex DFT coefficients, as well as for func-tions of the spectral amplitude. We will conclude this chapter with an overview of speech enhancement approaches using deep neural networks (DNNs).

## 12.1 Introduction

When a speech communication device is used in environments with high levels of ambi-ent noise, the noise picked up by the microphone will significantly impair the quality and the intelligibility of the transmitted speech signal. The quality degradations can be very annoying, especially in mobile communications where hands free devices are frequently used in noisy environments such as cars. Compared to a close talking microphone, the use of a hands free device can lower the signal-to-noise ratio (SNR) of the microphone signal by more than 20 dB. Thus, even when the noise levels are moderate, the SNR of a hands free microphone signal is rarely better than 25 dB. Traditional single-microphone speech enhancement systems, which process the signal without much temporal context and inde-pendently in a number of frequency bands, can improve the quality of speech signals and help to reduce listener fatigue. More recently, neural network-based approaches use the joint statistics of many time-frequency bins and are able to also improve the intelligibil-ity. All of these approaches are useful not only in hands-free communication but also in conjunction with speech coders, such as *code excited linear prediction* (CELP) or *mixed exci-tation linear prediction* (MELP) coders and hearing devices, including cochlear implants. Since they help to improve the quality of the estimated (spectral and temporal) parameters

*Digital Speech Transmission and Enhancement*, Second Edition. Peter Vary and Rainer Martin.
© 2024 John Wiley & Sons Ltd. Published 2024 by John Wiley & Sons Ltd.

noise reduction pre-processors may improve not only the quality [Kang, Fransen 1989] but also the intelligibility of the noisy speech signal [Collura 1999], [Chen et al. 2016]. Besides the basic requirement to deliver enhanced signal quality and intelligibility, the selection of the appropriate method will also depend on additional constraints, such as the maximum latency, the computational complexity, and the memory footprint. Furthermore, the naturalness of the remaining artifacts and the residual noise are of concern.

To illustrate the performance of single-channel systems, Figure 12.1 shows the time domain waveforms of a clean speech signal, a noisy signal, and two versions of an enhanced signal vs. the sampling index $k$. The noise is non-stationary car noise composed of relatively stationary motor and wind noise and highly transient noise bursts resulting

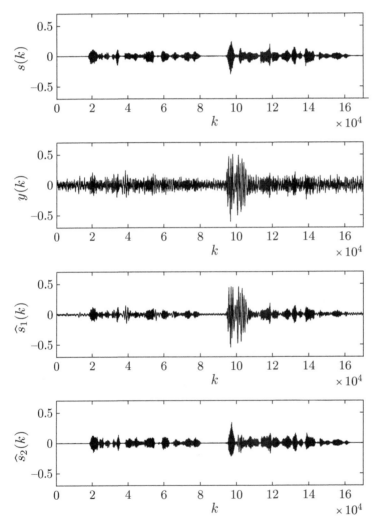

**Figure 12.1** Time domain waveforms of a clean (top), a noisy (second), and two enhanced speech signals. The waveforms in the third and bottom plots are the outputs of a traditional and a DNN-based enhancement system.

**Figure 12.2** The noise reduction filter.

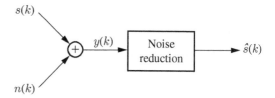

from potholes. The traditional single-channel noise reduction algorithm, which was used in the first example, significantly reduces the level of the slowly varying disturbing noise. However, due to the difficulty of tracking fast variations of the background noise, the short noise bursts around $k = 40\,000$ and $k = 100\,000$ are not removed. The second example uses a neural network that has been trained on a large number of speakers and noise examples. This method delivers an almost clean signal.

Given the large diversity of acoustic environments and noise reduction applications and the resulting, sometimes conflicting performance requirements for noise reduction algorithms, it is apparent that there cannot be only one single "optimal" algorithm. Hence, a large variety of algorithms have been developed, which have proved to be beneficial in certain noise environments or certain applications. Most of these more successful algorithms use statistical considerations for the computation of the enhanced signal. Most of them work in the frequency or some other transformation domain, and some of them also include models of human hearing [Tsoukalas et al. 1993], [Gustafsson et al. 1998], [Virag 1999]. So far, there are no international (ITU or ETSI) standards for noise reduction algorithms, although noise reduction algorithms have become part of speech coding and teleconferencing systems recently. In fact, a noise reduction pre-processor [Ramabadran et al. 1997] is part of the *enhanced variable rate codec* (EVRC, TIA/EIA IS-127) standard. Moreover, a set of minimum requirements for noise reduction pre-processors has been defined in conjunction with the ETSI/3GPP *adaptive multi-rate* (AMR) codec [3GPP TS 122.076 V5.0.0 2002], [3GPP TR 126.978, V4.0.0 2001].

In this chapter, we will give a general introduction to noise reduction methods and then discuss some of the more important topics in detail. While we will look especially at statistical models for the estimation of speech signals and their spectral coefficients, we note that many of these concepts are also useful in the context of the more recent deep learning approaches. Throughout this chapter, we will focus exclusively on additive noise, i.e., consider disturbed signals $y(k)$, which are a sum of the speech signal $s(k)$ and the noise signal $n(k)$, $y(k) = s(k) + n(k)$, as shown in Figure 12.2. Furthermore, $s(k)$ and $n(k)$ are assumed to be statistically independent, which also implies $E\{s(k)n(i)\} = 0 \; \forall \, k, i$. The enhanced signal will be denoted by $\hat{s}(k)$.

## 12.2 Linear MMSE Estimators

Linearly constrained minimum mean square error (MMSE) estimators play an important role in signal estimation theory as they are relatively easy to develop and to implement. Furthermore, they are optimal for Gaussian signals (see Section 5.11.4) in an unconstrained optimization scenario.

## 12.2.1 Non-causal IIR Wiener Filter

Our discussion is based on the signal model depicted in Figure 12.3. Adding the noise signal $n(k)$ to the undisturbed desired signal $s(k)$ yields a noisy signal $y(k)$. The enhanced signal $\hat{s}(k)$ is compared to a reference signal $d(k)$. The resulting error signal $e(k)$ is used to compute the filter coefficients $h(k)$. For the time being, we assume that all signals are wide sense stationary and zero mean random processes.

The optimal linear filter is, then, time invariant and can be characterized by its impulse response $h(k)$. In the general case, the filter with impulse response $h(k)$ is neither a causal nor a finite-impulse-response (FIR) filter. Thus, the output signal $\hat{s}(k)$ of the linear time-invariant infinite-impulse-response (IIR) filter is given by

$$\hat{s}(k) = \sum_{\kappa=-\infty}^{\infty} h(\kappa)y(k-\kappa). \tag{12.1}$$

For computing the impulse response of the optimal filter, we may define a target signal $d(k) = s(k)$ and minimize the mean square error

$$E\left\{e^2(k)\right\} = E\left\{(\hat{s}(k) - d(k))^2\right\} = E\left\{(\hat{s}(k) - s(k))^2\right\}. \tag{12.2}$$

The partial derivative of the mean square error $E\left\{e^2(k)\right\}$ with respect to a coefficient $h(i)$ leads to the condition

$$\frac{\partial E\left\{e^2(k)\right\}}{\partial h(i)} = \frac{\partial E\left\{\left(\sum_{\kappa=-\infty}^{\infty} h(\kappa)y(k-\kappa) - s(k)\right)^2\right\}}{\partial h(i)} = 0 \tag{12.3}$$

for all $i \in \mathbb{Z}$ which may be written as

$$\sum_{\kappa=-\infty}^{\infty} h(\kappa)\,\varphi_{yy}(i-\kappa) = \varphi_{ys}(i), \quad \forall i \in \mathbb{Z}. \tag{12.4}$$

$\varphi_{yy}(\lambda) = E\{y(k)y(k+\lambda)\}$ and $\varphi_{ys}(\lambda) = E\{y(k)s(k+\lambda)\}$ are the auto-correlation function of the noisy signal $y(k)$ and the cross-correlation function of the noisy signal $y(k)$ and the clean speech signal $s(k)$, respectively.

Equation (12.4) is recognized as a convolution in the correlation function domain and can therefore be easily solved in the Fourier domain using the relations $\varphi_{yy}(\lambda) \overset{F}{\circ\!\!-\!\!\bullet} \Phi_{yy}(e^{j\Omega})$ and

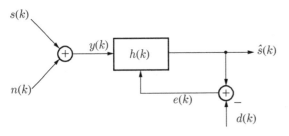

s(k)

**Figure 12.3** The linear filter problem for reducing additive noise.

$\varphi_{ys}(\lambda) \overset{\mathcal{F}}{\circ\!-\!\bullet} \Phi_{ys}(e^{j\Omega})$ with $\Phi_{yy}(e^{j\Omega})$ and $\Phi_{ys}(e^{j\Omega})$ denoting the auto-power and the cross-power spectral densities (see Section 5.8). For $\Phi_{yy}(e^{j\Omega}) \neq 0$ we obtain

$$H(e^{j\Omega}) = \frac{\Phi_{ys}(e^{j\Omega})}{\Phi_{yy}(e^{j\Omega})} \quad \text{and} \quad h(k) = \frac{1}{2\pi} \int_{-\pi}^{\pi} H(e^{j\Omega}) e^{j\Omega k} \, d\Omega. \tag{12.5}$$

For the above additive noise model with $\Phi_{yy}(e^{j\Omega}) = \Phi_{ss}(e^{j\Omega}) + \Phi_{nn}(e^{j\Omega})$, it follows that

$$H(e^{j\Omega}) = \frac{\Phi_{ss}(e^{j\Omega})}{\Phi_{ss}(e^{j\Omega}) + \Phi_{nn}(e^{j\Omega})} = \frac{\Phi_{ss}(e^{j\Omega})/\Phi_{nn}(e^{j\Omega})}{1 + \Phi_{ss}(e^{j\Omega})/\Phi_{nn}(e^{j\Omega})}, \tag{12.6}$$

where $\Phi_{ss}(e^{j\Omega})$ and $\Phi_{nn}(e^{j\Omega})$ are the power spectral densities of the clean speech and the noise signal, respectively. The estimated signal spectrum may then be written as

$$\hat{S}(e^{j\Omega}) = H(e^{j\Omega}) Y(e^{j\Omega}). \tag{12.7}$$

The non-causal IIR *Wiener filter* (12.6) evaluates the SNR $\Phi_{ss}(e^{j\Omega})/\Phi_{nn}(e^{j\Omega})$ at a given frequency $\Omega$. This is illustrated in Figure 12.4 where the power spectral densities of the clean speech and the noise are shown in the upper plot and the resulting frequency response of the Wiener filter in the lower plot. When the SNR is large at a given frequency $\Omega$, $H(e^{j\Omega})$ approaches unity and the corresponding frequency component will be passed on without being attenuated. For low SNR conditions, we have $\Phi_{ss}(e^{j\Omega}) \ll \Phi_{nn}(e^{j\Omega})$ and $H(e^{j\Omega}) \approx 0$. The corresponding frequency component of the input signal will be attenuated. Therefore, the noise reduction task will be most effectively accomplished if the speech signal and the noise do not occupy the same frequency bands. In the case of overlapping frequency bands, the noise reduction will also result in an attenuation of the desired speech signal.

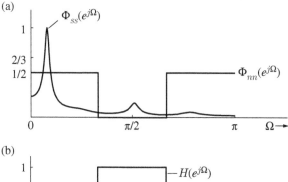

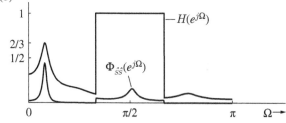

**Figure 12.4** Wiener filter: principle of operation. (a) Power spectral densities of the target signal and of the noise signal. (b) Resulting frequency response and the power spectral density of the output signal.

For the additive noise model the power spectral densities in (12.6) are real-valued and even symmetric. Therefore, the frequency response and the impulse response are also real-valued and even symmetric. Since the convolution with an even symmetric impulse response does not alter the phase spectrum of the input signal, the non-causal IIR Wiener filter is a "zero-phase" filter, an approximation of which is easily implemented in the discrete Fourier domain.

## 12.2.2  The FIR Wiener Filter

We now restrict the impulse response $h(k)$ of the optimal filter to be of finite and even order $N$, and to be causal, i.e.,

$$h(k) = \begin{cases} \text{arbitrary,} & 0 \le k \le N, \\ 0, & \text{otherwise.} \end{cases} \tag{12.8}$$

To obtain a linear-phase solution for the additive noise model, we use a delayed reference signal $d(k) = s(k - N/2)$ in the derivation of the optimal filter in Figure 12.3. From (12.4) we find immediately a set of $N + 1$ equations

$$\sum_{\kappa=0}^{N} h(\kappa) \varphi_{yy}(i - \kappa) = \varphi_{ss}(i - N/2), \quad i = 0, \ldots, N, \tag{12.9}$$

which can be stacked in vector/matrix notation as

$$\begin{pmatrix} \varphi_{yy}(0) & \varphi_{yy}(1) & \cdots & \varphi_{yy}(N) \\ \varphi_{yy}(1) & \varphi_{yy}(0) & \cdots & \varphi_{yy}(N-1) \\ \vdots & \vdots & \ddots & \vdots \\ \varphi_{yy}(N) & \varphi_{yy}(N-1) & \cdots & \varphi_{yy}(0) \end{pmatrix} \begin{pmatrix} h(0) \\ h(1) \\ \vdots \\ h(N) \end{pmatrix} = \begin{pmatrix} \varphi_{ss}(-N/2) \\ \varphi_{ss}(-N/2+1) \\ \vdots \\ \varphi_{ss}(N/2) \end{pmatrix}, \tag{12.10}$$

where we used $\varphi_{yy}(\lambda) = \varphi_{yy}(-\lambda)$. Equation (12.10) may be written as

$$\mathbf{R}_{yy}\,\mathbf{h} = \boldsymbol{\varphi}_{ss}, \tag{12.11}$$

where

$$\mathbf{R}_{yy} = \begin{pmatrix} \varphi_{yy}(0) & \varphi_{yy}(1) & \cdots & \varphi_{yy}(N) \\ \varphi_{yy}(1) & \varphi_{yy}(0) & \cdots & \varphi_{yy}(N-1) \\ \vdots & \vdots & \ddots & \vdots \\ \varphi_{yy}(N) & \varphi_{yy}(N-1) & \cdots & \varphi_{yy}(0) \end{pmatrix} \tag{12.12}$$

is the auto-correlation matrix and

$$\mathbf{h} = \begin{pmatrix} h(0) \\ h(1) \\ \vdots \\ h(N) \end{pmatrix} \quad \text{and} \quad \boldsymbol{\varphi}_{ss} = \begin{pmatrix} \varphi_{ss}(-N/2) \\ \varphi_{ss}(-N/2+1) \\ \vdots \\ \varphi_{ss}(N/2) \end{pmatrix} \tag{12.13}$$

are the coefficient vector and the auto-correlation vector, respectively. If $\mathbf{R}_{yy}$ is invertible, the coefficient vector is given by

$$\mathbf{h} = \mathbf{R}_{yy}^{-1}\,\boldsymbol{\varphi}_{ss}. \tag{12.14}$$

Since $\mathbf{R}_{yy}$ is a symmetric Toeplitz matrix, the system of equations in (12.11) can be efficiently solved using the Levinson–Durbin recursion. Also, we note that the correlation vector $\boldsymbol{\varphi}_{ss}$ is symmetric, and therefore the vector of filter coefficients,

$$h(N/2 - i) = h(N/2 + i), \qquad i \in \{-N/2, \dots, N/2\}, \tag{12.15}$$

is symmetric as well with respect to $i = N/2$. When the reference signal is delayed by $N/2$ samples, the solution to the optimal filter problem is a linear-phase FIR filter. Of course, the filter can also be optimized without a delay in the reference signal. In general, the resulting filter is still causal but does not have the linear-phase property in this case.

To account for the time-varying statistics of speech and noise signals the application of the Wiener filter to speech processing requires the use of short-term correlation functions. This can be done either by using a block processing approach or by approximating the Wiener filter with, for instance, a stochastic gradient algorithm such as the *normalized least-mean-square* (NLMS) algorithm [Haykin 1996]. In both cases the resulting filter will be time variant.

## 12.3 Speech Enhancement in the DFT Domain

The IIR Wiener filter is based on time domain linear filtering of stationary signals. However, the IIR solution to the optimization problem has led us quite naturally into the frequency domain. For stationary signals, approximations to the IIR Wiener filter may easily be realized in the frequency domain by using either Fourier transform or filter bank techniques. When the input signal is non-stationary, however, the filter coefficients must be continuously adapted. The filter will then be linear only on short, quasi-stationary segments of the signal.

Before we investigate these implementation aspects in detail, we would like to approach the noise reduction problem from a different point of view and treat it directly as an estimation problem of short, quasi-stationary speech segments (or speech frames) in the discrete Fourier transform (DFT) domain. We compute DFT frames using a perfect reconstruction analysis–synthesis system (see Section 3.7) as shown in Figure 12.5. The frame-based processing approach segments the incoming noisy signal into short frames of typically 5–30 ms duration. Each of these frames is transformed into the DFT domain, enhanced, inverse DFT transformed, and added to the previously processed signal with some overlap to smooth out discontinuities at the frame boundaries. If we assume that our analysis–synthesis system is sufficiently accurate when the modified frames are overlap-added in the construction of the enhanced signal [Griffin, Lim 1984], we may focus on the enhancement of a *single* quasi-stationary frame of speech.

To this end, we define signal vectors of size $M$ for the noisy speech $y(k)$, the clean speech $s(k)$, and the noise $n(k)$,

$$\mathbf{y}(k) = (y(k - M + 1), y(k - M + 2), \dots, y(k))^T, \tag{12.16}$$

$$\mathbf{s}(k) = (s(k - M + 1), s(k - M + 2), \dots, s(k))^T, \tag{12.17}$$

$$\mathbf{n}(k) = (n(k - M + 1), n(k - M + 2), \dots, n(k))^T. \tag{12.18}$$

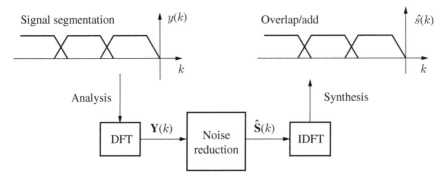

**Figure 12.5** DFT domain implementation of the noise reduction filter.

When $k$ is the index of the current sample, each of these vectors holds a frame of the $M$ most recent signal samples. Prior to computing the DFT, these vectors are weighted with an analysis window $\mathbf{w} = (w(0), w(1), \dots, w(M-1))^T$,

$$\mathbf{Y}(k) = \text{DFT}\{\mathbf{w} \odot \mathbf{y}(k)\}, \tag{12.19}$$

$$\mathbf{S}(k) = \text{DFT}\{\mathbf{w} \odot \mathbf{s}(k)\}, \tag{12.20}$$

$$\mathbf{N}(k) = \text{DFT}\{\mathbf{w} \odot \mathbf{n}(k)\}, \tag{12.21}$$

where $\odot$ denotes an element-by-element multiplication of two vectors or matrices. The DFT domain vectors at sample index $k$ are written in terms of their frequency components as

$$\mathbf{Y}(k) = (Y_0(k), \dots, Y_\mu(k), \dots, Y_{M-1}(k))^T, \tag{12.22}$$

$$\mathbf{S}(k) = (S_0(k), \dots, S_\mu(k), \dots, S_{M-1}(k))^T, \tag{12.23}$$

$$\mathbf{N}(k) = (N_0(k), \dots, N_\mu(k), \dots, N_{M-1}(k))^T. \tag{12.24}$$

### 12.3.1 The Wiener Filter Revisited

The correspondence between time domain convolution and Fourier domain multiplication suggests the definition of the DFT of the enhanced signal frame $\hat{\mathbf{S}}(k) = (\hat{S}_0(k), \hat{S}_1(k), \dots, \hat{S}_{M-1}(k))^T$ as the result of an elementwise multiplication

$$\hat{\mathbf{S}}(k) = \mathbf{H}(k) \odot \mathbf{Y}(k), \tag{12.25}$$

of a weight vector

$$\mathbf{H}(k) = (H_0(k), H_1(k), \dots, H_{M-1}(k))^T, \tag{12.26}$$

and the DFT vector $\mathbf{Y}(k)$. Assuming that the DFT coefficients are (asymptotically) independent, we may minimize the mean square error

$$\text{E}\left\{\left|S_\mu(k) - \hat{S}_\mu(k)\right|^2\right\}, \tag{12.27}$$

independently for each frequency bin $\mu$. The partial derivative of

$$\text{E}\left\{|S_\mu(k) - \hat{S}_\mu(k)|^2\right\} = \text{E}\left\{\left(S_\mu(k) - H_\mu(k)Y_\mu(k)\right)\left(S_\mu(k) - H_\mu(k)Y_\mu(k)\right)^*\right\},$$

with respect to the real part of $H_\mu(k)$ yields the condition

$$\frac{\partial \mathrm{E}\left\{\left|S_\mu(k) - \hat{S}_\mu(k)\right|^2\right\}}{\partial \mathrm{Re}\left\{H_\mu(k)\right\}} = 0 \tag{12.28}$$

and hence, with $\mathrm{E}\left\{S_\mu(k)Y_\mu^*(k)\right\} = \mathrm{E}\left\{S_\mu^*(k)Y_\mu(k)\right\} = \mathrm{E}\left\{|S_\mu(k)|^2\right\}$,

$$\mathrm{Re}\{H_\mu(k)\} = \frac{\mathrm{E}\left\{|S_\mu(k)|^2\right\}}{\mathrm{E}\left\{|Y_\mu(k)|^2\right\}} = \frac{\mathrm{E}\left\{|S_\mu(k)|^2\right\}}{\mathrm{E}\left\{|S_\mu(k)|^2\right\} + \mathrm{E}\left\{|N_\mu(k)|^2\right\}}. \tag{12.29}$$

For the imaginary part we obtain from

$$\frac{\partial \mathrm{E}\left\{|S_\mu(k) - \hat{S}_\mu(k)|^2\right\}}{\partial \mathrm{Im}\left\{H_\mu(k)\right\}} = 0, \tag{12.30}$$

the result

$$\mathrm{Im}\left\{H_\mu(k)\right\} = 0. \tag{12.31}$$

Using the abbreviations $\sigma_{S,\mu}^2(k) = \mathrm{E}\left\{|S_\mu(k)|^2\right\}$ and $\sigma_{N,\mu}^2(k) = \mathrm{E}\left\{|N_\mu(k)|^2\right\}$ we therefore obtain

$$H_\mu(k) = \frac{\mathrm{E}\left\{|S_\mu(k)|^2\right\}}{\mathrm{E}\left\{|S_\mu(k)|^2\right\} + \mathrm{E}\left\{|N_\mu(k)|^2\right\}} = \frac{\sigma_{S,\mu}^2(k)}{\sigma_{S,\mu}^2(k) + \sigma_{N,\mu}^2(k)}, \tag{12.32}$$

as a real valued weight, which does not modify the phase of the noisy coefficients $Y_\mu(k)$. $H_\mu(k)$ is defined in terms of statistical expectations as a function of time and is therefore in general time varying. There are numerous ways (non-parametric or model based) to compute approximations to these expectations. In general, some averaging (over time and/ or frequency) is necessary. For a single microphone system, the most prominent estimation procedure is based on time-averaged modified periodograms [Welch 1967]. In multi-microphone systems, we might also smooth over microphone channels and thus reduce the amount of smoothing over time.

It is quite instructive to juxtapose this result to the Wiener filter in (12.6). For stationary signals, $\mathrm{E}\left\{|S_\mu(k)|^2\right\}$ and $\mathrm{E}\left\{|N_\mu(k)|^2\right\}$ are estimates of the power spectra $\Phi_{ss}(e^{j\Omega_\mu})$ and $\Phi_{nn}(e^{j\Omega_\mu})$. For a properly normalized window $\mathbf{w}$, these estimates are asymptotically ($M \to \infty$) unbiased. On the one hand, (12.32) is therefore closely related to the Wiener filter. On the other hand, it is conceptually quite different from the Wiener filter as we did not start from a *time domain linear* filtering problem. While a frequency domain approximation of (12.6) implies the implementation of a linear filter, e.g., by means of a fast convolution, the estimation of DFT coefficients as in (12.25) and (12.32) leads to a cyclic convolution in the time domain. However, there is no principal reason why the latter should not work and the cyclic effects known from linear theory must not be interpreted as a disturbance in case of optimal frequency domain estimators. However, we should use a perfect reconstruction overlap-add scheme, tapered analysis and synthesis windows, and sufficient overlap between frames to control estimation errors at the frame edges [Griffin, Lim 1984].

In Figure 12.6, we illustrate the principle of the DFT-based noise reduction algorithm. The magnitude squared DFT coefficients of a noisy, voiced speech sound, and the estimated

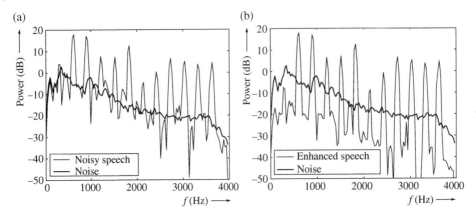

**Figure 12.6** Principle of DFT-based noise reduction. (a) Short-time spectrum of the noisy signal and the estimated noise PS. (b) Short-time spectrum of the enhanced signal and the estimated noise PS.

noise *power spectrum* (PS) are shown in the left hand plot. The right hand plot depicts the magnitude squared DFT coefficients of the enhanced speech sound, as well as the estimated noise PS. We observe that the harmonic (high SNR) peaks of the speech sound are well reproduced, while the valleys in between these sounds, where the noise is predominant, are attenuated. As a result the global SNR of the speech sound is improved.

Despite the apparently straightforward solution to the estimation problem, one critical question remains: (12.6) as well as (12.32) require knowledge of the PS of the clean speech or of the SNR of the noisy signal at a given frequency $\Omega_\mu$. However, neither the clean speech PS nor the SNR are readily available. To arrive at more practical solutions, we discuss another (however closely related) approach to noise reduction, which is known as *spectral subtraction*.

### 12.3.2 Spectral Subtraction

The basic idea of spectral subtraction [Boll 1979] and related proposals [Berouti et al. 1979], [Preuss 1979] is to subtract an estimate of the noise floor from an estimate of the spectrum of the noisy signal. Since the speech signal is non-stationary, this has to be done on a short-time basis, preferably using a DFT-based analysis–synthesis system.

Using the frame-wise processing approach of Section 12.3 and an appropriately normalized analysis window $\mathbf{w}$, the power of the clean speech signal at the discrete frequencies $\Omega_\mu = \frac{\mu 2\pi}{M}$ is given by

$$
\begin{aligned}
\mathrm{E}\left\{|S_\mu(k)|^2\right\} &= \mathrm{E}\left\{|Y_\mu(k)|^2\right\} - \mathrm{E}\left\{|N_\mu(k)|^2\right\} \\
&= \mathrm{E}\left\{|Y_\mu(k)|^2\right\} \left[1 - \frac{\mathrm{E}\left\{|N_\mu(k)|^2\right\}}{\mathrm{E}\left\{|Y_\mu(k)|^2\right\}}\right] \\
&= \mathrm{E}\left\{|Y_\mu(k)|^2\right\} |\tilde{H}_\mu(k)|^2 \ .
\end{aligned}
\tag{12.33}
$$

Since a sliding window DFT may be interpreted as an analysis filter bank (see Section 4.1.6), $\mathrm{E}\left\{|S_\mu(k)|^2\right\}$ represents the power of a complex-valued subband signal $S_\mu(k)$ for any fixed $\mu$.

In analogy to the relation of input and output power in linear systems [Papoulis, Unnikrishna Pillai 2001], the spectral subtraction method may be interpreted as a time-variant filter with magnitude frequency response

$$\tilde{H}_\mu(k) = \sqrt{1 - \frac{\mathrm{E}\left\{|N_\mu(k)|^2\right\}}{\mathrm{E}\left\{|Y_\mu(k)|^2\right\}}} = \sqrt{\frac{\mathrm{E}\left\{|S_\mu(k)|^2\right\}}{\mathrm{E}\left\{|Y_\mu(k)|^2\right\}}} = \sqrt{H_\mu(k)} \;, \tag{12.34}$$

which is the square root of (12.32). Since we are subtracting in the PS domain, this approach is called *power subtraction*. This approach has been further generalized using a variety of nonlinear transformations [Vary, Martin 2006].

Thus, the formulation of the subtraction approach in terms of a multiplicative spectral gain function not only unifies the various noise reduction approaches, but is also helpful for implementing these algorithms since $0 \leq |\tilde{H}_\mu(k)| \leq 1$ is a normalized quantity.

When the spectral subtraction in (12.33) is used in conjunction with the Wiener filter (12.32), we obtain

$$H_\mu(k) = \frac{\mathrm{E}\left\{|Y_\mu(k)|^2\right\} - \mathrm{E}\left\{|N_\mu(k)|^2\right\}}{\mathrm{E}\left\{|Y_\mu(k)|^2\right\}} = 1 - \frac{\mathrm{E}\left\{|N_\mu(k)|^2\right\}}{\mathrm{E}\left\{|Y_\mu(k)|^2\right\}}. \tag{12.35}$$

The PS of the noise, which is a necessary ingredient of these spectral weighting rules, can be estimated using the *minimum statistics* approach [Martin 2001], soft-decision and MMSE estimators [Sohn, Sung 1998], [Gerkmann, Hendriks 2012], or using a *voice activity detector* (VAD) [McAulay, Malpass 1980].

The "pure" spectral subtraction approach is not very robust with respect to estimation errors in the spectral gain function, especially in frequency bins containing noise only. Practical implementations therefore suffer from fluctuating residual noise, which is commonly known as "musical noise."

### 12.3.3 Estimation of the A Priori SNR

In Section 12.3.2, we have seen that the spectral subtraction technique can be used on its own or in conjunction with the Wiener filter. In the latter case, it supplies an estimate for the unknown power spectrum of the clean speech. Another approach to the realization of the Wiener filter [Scalart, Vieira Filho 1996] is to express it in terms of the a priori SNR

$$\eta_\mu(k) = \frac{\mathrm{E}\left\{|S_\mu(k)|^2\right\}}{\mathrm{E}\left\{|N_\mu(k)|^2\right\}} = \frac{\sigma_{S,\mu}^2(k)}{\sigma_{N,\mu}^2(k)}, \tag{12.36}$$

and to estimate the a priori SNR instead of the clean speech PS. Since for most noise reduction applications the a priori SNR ranges over an interval of $-20$ to $30\,\mathrm{dB}$ and is invariant with respect to the signal scaling, it is easier to deal with the SNR in (e.g., fixed point) implementations than using power spectra. Another reason for using the a priori SNR lies in the availability of a simple yet powerful estimation algorithm, which will be discussed below.

The estimation of the a priori SNR is based on the a posteriori SNR $\gamma_\mu(k)$, which is defined as the ratio of the periodogram of the noisy signal and the noise power

$$\gamma_\mu(k) = \frac{|Y_\mu(k)|^2}{\sigma_{N,\mu}^2(k)}. \tag{12.37}$$

If an estimate of the noise PS is available, the a posteriori SNR is easily measurable. The a priori SNR can now be expressed as

$$\eta_\mu(k) = \frac{\sigma^2_{S,\mu}(k)}{\sigma^2_{N,\mu}(k)} = \mathrm{E}\left\{\gamma_\mu(k) - 1\right\}, \tag{12.38}$$

where the "decision-directed" approach [Ephraim, Malah 1984] as shown in the following section may be used to estimate $\eta_\mu(k)$. To apply this approach to the estimation of the a priori SNR at time $k + r$, an estimate $|\widehat{S_\mu(k)}|$ of the clean speech amplitudes $|S_\mu(k)|$ at time $k$ must be available. Furthermore, we assume that $|S_\mu(k + r)| \approx |S_\mu(k)|$, which holds for quasi-stationary speech sounds, but is less valid for transient sounds.

### 12.3.3.1  Decision-Directed Approach

The estimated a priori SNR $\hat{\eta}_\mu(k + r)$ of the frame at $k + r$ is computed as a linear combination of a SNR estimate based on the signal frame at time $k$ and the instantaneous realization $\gamma_\mu(k + r) - 1$ of (12.38),

$$\hat{\eta}_\mu(k + r) = \alpha_\eta \frac{|\widehat{S_\mu(k)}|^2}{\sigma^2_{N,\mu}(k)} + (1 - \alpha_\eta) \max\left(\gamma_\mu(k + r) - 1, 0\right), \tag{12.39}$$

where $\alpha_\eta$ depends on the frame rate and is typically in the range $0.9 \le \alpha_\eta \le 0.99$. We note that the instantaneous estimate is identical to the ML estimate $\hat{\eta}_{\mu,\mathrm{ML}}(k + r) = \gamma_\mu(k + r) - 1$ under a complex Gaussian model. To show this, we note that magnitude squared DFT coefficients are exponentially distributed (see (5.137)). Then, we may normalize $R^2_\mu = |Y_\mu(k)|^2$ on $\sigma^2_{N,\mu}(k)$ and use (5.137) to obtain the PDF of the a posteriori SNR for $\gamma_\mu(k) \ge 0$ as

$$p_{\gamma_\mu}(u \mid \eta_\mu(k)) = \frac{\sigma^2_{N,\mu}(k)}{\sigma^2_{S,\mu}(k) + \sigma^2_{N,\mu}(k)} \exp\left(-\frac{u\,\sigma^2_{N,\mu}(k)}{\sigma^2_{S,\mu}(k) + \sigma^2_{N,\mu}(k)}\right)$$

$$= \frac{1}{1 + \eta_\mu(k)} \exp\left(-\frac{u}{1 + \eta_\mu(k)}\right), \quad u \ge 0, \tag{12.40}$$

which is parametrized by the a priori SNR $\eta_\mu(k)$. The maximization of this PDF w.r.t. $\eta_\mu(k)$ then leads to the above ML estimate $\hat{\eta}_{\mu,\mathrm{ML}}(k + r) = \gamma_\mu(k + r) - 1$.

It has frequently been argued [Cappé 1994], [Scalart, Vieira Filho 1996] that the decision-directed estimation procedure contributes considerably to the subjective quality of the enhanced speech, especially to the reduction of "musical noise." This has been further analyzed in Breithaupt, Martin [2011].

### 12.3.3.2  Smoothing in the Cepstrum Domain

Standard measures to avoid musical noise are based on smoothing estimated quantities and on limiting the amount of noise reduction such that single spurious peaks are less likely to occur. Further improvements to the SNR estimation procedure can be obtained with an adaptive smoothing process across frequency bins. This process is implemented in the cepstrum domain and hence requires two additional spectral transformations. Smoothing in the cepstrum domain has been shown to be very effective to suppress fast random fluctuations. We therefore briefly summarize this method [Breithaupt et al. 2007], [Breithaupt et al. 2008a], [Breithaupt, Martin 2008].

Starting with an ML estimate of the speech PS

$$\hat{\sigma}^2_{S,\mu,\text{ML}}(k) = \hat{\eta}_{\mu,\text{ML}}(k) \cdot \text{E}\left\{|N_\mu(k)|^2\right\} \tag{12.41}$$

we obtain its (real-valued) cepstrum as

$$c_s(q,k) = \sum_{\mu=0}^{M-1} \ln\left(\hat{\sigma}^2_{S,\mu,\text{ML}}(k)\right) e^{j2\pi\mu q/M}.$$

with $q = 0, \ldots, M-1$ denoting the cepstral bin index and $\ln(\cdot)$ denoting the natural logarithm. Now, we apply a first-order smoothing recursion to each cepstral bin

$$\bar{c}_s(q,k) = \alpha_c(q,k)\bar{c}_s(q,k-1) + \left(1 - \alpha_c(q,k)\right) c_s(q,k), \tag{12.42}$$

where we adjust the smoothing parameter $\alpha_c(q,k)$ such that little smoothing is applied to the coarse spectral features and more smoothing is applied to the fine and rapidly fluctuating spectral details. In the cepstrum domain this can be easily achieved as the coarse spectral structure is mapped onto bins with low $q$ indices and the bins that represent the fundamental frequency. The latter bins are found via a peak-picking procedure in the cepstrum domain. After an additional bias correction $c_B$, which is related to Euler's constant [Ephraim, Rahim 1998], [Gerkmann, Martin 2009], we obtain the smoothed speech PS as

$$\hat{\sigma}^2_{S,\mu,\text{CEPS}}(k) = \exp\left(\sum_{\mu=0}^{M-1} \bar{c}_s(q,k)e^{-j2\pi\mu q/M} + c_B\right) \tag{12.43}$$

which may then be used to compute the a priori SNR. We thus achieve a reduction of the variance of the residual noise with negligible impact on the speech signal as well as a preservation of the harmonic spectral structure of voiced speech. An advantage of this procedure is that weak harmonics are restored. This idea has been further explored in Elshamy et al. [2018].

### 12.3.4 Quality and Intelligibility Evaluation

The most common methods for the assessment of the enhanced signal are either based on listening experiments or on instrumental measures. In both cases we are interested to evaluate

- the distortions and the intelligibility of the enhanced speech signal,
- the overall reduction of the noise power level, and
- the naturalness of the residual noise signal including "musical noise" and other processing artifacts.

Prior to investing into extensive listening experiments, it is common practice to first evaluate these criteria in terms of instrumental measures, the most prominent of which are

- the SNR and frequency-weighted segmental SNR improvement [Tribolet et al. 1978],
- the perceptual evaluation of speech quality (PESQ) [ITU-T Rec. P.862 2001] and the more recent perceptual objective listening quality (POLQA) prediction [ITU-T Rec. P.863 2018] resulting in an estimated *mean opinion score* (MOS),

- the short-time objective intelligibility (STOI) measure [Taal et al. 2011] and its variants and extensions, for instance using information-theoretic principles [Taghia et al. 2012], [Taghia, Martin 2014], with improved robustness (ESTOI) [Jensen, Taal 2016], or for binaural signals [Andersen et al. 2016],
- composite measures [Tribolet et al. 1978] such as COVL [Hu, Loizou 2008] predicting ITU-T P.835 overall quality scores [ITU-T Rec. P.835 2003], and corresponding measures CSIG and CBAK for predicting the speech signal distortion and the intrusiveness of background noise,
- and measures specifically designed for source separation applications such as the signal-to-interference ratio (SIR), the signal-to-artifacts ratio (SAR) [Vincent et al. 2006], and the scale-invariant signal-to-distortion ratio (SI-SDR) [Le Roux et al. 2018].

Some measures have also been extended to model speech reception of hearing-impaired listeners, see e.g., [Falk et al. 2015].

Once the benefits of a processing method have been demonstrated, formal listening experiments may be performed. Often, these tests aim at the computation of a MOS as described in the ITU-T recommendation P.800 [ITU-T Rec. P.800 1996], which defines different rating procedures. In the case of an absolute category rating (ACR) test listeners will assess the signal's quality in terms of five categories "excellent," "good," "fair," "poor," and "bad." Alternatively, also comparison category rating (CCR) tests may be used. Here, the assessment is relative to a high-quality reference signal in terms of the categories "much better," "better," "slightly better," "about the same," "slightly worse," "worse," and "much worse." The aggregated test data then results in a (C)MOS-*listening quality subjective* (LQS) score. Furthermore, factors influencing results should be identified during the design of the listening experiment and differences in the MOS scores should be analyzed in order to assess their statistical significance. Among other factors, test results may depend on

- selection of listeners (age, hearing abilities and experience, cultural differences) and the test instructions,
- test materials (language, speaker characteristics), and the
- method of sound reproduction including reproduction sound pressure level.

In the context of traditional methods and applications where a full removal of background noise is not possible or desired, we are also concerned about the naturalness of the background noise. Here, the term "musical noise" describes the randomly fluctuating noise artifacts, which are frequently observed in noise-reduced signals. It is especially noticeable for the simple spectral weighting methods as outlined in Section 12.3.2. The musical noise phenomenon can be explained by estimation errors in the frequency domain, which lead to spurious peaks in the spectral representation of the enhanced signal. When the enhanced signal is reconstructed via IFFT and overlap-add, these peaks correspond to a sinusoidal excitation whose frequency varies randomly from frame to frame.

Figure 12.7 depicts the time evolution of the short-term magnitude spectrum of a clean (a), a noisy (b), and the corresponding enhanced signal (c) [Vary 1985]. The speech signal is disturbed by white Gaussian noise and three stationary harmonic tones at 1, 2, and 3 kHz. The enhanced spectrum is computed using spectral subtraction. Clearly, the stationary tones can be completely removed. Also, the level of the white noise is significantly reduced. However, some of the low-power speech components are lost and random spectral peaks, especially during the speech pause, remain.

(a)

(b)

(c)

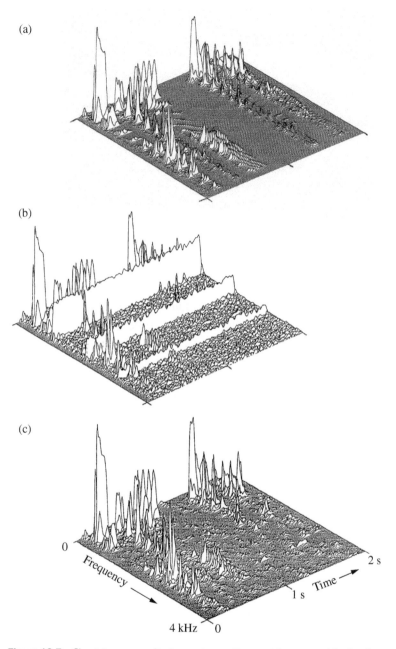

0

Frequency →

4 kHz     0

2 s

1 s     Time →

**Figure 12.7** Short-term magnitude spectra vs. time and frequency (a) of a clean speech signal, (b) of the clean signal with additive white noise and harmonic tones, and (c) of the enhanced signal using spectral subtraction.

A statistical evaluation of these estimation errors is given in [Vary 1985]. Quite a substantial effort in speech enhancement research has been spent on techniques, which avoid the "musical noise" phenomenon. We will briefly describe commonly used countermeasures.

### 12.3.4.1 Noise Oversubtraction

The subtraction of the average PS of the noise from the time-varying magnitude squared DFT coefficients of the noisy signal leads to residual spectral peaks, which in turn result in the "musical noise" phenomenon. The musical noise is especially disturbing during speech pauses. The noise power overestimation increases the estimated noise PS in order to reduce the amplitude of these random spectral peaks after subtraction [Berouti et al. 1979]. Furthermore, a "spectral floor" is applied to the resultant spectral coefficients, preventing them from falling below a preset minimum level. Hence, we obtain spectral components which fluctuate less during speech pauses. In practice, the estimated noise power level is increased by a factor of $1 \leq O_S \leq 2$, or a maximum of 3 dB, see [Vary, Martin 2006] for more details.

### 12.3.4.2 Spectral Floor

The spectral floor imposes a lower limit $\beta \, E \left\{ |N_\mu(k)|^2 \right\}$ on the magnitude squared DFT coefficients of the enhanced signal. For the spectral gain functions discussed so far, $\beta$ cannot be much smaller than 0.1 without giving rise to noticeable "musical noise." On the other hand, it cannot be much larger without rendering the noise reduction algorithm useless.

### 12.3.4.3 Limitation of the A Priori SNR

[Cappé 1994] and [Malah et al. 1999] proposed to apply a lower limit to the estimated a priori SNR in order to reduce the annoying musical tones. The lower limit has the greatest effect on frequency bins, which do *not* contain speech. Thus, it is of great importance for the perceived quality during speech pauses but also for the overall shape of the speech spectrum during speech activity. The limit on the a priori SNR limits the maximum attenuation. This can be easily seen when using the Wiener filter but also holds for the more complicated non-linear estimators [Martin et al. 2000]. For the Wiener filter in (12.6) and low SNR conditions, we obtain

$$H(e^{j\Omega}) = \frac{\sigma_{S,\mu}^2(k)/\sigma_{N,\mu}^2(k)}{1 + \sigma_{S,\mu}^2(k)/\sigma_{N,\mu}^2(k)} \approx \frac{\sigma_{S,\mu}^2(k)}{\sigma_{N,\mu}^2(k)}. \tag{12.44}$$

Thus, by applying a lower bound to the a priori SNR $\eta_\mu(k)$, the maximum attenuation of the filter is limited.

### 12.3.4.4 Adaptive Smoothing of the Spectral Gain

Standard measures to avoid musical noise are based on smoothing estimated quantities and on limiting the amount of noise reduction such that single spurious peaks are less likely to occur. Smoothing in the cepstrum domain has been shown to be very effective to suppress fast random fluctuations [Breithaupt et al. 2007]. This cepstrum smoothing method is in fact similar to the SNR estimation approach as outlined in Section 12.3.3, and it does not introduce additional algorithmic latency. However, since gain functions are confined to the range [0, 1] their dynamic range is much smaller than that of the a priori SNR. It is therefore not necessary to use a logarithmic compression in the smoothing process.

The transformation to the cepstral domain can be entirely avoided if the smoothing process is implemented directly in the spectral domain. In Esch, Vary [2009], the authors propose a moving average across the frequency bins of the gain function where the width of the smoothing window is dependent on the frame SNR. While this method is computationally more efficient, it does not provide smoothing in high SNR regions during speech activity.

### 12.3.5  Spectral Analysis/Synthesis for Speech Enhancement

The noise reduction methods presented so far were developed for spectral domain processing. Besides the DFT, other spectral transforms as well as filter banks are in principle suited for this task. Since spectral transforms and filter banks are discussed in detail in Chapters 3 and 4, we summarize here issues which are specifically related to noise reduction.

One of the most important questions to be answered concerns the spectral resolution which should be provided by the spectral transform. The design can be guided by the characteristics of the source signal and/or the properties of the receiver, the human ear, or a speech recognizer. While the former approach suggests a high and uniform resolution up to 3–4 kHz in order to separate the distinct peaks of voiced speech harmonics and to attenuate the noise in between the peaks, the latter suggests less resolution, especially at higher frequencies. Both approaches have been advocated in the literature: mostly DFT-based systems when a uniform resolution is desired (see Section 3.7) and filter banks (see Chapter 4) for non-uniform resolution.

Thus, filter bank implementations are attractive as they can be adapted to the spectral and the temporal resolution of the human ear. Also, there are many applications, such as hearing aids, where the computational complexity is of utmost importance. Filter banks with non-uniform spectral resolution allow a reduction in the number of channels without sacrificing the quality of the enhanced signal. In principle, the filter bank can be realized by discrete filters or by polyphase filter banks with uniform or non-uniform spectral resolution [Doblinger 1991], [Doblinger, Zeitlhofer 1996], [Kappelan et al. 1996], [Gülzow, Engelsberg 1998], [Gülzow et al. 2003], [Vary 2005]. Chapter 4 discusses typical realizations in detail.

## 12.4  Optimal Non-linear Estimators

We return to the issue of how to estimate the clean speech DFT coefficients given the noisy coefficients. In contrast to Section 12.2, we now aim at developing optimal *non-linear* estimators. Unlike the optimal linear estimators, these estimators require knowledge of the probability density functions of the speech and noise DFT coefficients, which may be acquired from training data. Similar to the exposition in Section 12.3, we do not assume stationary signals but develop estimators for a single, quasi-stationary frame of speech. To enhance the readability of the derivations, we will drop the (temporal) frame index $\lambda$ and denote all quantities as functions of the frequency bin index $\mu$ only. Thus, for the signal frame under consideration, the DFT coefficients of the clean speech and the noisy signal will be denoted by

$$S_\mu = A_\mu e^{j\alpha_\mu} \quad \text{and} \quad Y_\mu = R_\mu e^{j\theta_\mu}, \tag{12.45}$$

respectively.

Obviously, an ideal non-linear noise reduction filter could be defined using a *complex-valued* spectral gain function

$$\mathcal{H}_\mu = \frac{S_\mu}{Y_\mu} = \frac{S_\mu Y_\mu^*}{|Y_\mu|^2}, \tag{12.46}$$

or an ideal *amplitude* gain

$$\mathcal{H}_\mu^A = \frac{|S_\mu|}{|Y_\mu|}. \tag{12.47}$$

In the context of deep neural network (DNN)-based speech enhancement systems these filters are also known as the *complex ideal ratio mask* and the *spectral magnitude mask* [Wang et al. 2014], [Williamson et al. 2016], respectively. Then, given estimates $\hat{\mathcal{H}}_\mu$ and $\hat{\mathcal{H}}_\mu^A$ of these filters, we may estimate the spectrum of the enhanced signal as

$$\hat{S}_\mu = \hat{\mathcal{H}}_\mu Y_\mu \quad \text{and} \quad \hat{S}_\mu = \hat{\mathcal{H}}_\mu^A Y_\mu. \tag{12.48}$$

These gain functions constitute instantaneous, non-linear estimators, the implementation of which requires additional prior knowledge as $S_\mu$ or $|S_\mu|$ is not known. While such estimators may be implemented on the basis of traditional low-dimensional statistical models, impressive progress has been achieved recently in conjunction with DNNs. Nevertheless, we start out with approximations, which are derived on the basis of low-dimensional statistical models, as these do not require extensive training data.

To this end, we write the power of the $\mu$-th spectral component as

$$\begin{aligned}
\sigma_{S,\mu}^2 &= \mathrm{E}\left\{|S_\mu|^2\right\} = \mathrm{E}\left\{A_\mu^2\right\}, \\
\sigma_{N,\mu}^2 &= \mathrm{E}\left\{|N_\mu|^2\right\}, \\
\sigma_{Y,\mu}^2 &= \mathrm{E}\left\{|Y_\mu|^2\right\} = \mathrm{E}\left\{R_\mu^2\right\} = \sigma_{S,\mu}^2 + \sigma_{N,\mu}^2.
\end{aligned} \tag{12.49}$$

Most of the following derivations employ the complex Gaussian model for the PDF of the DFT coefficients of speech and noise signals, as outlined in Section 5.10. This model is valid when the DFT length is significantly larger than the span of correlation in the signal. However, we will also discuss solutions based on super-Gaussian distributions where appropriate.

Although we will primarily discuss estimators, which are optimal with respect to the MMSE, we will begin with ML estimation. We show that for the complex Gaussian model the maximum likelihood estimate is closely related to the spectral subtraction technique.

### 12.4.1 Maximum Likelihood Estimation

The ML estimate of the speech power $\sigma_{S,\mu}^2$ in the $\mu$-th DFT bin maximizes the joint probability density of the observed spectral coefficient $Y_\mu$ with respect to $\sigma_{S,\mu}^2$. The joint probability function conditioned on the unknown parameter $\sigma_{S,\mu}^2$ is given by (5.133), i.e.,

$$p(\mathrm{Re}\{Y_\mu\}, \mathrm{Im}\{Y_\mu\} \mid \sigma_{S,\mu}^2) = \frac{1}{\pi(\sigma_{S,\mu}^2 + \sigma_{N,\mu}^2)} \exp\left(-\frac{|Y_\mu|^2}{\sigma_{S,\mu}^2 + \sigma_{N,\mu}^2}\right). \tag{12.50}$$

If we set the first derivative of (12.50) with respect to $\sigma_{S,\mu}^2$ to zero, we obtain the estimated speech power [McAulay, Malpass 1980],

$$\hat{\sigma}_{S,\mu,\mathrm{ML}}^2 = |Y_\mu|^2 - \sigma_{N,\mu}^2 = R_\mu^2 - \sigma_{N,\mu}^2. \tag{12.51}$$

Since the second derivative is negative this is indeed a maximum. The ML estimator is an unbiased estimator of $\sigma_{S,\mu}^2$ since the expected value of $\hat{\sigma}_{S,\mu,\mathrm{ML}}^2$ is $\mathrm{E}\{\hat{\sigma}_{S,\mu,\mathrm{ML}}^2\} = \sigma_{Y,\mu}^2 - \sigma_{N,\mu}^2 = \sigma_{S,\mu}^2$. In a practical application of this estimator, we must ensure, however, that the estimate of the variance is always non-negative. Using the square-root of the above relation and the phase of the noisy input, an estimate of the spectral coefficient of clean speech is given by

$$\hat{S}_\mu = \sqrt{1 - \frac{\sigma_{N,\mu}^2}{|Y_\mu|^2}} |Y_\mu| e^{j\theta_\mu} = \sqrt{1 - \frac{\sigma_{N,\mu}^2}{|Y_\mu|^2}} Y_\mu = G_{\mu,\mathrm{ML}} Y_\mu, \tag{12.52}$$

where, again, we must make sure that the argument of the square root is non-negative. Note that although we have written the result as a product of the noisy coefficient and a gain function, the estimator is non-linear. The gain function depends on the DFT coefficients of the noisy signal. $|Y_\mu|^2$ in the denominator may be smoothed over frequency or, if the signal is short-time stationary, over time. Using (12.49), Eq. (12.52) is recognized as an approximation to the power subtraction rule (12.34).

Similarly, (12.51) may be used in conjunction with the MMSE filter (12.32). Replacing the unknown speech power by its ML estimate, we obtain

$$\hat{S}_\mu = \frac{|Y_\mu|^2 - \sigma_{N,\mu}^2}{\mathrm{E}\{|Y_\mu|^2\}} Y_\mu = G_{\mu,\mathrm{WML}} Y_\mu, \tag{12.53}$$

where the gain function in (12.53) is, after smoothing the numerator, an approximation to (12.32).

An alternative solution to the ML signal estimation problem can be based on the conditional joint density (5.139) of noisy spectral coefficients, given the spectral amplitude $A_\mu$ [McAulay, Malpass 1980]. In this case, the clean speech spectral amplitude $A_\mu$ is the unknown parameter and the clean speech phase $\alpha_\mu$ is assumed to be uniformly distributed between 0 and $2\pi$. The ML estimate of the spectral amplitude is obtained from (5.139) by averaging over the phase. More specifically, we maximize

$$p(Y_\mu \mid A_\mu) = \int_0^{2\pi} p(Y_\mu \mid A_\mu, \alpha_\mu) p(\alpha_\mu) \, d\alpha_\mu$$

$$= \frac{1}{\pi \sigma_{N,\mu}^2} \exp\left(-\frac{|Y_\mu|^2 + A_\mu^2}{\sigma_{N,\mu}^2}\right) \frac{1}{2\pi} \int_0^{2\pi} \exp\left(\frac{2A_\mu \mathrm{Re}\{Y_\mu e^{-j\alpha_\mu}\}}{\sigma_{N,\mu}^2}\right) d\alpha_\mu,$$

with respect to $A_\mu$. The integral in the above equation is known as the modified Bessel function of the first kind. For $\frac{2A_\mu |Y_\mu|}{\sigma_{N,\mu}^2} \geq 3$ it can be approximated by

$$\frac{1}{2\pi} \int_0^{2\pi} \exp\left(\frac{2A_\mu \mathrm{Re}\{Y_\mu e^{-j\alpha_\mu}\}}{\sigma_{N,\mu}^2}\right) d\alpha_\mu \approx \frac{1}{\sqrt{2\pi \frac{2A_\mu |Y_\mu|}{\sigma_{N,\mu}^2}}} \exp\left(\frac{2A_\mu |Y_\mu|}{\sigma_{N,\mu}^2}\right).$$

Differentiation of

$$p(Y_\mu|A_\mu) \approx \frac{1}{2\pi\sigma_{N,\mu}\sqrt{\pi A_\mu|Y_\mu|}} \exp\left(-\frac{(A_\mu - |Y_\mu|)^2}{\sigma_{N,\mu}^2}\right), \tag{12.54}$$

with respect to $A_\mu$ leads to the approximate ML estimate of the spectral magnitude

$$\hat{A}_\mu = \frac{|Y_\mu|}{2}\left(1 \pm \sqrt{1 - \frac{\sigma_{N,\mu}^2}{R_\mu^2}}\right). \tag{12.55}$$

Retaining the phase of the noisy signal, we obtain for the complex enhanced coefficient

$$\hat{S}_\mu = \left(0.5 + 0.5\sqrt{1 - \frac{\sigma_{N,\mu}^2}{R_\mu^2}}\right)Y_\mu = G_{\mu,\text{MML}}Y_\mu, \tag{12.56}$$

where, again, we have to make sure that the argument of the square root is non-negative. Note that this spectral gain function provides for a maximum of only 6 dB of noise reduction whereas other solutions, for instance (12.52), may provide much larger noise attenuation.

### 12.4.2 Maximum A Posteriori Estimation

The maximum a posteriori estimator finds the clean speech coefficients, which maximize $p(S_\mu \mid Y_\mu)$. It allows explicit modeling of the a priori density of the speech coefficients and noise, since, using Bayes' theorem, we might as well maximize

$$p(S_\mu \mid Y_\mu)p(Y_\mu) = p(Y_\mu \mid S_\mu)p(S_\mu). \tag{12.57}$$

For the Gaussian signal model, it is easily verified that the MAP estimator is identical to the Wiener filter (12.32).

More interesting solutions arise when we estimate the magnitude and phase, i.e., solve

$$p(A_\mu, \alpha_\mu \mid Y_\mu)p(Y_\mu) = p(Y_\mu \mid A_\mu, \alpha_\mu)p(A_\mu, \alpha_\mu)$$
$$= \frac{2A_\mu}{2\pi^2\sigma_{N,\mu}^2\sigma_{S,\mu}^2}\exp\left(-\frac{|Y_\mu - A_\mu e^{j\alpha_\mu}|^2}{\sigma_{N,\mu}^2}\right)\exp\left(-\frac{A_\mu^2}{\sigma_{S,\mu}^2}\right).$$

The MAP estimate of the clean speech phase $\alpha_\mu$ is simply the noisy phase $\theta_\mu$ [Wolfe, Godsill 2001], whereas the MAP estimate of the amplitude $A_\mu$ yields

$$\hat{A}_\mu = \frac{\eta_\mu + \sqrt{\eta_\mu^2 + 2(1 + \eta_\mu)\frac{\eta_\mu}{\gamma_\mu}}}{2(1 + \eta_\mu)}R_\mu. \tag{12.58}$$

MAP estimation procedures have also been developed in the context of super-Gaussian speech models (see Section 5.10.3) [Lotter, Vary 2003], [Lotter, Vary 2005], [Dat et al. 2005].

### 12.4.3 MMSE Estimation

In this section, we will formulate and discuss a number of well known solutions to the MMSE estimation problem.

### 12.4.3.1 MMSE Estimation of Complex Coefficients

We will begin with the MMSE estimation of DFT coefficients under the assumption that the real and the imaginary parts of the DFT coefficients are statistically independent and Gaussian distributed. Then, the MMSE estimate of the complex coefficients separates into two independent estimators for the real and the imaginary parts, i.e.,

$$\hat{S}_\mu = \mathrm{E}\left\{S_\mu \mid Y_\mu\right\}$$

$$= \int_{-\infty}^{\infty} \mathrm{Re}\left\{S_\mu\right\} p\left(\mathrm{Re}\left\{S_\mu\right\} \mid \mathrm{Re}\left\{Y_\mu\right\}\right) \mathrm{dRe}\left\{S_\mu\right\}$$

$$+ j \int_{-\infty}^{\infty} \mathrm{Im}\left\{S_\mu\right\} p\left(\mathrm{Im}\left\{S_\mu\right\} \mid \mathrm{Im}\left\{Y_\mu\right\}\right) \mathrm{dIm}\left\{S_\mu\right\}$$

$$= \mathrm{E}\left\{\mathrm{Re}\{S_\mu\} \mid \mathrm{Re}\{Y_\mu\}\right\} + j\,\mathrm{E}\left\{\mathrm{Im}\{S_\mu\} \mid \mathrm{Im}\{Y_\mu\}\right\}. \tag{12.59}$$

Using a Gaussian density for the clean speech coefficients and a conditional Gaussian density for the noisy speech coefficients given the clean speech coefficients as defined in (5.131) and (5.138), we immediately find the estimate of the real part as in (5.169), i.e.,

$$\mathrm{E}\left\{\mathrm{Re}\{S_\mu\} \mid \mathrm{Re}\{Y_\mu\}\right\} = \frac{\sigma_{S,\mu}^2}{\sigma_{S,\mu}^2 + \sigma_{N,\mu}^2}\,\mathrm{Re}\{Y_\mu\}. \tag{12.60}$$

The combination of (12.60) and the analog solution for the imaginary part yields

$$\mathrm{E}\left\{S_\mu \mid Y_\mu\right\} = \frac{\sigma_{S,\mu}^2}{\sigma_{S,\mu}^2 + \sigma_{N,\mu}^2}\left(\mathrm{Re}\{Y_\mu\} + j\,\mathrm{Im}\{Y_\mu\}\right) = \frac{\sigma_{S,\mu}^2}{\sigma_{S,\mu}^2 + \sigma_{N,\mu}^2}\,Y_\mu. \tag{12.61}$$

For the Gaussian signal model the optimal filter is therefore identical to the Wiener filter and thus a zero-phase filter. The phase of the noisy signal is (except for a possible overall delay of the signal due to spectral analysis) not modified. For super-Gaussian speech models, estimators for the real and the imaginary parts can be developed as well [Martin 2002], [Martin 2005b], [Martin 2005c], [Martin, Breithaupt 2003]. In contrast to the estimators based on Gaussian speech models, they also lead to a modification of the short-time phase. It has been shown, however, that in general the independence of the real and imaginary parts is not strictly observed and the independence of the magnitude and phase provides a more valid model. Under some mild assumptions and regardless of the actual distribution of the amplitude of speech DFT coefficients, the MMSE spectral gain function is always real-valued and the estimate of the speech phase equals the noisy phase [Erkelens et al. 2008].

### 12.4.3.2 MMSE Amplitude Estimation

We now turn to the estimation of the spectral amplitude $A_\mu$ and functions thereof. This approach has the obvious advantage that only a single scalar quantity needs to be estimated. In most cases, the amplitude estimate is then combined with the phase of the noisy signal to yield the complex enhanced coefficients. Amplitude estimation and its variants have become popular with the landmark works of Ephraim and Malah [Ephraim, Malah 1984], [Ephraim, Malah 1985] and are widely used. While these original works relied on the Gaussian model we will present a more general formulation that includes a parametrized model for the PDF of the amplitude as well as an adjustable exponent. We aim to find a solution to

$$\hat{A}_\mu = \underset{\tilde{A}_\mu}{\mathrm{argmin}} \ \mathrm{E}\left\{ \left( e_\beta(A_\mu, \tilde{A}_\mu) \right)^2 \ \middle| \ Y_\mu, \sigma_{N,\mu}^2, \eta_\mu \right\}, \tag{12.62}$$

where we use a parametric error function

$$e_\beta(A_\mu, \tilde{A}_\mu) = (A_\mu)^\beta - (\tilde{A}_\mu)^\beta, \tag{12.63}$$

with an exponent $\beta$ and a parametric $\chi$-distribution [Johnson et al. 1994] for the speech amplitudes

$$p(A_\mu) = \frac{2}{\Gamma(\delta)} \left( \frac{\delta}{\sigma_{s,\mu}^2} \right)^\delta A_\mu^{2\delta-1} \exp\left( -\frac{\delta}{\sigma_{s,\mu}^2} A_\mu^2 \right). \tag{12.64}$$

We denote $\beta$ and $\delta$ as compression parameter and shape parameter, respectively [Breithaupt et al. 2008b]. The effect of the compression parameter is illustrated in Figure 12.8 for $\beta < 1$: deviations $\Delta A_\mu = A_\mu - \hat{A}_\mu$ at large amplitudes $A_\mu$ result in smaller errors $e_\beta(A_\mu, \hat{A}_\mu)$ than similar deviations at small amplitudes. Thus, the variance of the estimation errors receives a larger weight for weak speech sounds and background noise. Note, that the case $\beta > 1$ is also of interest: For $\beta = 2$ the estimator will approximate the power of the spectral coefficient.

The impact of the shape parameter on the log-density of the spectral magnitude is illustrated in Figure 12.9. For the Gaussian model we have $\delta = 1$, which results in a Rayleigh PDF for the spectral amplitudes. For $\delta < 1$ we obtain PDFs for the amplitudes that correspond to a super-Gaussian distribution of the spectral coefficients.

Given these two parameters we now find [Breithaupt et al. 2008b] the analytic solution to the estimation problem (12.62), i.e.,

$$\hat{A}_\mu = \sqrt{\frac{\eta_\mu}{\delta + \eta_\mu}} \left[ \frac{\Gamma(\delta + \frac{\beta}{2})}{\Gamma(\delta)} \frac{{}_1F_1\left(1 - \delta - \frac{\beta}{2}, 1; -v_\mu\right)}{{}_1F_1\left(1 - \delta, 1; -v_\mu\right)} \right]^{\frac{1}{\beta}} \sqrt{\sigma_{N,\mu}^2}, \tag{12.65}$$

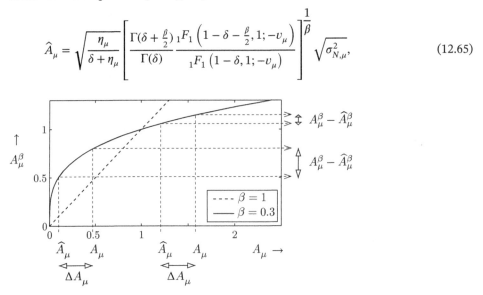

**Figure 12.8** Compression of the approximation error for $\beta = 1$ and $\beta = 0.3$.

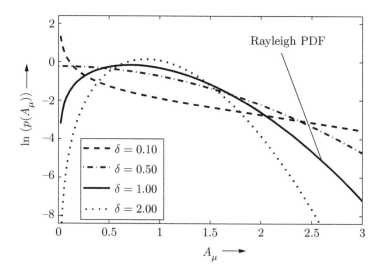

**Figure 12.9** Impact of the shape parameter $\delta$ on the log-density of a $\chi$-distributed variable [Breithaupt 2008].

where $_1F_1(a, c; x)$ denotes a confluent hypergeometric function. $\gamma_\mu = |Y_\mu|^2/\sigma_{N,\mu}^2$ and $\eta_\mu$ are the a posteriori SNR and the a priori SNR as before. Furthermore, we define

$$v_\mu = \frac{\eta_\mu}{1 + \eta_\mu}\gamma_\mu. \tag{12.66}$$

The solution (12.65) is valid for $\delta > 0$ and $\delta + \beta/2 > 0$ with $\beta \neq 0$. Using the phase of the noisy spectral coefficient we may then write

$$\hat{S}_\mu = \hat{A}_\mu \frac{Y_\mu}{R_\mu} = G_\mu Y_\mu. \tag{12.67}$$

Parameter combinations that can be realized with (12.65) are indicated within the light-shaded area in Figure 12.10. The above estimator includes a number of well known estimation error criteria and solutions, most notably

$\beta = 1, \delta = 1$: MMSE short-time spectral amplitude (MMSE-STSA) estimator [Ephraim, Malah 1984]:

$$\hat{A}_\mu = \frac{R_\mu}{\gamma_\mu} \sqrt{v_\mu} \, \Gamma(1.5) \, _1F_1(-0.5, 1; -v_\mu), \tag{12.68}$$

$\beta \to 0, \delta = 1$: MMSE log-spectral amplitude (MMSE-LSA) estimator [Ephraim, Malah 1985]

$$\hat{A}_\mu = \exp\left(\mathrm{E}\left\{\ln(A_\mu)|Y_\mu\right\}\right)$$

$$= \frac{\xi_\mu}{1 + \xi_\mu} \exp\left(\frac{1}{2}\int_{v_\mu}^\infty \frac{\exp\{-t\}}{t}\,dt\right)|Y_\mu|, \tag{12.69}$$

where for a practical implementation, the exponential integral function in (12.69) can be tabulated as a function of $v_\mu$.

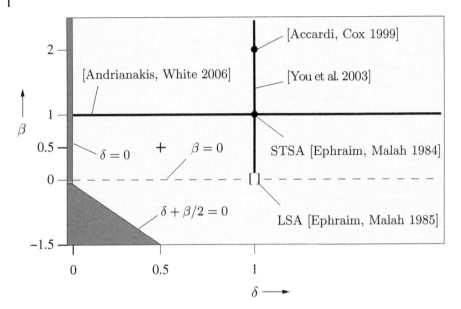

**Figure 12.10** Parameter space of the general MMSE estimator (12.65) and related solutions. A parameter combination that has been found to perform well is indicated by the cross. Source: Adapted from [Breithaupt, Martin 2011] © 2011 IEEE.

$\beta = 2, \delta = 1$:  MMSE magnitude-squared estimator [Accardi, Cox 1999]

$$\hat{A}_\mu^2 = \left( \frac{\sigma_{S,\mu}^2}{\sigma_{S,\mu}^2 + \sigma_{N,\mu}^2} \right)^2 |Y_\mu|^2 + \frac{\sigma_{S,\mu}^2 \sigma_{N,\mu}^2}{\sigma_{S,\mu}^2 + \sigma_{N,\mu}^2}, \tag{12.70}$$

$\delta = 1$:  $\beta$-order MMSE spectral amplitude estimator [You et al. 2003]

$$\hat{A}_\mu = \sqrt{\frac{\eta_\mu}{\delta + \eta_\mu}} \left[ \Gamma\left(1 + \frac{\beta}{2}\right) {}_1F_1\left(-\frac{\beta}{2}, 1; -v_\mu\right) \right]^{\frac{1}{\beta}} \sqrt{\sigma_{N,\mu}^2} \tag{12.71}$$

$\beta = 1$:  MMSE estimators with $\chi$-speech prior [Andrianakis, White 2006], [Erkelens et al. 2007]

$$\hat{A}_\mu = \sqrt{\frac{\eta_\mu}{\delta + \eta_\mu}} \left[ \frac{\Gamma(\delta + 0.5)}{\Gamma(\delta)} \frac{{}_1F_1\left(0.5 - \delta, 1; -v_\mu\right)}{{}_1F_1\left(1 - \delta, 1; -v_\mu\right)} \right]^{\frac{1}{\beta}} \sqrt{\sigma_{N,\mu}^2}. \tag{12.72}$$

All of the above estimators require knowledge of the speech power and noise power or corresponding SNR values, which requires additional estimation procedures such as (12.39). Finally, we consider the input–output characteristic of the general estimator (see Figure 12.11). To this end, we normalize the input and output variables

$$\frac{\hat{A}_\mu}{\sqrt{\sigma_{N,\mu}^2}} = f(\sqrt{\gamma_\mu}) = f\left( \frac{|Y_\mu|}{\sqrt{\sigma_{N,\mu}^2}} \right), \tag{12.73}$$

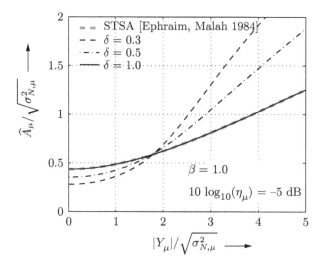

**Figure 12.11** Input–output characteristics for shape parameters $\delta \in \{0.3, 0.5, 1.0\}$ without compression ($\beta = 1$) and $-5\,\text{dB}$ a priori SNR $\eta_\mu$.

and note that for large normalized input values we obtain the following asymptotic behavior independently of $\beta$

$$\left.\frac{\hat{A}_\mu}{\sqrt{\sigma_{N,\mu}^2}}\right|_{v_\mu \gg 1} = \frac{\eta_\mu}{\delta + \eta_\mu} \sqrt{\gamma_\mu}, \tag{12.74}$$

while for small input values we obtain

$$\left.\frac{\hat{A}_\mu}{\sqrt{\sigma_{N,\mu}^2}}\right|_{Y_\mu = 0} = \sqrt{\frac{\eta_\mu}{\delta + \eta_\mu}} \left[\frac{\Gamma(\delta + \frac{\beta}{2})}{\Gamma(\delta)}\right]^{1/\beta}. \tag{12.75}$$

Thus, small values of $\delta$ lead to less clipping of speech onsets and more noise reduction but also to more large random outliers (*musical noise*). To conclude this section, we note that these estimators have been extended to explicitly account for information on the clean speech phase with or without phase uncertainty, e.g., [Gerkmann, Krawczyk 2013] and [Mowlaee, Kulmer 2015]. Furthermore, while the above estimators model the PS of speech and noise as deterministic quantities, also stochastic models have been explored such that the PS and its estimate are both treated as random variables. PS uncertainty is considered for the speech target in [Enzner, Thüne 2017], [Krawczyk-Becker, Gerkmann 2018], and for both the speech and noise signal in [Kim et al. 2022]. These methods improve the performance in the presence of estimation errors and thus provide more robust results.

## 12.5 Joint Optimum Detection and Estimation of Speech

The optimal estimators of Sections 12.2 and 12.4 implicitly assume that speech is present in the noisy signal. This is, of course, not always the case and the estimator for clean speech conditioned on the presence of speech is not optimal when speech is absent. Also, to improve the subjective listening quality, it can be advantageous to explicitly distinguish the two cases "speech present" and "speech absent." For example, the listening quality can be improved when the speech enhancement algorithm applies a constant, frequency-independent attenuation $G_{\min}$ to the noisy signal during speech pauses [Yang 1993], [Malah et al. 1999]. The MMSE estimator can be extended to account for speech presence or absence. Below, the resulting joint optimum detection and estimation problem [Middleton, Esposito 1968], as outlined in Section 5.11.5, is solved in the context of amplitude estimation.

To this end, we introduce the two hypotheses of "speech is present in DFT bin $\mu$" and "speech is absent in DFT bin $\mu$" and denote these two hypotheses by $H_{\mu}^{(1)}$ and $H_{\mu}^{(0)}$, respectively. All statistical quantities will now be conditioned on these hypotheses. The two hypotheses $H_{\mu}^{(1)}$ and $H_{\mu}^{(0)}$ can be stated in terms of the DFT coefficients as

$$H_{\mu}^{(0)} : Y_{\mu} = N_{\mu} \quad \text{and} \tag{12.76}$$

$$H_{\mu}^{(1)} : Y_{\mu} = S_{\mu} + N_{\mu}. \tag{12.77}$$

We denote the a priori probability that speech is present and the a priori probability that speech is absent in bin $\mu$ with $p_{\mu} = P(H_{\mu}^{(1)})$ and $q_{\mu} = P(H_{\mu}^{(0)}) = 1 - p_{\mu}$, respectively.

In the case of quadratic cost functions, the joint optimal detector and estimator is a linear combination of two terms as in (5.176). For example, the estimator for the spectral magnitudes may be decomposed into

$$\hat{A}_{\mu} = G_{\mu}^{(1)} \, \mathrm{E}\left\{ A_{\mu} \mid Y_{\mu}, H_{\mu}^{(1)} \right\} + G_{\mu}^{(0)} \, \mathrm{E}\left\{ A_{\mu} \mid Y_{\mu}, H_{\mu}^{(0)} \right\}, \tag{12.78}$$

where $\mathrm{E}\left\{ A_{\mu} \mid Y_{\mu}, H_{\mu}^{(1)} \right\}$ and $\mathrm{E}\left\{ A_{\mu} \mid Y_{\mu}, H_{\mu}^{(0)} \right\}$ are the optimal estimators under the hypotheses $H_{\mu}^{(1)}$ and $H_{\mu}^{(0)}$, respectively, and the multiplicative weighting factors are given by

$$G_{\mu}^{(1)} = \frac{\Lambda_{\mu}(Y_{\mu})}{1 + \Lambda_{\mu}(Y_{\mu})} = \frac{p_{\mu} p(Y_{\mu} \mid H_{\mu}^{(1)})}{q_{\mu} p(Y_{\mu} \mid H_{\mu}^{(0)}) + p_{\mu} p(Y_{\mu} \mid H_{\mu}^{(1)})}, \tag{12.79}$$

$$G_{\mu}^{(0)} = \frac{1}{1 + \Lambda_{\mu}(Y_{\mu})} = \frac{q_{\mu} p(Y_{\mu}(\lambda) \mid H_{\mu}^{(0)}(\lambda))}{q_{\mu} p(Y_{\mu} \mid H_{\mu}^{(0)}) + p_{\mu} p(Y_{\mu} \mid H_{\mu}^{(1)})}. \tag{12.80}$$

These *soft-decision* weights are functions of the generalized likelihood ratio

$$\Lambda_{\mu}(Y_{\mu}) = \frac{p(Y_{\mu} \mid H_{\mu}^{(1)}) p_{\mu}}{p(Y_{\mu} \mid H_{\mu}^{(0)}) q_{\mu}}. \tag{12.81}$$

In the case of speech absence, we strive for a frequency-independent attenuation $\mathrm{E}\left\{ A_{\mu} \mid Y_{\mu}, H_{\mu}^{(0)} \right\} = G_{\min} R_{\mu}$ of the noisy input magnitude. Thus, we may simplify the above estimator, i.e.,

$$\hat{A}_{\mu} = \frac{\Lambda_{\mu}(Y_{\mu})}{1 + \Lambda_{\mu}(Y_{\mu})} \mathrm{E}\left\{ A_{\mu} \mid Y_{\mu}, H_{\mu}^{(1)} \right\} + \frac{1}{1 + \Lambda_{\mu}(Y_{\mu})} G_{\min} R_{\mu}. \tag{12.82}$$

Another solution may be obtained when the conditional density $p(Y_\mu \mid A_\mu, \alpha_\mu)$ is replaced by the conditional density of the magnitude $p(R_\mu \mid A_\mu, \alpha_\mu)$. A completely analogous derivation leads to the soft-decision estimation as proposed in [McAulay, Malpass 1980],

$$\hat{A}_\mu = \mathrm{E}\left\{A_\mu \mid R_\mu, H_\mu^{(1)}\right\} P(H_\mu^{(1)} \mid R_\mu) + \mathrm{E}\left\{A_\mu \mid R_\mu, H_\mu^{(0)}\right\} P(H_\mu^{(0)} \mid R_\mu), \qquad (12.83)$$

where now also the a posteriori probabilities of speech presence and absence, $P(H_\mu^{(1)} \mid R_\mu)$ and $P(H_\mu^{(0)} \mid R_\mu)$, come into play. In [McAulay, Malpass 1980], $\mathrm{E}\left\{A_\mu \mid R_\mu, H_\mu^{(1)}\right\}$ is replaced by the ML estimate (12.56), $\mathrm{E}\left\{A_\mu \mid R_\mu, H_\mu^{(0)}\right\}$ is set to zero, and the likelihood ratio $\Lambda_\mu$ is based on the conditional PDF (5.142):

$$\Lambda_\mu(R_\mu) = \frac{p(R_\mu \mid H_\mu^{(1)}) p_\mu}{p(R_\mu \mid H_\mu^{(0)}) q_\mu}. \qquad (12.84)$$

We note that the soft-decision approach can be used in conjunction with other estimators, e.g., the MMSE-LSA estimator, and other cost functions as well. In general, however, the factoring property of (12.78) will be lost.

As a final example, we consider the estimation of $\ln(A_\mu)$ and obtain

$$\hat{A}_\mu = \exp\left(\mathrm{E}\left\{\ln(A_\mu) \mid Y_\mu, H_\mu^{(1)}\right\}\right)^{\frac{\Lambda_\mu(Y_\mu)}{1+\Lambda_\mu(Y_\mu)}} (G_{\min} R_\mu)^{\frac{1}{1+\Lambda_\mu(Y_\mu)}}, \qquad (12.85)$$

where we make use of the MMSE-LSA estimator as outlined in Section 12.4.3.2 and the fixed attenuation $G_{\min}$ during speech pause. This estimator [Cohen, Berdugo 2001] results in larger improvements in the segmental SNR than the estimator in (12.82) but also in more speech distortions. This less desirable behavior (see [Ephraim, Malah 1985]) is improved when the soft-decision modifier is multiplicatively combined with the MMSE-LSA estimator [Malah et al. 1999],

$$\hat{A}_\mu = \frac{\Lambda_\mu(Y_\mu)}{1 + \Lambda_\mu(Y_\mu)} \exp\left(\mathrm{E}\left\{\ln(A_\mu) \mid Y_\mu, H_\mu^{(1)}\right\}\right). \qquad (12.86)$$

This estimator achieves less noise reduction than the estimator in (12.85).

## 12.6 Computation of Likelihood Ratios

For the Gaussian model, the PDFs of $Y_\mu$ conditioned on the hypotheses $H_\mu^{(0)}$ or $H_\mu^{(1)}$ follow from (5.133),

$$p(Y_\mu \mid H_\mu^{(0)}) = \frac{1}{\pi \sigma_{N,\mu}^2} \exp\left(-\frac{R_\mu^2}{\sigma_{N,\mu}^2}\right) \qquad (12.87)$$

and

$$p(Y_\mu \mid H_\mu^{(1)}) = \frac{1}{\pi \left(\sigma_{N,\mu}^2 + \mathrm{E}\left\{A_\mu^2 \mid H_\mu^{(1)}\right\}\right)} \exp\left(-\frac{R_\mu^2}{\sigma_{N,\mu}^2 + \mathrm{E}\left\{A_\mu^2 \mid H_\mu^{(1)}\right\}}\right).$$

The PDF of complex Gaussians depends only on the magnitude $R_\mu$ and not on the phase. The expectation of the speech power, $E\left\{A_\mu^2 \mid H_\mu^{(1)}\right\}$, is now explicitly conditioned on the presence of speech and thus excludes speech pauses. The likelihood ratio is then given by

$$
\Lambda_\mu = \frac{1 - q_\mu}{q_\mu} \frac{\sigma_{N,\mu}^2}{\sigma_{N,\mu}^2 + E\left\{A_\mu^2 \mid H_\mu^{(1)}\right\}}
$$

$$
\cdot \exp\left(-\frac{R_\mu^2}{\sigma_{N,\mu}^2 + E\left\{A_\mu^2 \mid H_\mu^{(1)}\right\}} + \frac{R_\mu^2}{\sigma_{N,\mu}^2}\right)
$$

$$
= \frac{1 - q_\mu}{q_\mu} \frac{1}{1 + \xi_\mu} \exp\left(\gamma_\mu \frac{\xi_\mu}{1 + \xi_\mu}\right), \tag{12.88}
$$

where $\xi_\mu$ is the a priori SNR conditioned on the presence of speech

$$
\xi_\mu = \frac{E\left\{A_\mu^2 \mid H_\mu^{(1)}\right\}}{\sigma_{N,\mu}^2}. \tag{12.89}
$$

According to [Ephraim, Malah 1984], $\xi_\mu$ can be expressed as a function of the unconditional expectation $E\left\{A_\mu^2\right\}$

$$
\frac{E\left\{A_\mu^2 \mid H_\mu^{(1)}\right\}}{\sigma_{N,\mu}^2} = \frac{E\left\{A_\mu^2\right\}}{\sigma_{N,\mu}^2} \frac{1}{1 - q_\mu} = \eta_\mu \frac{1}{1 - q_\mu}, \tag{12.90}
$$

with

$$
\eta_\mu = \frac{E\left\{A_\mu^2\right\}}{\sigma_{N,\mu}^2}. \tag{12.91}
$$

Other solutions are discussed in [Cohen, Berdugo 2001].

## 12.7 Estimation of the A Priori and A Posteriori Probabilities of Speech Presence

The computation of the likelihood ratio requires knowledge of the a priori probability $p_\mu$ of speech presence in each frequency bin $\mu$. These probabilities do not just reflect the proportion of speech segments to speech pauses. They should also take into account that, during voiced speech, most of the speech energy is concentrated in the frequency bins, which correspond to the speech harmonics. The frequency bins in between the speech harmonics contain mostly noise. Consequently, there has been some debate as to which value would be most appropriate for $p_\mu$. In the literature a priori probabilities for speech presence range between 0.5 and 0.8 [McAulay, Malpass 1980], [Ephraim, Malah 1984].

However, a fixed a priori probability $p_\mu$ can only be a compromise since the location of the speech harmonics varies with the fundamental frequency of the speaker. Tracking the a priori speech presence probability (SPP) individually in all frequency bins and furthermore the a posteriori probability $P(H_\mu^{(1)} \mid R_\mu)$ should therefore result in improved performance. These task may be solved via the estimation procedures, which we outline below.

### 12.7.1 Estimation of the A Priori Probability

The a priori probability for speech presence can be obtained from a test on the a priori SNR $\eta_\mu$ where the a priori SNR is compared to a preset threshold $\eta_{min}$. Since the a priori SNR parameterizes the statistics of the a posteriori SNR in terms of the exponential density, a test can be devised, which relies exclusively on the a posteriori SNR [Malah et al. 1999]. In this test, we compare the a posteriori SNR to a threshold $\gamma_q$. When speech is present, the decisions are smoothed over time for each frequency bin using a first-order recursive system

$$\hat{p}_\mu(\lambda) = \alpha_p \hat{p}_\mu(\lambda - 1) + (1 - \alpha_p) I_\mu(\lambda), \tag{12.92}$$

where $I_\mu(\lambda)$ denotes an index function with

$$I_\mu(\lambda) = \begin{cases} 1, & \gamma_\mu(\lambda) > \gamma_q, \\ 0, & \gamma_\mu(\lambda) \leq \gamma_q. \end{cases} \tag{12.93}$$

During speech pauses, the probability $p_\mu$ is set to a fixed value $p_\mu \gtrsim 0$. Good results are obtained using $\gamma_q = 0.8$ and $\alpha_q = 0.95$ [Malah et al. 1999].

### 12.7.2 A Posteriori Speech Presence Probability Estimation

The a posteriori SPP can be obtained from Bayes' theorem as

$$P(H_\mu^{(1)} \mid R_\mu) = \frac{p(R_\mu \mid H_\mu^{(1)}) p_\mu}{p(R_\mu \mid H_\mu^{(1)}) p_\mu + p(R_\mu \mid H_\mu^{(0)}) q_\mu} = \frac{\Lambda_\mu(R_\mu)}{1 + \Lambda_\mu(R_\mu)}$$

$$= \frac{p(R_\mu \mid H_\mu^{(1)})}{p(R_\mu \mid H_\mu^{(1)}) + p(R_\mu \mid H_\mu^{(0)})}, \tag{12.94}$$

where the last identity assumes equal a priori probabilities, $p_\mu = q_\mu = 0.5$. Substituting the conditional densities, we obtain the SPP [McAulay, Malpass 1980]

$$P(H_\mu^{(1)} \mid R_\mu) = \frac{\exp(-\xi_\mu) I_0(2\sqrt{\xi_\mu \gamma_\mu})}{1 + \exp(-\xi_\mu) I_0(2\sqrt{\xi_\mu \gamma_\mu})}. \tag{12.95}$$

This estimate may be recursively smoothed to approximate the a priori probability of speech presence

$$\hat{p}_\mu(\lambda) = (1 - \beta_p) \hat{p}_\mu(\lambda - 1) + \beta_p P(H_\mu^{(1)} \mid R_\mu). \tag{12.96}$$

A similar SPP estimator may be formulated [Ephraim, Malah 1984] using the complex Gaussian model and substituting the likelihood ratio (12.88) into

$$P(H_\mu^{(1)} \mid Y_\mu) = \frac{\Lambda_\mu(Y_\mu)}{1 + \Lambda_\mu(Y_\mu)} = \left(1 + \frac{q_\mu}{1 - q_\mu} (1 + \xi_\mu) \exp\left(-\gamma_\mu \frac{\xi_\mu}{1 + \xi_\mu}\right)\right)^{-1}. \tag{12.97}$$

Figure 12.12 plots the SPP $P(H_\mu^{(1)} \mid Y_\mu)$ as a function of $\xi_\mu$ and $\gamma_\mu$. For low a priori SNR, the probability of speech presence is confined to a rather small interval around 0.5. Only for larger SNRs we observe the full range of probability values. Both (12.95) and (12.97) require estimates of the a priori and a posteriori SNRs. In practical systems, SPP estimation and SNR estimation are therefore often connected in a recursive loop.

### 12.7.3 SPP Estimation Using a Fixed SNR Prior

The estimation procedures detailed above depend on the local a priori SNR, which needs to be estimated, e.g., using the decision-directed approach in (12.39). In low SNR conditions, the estimated a priori SNR will attain small values and, with $q_\mu = 0.5$, the probabilities $P(H_\mu^{(0)}|Y_\mu)$ and $P(H_\mu^{(1)}|Y_\mu)$ both approach 0.5. This will in turn render the discrimination between speech absence and speech presence unreliable and, as SPP and SNR estimation are often connected in a recursive loop, will also have a detrimental effect on SNR estimation.

Therefore, it is beneficial to decouple SNR estimation and SPP estimation in (12.97) via a pre-selected *fixed* a priori SNR $\xi_\mu$ and a *fixed* a priori probability $q_\mu$ [Gerkmann et al. 2008]. The pre-selected $\xi_{\mu,\mathrm{fix}}$ should reflect the local SNR that a typical speech sound would have if speech were present in the considered time-frequency bin. The fixed SNR should be carefully chosen: if it is too high, weak speech components are not recognized, if it is too low, random fluctuations will disturb the likelihood ratio. The effect of a fixed a priori SNR can be analyzed in Figure 12.12, where a vertical dashed line marks a fixed SNR of 12 dB. Then, the SPP is a function of the a posteriori SNR only. When the fixed a priori SNR is sufficiently large, the a posteriori SNR is mapped on almost the full range [0, 1], which is easily verified by moving along the dashed vertical in Figure 12.12. A specific optimal value for $\xi_{\mu,\mathrm{fix}}$ may be found via minimizing a cost function which takes the cost of missed speech detections and false hits into account. Averaging the a posteriori SNR and the a priori SNR across time-frequency bins in the neighborhood of the bin under consideration further improves this process [Gerkmann et al. 2008].

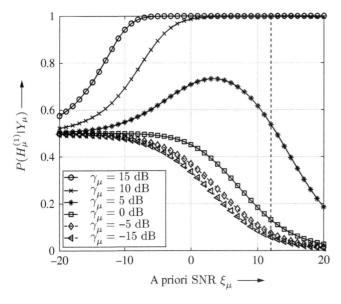

**Figure 12.12** Speech presence probability $P(H_\mu^{(1)}|Y_\mu)$ as a function of a priori SNR $\xi_\mu$ and a posteriori SNR $\gamma_\mu$. The dashed vertical line marks an exemplary fixed SNR value of 12 dB.

## 12.8  VAD and Noise Estimation Techniques

All noise suppression methods discussed in the previous sections require knowledge of the PS of the disturbing noise. Noise power estimation is therefore an important component of speech enhancement algorithms. The noise PS is in general time varying and not known a priori. It must be estimated and updated during the execution of the noise reduction algorithm. When the noise is non-stationary, it is not sufficient to sample the noise in a speech pause prior to a speech phrase and to keep this estimate fixed during speech activity. On the contrary, we must frequently update the noise estimate in order to track the noise PS with sufficient accuracy.

VADs are also used for controlling *discontinuous transmission* (DTX) in mobile voice communication systems [Freeman et al. 1989], [Srinivasan, Gersho 1993], [3GPP TS 126.094, V4.0.0 2001], [ETSI TS 126.445 V14.2.0 2018, Section 5.1.12] and for detecting speech spurts in speech recognition applications. As we will see, a single algorithm will not be perfect for all applications. The VAD algorithm needs to be optimized for the specific task.

In many applications, however, we need an estimate of the noise floor rather than binary voice activity decisions. Methods based on "soft" estimation criteria are therefore preferable since they provide means for an update of the noise power during speech activity. Besides VAD-based methods, we therefore also consider techniques, which are based on the probability of speech presence and soft-decision updates [Yu 2009], [Gerkmann, Hendriks 2012], and on minimum power tracking, also known as *minimum statistics* [Martin 1994], [Martin 2001].

### 12.8.1  Voice Activity Detection

Over the years, many algorithms for voice activity detection have been proposed, e.g., based on likelihood ratios [Sohn et al. 1999], based on histogram modeling via a two-component GMM [Van Compernolle 1989], and using likelihood ratios with multiple observations [Ramiréz et al. 2005]. A VAD can be described by means of a finite state machine with at least two states, "speech is present" and "speech is absent." Sometimes, additional states are used to cope with transitions between these two basic states [Breithaupt, Martin 2006].

When speech is absent, the noise PS can be estimated using a first-order recursive system with $0 < \alpha < 1$,

$$\hat{\sigma}^2_{Y,\mu}(\lambda) = \alpha \hat{\sigma}^2_{Y,\mu}(\lambda - 1) + (1 - \alpha)|Y_\mu(\lambda)|^2. \tag{12.98}$$

The key to the successful design of a VAD is to use features of the speech signal, which either are – at least in principle – independent of the noise power level or can be normalized on the noise power. Much ingenuity is required to find the most appropriate set of features, see, e.g., [Graf 2015] for an overview. Once these features have been selected, the problem can be treated with classical detection theory, or, alternatively, a data-driven model may be used. Furthermore, DNNs have improved the performance of VAD recently, e.g., [Tashev, Mirsamadi 2016] and [Heitkämper et al. 2020].

A comparison of standard VAD algorithms is presented, for example, in [Beritelli et al. 2001]. A simple approach is to use an estimated SNR [Martin 1993] and to compare the estimated SNR to one or several fixed thresholds. This might work well for high SNR conditions, but for low SNR conditions we will encounter a substantial amount of erroneous detections. Typical VAD implementations therefore use several features. Some of the more common methods are discussed below.

### 12.8.1.1 Detectors Based on the Subband SNR

An SNR-based VAD may be made more robust by computing the instantaneous SNR in frequency subbands and by averaging the SNR over all of these bands. The spectral analysis may be realized as a filter bank with uniform or non-uniform resolution or by a sliding window DFT. We will now examine this concept in greater detail and consider a detection algorithm based on the a posteriori SNR of DFT coefficients. To develop the detection algorithm, we introduce two hypotheses

$$H^{(0)}(\lambda) : \quad \text{speech is absent in signal frame } \lambda, \text{ and}$$

$$H^{(1)}(\lambda) : \quad \text{speech is present in signal frame } \lambda.$$

Assuming Gaussian PDFs of equal variance for the real and the imaginary parts of the complex $M$-point DFT coefficients and mutual statistical independence of all DFT bins, we may write the joint PDF for a vector $\mathbf{Y}(\lambda)$ of $M/2 - 1$ complex noisy DFT coefficients when speech is present as

$$p_{\mathbf{Y}(\lambda)|H^{(1)}(\lambda)} \left( Y_1(\lambda), \ldots, Y_{M/2-1}(\lambda) \mid H^{(1)}(\lambda) \right)$$

$$= \prod_{\mu=1}^{M/2-1} \frac{1}{\pi \left( \sigma_{N,\mu}^2 + \sigma_{S,\mu}^2 \right)} \exp\left( -\frac{|Y_\mu(\lambda)|^2}{\sigma_{N,\mu}^2 + \sigma_{S,\mu}^2} \right). \tag{12.99}$$

The DC and Nyquist frequency bins are not included in the product. $\sigma_{S,\mu}^2$ and $\sigma_{N,\mu}^2$ denote the power of the speech and the noise coefficients in the $\mu$-th frequency bin, respectively. When speech is absent, we have

$$p_{\mathbf{Y}(\lambda)|H^{(0)}(\lambda)} \left( Y_1(\lambda), \ldots, Y_{M/2-1}(\lambda) \mid H^{(0)}(\lambda) \right)$$

$$= \prod_{\mu=1}^{M/2-1} \frac{1}{\pi \sigma_{N,\mu}^2} \exp\left( -\frac{|Y_\mu(\lambda)|^2}{\sigma_{N,\mu}^2} \right). \tag{12.100}$$

Speech can now be detected by comparing the log-likelihood ratio [Van Trees 1968], [Sohn et al. 1999]

$$\Lambda\left(\mathbf{Y}(\lambda)\right) = \frac{1}{M/2 - 1} \log\left( \frac{p_{\mathbf{Y}(\lambda)|H^{(1)}(\lambda)}(Y_1(\lambda), \ldots, Y_{M/2-1}(\lambda) \mid H^{(1)}(\lambda))}{p_{\mathbf{Y}(\lambda)|H^{(0)}(\lambda)}(Y_1(\lambda), \ldots, Y_{M/2-1}(\lambda) \mid H^{(0)}(\lambda))} \right)$$

$$= \frac{1}{M/2 - 1} \sum_{\mu=1}^{M/2-1} \left( \frac{|Y_\mu(\lambda)|^2}{\sigma_{N,\mu}^2} - \frac{|Y_\mu(\lambda)|^2}{\sigma_{N,\mu}^2 + \sigma_{S,\mu}^2} - \log\left( \frac{\sigma_{N,\mu}^2 + \sigma_{S,\mu}^2}{\sigma_{N,\mu}^2} \right) \right),$$

to a threshold $\mathfrak{L}_{\text{thr}}$

$$\Lambda\left(\mathbf{Y}(\lambda)\right) \underset{H^{(1)}(\lambda)}{\overset{H^{(0)}(\lambda)}{\lessgtr}} \mathfrak{L}_{\text{thr}}. \tag{12.101}$$

The test can be further simplified if the unknown speech variances $\sigma_{S,\mu}^2$ are replaced by their ML estimates (12.51)

$$\hat{\sigma}_{S,\mu,ML}^2 = |Y_\mu(\lambda)|^2 - \sigma_{N,\mu}^2. \tag{12.102}$$

In this case we obtain

$$\Lambda(\mathbf{Y}(\lambda)) = \frac{1}{M/2-1} \sum_{\mu=1}^{M/2-1} \left( \frac{|Y_\mu(\lambda)|^2}{\sigma_{N,\mu}^2} - \log\left( \frac{|Y_\mu(\lambda)|^2}{\sigma_{N,\mu}^2} \right) - 1 \right)$$

$$= \frac{1}{M/2-1} \sum_{\mu=1}^{M/2-1} \left( \gamma_\mu(\lambda) - \log\left( \gamma_\mu(\lambda) \right) - 1 \right) \tag{12.103}$$

$$\Lambda(\mathbf{Y}(\lambda)) \underset{H^{(1)}(\lambda)}{\overset{H^{(0)}(\lambda)}{\lessgtr}} \mathfrak{L}_{\text{thr}}, \tag{12.104}$$

which is recognized as a discrete approximation of the Itakura–Saito distortion measure between the magnitude squared signal spectrum $|Y_\mu(\lambda)|^2$ and the noise PS $\sigma_{N,\mu}^2$ [Markel, Gray, 1976, chapter 6].

If we retain only the first term in the summation, the detection test simplifies to a test on the average a posteriori SNR (12.37). In fact, as observed by several authors [Häkkinen, Väänänen 1993], [Malah et al. 1999], the a posteriori SNR $\gamma_\mu$ can be used to build reliable VADs. Since $\gamma_\mu$ is normalized on the noise PS, its expectation is equal to unity during speech pause. A VAD is therefore obtained by comparing the average a posteriori SNR $\bar{\gamma}(\lambda)$ to a fixed threshold $\gamma_{\text{thr}}$,

$$\bar{\gamma}(\lambda) = \frac{1}{M/2-1} \sum_{\mu=1}^{M/2-1} \gamma_\mu(\lambda) \underset{H^{(1)}(\lambda)}{\overset{H^{(0)}(\lambda)}{\lessgtr}} \gamma_{\text{thr}}. \tag{12.105}$$

For the Gaussian model, the variance of $\gamma_\mu(\lambda)$ is equal to one during speech pause. If we assume that all frequency bins are mutually independent, the variance of the average a posteriori SNR is, during speech pause, given by

$$\text{var}\{\bar{\gamma}\} = \frac{1}{M/2-1}. \tag{12.106}$$

As a consequence, the speech detection threshold can be set to $\gamma_{\text{thr}} = 1 + a\sqrt{\text{var}\{\bar{\gamma}\}}$ where $a$ is in the range $2 \le a \le 4$. The examples in Figure 12.13 show the typical performance of this detector for a high SNR and a low SNR signal. The noise in this experiment is computer-generated white Gaussian noise. As we can see, this detector works almost perfectly for high SNR conditions. When the SNR is low, speech activity is not always properly indicated. For low SNR conditions, a more advanced algorithm is required. We note that this approach may be varied in several ways. For example, [Yang 1993] employs a test

$$\frac{1}{M/2-1} \sum_{\mu=1}^{M/2-1} \max\left( \frac{R_\mu^2 - \sigma_{N,\mu}^2}{R_\mu^2}, 0 \right) \underset{H^{(1)}(\lambda)}{\overset{H^{(0)}(\lambda)}{\lessgtr}} \mathfrak{L}_{\text{thr}}. \tag{12.107}$$

### 12.8.2 Noise Power Estimation Based on Minimum Statistics

In contrast to noise estimation based on voice activity detection, the *minimum statistics* algorithm does not use any explicit threshold to distinguish between speech activity

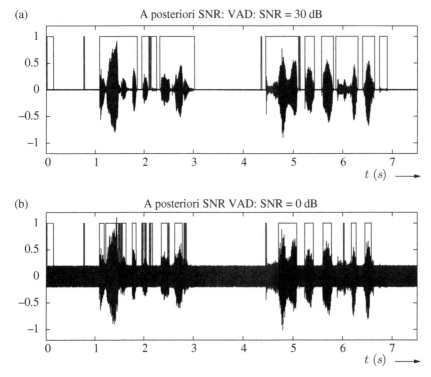

**Figure 12.13** Speech signals and VAD decision for the algorithm based on the a posteriori SNR (12.105) (a) 30 dB segmental SNR and (b) 0 dB segmental SNR.

and speech pause and is therefore more closely related to soft-decision methods than to traditional voice activity detection. Similar to soft-decision methods, it can also update the estimated noise PS during speech activity. Speech enhancement based on minimum statistics has been proposed in [Martin 1994] and improved in [Martin 2001].

The minimum statistics method rests on two conditions, namely that the speech and the disturbing noise are statistically independent and that the power of a noisy speech signal frequently decays to the power level of the disturbing noise. It is therefore possible to derive an accurate noise PS estimate by tracking the minimum of the (smoothed) noisy signal PS. Since the minimum is smaller than or equal to the mean, the minimum tracking method requires a bias correction. It turns out that the bias is a function of the variance of the smoothed signal PS and as such depends on the smoothing parameter of the PS estimator.

Given the first-order smoothing recursion,

$$\hat{\sigma}^2_{Y,\mu}(\lambda) = \alpha_\mu(\lambda) \cdot \hat{\sigma}^2_{Y,\mu}(\lambda - 1) + (1 - \alpha_\mu(\lambda)) \cdot |Y_\mu(\lambda)|^2, \tag{12.108}$$

we may use a time- and frequency-dependent smoothing parameter $\alpha_\mu(\lambda)$ to track the variations in the input PS [Martin, Lotter 2001], [Martin 2001], where for speech pause, we want $\hat{\sigma}^2_{Y,\mu}(\lambda)$ to be as close as possible to the noise PS $\sigma^2_{N,\mu}(\lambda)$. Then, we track the minimum noise floor $\sigma^2_{\min,\mu}(\lambda)$ using a sliding window of $D$ consecutive frames and apply a bias correction factor $B_{\min,\mu}\left(D, Q_{eq,\mu}(\lambda)\right)$ to arrive at an *unbiased* estimator of the noise PS $\sigma^2_{N,\mu}(\lambda)$,

$$\hat{\sigma}^2_{N,\mu}(\lambda) = B_{\min,\mu}\left(D, Q_{eq,\mu}(\lambda)\right)\sigma^2_{\min,\mu}(\lambda). \tag{12.109}$$

Besides the length $D$ of the sliding minimum search window, the unbiased estimator requires knowledge of the normalized variance

$$\frac{1}{Q_{eq,\mu}(\lambda)} = \frac{\text{var}\{\hat{\sigma}^2_{Y,\mu}(\lambda)\}}{2\,\sigma^4_{N,\mu}(\lambda)} \tag{12.110}$$

of the smoothed PS estimate $\hat{\sigma}^2_{Y,\mu}(\lambda)$ at any given time and frequency index. The appropriate bias compensation factor is then obtained via a lookup table or approximate functional expressions [Martin 2001], [Martin 2005a].

To estimate the variance of the smoothed power spectrum estimate $\hat{\sigma}^2_{Y,\mu}(\lambda)$, we use a first-order smoothing recursion for the approximation of the first moment, $E\{\hat{\sigma}^2_{Y,\mu}(\lambda)\}$, and the second moment, $E\{(\hat{\sigma}^2_{Y,\mu}(\lambda))^2\}$, of $\hat{\sigma}^2_{Y,\mu}(\lambda)$,

$$\overline{P}_\mu(\lambda) = \beta_\mu(\lambda)\,\overline{P}_\mu(\lambda-1) + (1 - \beta_\mu(\lambda))\,\hat{\sigma}^2_{Y,\mu}(\lambda)$$

$$\overline{P^2}_\mu(\lambda) = \beta_\mu(\lambda)\,\overline{P^2}_\mu(\lambda-1) + (1 - \beta_\mu(\lambda))\left(\hat{\sigma}^2_{Y,\mu}(\lambda)\right)^2 \tag{12.111}$$

$$\widehat{\text{var}}\{\hat{\sigma}^2_{Y,\mu}(\lambda)\} = \overline{P^2}_\mu(\lambda) - \overline{P}^2_\mu(\lambda),$$

where $\overline{P}_\mu(\lambda)$ and $\overline{P^2}_\mu(\lambda)$ denote the estimated first and second moments, respectively. Good results are obtained by choosing the smoothing parameter $\beta_\mu(\lambda) = \alpha^2_\mu(\lambda)$ and by limiting $\beta_\mu(\lambda)$ to values less than or equal to 0.8 [Martin 2001]. Finally, $1/Q_{eq,\mu}(\lambda)$ is estimated by

$$\frac{1}{Q_{eq,\mu}(\lambda)} \approx \frac{\widehat{\text{var}}\{\hat{\sigma}^2_{Y,\mu}(\lambda)\}}{2\,\hat{\sigma}^4_{N,\mu}(\lambda-1)}, \tag{12.112}$$

and this estimate is limited to a maximum of 0.5 corresponding to $Q_{eq,\mu}(\lambda) = 2$. Since an increasing noise power can be tracked only with some delay, the minimum statistics estimator has a tendency to underestimate non-stationary noise.

As an example Figure 12.14 plots $|Y_\mu(\lambda)|^2$, the smoothed power $\hat{\sigma}^2_{Y,\mu}(\lambda)$, the noise estimate $\hat{\sigma}^2_{N,\mu}(\lambda)$ and the time-varying smoothing parameter $\alpha_\mu(\lambda)$. We see that the time varying smoothing parameter allows the estimated signal power to closely follow the variations of the speech signal. During speech pauses the noise is well smoothed. The bias compensation works very well, as the smoothed power and the estimated noise power follow each other closely during speech pauses. We also note that the noise power spectrum estimate is updated during speech activity. This is an advantage of the minimum statistics approach w.r.t. VAD-based methods.

Although the minimum statistics approach was originally developed for a sampling rate of $f_s = 8000$ Hz and a frame advance of 128 samples, it can easily be adapted to other sampling rates and frame advance schemes. The length $D$ of the minimum search window must be set proportional to the frame rate. For a given sampling rate $f_s$ and a frame advance $r$, the duration of the time window for minimum search, $D \cdot r/f_s$, should be equal to approximately 1.5 s. The minimum search itself is efficiently implemented by subdividing

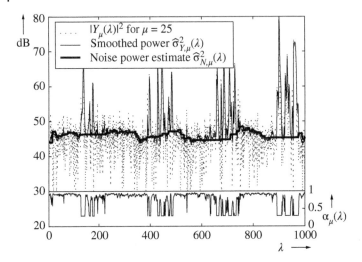

**Figure 12.14** $|Y_\mu(\lambda)|^2$, smoothed power $\hat{\sigma}^2_{Y,\mu}(\lambda)$ (12.108) and noise estimate $\hat{\sigma}^2_{N,\mu}(\lambda)$ (12.109) for a noisy speech signal (vehicular noise, sampling rate 8 kHz, DFT length 256) and a single frequency bin $\mu = 25$. The time-varying smoothing parameter $\alpha_\mu(\lambda)$ is shown in the lower inset graph. Source: [Martin 2001]; © 2001 IEEE.

the search window of length $D$ into subwindows. This, as well as a method to improve tracking of non-stationary noise, is explained in greater detail in [Martin 2001].

### 12.8.3  Noise Estimation Using a Soft-Decision Detector

Using a Gaussian model as outlined in Section 12.8.1.1, the probability of speech presence or absence can be determined. These probabilities can also be used to estimate and/or update the background noise power. A noise estimation algorithm based on these probabilities was proposed in [Sohn, Sung 1998] for the purpose of robust voice activity detection. This noise estimate can also be used directly for speech enhancement without employing a VAD [Beaugeant 1999]. We will briefly outline the method.

Under the assumption that DFT coefficients are mutually independent, the MMSE estimate $\hat{\sigma}^2_{N,\mu}(\lambda) = E\left\{\sigma^2_{N,\mu}(\lambda) \mid Y_\mu(\lambda)\right\}$ of the background noise power can be written as

$$\hat{\sigma}^2_{N,\mu}(\lambda) = E\left\{\sigma^2_{N,\mu}(\lambda) \mid Y_\mu(\lambda), H^{(0)}_\mu(\lambda)\right\} P\left(H^{(0)}_\mu(\lambda) \mid Y_\mu(\lambda)\right)$$
$$+ E\left\{\sigma^2_{N,\mu}(\lambda) \mid Y_\mu(\lambda), H^{(1)}_\mu(\lambda)\right\} P\left(H^{(1)}_\mu(\lambda) \mid Y_\mu(\lambda)\right), \tag{12.113}$$

where we now define the hypotheses $H^{(0)}_\mu(\lambda)$ and $H^{(1)}_\mu(\lambda)$ for speech absence and presence, respectively, individually for each frequency bin. In analogy to Section 12.5, the probabilities $P(H^{(0)}_\mu(\lambda) \mid Y_\mu(\lambda))$ and $P(H^{(1)}_\mu(\lambda) \mid Y_\mu(\lambda))$ can be written as a function of the generalized likelihood ratio

$$\Lambda_\mu(\lambda) = \frac{p(Y_\mu(\lambda) \mid H^{(1)}_\mu(\lambda))P(H^{(1)}_\mu(\lambda))}{p(Y_\mu(\lambda) \mid H^{(0)}_\mu(\lambda))P(H^{(0)}_\mu(\lambda))}, \tag{12.114}$$

i.e.,

$$P(H_\mu^{(0)}(\lambda) \mid Y_\mu(\lambda)) = \frac{1}{1 + \Lambda_\mu(\lambda)} \tag{12.115}$$

and

$$P(H_\mu^{(1)}(\lambda) \mid Y_\mu(\lambda)) = \frac{\Lambda_\mu(\lambda)}{1 + \Lambda_\mu(\lambda)}. \tag{12.116}$$

The evaluation of the expectations in (12.113) is in general difficult. Therefore, simplified estimators are used. During speech activity the noise power estimate is not updated and the estimate is replaced by the estimate of the previous frame

$$E\left\{\sigma_{N,\mu}^2(\lambda) \mid H_\mu^{(1)}(\lambda)\right\} \approx \hat\sigma_{N,\mu}^2(\lambda - 1). \tag{12.117}$$

During speech pause the squared magnitude $|Y_\mu(k)|^2$ of the current frame is used as an estimate of the noise power, i.e., $E\left\{\sigma_{N,\mu}^2(k) \mid H_\mu^{(0)}(\lambda)\right\} \approx |Y_\mu(k)|^2$. The estimator based on the probabilities of speech presence and absence can therefore be written as [Sohn, Sung 1998]

$$\hat\sigma_{N,\mu}^2(\lambda) = \frac{1}{1 + \Lambda_\mu(\lambda)} |Y_\mu(\lambda)|^2 + \frac{\Lambda_\mu(\lambda)}{1 + \Lambda_\mu(\lambda)} \hat\sigma_{N,\mu}^2(\lambda - 1), \tag{12.118}$$

which is recognized as a first-order recursive (smoothing) filter with a time-varying and frequency-dependent smoothing parameter. The likelihood ratio can be estimated as outlined in Section 12.5 or be replaced by a frequency-independent average.

### 12.8.4 Noise Power Tracking Based on Minimum Mean Square Error Estimation

Robust noise power tracking approaches require some form of non-linear decoupling of speech and noise components using, for example, order statistics or, alternatively, non-linear estimators in conjunction with a non-linear dynamic control of the tracking loop [Yu 2009], [Hendriks et al. 2010], [Gerkmann 2011], [Gerkmann, Hendriks 2012]. While the minimum-statistics approach quite reliably prevents speech leaking into the noise power estimate it also comes along with a tracking delay of one to two seconds. It is therefore less suitable for rapidly varying noise types.

The tracking speed is significantly improved when the non-linear separation between speech and noise is based on an instantaneous non-linear MMSE estimator of the noise periodogram and subsequent smoothing. This approach has been investigated in conjunction with a fixed smoothing parameter in [Yu 2009], [Hendriks et al. 2010] and using the SPP as a dynamic smoothing parameter in [Sohn, Sung 1998], [Gerkmann 2011], and [Gerkmann, Hendriks 2012]. We will briefly outline the latter approach.

In analogy to the MMSE estimator of the magnitude-squared speech components (12.70) the corresponding noise periodogram estimator is given by [Yu 2009] and [Hendriks et al. 2010] as

$$E\left\{|N_\mu|^2 \mid Y_\mu, H_\mu^{(1)}\right\} = \left(\frac{\sigma_{N,\mu}^2}{\sigma_{S,\mu}^2 + \sigma_{N,\mu}^2}\right)^2 |Y_\mu|^2 + \frac{\sigma_{S,\mu}^2 \sigma_{N,\mu}^2}{\sigma_{S,\mu}^2 + \sigma_{N,\mu}^2}, \tag{12.119}$$

which is in principle an unbiased estimator as its expectation is equal to $\sigma^2_{N,\mu}$. However, it has also been shown that this estimator is biased when the estimated speech and noise powers, which are required for its implementation, differ from their true values. Therefore, the authors in [Yu 2009] and [Hendriks et al. 2010] analyze this bias and provide a bias correction. Both these works use a fixed smoothing parameter and additional countermeasures to avoid stagnation in the tracking loop.

Alternatively and in analogy to (5.176) and the exposition in Section 12.5 the estimator might be written as a joint estimation and detection problem

$$
\mathrm{E}\left\{|N_\mu(\lambda)|^2 \mid Y_\mu(\lambda)\right\} = \mathrm{E}\left\{|N_\mu(\lambda)|^2 \mid Y_\mu(\lambda), H^{(1)}_\mu(\lambda)\right\} P(H^{(1)}_\mu(\lambda) \mid Y_\mu(\lambda))
$$
$$
+ \mathrm{E}\left\{|N_\mu(\lambda)|^2 \mid Y_\mu(\lambda), H^{(0)}_\mu(\lambda)\right\} P(H^{(0)}_\mu(\lambda) \mid Y_\mu(\lambda)), \quad (12.120)
$$

which naturally incorporates the probability of speech presence $P(H^{(1)}_\mu(\lambda))$ in the estimation procedure for $\mathrm{E}\left\{|N_\mu(\lambda)|^2 \mid Y_\mu(\lambda)\right\}$.

For speech activity, an unbiased estimator of noise power may be approximated by the power estimate of the previous frame

$$
\mathrm{E}\left\{|N_\mu(\lambda)|^2 \mid Y_\mu(\lambda), H^{(1)}_\mu(\lambda)\right\} \approx \hat{\sigma}^2_{N,\mu}(\lambda - 1), \quad (12.121)
$$

and for speech absence we have

$$
\mathrm{E}\left\{|N_\mu(\lambda)|^2 \mid Y_\mu(\lambda), H^{(0)}_\mu(\lambda)\right\} = |Y_\mu(\lambda)|^2. \quad (12.122)
$$

The computation of the SPP estimation starts with the generalized likelihood ratio

$$
\Lambda_\mu(Y_\mu(\lambda)) = \frac{p(Y_\mu \mid H^{(1)}_\mu)(\lambda)\, p_\mu}{p(Y_\mu \mid H^{(0)}_\mu)(\lambda)\, q_\mu}, \quad (12.123)
$$

where $p_\mu$ and $q_\mu$ denote the a priori probabilities of speech presence and absence. The SPP is then given by

$$
P(H^{(1)}_\mu(\lambda) \mid Y_\mu(\lambda)) = \frac{\Lambda_\mu(Y_\mu(\lambda))}{1 + \Lambda_\mu(Y_\mu(\lambda))}. \quad (12.124)
$$

Authors in [Gerkmann 2011] and [Gerkmann, Hendriks 2012] use the complex Gaussian distribution model as in (12.88) with $q_\mu = (1 - p_\mu) = 0.5$ and a fixed, optimized a priori SNR of $\xi_\mu = 15\,\mathrm{dB}$ [Gerkmann et al. 2008]. The latter feature is indeed important as it decouples the a priori SNR estimate from the SPP computation and constitutes the difference w.r.t. earlier work [Sohn, Sung 1998]. After additional first-order smoothing of the SPP with a fixed smoothing parameter $\alpha_{\mathrm{SPP}}$ a preliminary noise power estimate is obtained,

$$
\tilde{\sigma}^2_{N,\mu}(\lambda) = P(H^{(1)}_\mu(\lambda) \mid Y_\mu(\lambda))\hat{\sigma}^2_{N,\mu}(\lambda - 1) + \left(1 - P(H^{(1)}_\mu(\lambda) \mid Y_\mu(\lambda))\right)|Y_\mu(\lambda)|^2. \quad (12.125)
$$

Furthermore, whenever the smoothed SPP is larger than 0.99, the estimated SPP is constrained to values smaller than 0.99 to avoid stagnation. Then, the output of another first-order recursive system with input $\tilde{\sigma}^2_{N,\mu}(\lambda)$ and a fixed smoothing parameter $\alpha_{\mathrm{PS}}$ delivers the final estimate $\hat{\sigma}^2_{N,\mu}(\lambda)$.

Finally, we like to mention that for all methods, parameters and additional measures need to be carefully tuned on larger data sets (e.g., in cases where the spectral resolution and the frame rate differ from the original settings) in order to achieve the desired tracking performance and to minimize speech leakage.

### 12.8.5  Evaluation of Noise Power Trackers

The evaluation of a noise PS tracker may follow different strategies. While in the initial stages of development, direct measurements of estimation errors are most useful, the tracker will eventually function in the context of a speech enhancement system. Embedded into a speech enhancement system, the tracker will, in many cases, interact with other estimators or gain functions. Therefore, not only speech quality and speech intelligibility measures (such as PESQ and STOI or mutual-information-based measures [Taghia, Martin 2014]) but also listening experiments are very helpful in the assessment of the overall system and the tracker in particular.

Yet, for a comparison of different versions in the early stages of a development, it is interesting to evaluate the estimation error, and more specifically, the over- and under-estimation errors. A particular noise PS tracker may have a tendency to underestimate the true noise PS. This tracker is hence likely to generate a higher level of residual noise in the processed signal. Estimators, which allow the tracking of rapidly varying noise power levels might also have a tendency to overestimate the true noise power level and therefore might lead to an attenuation of the target signal. Furthermore, trackers with irregular fluctuations in the estimate are likely to produce unnatural artifacts at the output of the enhancement system. These general aspects can be assessed via the log-error measure [Hendriks et al. 2008] and the log-error variance [Taghia et al. 2011]. The log-error distortion measure is defined as

$$\text{LogErr} = \frac{1}{NF} \sum_{\lambda=0}^{N-1} \sum_{\mu=0}^{F-1} \left| 10 \log \left( \frac{\hat{\sigma}_{N,\mu}^2(\lambda)}{\sigma_{N,\mu}^2(\lambda)} \right) \right|, \tag{12.126}$$

where $N$ and $F$ define the overall number of time and frequency bins, respectively. Furthermore, [Gerkmann, Hendriks 2012] also proposed to compute separate distortion measures $\text{LogErr}_u$ and $\text{LogErr}_o$ for errors related to underestimation and overestimation as these error types cause different perceptual effects, as discussed above. These measures are defined as

$$\text{LogErr}_u = \frac{1}{NF} \sum_{\lambda=0}^{N-1} \sum_{\mu=0}^{F-1} \left| \min \left( 0, 10 \log \left( \frac{\hat{\sigma}_{N,\mu}^2(\lambda)}{\sigma_{N,\mu}^2(\lambda)} \right) \right) \right| \tag{12.127}$$

and

$$\text{LogErr}_o = \frac{1}{NF} \sum_{\lambda=0}^{N-1} \sum_{\mu=0}^{F-1} \left| \max \left( 0, 10 \log \left( \frac{\hat{\sigma}_{N,\mu}^2(\lambda)}{\sigma_{N,\mu}^2(\lambda)} \right) \right) \right|. \tag{12.128}$$

[Gerkmann, Hendriks 2012] also show that for babble noise and traffic noise, the rapidly-tracking MMSE estimators significantly reduce the errors associated with noise power underestimation in comparison to the minimum-statistics method but increase the errors associated with noise power overestimation.

Thus, in the context of speech enhancement, fast tracking methods may lead to additional speech distortions. A comparison of several state-of-the-art methods with respect to the above measures is provided, e.g., by [Taghia et al. 2011] and [Yong, Nordholm 2016], where in the latter work further refinements of the above method are described as well.

## 12.9  Noise Reduction with Deep Neural Networks

In recent years, DNNs have been successfully deployed also in the field of single-channel noise reduction. The significant performance improvements achieved through DNN-based systems do not only stem from the use of improved multi-variate signal models but also from DNN structures that are specifically geared toward the separation of independent audio sources and the generation of speech signals from embedded representations. Besides discriminative models, which aim at a functional mapping from input features to the desired target, also generative models which learn the statistical distribution of the target signal have been investigated. These include generative adversarial approaches (e.g., SEGAN [Pascual et al. 2017]) and the more recent diffusion methods, which iteratively synthesize the noise-free output signal via a large-scale generative network [Welker et al. 2022]. Most remarkably, also single-channel speaker separation has seen significant breakthroughs in the past decade as DNNs can be trained to separate speakers in an embedded space and to derive separating masks from these embeddings, see e.g., [Hershey et al. 2016], [Wang, Chen 2018], and [Subakan et al. 2021].

The discriminative approaches for speech enhancement, which will be discussed with some more detail below, aim at a variety of training targets [Wang et al. 2014], [Nicolson, Paliwal 2021]. They include the estimation of binary spectral masks, the estimation of spectral magnitudes and phases, the estimation of SNR, the estimation of real and imaginary parts of the clean-speech spectrum, and the direct estimation of the time domain signal, e.g., [Luo, Mesgarani 2019] which features a fully convolutional end-to-end audio separation network (Conv-TasNet) for speech separation. To this date, many different methods have been proposed and developments toward computationally efficient implementations is still ongoing.

An obvious advantage of DNNs resides in the use of extended multi-variate context in the input data (sometimes called "receptive field") for the estimation of the output target. While traditional methods almost always rely on decorrelating transformations and processing schemes that act mostly independent in frequency sub-bands, data-driven DNN-based methods may capitalize on dependencies across frequency bins and time frames. As argued in Section 5.10.3, the assumption of independent frequency bins is not well justified when using short analysis frames. DNN-based methods consider dependencies across frequency bins via high-dimensional vectors of spectral features while temporal context is often taken care of via recurrent networks or time-delay networks (TDNNs), and their variants.

A general block diagram of a DFT-based processing system is depicted in Figure 12.15. In the horizontal signal path, we buffer DFT frames with the option to include a look-ahead of $L$ frames or to use a succession of past DFT frames in the estimation process. To capture the context of successive spectra, also the input feature set relies on a larger number of buffered past frames and possibly additional features, which are extracted from the time domain signal. The DNN then uses these features to estimate a target gain function, which is applied to $Y_\mu(\lambda - L)$.

The DNN may learn the transformation from noisy features to a variety of target representations where the context, both in frequency and time, may be exploited. The implementation of such data-driven noise reduction approaches requires a number of essential ingredients. They are briefly discussed as follows:

### 12.9.1 Processing Model

Typically, the network for estimating the target quantity may be composed of feed-forward, time-delay, convolutional, or recurrent neural network layers (see Section 5.12), possibly with additional elements such as self-attention networks [Vaswani et al. 2017]. Often, the layers are arranged in an encoder–decoder structure or in an U-net structure [Ronneberger et al. 2015] and the specific layouts also depend on the estimation target. When, for instance a spectral mask (gain function) is estimated on a frame-by-frame basis, as shown in Figure 12.15, the estimated clean speech spectrum is obtained via a multiplicative application of the mask $H_\mu(\lambda)$

$$\widehat{S}_{\mu-L}(\lambda) = H_\mu(\lambda)\, Y_\mu(\lambda - L), \tag{12.129}$$

where $L$ denotes the number of look-ahead frames. It appears to be a consensus that the spectral mask should also include phase modifications, and in this case it is also beneficial to implement complex-valued operations in all layers [Hu et al. 2020]. It has been observed that simply estimating real and imaginary parts for the construction of a complex mask may not be most efficient. Alternative models provide separate estimation paths for the amplitude and phase information, albeit with interactions between the two [Yin et al. 2020]. Also, the estimation of complex filter functions using data across time and frequency have been introduced to DNN-based processing models. These, are also known as deep filtering [Mack, Habets 2020], [Schröter et al. 2022] and are formulated as

$$\widehat{S}_\mu(\lambda) = \mathbf{H}_\mu^H(\lambda)\, \mathbf{Y}_\mu(\lambda), \tag{12.130}$$

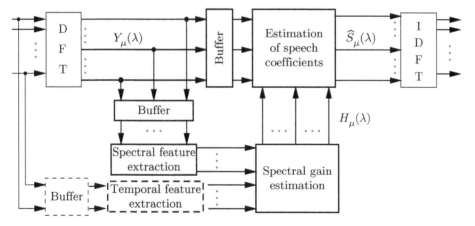

**Figure 12.15** A general diagram of a DFT-based noise reduction system using frequency domain input and optional time domain features, excluding segmentation and overlap-add operations. On the one hand, buffering of DFT frames enlarges the receptive field for feature extraction and on the other hand, it allows a non-causal operation or "deep filtering" operation in the signal path.

where $\bullet^H$ denotes the Hermitian operator. Here, we employ a complex-valued filter vector $\mathbf{H}_\mu(\lambda)$ on a vector $\mathbf{Y}_\mu(\lambda)$ of corresponding size in the time-frequency domain. The latter vector may hold a rectangular patch of time-frequency data or a sparsified version thereof. It has been observed that for any specific time-frequency bin $(\mu, \lambda)$, mostly the temporal context of this bin and the frequency context of its adjacent bins are of importance [Schröter et al. 2022]. The data in $\mathbf{Y}_\mu(\lambda)$ may be purely causal or may include look-ahead frames. Note that temporal context had also been considered in traditional methods, most notably in the context of single-channel minimum variance distortionless response (MVDR) approaches [Benesty, Huang 2011], [Schasse, Martin 2014], which provide less distorted target signals than, e.g., a Wiener filter. To reduce the overall computational complexity and latency it has also been found helpful to exploit knowledge about the speech production process such that the envelope information and the excitation is processed in dedicated (however linked) networks [Valin et al. 2020], [Schröter et al. 2022]. Moreover, improved performance has been demonstrated by networks, which first estimate the spectral magnitudes and then recover the complex-valued spectrum in a second stage [Li et al. 2021a].

### 12.9.2 Estimation Targets

A variety of estimation targets have been successfully used. These are explored, e.g., in the context of spectral amplitude features in [Nicolson, Paliwal 2021] and for speaker separation in Wang et al. [2014], Wang, Chen [2018]. Prominent options are

- compressed spectral amplitudes, log-PS, (compressed) complex spectra, or real and imaginary parts of the speech spectrum $S_\mu(\lambda)$,
- the noise PS [Li et al. 2019], [Zhang et al. 2020],
- the a priori SNR $\eta_\mu(\lambda)$ and/or a posteriori SNR $\gamma_\mu(\lambda)$ [Nicolson, Paliwal 2019], [Rehr, Gerkmann 2021], [Kim et al. 2022],
- the SPP [Tu et al. 2019], [Kim et al. 2022],
- the *ideal ratio mask*

$$H_\mu^\beta(\lambda) = \left( \frac{|S_\mu(\lambda)|^2}{|S_\mu(\lambda)|^2 + |N_\mu(\lambda)|^2} \right)^\beta$$

where the exponent $\beta$ may be further tuned, and
- the *complex ideal ratio mask* [Williamson et al. 2016b]

$$
\begin{aligned}
H_\mu(\lambda) &= \frac{S_\mu(\lambda)}{Y_\mu(\lambda)} = \frac{S_\mu(\lambda)Y_\mu^*(\lambda)}{\left|Y_\mu(\lambda)\right|^2} \\
&= \frac{\mathrm{Re}\{S_\mu(\lambda)\}\mathrm{Re}\{Y_\mu(\lambda)\} + \mathrm{Im}\{S_\mu(\lambda)\}\mathrm{Im}\{Y_\mu(\lambda)\}}{|Y_\mu(\lambda)|^2} \\
&\quad + j\frac{\mathrm{Im}\{S_\mu(\lambda)\}\mathrm{Re}\{Y_\mu(\lambda)\} - \mathrm{Re}\{S_\mu(\lambda)\}\mathrm{Im}\{Y_\mu(\lambda)\}}{|Y_\mu(\lambda)|^2} \\
&= \mathrm{Re}\{H_\mu(\lambda)\} + j\mathrm{Im}\{H_\mu(\lambda)\}.
\end{aligned}
$$

### 12.9.3 Loss Function

Loss functions comprise most often the mean square error in the time or frequency domain, the mean absolute error, or compressed versions (similar to 12.62) thereof. It has also been shown that loss functions composed of several criteria outperform single error criteria and may also include perceptual measures. Most notably, it is useful to also consider phase errors in the loss function. A typical composite cost function that takes compressed amplitudes as well as phases in the frequency domain into account is given by Braun et al. [2021]

$$
\mathcal{L} = \alpha \sum_{\mu,\lambda} \left| |S_\mu(\lambda)|^\beta \frac{S_\mu(\lambda)}{|S_\mu(\lambda)|} - |\widehat{S}_\mu(\lambda)|^\beta \frac{\widehat{S}_\mu(\lambda)}{|\widehat{S}_\mu(\lambda)|} \right|^2 + (1-\alpha) \sum_{\mu,\lambda} \left| |S_\mu(\lambda)|^\beta - |\widehat{S}_\mu(\lambda)|^\beta \right|^2,
$$

(12.131)

with, e.g., $\beta = 0.3$ and $\alpha = 0.3$ and $\widehat{S}_\mu(\lambda)$ denoting the estimated target. A composite loss considering time-domain and frequency-domain errors is proposed, for instance, in [Isik et al. 2020], which also allows to assign different weights to underestimation and overestimation errors in the frequency domain.

When the estimation target are (level-dependent) signal spectra $S_\mu(\lambda)$ (instead of gain functions), it is of advantage to normalize the amplitudes on the standard deviation $\sigma_S$ across each training utterance, i.e., $\widetilde{S}_\mu(\lambda) = \frac{S_\mu(\lambda)}{\sigma_S}$ and $\widetilde{Y}_\mu(\lambda) = \frac{Y_\mu(\lambda)}{\sigma_S}$, where speech pauses are discarded in the computation of the standard deviation of $S_\mu(\lambda)$. The normalized target $\widetilde{S}_\mu(\lambda)$ then replaces the corresponding original quantities $S_\mu(\lambda)$ in the loss function.

### 12.9.4 Input Features

The network input for the estimation (inference) process often constitutes log-mel band energies, the amplitudes or real and imaginary parts of DFT coefficients or the output of an auditory filterbank. Input features have been studied in a variety of works, e.g., for SNR estimation in Li et al. [2021b]. It turns out that the best set of features depends on the estimation target and also on the network structure. Recently, it has been noted that additional frequency-positional codes, concatenated to the feature sets, improve the enhancement procedure [Isik et al. 2020].

### 12.9.5 Data Sets

Large training, validation, and test data sets with sufficiently large diversity are a crucial ingredient. Big networks require big data sets.

A simple DNN processing model is shown in Figure 12.16 where the input features enter a succession of convolutional and recursive network layers, e.g., [Braun et al. 2021]. Through decimation of the input data matrix, the dimensions are successively reduced whereas the use of a multitude of filter kernels ("channels") in each convolutional layer provides a multifaceted signal representation. The output will thus be forced to focus on the most salient features of the speech signal and reject noise and interference. Then, the temporal

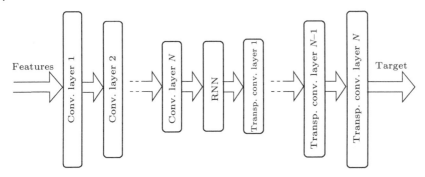

**Figure 12.16** A DNN structure for the estimation of a specific target (spectra or mask functions) using convolutional layers in the encoder and the decoder and a recursive layer in between. Often, skip connections are employed between the corresponding layers of the encoder and the decoder (e.g., [Braun et al. 2021]).

coherence of the short-time spectra is modeled via a recursive network composed of one or multiple long short-term memory (LSTM) units or gated recursive units (GRU) [Erdogan et al. 2015], [Sun et al. 2017], [Braun et al. 2021]. Thus, the recursive layers situated in the bottleneck of the network structure introduce additional temporal memory. Finally, a succession of transposed convolutional layers interpolates the output of the RNN layer(s) to the desired target dimensions.

Figure 12.17 depicts a typical block diagram of a setup for network training. To also include overlap-add reconstruction errors in the loss function, the short-time Fourier transform (STFT) of the reconstructed signal is optionally fed back to the training module.

Compared to low-dimensional traditional estimation and processing models as discussed in the first part of this chapter, pre-trained DNN methods exhibit different behavior: On the

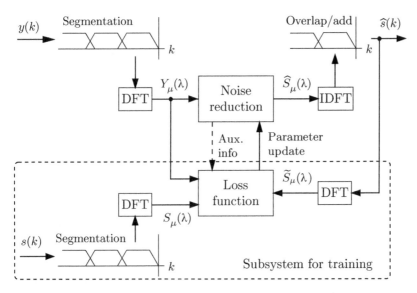

**Figure 12.17** Overview on training procedure for STFT-based systems taking the consistency constraint into account.

one hand, it has been observed early on in the development of DNN-based noise reduction algorithms [Xu et al. 2014], [Xu et al. 2015] that pre-trained deep networks suffer much less from fast fluctuating residual noise (*musical noise*) than traditional adaptive methods. On the other hand, the adaptation to high-dimensional input data and the generalization to unseen noise examples requires large and diverse data sets in the training step. As a result, significant computational resources are necessary, at least for network training.

In order to achieve low-latency operation and a lower computational complexity several works have also considered combinations of traditional and DNN-based algorithmic components. The *Deep Xi* framework [Nicolson, Paliwal 2019] uses a recurrent neural network to estimate the a priori SNR, which is known to be a crucial quantity for the computation of many traditional gain functions, including the Wiener filter. The method has been further refined via a non-recursive temporal convolutional network (TCN) and has been also applied to noise PS estimation [Zhang et al. 2020]. In Kim et al. [2022], the method has been extended to estimate both the noise and the speech PS using an augmented convolutional transformer (*conformer*) to improve the estimation of the a priori and the a posteriori SNR and thus the performance of the resulting speech enhancement system (using the MMSE-LSA gain function) further. Essentially, these method are capable of eliminating the tracking delay of traditional estimation methods. DNN-based methods have also been combined with elements from speech coders. The *PerceptNet* [Valin et al. 2021] enhances speech features related to the spectral envelope (via a post-filter) and the speech excitation via a comb filter, which is adapted to the fundamental frequency. This approach achieves a low computational complexity even on full-band signals.

# References

3GPP TR 126.978, V4.0.0 (2001). Universal Mobile Telecommunications System (UMTS); Results of the AMR Noise Suppression Selection Phase.

3GPP TS 122.076 V5.0.0 (2002). Digital Cellular Telecommunications System (Phase 2+); Universal Mobile Telecommunications System (UMTS); Noise Suppression for the AMR Codec; Service Description.

3GPP TS 126.094, V4.0.0 (2001). Universal Mobile Telecommunications System (UMTS); Mandatory Speech Codec Speech Processing Functions, AMR Speech Codec; Voice Activity Detector (VAD).

Accardi, A. J.; Cox, R. (1999). A Modular Approach to Speech Enhancement with an Application to Speech Coding, *Proceedings of the IEEE International Conference on Acoustics, Speech, and Signal Processing (ICASSP)*, Phoenix, Arizona, USA, vol. 1, pp. 201–204.

Andersen, A. H.; de Haan, J. M.; Tan, Z.-H.; Jensen, J. (2016). Predicting the Intelligibility of Noisy and Nonlinearly Processed Binaural Speech, *IEEE/ACM Transactions on Audio, Speech, and Language Processing*, vol. 24, no. 11, pp. 1908–1920.

Andrianakis, I.; White, P. (2006). MMSE Speech Spectral Amplitude Estimators With Chi and Gamma Speech Priors, *2006 IEEE International Conference on Acoustics Speech and Signal Processing Proceedings*, vol. 3, pp. III–III.

Beaugeant, C. (1999). *Réduction de Bruit et Contrôle d'Echo pour les Applications Radiomobiles*, PhD thesis, University of Rennes 1.

Benesty, J.; Huang, Y. (2011). A Single-Channel Noise Reduction MVDR Filter, *2011 IEEE International Conference on Acoustics, Speech and Signal Processing (ICASSP)*, pp. 273–276.

Beritelli, F.; Casale, S.; Ruggeri, G. (2001). Performance Evaluation and Comparison of ITU-T/ETSI Voice Activity Detectors, *Proceedings of the IEEE International Conference on Acoustics, Speech, and Signal Processing (ICASSP)*, Salt Lake City, Utah, USA, pp. 1425–1428.

Berouti, M.; Schwartz, R.; Makhoul, J. (1979). Enhancement of Speech Corrupted by Acoustic Noise, *Proceedings of the IEEE International Conference on Acoustics, Speech, and Signal Processing (ICASSP)*, Tulsa, Oklahoma, USA, pp. 208–211.

Boll, S. F. (1979). Suppression of Acoustic Noise in Speech Using Spectral Subtraction, *IEEE Transactions on Acoustics, Speech and Signal Processing*, vol. 27, pp. 113–120.

Braun, S.; Gamper, H.; Reddy, C. K.; Tashev, I. (2021). Towards Efficient Models for Real-Time Deep Noise Suppression, *ICASSP 2021 - 2021 IEEE International Conference on Acoustics, Speech and Signal Processing (ICASSP)*, pp. 656–660.

Breithaupt, C. (2008). *Noise Reduction Algorithms for Speech Communications – Statistical Analysis and Improved Estimation Procedures*, PhD thesis. *Institute of Communication Acoustics*, Ruhr-Universität Bochum.

Breithaupt, C.; Martin, R. (2006). Voice Activity Detection in the DFT Domain Based on a Parametric Noise Model, *Proceedings of International Workshop on Acoustic Echo and Noise Control (IWAENC)*, Paris, pp. 1–4.

Breithaupt, C.; Martin, R. (2008). Noise Reduction – Statistical Analysis and Control of Musical Noise, *in* R. Martin; U. Heute; C. Antweiler (eds.), *Advances in Digital Speech Transmission*, John Wiley & Sons, Ltd., Chichester, pp. 107–133.

Breithaupt, C.; Martin, R. (2011). Analysis of the Decision-Directed SNR Estimator for Speech Enhancement With Respect to Low-SNR and Transient Conditions, *IEEE Transactions on Audio, Speech, and Language Processing*, vol. 19, no. 2, pp. 277–289.

Breithaupt, C.; Gerkmann, T.; Martin, R. (2007). Cepstral Smoothing of Spectral Filter Gains for Speech Enhancement Without Musical Noise, *IEEE Signal Processing Letters*, vol. 14, no. 12, pp. 1036–1039.

Breithaupt, C.; Gerkmann, T.; Martin, R. (2008a). A Novel a Priori SNR Estimation Approach Based on Selective Cepstro-Temporal Smoothing, *2008 IEEE International Conference on Acoustics, Speech and Signal Processing*, pp. 4897–4900.

Breithaupt, C.; Krawczyk, M.; Martin, R. (2008b). Parameterized MMSE Spectral Magnitude Estimation for the Enhancement of Noisy Speech, *2008 IEEE International Conference on Acoustics, Speech and Signal Processing*, pp. 4037–4040.

Cappé, O. (1994). Elimination of the Musical Noise Phenomenon with the Ephraim and Malah Noise Suppressor, *IEEE Transactions on Speech and Audio Processing*, vol. 2, no. 2, pp. 345–349.

Chen, J.; Wang, Y.; Yoho, S. E.; Wang, D.; Healy, E. W. (2016). Large-Scale Training to Increase Speech Intelligibility for Hearing-Impaired Listeners in Novel Noises, *The Journal of the Acoustical Society of America*, vol. 139, no. 5, pp. 2604–2612.

Cohen, I.; Berdugo, B. (2001). Speech Enhancement for Non-Stationary Noise Environments, *Signal Processing*, vol. 81, pp. 2403–2418.

Collura, J. S. (1999). Speech Enhancement and Coding in Harsh Acoustic Noise Environments, *Proceedings of the IEEE Workshop on Speech Coding*, pp. 162–164.

Dat, T. H.; Takeda, K.; Itakura, F. (2005). Generalized Gamma Modeling of Speech and its Online Estimation for Speech Enhancement, *Proceedings of the IEEE International Conference on Acoustics, Speech, and Signal Processing (ICASSP)*, Philadelphia, Pennsylvania, USA, vol. IV, pp. 181–184.

Doblinger, G. (1991). An Efficient Algorithm for Uniform and Nonuniform Digital Filter Banks, *IEEE International Symposium on Circuits and Systems (ISCAS)*, Raffles City, Singapore, pp. 646–649.

Doblinger, G.; Zeitlhofer, T. (1996). Improved Design of Uniform and Nonuniform Modulated Filter Banks, *Proceedings of the IEEE Nordic Signal Processing Symposium (NORSIG)*, Helsinki, Finland, pp. 327–330.

Elshamy, S.; Madhu, N.; Tirry, W.; Fingscheidt, T. (2018). DNN-Supported Speech Enhancement With Cepstral Estimation of Both Excitation and Envelope, *IEEE/ACM Transactions on Audio, Speech, and Language Processing*, vol. 26, no. 12, pp. 2460–2474.

Enzner, G.; Thüne, P. (2017). Robust MMSE Filtering for Single-Microphone Speech Enhancement, *2017 IEEE International Conference on Acoustics, Speech and Signal Processing (ICASSP)*, pp. 4009–4013.

Ephraim, Y.; Malah, D. (1984). Speech Enhancement Using a Minimum Mean-Square Error Short-Time Spectral Amplitude Estimator, *IEEE Transactions on Acoustics, Speech and Signal Processing*, vol. 32, no. 6, pp. 1109–1121.

Ephraim, Y.; Malah, D. (1985). Speech Enhancement Using a Minimum Mean-Square Error Log-Spectral Amplitude Estimator, *IEEE Transactions on Acoustics, Speech and Signal Processing*, vol. 33, no. 2, pp. 443–445.

Ephraim, Y.; Rahim, M. (1998). On Second Order Statistics and Linear Estimation of Cepstral Coefficients, *Proceedings of the 1998 IEEE International Conference on Acoustics, Speech and Signal Processing, ICASSP '98 (Cat. No.98CH36181)*, vol. 2, pp. 965–968.

Erdogan, H.; Hershey, J. R.; Watanabe, S.; Le Roux, J. (2015). Phase-sensitive and Recognition-boosted Speech Separation Using Deep Recurrent Neural Networks, *2015 IEEE International Conference on Acoustics, Speech and Signal Processing (ICASSP)*, pp. 708–712.

Erkelens, J. S.; Hendriks, R. C.; Heusdens, R.; Jensen, J. (2007). Minimum Mean-Square Error Estimation of Discrete Fourier Coefficients With Generalized Gamma Priors, *IEEE Transactions on Audio, Speech, and Language Processing*, vol. 15, no. 6, pp. 1741–1752.

Erkelens, J. S.; Hendriks, R. C.; Heusdens, R. (2008). On the Estimation of Complex Speech DFT Coefficients Without Assuming Independent Real and Imaginary Parts, *IEEE Signal Processing Letters*, vol. 15, pp. 213–216.

Esch, T.; Vary, P. (2009). Efficient Musical Noise Suppression for Speech Enhancement System, *2009 IEEE International Conference on Acoustics, Speech and Signal Processing*, pp. 4409–4412.

ETSI TS 126.445 V14.2.0 (2018). Universal Mobile Telecommunications System (UMTS); LTE; Codec for Enhanced Voice Services (EVS); Detailed Algorithmic Description.

Falk, T. H.; Parsa, V.; Santos, J. F.; Arehart, K.; Hazrati, O.; Huber, R.; Kates, J. M.; Scollie, S. (2015). Objective Quality and Intelligibility Prediction for Users of Assistive Listening Devices: Advantages and Limitations of Existing Tools, *IEEE Signal Processing Magazine*, vol. 32, no. 2, pp. 114–124.

Freeman, D. K.; Cosier, G.; Southcott, C. B.; Boyd, I. (1989). The Voice Activity Detector for the Pan-European Digital Cellular Mobile Telephone Service, *Proceedings of the IEEE*

*International Conference on Acoustics, Speech, and Signal Processing (ICASSP)*, Glasgow, Schottland, pp. 369–372.

Gerkmann, T.; Hendriks, R. C. (2011). Noise Power Estimation Based on the Probability of Speech Presence, *2011 IEEE Workshop on Applications of Signal Processing to Audio and Acoustics (WASPAA)*, pp. 145–148.

Gerkmann, T.; Hendriks, R. C. (2012). Unbiased MMSE-Based Noise Power Estimation With Low Complexity and Low Tracking Delay, *IEEE Transactions on Audio, Speech, and Language Processing*, vol. 20, no. 4, pp. 1383–1393.

Gerkmann, T.; Krawczyk, M. (2013). MMSE-Optimal Spectral Amplitude Estimation Given the STFT-Phase, *IEEE Signal Processing Letters*, vol. 20, no. 2, pp. 129–132.

Gerkmann, T.; Martin, R. (2009). On the Statistics of Spectral Amplitudes After Variance Reduction by Temporal Cepstrum Smoothing and Cepstral Nulling, *IEEE Transactions on Signal Processing*, vol. 57, no. 11, pp. 4165–4174.

Gerkmann, T.; Breithaupt, C.; Martin, R. (2008). Improved a Posteriori Speech Presence Probability Estimation Based on a Likelihood Ratio with Fixed Priors, *IEEE Transactions on Audio, Speech, and Language Processing*, vol. 16, no. 5, pp. 910–919.

Graf, S.; Herbig, T.; Buck, M.; Schmidt, G. (2015). Features for Voice Activity Detection: A Comparative Analysis, *EURASIP Journal on Advances in Signal Processing*, vol. 2015, p. 91.

Griffin, D. W.; Lim, J. S. (1984). Signal Estimation from Modified Short-Time Fourier Transform, *IEEE Transactions on Acoustics, Speech and Signal Processing*, vol. 32, no. 2, pp. 236–243.

Gülzow, T.; Engelsberg, A. (1998). Comparison of a Discrete Wavelet Transformation and a Nonuniform Polyphase Filterbank Applied to Spectral Subtraction Speech Enhancement, *Signal Processing*, vol. 64, no. 1, pp. 5–19.

Gülzow, T.; Ludwig, T.; Heute, U. (2003). Spectral-Substraction Speech Enhancement in Multirate Systems with and without Non-uniform and Adaptive Bandwidths, *Signal Processing*, vol. 83, pp. 1613–1631.

Gustafsson, S.; Jax, P.; Vary, P. (1998). A Novel Psychoacoustically Motivated Audio Enhancement Algorithm Preserving Background Noise Characteristics, *Proceedings of the IEEE International Conference on Acoustics, Speech, and Signal Processing (ICASSP)*, Seattle, Washington, USA, pp. 397–400.

Häkkinen, J.; Väänänen, M. (1993). Background Noise Suppressor for a Car Hands-Free Microphone, *Proceedings of the International Conference on Signal Processing Applications and Techniques (ICSPAT)*, Santa Clara, California, USA, pp. 300–307.

Haykin, S. (1996). *Adaptive Filter Theory*, 3rd edn, Prentice Hall, Englewood Cliffs, New Jersey.

Heitkämper, J.; Schmalenströer, J.; Häb-Umbach, R. (2020). Statistical and Neural Network Based Speech Activity Detection in Non-Stationary Acoustic Environments, *Proceedings of INTERSPEECH 2020*, pp. 2597–2601.

Hendriks, R. C.; Jensen, J.; Heusdens, R. (2008). Noise Tracking Using DFT Domain Subspace Decompositions, *IEEE Transactions on Audio, Speech, and Language Processing*, vol. 16, no. 3, pp. 541–553.

Hendriks, R. C.; Heusdens, R.; Jensen, J. (2010). MMSE Based Noise PSD Tracking with Low Complexity, *2010 IEEE International Conference on Acoustics, Speech and Signal Processing*, pp. 4266–4269.

Hershey, J. R.; Chen, Z.; Le Roux, J.; Watanabe, S. (2016). Deep Clustering: Discriminative Embeddings for Segmentation and Separation, *2016 IEEE International Conference on Acoustics, Speech and Signal Processing (ICASSP)*, pp. 31–35.

Hu, Y.; Loizou, P. C. (2008). Evaluation of Objective Quality Measures for Speech Enhancement, *IEEE Transactions on Audio, Speech, and Language Processing*, vol. 16, no. 1, pp. 229–238.

Hu, Y.; Liu, Y.; Lv, S.; Xing, M.; Zhang, S.; Fu, Y.; Wu, J.; Zhang, B.; Xie, L. (2020). DCCRN: Deep Complex Convolution Recurrent Network for Phase-Aware Speech Enhancement, *Proceedings of INTERSPEECH 2020*, pp. 2472–2476.

Isik, U.; Giri, R.; Phansalkar, N.; Valin, J.-M.; Helwani, K.; Krishnaswamy, A. (2020). PoCoNet: Better Speech Enhancement with Frequency-Positional Embeddings, Semi-Supervised Conversational Data, and Biased Loss, *Proceedings of INTERSPEECH 2020*, pp. 2487–2491.

ITU-T Rec. P.800 (1996). Methods for Subjective Determination of Transmission Quality, *Series P: Recommendation P.800*, International Telecommunication Union (ITU).

ITU-T Rec. P.835 (2003). Subjective Test Methodology for Evaluating Speech Communication Systems that Include Noise Suppression Algorithm, *Series P: Recommendation P.835*, vol. Telephone Transmission Quality, Telephone Installations, Local Line Networks, Methods for objective and subjective assessment of quality, International Telecommunication Union (ITU).

ITU-T Rec. P.862 (2001). Perceptual Evaluation of Speech Quality (PESQ): An Objective Method for End-to-End Speech Quality Assessment of Narrow-Band Telephone Networks and Speech Codecs, *Series P: Recommendation P.862*, International Telecommunication Union (ITU).

ITU-T Rec. P.863 (2018). Perceptual objective listening quality prediction, *Series P: Recommendation P.863*, International Telecommunication Union (ITU).

Jensen, J.; Taal, C. H. (2016). An Algorithm for Predicting the Intelligibility of Speech Masked by Modulated Noise Maskers, *IEEE/ACM Transactions on Audio, Speech, and Language Processing*, vol. 24, no. 11, pp. 2009–2022.

Johnson, N. L.; Kotz, S.; Balakrishnan, N. (1994). *Continuous Univariate Distributions*, John Wiley & Sons, Ltd., Chichester.

Kang, G. S.; Fransen, L. J. (1989). Quality Improvements of LPC-Processed Noisy Speech by Using Spectral Subtraction, *IEEE Transactions on Acoustics, Speech and Signal Processing*, vol. 37, no. 6, pp. 939–942.

Kappelan, M.; Strauß, B.; Vary, P. (1996). Flexible Nonuniform Filter Banks Using Allpass Transformation of Multiple Order, *Proceedings of Signal Processing VIII - Theories and Applications*, Trieste, Italy, pp. 1745–1748.

Kim, M.; Song, H.; Cheong, S.; Shin, J. W. (2022). iDeepMMSE: An Improved Deep Learning Approach to MMSE Speech and Noise Power Spectrum Estimation for Speech Enhancement, *Proceedings of INTERSPEECH 2022*, pp. 181–185.

Krawczyk-Becker, M.; Gerkmann, T. (2018). On Speech Enhancement Under PSD Uncertainty, *IEEE/ACM Transactions on Audio, Speech, and Language Processing*, vol. 26, no. 6, pp. 1144–1153.

Le Roux, J.; Wisdom, S.; Erdogan, H.; Hershey, J. R. (2018). SDR - Half-baked or Well Done?, *CoRR*, vol. abs/1811.02508.

Li, X.; Leglaive, S.; Girin, L.; Horaud, R. (2019). Audio-Noise Power Spectral Density Estimation Using Long Short-Term Memory, *IEEE Signal Processing Letters*, vol. 26, no. 6, pp. 918–922.

Li, A.; Liu, W.; Luo, X.; Zheng, C.; Li, X. (2021a). ICASSP 2021 Deep Noise Suppression Challenge: Decoupling Magnitude and Phase Optimization with a Two-Stage Deep Network, *ICASSP 2021 - 2021 IEEE International Conference on Acoustics, Speech and Signal Processing (ICASSP)*, pp. 6628–6632.

Li, H.; Wang, D.; Zhang, X.; Gao, G. (2021b). Recurrent Neural Networks and Acoustic Features for Frame-Level Signal-to-Noise Ratio Estimation, *IEEE/ACM Transactions on Audio, Speech, and Language Processing*, vol. 29, pp. 2878–2887.

Lotter, T.; Vary, P. (2003). Noise Reduction by Maximum a Posteriori Spectral Amplitude Estimation with Super-Gaussian Speech Modeling, *Proceedings of the International Workshop on Acoustic Echo and Noise Control (IWAENC)*, Kyoto, Japan, pp. 83–86.

Lotter, T.; Vary, P. (2005). Speech Enhancement by MAP Spectral Amplitude Estimation Using a Super-Gaussian Speech Model, *EURASIP Journal on Applied Signal Processing*, vol. 2005, no. 7, pp. 1110–1126.

Luo, Y.; Mesgarani, N. (2019). Conv-TasNet: Surpassing Ideal Time-Frequency Magnitude Masking for Speech Separation, *IEEE/ACM Transactions on Audio, Speech, and Language Processing*, vol. 27, no. 8, pp. 1256–1266.

Mack, W.; Habets, E. A. P. (2020). Deep Filtering: Signal Extraction and Reconstruction Using Complex Time-Frequency Filters, *IEEE Signal Processing Letters*, vol. 27, pp. 61–65.

Malah, D.; Cox, R. V.; Accardi, A. J. (1999). Tracking Speech-Presence Uncertainty to Improve Speech Enhancement in Non-Stationary Noise Environments, *Proceedings of the IEEE International Conference on Acoustics, Speech, and Signal Processing (ICASSP)*, Phoenix, Arizona, USA, pp. 789–792.

Markel, J. D.; Gray, A. H. (1976). *Linear Prediction of Speech*, Springer-Verlag, Berlin, Heidelberg, New York.

Martin, R. (1993). An Efficient Algorithm to Estimate the Instantaneous SNR of Speech Signals, *Proceedings of the European Conference on Speech Communication and Technology (EUROSPEECH)*, Berlin, Germany, pp. 1093–1096.

Martin, R. (1994). Spectral Subtraction Based on Minimum Statistics, *Proceedings of the European Signal Processing Conference (EUSIPCO)*, Edinburgh, Scotland, pp. 1182–1185.

Martin, R. (2001). Noise Power Spectral Density Estimation Based on Optimal Smoothing and Minimum Statistics, *IEEE Transactions on Speech and Audio Processing*, vol. 9, no. 5, pp. 504–512.

Martin, R. (2002). Speech Enhancement Using MMSE Short Time Spectral Estimation with Gamma Distributed Speech Priors, *Proceedings of the IEEE International Conference on Acoustics, Speech, and Signal Processing (ICASSP)*, Orlando, Florida, USA, vol. I, pp. 253–256.

Martin, R. (2005a). Bias Compensation Methods for Minimum Statistics Noise Power Spectral Density Estimation, *Signal Processing*, vol. 86, no. 6, pp. 1215–1229. Special Issue on Speech and Audio Processing (to appear).

Martin, R. (2005b). Speech Enhancement based on Minimum Mean Square Error Estimation and Supergaussian Priors, *IEEE Transactions on Speech and Audio Processing*, vol. 13, no. 5, pp. 845–856.

Martin, R. (2005c). Statistical Methods for the Enhancement of Noisy Speech, *in* J. Benesty; S. Makino; J. Chen (eds.), *Speech Enhancement*, Springer-Verlag, Berlin, Heidelberg, New York.

Martin, R.; Breithaupt, C. (2003). Speech Enhancement in the DFT Domain Using Laplacian Speech Priors, *Proceedings of the International Workshop on Acoustic Echo and Noise Control (IWAENC)*, Kyoto, Japan, pp. 87–90.

Martin, R.; Lotter, T. (2001). Optimal Recursive Smoothing of Non-Stationary Periodograms, *Proceedings of the International Workshop on Acoustic Echo and Noise Control (IWAENC)*, Darmstadt, Germany, pp. 167–170.

Martin, R.; Wittke, J.; Jax, P. (2000). Optimized Estimation of Spectral Parameters for the Coding of Noisy Speech, *Proceedings of the IEEE International Conference on Acoustics, Speech, and Signal Processing (ICASSP)*, Istanbul, Turkey, vol. III, pp. 1479–1482.

McAulay, R. J.; Malpass, M. L. (1980). Speech Enhancement Using a Soft-Decision Noise Suppression Filter, *IEEE Transactions on Acoustics, Speech and Signal Processing*, vol. 28, no. 2, pp. 137–145.

Middleton, D.; Esposito, R. (1968). Simultaneous Optimum Detection and Estimation of Signals in Noise, *IEEE Transactions on Information Theory*, vol. 14, no. 3, pp. 434–444.

Mowlaee, P.; Kulmer, J. (2015). Harmonic Phase Estimation in Single-Channel Speech Enhancement Using Phase Decomposition and SNR Information, *IEEE/ACM Transactions on Audio, Speech, and Language Processing*, vol. 23, no. 9, pp. 1521–1532.

Nicolson, A.; Paliwal, K. K. (2019). Deep Learning for Minimum Mean-Square Error Approaches to Speech Enhancement, *Speech Communication*, vol. 111, no. C, pp. 44–55.

Nicolson, A.; Paliwal, K. K. (2021). On Training Targets for Deep Learning Approaches to Clean Speech Magnitude Spectrum Estimation, *The Journal of the Acoustical Society of America*, vol. 149, no. 5, pp. 3273–3293.

Papoulis, A.; Unnikrishna Pillai, S. (2001). *Probability, Random Variables, and Stochastic Processes*, 4th edn, McGraw-Hill, New York.

Pascual, S.; Bonafonte, A.; Serrà, J. (2017). SEGAN: Speech Enhancement Generative Adversarial Network, *Proceedings of INTERSPEECH 2017*, pp. 3642–3646.

Preuss, R. D. (1979). A Frequency Domain Noise Cancelling Preprocessor for Narrowband Speech Communication Systems, *Proceedings of the IEEE International Conference on Acoustics, Speech, and Signal Processing (ICASSP)*, Washington, District of Columbia, USA, pp. 212–215.

Ramabadran, T. V.; Ashley, J. P.; McLaughlin, M. J. (1997). Background Noise Suppression for Speech Enhancement and Coding, *Proceedings of the IEEE Workshop on Speech Coding*, Pocono Manor, Pennsylvania, USA, pp. 43–44.

Ramiréz, J.; Segura, J.; Benítez, C.; García, L.; Rubio, A. (2005). Statistical Voice Activity Detection Using a Multiple Observation Likelihood Ratio Test, *IEEE Signal Processing Letters*, vol. 12, no. 10, pp. 689–692.

Rehr, R.; Gerkmann, T. (2021). SNR-Based Features and Diverse Training Data for Robust DNN-Based Speech Enhancement, *IEEE/ACM Transactions on Audio, Speech, and Language Processing*, vol. 29, pp. 1937–1949.

Ronneberger, O.; Fischer, P.; Brox, T. (2015). U-Net: Convolutional Networks for Biomedical Image Segmentation, *in* N. Navab; J. Hornegger; W. Wells; A. Frangi (eds.), *Medical Image Computing and Computer-Assisted Intervention - MICCAI 2015. MICCAI 2015*. Lecture Notes in Computer Science, vol. 9351. Springer, Cham, pp. 234–241.

Scalart, P.; Vieira Filho, J. (1996). Speech Enhancement Based on A Priori Signal to Noise Estimation, *Proceedings of the IEEE International Conference on Acoustics, Speech, and Signal Processing (ICASSP)*, Atlanta, Georgia, USA, pp. 629–632.

Schasse, A.; Martin, R. (2014). Estimation of Subband Speech Correlations for Noise Reduction via MVDR Processing, *IEEE/ACM Transactions on Audio, Speech, and Language Processing*, vol. 22, no. 9, pp. 1355–1365.

Schröter, H.; Rosenkranz, T.; Escalante-B, A.-N.; Maier, A. (2022). Low Latency Speech Enhancement for Hearing Aids Using Deep Filtering, *IEEE/ACM Transactions on Audio, Speech, and Language Processing*, vol. 30, pp. 2716–2728.

Sohn, J.; Sung, W. (1998). A Voice Activity Detector Employing Soft Decision Based Noise Spectrum Adaptation, *Proceedings of the IEEE International Conference on Acoustics, Speech, and Signal Processing (ICASSP)*, Seattle, Washington, USA, vol. 1, pp. 365–368.

Sohn, J.; Kim, N. S.; Sung, W. (1999). A Statistical Model-Based Voice Activity Detector, *Signal Processing Letters*, vol. 6, no. 1, pp. 1–3.

Srinivasan, K.; Gersho, A. (1993). Voice Activity Detection for Cellular Networks, *Proceedings of the IEEE Workshop on Speech Coding*, St. Jovite, Canada, pp. 85–86.

Subakan, C.; Ravanelli, M.; Cornell, S.; Bronzi, M.; Zhong, J. (2021). Attention Is All You Need In Speech Separation, *IEEE International Conference on Acoustics, Speech and Signal Processing (ICASSP)*, pp. 21–25.

Sun, L.; Du, J.; Dai, L.-R.; Lee, C.-H. (2017). Multiple-target Deep Learning for LSTM-RNN Based Speech Enhancement, *2017 Hands-free Speech Communications and Microphone Arrays (HSCMA)*, pp. 136–140.

Taal, C. H.; Hendriks, R. C.; Heusdens, R.; Jensen, J. (2011). An Algorithm for Intelligibility Prediction of Time-Frequency Weighted Noisy Speech, *IEEE Transactions on Audio, Speech, and Language Processing*, vol. 19, no. 7, pp. 2125–2136.

Taghia, J.; Martin, R. (2014). Objective Intelligibility Measures Based on Mutual Information for Speech Subjected to Speech Enhancement Processing, vol. 22, no. 1, pp. 6–16.

Taghia, J.; Taghia, J.; Mohammadiha, N.; Sang, J.; Bouse, V.; Martin, R. (2011). An Evaluation of Noise Power Spectral Density Estimation Algorithms in Adverse Acoustic Environments, *Proceedings of the IEEE International Conference on Acoustics, Speech, and Signal Processing (ICASSP)*, pp. 4640–4643.

Taghia, J.; Martin, R.; Hendriks, R. C. (2012). On Mutual Information as a Measure of Speech Intelligibility, *2012 IEEE International Conference on Acoustics, Speech and Signal Processing (ICASSP)*, pp. 65–68.

Tashev, I.; Mirsamadi, S. (2016). DNN-Based Causal Voice Activity Detector.

Tribolet, J.; Noll, P.; McDermott, B.; Crochiere, R. (1978). A Study of Complexity and Quality of Speech Waveform Coders, *ICASSP '78. IEEE International Conference on Acoustics, Speech, and Signal Processing*, vol. 3, pp. 586–590.

Tsoukalas, D.; Paraskevas, M.; Mourjopoulos, J. (1993). Speech Enhancement using Psychoacoustic Criteria, *Proceedings of the IEEE International Conference on Acoustics, Speech, and Signal Processing (ICASSP)*, pp. 359–362.

Tu, Y.-H.; Du, J.; Lee, C.-H. (2019). DNN Training Based on Classic Gain Function for Single-channel Speech Enhancement and Recognition, *ICASSP 2019 - 2019 IEEE International Conference on Acoustics, Speech and Signal Processing (ICASSP)*, pp. 910–914.

Valin, J.; Isik, U.; Phansalkar, N.; Giri, R.; Helwani, K.; Krishnaswamy, A. (2020). A Perceptually-Motivated Approach for Low-Complexity, Real-Time Enhancement of Fullband Speech, *in* H. Meng; B. Xu; T. F. Zheng (eds.), *Interspeech 2020, 21st Annual Conference of the International Speech Communication Association, Virtual Event*, Shanghai, China, 25–29 October 2020, ISCA, pp. 2482–2486.

Valin, J.-M.; Tenneti, S.; Helwani, K.; Isik, U.; Krishnaswamy, A. (2021). Low-Complexity, Real-Time Joint Neural Echo Control and Speech Enhancement Based on Percepnet, *ICASSP 2021 - 2021 IEEE International Conference on Acoustics, Speech and Signal Processing (ICASSP)*, pp. 7133–7137.

Van Compernolle, D. (1989). Noise Adaptation in a Hidden Markov Model Speech Recognition System, *Computer Speech and Language*, vol. 3, pp. 151–167.

Van Trees, H. L. (1968). *Detection, Estimation, and Modulation Theory*, MIT Press, Cambridge, Massachusetts.

Vary, P. (1985). Noise Suppression by Spectral Magnitude Estimation - Mechanism and Theoretical Limits, *Signal Processing*, vol. 8, pp. 387–400.

Vary, P. (2005). An Adaptive Filterbank Equalizer for Speech Enhancement, *Signal Processing*, vol. 86, no. 6, pp. 1206–1214. Special Issue on Speech and Audio Processing (to appear).

Vary, P.; Martin, R. (2006). *Digital Speech Transmission*, John Wiley & Sons, Ltd.

Vaswani, A.; Shazeer, N.; Parmar, N.; Uszkoreit, J.; Jones, L.; Gomez, A.; Kaiser, L.; Polosukhin, I. (2017). Attention Is All You Need, *31st Conference on Neural Information Processing Systems (NIPS 2017)*.

Vincent, E.; Gribonval, R.; Fevotte, C. (2006). Performance Measurement in Blind Audio Source Separation, *IEEE Transactions on Audio, Speech, and Language Processing*, vol. 14, no. 4, pp. 1462–1469.

Virag, N. (1999). Single Channel Speech Enhancement Based on Masking Properties of the Human Auditory System, *IEEE Transactions on Speech and Audio Processing*, vol. 7, no. 2, pp. 126–137.

Wang, D.; Chen, J. (2018). Supervised Speech Separation Based on Deep Learning: An Overview, *IEEE/ACM Transactions on Audio, Speech, and Language Processing*, vol. 26, no. 10, pp. 1702–1726.

Wang, Y.; Narayanan, A.; Wang, D. (2014). On Training Targets for Supervised Speech Separation, *IEEE/ACM Transactions on Audio, Speech, and Language Processing*, vol. 22, no. 12, pp. 1849–1858.

Welch, P. D. (1967). The Use of Fast Fourier Transform for the Estimation of Power Spectra: A Method Based on Time Averaging Over Short, Modified Periodograms, *IEEE Transactions on Audio and Electroacoustics*, vol. 15, no. 2, pp. 70–73.

Welker, S.; Richter, J.; Gerkmann, T. (2022). Speech Enhancement with Score-Based Generative Models in the Complex STFT Domain, *Proceedings of INTERSPEECH 2022*, pp. 2928–2932.

Williamson, D. S.; Wang, Y.; Wang, D. (2016a). Complex Ratio Masking for Joint Enhancement of Magnitude and Phase, *2016 IEEE International Conference on Acoustics, Speech and Signal Processing (ICASSP)*, pp. 5220–5224.

Williamson, D. S.; Wang, Y.; Wang, D. (2016b). Complex Ratio Masking for Monaural Speech Separation, *IEEE/ACM Transactions on Audio, Speech, and Language Processing*, vol. 24, no. 3, pp. 483–492.

Wolfe, P. J.; Godsill, S. J. (2001). Simple Alternatives to the Ephraim and Malah Suppression Rule for Speech Enhancement, *Proceedings of the Eleventh IEEE Workshop on Statistical Signal Processing*, Singapore, vol. II, pp. 496–499.

Xu, Y.; Du, J.; Dai, L.-R.; Lee, C.-H. (2014). An Experimental Study on Speech Enhancement Based on Deep Neural Networks, *IEEE Signal Processing Letters*, vol. 21, no. 1, pp. 65–68.

Xu, Y.; Du, J.; Dai, L.-R.; Lee, C.-H. (2015). A Regression Approach to Speech Enhancement Based on Deep Neural Networks, *IEEE/ACM Transactions on Audio, Speech, and Language Processing*, vol. 23, no. 1, pp. 7–19.

Yang, J. (1993). Frequency Domain Noise Suppression Approaches in Mobile Telephone Systems, *Proceedings of the IEEE International Conference on Acoustics, Speech, and Signal Processing (ICASSP)*, Minneapolis, Minnesota, USA, pp. 363–366.

Yin, D.; Luo, C.; Xiong, Z.; Zeng, W. (2020). PHASEN: A Phase-and-Harmonics-Aware Speech Enhancement Network, *The Thirty-Fourth AAAI Conference on Artificial Intelligence, AAAI 2020, The Thirty-Second Innovative Applications of Artificial Intelligence Conference, IAAI 2020, The Tenth AAAI Symposium on Educational Advances in Artificial Intelligence, EAAI 2020, New York, NY, USA, February 7–12, 2020*, AAAI Press, pp. 9458–9465.

Yong, P. C.; Nordholm, S. (2016). An Improved Soft Decision Based Noise Power Estimation Employing Adaptive Prior and Conditional Smoothing, *2016 IEEE International Workshop on Acoustic Signal Enhancement (IWAENC)*.

You, C.; Koh, S.; Rahardja, S. (2003). Adaptive $\beta$-Order MMSE Estimation for Speech Enhancement, *2003 IEEE International Conference on Acoustics, Speech, and Signal Processing, 2003. Proceedings. (ICASSP '03).*, vol. 1, pp. I–I.

Yu, R. (2009). A Low-Complexity Noise Estimation Algorithm Based on Smoothing of Noise Power Estimation and Estimation Bias Correction, *2009 IEEE International Conference on Acoustics, Speech and Signal Processing*, pp. 4421–4424.

Zhang, Q.; Nicolson, A.; Wang, M.; Paliwal, K. K.; Wang, C. (2020). DeepMMSE: A Deep Learning Approach to MMSE-Based Noise Power Spectral Density Estimation, *IEEE/ACM Transactions on Audio, Speech, and Language Processing*, vol. 28, pp. 1404–1415.

# 13

# Dual-Channel Noise and Reverberation Reduction

Single-microphone speech enhancement algorithms are favored in many applications because they are relatively easy to apply. For instance, cloud-based teleconferencing services do not have direct access to the acoustic environment of the user and most often have to rely on a single (mono) audio signal. However, for applications embedded in smart phones or hearing devices, access to multiple microphones is frequently foreseen. For such devices, further performance improvements are possible and dual-channel (and multi-channel, see Chapter 15) approaches may become the method of choice. Using more than one microphone channel, the performance of noise and echo reduction algorithms can be expected to improve, as the spatial characteristics of the sound field can be exploited, e.g., for the estimation of a priori unknown statistical quantities or for voice activity detection.

Figure 13.1 depicts two important application domains for dual-channel approaches. On the left-hand side of this figure, we consider a smart phone in the near-field of a speaker. The smart phone is in hands-held mode and receives the voice signal via a primary microphone at its lower end and a secondary microphone at its upper end. Due to the sound propagation, we find a significant sound pressure level difference at the two microphones as well as a time delay difference. Both features may be exploited in an adaptive enhancement algorithm. On the right-hand side, we depict the binaural configuration of two microphones as it is used, e.g., in hearing aids. Now, the two microphones are typically in the far-field of the target source and often surrounded by several acoustic noise sources and reverberation. Moreover, the head of the listener acts as an acoustic obstacle and induces distinct direction dependent level and time delay differences. In what follows, we will now include adaptive filters in our consideration and explore their utility in the context of these acoustic scenarios.

## 13.1 Dual-Channel Wiener Filter

We begin our discussion of dual-channel approaches with the most basic scenario and an analysis in terms of second-order statistics. Figure 13.2 depicts two microphones and a single, possibly adaptive filter with impulse response $h(k)$. This system differs from

(a) (b)

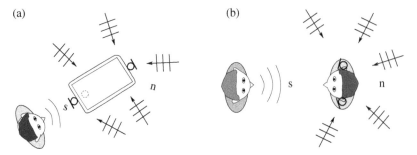

**Figure 13.1** Prominent applications of dual-channel speech enhancement: (a) smart phone in the acoustic near-field of its user and (b) binaural microphone configuration for hearing aids.

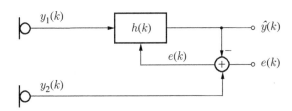

**Figure 13.2** Dual-channel noise reduction system.

Figure 12.3 in that a second microphone is added and this second signal $y_2(k)$ is taken as the target signal $d(k)$ in Figure 12.3. Furthermore, we also provide the error signal $e(k)$ as an additional output.

The computation of the impulse response $h(k)$ of the non-causal infinite impulse response (IIR) Wiener filter for the dual-channel case is analogous to the derivation in Section 12.2.1 where we now assume that the desired signal is simply derived from the secondary microphone input. For wide sense stationary signals, the minimization of

$$E\left\{e^2(k)\right\} = E\left\{(y_2(k) - \hat{y}(k))^2\right\} \tag{13.1}$$

in the mean-square sense with

$$\hat{y}(k) = \sum_{i=-\infty}^{\infty} h(i)y_1(k - i) \tag{13.2}$$

results in the necessary (orthogonality) condition

$$E\left\{(y_2(k) - \hat{y}(k))\,\hat{y}(k)\right\} \overset{!}{=} 0, \tag{13.3}$$

and in the frequency response (see Section 12.2.1)

$$H(e^{j\Omega}) = \frac{\Phi_{y_1 y_2}(e^{j\Omega})}{\Phi_{y_1 y_1}(e^{j\Omega})} \tag{13.4}$$

of the optimal filter. In the general case of additive speech and noise signals,

$$y_1(k) = s_1(k) + n_1(k)$$
$$y_2(k) = s_2(k) + n_2(k), \tag{13.5}$$

and when the speech and the noise signals are statistically independent, the minimum mean square error (MMSE) IIR filter is given by

$$H(e^{j\Omega}) = \frac{\Phi_{y_1 y_2}(e^{j\Omega})}{\Phi_{y_1 y_1}(e^{j\Omega})} = \frac{\Phi_{s_1 s_2}(e^{j\Omega}) + \Phi_{n_1 n_2}(e^{j\Omega})}{\Phi_{s_1 s_1}(e^{j\Omega}) + \Phi_{n_1 n_1}(e^{j\Omega})}. \tag{13.6}$$

The frequency response of the optimal linear filter may be decomposed into two independent optimal filters $H_s(e^{j\Omega})$ and $H_n(e^{j\Omega})$ for the estimation of the speech signal $s_2(k)$ and the noise signal $n_2(k)$, respectively,

$$H(e^{j\Omega}) = \frac{\Phi_{s_1 s_2}(e^{j\Omega})}{\Phi_{y_1 y_1}(e^{j\Omega})} + \frac{\Phi_{n_1 n_2}(e^{j\Omega})}{\Phi_{y_1 y_1}(e^{j\Omega})} = H_s(e^{j\Omega}) + H_n(e^{j\Omega}). \tag{13.7}$$

At the output of the optimal filter, we obtain the components of $y_1(k)$, which are correlated with the second channel $y_2(k)$, regardless whether they are speech or noise. Uncorrelated components are suppressed. Depending on the correlation properties of the speech and the noise signals, the optimal filter will act primarily as either a noise or a speech estimator.

For the optimal estimate $\hat{y}_{\text{opt}}(k)$, condition (13.3) holds and thus the minimum error $E\{e^2(k)\}_{|\text{min}}$ is given by

$$\begin{aligned}
E\{e^2(k)\}_{|\text{min}} &= E\left\{ \left( y_2(k) - \hat{y}_{\text{opt}}(k) \right)^2 \right\} \\
&= E\left\{ \left( y_2(k) - \hat{y}_{\text{opt}}(k) \right) y_2(k) \right\} \\
&= \varphi_{y_2 y_2}(0) - \sum_{i=-\infty}^{\infty} h_{\text{opt}}(i)\varphi_{y_1 y_2}(i) \tag{13.8} \\
&= \frac{1}{2\pi} \int_{-\pi}^{\pi} \Phi_{y_2 y_2}(e^{j\Omega}) d\Omega - \frac{1}{2\pi} \int_{-\pi}^{\pi} H(e^{j\Omega}) \Phi_{y_1 y_2}^*(e^{j\Omega}) d\Omega,
\end{aligned}$$

where Parseval's theorem (Table 3.2) was used in the last equality and $*$ denotes a complex-conjugate variable. The cross-power spectrum $\Phi_{y_1 y_2}(e^{j\Omega})$ is the Fourier transform of $\varphi_{y_1 y_2}(\ell) = E\left\{ (y_1(k)y_2(k+\ell)) \right\}$. Using (13.4) in (13.8) and merging both integrals, we obtain

$$\begin{aligned}
E\{e^2(k)\}_{|\text{min}} &= \frac{1}{2\pi} \int_{-\pi}^{\pi} \Phi_{y_2 y_2}(e^{j\Omega}) \left( 1 - \frac{\Phi_{y_1 y_2}(e^{j\Omega})\Phi_{y_1 y_2}^*(e^{j\Omega})}{\Phi_{y_1 y_1}(e^{j\Omega})\Phi_{y_2 y_2}(e^{j\Omega})} \right) d\Omega \\
&= \frac{1}{2\pi} \int_{-\pi}^{\pi} \Phi_{y_2 y_2}(e^{j\Omega}) \left( 1 - \left| \gamma_{y_1 y_2}(e^{j\Omega}) \right|^2 \right) d\Omega, \tag{13.9}
\end{aligned}$$

where $\left| \gamma_{y_1 y_2}(e^{j\Omega}) \right|^2$ denotes the *magnitude-squared coherence* (MSC) function [Bendat, Piersol 1966], [Carter 1987] of the two microphone signals, i.e.,

$$\left| \gamma_{y_1 y_2}(e^{j\Omega}) \right|^2 = \frac{|\Phi_{y_1 y_2}(e^{j\Omega})|^2}{\Phi_{y_1 y_1}(e^{j\Omega})\Phi_{y_2 y_2}(e^{j\Omega})}. \tag{13.10}$$

The MSC constitutes a normalized, frequency-dependent measure of correlation with

$$0 \leq \left| \gamma_{y_1 y_2}(e^{j\Omega}) \right|^2 \leq 1, \tag{13.11}$$

and is frequently used to characterize spatial properties of sound fields. It indicates the linear relation between the two signals and, according to (13.9), gives an indication of how

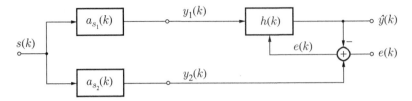

**Figure 13.3** Single-source signal model.

effective the Wiener filter is. Obviously, the effectiveness of the Wiener filter depends on the signal model, which the two microphone signals obey. In what follows, we consider three special cases:

- When the two microphone signals $y_1(k)$ and $y_2(k)$ are uncorrelated, i.e., $E\{y_1(k) y_2(\ell)\} = 0$, $\forall k, \ell$, the MSC is equal to zero and the error power is equal to the power of the signal $y_2(k)$. The impulse response of the optimal filter is identical to zero.
- When the two microphone signals are linearly related, the MSC is equal to one. In Figure 13.3, the two microphone signals originate from a single-source signal $s(k)$, i.e., $y_1(k) = a_{s_1}(k) * s(k)$ and $y_2(k) = a_{s_2}(k) * s(k)$. For this signal model, we obtain with (see Section 5.8.4)

$$\Phi_{y_1 y_2}(e^{j\Omega}) = A^*_{s_1}(e^{j\Omega}) A_{s_2}(e^{j\Omega}) \Phi_{ss}(e^{j\Omega})$$

$$\Phi_{y_1 y_1}(e^{j\Omega}) = \left| A_{s_1}(e^{j\Omega}) \right|^2 \Phi_{ss}(e^{j\Omega}) \tag{13.12}$$

$$\Phi_{y_2 y_2}(e^{j\Omega}) = \left| A_{s_2}(e^{j\Omega}) \right|^2 \Phi_{ss}(e^{j\Omega}),$$

and (13.10) the MSC $\left| \gamma_{y_1 y_2}(e^{j\Omega}) \right|^2 = 1$. In general, it can be shown that the MSC is invariant with respect to linear transformations [Bendat, Piersol 1966].

- In many practical cases, the correlation and, thus, the coherence vary with frequency. This situation occurs when linearly related signals are disturbed by uncorrelated noise, or, when spatially distributed, mutually uncorrelated sources contribute to the microphone signals. This frequently encountered scenario is considered in more detail in Section 13.2.

## 13.2 The Ideal Diffuse Sound Field and Its Coherence

The coherence of two analog microphone signals $y_1(t)$ and $y_2(t)$ is defined as

$$\gamma_{y_1 y_2}(f) = \frac{\Phi_{y_1 y_2}(f)}{\sqrt{\Phi_{y_1 y_1}(f) \Phi_{y_2 y_2}(f)}}, \tag{13.13}$$

where $\Phi_{y_1 y_2}(f)$ denotes the cross-power spectrum and $\Phi_{y_1 y_1}(f)$ and $\Phi_{y_2 y_2}(f)$ the auto-power spectra at a given frequency $f$.

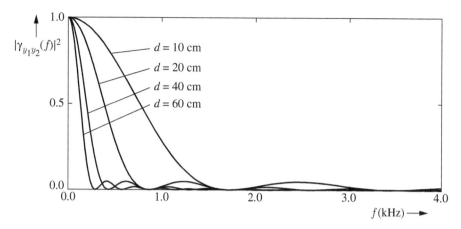

**Figure 13.4** MSC of two microphone signals in an isotropic sound field for omnidirectional microphones and a distance $d = 0.1$ m, 0.2 m, 0.4 m, and 0.6 m.

An interesting special case is the *ideal diffuse sound field*. In such a *spherically-isotropic* sound field, the coherence function of two microphone signals $y_1(t)$ and $y_2(t)$ follows the closed-form solution [Kuttruff 1990]

$$\gamma_{y_1 y_2}(f) = \begin{cases} \dfrac{\sin(2 \pi f\, d\, c^{-1})}{2 \pi f\, d\, c^{-1}}, & f > 0, \\ 1, & f = 0, \end{cases} \tag{13.14}$$

where $d$ and $c$ denote the distance between the *omnidirectional* microphones and the speed of sound, respectively. Using (13.14), Figure 13.4 plots the MSC $|\gamma_{y_1 y_2}(f)|^2$ as a function of frequency $f$ for several inter-microphone distances $d$. The MSC of the ideal diffuse sound field attains its first zero at $f_c = \frac{c}{2d}$. Hence, for frequencies above $f_c = \frac{c}{2d}$, very little correlation is observed.

Furthermore, it is interesting to note that while an ideal diffuse sound field results in an MSC according to (13.14) the converse is not necessarily true [Dämmig 1957]: sound fields, which are not ideally diffuse may also exhibit an MSC as in (13.14).

As an example, Figure 13.5 shows the estimated MSC for stationary office noise and two omnidirectional microphones. Especially for low frequencies, the estimated coherence matches the MSC of the ideal diffuse sound field quite well. By contrast, for signals, which originate from a single source, a high degree of coherence is observed. Figure 13.6 depicts the power spectrum (a) and the estimated coherence (b) of two signals, which originate from a speaker in a car compartment. For most frequencies, an MSC above 0.9 is observed. The distance between the two hypercardioid microphones is $d = 20$ cm.

It is common practice to estimate the coherence on the basis of magnitude squared Fourier coefficients or the periodogram. The estimation of the coherence function necessarily requires averaging [Carter et al. 1973]. If we use the (cross-)periodogram without averaging as an approximation to the (cross-)power spectrum, we have

$$\left| \gamma_{y_1 y_2, \mu}(k) \right|^2 = \frac{|Y_{1,\mu}(k) Y_{2,\mu}^*(k)|^2}{|Y_{1,\mu}(k)|^2\, |Y_{2,\mu}(k)|^2} = 1 \tag{13.15}$$

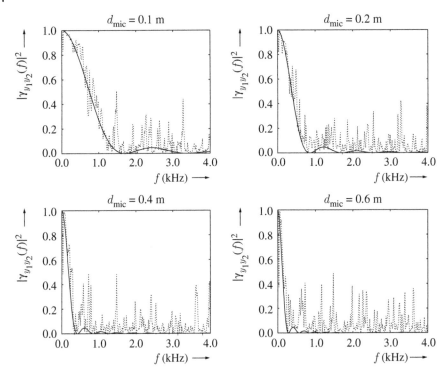

**Figure 13.5** MSC of the ideal diffuse sound field and estimated MSC of office noise for omnidirectional microphones and a distance $d = 0.1$ m, 0.2 m, 0.4 m and 0.6 m [Martin 1995].

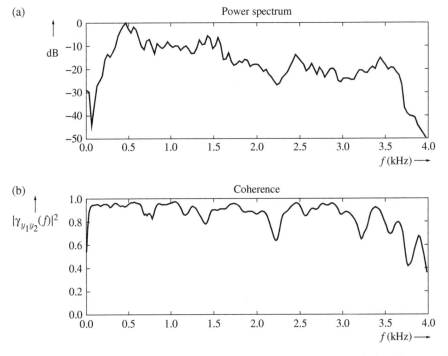

**Figure 13.6** Average power spectrum (a) and measured MSC (b) of a speech signal in a reverberant car compartment. The distance between the hypercardioid microphones is 0.2 m [Martin 1995].

for frequency bin $\mu$ independent of the actual correlation properties. When reverberation comes into play, the estimation of the MSC using the discrete Fourier transform (DFT) is not trivial, even when the signals are stationary and when sufficient averaging is applied. Long reverberation tails combined with DFT-based block processing might introduce a substantial bias in the estimated MSC [Martin, 1995, chapter 2]. Then, even when the wave field is generated by a single point-like sound source, reverberation tails exceeding the DFT length will reduce the coherence to values below unity.

The coherence of two microphone signals depends on the sound field and the geometric configuration and directional characteristics of the microphones. Closed-form and numerical solutions have been computed for the spherically isotropic and the cylindric-isotropic noise field and for a variety of directional microphones [Elko 2001]. Furthermore, in some applications the microphones are not positioned in free-field but are attached to the human head, which modifies wave propagation through diffraction and scattering. In this case, the coherence can be either measured (e.g., in a reverberant chamber), analytically computed using an acoustic model, or estimated from a set of given *head-related transfer functions* (HRTFs). For a binaural configuration of microphones, i.e., microphones positioned at the left and the right ear, measurements of the coherence function have been presented in [Lindevald, Benade 1986]. The authors propose to approximate their data, which has been measured at the entrances of the ear canals of a human head in a large reverberant lecture hall as

$$\gamma_{y_1 y_2}(f) = \begin{cases} \dfrac{\sin(\pi f / f_{c1})}{\pi f / f_{c1}} \dfrac{1}{\sqrt{1 + (f/f_{c2})^4}} & f > 0 \\ 1 & f = 0, \end{cases} \tag{13.16}$$

with $f_{c1} \approx 520\,\text{Hz}$ and $f_{c2} \approx 728\,\text{Hz}$. They also provide an analytical computation of the coherence when the head is approximated by an acoustically hard sphere. The binaural measurement has been replicated in an echoic chamber [Borß, 2011, chapter 4] and approximated in a slightly simpler form as

$$\gamma_{y_1 y_2}(f) = \begin{cases} \dfrac{\sin(\pi f / f_{c1})}{\pi f / f_{c1}} \max\left(0,\ 1 - \dfrac{f}{f_{c2}}\right), & f > 0, \\ 1, & f = 0, \end{cases} \tag{13.17}$$

with $f_{c1} = 550\,\text{Hz}$ and $f_{c2} = 2700\,\text{Hz}$. Relating the latter approximation to the free-field model in (13.14) we find that the frequency of the first zero $f_{c1} = 550\,\text{Hz}$ corresponds to an effective microphone distance of $d = \frac{c}{2 f_{c1}} \approx 0.31\,\text{m}$, which is about twice as large as the true ear distance. Furthermore, a semi-analytical model for the computation of the binaural coherence with additional measurements is presented in [Jeub et al. 2011]. which also confirms the position of the first zero of the MSC.

In Figure 13.7, we juxtapose the free-field MSC for a microphone distance of $d = 0.15\,\text{m}$ and the two above approximations. We note that the central lobe of the MSC is much more narrow in the binaural configuration.

As the effectiveness of the dual-channel Wiener filter depends much on the correlation between the microphone signals, we will now consider several applications and the corresponding signal models. We distinguish two basic methods: *noise cancellation* based on a

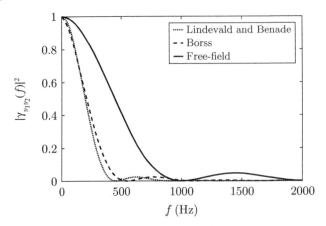

**Figure 13.7** Free-field MSC of the ideal diffuse sound field (microphone distance $d = 0.15$ m) and approximated MSC of reverberant diffuse sound fields for omnidirectional microphones attached to a human head.

noise-only reference signal and *noise reduction* based on a symmetric dual-channel signal model. In both cases, the MSCs of speech and noise are the key to analyzing and understanding the performance of these algorithms.

## 13.3 Noise Cancellation

The noise cancellation technique can be applied when the noise is coherently received by the two microphones. This is the case, when the noise signals are linear transforms of a single noise source as shown in Figure 13.8. We assume that the desired speech signal $s(k)$ is disturbed by additive noise, which originates from a noise source $n(k)$ via the impulse response $a_{n_2}(k)$ of a linear system. The noise and the speech signals are picked up by the second microphone. Therefore, $y_2(k) = a_{s_2}(k) * s(k) + a_{n_2}(k) * n(k)$. Furthermore, a noise-only reference is available, which originates from the same noise signal $n(k)$ and is picked up at the first microphone as $y_1(k) = a_{n_1}(k) * n(k)$. This noise reference signal is free of the speech signal. Assuming statistical independence of speech and noise, the Wiener filter is then used to estimate the noise, which disturbs the desired signal $s(k)$.

For this signal model we find that (also see Section 5.8.4)

$$\Phi_{y_1 y_2}(e^{j\Omega}) = A_{n_1}^*(e^{j\Omega}) A_{n_2}(e^{j\Omega}) \Phi_{nn}(e^{j\Omega}), \tag{13.18a}$$

$$\Phi_{y_1 y_1}(e^{j\Omega}) = \left|A_{n_1}(e^{j\Omega})\right|^2 \Phi_{nn}(e^{j\Omega}), \tag{13.18b}$$

$$\Phi_{y_2 y_2}(e^{j\Omega}) = \left|A_{s_2}(e^{j\Omega})\right|^2 \Phi_{ss}(e^{j\Omega}) + \left|A_{n_2}(e^{j\Omega})\right|^2 \Phi_{nn}(e^{j\Omega}), \tag{13.18c}$$

and therefore with (13.10)

$$\left|\gamma_{y_1 y_2}(e^{j\Omega})\right|^2 \tag{13.19}$$

$$= \frac{\left|A_{n_1}^*(e^{j\Omega}) A_{n_2}(e^{j\Omega}) \Phi_{nn}(e^{j\Omega})\right|^2}{\left|A_{n_1}(e^{j\Omega})\right|^2 \Phi_{nn}(e^{j\Omega}) \left(\left|A_{s_2}(e^{j\Omega})\right|^2 \Phi_{ss}(e^{j\Omega}) + \left|A_{n_2}(e^{j\Omega})\right|^2 \Phi_{nn}(e^{j\Omega})\right)}$$

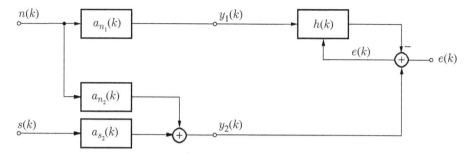

**Figure 13.8** Signal model for noise cancellation.

and

$$H(e^{j\Omega}) = \frac{A_{n_1}^*(e^{j\Omega})A_{n_2}(e^{j\Omega})}{\left|A_{n_1}(e^{j\Omega})\right|^2} = \frac{A_{n_2}(e^{j\Omega})}{A_{n_1}(e^{j\Omega})}. \tag{13.20}$$

The filter $h(k)$ estimates the *relative transfer function* (RTF, see [Gannot et al. 2001]) of the second channel with the first channel being used for normalization. As the output of the Wiener filter is the estimated noise of the second channel, the error signal $e(k)$ is the desired output containing the speech signal

$$e(k) = y_2(k) - h(k) * y_1(k)$$

$$= a_{s_2}(k) * s(k) + \left( a_{n_2}(k) - a_{n_1}(k) * h(k) \right) * n(k). \tag{13.21}$$

Thus, using the non-causal IIR Wiener filter, perfect noise cancellation is possible if $A_{n_1}(e^{j\Omega}) \neq 0$. In a practical implementation using a finite impulse response (FIR) filter, a causal approximation to (13.20) must be used, which may lead to less noise reduction. Furthermore, we notice from (13.19) that the speech signal reduces the coherence of the microphone signals and therefore increases the error. Thus, in the noise cancellation application, the speech signal disturbs the filter adaptation! In an adaptive implementation of the noise canceller using, e.g., the least-mean square (LMS) algorithm [Widrow et al. 1975], it is therefore advisable to adapt the noise estimation filter only, when little or no speech is present.

For the noise cancellation application, we define with (13.9) the normalized error power

$$R = -10 \log_{10} \frac{\int_{-\pi}^{\pi} \Phi_{y_2 y_2}(e^{j\Omega}) \left( 1 - \left| \gamma_{y_1 y_2}(e^{j\Omega}) \right|^2 \right) d\Omega}{\int_{-\pi}^{\pi} \Phi_{y_2 y_2}(e^{j\Omega}) d\Omega} \tag{13.22}$$

as a measure of performance. For example, a frequency-independent MSC of the noise signals of 0.9 provides a noise power reduction of 10 dB.

A successful application of the noise cancellation technique is acoustic echo cancellation (see Chapter 14). In acoustic echo cancellation, the disturbing loudspeaker signal is available as a digital signal and can be fed directly into the canceller. Then, without local noise, a high degree of coherence is achieved. In the context of reducing additive acoustic noise, however, the requirements of the signal model in Figure 13.8 are hard to fulfill. In the interior of a car, for example, we frequently encounter distributed noise sources and diffuse noise fields. Then, the microphones must be sufficiently close to achieve correlation over a large range of frequencies. This, however, leads inevitably to leakage of the speech signal into the reference channel [Armbrüster et al. 1986], [Degan, Prati 1988]. Therefore, it is difficult to obtain a noise-only reference, which is free of the desired speech signal $s(k)$. In the diffuse noise field, the noise cancellation approach can work only for low frequencies and when very little speech leaks into the reference microphone.

A decoupling of the two microphones with respect to the speech signal can be achieved by additional means, e.g., by using a facemask in an aircraft cockpit [Harrison et al. 1986], or by using a vibration sensor, which is immune to air-borne speech sounds. In the former application, the information bearing signal $y_2(k)$ is picked up inside the mask while the noise reference, i.e., the input to the Wiener filter, is picked up outside.

### 13.3.1 Implementation of the Adaptive Noise Canceller

In practice, the filter $h(k)$ must be causal and also adaptive, since $a_{n_1}(k)$ and $a_{n_2}(k)$ are in general not fixed but time varying. A standard solution to the adaptation problem is to use an FIR filter $h(k)$ of order $N$ and either a block adaptation of the coefficient vector according to

$$\mathbf{R}_{y_1 y_1} \, \mathbf{h} = \boldsymbol{\varphi}_{y_1 y_2}, \tag{13.23}$$

or the *normalized least-mean square* (NLMS) algorithm (see Section 14.5) for an iterative coefficient update [Widrow et al. 1975],

$$\mathbf{h}(k+1) = \mathbf{h}(k) + \beta(k)\, e(k)\, \mathbf{y}_1(k). \tag{13.24}$$

Here, $\beta(k)$ denotes a possibly time-varying stepsize parameter. Since the speech signal of the primary channel disturbs the adaptation of the noise cancellation filter, the adaptation must be slowed down or completely stopped whenever the speaker becomes active. As discussed in depth in Chapter 14, this is quite analogous to the echo cancellation problem where near-end speech might disturb the adaptation of the canceller. In the presence of speech, too large a stepsize will lead to distortions of the estimated noise. Since these distortions are correlated with the speech signal $s(k)$ they will be perceived as distortions of the speech signal. Too small a stepsize will slow down the adaptation of the adaptive filter. Thus, a good balance between averaging and tracking is desirable.

Furthermore, adaptive noise cancellers have been devised, which exploit the short-term periodic structure of voiced speech [Sambur 1978]. Other approaches employ a cascade of cancellers to remove the speech in the reference signal and to remove the noise in the output signal [Faucon et al. 1989], or combine the two-microphone canceller with single channel approaches [Kroschel, Linhard 1988], [Gustafsson et al. 1999].

The noise cancellation and suppression schemes presented in this chapter are also useful in conjunction with beamforming microphone arrays (see Chapter 15). In fact, the (multi-channel) adaptive noise canceler is an important component of the *generalized sidelobe canceller*. In case of an array with only two microphones the axis of which is oriented forward toward the desired source, a backwards-directed beam with a null toward the (forward) target source can be used to derive an estimate of the ambient noise.

In an alternative approach, which has been devised for use in smart phones with (at least) two microphones [Jeub et al. 2012], the authors exploit acoustic near-field effects. In this specific source-microphone configuration, the source is close to the communication device with its microphones. It is then assumed that the target speech level is at least 10 dB higher in the primary microphone as compared to a secondary microphone at the opposite side of the device (see Figure 13.1a). Under the additive noise model considered in (13.28) and Figure 13.9 we may write the power spectra of both microphone signals as

$$\Phi_{y_1 y_1}(e^{j\Omega}) = |A_{s_1}(e^{j\Omega})|^2 \Phi_{ss}(e^{j\Omega}) + \Phi_{n_1 n_1}(e^{j\Omega})$$
$$\Phi_{y_2 y_2}(e^{j\Omega}) = |A_{s_2}(e^{j\Omega})|^2 \Phi_{ss}(e^{j\Omega}) + \Phi_{n_2 n_2}(e^{j\Omega}),$$
$$\tag{13.25}$$

and the power difference [Yousefian et al. 2009] as

$$\Phi_{y_1 y_1}(e^{j\Omega}) - \Phi_{y_2 y_2}(e^{j\Omega}) = |A_{s_1}(e^{j\Omega})|^2 \Phi_{ss}(e^{j\Omega}) \left(1 - \frac{|A_{s_2}(e^{j\Omega})|^2}{|A_{s_1}(e^{j\Omega})|^2}\right)$$
$$+ \Phi_{n_1 n_1}(e^{j\Omega}) - \Phi_{n_2 n_2}(e^{j\Omega}) \tag{13.26}$$
$$= |A_{s_1}(e^{j\Omega})|^2 \Phi_{ss}(e^{j\Omega}) \left(1 - |R_{s_1 s_2}(e^{j\Omega})|^2\right) + \Phi_{\Delta n}(e^{j\Omega}),$$

where $|R_{s_1 s_2}(e^{j\Omega})|^2 > 0$ is the magnitude-squared RTF between the two microphones. Under the assumption of a homogeneous noise field we have $\Phi_{\Delta n}(e^{j\Omega}) \approx 0$. The power difference may then be used to construct a Wiener-like noise reduction filter [Yousefian et al. 2009].

Alternatively, we may normalize the power difference [Jeub et al. 2012]

$$\Delta\Phi_{\mathrm{PLDNE}}(e^{j\Omega}) = \left|\frac{\Phi_{y_1 y_1}(e^{j\Omega}) - \Phi_{y_2 y_2}(e^{j\Omega})}{\Phi_{y_1 y_1}(e^{j\Omega}) + \Phi_{y_2 y_2}(e^{j\Omega})}\right|$$
$$= \left|\frac{|A_{s_1}(e^{j\Omega})|^2 \Phi_{ss}(e^{j\Omega}) \left(1 - |R_{s_1 s_2}(e^{j\Omega})|^2\right) + \Phi_{\Delta n}(e^{j\Omega})}{|A_{s_1}(e^{j\Omega})|^2 \Phi_{ss}(e^{j\Omega}) \left(1 + |R_{s_1 s_2}(e^{j\Omega})|^2\right) + \Phi_{n_1 n_1}(e^{j\Omega}) + \Phi_{n_2 n_2}(e^{j\Omega})}\right|$$
$$\tag{13.27}$$

as this quantity also provides valuable information on the activity of the target speech signal and can be used to control a recursive noise power estimator. When the target speech source is not active $\Delta\Phi_{\mathrm{PLDNE}}(e^{j\Omega})$ will be close to zero. Then, the primary microphone may be used for the estimation of the noise power. When the power difference is larger than a minimum value, the secondary microphone will deliver a suitable noise power estimate while for large differences the estimation can be halted to avoid speech leakage.

## 13.4 Noise Reduction

In many speech communication scenarios, it will not be possible to prevent the speech signal from leaking into the microphone signal $y_1(k)$. Thus, the speech signal will also be estimated by the adaptive filter and canceled to some extent. Also, the noise originates, in general, not from a single acoustic point source but from several spatially distributed sources. Hence, the correlation between the noise components of both channels will be reduced. A typical example where this kind of problem prevails is speech pickup with two microphones in the ideal diffuse noise field. Diffuse noise fields arise in reverberant environments and are thus quite common in speech communication applications.

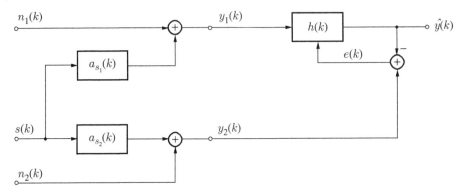

**Figure 13.9** Signal model for noise reduction.

### 13.4.1 Principle of Dual-Channel Noise Reduction

We therefore consider a scenario where both ambient noise and speech are picked up by the microphones as shown in Figure 13.9. We assume that the speech signal originates from a point source and that the noise components in the two microphone signals exhibit a low degree of correlation. These requirements can be fulfilled in an ideal diffuse noise field, where above a cut-off frequency $f_c = c/(2d)$ the correlation of the noise signals is close to zero. The microphone signals can be therefore written as

$$y_1(k) = a_{s_1}(k) * s(k) + n_1(k)$$
$$y_2(k) = a_{s_2}(k) * s(k) + n_2(k) .$$

(13.28)

As before, we assume that the speech signals are not correlated with the noise signals. Thus, the linearly constrained MMSE IIR filter (13.4) yields

$$H(e^{j\Omega}) = \frac{\Phi_{y_1 y_2}(e^{j\Omega})}{\Phi_{y_1 y_1}(e^{j\Omega})} = \frac{A_{s_1}^*(e^{j\Omega})A_{s_2}(e^{j\Omega})\Phi_{ss}(e^{j\Omega}) + \Phi_{n_1 n_2}(e^{j\Omega})}{\left|A_{s_1}(e^{j\Omega})\right|^2 \Phi_{ss}(e^{j\Omega}) + \Phi_{n_1 n_1}(e^{j\Omega})},$$

(13.29)

where the numerator follows from (5.125). In this case, the output signal of the dual-channel noise reduction system is the filtered signal $\hat{y}(k)$. When the noise is uncorrelated above a cut-off frequency $f_c = \frac{c}{2d}$, i.e.,

$$\Phi_{n_1 n_2}(e^{j\Omega}) \approx 0$$

(13.30)

and

$$A_{s_1}(e^{j\Omega}) \approx A_{s_2}(e^{j\Omega}),$$

(13.31)

the frequency response of the optimal filter (13.29) approaches the frequency response of the single-channel Wiener filter for the estimation of the speech components in $y_1(k)$,

$$H(e^{j\Omega}) = \frac{A_{s_1}^*(e^{j\Omega})A_{s_2}(e^{j\Omega})\Phi_{ss}(e^{j\Omega})}{\left|A_{s_1}(e^{j\Omega})\right|^2 \Phi_{ss}(e^{j\Omega}) + \Phi_{n_1 n_1}(e^{j\Omega})}$$

$$= \frac{A_{s_2}(e^{j\Omega})}{A_{s_1}(e^{j\Omega})} \frac{\left|A_{s_1}(e^{j\Omega})\right|^2 \Phi_{ss}(e^{j\Omega})}{\left|A_{s_1}(e^{j\Omega})\right|^2 \Phi_{ss}(e^{j\Omega}) + \Phi_{n_1 n_1}(e^{j\Omega})}.$$

(13.32)

In contrast to the single-channel Wiener filter, no a priori knowledge about the clean speech power spectrum is required. The filter is computed using the microphone signals only. However, for the above assumption to hold, we must place the microphones within the *critical distance* of the speech source. Within the critical distance, the direct sound path is dominant, and the random fluctuations in the reverberant tails of the impulse responses $a_{s_1}(k)$ and $a_{s_2}(k)$ are of less importance. Finally, we note that for the above noise reduction constellation the error signal $e(k)$ is composed of noise terms only. The adaptive filter (13.32) applies a relative frequency response $\frac{A_{s2}(e^{j\Omega})}{A_{s1}(e^{j\Omega})}$ (or RTF) to align the estimated speech components of the first channel with the speech components in the second input channel. When these two components match, only the combined noise terms remain in the error signal $e(k)$. This is the basis of the equalization–cancellation (EC) approach discussed below.

### 13.4.2 Binaural Equalization–Cancellation and Common Gain Noise Reduction

The above approaches are useful in mobile telephony when two microphones are available and a single processed signal is to be transmitted to the far-end side. However, in binaural speech communication devices such as hearing aids or smart headphones ("hearables" or "earbuds,"), the two input signals are picked up in a binaural configuration and the output signals $\hat{s}_l(k)$ and $\hat{s}_r(k)$ are delivered to the left and the right ear, respectively. Figure 13.10 depicts a frequency-domain implementation of a typical example of a system with two input and two output channels as described, e.g., in [Azarpour et al. 2014]. Here, we make use of two adaptive filters $H_1$ and $H_2$ and a causality delay $T_c$ to cancel the target speech signal and to derive noise components $N_1$ and $N_2$. Based on the power spectra of these components, spectral gain functions for noise reduction are computed. This is in fact reminiscent of a null-steering beamformer (see Section 15.4.4) and, interestingly, this principle (in a simpler time-domain form) has also been introduced to explain binaural masking effects in the human auditory system [Durlach 1963].

An important feature of binaural speech processing systems as implemented in hearing aids is the preservation of binaural cues. It is highly desirable to preserve the interaural

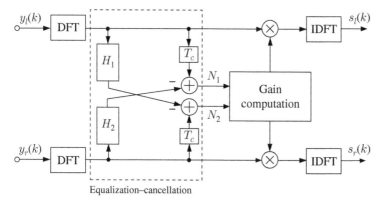

**Figure 13.10**  Dual-channel noise power estimation using the equalization–cancellation principle with a subsequent computation of spectral gains for noise suppression.

time difference/interaural phase difference (ITD/IPD) and interaural level difference (ILD) such that the auditory localization of sound sources is not impaired. Binaural localization cues may be restored by remixing the output signals with the unprocessed input signals according to

$$\hat{s}_l(k) = (1 - \alpha) \cdot s_l(k) + \alpha \cdot y_l(k), \tag{13.33a}$$

$$\hat{s}_r(k) = (1 - \alpha) \cdot s_r(k) + \alpha \cdot y_r(k). \tag{13.33b}$$

Obviously, this solution entails a tradeoff between maximum noise suppression and spatial fidelity. It is therefore worthwhile to look for more advanced solutions, which make use of the same spectral gain in both output channels, thereby preserving interaural time and level differences explicitly.

The combination of the EC principle with a noise reduction filter has been proposed in [Li et al. 2011] and [Enzner et al. 2016], where the EC stage estimates the interference and a cue-preserving common Wiener filter removes noise from both input channels. Since the EC stage eliminates the target speech signal, the system is able to track also non-stationary interference.

To briefly sketch the derivation of this common-gain binaural Wiener filter [Enzner et al. 2016], we define the desired output signals of the left and the right side, $s_l(k)$ $s_r(k)$ and consider the combined mean square error (MSE) of the left and the right output signals

$$\mathrm{E}\left\{e_l^2(k)\right\} + \mathrm{E}\left\{e_r^2(k)\right\} = \mathrm{E}\left\{(\hat{y}_l(k) - s_l(k))^2\right\} + \mathrm{E}\left\{(\hat{y}_r(k) - s_r(k))^2\right\}. \tag{13.34}$$

As we like to compute a common impulse response $g(k)$ for both sides, we have

$$\hat{y}_l(k) = g(k) * y_l(k), \tag{13.35}$$

$$\hat{y}_r(k) = g(k) * y_r(k). \tag{13.36}$$

Similar to the single-channel Wiener filter, the solution of the IIR Wiener filter is computed in the frequency domain and yields under the assumption of additive and independent noise

$$\Phi_{s_l y_l}(e^{j\Omega}) = \Phi_{s_l s_l}(e^{j\Omega}), \tag{13.37a}$$

$$\Phi_{s_r y_r}(e^{j\Omega}) = \Phi_{s_r s_r}(e^{j\Omega}), \tag{13.37b}$$

the frequency response

$$G(e^{j\Omega}) = \frac{\Phi_{s_l y_l}(e^{j\Omega}) + \Phi_{s_r y_r}(e^{j\Omega})}{\Phi_{y_l y_l}(e^{j\Omega}) + \Phi_{y_r y_r}(e^{j\Omega})} = \frac{\Phi_{s_l s_l}(e^{j\Omega}) + \Phi_{s_r s_r}(e^{j\Omega})}{\Phi_{y_l y_l}(e^{j\Omega}) + \Phi_{y_r y_r}(e^{j\Omega})}. \tag{13.38}$$

Thus, the optimal solution comprises averages of the power spectra in the numerators and denominators of individual single-channel Wiener filter frequency responses. It is shown in [Enzner et al. 2016] that this gain function results in a better balance between the left and the right ear than solutions with channel-specific or heuristic common gains. The target signal blocking via EC and variations thereof have been thoroughly investigated in [Azarpour, Enzner 2017] in conjunction with the common gain approach.

Note that these approaches are also related to [Lotter, Vary 2006], which estimates the target signal power via a superdirective beamformer and computes a common gain for

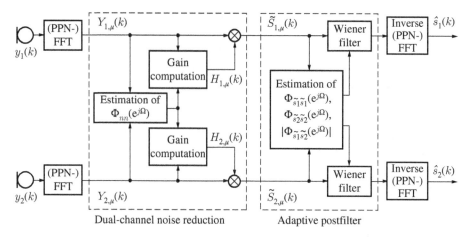

**Figure 13.11** Binaural noise reduction system in the DFT domain using a two-channel noise power spectrum estimate and an adaptive Wiener postfilter (PPN = Poly-Phase Network for high-resolution spectral analysis) [Dörbecker, Ernst 1996].

the left and the right output channel. Similar to the methods in [Azarpour, Enzner 2017], the computation of the optimal beamformer for target extraction requires assumptions about the spatial covariance of the noise field. In many cases this is modeled as an ideal diffuse noise field.

### 13.4.3 Combined Single- and Dual-Channel Noise Reduction

It is furthermore interesting to explore combinations of single-and dual-channel approaches. An early example of such a system is given in [Dörbecker, Ernst 1996]. This system has been designed to deliver two output channels for use in a binaural hearing aid and is depicted in Figure 13.11. Here, the single-channel noise reduction subsystem as well as the Wiener postfilters use cross-correlation information of the two microphone channels. First, the dual-channel noise power estimate is used in conjunction with single-channel log-MMSE gain functions [Ephraim, Malah 1985] in each of the two microphone channels. In the second stage, a frequency-domain adaptive postfilter is employed, which implements the dual-channel noise reduction principle as outlined in Section 13.4.1. While the first stage computes two gain functions, the second stage uses the same filter in both the left and right output channels. This common Wiener filter has been computed taking either the mean or the minimum of the per-channel filters.

## 13.5 Dual-Channel Dereverberation

It is interesting to note that the above dual-channel noise reduction approach can also be employed for the suppression of reverberation components in microphone signals. Single and multi-channel methods for dereverberation were summarized, including evaluation procedures and results, in conjunction with the *Reverb Challenge* [Kinoshita et al. 2016].

For the application of dual-channel methods, we note that late reverberation is dominated by a huge number of sound reflections arriving from a multitude of directions. Therefore, the coherence of the received signals decays with increasing reverberation levels and approaches the coherence of the ideal diffuse sound field. This effect has been measured and discussed, e.g., in [Martin, Vary 1994].

First proposals utilizing this property reach back to [Allen et al. 1977] and even beyond, as this method had been originally devised using analog electronics [Danilenko 1968], [Simmer et al. 2001]. In the first digital realization of this principle, [Allen et al. 1977] propose a dual-channel system that first performs a decomposition of the two broadband audio signals into frequency subbands and, after a phase-aligned combination of the two channels, uses the MSC function (and variations thereof) as a gain function on the combined signal. A block diagram of this system, extended to two output channels for binaural listening, is depicted in Figure 13.12. The analysis and synthesis stages may be implemented via filter banks or the STFT/iSTFT. The frequency responses of the alignment filters $A_\mu(k)$ are given by

$$A_\mu(k) = \frac{Y_{2,\mu}(k)Y_{1,\mu}^*(k)}{|Y_{1,\mu}(k)Y_{2,\mu}(k)|},$$ (13.39)

where $Y_{\kappa,\mu}(k)$ denotes the complex signal at time $k$ and frequency bin $\mu$ of the $\kappa$-th channel. Then, the combined and processed subband signals in each output channel are given by

$$\hat{S}_{1,\mu}(k) = \frac{\hat{K}_\mu(k)}{2}\left(Y_{1,\mu}(k) + A_\mu^*(k)Y_{2,\mu}(k)\right)$$
$$= \frac{\hat{K}_\mu(k)}{2}\left(|Y_{1,\mu}(k)| + |Y_{2,\mu}(k)|\right)\frac{Y_{1,\mu}(k)}{|Y_{1,\mu}(k)|},$$ (13.40)

$$\hat{S}_{2,\mu}(k) = \frac{\hat{K}_\mu(k)}{2}\left(Y_{2,\mu}(k) + A_\mu(k)Y_{1,\mu}(k)\right)$$
$$= \frac{\hat{K}_\mu(k)}{2}\left(|Y_{1,\mu}(k)| + |Y_{2,\mu}(k)|\right)\frac{Y_{2,\mu}(k)}{|Y_{2,\mu}(k)|}.$$ (13.41)

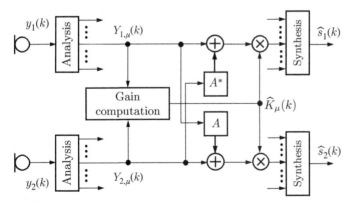

**Figure 13.12** Dual-channel dereverberation using spectral gains based on the coherence function.

The gain function $\hat{K}_\mu(k)$ may approximate the MSC, i.e., $\hat{K}_\mu(k) \approx \left| \gamma_{y_1 y_2, \mu}(k) \right|^2$ where statistical expectations are replaced by first-order recursive systems with smoothing parameter $\alpha$,

$$P_{Y_{\kappa,\mu} Y_{\lambda,\mu}}(k) = \alpha P_{Y_{\kappa,\mu} Y_{\lambda,\mu}}(k-r) + (1-\alpha) Y_{\kappa,\mu}(k-r) Y^*_{\lambda,\mu}(k-r), \tag{13.42}$$

$r$ denotes the frame shift and $\kappa, \lambda \in \{1,2\}$ the channel indices. Thus, the estimated gain function is given by

$$\hat{K}_\mu(k) = \frac{|P_{Y_{1,\mu} Y_{2,\mu}}(k)|^2}{P_{Y_{1,\mu} Y_{1,\mu}}(k) P_{Y_{2,\mu} Y_{2,\mu}}(k)}. \tag{13.43}$$

A substantial reduction of reverberation is achieved, as demonstrated in Figure 13.13. The parameter $\alpha$ must strike a balance between sufficient averaging for the estimation of a proper coherence function (see (13.15)) and a fast adaptation to onsets and reverberation tails. While $\alpha$ close to zero will result in gain functions close to one, $\alpha$ close to one may increase the perceived reverberance and is thus also not desirable. In an STFT-based

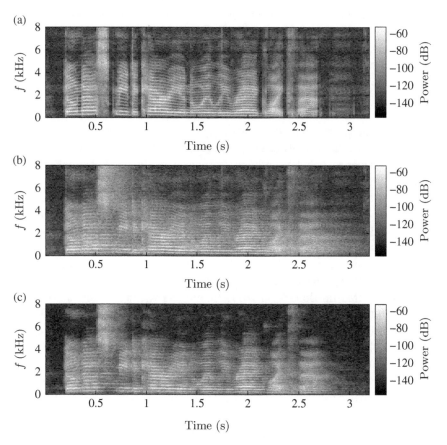

**Figure 13.13** Dual-channel dereverberation using the coherence function: (a) spectrogram of clean speech signal, (b) spectrogram of artificially reverberated speech signal using binaural room impulse responses [Jeub et al. 2009] of a lecture room with $T_{60} = 700$ ms, (c) spectrogram of the enhanced speech signal.

implementation, it is therefore advised to use small frame shifts such that sufficient local smoothing can be achieved without blurring speech sounds.

The basic method has been further refined and combined with dual-channel noise reduction methods in several works. In [Kollmeier et al. 1993], the authors evaluate a real-time implementation in conjunction with dynamic multi-band compression for use in binaural hearing aids. They stress the importance of striking a balance between interference suppression, target signal distortion, and processing artifacts. [Jeub et al. 2010] present a two-stage combination of a single-channel dereverberation method and a second coherence-based dereverberation stage. While the first stage uses a statistical model of the sound energy decay in reverberant rooms, the second stage is based on a model of binaural coherence [Jeub et al. 2011] and thus exploits the low coherence of the residual reverberation and the high coherence of the direct sound components. Both stages use a common gain in both binaural channels in order to preserve binaural cues. The common gain method is also applied in [Westermann et al. 2013] where the authors propose and evaluate an adaptive non-linear (sigmoidal) coherence-to-gain mapping for speech dereverberation. Methods to reduce unnatural fluctuations and processing artifacts via cepstrum smoothing have been explored in [Gerkmann 2011] and [Martin et al. 2015].

## 13.6 Methods Based on Deep Learning

The principles outlined in this chapter may also be used in the context of deep neural networks (DNNs) and data-driven methods. However, given the recent success of DNN-based single-channel methods, the benefits of using a secondary microphone may be less pronounced. As a consequence, only few authors have considered the dual-channel configuration and have explored novel processing strategies in this context.

The estimation of the noise power spectrum using log-mel features derived from the two microphone signals of a smart phone and a deep feed-forward neural network has been explored in [López-Espejo et al. 2016]. Here, the dual-channel configuration allows the exploitation of power level differences between the input channels. The method aims to find a non-linear mapping between the noisy speech of the two input signals and the log-mel features of the noise in the primary microphone channel. It has been evaluated in the context of an ASR task. In [Martín-Doñas et al. 2017], the authors extend this work and explore the use of both feed-forward and recurrent network structures for use in dual-microphones smart phones. In this feature-based approach, DFT-based spectral log-power features are computed for both channels. Features from both channels are stacked and are subject to a joint mean and variance normalization, thus also preserving power level differences. Then, during the training stage, a mapping from these features to the clean speech features is computed. The output signal is reconstructed using the enhanced log-power spectrum and the noisy phase of the primary microphone.

In [Tan et al. 2019], the authors propose to use a convolutional recurrent network (CRN) to predict a *phase-sensitive* time-frequency mask. This mask is then applied to the magnitude spectrum of the primary microphone signal. The noisy phase is used for the reconstruction of the time-domain signal. The input feature matrix consists not only of the magnitude spectra of both channels but also of the magnitude of the spectral sum and spectral difference of both channels ("inter-channel features"). Thus, as the sum and the difference are computed on complex spectra, the method implicitly considers phase information in these magnitude features. However, it is shown in [Tan et al. 2021], using a densely connected CRN that inter-channel features do not provide an advantage when complex-valued spectra of both channels are used as input features and the network is configured to produce a direct complex spectral mapping. In line with recent developments in single-channel noise reduction, fully complex-valued computations on complex-valued features appear to be the method of choice. The above and other works also focus on the reduction of the computational complexity and algorithmic latency as this is a recurring requirement for communication devices with a small computational footprint.

## References

Allen, J. B.; Berkley, D. A.; Blauert, J. (1977). Multimicrophone Signal-Processing Technique to Remove Room Reverberation from Speech Signals, *Journal of the Acoustical Society of America*, vol. 62, no. 4, pp. 912–915.

Armbrüster, W.; Czarnach, R.; Vary, P. (1986). Adaptive Noise Cancellation with Reference Input - Possible Applications and Theoretical Limits, *Proceedings of the European Signal Processing Conference (EUSIPCO)*, pp. 391–394.

Azarpour, M.; Enzner, G. (2017). Binaural Noise Reduction via Cue-Preserving MMSE Filter and Adaptive-Blocking-Based Noise PSD Estimation, *EURASIP Journal on Advances in Signal Processing*, vol. 2017, pp. 1–17.

Azarpour, M.; Enzner, G.; Martin, R. (2014). Binaural Noise PSD Estimation for Binaural Speech Enhancement, *Proceedings of the IEEE International Conference on Acoustics, Speech and Signal Processing (ICASSP)*, pp. 7068–7072.

Bendat, J. S.; Piersol, A. G. (1966). *Measurement and Analysis of Random Data*, John Wiley & Sons, Ltd., Chichester.

Borß, C. (2011). *An Improved Parametric Model for the Design of Virtual Acoustics and its Applications*, PhD thesis, Ruhr-Universität Bochum, Institute of Communication Acoustics.

Carter, G. C. (1987). Coherence and Time Delay Estimation, *Proceedings of the IEEE*, vol. 75, no. 2, pp. 236–255.

Carter, G. C.; Knapp, C. H.; Nuttall, A. H. (1973). Estimation of the Magnitude-Squared Coherence Function via Overlapped Fast Fourier Transform Processing, *IEEE Transactions on Audio and Electroacoustics*, vol. 21, no. 4, pp. 337–344.

Dämmig, P. (1957). Zur Messung der Diffusität von Schallfeldern durch Korrelation, *Acustica*, vol. 7, p. 387 (in German).

Danilenko, L. (1968). *Binaurales Hören im nichtstationären diffusen Schallfeld*, PhD thesis, RWTH Aachen University (in German).

Degan, N. D.; Prati, C. (1988). Acoustic Noise Analysis and Speech Enhancement Techniques for Mobile Radio Applications, *Signal Processing*, vol. 15, no. 1, pp. 43–56.

Dörbecker, M.; Ernst, S. (1996). Combination of Two-channel Spectral Subtraction and Adaptive Wiener Post-Filtering for Noise Reduction and Dereverberation, *Proceedings of the European Signal Processing Conference (EUSIPCO)*, pp. 995–998.

Durlach, N. I. (1963). Equalization and Cancellation Theory of Binaural Masking?Level Differences, *Journal of the Acoustical Society of America*, vol. 35, no. 8, pp. 1206–1218.

Elko, G. (2001). Spatial Coherence Function for Differential Microphones in Isotropic Noise Fields, *in* M. Brandstein; D. Ward (eds.), *Microphone Arrays*, Springer-Verlag, Berlin, Heidelberg, New York.

Enzner, G.; Azarpour, M.; Siska, J. (2016). Cue-preserving MMSE Filter for Binaural Speech Enhancement, *2016 IEEE International Workshop on Acoustic Signal Enhancement (IWAENC)*, pp. 1–5.

Ephraim, Y.; Malah, D. (1985). Speech Enhancement Using a Minimum Mean-Square Error Log-Spectral Amplitude Estimator, *IEEE Transactions on Acoustics, Speech and Signal Processing*, vol. 33, no. 2, pp. 443–445.

Faucon, G.; Mezalek, S. T.; Le Bouquin, R. (1989). Study and Comparison of Three Structures for Enhancement of Noisy Speech, *Proceedings of the IEEE International Conference on Acoustics, Speech, and Signal Processing (ICASSP)*, Glasgow, Scotland, pp. 385–388.

Gannot, S.; Burshtein, D.; Weinstein, E. (2001). Signal Enhancement using Beamforming and Nonstationarity with Applications to Speech, *IEEE Transactions on Signal Processing*, vol. 49, no. 8, pp. 1614–1626.

Gerkmann, T. (2011). Cepstral Weighting for Speech Dereverberation Without Musical Noise, *2011 19th European Signal Processing Conference*, pp. 2309–2313.

Gustafsson, H.; Nordholm, S.; Claesson, I. (1999). Spectral Subtraction Using Dual Microphones, *Proceedings of the International Workshop on Acoustic Echo and Noise Control (IWAENC)*, Pocono Manor, Pennsylvania, USA, pp. 60–63.

Harrison, W. A.; Lim, J. S.; Singer, E. (1986). A New Application of Adaptive Noise Cancellation, *IEEE Transactions on Acoustics, Speech and Signal Processing*, vol. 34, no. 1, pp. 21–27.

Jeub, M.; Schäfer, M.; Vary, P. (2009). A Binaural Room Impulse Response Database for the Evaluation of Dereverberation Algorithms, *Proceedings of International Conference on Digital Signal Processing (DSP)*, Santorini, Greece, pp. 1–4.

Jeub, M.; Schafer, M.; Esch, T.; Vary, P. (2010). Model-Based Dereverberation Preserving Binaural Cues, *IEEE Transactions on Audio, Speech, and Language Processing*, vol. 18, no. 7, pp. 1732–1745.

Jeub, M.; Dörbecker, M.; Vary, P. (2011). A Semi-Analytical Model for the Binaural Coherence of Noise Fields, *IEEE Signal Processing Letters*, vol. 18, no. 3, pp. 197–200.

Jeub, M.; Herglotz, C.; Nelke, C.; Beaugeant, C.; Vary, P. (2012). Noise Reduction for Dual-microphone Mobile Phones Exploiting Power Level Differences, *2012 IEEE International Conference on Acoustics, Speech and Signal Processing (ICASSP)*, pp. 1693–1696.

Kinoshita, K.; Delcroix, M.; Gannot, S.; Habets, E. A. P.; Haeb-Umbach, R.; Kellermann, W.; Leutnant, V.; Maas, R.; Nakatani, T.; Raj, B.; Sehr, A.; Yoshioka, T. (2016). A Summary of the REVERB Challenge: State-of-the-art and Remaining Challenges in Reverberant Speech Processing Research, *EURASIP Journal on Advances in Signal Processing*, vol. 2016, pp. 1–19.

Kollmeier, B.; Peissig, J.; Hohmann, V. (1993). Real-Time Multiband Dynamic Compression and Noise Reduction for Binaural Hearing Aids, *Journal of Rehabilitation Research and Development*, vol. 30, no. 1, pp. 82–94.

Kroschel, K.; Linhard, K. (1988). Combined Methods for Adaptive Noise Cancellation, *Proceedings of the European Signal Processing Conference (EUSIPCO)*, Grenoble, France, pp. 411–414.

Kuttruff, H. (1990). *Room Acoustics*, 3rd edn, Applied Science Publishers, Barking.

Li, J.; Sakamoto, S.; Hongo, S.; Akagi, M.; Suzuki, Y. (2011). Two-Stage Binaural Speech Enhancement with Wiener Filter for High-Quality Speech Communication, *Speech Communication*, vol. 53, no. 5, pp. 677–689. Perceptual and Statistical Audition.

Lindevald, I.; Benade, A. (1986). Two-Ear Correlation in the Statistical Sound Fields of Rooms, *Journal of the Acoustical Society of America*, vol. 80, no. 2, pp. 661–664.

López-Espejo, I.; Peinado, A. M.; Gomez, A. M.; Martín-Doñas, J. M. (2016). Deep Neural Network-Based Noise Estimation for Robust ASR in Dual-Microphone Smartphones, *International Conference on Advances in Speech and Language Technologies for Iberian Languages*, Springer-Verlag, pp. 117–127.

Lotter, T.; Vary, P. (2006). Dual-Channel Speech Enhancement by Superdirective Beamforming, *EURASIP Journal on Applied Signal Processing*, vol. 2006, pp. 1–14.

Martín-Doñas, J. M.; Gomez, A. M.; López-Espejo, I.; Peinado, A. M. (2017). Dual-Channel DNN-Based Speech Enhancement for Smartphones, *IEEE 19th International Workshop on Multimedia Signal Processing (MMSP)*, pp. 1–6.

Martin, R. (1995). *Hands-free Telephones Based on Multi-Channel Echo Cancellation and Noise Reduction*, PhD thesis. *Aachener Beiträge zu digitalen Nachrichtensystemen*, vol. 3, P. Vary (ed.), RWTH Aachen University (in German).

Martin, R.; Vary, P. (1994). Combined Acoustic Echo Cancellation, Dereverberation, and Noise Reduction: A Two Microphone Approach, *Annales des Télécommunications*, vol. 7–8, pp. 429–438.

Martin, R.; Azarpour, M.; Enzner, G. (2015). Binaural Speech Enhancement with Instantaneous Coherence Smoothing using the Cepstral Correlation Coefficient, *Proceedings of the IEEE International Conference on Acoustics, Speech and Signal Processing (ICASSP)*, pp. 111–115.

Sambur, M. R. (1978). Adaptive Noise Canceling for Speech Signals, *IEEE Transactions on Acoustics, Speech and Signal Processing*, vol. 26, no. 5, pp. 419–423.

Simmer, K.; Bitzer, J.; Marro, C. (2001). Post-filtering Techniques, *in* M. Brandstein; D. Ward (eds.), *Microphone Arrays*, Springer-Verlag, Berlin, Heidelberg, New York.

Tan, K.; Zhang, X.; Wang, D. (2019). Real-Time Speech Enhancement Using an Efficient Convolutional Recurrent Network for Dual-microphone Mobile Phones in Close-talk Scenarios, *Proceedings of the IEEE International Conference on Acoustics, Speech and Signal Processing (ICASSP)*, pp. 5751–5755.

Tan, K.; Zhang, X.; Wang, D. (2021). Deep Learning Based Real-Time Speech Enhancement for Dual-Microphone Mobile Phones, *IEEE/ACM Transactions on Audio, Speech, and Language Processing*, vol. 29, pp. 1853–1863.

Westermann, A.; Buchholz, J. M.; Dau, T. (2013). Binaural Dereverberation Based on Interaural Coherence Histograms, *The Journal of the Acoustical Society of America*, vol. 133, no. 5, pp. 2767–2777.

Widrow, B.; Glover, J. R.; McCool, J. M.; Kaunitz, J.; Williams, C. S.; Hearn, R. H.; Zeidler, J. R.; Dong, E.; Goodlin, R. C. (1975). Adaptive Noise Cancelling: Principles and Applications, *Proceedings of the IEEE*, vol. 63, no. 12, pp. 1692–1716.

Yousefian, N.; Akbari, A.; Rahmani, M. (2009). Using Power Level Difference for Near Field Dual-microphone Speech Enhancement, *Applied Acoustics*, vol. 70, no. 11–12, pp. 1412–1421.

# 14

## Acoustic Echo Control

In this chapter, we discuss algorithms for feedback control in handsfree voice communication systems, which use loudspeaker(s) and microphone(s) instead of a telephone handset. At the center of our discussion are adaptive algorithms for acoustic echo cancellation. In principle, the acoustic echo canceller can remove the annoying echo, i.e., the crosstalk from the loudspeaker to the microphone without distorting the near-end speech. Acoustic echo cancellation (AEC) is therefore an essential component of high-quality full-duplex handsfree communication devices. AEC is quite a challenging application of adaptive filters and other signal processing techniques. We present and discuss the most important algorithms for the adaptation of acoustic echo cancellers in the time and frequency domain, as well as additional measures for echo control.

## 14.1 The Echo Control Problem

Many speech communication devices offer a so-called handsfree mode. For this, a loudspeaker and a microphone are used instead of a hand-held telephone set, in order to increase user comfort or for safety reasons, e.g., while driving a car. Applications of handsfree systems comprise not only car phones and smart phones, but also multimedia systems with speech input, human–machine interfaces, or teleconferencing facilities. Even in communication headsets, AEC plays a role, especially for in-ear devices. Due to the small distance between the loudspeaker and the microphone, crosstalk occurs which has to be eliminated.

The basic set-up of a handsfree communication system is illustrated in Figure 14.1. The echo control problem arises as a consequence of the acoustic coupling between the loudspeaker and the microphone. The microphone picks up not only the desired signal $s$ of the near-end speaker but also undesirable background noise $n$, and in particular the signal of the far-end speaker, denoted $\tilde{x}$, which is received via the electro-acoustic transmission path from the loudspeaker to the microphone. In reverberant rooms, the signal $\tilde{x}$ includes multiple acoustic reflections. It is commonly called the acoustic echo signal (in distinction from the electric line echoes of the telephone network).

In digital networks, the far-end speaker's signal $x(k)$ and the near-end signal $y(k)$ are available separately and are transmitted over separate physical or logical channels. A special situation exists if an analog telephone set is connected by a 2-wire cable or subscriber

*Digital Speech Transmission and Enhancement*, Second Edition. Peter Vary and Rainer Martin.
© 2024 John Wiley & Sons Ltd. Published 2024 by John Wiley & Sons Ltd.

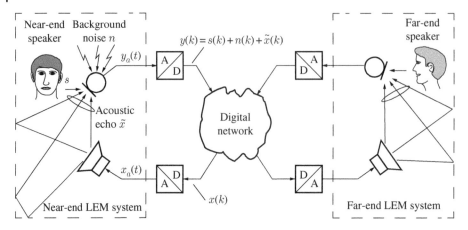

**Figure 14.1** Loudspeaker–enclosure–microphone (LEM) system of a handsfree telephone with digital input signal $x(k)$ and output signal $y(k)$.

line to an analog port of a digital network, e.g., a router or a local exchange. Then, the directional separation of the analog signals $x_a(t)$ and $y_a(t)$ is achieved in the telephone set by using directional filters, also called hybrid coils or bridge transformers. A telephone hybrid converts, the 2+2=4 wires of the loudspeaker and the microphone-signal from and to the 2-wire connection. A hybrid is also needed at the 2-wire analog port of the network (e.g., cable to a router or subscriber line to a local telephone exchange).

As hybrids do not provide perfect separation, some electric crosstalk from $x_a(t)$ to $y_a(t)$ and $y_a(t)$ to $x_a(t)$ occurs, which is a further source of voice echos.

In the following, we do not need to differentiate between the acoustic or analog signals and their digital counterparts. Only the discrete-time signals $x(k), s(k), n(k)$, etc., i.e., digital versions of band-limited analog signals need to be considered for signal processing.

The digital microphone signal $y(k)$ of the handsfree device in Figure 14.1 is thus given by

$$y(k) = s(k) + n(k) + \tilde{x}(k). \tag{14.1}$$

The task of acoustic echo cancellation is to prevent the echo signal $\tilde{x}(k)$ from being fed back to the far-end user. Thus, the stability of the electro-acoustic loop is ensured, even when the far-end terminal is a handsfree telephone as well.

Without echo cancellation, the far-end user receives an echo of her own voice. As the round-trip time (RTT) echo delay (which equals twice the one way transmission delay) might be large, the echo is not masked by the far-end user's own voice and thus may make it difficult for the user to speak.

The RTT is perceivable and sometimes even annoying, especially in cellular and in all-IP networks. The latency is due to several reasons such as source coding with a certain frame length, error protection with interleaving, modulation, radio propagation, and last but not least packet transmission with jitter buffers at the receiving end.

For example, in 2G cellular network with "circuit switched" transmission we have an RTT of approx. 180 ms. In voice-over-IP (VoIP) landline, satellite or 4G/5G cellular systems the RTT can take several 10 ms up to 300 ms.

Simple solutions to the echo control problem employ voice-controlled *echo suppressors,* which consist of a variable attenuator in both the transmit and the receive branch. Depending on the speech activity of the two speakers, complementary attenuation is applied to the transmit and the receive branch such that the total round trip attenuation in the echo path is, e.g., 40 dB.

This principle can be easily realized using analog or digital technology. A digital solution is shown in Figure 14.2. When only one of the two speakers is active (*single talk*) the echo suppressor can achieve a high level of echo suppression. However, as the variable attenuation factors should fulfill the condition

$$- \left( 20 \lg a_x(k) + 20 \lg a_y(k) \right) = 40 \, \text{dB}, \tag{14.2}$$

simultaneous communication of the two users (*double talk*) is possible only to a limited extent. This limitation can be circumvented by applying *echo cancellation* instead as depicted in Figure 14.3. The basic idea of acoustic echo cancellation is to model the electro-acoustic echo path (the loudspeaker–enclosure–microphone system, LEM system) as a linear system and to identify the possibly time-varying impulse response $\mathbf{g}(k)$ of this system by means of an adaptive transversal (FIR) filter $\mathbf{h}(k)$ with the time-variant coefficient vector of length $m$

$$\mathbf{h}(k) = \left( h_0(k), h_1(k), \ldots, h_{m-1}(k) \right)^T. \tag{14.3}$$

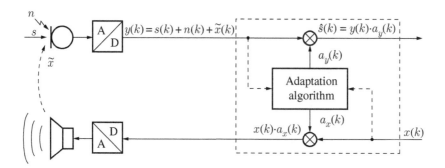

**Figure 14.2** Loudspeaking telephone with a voice-controlled echo suppressor.

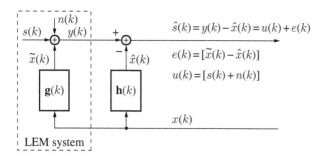

**Figure 14.3** Discrete-time model of a handsfree telephone with echo cancellation.

Thus, the echo signal $\tilde{x}(k)$ captured with the microphone can be estimated as

$$\hat{x}(k) = \mathbf{h}^T(k)\mathbf{x}(k). \tag{14.4}$$

The estimated echo, $\hat{x}(k)$, is then subtracted from the microphone signal $y(k)$.

Due to the frequency band limitation of the microphone signal, the electro-acoustic transmission of the far-end speaker's signal $x(k)$ via the LEM system can be described as a discrete-time, linear system with the causal impulse response

$$\mathbf{g}(k) = \big(g_0(k), g_1(k), g_2(k), \ldots\big)^T. \tag{14.5}$$

Because of the physical characteristics of the LEM system, the length of this impulse response is in principle infinite (IIR).

Two typical impulse responses of LEM systems are shown in Figure 14.4. While the initial part of the LEM impulse response is dominated by the direct path from the loudspeaker to the microphone and by distinct peaks of early reflections, the reverberation manifests itself in a large number of decaying impulses, which are best described by statistical models.

The exponential decay of the late reverberation components is usually characterized by a time constant, the *reverberation time* $T_H$. It describes the time span during which, according to an exponential law, the energy density drops by 60 dB of its initial value after turning off the sound source. Using *Sabine's reverberation formula* (e.g., [Kuttruff 2017]), the reverberation time can be approximately determined as a function of the room volume $V$, the wall areas $A_i$ with absorption coefficients $\alpha_i$ and the sound velocity $c$ as

$$T_H = \frac{24 \; \ln(10) \; V}{c \; \sum_i A_i \, \alpha_i}. \tag{14.6}$$

The reverberation time in a car with an acoustically relevant volume of, e.g., $V = 1.3\,\text{m}^3$ amounts to about $T_H = 0.065\,\text{s}$; in an office room with $V = 100\,\text{m}^3$ it can be much larger, e.g., $T_H = 0.7\,\text{s}$ [Martin 1995].

The large number of quasi-random reflections in a reverberant enclosure also implies that any accurate model of the LEM system requires many degrees of freedom. Because of the exponential decay of LEM responses and because of amplifier and quantization noise in the

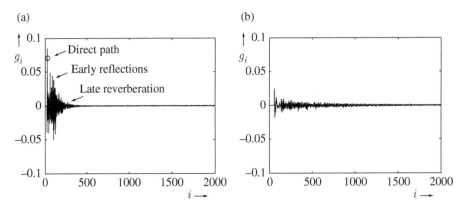

**Figure 14.4** Measured impulse responses of LEM systems at $f_s = 8$ kHz. (a) Car and (b) office room.

microphone signal, the infinite impulse response $\mathbf{g}(k)$ (14.5) can be truncated for modeling purposes to a finite number of $m'$ coefficients, provided that $m'$ is chosen sufficiently large. The LEM system is thus modeled as a transversal (FIR) filter of order $m' - 1$ with impulse response

$$\mathbf{g}(k) = \left( g_0(k), g_1(k), \ldots, g_{m'-1}(k) \right)^T, \tag{14.7}$$

and with input signal $x(k)$. The impulse response is in general time-variant as it is influenced by movements of the near-end speaker and by other changes of the acoustic environment such as temperature, causing a change of the speed of sound.

The underlying signal processing problem therefore consists in identifying the time-varying system impulse response $\mathbf{g}(k)$. If the impulse responses $\mathbf{g}$ and $\mathbf{h}$ match exactly and $m = m'$ is sufficiently large, the echo signal will be eliminated from the transmit branch. In practice, the impulse response $\mathbf{h}(k)$ of the cancellation filter is adapted using an iterative algorithm. One of the most prominent adaptation algorithms is the *normalized least-mean-square* (NLMS) algorithm, which has gained widespread acceptance due to its simple adaptation rule and stability. Tracking of fast-changing impulse responses is, however, not easily accomplished [van de Kerkhof, Kitzen 1992]. Fast-converging time-domain algorithms and efficient frequency-domain implementations of adaptive filtering concepts, are therefore also of significant interest. A vast amount of literature deals with these adaptive algorithms and their application to echo cancellation. Excellent surveys are given in, for example, [Breining et al. 1999], [Gay, Benesty 2000], [Benesty et al. 2001], [Hänsler, Schmidt 2004], and [Enzner et al. 2014].

To fully appreciate the difficulties of the echo cancellation problem, an estimate of the required order of the cancellation filter will be derived. In Figure 14.5, an idealized LEM system with two sound propagation paths is depicted: one direct path leads from the loudspeaker to the microphone via the distance $d$, and one indirect path with two reflections via the total distance of $d_1 + d_2 + d_3$. The length of the impulse response $\mathbf{h}$ of the cancellation filter should cover the corresponding propagation delay $\tau$ of the acoustic signal.

With a sampling rate $f_s$ and a sound velocity $c$, the minimal required filter order amounts to

$$m - 1 \geq \frac{\sum_i d_i}{c} \cdot f_s \approx T_H \cdot f_s. \tag{14.8}$$

For $c \approx 343$ m/s and a distance of $d = 20$ cm between loudspeaker and microphone, the direct sound path could be canceled using

$$m = \text{round} \left( \frac{d}{c} \cdot f_s \right), \tag{14.9}$$

**Figure 14.5** Sound propagation paths in the LEM system.

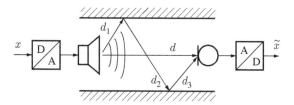

coefficients, i.e., only $m = 6$ in case of a sampling rate of $f_s = 8\,\text{kHz}$ and $m = 12$ in case of $f_s = 16\,\text{kHz}$.

However, the cancellation of the indirect sound, which is due to multiple reflections requires significantly longer impulse responses. The average total distance can be obtained from the mean free path lengths $\bar{d}_i$ between reflections [Kuttruff 2017].

The cumulative mean free path length of

$$\bar{d} = \sum_i \bar{d}_i = c \cdot T_H, \tag{14.10}$$

amounts in a car to, e.g., $\bar{d} = 343\,\text{m/s} \cdot 0.065\,\text{s} \approx 22.3\,\text{m}$ and in reverberating office rooms with $0.2\,\text{s} \le T_H \le 1.0\,\text{s}$ to $68.6\,\text{m} \le \bar{d} \le 343\,\text{m}$.

At a sampling rate of $f_s = 8\,\text{kHz}$ the required number of filter coefficients would amount in a car to $m \approx 500$ and in the office room to $m \approx 1600...8000$.

If the sampling rate is increased, e.g., for wideband speech transmission with $f_s = 16\,\text{kHz}$, the number of coefficients has to be doubled.

The computational complexity of a time-domain FIR filter in terms of *Mega operations per second* (MOPS), i.e., multiply add operations, is for $f_s = 16\,\text{kHz}$ in an environment requiring $m = 1000 ... 16000$ coefficients about

$$m \cdot f_s = 16 ... 256\,\text{MOPS}. \tag{14.11}$$

The computational complexity for the adaptation of the coefficients must be added, which increases (14.11) by a factor of at least 2–3. The high complexity of a full-size canceller, the required fast adaptation and the difficulty of adapting a canceler in the presence of acoustic noise and double talk usually requires implementations with significantly fewer canceler coefficients. Thus, the echo cannot be fully cancelled and additional measures, as discussed in Section 14.2, are necessary to suppress the residual echo.

## 14.2 Echo Cancellation and Postprocessing

As the acoustic environment may vary with time, the echo canceller $\mathbf{h}(k)$ has to be implemented as an adaptive filter. Various adaptation algorithms will be described in Sections 14.5–14.11. The echo path $\mathbf{g}(k)$ is altered when acoustic reflections change, e.g., by movements of the user or by opening a car window. As adaptation takes some time, the cancellation filter may not deliver sufficient echo suppression during this transition period. Some annoying residual echo remains at least for a while, which requires additional measures for echo reduction.

Such measures are also advisable if, due to complexity limitations in terms of available processing power and memory, the impulse response of the cancellation filter $\mathbf{h}(k)$ is noticeably shorter than the actual impulse response $\mathbf{g}(k)$ of the LEM system.

As late echoes are more audible and more annoying than early echoes or reflections, the required amount of echo suppression depends on the signal dispersion in the acoustic path as well on the transmission delay caused by the network. In case of insufficient echo suppression, the user at the far-end perceives the residual echo of his or her own voice with the possibly annoying round trip delay. Quality standards for handsfree devices have been established in international recommendations as discussed in Section 14.3.

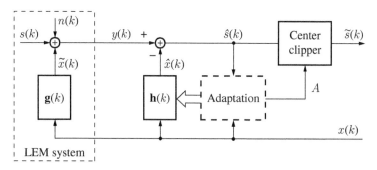

**Figure 14.6** Echo canceller with postprocessing for residual echo suppression by center clipping as described by (14.13).

### 14.2.1 Echo Canceller with Center Clipper

A low-level residual echo signal

$$e(k) = \tilde{x}(k) - \hat{x}(k), \tag{14.12}$$

might be audible at the far-end during speech pauses of the near-end speaker, especially in case of a long transmission delay.

In such a situation, the residual echo signal can be effectively suppressed with a non-linear *center clipping device,* as depicted in Figure 14.6.

The task of a center clipper is to suppress signal portions with small amplitudes.

The output signal $\hat{s}(k)$ of the echo canceller is processed by a center clipper as described by either

$$\tilde{s}(k) = \begin{cases} \hat{s}(k) - A & \hat{s}(k) > +A \\ 0 & |\hat{s}(k)| \leq +A \\ \hat{s}(k) + A & \hat{s}(k) < -A, \end{cases} \tag{14.13}$$

or alternatively by

$$\tilde{s}(k) = \begin{cases} \hat{s}(k) & \hat{s}(k) > +A \\ 0 & |\hat{s}(k)| \leq +A \\ \hat{s}(k) & \hat{s}(k) < -A. \end{cases} \tag{14.14}$$

and as illustrated in Figure 14.7. Both variants provide similarly good results. The threshold value $A$, which can also be adapted, should be as large as necessary, but as small as possible.

### 14.2.2 Echo Canceller with Voice-Controlled Soft-Switching

The center clipper can suppress, without perceptible distortions of $s(k)$, only low-level residual echoes below a small threshold $A$. However, during the initialization phase of the cancellation filter, and after sudden changes in the LEM impulse response, the amplitudes of the residual echoes might be temporarily much larger than the threshold $A$ of the center clipper.

(a) (b)

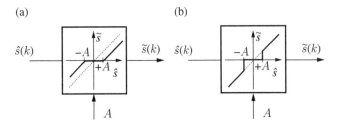

**Figure 14.7** Center clipper characteristics. (a) Equation (14.13) and (b) Equation (14.14).

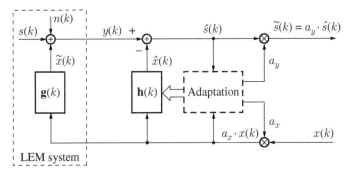

**Figure 14.8** Echo canceler with voice-controlled soft-switching supporting the initialization phase and suppressing the residual echo.

Better attenuation of the residual echo $e(k) = \tilde{x}(k) - \hat{x}(k)$ can be provided by combining the echo canceller with the voice-controlled soft-switch of Figure 14.2, e.g., [Armbrüster 1988], as shown in Figure 14.8.

If the echo canceler does not produce any loop attenuation at the start of the adaptation, the two factors are set, e.g., to $20\ \lg(a_x) = 20\ \lg(a_y) = -20\,\text{dB}$, corresponding to a loop attenuation of 40 dB. Then, the far-end speaker starts talking. In this early stage of adaptation, the receive branch is adjusted to an attenuation of 0 dB ($a_x = 1$) and the transmit branch to an attenuation of 40 dB ($a_y = 0.01$). Full-duplex double talk is not yet possible, because of a receive gain of, e.g., $a_x = 1$ and a transmit gain $a_y \approx 0$ or vice versa. This concept is also called *level weighing*. The initial span of the compound attenuation of e.g., $20\ \lg(a_x) + 20\ \lg(a_y) = 40\,\text{dB}$ is gradually reduced, as the cancellation filter $\mathbf{h}(k)$ is adapted for taking over the cancellation of the echo signal $\tilde{x}(k)$.

After a change in the impulse response of the LEM, the echo canceler typically achieves an effective reduction $20\ \lg(a_c)$ of less than 40 dB. When this condition is detected, the voice-controlled soft-switching becomes active again, with the required additional reduction of $(40 - 20\ \lg(a_c))\,\text{dB}$ half in the transmit and half in the receive branch.

A further variation is to apply voice-controlled soft-switching individually in subbands of a filter bank or a filter bank equalizer.

### 14.2.3 Echo Canceller with Adaptive Postfilter

Instead of residual echo suppression by center clipping (Figure 14.6) or by soft-switching (Figure 14.8), an adaptive postfilter $\mathbf{c}(k)$ can be used in the transmit branch, as shown in Figure 14.9, e.g., [Faucon, Le Bouquin-Jeannès 1995], [Martin, Altenhöner 1995], [Ayad

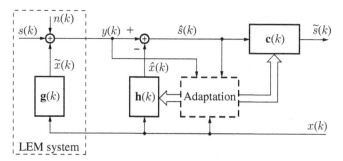

**Figure 14.9** Combination of an echo canceler with an adaptive filter for a frequency selective attenuation of the echo.

et al. 1996], [Martin, Vary 1996], [Le Bouquin-Jeannès et al. 2001], and [Hänsler, Schmidt 2004]. This is a transversal filter which, in contrast to soft switching, provides frequency selective attenuation. The spectrum of the signal $\hat{s}(k)$ is weighted as a function of frequency and time, depending on the instantaneous, spectral shape of the residual echo $e(k)$, the desired signal $s(k)$, and the background noise $n(k)$. In this, the psychoacoustic effect of masking is exploited. The residual echo is suppressed when it has significantly more power in a given frequency band than the near-end signals, i.e., when it is audible and not masked by the near-end signal. In the speech pauses of the far-end speaker, the signal $\hat{s}(k)$ is not influenced by the postfilter. The time-variant impulse response $\mathbf{c}(k)$ can be adjusted in the time domain or in the frequency domain, in analogy to single-channel noise reduction systems, e.g., [Martin, Gustafsson 1996] and [Gustafsson et al. 1998] (see also Chapter 12). Alternatively, DNN-based methods can be used for adjusting the postfilter, e.g., [Haubner, Kellermann 2022].

This widely used method achieves a significant additional reduction in the residual echo. Therefore, the order of the echo canceller can be greatly reduced, thus leading to very efficient implementations. In fact, the combined echo cancellation and postfiltering approach opens up a twofold perspective on the design of echo control systems: a relative short canceller can be used to guarantee *stability* of the transmission loop, while the *audibility* of echoes can be tackled by a less expensive postfilter. During double talk, the latter device needs only to achieve a less strong reduction of the residual echo since much of the residual echo is masked by the near-end speech and stability is provided by the echo canceller.

## 14.3 Evaluation Criteria

Since the echo problem has a profound effect on communication quality, numerous international recommendations and standards have been established to guide the development of handsfree devices. The required echo suppression depends on the signal delay of the transmission path and on the double talk condition. The actual requirements are frequently specified in terms of the *talker echo loudness rating* (TELR), i.e., the level difference between the original far-end voice and the resulting echo. The TELR, which is deemed acceptable during single talk is given by 35 dB and 52 dB for transmission delays of 30 ms and 200 ms, respectively. For transmission delays of more than 200 ms, even more echo attenuation is necessary. During double talk, some of the echo is masked by the near-end signal

[Gilloire 1994]. In this case, good listening quality is achieved when the single talk TELR requirement is lowered by no more than 4 dB [ITU-T G.131 2003].

The required high level of echo suppression, particularly needed for long transmission delays, can in practice only be achieved with an echo canceller in conjunction with additional measures such as a center clipper (Section 14.2.1), voice-controlled soft-switching [Armbrüster 1988] (Section 14.2.2), or an echo reduction postfilter [Martin, Gustafsson 1996] (Section 14.2.3). However, the contribution of an echo suppressor to the required TELR has a profound effect on the double talk capability of the handsfree terminal. Thus, handsfree terminals are further categorized based on the attenuation range, which is applied to the near-end signal by the echo suppressor [ITU-T P.340 2000]. Full duplex capability, for instance, is achieved when the contribution of the echo suppressor to the required TELR entails less than 3 dB of near-end signal attenuation.

Since the TELR measure comprises the entire electro-acoustic echo loop, including the transmission network and the handsfree terminal, it is often not practical for a fair comparison of echo control devices. For an instrumental evaluation of echo cancellation and echo suppression algorithms, two instrumental criteria, which characterize the system identification error and the level of echo reduction, are therefore commonly used.

### 14.3.1 System Distance

The electro-acoustic LEM path is modeled as an FIR filter $\mathbf{g}(k)$ of length $m'$ which is approximated by the FIR impulse response $\mathbf{h}(k)$ of the echo cancellation filter of length $m$. In case of matching dimensions with $m = m'$ the distance vector

$$\mathbf{d}(k) = \mathbf{g}(k) - \mathbf{h}(k), \tag{14.15}$$

of the impulse responses allows us to define the relative system distance

$$D(k) = \frac{||\mathbf{d}(k)||^2}{||\mathbf{g}(k)||^2}, \tag{14.16}$$

as a performance measure for system identification. $||\mathbf{d}(k)||^2 = \mathbf{d}^T(k)\mathbf{d}(k)$ denotes the squared vector norm. Generally, the logarithmic distance $10 \lg(D(k))$ (in dB) is used. If the coefficient vectors $\mathbf{g}(k)$ and $\mathbf{h}(k)$ differ in length, the shorter vector is padded with zero values. According to definition (14.16), the system distance $D(k)$ does not depend directly on the signal $x(k)$. However, since the cancellation filter is adapted using the signal $x(k)$, the system distance obtained at time instant $k$ depends in fact on the particular excitation $x(k)$. Since the impulse responses of real LEM systems are generally not known, the system distance as defined in (14.16) is primarily an important evaluation criterion for off-line simulations with given impulse responses $\mathbf{g}(k)$.

### 14.3.2 Echo Return Loss Enhancement

One criterion, which is more closely related to the subjectively perceived performance is the achievable reduction of the power of the echo signal $\tilde{x}(k)$. The corresponding measure is called echo return loss enhancement (ERLE) and is defined as

$$\frac{\mathrm{ERLE}(k)}{\mathrm{dB}} = 10 \lg \left( \frac{\mathrm{E}\{\tilde{x}^2(k)\}}{\mathrm{E}\{(\tilde{x}(k) - \hat{x}(k))^2\}} \right). \tag{14.17}$$

This criterion depends on the echo signal $\tilde{x}(k)$ and its estimate $\hat{x}(k)$. The ensemble average $\mathrm{E}\{\cdot\}$ provides a measure of echo reduction for every time instant $k$. In practice, the expected values $\mathrm{E}\{\cdot\}$ in (14.17) are replaced by short-term estimates.

A small system distance $D(k)$ implies a high level of ERLE($k$). The inverse conclusion does not hold, which is easily demonstrated with a specific excitation signal $x(k)$, consisting of one or several sinusoidal components. To achieve a high level of ERLE, the frequency response of the cancellation filter must match the frequency response of the LEM system only for the frequencies, which are actually excited by $x(k)$. Deviations at other frequencies do not have an impact on the measured ERLE($k$).

In a real system, the *residual echo*

$$e(k) = \tilde{x}(k) - \hat{x}(k), \tag{14.18}$$

is only accessible for $s(k) = 0$ (cf. Figure 14.3), i.e., during speech pauses of the near-end speaker (*single talk*), and only if there is no background noise, i.e., for $n(k) = 0$. In simulation experiments, however, $e(k)$ can be calculated if the signals $s(k)$ and $n(k)$ are available as separate files, which can be processed separately so that the observation of the ERLE over time is possible in double talk and additive background noise conditions as well.

Besides ERLE($k$) and the system distance $D(k)$, measures for near-end speech quality and for double talk capability are used to characterize a handsfree device. Since modern handsfree devices are neither time invariant nor linear, special measurement procedures are required. An example based on *composite source signals* is given in [Gierlich 1992]. The standard procedures are summarized in [Carini 2001], [Gierlich, Kettler 2005], and [Kettler, Gierlich 2008].

## 14.4 The Wiener Solution

To lay the ground for the development of adaptive algorithms we briefly review the minimum mean square error solution (Wiener filter) to the echo cancellation problem.

The starting point is the minimization of the mean square error (MSE) of the residual echo $e(k)$ according to Figure 14.3

$$\mathrm{E}\left\{e^2(k)\right\} = \mathrm{E}\left\{\left(\tilde{x}(k) - \mathbf{h}^T(k)\mathbf{x}(k)\right)^2\right\}, \tag{14.19}$$

with the excitation vector

$$\mathbf{x}(k) = (x(k), x(k-1), \ldots, x(k-m+1))^T, \tag{14.20}$$

and the coefficient vector (tap weight vector)

$$\mathbf{h}(k) = \left(h_0(k), h_1(k), \ldots, h_{m-1}(k)\right)^T. \tag{14.21}$$

The partial derivation of (14.19) w.r.t. the unknown impulse response vector $\mathbf{h}(k)$ results in

$$\frac{\partial \mathrm{E}\{e^2(k)\}}{\partial \mathbf{h}(k)} = \mathrm{E}\{-2 \cdot [\tilde{x}(k) - \mathbf{h}^T(k)\mathbf{x}(k)] \cdot \mathbf{x}(k)\} \tag{14.22a}$$

$$= -2\,\mathrm{E}\{\tilde{x}(k) \cdot \mathbf{x}(k)\} + 2\,\mathrm{E}\{\mathbf{x}(k) \cdot \mathbf{x}^T(k) \cdot \mathbf{h}(k)\} \overset{!}{=} 0. \tag{14.22b}$$

For wide sense stationary input signals and a *time invariant* LEM system response $g(k) = g$, the general solution for a *time invariant* FIR Wiener filter $h(k) = h_W$ follows as (cf. also Section 5.11.1)

$$h_W = E\{x(k)x^T(k)\}^{-1} \cdot E\{\tilde{x}(k)x(k)\} = R_{xx}^{-1} \cdot E\{\tilde{x}(k)x(k)\} \tag{14.23}$$

provided that the $m \times m$ stationary auto-correlation matrix $R_{xx} = E\{x(k)x^T(k)\}$ is invertible.

In the echo cancellation context with $m \le m'$ and $\tilde{x}(k) = \sum_{i=0}^{m'-1} g_i x(k-i)$, the cross-correlation vector $E\{\tilde{x}(k)x(k)\}$ may be written as

$$E\{\tilde{x}(k)x(k)\} = \begin{pmatrix} \sum_{i=0}^{m'-1} g_i \varphi_{xx}(i) \\ \sum_{i=0}^{m'-1} g_i \varphi_{xx}(i-1) \\ \vdots \\ \sum_{i=0}^{m'-1} g_i \varphi_{xx}(i-m+1) \end{pmatrix} \tag{14.24}$$

$$= \begin{pmatrix} \varphi_{xx}(0) & \varphi_{xx}(1) & \cdots & \varphi_{xx}(m'-1) \\ \varphi_{xx}(1) & \varphi_{xx}(0) & \cdots & \varphi_{xx}(m'-2) \\ \vdots & \vdots & & \vdots \\ \varphi_{xx}(m-1) & \varphi_{xx}(m-2) & \cdots & \varphi_{xx}(m'-m) \end{pmatrix} \begin{pmatrix} g_0 \\ g_1 \\ \vdots \\ g_{m'-1} \end{pmatrix}$$

$$= R_{xx} \cdot g$$

with $\varphi_{xx}(i) = E\{x(k)x(k+i)\}$.

Combining (14.23) and (14.24) we conclude that for $m = m'$ a perfect identification $h_W = g$ is in principle possible. For $m < m'$ the FIR Wiener filter identifies the first $m$ coefficients without error if the auto-correlation $\varphi_{xx}(i)$ is equal to zero for $1 \le i \le m'$, i.e., for a white noise excitation signal $x(k)$. If the auto-correlation does not vanish for $1 \le i \le m'$ and $m < m'$ the estimate $h_W$ is biased.

The Wiener filter provides a solution for stationary signals. In general, the statistics of the excitation signal as well as the target impulse response $g(k)$ are time-varying. Therefore, adaptive algorithms are of great importance and will be studied in more detail below.

## 14.5 The LMS and NLMS Algorithms

### 14.5.1 Derivation and Basic Properties

In analogy to the iterative adaptation of a linear predictor by means of the *least-mean-square* (LMS) algorithm (see Section 6.4.2) we derive a rule for adjusting the cancellation filter step-by-step. The gradient of the MSE (14.19) results in the vector

$$\nabla(k) = \frac{\partial E\{e^2(k)\}}{\partial h(k)} \tag{14.25a}$$

$$= 2\,\mathrm{E}\left\{ e(k)\,\frac{\partial e(k)}{\partial \mathbf{h}(k)} \right\} \tag{14.25b}$$

$$= -2\,\mathrm{E}\{e(k)\mathbf{x}(k)\}. \tag{14.25c}$$

In order to decrease the error, the tap weights $h_i(k)$ must be adapted in the direction of the negative gradient. The fundamental idea of the LMS algorithm is to replace the mean gradient by the instantaneous gradient

$$\hat{\boldsymbol{\nabla}}(k) = -2\,e(k)\mathbf{x}(k). \tag{14.26}$$

The resulting *stochastic gradient descent* algorithm may be written as

$$\mathbf{h}(k+1) = \mathbf{h}(k) + \beta(k)\,e(k)\mathbf{x}(k), \tag{14.27}$$

with the effective (and in general time-variant) stepsize parameter $\beta(k)$ (cf. (6.81-b)).

It can be shown that the LMS algorithm converges in the mean square sense if and only if the stepsize parameter $\beta(k)$ satisfies

$$0 < \beta(k) < \frac{2}{\lambda_{\max}}, \tag{14.28}$$

where $\lambda_{\max}$ is the largest eigenvalue of the input correlation matrix $\mathbf{R}_{xx}$ [Haykin 1996].

Viewed as a dynamic system, the speed of convergence of the LMS algorithm depends on the stepsize as well as on the characteristic modes as described by the eigenvalues of the correlation matrix. For a small stepsize parameter $\beta$, we might describe the mean trajectory of the squared error $e^2(k)$ ("learning curve") by a single exponential with time constant [Haykin 1996]

$$\tau_{av} \approx \frac{1}{2\beta\frac{1}{m}\sum\limits_{i=1}^{m}\lambda_i}. \tag{14.29}$$

Since the stepsize must take the largest eigenvalue into account it follows that a large eigenvalue spread leads to a slow convergence. It is also obvious that for non-stationary signals the stepsize must be adapted to the time-varying eigenstructure of the input signal. Since $\sum\limits_{i=1}^{m}\lambda_i = \mathrm{trace}(\mathbf{R}_{xx})$, a conservative upper bound for the stepsize is given by

$$0 < \beta(k) < \frac{2}{\mathrm{trace}(\mathbf{R}_{xx})} = \frac{2}{m\sigma_x^2}. \tag{14.30}$$

If $\beta(k)$ is chosen proportional to $\mathrm{trace}(\mathbf{R}_{xx})^{-1}$,

$$\beta(k) = \alpha \cdot \mathrm{trace}(\mathbf{R}_{xx})^{-1}, \tag{14.31}$$

the mean time constant is given by

$$\tau_{av} \approx \frac{m}{2\alpha}, \tag{14.32}$$

which is a first indication that long adaptive filters converge less rapidly than short filters.

Since in a real system the residual echo signal $e(k)$ cannot be isolated, the microphone signal

$$\hat{s}(k) = s(k) + n(k) + e(k), \tag{14.33}$$

is used instead of $e(k)$, leading to the practical LMS adaptation

$$\mathbf{h}(k+1) = \mathbf{h}(k) + \beta(k)\,\hat{s}(k)\,\mathbf{x}(k)$$
$$= \mathbf{h}(k) + \beta(k)\,e(k)\,\mathbf{x}(k) + \beta(k)\,(s(k) + n(k))\,\mathbf{x}(k). \tag{14.34}$$

The near-end speaker's signal $s(k)$ and the background noise $n(k)$ must be considered as interfering signals for the adaptation process. Because of short-time correlations between $s(k) + n(k)$ and $x(k)$ the minimization of the power of $e(k)$ might be severely disturbed by the near signals. Consequently, to avoid misalignment of the canceller the adaptation must be frozen (or at least slowed down) as soon as the near-end speaker becomes active. This can be achieved by an adaptive stepsize control mechanism, which will be discussed in more detail in Section 14.6.4.

The *NLMS* algorithm may be developed as a modification of the LMS algorithm using the normalized time-varying stepsize

$$\beta(k) = \frac{\alpha(k)}{||\mathbf{x}(k)||^2} = \frac{\alpha(k)}{\mathbf{x}^T(k)\mathbf{x}(k)}. \tag{14.35}$$

It can be shown that the NLMS converges in the mean square for [Haykin 1996]

$$0 < \alpha(k) < 2. \tag{14.36}$$

Moreover, the specific properties of the normalized coefficient update

$$\mathbf{h}(k+1) - \mathbf{h}(k) = \frac{\alpha(k)}{\mathbf{x}^T(k)\mathbf{x}(k)}\,e(k)\mathbf{x}(k), \tag{14.37}$$

allow an interpretation of the NLMS algorithm and its adaptation in terms of geometric projections. In contrast to the LMS algorithm, the NLMS algorithm with (14.36) is stable not only in the mean but also deterministically in each iteration [Rupp 1993], [Slock 1993]. This will be analyzed in more detail in Section 14.7. In Section 14.6, we will first consider the performance of the LMS and NLMS algorithms in the context of the echo cancellation application.

## 14.6 Convergence Analysis and Control of the LMS Algorithm

We now return to the LMS algorithm for an in-depth analysis of its convergence behavior. To guarantee stability for stationary signals, we will use a normalized stepsize

$$\beta(k) = \frac{\alpha}{m\,\sigma_x^2}, \tag{14.38}$$

with $0 < \alpha < 2$ and

$$\sigma_x^2 = E\{x^2(k)\}. \tag{14.39}$$

Since for wide sense stationary signals and large $m$ we have

$$||\mathbf{x}(k)||^2 \approx m\,\sigma_x^2, \tag{14.40}$$

the results of this analysis will approximately hold also for the NLMS algorithm, which we will use in the following evaluation. We will also show how the convergence analysis leads to the design of optimal stepsize parameters.

### 14.6.1 Convergence in the Absence of Interference

In this section, we analyze the convergence behavior for stationary noise-like excitation signals $x(k)$ in the absence of interference, i.e., for $s(k) = 0$ and $n(k) = 0$. The impulse response of the LEM system is assumed to be time invariant, i.e., $\mathbf{g}(k) = \mathbf{g}$, and of same length as the response of the echo canceller, i.e., $m = m'$.

When the adaptive filter is excited with a zero mean, white noise signal $x(k)$ of variance $\sigma_x^2$, and when the vectors $\mathbf{d}(k)$ and $\mathbf{x}(k)$ are assumed to be statistically independent we have

$$E\left\{e^2(k)\right\} = E\left\{\left(\mathbf{d}^T(k)\mathbf{x}(k)\right)^2\right\} \tag{14.41a}$$

$$= \sigma_x^2 E\left\{||\mathbf{d}(k)||^2\right\}, \tag{14.41b}$$

where

$$\mathbf{d}(k) = \mathbf{g} - \mathbf{h}(k) \tag{14.41c}$$

denotes the distance vector.

With (14.38) and (14.40) we get

$$\mathbf{d}(k+1) = \mathbf{g} - \mathbf{h}(k+1) \tag{14.42a}$$

$$= \mathbf{g} - \left(\mathbf{h}(k) + \frac{\alpha}{m\,\sigma_x^2}e(k)\mathbf{x}(k)\right) \tag{14.42b}$$

$$= \mathbf{d}(k) - \frac{\alpha}{m\,\sigma_x^2}e(k)\mathbf{x}(k). \tag{14.42c}$$

Thus, the evolution of the squared norm of the distance vector can be formulated as

$$E\left\{||\mathbf{d}(k+1)||^2\right\} \approx E\left\{||\mathbf{d}(k)||^2\right\} - E\left\{e^2(k)\right\}\frac{\alpha}{m\,\sigma_x^2}(2-\alpha) \tag{14.43a}$$

$$= E\left\{||\mathbf{d}(k)||^2\right\}\left(1 - \frac{\alpha}{m}(2-\alpha)\right). \tag{14.43b}$$

For $0 < \alpha < 2$ the ensemble average of the system distance decreases in each iteration step. Choosing the initial vector $\mathbf{h}(k = 0) = \mathbf{0}$, the average system distance at the beginning of the recursion equals

$$E\{||\mathbf{d}(0)||^2\} = ||\mathbf{g}||^2,$$

and (14.43b) can also be expressed for $k = 0, 1, 2, \dots$ as

$$E\left\{||\mathbf{d}(k)||^2\right\} = ||\mathbf{g}||^2\left(1 - \frac{\alpha}{m}(2-\alpha)\right)^k. \tag{14.44}$$

This relation is illustrated in Figure 14.10 for different values of $\alpha$ and the measured LEM car impulse response of Figure 14.4a. With the definition of the relative system distance (14.16), we obtain

$$\frac{E\left\{||\mathbf{d}(k)||^2\right\}}{||\mathbf{g}||^2} = E\left\{D(k)\right\}.$$

As shown below, the fastest *mean* convergence is attained for $\alpha = 1$ (cf. Section 14.6.4).

Next, the system excited with a colored noise signal $x(k)$ will be examined. We determine to which extent the convergence is influenced by a correlation of adjacent samples of $x(k)$.

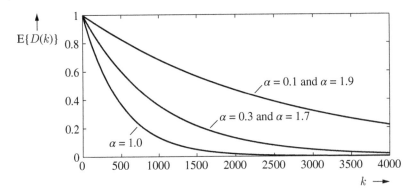

**Figure 14.10** Convergence behavior of the LMS algorithm excited with white noise, LEM impulse response measured in a car as shown in Figure 14.4a, $m = m' = 500$.

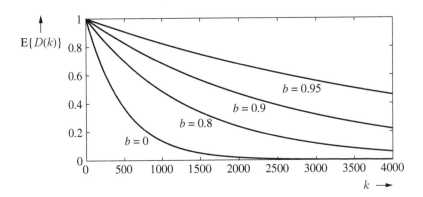

**Figure 14.11** Convergence behavior of the LMS algorithm excited with colored noise ($m = m' = 500$, $\alpha = 1$).

This investigation delivers first qualitative conclusions on the efficiency of an NLMS-driven echo cancellation filter excited with speech signals.

The excitation signal will be derived from a white noise process $u(k)$ of power $\sigma_u^2$, by filtering with a first-order, recursive filter (first-order Markov process):

$$x(k) = b \cdot x(k-1) + u(k); \quad 0 \le b < 1 \tag{14.45}$$

with

$$\sigma_x^2 = \frac{1}{1-b^2} \cdot \sigma_u^2. \tag{14.46}$$

In analogy to (14.44) we find for $\alpha = 1$

$$E\left\{||\mathbf{d}(k)||^2\right\} = ||\mathbf{g}||^2 \left(1 - \frac{1-b^2}{m}\right)^k, \quad k = 0, 1, 2, \dots. \tag{14.47}$$

Figure 14.11 shows the corresponding convergence behavior for different values of the parameter $b$. The learning curve for $b = 0$ is identical to the one in Figure 14.10 for a stepsize $\alpha = 1$ according to (14.31).

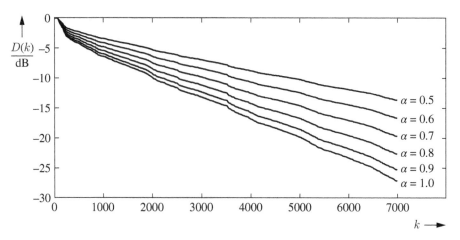

**Figure 14.12** System distance with different stepsizes $\alpha$ in absence of interference (i.e., $s(k) = 0$, $n(k) = 0$); $m = m' = 500$, $x(k)$ = colored noise ($b = 0.8$).

Interestingly, the convergence behavior is depending on $\sigma_x^2$ and the correlation. With increasing correlation, i.e., increasing $b$, the convergence speed of the algorithm decreases. This leads to the conclusion that LMS- or NLMS-based echo cancellers will show poor adaptation performance when excited with speech signals. As outlined in Section 14.5 the eigenvalue spread of the correlation matrix of the excitation vector $\mathbf{x}(k)$ has a profound effect on the speed of adaptation [Haykin 1996].

The above analysis yields the mean convergence as a function of the discrete time index $k$ in the sense of an ensemble average. The result of a simulation using the NLMS algorithm with colored noise is depicted for $b = 0.8$ in Figure 14.12. It shows the logarithmic system distance $D(k)$ vs. time for different values of the stepsize $\alpha$. In general, the time evolution is in line with (14.47). However, deviations from the analytical result (14.47) can be observed for $k < m$. This is primarily caused by the approximations (14.38) and (14.41b) which are affected by the initialization of the coefficient and the excitation vectors.

The result of simulations with speech, colored, and white noise is presented in Figure 14.13. The general conclusions from (14.47) for colored noise can be confirmed. The relatively poor convergence behavior for correlated noise and for speech signals can be clearly observed.

### 14.6.2 Convergence in the Presence of Interference

It was stated earlier that in a real system, the true instantaneous residual echo $e(k) = \tilde{x}(k) - \hat{x}(k)$ is not accessible and therefore the NLMS algorithm must use the signal

$$\hat{s}(k) = s(k) + n(k) + e(k), \tag{14.48}$$

which comprises the speech signal $s(k)$ of the near-end speaker and the background noise $n(k)$.

As a result, the steady-state echo reduction

$$\mathrm{ERLE}_\infty = \lim_{k \to \infty} \mathrm{E}\left\{\mathrm{ERLE}(k)\right\}, \tag{14.49}$$

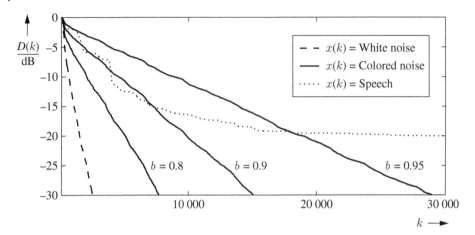

**Figure 14.13** System distance with different input signals $x(k)$ in absence of interference (i.e., $s(k) = 0$, $n(k) = 0$); $m = m' = 500$, $\alpha = 1$.

turns out to be lower, while the attainable system distance

$$D_\infty = \lim_{k \to \infty} E\{D(k)\}, \tag{14.50}$$

proves to be larger.

A closed-form analytical analysis of the interfering factors is only possible for special cases such as a system excitation with a stationary white noise signal $x(k)$ of power $\sigma_x^2$.

For the analysis below, the signal of the near-end speaker is set to zero ($s(k) = 0$, i.e., speech pause). Furthermore, an interference $n(k)$ with

$$E\{n^2(k)\} = \sigma_n^2, \tag{14.51}$$

is assumed, which is statistically independent of the signal $x(k)$, leading to

$$E\{x(k) \cdot n(\ell)\} = 0 \quad \text{and} \quad E\{e(k) \cdot n(\ell)\} = 0. \tag{14.52}$$

With these assumptions the LMS algorithm with normalized stepsize can be written as

$$\mathbf{h}(k+1) = \mathbf{h}(k) + \frac{\alpha}{m\,\sigma_x^2}\,(e(k) + n(k))\,\mathbf{x}(k). \tag{14.53}$$

In analogy to (14.43a), the mean system distance is given by

$$E\{||\mathbf{d}(k+1)||^2\} = E\{||\mathbf{d}(k)||^2\} - E\{e^2(k)\}\,\frac{\alpha}{m\,\sigma_x^2}(2 - \alpha) + \frac{\alpha^2}{m\,\sigma_x^2}E\{n^2(k)\}. \tag{14.54a}$$

Using again (14.41b) and (14.51), this equation can be modified to

$$E\{||\mathbf{d}(k+1)||^2\} = E\{||\mathbf{d}(k)||^2\}\left(1 - \frac{\alpha}{m}(2 - \alpha)\right) + \frac{\alpha^2}{m\,\sigma_x^2}\sigma_n^2. \tag{14.54b}$$

In comparison to the case without interference (14.43b), an additional constant term has been added. This implies that with increasing $k$ and for constant $\alpha$ the system distance cannot become arbitrarily small.

The steady-state solution of (14.54b) is obtained for

$$\lim_{k\to\infty} E\left\{||\mathbf{d}(k+1)||^2\right\} = \lim_{k\to\infty} E\left\{||\mathbf{d}(k)||^2\right\} = ||\mathbf{d}_\infty||^2, \tag{14.55}$$

and is given by

$$||\mathbf{d}_\infty||^2 = \frac{\alpha}{2-\alpha} \frac{\sigma_n^2}{\sigma_x^2}. \tag{14.56}$$

With the stated assumptions, the power of the echo signal $\tilde{x}(k)$ can be expressed as

$$\sigma_{\tilde{x}}^2 = ||\mathbf{g}||^2 \sigma_x^2. \tag{14.57}$$

Therefore, the mean steady-state relative system distance results in

$$\lim_{k\to\infty} E\{D(k)\} = D_\infty = \frac{||\mathbf{d}_\infty||^2}{||\mathbf{g}||^2} \tag{14.58a}$$

$$= \frac{\alpha}{2-\alpha} \frac{\sigma_n^2}{\sigma_{\tilde{x}}^2}. \tag{14.58b}$$

In the special case of $\alpha = 1$, the achievable system distance is equivalent to the power ratio between the noise signal $n(k)$ and the echo signal $\tilde{x}(k)$ at the microphone. For $\alpha < 1$, the system distance can be improved at the expense of a slower convergence. Figure 14.14 shows the result of a simulation of the NLMS algorithm for different stepsize parameters $\alpha$. The theoretical limit according to (14.58b) is indicated by dashed lines.

Note that in the above derivation of the steady-state performance for $s(k) = 0$, only the statistical independence of the echo signal $\tilde{x}(k)$ with respect to the noise $n(k)$ has been assumed.

In the special case of a white excitation signal $x(k)$, a further interesting conclusion can be drawn from (14.58b). With

$$\frac{1}{D_\infty} = \frac{||\mathbf{g}||^2}{||\mathbf{d}_\infty||^2} = \frac{||\mathbf{g}||^2}{||\mathbf{d}_\infty||^2} \cdot \frac{\sigma_x^2}{\sigma_x^2} = \frac{\sigma_{\tilde{x}}^2}{E\{(\tilde{x}-\hat{x})^2\}}, \tag{14.59}$$

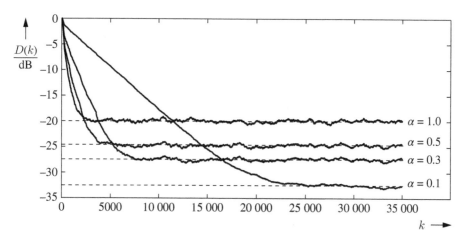

**Figure 14.14** System distance for different stepsizes $\alpha$ in the presence of background noise with $10 \lg \left(E\{n^2(k)\}/E\{\tilde{x}^2(k)\}\right) = -20$ dB; $x(k), n(k) =$ white noise, $s(k) = 0$, $m = m' = 500$ [Antweiler 1995].

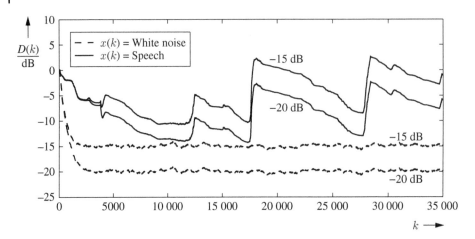

**Figure 14.15** System distance for different excitation signals $x(k)$ in the presence of background noise with 10 lg $\left(E\{n^2(k)\}/E\{\tilde{x}^2(k)\}\right) = -15$ dB or $-20$ dB, respectively; $n(k) =$ white noise, $s(k) = 0$, $m = m' = 500$, $\alpha = 1.0$.

we derive from (14.59) with (14.58b) the logarithmic echo return loss enhancement (ERLE) (14.17) for $\alpha = 1$

$$\frac{\text{ERLE}(k)}{\text{dB}} = 10 \lg\left(\frac{\sigma_{\tilde{x}}^2(k)}{E\{(\tilde{x} - \hat{x})^2\}}\right) = 10 \lg\left(\frac{\sigma_{\tilde{x}}^2}{\sigma_n^2}\right). \tag{14.60}$$

The conclusion is that both the achievable system distance (14.58b) and the logarithmic ERLE (14.60) are limited by the echo-to-noise ratio.

The results of simulation examples are depicted for $\alpha = 1$, i.e. $\frac{2-\alpha}{\alpha} = 1$, in Figure 14.15 for different excitation signal.

### 14.6.3 Filter Order of the Echo Canceller

Neglecting the complexity for computing the stepsize parameter, the LMS algorithm requires $2m$ multiply accumulate operations per sample, i.e., $m$ operations for the adaptation of the coefficients and $m$ for the filtering, respectively. The order of the cancellation filter should therefore be kept as small as possible. This requirement is also supported by the speed of convergence, which slows down with increasing filter order as seen in (14.44).

A restriction in the length of the impulse response of the cancellation filter inevitably leads to a limitation of the attainable system distance. This limitation can easily be estimated as follows. With an ideal match between the impulse response of the compensation filter $\mathbf{h}$ and the first $m$ values of the LEM impulse response $\mathbf{g}$, the best possible system distance is given by

$$D_{opt} = \frac{||\mathbf{h} - \mathbf{g}||^2}{||\mathbf{g}||^2} = \frac{\sum\limits_{i=m}^{\infty} g_i^2}{\sum\limits_{i=0}^{\infty} g_i^2}. \tag{14.61}$$

By measuring the impulse response of any LEM configuration, the required minimum order of the cancellation filter can thus be determined in advance. For the LEM impulse

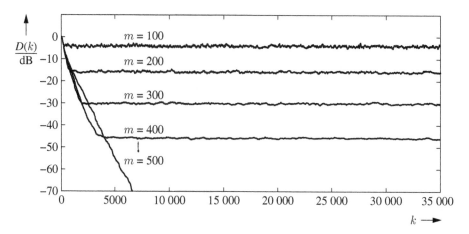

**Figure 14.16** System distance for different filter lengths $m$ in absence of interference (i.e., $s(k) = 0, n(k) = 0$); $x(k) =$ white noise, $m' = 500, \alpha = 1.0, f_s = 8\,\text{kHz}$.

response measured in a car (see Figure 14.4a) at a sampling rate of $f_s = 8\,\text{kHz}$, Figure 14.16 clearly shows the limitation of the system distance caused by a limited filter order.

### 14.6.4 Stepsize Parameter

So far, the normalized stepsize parameter $\alpha$ was assumed to be constant. Below, an optimal and possibly adaptive adjustment of this parameter will be investigated. The optimization is based on the maximum improvement of the mean system distance from one instant to the next.

First, a system excited with a white noise signal $x(k)$ is considered in the absence of interference. The insertion of a now time-variant (but deterministic) stepsize parameter $\alpha(k)$ into (14.43a) results for every arbitrary but fixed time index $k$ in

$$\text{E}\left\{||\mathbf{d}(k+1)||^2\right\} \approx \text{E}\left\{||\mathbf{d}(k)||^2\right\} - \text{E}\left\{e^2(k)\right\} \frac{\alpha(k)}{m\,\sigma_x^2} (2 - \alpha(k)). \tag{14.62}$$

Thus, the difference of the mean system distances at time instant $k$ and time instant $k+1$ is given by

$$\Delta_{\text{E}}^2(k) = \text{E}\left\{||\mathbf{d}(k)||^2\right\} - \text{E}\{||\mathbf{d}(k+1)||^2\}$$

$$\approx \text{E}\left\{e^2(k)\right\} \frac{\alpha(k)}{m\,\sigma_x^2} (2 - \alpha(k)). \tag{14.63}$$

In the admissible range of the stepsize parameter

$$0 < \alpha(k) < 2, \tag{14.64}$$

the quantity $\Delta_{\text{E}}^2(k)$ only attains positive values. It is a quadratic function of $\alpha(k)$.
From the condition

$$\frac{\partial \Delta_{\text{E}}^2(k)}{\partial \alpha(k)} \overset{!}{=} 0, \tag{14.65}$$

the optimal stepsize for each time index $k$ results in a constant stepsize parameter

$$\alpha(k) = \alpha = 1, \tag{14.66}$$

which confirms the previous results.

The above is also reflected in the temporal evolution of the average system distance in Figure 14.10. In the presence of interference (double talk, background noise), the residual echo $e(k)$ is not directly accessible.

When $e(k)$ is replaced by $\hat{s}(k) = e(k) + s(k) + n(k)$ in the adaptation rule for the LMS algorithm according to

$$\mathbf{h}(k+1) = \mathbf{h}(k) + \frac{\alpha(k)}{||\mathbf{x}(k)||^2}\,\hat{s}(k)\,\mathbf{x}(k), \tag{14.67}$$

the difference between the mean system distances at time instants $k$ and $k+1$ follows in analogy to (14.54a)

$$\Delta_E^2(k) = \mathrm{E}\left\{e^2(k)\right\}\frac{\alpha(k)}{m\,\sigma_x^2}\,(2 - \alpha(k)) - \frac{\alpha^2(k)}{m\,\sigma_x^2}\,\left(\mathrm{E}\{s^2(k)\} + \mathrm{E}\left\{n^2(k)\right\}\right). \tag{14.68}$$

In order to obtain the best possible stepsize, the condition

$$\frac{\partial \Delta_E^2(k)}{\partial \alpha(k)} = 2\,\frac{\mathrm{E}\{e^2(k)\}}{m\,\sigma_x^2} - 2\,\alpha(k)\,\frac{\mathrm{E}\{\hat{s}^2(k)\}}{m\,\sigma_x^2} \stackrel{!}{=} 0, \tag{14.69}$$

with

$$\mathrm{E}\{\hat{s}^2(k)\} = \mathrm{E}\{e^2(k)\} + \mathrm{E}\{s^2(k)\} + \mathrm{E}\{n^2(k)\}$$

must be fulfilled. The general solution is [Schultheiß 1988] and [Mader et al. 2000]

$$\alpha_{opt}(k) = \frac{\mathrm{E}\left\{e^2(k)\right\}}{\mathrm{E}\left\{\hat{s}^2(k)\right\}}. \tag{14.70}$$

In this, the interference-free case with $e(k) = \hat{s}(k)$ is included as the special case $\alpha(k) = 1$.

The optimal stepsize can also be written as

$$\alpha_{opt}(k) = \frac{\mathrm{E}\{e^2(k)\}}{\mathrm{E}\{\hat{s}^2(k)\}} = \frac{\mathrm{E}\{e^2(k)\}}{\mathrm{E}\{s^2(k)\} + \mathrm{E}\{n^2(k)\} + \mathrm{E}\{e^2(k)\}} \tag{14.71a}$$

$$= \frac{1}{1 + \dfrac{\mathrm{E}\{s^2(k)\} + \mathrm{E}\{n^2(k)\}}{\mathrm{E}\{e^2(k)\}}} \leq 1. \tag{14.71b}$$

We see from (14.71b) that during double talk phases the stepsize is reduced. By reducing the stepsize, the signal-dependent bias of the echo canceller is minimized and thus distortions of the transmitted signal are avoided. During double talk, the adaptation need not be explicitly stopped so that slow improvements of the state of convergence are still possible. When the echo canceller has converged, the power of the residual echo $e(k)$ is in general significantly smaller than the power of the near-end speech signal $s(k)$ or the near-end noise signal $n(k)$. In this case, the stepsize parameter is much smaller than unity, which in turn is a prerequisite to maintain a small system distance (see (14.58b)).

Since the residual echo $e(k)$ does not exist as an isolated signal, the stepsize parameter $\alpha_{opt}(k)$ can only be approximated. In [Yamamoto, Kitayama 1982] and [Schultheiß 1988] a solution, which is based on an estimation of the instantaneous system distance is proposed.

A delay of $m_0$ samples is applied to the microphone signal. This delays the effective LEM impulse response by $m_0$ samples as well, i.e., it introduces $m_0$ leading zero values. The $m_0$ leading taps of the echo canceller can now be used to estimate the system distance and hence the residual echo power.

This procedure is outlined in [Schultheiß 1988] where $m_0 = 20$ delay coefficients are used. Additional measures, however, are necessary to detect a change in the LEM response, otherwise the adaptation will freeze [Mader et al. 2000], [Hänsler, Schmidt 2004].

## 14.7 Geometric Projection Interpretation of the NLMS Algorithm

The operation of the NLMS algorithm can be interpreted in terms of a vector space representation of the individual adaptation step [Claasen, Mecklenbräuker 1981], [Sommen, van Valburg 1989]. However, before we develop this "geometric" approach to the echo cancellation problem we introduce the concept of orthogonal vectors and spaces in $\mathbb{R}^m$ (see, e.g., [Debnath, Mikusinski 1999]).

Two column vectors $\mathbf{d}_1(k)$ and $\mathbf{d}_2(k)$ are said to be orthogonal if their inner product equals zero, i.e., $\mathbf{d}_1^T(k)\mathbf{d}_2(k) = 0$. By the same token, a vector $\mathbf{d}(k)$ is said to be orthogonal to a subspace $\mathbf{S}_x$ if $\mathbf{d}^T(k)\mathbf{x}(k) = 0$ for every $\mathbf{x}(k) \in \mathbf{S}_x$. The set of all vectors orthogonal to $\mathbf{S}_x$ are called the orthogonal complement of $\mathbf{S}_x$ and is denoted by $\mathbf{S}^\perp$.

Every element $\mathbf{d}(k) \in \mathbb{R}^m$ thus has a unique decomposition in the form

$$\mathbf{d}(k) = \mathbf{d}^\|(k) + \mathbf{d}^\perp(k), \tag{14.72}$$

where $\mathbf{d}^\|(k)$ is an element of a subspace $\mathbf{S}_x$ of $\mathbb{R}^m$ and $\mathbf{d}^\perp(k)$ an element of the orthogonal complement $\mathbf{S}^\perp$ of $\mathbf{S}_x$. $\mathbf{d}^\|(k)$ is then called the projection of $\mathbf{d}(k)$ onto $\mathbf{S}_x$ and $\mathbf{P}_x$, with $\mathbf{P}_x\mathbf{d}(k) = \mathbf{d}^\|(k)$ being the associated projection operator.

If $\mathbf{S}_x$ is a subspace and $\mathbf{x}(k) \in \mathbf{S}_x$, then the projection of $\mathbf{d}(k)$ onto $\mathbf{S}_x$ is given by

$$\mathbf{d}^\|(k) = \mathbf{P}_x(k)\mathbf{d}(k) = \frac{\mathbf{x}(k)}{||\mathbf{x}(k)||} \cdot \frac{\mathbf{x}^T(k)}{||\mathbf{x}(k)||} \mathbf{d}(k), \tag{14.73}$$

where the scalar term $\mathbf{x}^T(k)\mathbf{d}(k)/||\mathbf{x}(k)||$ quantifies the length of the projection of $\mathbf{d}(k)$ onto $\mathbf{x}(k)$ and $\mathbf{x}(k)/||\mathbf{x}(k)||$ is the unit vector in the direction of $\mathbf{x}(k)$.

From (14.73) it can be concluded that the projection operator becomes an $m \times m$ time-variable matrix because of its dependency on $\mathbf{x}(k)$

$$\mathbf{P}_x(k) = \frac{\mathbf{x}(k)\mathbf{x}^T(k)}{||\mathbf{x}(k)|| \, ||\mathbf{x}(k)||}. \tag{14.74}$$

It can easily be shown that the projection operator (14.74) satisfies the following necessary conditions (for ease of notation, dependency on time index $k$ is omitted in the following Eqs. (14.75) ... (14.77)),

1. Self mapping:

$$\mathbf{P}_x\mathbf{x} = \mathbf{x}. \tag{14.75}$$

2. Idempotence

$$\mathbf{P}_x \mathbf{P}_x = \mathbf{P}_x, \tag{14.76}$$

which leads to

$$\mathbf{d} = \mathbf{d}^{\parallel} + \mathbf{d}^{\perp} = \mathbf{P}_x \mathbf{d} + \mathbf{d}^{\perp}$$
$$\mathbf{P}_x \mathbf{d} = \mathbf{P}_x [\mathbf{P}_x \mathbf{d} + \mathbf{d}^{\perp}] = \mathbf{P}_x \mathbf{P}_x \mathbf{d} + 0$$

3. Self-adjointness

$$\mathbf{P}_x^T = \mathbf{P}_x \tag{14.77}$$

or, equivalently,

$$\left(\mathbf{P}_x \mathbf{d}_1\right)^T \mathbf{d}_2 = \mathbf{d}_1^{\parallel T} \mathbf{d}_2 = \mathbf{d}_1^T \mathbf{P}_x^T \mathbf{d}_2$$
$$= \mathbf{d}_1^T \mathbf{P}_x \mathbf{d}_2 = \mathbf{d}_1^T \mathbf{d}_2^{\parallel}$$

for any $\mathbf{d}_1, \mathbf{d}_2 \in \mathbb{R}^m$.

If we return to the NLMS-adapted echo canceller and consider the distance vector

$$\mathbf{d}(k) = \mathbf{g} - \mathbf{h}(k), \tag{14.78}$$

the following equation holds for the NLMS update (14.37) in an environment with no inter-ference, i.e., $s(k) + n(k) = 0$, and $e(k) = \mathbf{d}^T(k)\mathbf{x}(k)$,

$$\mathbf{d}(k+1) = \mathbf{g} - \mathbf{h}(k+1) \tag{14.79a}$$

$$= \mathbf{g} - \left(\mathbf{h}(k) + \alpha \frac{e(k)}{||\mathbf{x}(k)||^2} \mathbf{x}(k)\right) \tag{14.79b}$$

$$= \mathbf{d}(k) - \alpha \frac{\mathbf{d}^T(k)\mathbf{x}(k)}{||\mathbf{x}(k)||^2} \mathbf{x}(k) \tag{14.79c}$$

$$= \mathbf{d}(k) - \alpha \frac{\mathbf{x}(k)\mathbf{x}^T(k)}{||\mathbf{x}(k)|| \, ||\mathbf{x}(k)||} \mathbf{d}(k) \tag{14.79d}$$

$$= \mathbf{d}(k) - \alpha \, \mathbf{d}^{\parallel}(k) \tag{14.79e}$$

$$= (\mathbf{I} - \alpha \mathbf{P}_x(k)) \, \mathbf{d}(k) \tag{14.79f}$$

with $\mathbf{I}$ denoting the $m \times m$ unity matrix.

Note that the term $\mathbf{d}^T(k)\mathbf{x}(k) = \mathbf{x}^T(k)\mathbf{d}(k)$ in (14.79c) and (14.79d) quantifies a *scalar* sample of the residual echo $e(k) = \tilde{x}(k) - \hat{x}(k)$ according to (14.18).

The correction vector $\mathbf{d}^{\parallel}(k)$ can therefore be interpreted as an orthogonal projection of the distance vector $\mathbf{d}(k)$ onto the signal vector $\mathbf{x}(k)$ and

$$\mathbf{P}_x(k) = \frac{\mathbf{x}(k)\mathbf{x}^T(k)}{||\mathbf{x}(k)||^2}, \tag{14.80}$$

is the corresponding projection operator. It is easily verified that (14.80) fulfills both necessary conditions (14.76) and (14.77). This concept is illustrated in Figure 14.17 for $m = 2$.

The length of the distance vector $\mathbf{d}(k)$ is reduced by subtracting the component $\alpha \, \mathbf{d}^{\parallel}(k)$, which is parallel to the signal vector $\mathbf{x}(k)$. Provided that the stepsize parameter is within the range

$$0 < \alpha < 2, \tag{14.81}$$

**Figure 14.17** Geometric interpretation of the NLMS algorithm, $m = 2$ [Antweiler 1995].

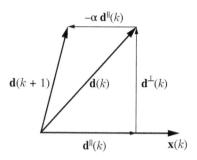

the length of the distance vector $\mathbf{d}(k)$ always decreases. The convergence condition (14.36) is thus confirmed. Obviously, the update of $\mathbf{d}(k)$ is restricted to the direction of $\mathbf{x}(k)$. For $\alpha < 1$ the projection does not fully eliminate the components in the space spanned by $\mathbf{x}(k)$ but amounts to a *relaxed* projection.

For $\alpha = 1$, we find $\mathbf{d}(k+1) = (\mathbf{I} - \mathbf{P}_x(k))\,\mathbf{d}(k)$ where $\mathbf{I} - \mathbf{P}_x(k) = \mathbf{P}_x^{\perp}(k)$ may be interpreted as an operator for projecting $\mathbf{d}(k)$ onto the orthogonal complement of $\mathbf{x}(k)$, since

$$\mathbf{P}_x(k)(\mathbf{I} - \mathbf{P}_x(k)) = \mathbf{P}_x(k) - \mathbf{P}_x(k)\mathbf{P}_x(k) = \mathbf{P}_x(k) - \mathbf{P}_x(k) = 0. \tag{14.82}$$

A relative system distance $D(k) = 0$ according (14.16) can only be obtained if, during the adaptation phase, the excitation signal vector points toward all directions in the $m$-dimensional vector space. This interpretation explains the fact that the most rapid convergence is achieved for a *perfect sequence* or a *perfect sweep* excitation signal [Antweiler, Dörbecker 1994], [Antweiler, Antweiler 1995], [Lüke, Schotten 1995] (also see Section 14.10.5).

## 14.8 The Affine Projection Algorithm

The geometric interpretation leads to an interesting generalization of the NLMS algorithm, namely the *affine projection* (AP) algorithm [Ozeki, Umeda 1984].

For $s(k) = n(k) = 0$ and $m = m'$ the NLMS filter coefficient update can be written in terms of the projection operator $\mathbf{P}_x(k)$ as

$$\mathbf{h}(k+1) = \mathbf{h}(k) + \alpha\,\mathbf{P}_x(k)\,\mathbf{d}(k) \tag{14.83a}$$

$$= \mathbf{h}(k) + \alpha\,\mathbf{P}_x(k)\,[\mathbf{g}(k) - \mathbf{h}(k)] \tag{14.83b}$$

$$= [\mathbf{I} - \alpha\,\mathbf{P}_x(k)]\,\mathbf{h}(k) + \alpha\,\mathbf{P}_x(k)\,\mathbf{g}(k). \tag{14.83c}$$

For $\alpha = 1$ we now have an AP

$$\mathbf{h}(k+1) = \mathbf{P}_x^{\perp}(k)\,\mathbf{h}(k) + \mathbf{g}^{\parallel}(k). \tag{14.84}$$

The updated filter coefficient vector $\mathbf{h}(k+1)$ consists of two components. The term $\mathbf{P}_x^{\perp}(k)\,\mathbf{h}(k)$ eliminates the portion of $\mathbf{h}(k)$ in the direction of $\mathbf{x}(k)$, while $\mathbf{g}^{\parallel}(k)$ quantifies the portion of the true impulse response $\mathbf{g}(k)$ in the direction of $\mathbf{x}(k)$.

The AP algorithm generalizes this idea. In each iteration, the AP algorithm reduces the system distance not just in one but in several directions. Hence, we now use the projection onto the space spanned by a sequence of $p < m$ successive vectors of the input signal

$$\mathbf{X}_p(k) = (\mathbf{x}(k), \mathbf{x}(k-1), \ldots, \mathbf{x}(k-p+1)). \tag{14.85}$$

If we consider $\mathbf{X}_p(k)$ to be a set of $p$ non-orthogonal basis vectors for a subspace of $\mathbb{R}^m$, the inner product vector $\mathbf{X}_p^T(k)\,\mathbf{d}(k)$, in conjunction with an appropriate normalization, yields the components of $\mathbf{d}(k)$ with respect to this subspace basis. Therefore, the projection operator onto this subspace, $\mathbf{P}_{X_p}(k)$, may be written in analogy to (14.74) as $\mathbf{P}_{X_p}(k) = \mathbf{X}_p(k)\,\mathbf{A}\,\mathbf{X}_p^T(k)$ where the normalization matrix $\mathbf{A}$ is to be determined. Also, we note that the projection operator has to satisfy

$$\mathbf{P}_{X_p}(k)\,\mathbf{X}_p(k) = \mathbf{X}_p(k). \tag{14.86}$$

With the above observations the operator can be constructed such that

$$\mathbf{X}_p(k)\,\mathbf{A}\,\mathbf{X}_p^T(k)\,\mathbf{X}_p(k) = \mathbf{X}_p(k), \tag{14.87}$$

where $\mathbf{A}$ is now identified as

$$\mathbf{A} = [\mathbf{X}_p^T(k)\,\mathbf{X}_p(k)]^{-1}. \tag{14.88}$$

We obtain

$$\mathbf{P}_{X_p}(k) = \mathbf{X}_p(k)\,[\mathbf{X}_p^T(k)\,\mathbf{X}_p(k)]^{-1}\,\mathbf{X}_p^T(k). \tag{14.89}$$

It is easily verified that $\mathbf{P}_{X_p}(k)$ is indeed a projection operator, i.e.,

$$\mathbf{P}_{X_p}(k)\,\mathbf{P}_{X_p}(k) = \mathbf{P}_{X_p}(k) \text{ and} \tag{14.90}$$

$$(\mathbf{P}_{X_p}(k)\,\mathbf{d}_1(k))^T\,\mathbf{d}_2(k) = \mathbf{d}_1^T(k)\,\mathbf{P}_{X_p}(k)\,\mathbf{d}_2(k), \tag{14.91}$$

with

$$\mathbf{P}_{X_p}^T(k) = \mathbf{P}_{X_p}(k) \tag{14.92}$$

since $\mathbf{X}_p^T(k)\,\mathbf{X}_p(k)$ and its inverse are symmetric matrices.

In the noise-free case the AP algorithm is given by

$$\mathbf{h}(k+1) = \mathbf{h}(k) + \alpha\,\mathbf{X}_p(k)\,[\mathbf{X}_p^T(k)\mathbf{X}_p(k)]^{-1}\,\mathbf{X}_p^T(k)\,\mathbf{d}(k). \tag{14.93}$$

In the general case with a near-end speech signal $s(k)$ and a background noise $n(k)$ according to $\hat{s}(k) = \tilde{x}(k) - \hat{x}(k) + s(k) + n(k) = e(k) + s(k) + n(k)$ we define

$$\hat{\mathbf{s}}_p(k) = (\hat{s}(k), \hat{s}(k-1), \ldots, \hat{s}(k-p+1))^T, \tag{14.94}$$

with (see also notation in Figure 14.3)

$$\hat{\mathbf{s}}_p(k) = \mathbf{y}_p(k) - \hat{\mathbf{x}}_p(k) \tag{14.95a}$$
$$= \mathbf{y}_p(k) - \mathbf{X}_p^T(k)\,\mathbf{h}(k)$$
$$= \mathbf{s}_p(k) + \mathbf{n}_p(k) + \tilde{\mathbf{x}}_p(k) - \mathbf{X}_p^T(k)\,\mathbf{h}(k). \tag{14.95b}$$

With the introduction of a regularization parameter $\delta$ for decreasing the condition number of the matrix before inversion, we obtain as a generalization of (14.83a)

$$\mathbf{h}(k+1) = \mathbf{h}(k) + \alpha\,\mathbf{X}_p(k)\,[\mathbf{X}_p^T(k)\mathbf{X}_p(k) + \delta\,\mathbf{I}]^{-1}\,\hat{\mathbf{s}}_p(k). \tag{14.96}$$

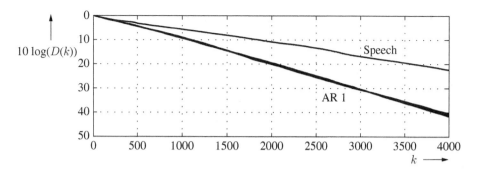

**Figure 14.18** Convergence behavior of the AP algorithm with $p = 10$ and $\alpha = 1$ excited with colored noise (AR 1; $b = 0, 0.8, 0.9, 0.95$) and speech. Car LEM impulse response, $m = m' = 500$.

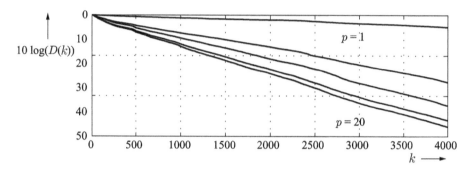

**Figure 14.19** Convergence behavior of the AP algorithm with $\alpha = 1$ and $p = 1, 5, 10, 15, 20$ excited with speech ($m = m' = 500$).

The regularization parameter $\delta$ is necessary since $\mathbf{X}_p^T(k)\mathbf{X}_p(k)$ is a rank deficient matrix for $p \to m$. Optimal regularization parameters are considered in [Myllylä, Schmidt 2002]. In the noise-free case, fastest convergence is achieved for $\alpha = 1$. However, in the presence of near-end speech and noise $\alpha \ll 1$ results in better convergence. Figure 14.18 depicts the system distance for various far-end signals and $p = 10$. Also, in the case of a first-order Markov process (AR 1, with AR coefficient $b$ in Figure 14.18) convergence does not depend on the signal correlation. In the case of speech signals the AP algorithm leads to much faster convergence than the NLMS algorithm.

Figure 14.19 shows the impact of the projection order $p$ on the convergence for speech signals and $\alpha = 1$. It is seen that a larger projection order is beneficial. Most of the gains are achieved, however, for a projection order of $p = 10$–$20$.

Furthermore, fast implementations of the AP algorithm exist [Gay, Tavathia 1995], [Tanaka et al. 1995], which make the AP algorithm attractive for real-time applications. For a robust implementation of these algorithms see, for example, [Myllylä 2001]. A variation of the AP algorithm with exponentially weighted stepsize parameters is outlined in [Makino, Kaneda 1992]. In this algorithm an individual stepsize is assigned to each canceller coefficient.

## 14.9   Least-Squares and Recursive Least-Squares Algorithms

The Wiener solution requires knowledge about first- and second-order statistics such as auto- and cross-correlation functions. These quantities must be estimated from the available data. In this section, we will discuss a family of algorithms, which relies directly on the minimization of the observed error of a block of data. We begin our presentation with the *weighted least-squares algorithm*, which is, in the context of acoustic echo cancellation, more of theoretical than of practical interest since it requires the inversion of a possibly large matrix.

However, a recursive approximation, the *recursive least-squares* (RLS) algorithm, avoids this inversion and is better suited for practical implementations.

### 14.9.1   The Weighted Least-Squares Algorithm

The *weighted least-squares approach* combines the actually measured adaptation error (including errors due to near-end speech and noise) of a block of data into a vector of dimensions $M \geq m$ and minimizes the weighted norm of this vector. An $m \times M$ data matrix is constructed from the $M$ column vectors $\mathbf{x}(k - i)$ each of dimension $m$

$$\mathbf{X}_M(k) = (\mathbf{x}(k), \mathbf{x}(k - 1), \dots, \mathbf{x}(k - M + 1)), \tag{14.97}$$

and a segment of the output signal

$$\hat{\mathbf{s}}_M(k) = (\hat{s}(k), \hat{s}(k - 1), \dots, \hat{s}(k - M + 1))^T, \tag{14.98}$$

of length $M$ is considered, which is given by

$$\hat{\mathbf{s}}_M(k) = \mathbf{y}_M(k) - \mathbf{X}_M^T(k)\,\mathbf{h}(k), \tag{14.99}$$

where $\mathbf{y}_M(k)$ denotes the vector of the last $M$ samples of the microphone signal

$$\mathbf{y}_M(k) = (y(k), y(k - 1), \dots, y(k - M + 1))^T. \tag{14.100}$$

The output vector $\hat{\mathbf{s}}_M(k)$, including the near end speech signal $s(k)$ and the near end noise signal $n(k)$ (see Figure 14.3) assumes the role of the error signal.

The weighted least-squares algorithm minimizes the weighted squared error norm

$$J_{LS}(k) = \hat{\mathbf{s}}_M^T(k)\,\mathbf{W}\,\hat{\mathbf{s}}_M(k), \tag{14.101}$$

where $\mathbf{W} = diag\{(w_{11}, w_{22}, \dots, w_{MM})^T\}$ is a positive-definite diagonal weighting matrix. The minimization of the error norm $J_{LS}$ with respect to the unknown coefficient vector $\mathbf{h}(k)$ requires the solution of

$$\frac{\partial J_{LS}(k)}{\partial \mathbf{h}(k)} = -2\,\mathbf{X}_M(k)\mathbf{W}\,\mathbf{y}_M(k) + 2\,[\mathbf{X}_M(k)\,\mathbf{W}\,\mathbf{X}_M^T(k)]\,\mathbf{h}(k) \overset{!}{=} \mathbf{0}. \tag{14.102}$$

If $\mathbf{X}_M(k)\,\mathbf{W}\,\mathbf{X}_M^T(k)$ is invertible (for which $M \geq m$ is a necessary and $rank(\mathbf{X}_M(k)) = m$ a sufficient condition), the solution is given by

$$\mathbf{h}_{LS}(k) = \left(\mathbf{X}_M(k)\,\mathbf{W}\,\mathbf{X}_M^T(k)\right)^{-1}\mathbf{X}_M(k)\,\mathbf{W}\,\mathbf{y}_M(k). \tag{14.103}$$

$\mathbf{X}_M(k)\mathbf{W}\mathbf{X}_M^T(k)$ and $\mathbf{X}_M(k)\mathbf{W}\mathbf{y}_M(k)$ are recognized as estimates of a weighted data auto-correlation matrix and weighted cross-correlation vector, respectively. For stationary signals and $M \to \infty$ the least-squares solution $\mathbf{h}_{LS}$ approaches the FIR Wiener solution.

It can be shown that if the near-end signal $s(k) + n(k)$ is zero mean and if $s(k) + n(k)$ and $x(k)$ are statistically independent $\mathbf{h}_{LS}$ is an unbiased estimate of $\mathbf{g}$, provided $m = m'$ holds.

With $\mathbf{W} = \mathbf{I}$ the least-squares solution is suited for stationary signals. However, in the context of acoustic echo cancellation where the excitation is speech an exponential weighting $w_{ii} = \lambda^{i-1}$ with $0 < \lambda < 1$ is much better suited. For large $m$ or $M$ the computational complexity of $O(M^3)$ associated with the matrix inversion prohibits the direct implementation of the weighted least-squares algorithm.

### 14.9.2 The RLS Algorithm

The computational effort for the least-squares algorithm can be significantly reduced when the inverse auto-correlation matrix is estimated in a recursive way. This leads to the *recursive least-squares* algorithm [Haykin 1996]. We note that

$$\mathbf{R}_{xx}(k) = \mathbf{X}_M(k)\,\mathbf{W}\,\mathbf{X}_M^T(k), \tag{14.104}$$

can also be written as

$$\mathbf{R}_{xx}(k) = \left(w_{11}\mathbf{x}(k), \ldots, w_{MM}\mathbf{x}(k - M + 1)\right) \cdot \begin{pmatrix} \mathbf{x}^T(k) \\ \mathbf{x}^T(k-1) \\ \vdots \\ \mathbf{x}^T(k-M+1) \end{pmatrix}$$

$$= \sum_{i=0}^{M-1} w_{i+1,i+1}\,\mathbf{x}(k-i)\mathbf{x}^T(k-i). \tag{14.105}$$

The estimated auto-correlation matrix is a sum over $M$ rank 1 matrices. With $w_{i+1,i+1} = \lambda^i$, $0 < \lambda < 1$ we obtain

$$\mathbf{R}_{xx}(k) = \sum_{i=0}^{M-1} \lambda^i \mathbf{x}(k-i)\mathbf{x}^T(k-i)$$

$$= \mathbf{x}(k)\mathbf{x}^T(k) + \sum_{i=1}^{M-1} \lambda^i \mathbf{x}(k-i)\mathbf{x}^T(k-i) \tag{14.106}$$

$$= \mathbf{x}(k)\mathbf{x}^T(k) + \lambda \sum_{i=0}^{M-2} \lambda^i \mathbf{x}(k-i-1)\mathbf{x}^T(k-i-1).$$

If $0 < \lambda < 1$ and $M$ is sufficiently large we may write

$$\mathbf{R}_{xx}(k) \approx \lambda\,\mathbf{R}_{xx}(k-1) + \mathbf{x}(k)\mathbf{x}^T(k), \tag{14.107}$$

because of

$$\mathbf{R}_{xx}(k-1) = \sum_{i=0}^{M-1} \lambda^i \mathbf{x}(k-i-1)\mathbf{x}^T(k-i-1), \tag{14.108}$$

and $\lambda^{M-1} \approx 0$.

In the same way, we obtain for the weighted cross-correlation vector

$$\boldsymbol{\varphi}_{xy}(k) = \mathbf{X}_M(k)\,\mathbf{W}\,\mathbf{y}_M(k)$$

$$= \sum_{i=0}^{M-1} \lambda^i \mathbf{x}(k-i)\,y(k-i) \tag{14.109}$$

$$\approx \lambda\,\boldsymbol{\varphi}_{xy}(k-1) + \mathbf{x}(k)\,y(k).$$

The inversion of $\mathbf{R}_{xx}(k)$ can be avoided if we make use of the matrix inversion lemma [Haykin 1996]

$$\mathbf{A} = \mathbf{B}^{-1} + \mathbf{C}\mathbf{D}^{-1}\mathbf{C}^T \Leftrightarrow \mathbf{A}^{-1} = \mathbf{B} - \mathbf{B}\mathbf{C}(\mathbf{D} + \mathbf{C}^T\mathbf{B}\mathbf{C})^{-1}\mathbf{C}^T\mathbf{B}. \tag{14.110}$$

We relate the quantities $\mathbf{A}$, $\mathbf{B}^{-1}$, $\mathbf{C}$, and $\mathbf{D}$ to the matrices in (14.107) as follows:

$$\mathbf{A} = \mathbf{R}_{xx}(k),\ \mathbf{B}^{-1} = \lambda\,\mathbf{R}_{xx}(k-1),\ \mathbf{C} = \mathbf{x}(k),\ \mathbf{D} = 1, \tag{14.111}$$

and obtain finally with (14.110) by using an equal sign in (14.107)

$$\mathbf{R}_{xx}^{-1}(k) = \lambda^{-1}\mathbf{R}_{xx}^{-1}(k-1) - \lambda^{-1}\mathbf{R}_{xx}^{-1}(k-1)\mathbf{x}(k)$$

$$\cdot (1 + \mathbf{x}^T(k)\,\lambda^{-1}\mathbf{R}_{xx}^{-1}(k-1)\mathbf{x}(k))^{-1} \cdot \mathbf{x}^T(k)\,\lambda^{-1}\mathbf{R}_{xx}^{-1}(k-1)$$

$$= \lambda^{-1}\mathbf{R}_{xx}^{-1}(k-1) - \frac{\lambda^{-2}\mathbf{R}_{xx}^{-1}(k-1)\mathbf{x}(k)\mathbf{x}^T(k)\,\mathbf{R}_{xx}^{-1}(k-1)}{1 + \lambda^{-1}\mathbf{x}^T(k)\,\mathbf{R}_{xx}^{-1}(k-1)\mathbf{x}(k)}$$

or

$$\mathbf{R}_{xx}^{-1}(k) = \lambda^{-1}\,\mathbf{R}_{xx}^{-1}(k-1) - \lambda^{-1}\,\mathbf{k}(k)\mathbf{x}^T(k)\,\mathbf{R}_{xx}^{-1}(k-1) \tag{14.112a}$$

$$= \lambda^{-1}\,\mathbf{R}_{xx}^{-1}(k-1) - \mathbf{Q}(k) \tag{14.112b}$$

where $\mathbf{Q}(k)$ is just an abbreviation for the corresponding term in (14.112a) and

$$\mathbf{k}(k) = \frac{\lambda^{-1}\,\mathbf{R}_{xx}^{-1}(k-1)\mathbf{x}(k)}{1 + \lambda^{-1}\,\mathbf{x}^T(k)\,\mathbf{R}_{xx}^{-1}(k-1)\mathbf{x}(k)} \tag{14.113a}$$

$$= \mathbf{R}_{xx}^{-1}(k)\mathbf{x}(k) \tag{14.113b}$$

is also known as the *Kalman gain* vector. The last equality can be verified by rearranging (14.113a)

$$\mathbf{k}(k) = \lambda^{-1}\,\mathbf{R}_{xx}^{-1}(k-1)\mathbf{x}(k) - \lambda^{-1}\,\mathbf{k}(k)\mathbf{x}^T(k)\,\mathbf{R}_{xx}^{-1}(k-1)\mathbf{x}(k) \tag{14.114a}$$

$$= \lambda^{-1}\,\mathbf{R}_{xx}^{-1}(k-1)\mathbf{x}(k) - \mathbf{Q}(k)\mathbf{x}(k) \tag{14.114b}$$

and by substituting $\mathbf{Q}(k)\mathbf{x}(k)$ at the right hand side of (14.112b). These equations can now be used to develop a recursive update equation for the filter coefficients. Using the least-squares solution (14.103), (14.104), and (14.109) we find

$$\mathbf{h}_{\mathrm{RLS}}(k) = \lambda\,\mathbf{R}_{xx}^{-1}(k)\,\boldsymbol{\varphi}_{xy}(k-1) + \mathbf{R}_{xx}^{-1}(k)\,\mathbf{x}(k)\,y(k), \tag{14.115}$$

and by replacing the correlation term $\mathbf{R}_{xx}^{-1}(k)$ in (14.115) using (14.112a), we finally find with (14.113b) the RLS recursion

$$\mathbf{h}_{\mathrm{RLS}}(k) = \mathbf{R}_{xx}^{-1}(k-1)\,\boldsymbol{\varphi}_{xy}(k-1) - \mathbf{k}(k)\mathbf{x}^T(k)\,\mathbf{R}_{xx}^{-1}(k-1)\,\boldsymbol{\varphi}_{xy}(k-1)$$
$$+ \mathbf{R}_{xx}^{-1}(k)\mathbf{x}(k)\,y(k) \tag{14.116a}$$

$$= \mathbf{h}_{\mathrm{RLS}}(k-1) - \mathbf{k}(k)\mathbf{x}^T(k)\,\mathbf{h}_{\mathrm{RLS}}(k-1) + \mathbf{k}(k)\,y(k) \tag{14.116b}$$

$$= \mathbf{h}_{\mathrm{RLS}}(k-1) + \mathbf{k}(k)\,[y(k) - \mathbf{x}^T(k)\,\mathbf{h}_{\mathrm{RLS}}(k-1)] \tag{14.116c}$$

$$= \mathbf{h}_{\mathrm{RLS}}(k-1) + \mathbf{k}(k)\,\hat{s}(k|k-1). \tag{14.116d}$$

Note that a new *conditional* notation $\hat{s}(k|k-1)$ is used in (14.116d) to indicate, that the far-end signal $x(k)$ (see Figure 14.3) is filtered at the time instant $k$ with the previous impulse response $\mathbf{h}_{\mathrm{RLS}}(k-1)$ and that the Kalman gain vector $\mathbf{k}(k)$ (14.113a) is updated using the previous estimate $\mathbf{R}_{xx}^{-1}(k-1)$ of the inverse of the correlation matrix and the actual state vector $\mathbf{x}(k)$.

As an example, Figure 14.20 depicts the system distance for several values of the forgetting parameter $\lambda$, for a speech excitation, and no noise. For $\lambda \approx 1$ the RLS converges much faster than the NLMS algorithm.

Numerical stability is a critical issue for the RLS, especially for the fast $O(M)$ versions of the algorithm. A numerically stable version of the RLS is proposed in e.g., [Slock, Kailath 1991]. Some guidelines and initialization procedures are outlined in [Breining et al. 1999].

To conclude, we note that the convergence of the RLS algorithm is independent of the auto-correlation properties of the input signal. Finally, it should be mentioned that related algorithms with convergence properties similar to the RLS algorithm have been proposed in the literature (see, e.g., [Petillon et al. 1994]).

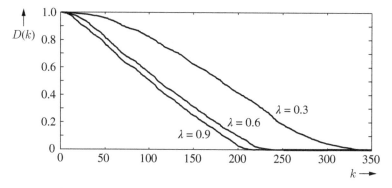

**Figure 14.20** Convergence behavior of the RLS algorithm excited with speech. Car LEM impulse response, $m = m' = 200$, $\alpha = 1$).

### 14.9.3 NLMS- and Kalman-Algorithm

In this section, the iterative adaptation of the echo canceller using the *LMS* and the *NLMS* algorithm (NLMS) of Sec. 14.5 will be revisited. Their relation with the Kalman algorithm [Kalman 1960] is investigated. It will be shown that the NLMS algorithm may be interpreted as a special case of the Kalman approach. The two algorithms differ w.r.t. optimization criteria and the underlying model assumptions, as follows:

- NLMS: The optimization criterion is the minimization of the *MSE* of the *residual echo signal e(k)*, which is related to the *ERLE* as defined in Section 14.3. No knowledge about the LEM model is included. A *time-invariant* finite LEM impulse response $\mathbf{g}(k)$ is assumed.
- Kalman: The optimization criterion is the minimization of the *mean square deviation* (MSD) between the LEM impulse response $\mathbf{g}(k)$ and the impulse response $\mathbf{h}(k)$ of the echo canceller, i.e., $E\{\mathbf{d}^T(k)\mathbf{d}(k)\} \overset{!}{=} min$. This criterion is related to the *system distance* as defined in Section 14.3. For the LEM system, a *time-variant* finite impulse response $\mathbf{g}(k)$ is assumed which follows a state-space model.

In the literature, the LEM impulse response is mostly considered to be constant (i.e., $\mathbf{g}(k) = \mathbf{g}_0$) or just varying slowly with time while the far-end signal $x(k)$ is modeled as a random process. In contrast, it is pointed out in [Enzner 2006] that actually the signal vector $\mathbf{x}(k)$ is known sample-by-sample and that the *variations* of the LEM impulse response are random by nature. Therefore, we consider the discrete-time model of echo cancellation in Figure 14.21a. The slowly time varying LEM impulse response $\mathbf{g}(k)$ is now modeled by a first order Markov chain, e.g., [Enzner 2006] and [Paleologu et al. 2013]

$$\mathbf{g}(k) = \gamma \cdot \mathbf{g}(k-1) + \eta \cdot \mathbf{q}(k) \tag{14.117}$$

$$\tilde{x}(k) = \mathbf{x}^T(k) \cdot \mathbf{g}(k) \tag{14.118}$$

as shown in Figure 14.21b, with

- constant *forgetting* factor $\gamma$,
- random white noise innovation $\mathbf{q}(k)$, $\mathcal{N}(\mathbf{0}, \sigma_q^2 \cdot \mathbf{I})$, $E\{\mathbf{q}\mathbf{q}^T\} = \sigma_q^2 \cdot \mathbf{I}$,
- on-off control switch $\eta \in \{0, 1\}$.

The time-invariant case is described with parameters $\gamma = 1$ and $\eta = 0$ while the time-variant model is given by $0 \ll \gamma < 1$ and $\eta = 1$.

Note: This is a special case of the linear state-space model. The more general approach uses time-variable matrices

$\mathbf{A}(k)$ : state transition matrix, instead of $\gamma$,

$\mathbf{B}(k)$ : input control matrix, instead of $\eta$,

$\mathbf{C}(k)$ : observation matrix, instead of $\mathbf{x}^T(k)$

according to

$$\mathbf{g}(k) = \mathbf{A}(k)\mathbf{g}(k-1) + \mathbf{B}(k)\mathbf{q}(\mathbf{k}),$$
$$\tilde{x}(k) = \mathbf{C}(k)\mathbf{g}(k).$$

**Figure 14.21** Model-based echo cancellation. (a) LEM system $\mathbf{g}(k)$ and echo canceller $\mathbf{h}(k)$ and (b) first order Markov model of the LEM system.

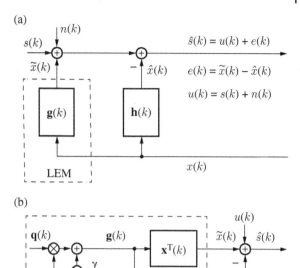

For the echo cancellation problem, the simpler model has proven to be a pragmatic solution, e.g., [Enzner 2006], [Paleologu et al. 2013], [Mandic et al. 2015], [Kühl et al. 2017], [Fabry et al. 2020], and [Kühl 2022].

The reasoning is that an exact state transition matrix is hard to specify. However, as typical acoustical LEM systems change very slowly, a first-order Markov model with a forgetting factor close to one, e.g., $\gamma = 0.999$ is adequate for tracing at least the slow variations of typical echo paths.

The predictable portion of $\mathbf{g}(k)$ is given by $\gamma \cdot \mathbf{g}(k-1)$ while the unpredictable components are modeled as a low-level white noise vector $\mathbf{q}(k)$. This predictability is included in the Kalman-type algorithms to improve the adaptation of $\mathbf{h}(k)$.

For the purpose of deriving the relations between the LMS type gradient descent algorithm and the more general Kalman-type state-space estimator, we need to introduce the logical distinction between the "first guess" or preliminary *updated* impulse response $\mathbf{h}^+(k)$, based on information available at time instant $k$ and the final impulse response $\mathbf{h}(k+1)$, which includes knowledge about the state-space model and is actually used at the next time instant $k+1$ for producing the sample $\hat{x}(k+1)$.

If at time instant $k$, a preliminary updated impulse response $\mathbf{h}^+(k)$ is estimated, e.g., by an NLMS-type algorithm, this first estimate can be improved by prediction, taking into account the Markov model as

$$\mathbf{h}(k+1) = \gamma \cdot \mathbf{h}^+(k). \tag{14.119}$$

Taking into account (14.117), this effects the evolution of the distance vector as follows:

$$\mathbf{d}^+(k) = \mathbf{g}(k) - \mathbf{h}^+(k) \tag{14.120a}$$

$$\mathbf{d}(k+1) = \mathbf{g}(k+1) - \mathbf{h}(k+1) \tag{14.120b}$$

$$= \gamma \cdot \mathbf{d}^+(k) + \eta \cdot \mathbf{q}(k+1). \tag{14.120c}$$

The impulse response $\mathbf{h}(k)$ of the echo canceller is adapted iteratively. Preferably, the adaptation is steered by the residual error signal $e(k)$. However, in a real system with near-end speech $s(k)$ and background noise $n(k)$, this *clean* error signal $e(k)$ is not accessible. Thus, the *disturbed error signal*, i.e., the output signal $\hat{s}(k)$ has to be used for the adaptation.

### 14.9.3.1 NLMS Algorithm

The starting point is the basic LMS algorithm of Section 14.5, i.e., the iterative solution according to the MSE criterion, using the (disturbed) error signal $\hat{s}(k) = e(k) + s(k) + n(k)$ for adaptation of $\mathbf{h}(k)$,

$$\mathbf{h}(k + 1) = \mathbf{h}^+(k) = \mathbf{h}(k) + \beta(k)\mathbf{x}(k)\hat{s}(k), \tag{14.121}$$

where $\beta(k)$ is the stepsize control parameter.

As the NLMS algorithm does not take into account the Markov model of the LEM impulse response, no distinction exists between the updated impulse response $\mathbf{h}^+(k)$ and the final version $\mathbf{h}(k + 1)$.

Preferably, the normalized version of the stepsize is chosen (NLMS)

$$\beta(k) = \frac{\alpha(k)}{||\mathbf{x}(k)||^2} = \frac{\alpha(k)}{\mathbf{x}^T(k)\mathbf{x}(k)} \quad \text{with } 0 < \alpha < 2. \tag{14.122}$$

The same scalar stepsize factor $\beta(k)$ is used for updating each element of the vector $\mathbf{h}(k)$.

### 14.9.3.2 Kalman Algorithm

As a generalization of (14.121), the scalar stepsize factor $\beta(k)$ is replaced by a time varying weighting matrix $\mathbf{K}(k)$ of dimension $m \times m$

$$\mathbf{h}^+(k) = \mathbf{h}(k) + \mathbf{K}(k)\mathbf{x}(k)\hat{s}(k) \tag{14.123a}$$

$$= \mathbf{h}(k) + \mathbf{w}(k)\hat{s}(k) \tag{14.123b}$$

$$= \mathbf{h}(k) + \mathbf{w}(k)[u(k) + \mathbf{x}^T(k)\mathbf{d}(k)], \tag{14.123c}$$

with column vectors

$$\mathbf{w}(k) = \mathbf{K}(k)\mathbf{x}(k)$$

$$\mathbf{d}(k) = \mathbf{g}(k) - \mathbf{h}(k)$$

and scalar samples

$$u(k) = s(k) + n(k).$$

$\mathbf{K}(k)$ is called the Kalman matrix and $\mathbf{w}(k)$ is the Kalman gain vector.

For ease of notation, the time index $k$ is omitted in the following derivations, i.e., we use

$$u = u(k) = s(k) + n(k), \quad \mathbf{h} = \mathbf{h}(k), \quad \mathbf{w} = \mathbf{w}(k), \quad \mathbf{d}^+ = \mathbf{d}^+(k).... \text{etc.}$$

Thus, the adaptation (14.123c) of the impulse response vector $\mathbf{h}(k)$ reads

$$\mathbf{h}^+ = \mathbf{h} + \mathbf{w}[\mathbf{x}^T\mathbf{d} + u], \tag{14.124}$$

and the updated distance vector can be formulated as

$$\mathbf{d}^+ = \mathbf{g} - \mathbf{h}^+ \tag{14.125a}$$

$$= \mathbf{g} - \mathbf{h} - \mathbf{w}[\mathbf{x}^T\mathbf{d} + u] \tag{14.125b}$$

$$= \mathbf{d} - \mathbf{w}[\mathbf{x}^T\mathbf{d} + u] \tag{14.125c}$$

$$= \mathbf{d} - \mathbf{w}\mathbf{x}^T\mathbf{d} - \mathbf{w}u, \tag{14.125d}$$

with the transposed version of $\mathbf{d}^+$

$$\mathbf{d}^{+T} = \mathbf{d}^T - \mathbf{d}^T\mathbf{x}\mathbf{w}^T - \mathbf{w}^T u. \tag{14.125e}$$

Next, we derive the optimal weight vector $\mathbf{w}$, and thus the optimal Kalman matrix $\mathbf{K}$, by minimization of the MSD in terms of the statistical expectation

$$\epsilon = \mathrm{E}\left\{||\mathbf{d}^+||^2\right\} \overset{!}{=} \min. \tag{14.126}$$

With (14.125e) and (14.125d), we obtain the following relation by exploiting the fact that $\mathbf{d}^T\mathbf{x} = \mathbf{x}^T\mathbf{d}$ delivers a scalar value:

$$\epsilon = \mathrm{E}\left\{\mathbf{d}^{+T}\mathbf{d}^+\right\}$$

$$= \mathrm{E}\left\{(\mathbf{d}^T - \mathbf{d}^T\mathbf{x}\mathbf{w}^T - \mathbf{w}^T u)(\mathbf{d} - \mathbf{w}\mathbf{x}^T\mathbf{d} - \mathbf{w}u)\right\} \tag{14.127a}$$

$$= \mathrm{E}\{\mathbf{d}^T\mathbf{d} - 2\mathbf{w}^T\mathbf{d}\mathbf{d}^T\mathbf{x} + \mathbf{w}^T\mathbf{w}\mathbf{x}^T\mathbf{d}\mathbf{d}^T\mathbf{x} + \mathbf{w}^T\mathbf{w}u^2 + $$
$$+ 2[\mathbf{w}^T\mathbf{w}\mathbf{x}^T\mathbf{d} - \mathbf{w}^T\mathbf{d}]u\}. \tag{14.127b}$$

In applying the expectation operation to (14.127b), we take into account that the current signal vector $\mathbf{x}$ is known and that the disturbing signal $u$ is zero mean, i.e., $\mathrm{E}\{u\} = 0$ with $\sigma_u^2 = \mathrm{E}\{u^2\}$.

Thus, (14.127b) results in

$$\epsilon = \mathrm{E}\{\mathbf{d}^T\mathbf{d}\} - 2\mathbf{w}^T\mathrm{E}\{\mathbf{d}\mathbf{d}^T\}\mathbf{x} + \mathbf{w}^T\mathbf{w}[\mathbf{x}^T\mathrm{E}\{\mathbf{d}\mathbf{d}^T\}\mathbf{x} + \sigma_u^2]. \tag{14.128}$$

With the error covariance matrix

$$\mathbf{P} = \mathrm{E}\{\mathbf{d}\mathbf{d}^T\}, \tag{14.129}$$

the optimality condition is given by partial derivation of (14.128) w.r.t. $\mathbf{w}$

$$\frac{\partial \epsilon}{\partial \mathbf{w}} = -2\mathbf{P}\mathbf{x} + 2\mathbf{w}[\mathbf{x}^T\mathbf{P}\mathbf{x} + \sigma_u^2] \overset{!}{=} \mathbf{0}. \tag{14.130}$$

Noting that the expression $[\mathbf{x}^T\mathbf{P}\mathbf{x} + \sigma_u^2]$ delivers a scalar quantity, this condition can be solved for the optimal weight vector $\mathbf{w}(k)$ and the Kalman matrix $\mathbf{K}(k)$ as

$$\mathbf{w}(k) = \frac{\mathbf{P}(k)}{\mathbf{x}^T(k)\mathbf{P}(k)\mathbf{x}(k) + \sigma_u^2}\mathbf{x}(k) \tag{14.131a}$$

$$= \mathbf{K}(k)\mathbf{x}(k). \tag{14.131b}$$

By using (14.125d), (14.125e), and (14.131a), a recursion for the updated covariance matrix

$$\mathbf{P}^+(k) = \mathrm{E}\left\{\mathbf{d}^+(k)\mathbf{d}^{+T}(k)\right\}, \tag{14.132}$$

can be evaluated as

$$\mathbf{P}^+(k) = \mathbf{P}(k) - \mathbf{w}(k)\mathbf{x}^T(k)\mathbf{P}(k) \tag{14.133a}$$

$$= \mathbf{P}(k) - \mathbf{K}(k)\mathbf{x}(k)\mathbf{x}^T(k)\mathbf{P}(k). \tag{14.133b}$$

This result can be checked easily by inserting (14.129) and (14.131a) in (14.133a).

Finally, the Markov model can be included in the calculation of $\mathbf{P}(k+1)$ for the next iteration step $k+1$ by using (14.120a) and (14.124)

$$\mathbf{P}(k+1) = \gamma^2 \mathbf{P}^+(k) + \sigma_q^2 \cdot \mathbf{I}. \tag{14.134}$$

### 14.9.3.3 Summary of Kalman Algorithm

Step 0: Initialization of impulse response $\mathbf{h}(k = 0)$ and distance covariance matrix $\mathbf{P}(k = 0)$ by using the initial distance vector $\mathbf{d}(k = 0) = \mathbf{g}(k = 0) - \mathbf{h}(k = 0)$ and calculation of $\hat{x}(k = 0)$ and $\hat{s}(k = 0)$.

Step 1: Computation of the optimal Kalman matrix according to (14.131a)

$$\mathbf{K}(k) = \frac{\mathbf{P}(k)}{\mathbf{x}(k)^T \mathbf{P}(k)\mathbf{x}(k) + \sigma_u^2},$$

with the propagated covariance matrix $\mathbf{P}(k)$ calculated with (14.134) in the previous iteration step $k - 1$.

Step 2: Update (14.124) of the impulse response $\mathbf{h}(k)$ of the echo canceller according to the discrete time model of Figure 14.3 using the estimate $\hat{s}(k)$ of near-end signal $s(k)$

$$\mathbf{h}^+(k) = \mathbf{h}(k) + \mathbf{K}(k)\mathbf{x}(k)\hat{s}(k) = \mathbf{h}(k) + \mathbf{w}(k)\hat{s}(k).$$

Step 3: Prediction (14.119) of the next state of the impulse response

$$\mathbf{h}(k+1) = \gamma\,\mathbf{h}^+(k).$$

Step 4: Update (14.133a) of the error covariance matrix

$$\mathbf{P}^+(k) = \mathbf{P}(k) - \mathbf{w}(k)\mathbf{x}^T(k)\mathbf{P}(k) = [\mathbf{I} - \mathbf{w}(k)\mathbf{x}^T(k)]\,\mathbf{P}(k).$$

Step 5: Propagation (14.134) of the distance covariance matrix for the next iteration $k + 1$

$$\mathbf{P}(k+1) = \gamma^2 \mathbf{P}^+(k) + \sigma_q^2 \cdot \mathbf{I}.$$

Step 6: Calculations of the next samples $\hat{x}(k + 1)$ and $\hat{s}(k + 1)$ using the latest impulse response $\mathbf{h}(k + 1)$. Repetition of steps 1 to 6 for $k \to k + 1$.

### 14.9.3.4 Remarks

1. The original NLMS algorithm with a scalar step size factor $\beta(k)$ is based on the *instantaneous* gradient of the squared residual echo signal $e^2(k)$ under the assumption of a time invariant LEM system. The corresponding stepsize factor (14.121) does not take into account the near-end speech signal $s(k)$ and the background noise $n(k)$ which both disturb the adaptation process. However, as shown in Section 14.6.4, the disturbing influence of $u(k) = s(k) + n(k)$ can be diminished by empirical adaptive stepsize control. The optimization criterion is the minimization of the *MSE* of the *residual echo signal*, which is related to the *ERLE*.

2. The filter adaptation rule of the Kalman algorithm (step 2) may be considered a generalization of NLMS algorithm (14.123b) by replacing the scalar stepsize factor $\beta(k)$ by the
   Kalman matrix $\mathbf{K}(k)$. The optimization criterion is the minimization of the *mean square
   deviation* (MSD) between the LEM impulse response $\mathbf{g}(k)$ and the impulse response $\mathbf{h}(k)$
   of the echo canceller, which is related to the *system distance*.
   The optimal Kalman matrix $\mathbf{K}(k)$ (step 1) includes systematically the power $\sigma_u^2$ of the
   disturbance $u(k)$ as well as the time varying Markov model of the LEM impulse response
   (steps 3,4,5). The elements of the resulting weighting matrix act as local stepsize factors
   for the individual elements of the impulse response vector $\mathbf{h}(k)$ as can be seen from step 2.
   These *local* stepsize factors have similar dependencies as the *global* optimum stepsize
   control factor $\alpha_{opt}(k)$ according to (14.71a).

3. The outlined derivations of the Kalman algorithm can be translated to the short-term
   frequency domain (DFT/FFT) e.g., [Enzner 2006], [Enzner 2008] and can favorably
   be combined with an adaptive postfilter, for reducing residual echo components after
   echo cancellation. This is also of special interest in case of a time varying LEM impulse
   response as the iterative adaptation algorithm of the echo canceller will not be able
   to trace perfectly the dynamic variations of the LEM impulse reponse. The inevitable
   residual echo signal can be tackled by a postfilter, which operates like a frequency
   domain noise suppression filter (see Chapter 12). The most crucial point is the joint
   control of the echo canceller and the postfilter. In [Enzner 2006] and [Enzner, Vary 2006]
   the optimal analytical solution is given, which is based on the idea of the state-space
   model (14.117) formulated in the frequency domain. An advantageous feature of this
   combined solution is that the postfilter also reduces the residual echo originating from
   a mismatch between the lengths of the impulse responses $\mathbf{g}(k)$ and $\mathbf{h}(k)$.

## 14.10  Block Processing and Frequency Domain Adaptive Filters

Depending on the acoustic environment, filters with hundreds to thousands of coefficients
are required for acoustic echo cancellation. When using the NLMS algorithm, a large filter
order is associated with slow convergence (see Figure 14.16). Furthermore, the computational effort increases linearly with the filter length $m$. Both problems can be alleviated to
a certain degree by frequency-domain processing.

The frequency domain approach reduces the computational complexity of the adaptive
filter by using the FFT for fast convolution and correlation (see Section 3.6) or by using filter
banks and multi-rate signal processing techniques. The improved convergence results from
the inherent decorrelation of the signal and the associated reduction of eigenvalue spread.
Among the many frequency domain approaches we will discuss the FFT-based *frequency
domain adaptive filter* (FDAF) in greater detail as it allows for a reduction in complexity
and a speed up of convergence.

As a starting point for deriving the FDAF algorithm, we consider the coefficient vector
update in the time domain according to the LMS algorithm

$$\mathbf{h}(k+1) = \mathbf{h}(k) + \beta\, e(k)\, \mathbf{x}(k), \tag{14.135}$$

with

$$e(k) = \tilde{x}(k) - \hat{x}(k) = \tilde{x}(k) - \mathbf{h}^T(k)\mathbf{x}(k), \tag{14.136}$$

and a constant stepsize parameter $\beta$. In each time step, $m$ MAC multiply accumulate operations are required for the filtering $\mathbf{h}^T(k)\mathbf{x}(k)$ in (14.136) and $m$ MAC operations for the coefficient vector update (14.135). The product $\beta \cdot e(k)$ and the residual error $e(k) = \tilde{x}(k) - \hat{x}(k)$ have to be computed only once per sample. Since $m \gg 1$, this contribution to the total computational effort can be neglected in the following complexity estimations. Thus, the computational effort is split almost evenly between adaptation and filtering. It seems obvious that the computational effort for filtering can markedly be reduced by using FFT-based fast convolution instead of time-domain convolution. As an intermediate step toward frequency domain methods, we consider the *block LMS* and the *exact block NLMS* algorithms.

## 14.10.1 Block LMS Algorithm

The *block LMS* algorithm [Clark et al. 1981] adjusts the filter coefficients not in each iteration, but only every $L$ samples. Hence, the signal $x(k)$ is convolved with an impulse response that is constant over a time interval of length $L \cdot T$. Also, $L$ residual error samples $e(k), e(k + 1), \dots, e(k + L - 1)$ are used to compute the new coefficient vector, which is valid from time instant $k + L$ on,

$$\mathbf{h}(k + L) = \mathbf{h}(k) + \beta \sum_{\lambda=0}^{L-1} e(k + \lambda)\mathbf{x}(k + \lambda). \tag{14.137a}$$

In analogy to (14.135) the coefficient vector update is now composed of the sum of the increments $\beta e(k + \lambda)\mathbf{x}(k + \lambda)$. The output signal therefore differs from the original LMS algorithm, where each residual error sample $e(k)$ is generated with a different set of filter coefficients.

We note that (14.137a) can also be written as

$$\mathbf{h}(k + L) = \mathbf{h}(k) + L\beta \frac{1}{L} \sum_{\lambda=0}^{L-1} e(k + \lambda)\mathbf{x}(k + \lambda) \tag{14.137b}$$

$$= \mathbf{h}(k) + \tilde{\beta}\,\hat{\mathbf{v}}_L(k + L - 1), \tag{14.137c}$$

which indicates that the original instantaneous gradient is replaced by a temporally smoothed *average* gradient

$$\hat{\mathbf{v}}_L(k + L - 1) = \frac{1}{L} \sum_{\lambda=0}^{L-1} e(k + \lambda)\,\mathbf{x}(k + \lambda), \tag{14.138}$$

and the effective stepsize is $\tilde{\beta} = L\beta$. Thus, for a stationary excitation the gradient is estimated more accurately. However, a slower convergence results, because, compared to the conventional LMS algorithm, the maximal stepsize parameter $\tilde{\beta}_{max}$ must be reduced by the factor $1/L$, in order to guarantee convergence (e.g., [Clark et al. 1981]).

For a white noise signal $x(k)$, both the conventional LMS algorithm and the block LMS algorithm converge with different speeds toward the Wiener solution.

It will be shown below, that the basic FDAF algorithm exhibits the reduced convergence behavior of the block LMS algorithm. But the speed of convergence can be significantly improved by using frequency- and time-dependent stepsize parameters.

The block LMS algorithm does not produce exactly the results of the LMS algorithm (14.27) with sample-by sample adaptation. It is, however, possible to develop mathematically exact block realizations of the LMS and the NLMS algorithms [Nitsch 1997].

In analogy to (14.137a) a block version of the NLMS algorithm with variable step size (14.35) may be written as

$$\mathbf{h}(k+L) = \mathbf{h}(k) + \sum_{\lambda=0}^{L-1} \alpha(k+\lambda) \frac{e_{block}(k+\lambda)\mathbf{x}(k+\lambda)}{||\mathbf{x}(k+\lambda)||^2}, \tag{14.139}$$

where $e_{block}(k+\lambda)$, $\lambda = 0, \ldots, L-1$, is computed with the temporarily fixed coefficient vector $\mathbf{h}(k)$

$$e_{block}(k+\lambda) = y(k+\lambda) - \mathbf{x}^T(k+\lambda)\mathbf{h}(k), \quad \lambda = 0, \ldots, L-1. \tag{14.140}$$

The difference between the block NLMS algorithm (14.139), (14.140), and the "sample-by-sample" NLMS algorithm according to (14.27) and (14.35) consists in the computation of the error signal. The "sample-by-sample" (sbs) algorithm produces the error signal

$$e_{sbs}(k+\lambda) = y(k+\lambda) - \mathbf{x}^T(k+\lambda)\mathbf{h}(k+\lambda), \tag{14.141}$$

with an impulse response $\mathbf{h}(k+\lambda)$ which is, in contrast to (14.140), not fixed over a period of $L$ samples. We may write the NLMS error as

$$e_{sbs}(k+\lambda) = e_{block}(k+\lambda) + \Delta e(k+\lambda), \tag{14.142}$$

as a sum of the block NLMS error $e_{block}(k+\lambda)$ and a correction term

$$\Delta e(k+\lambda) = \begin{cases} 0 & \lambda = 0 \\ -\sum_{i=1}^{\lambda} \frac{\alpha(k+i-1)}{||\mathbf{x}(k+i-1)||^2} e_{sbs}(k+i-1) \\ \quad \cdot \mathbf{x}^T(k+\lambda)\mathbf{x}(k+i-1) & \lambda = 1, \ldots, \end{cases} \tag{14.143}$$

which accounts for the intermediate coefficient vector updates.

Based on (14.143) a modified stepsize-error product can be derived for the adaptation (14.139) such that a mathematically exact block version of the NLMS algorithm is obtained, e.g., [Nitsch 1997] and [Vary, Martin 2006] (Section 13.9.2).

## 14.10.2 Frequency Domain Adaptive Filter (FDAF)

The frequency domain implementation of block adaptive algorithms comprises two steps. First, the convolution of the far-end signal $\mathbf{x}(k)$ with the filter impulse response $\mathbf{h}(k)$ of the echo canceller is implemented as a fast convolution using the overlap-save scheme (see Section 3.6.3). Secondly, the coefficient vector is adapted in the frequency domain. For large $m$, both steps result in significant computational savings.

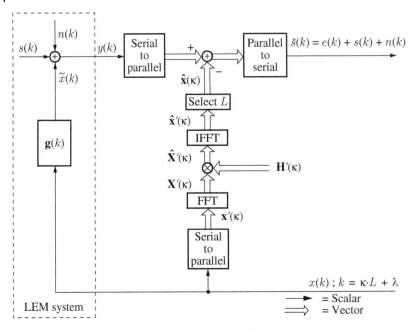

**Figure 14.22** Echo cancellation with fast convolution.

### 14.10.2.1 Fast Convolution and Overlap-Save

To develop the FDAF the time index $k$ is replaced by

$$k = \kappa L + \lambda; \qquad \kappa = 0, 1, 2, \dots; \qquad \lambda = 0, 1, \dots, L - 1, \tag{14.144}$$

where $\kappa$ is the block index and $\lambda$ is the sample index within the block. The convolution is performed in the frequency domain using the FFT and the *overlap-save* algorithm. In each block, $L$ new output samples $\hat{x}(k)$ are computed, while the coefficients of the filter are constant for at least $L$ time steps.

Figure 14.22 illustrates the basic structure. For an impulse response of order $m$, the length $M$ of the FFT must be chosen to accommodate $L$ valid output samples, therefore

$$M = m - 1 + L. \tag{14.145}$$

For better clarity, the effective time varying impulse response of the echo canceller, which is valid for the time period

$$\kappa \cdot L \le k \le (\kappa + 1) \cdot L - 1, \tag{14.146}$$

is now denoted as

$$h_i(\kappa \cdot L), \quad i = 0, 1, \dots, m - 1. \tag{14.147}$$

For the purpose of fast convolution, the impulse response of length $m$ is padded with zeros to a vector $\mathbf{h}'(\kappa)$ of length $M$ with elements

$$h_i'(\kappa) = \begin{cases} h_i(\kappa \cdot L) & i = 0, 1, \dots, m - 1 \\ 0 & i = m, m + 1, \dots, M - 1. \end{cases} \tag{14.148}$$

The input vector $\mathbf{x}'(\kappa)$ consists of the $M$ elements

$$\mathbf{x}'(\kappa) = (x'_0(\kappa), x'_1(\kappa), \ldots, x'_{M-1}(\kappa))^T$$
$$= (x(\kappa L - m + 1), \ldots, x(\kappa L), x(\kappa L + 1), \ldots, x(\kappa L + L - 1))^T \qquad (14.149)$$

and is also fed into an FFT of length $M$. In contrast to the definition (14.20) of the state vector of the time domain echo canceller, the input vector $\mathbf{x}'(\kappa)$ contains the samples in the order of increasing sampling indices.

The transformation of the coefficient vector

$$\mathbf{h}'(\kappa) = \left(h'_0(\kappa), h'_1(\kappa), \ldots, h'_{M-1}(\kappa)\right)^T$$

is written as

$$\mathbf{H}'(\kappa) = \mathrm{FFT}\{\mathbf{h}'(\kappa)\},$$

with

$$\mathbf{H}'(\kappa) = \left(H'_0(\kappa), H'_1(\kappa), \ldots, H'_{M-1}(\kappa)\right)^T.$$

The transformed vectors $\mathbf{H}'(\kappa)$ and $\mathbf{X}'(\kappa)$ are multiplied element by element, i.e.,

$$\hat{X}'_\mu(\kappa) = H'_\mu(\kappa) \cdot X'_\mu(\kappa). \qquad (14.150)$$

The resulting vector $\hat{\mathbf{X}}'(\kappa)$, after inverse transformation, yields the vector $\hat{\mathbf{x}}'(\kappa)$ of length $M$,

$$\hat{\mathbf{x}}'(\kappa) = (\hat{x}'_0(\kappa), \hat{x}'_1(\kappa), \ldots, \hat{x}'_{M-1}(\kappa))^T, \qquad (14.151)$$

which is equivalent to a cyclic convolution (see Figure 14.23) of the vectors $\mathbf{x}'(k)$ and $\mathbf{h}'(k)$. Consequently, the first $m - 1$ values $\hat{x}'_i(\kappa)$, $i = 0, 1, \ldots, m - 2$, are affected by cyclic effects. The selection of $L$ valid output samples may be described by an elementwise multiplication with window $\mathbf{w}$, where

$$w_i = \begin{cases} 0 & i = 0, 1, \ldots, m - 2 \\ 1 & i = m - 1, m, \ldots, M - 1, \end{cases} \qquad (14.152)$$

are the components of $\mathbf{w}$. The $L$ valid output values,

$$\hat{x}(\kappa \cdot L + \lambda) = \hat{x}'_{m-1+\lambda}(\kappa), \quad \lambda = 0, 1, \ldots, L - 1, \qquad (14.153)$$

correspond to the result of the linear convolution. Finally, the residual echo (undisturbed compensation error) is obtained from

$$e(\kappa \cdot L + \lambda) = \tilde{x}(\kappa \cdot L + \lambda) - \hat{x}(\kappa \cdot L + \lambda), \qquad (14.154)$$

whereas in real applications only the disturbed error

$$\hat{s}(\kappa \cdot L + \lambda) = y(\kappa \cdot L + \lambda) - \hat{x}(\kappa \cdot L + \lambda), \qquad (14.155)$$

is available instead of $e(\kappa \cdot L + \lambda)$. The serial-to-parallel operation comprises buffering of $L$ successive signal samples. $L$ samples must be collected before the next block of the transmit signal $\hat{s}(k)$ can be computed. The block processing therefore introduces a delay of $L - 1$ samples. However, at the expense of increased computational complexity, $L$ can be chosen to be much smaller than the transform length $M$. The FDAF concept therefore provides a flexible framework for balancing algorithmic delay and computational complexity.

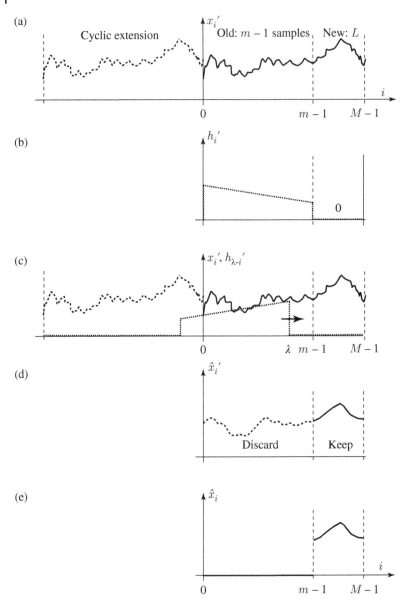

**Figure 14.23** Illustration of cyclic convolution with $M = m - 1 + L$. (a) Input vector $\mathbf{x}'(\kappa)$ with cyclic extension, (b) coefficient vector $\mathbf{h}'(\kappa)$, (c) cyclic relations in the time domain, (d) result of cyclic convolution, and (e) selection of $L$ valid samples.

However, when $m$ is large and when a small algorithmic delay $L$ is required, the computational complexity might not be acceptable. In [Sommen 1989] and [Soo, Pang 1990] a partitioning of the coefficient vector, i.e., a distribution of the long impulse response onto several partial filters with shorter impulse responses, is proposed to deal with this problem. For $Q$ partitions the *partitioned block frequency domain adaptive filter* (PBFDAF) is

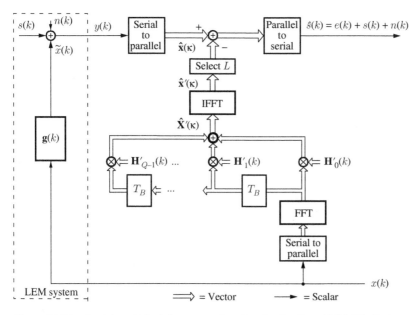

**Figure 14.24** Partitioned block frequency domain adaptive filter (PBFDAF). $T_B$ denotes a unit delay of frequency domain vectors.

illustrated in Figure 14.24. Instead of FFTs of length $m + L - 1$ this scheme uses FFTs of length $m/Q + L - 1$.

### 14.10.2.2 FLMS Algorithm

In the next step, the adaptation of the coefficients is transformed into the frequency domain as well. This leads to the *fast LMS* (FLMS) algorithm [Ferrara 1980], [Clark et al. 1983]. The benefits of an adaptation in the frequency domain are an additional reduction of the computational effort and an improvement of the convergence behavior using a time- *and* frequency-dependent stepsize control.

For this purpose, the adaptation rule of the block LMS algorithm according to (14.137c) is inspected again. As before, we develop this algorithm on the basis of the undisturbed error signal. The coefficient update Eq.(14.137b) is now written in terms of the block index $\kappa$,

$$\mathbf{h}'(\kappa + 1) = \mathbf{h}'(\kappa) + \beta \, \hat{\mathbf{v}}'(\kappa), \tag{14.156}$$

or more generally as

$$\mathbf{h}'(\kappa + 1) = \mathbf{h}'(\kappa) + \Delta \mathbf{h}'(\kappa), \tag{14.157}$$

where in the case of a constant stepsize $\beta$ the coefficient update vector $\Delta \mathbf{h}'(\kappa)$ is computed as $\Delta \mathbf{h}'(\kappa) = \beta \, \hat{\mathbf{v}}'(\kappa)$. The components $\hat{v}'_i(\kappa)$ of the gradient vector $\hat{\mathbf{v}}'(\kappa)$ are given by

$$\hat{v}'_i(\kappa) = \begin{cases} \displaystyle\sum_{\lambda=0}^{L-1} e(\kappa \cdot L + \lambda) x(\kappa \cdot L + \lambda - i) & i = 0, 1, \ldots, m - 1 \\[4mm] 0 & i = m, m + 1, \ldots, M - 1. \end{cases} \tag{14.158}$$

When the discrete Fourier transform is applied to both sides of (14.157), we obtain

$$\mathbf{H}'(\kappa + 1) = \mathbf{H}'(\kappa) + \Delta\mathbf{H}'(\kappa). \tag{14.159}$$

Thus, the update loop can be implemented in the frequency domain as shown in Figure 14.25a. In the simple case of a constant stepsize $\beta$ we have $\Delta\mathbf{H}'(\kappa) = \beta\hat{\mathbf{V}}'(\kappa)$ where in principle the frequency domain gradient vector $\hat{\mathbf{V}}'(\kappa)$ could be obtained via the FFT of the time domain gradient vector $\hat{\mathbf{v}}'(\kappa)$.

However, the computational complexity can be further reduced when the update vector $\Delta\mathbf{H}'(\kappa)$ is computed in the frequency domain as well. As a function of $i$, the individual component $\hat{v}'_i(\kappa)$ according to (14.158) for fixed $\kappa$ can be interpreted as a correlation between a segment of the residual error signal $e(k)$ of length $L$ and the signal $x(k)$. In analogy to the fast convolution, this correlation can be implemented as a fast correlation using the FFT.

In order to employ the FFT, the signal vectors must be constructed such that the cyclic convolution yields the components $\hat{v}'_i(\kappa)$ of the gradient vector as in (14.158). This can be achieved by padding $\mathbf{e}(\kappa) = (e(\kappa L), \dots, e(\kappa L + L - 1))^T$ with $m - 1$ leading zeros

$$e'_i(\kappa) = \begin{cases} 0 & i = 0, 1, \dots, m - 2 \\ e(\kappa \cdot L + i - m + 1) & i = m - 1, m, \dots, M - 1, \end{cases} \tag{14.160}$$

and by using the vectors $\mathbf{x}'(\kappa)$ and $\mathbf{e}'(\kappa)$ as inputs to the fast correlation. If we compute the FFTs $\mathbf{E}'(\kappa)$ and $\mathbf{X}'(\kappa)$ of vectors $\mathbf{e}'(\kappa)$ and $\mathbf{x}'(\kappa)$, respectively, the componentwise multiplication of the vectors $\mathbf{E}'(\kappa)$ and $\mathbf{X}'^*(\kappa)$,

$$\hat{V}'_\mu(\kappa) = E'_\mu(\kappa)X'^*_\mu(\kappa), \quad \mu = 0, \dots, M - 1, \tag{14.161}$$

and a subsequent inverse transform produce the desired values $\hat{v}'_i(\kappa)$ in the first $m$ elements of the inverse of $\tilde{\mathbf{V}}'(\kappa)$.

The desired gradient vector $\hat{\mathbf{v}}'$ is thus generated by applying a (gradient) constraint $\tilde{\mathbf{w}}$ to IDFT$\{\tilde{\mathbf{V}}(\kappa)\}$, which sets the last $L - 1$ samples to zero. The constraint window $\tilde{\mathbf{w}}$ is given

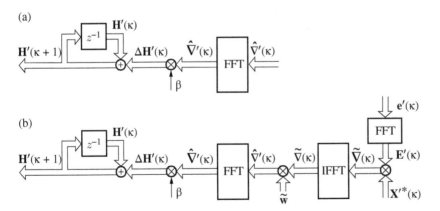

**Figure 14.25** Derivation of the adaptation rule of the FLMS algorithm. (a) Adaptation of the coefficient vector in the frequency domain, and (b) fast computation of the gradient vector.

by

$$\tilde{w}_i = \begin{cases} 1 & i = 0, 1, \dots, m-1 \\ 0 & i = m, m+1, \dots, M-1, \end{cases}$$ (14.162)

and the resulting FLMS algorithm is illustrated in Figure 14.25b. Including the two transformations for fast convolution (Figure 14.22), a total of five transformations of length $M$
are performed in order to determine $L$ values of the estimated echo signal $\hat{x}(k)$. The computational complexity of the constraint FLMS can then be estimated as follows.

The FFT of a real-valued sequence of length $M$ requires about $(M/4)\,\mathrm{ld}(M/2)$ complex
arithmetic operations. A total of five transformations are needed. For real-valued signals
the symmetry of the DFT vectors can be exploited. Therefore, approximately $4 \cdot M/2 = 2M$
complex operations are performed to multiply the components of $\mathbf{E}'(\kappa)$ and $\mathbf{X}'^*(\kappa)$, to multiply the components of $\mathbf{X}'(\kappa)$ and $\mathbf{H}'(\kappa)$, and to adapt the coefficient vector. If we equate
one complex operation with four real operations, the computational complexity for one
block of $L$ samples is given by $5M\,\mathrm{ld}(M/2) + 4 \cdot 2M$. Per output sample, we thus obtain a
computational complexity of

$$c = \frac{4 \cdot 5\frac{M}{4}\,\mathrm{ld}(M/2) + 4 \cdot 2M}{L}, \qquad M = m - 1 + L.$$

For $m = 500$, Figure 14.26 plots $c$ as a function of $0 \le L \le m/2$. Clearly, $c$ attains its minimum for the maximum value of $L$. Fortunately, the computational complexity exhibits a
sharp decline for increasing $L$. For example, for $m = 500$, we already achieve a significant
complexity reduction for $L > 40$.

Furthermore, the complexity of the FLMS may be related to the complexity of the time
domain LMS algorithm requiring $2m$ operations per sample. The relative complexity is thus
given for $M = m + L - 1 \approx m + L$ by

$$\rho = \frac{5M\,\mathrm{ld}(M/2) + 8M}{2mL}.$$ (14.163)

For $m = L$, Table 14.1 gives typical values. For increasing $L$, and a corresponding increase
of the block delay, $\rho$ decreases significantly.

In comparison to the LMS algorithm, the computational complexity can be significantly
reduced for large $L$. In contrast, the required memory capacity increases, because even with

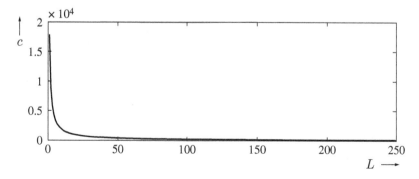

**Figure 14.26** Computational complexity of FLMS per output sample vs. block length $L$.

**Table 14.1** Relative computational complexity of the FLMS algorithm in comparison to the LMS algorithm ($M = 2\,L$; $m = L$), according to [Ferrara 1985].

| L | 16 | 32 | 64 | 256 | 1024 |
|---|------|------|------|------|-------|
| $\rho$ | 1.75 | 1.03 | 0.59 | 0.19 | 0.057 |

a skillful organization of the processes needed for an *in-place* FFT computation several vectors of the dimension $M$ must be stored.

The computational complexity can be reduced if the gradient constraint is neglected. Compared to the constrained FLMS, the unconstrained FLMS [Mansour, Gray 1982] saves two FFT/IFFT operations. However, in this case, the coefficient update must rely on a circular convolution. Therefore, the algorithm does not approach the Wiener solution in steady state. A biased filter coefficient vector results and convergence deteriorates for $L \approx m$ [Haykin 1996]. Furthermore, the *partitioned* FDAF does not converge well without the gradient constraint. The *soft-partitioned* FDAF [Enzner, Vary 2003a] provides a good compromise between computational complexity, speed of convergence, and roundoff noise in fix-point implementations.

### 14.10.2.3 Improved Stepsize Control

In addition to computational advantages, the frequency domain approach provides the possibility to control the stepsize parameter $\beta$ as a function of frequency and time, in order to positively influence the convergence behavior during double talk phases ($s(k) \neq 0$), for additive interference ($n(k) \neq 0$), and for time variance of the LEM impulse response. It can be shown [Nitsch 2000] that by minimizing the average convergence state the optimal stepsize is given as a function of frequency by

$$\beta_\mu(\kappa) = \frac{\alpha_\mu(\kappa)}{\mathrm{E}\left\{|X_\mu(\kappa)|^2\right\}} = \frac{\mathrm{E}\left\{|E_\mu(\kappa)|^2\right\}}{\mathrm{E}\left\{|\hat{S}_\mu(\kappa)|^2\right\}}\,\frac{1}{\mathrm{E}\left\{|X_\mu(\kappa)|^2\right\}}. \tag{14.164}$$

The power spectra $\mathrm{E}\left\{|\hat{S}_\mu(\kappa)|^2\right\}$ and $\mathrm{E}\{|X_\mu(\kappa)|^2\}$ are determined from the signals $\hat{s}(k)$ and $x(k)$, respectively. The residual echo power spectrum $\mathrm{E}\left\{|E_\mu(\kappa)|^2\right\}$ or, with $\mathrm{E}\left\{|E_\mu(\kappa)|^2\right\} = |D_\mu(\kappa)|^2\mathrm{E}\left\{|X_\mu(\kappa)|^2\right\}$, the convergence state $|D_\mu(\kappa)|^2$ of the adaptive filter must be estimated as outlined, for example, in [Mader et al. 2000], [Enzner et al. 2002], and [Hänsler, Schmidt 2004]. A simple and robust method for the continuous estimation of the convergence state, which does not need double talk detection mechanisms, is proposed in [Enzner, Vary 2003b].

### 14.10.3 Subband Acoustic Echo Cancellation

Besides the FDAF approach, subband acoustic echo cancellation is another widely used frequency domain method. In subband acoustic echo cancellation, we use a digital filter bank (e.g., QMF or PPN, see Chapter 4) instead of the DFT or FFT e.g., [Kellermann

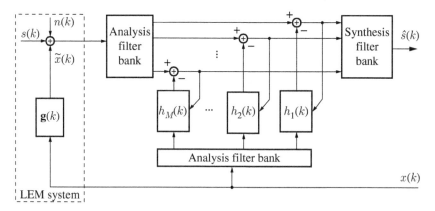

**Figure 14.27** Subband acoustic echo cancellation.

1985], [Kellermann 1988]. The far-end speaker's signal $x(k)$ and the microphone signal $y(k)$ are separated by a filter bank into $M$ subband signals with a reduced sampling rate (Figure 14.27). In each subband an individual echo canceller is utilized, which due to the downsampling has a correspondingly shortened impulse response in comparison to the full-band echo canceller. The individual cancellation filters are adapted with, for instance, the NLMS algorithm. By means of a synthesis filter bank, the compensated subband signals are interpolated and superimposed. As the spectrum of each individual subband signal is relatively flat within the respective frequency band, a favorable convergence behavior results.

In practical applications, *oversampled* filter banks are preferred, e.g., [Kellermann 1985]. Due to the inevitable spectral overlaps of adjacent channels of the critically sampled filter bank, the sampling rate reduction must be selected to be smaller than $M$, e.g., $r_0 = M/2$. Compared the critically sampled filter bank, this increases the effort for the subband cancellers, whose impulse responses then contain approximately $m' = m/r_0 = 2m/M$ coefficients. We refer to the literature [Kellermann 1985], [Kellermann 1989], and [Shynk 1992] for further details. Critically sampled filter banks and IIR filter banks are possible as well but then additional cross-channel filters [Gilloire, Vetterli 1992] or notch filters [Naylor et al. 1998] must be employed to cancel aliasing components.

### 14.10.4 Echo Canceller with Adaptive Postfilter in the Frequency Domain

The adaptive postfilter can be realized in the frequency domain as well, and can then be combined with measures for noise reduction [Faucon, Le Bouquin-Jeannès 1995], [Ayad et al. 1996], [Martin, Vary 1996], [Le Bouquin-Jeannès et al. 2001], [Enzner, Vary 2003b], [Enzner, Vary 2006], [Hänsler, Schmidt 2004], [Enzner 2006]. Also, the frequency domain implementation allows to compute psychoacoustic masking thresholds and thus provides means for a more accurate computation of filter weights. A psychoacoustically motivated approach along these lines has been proposed in [Gustafsson et al. 2002].

In principle, any of the well-known noise reduction methods can also be employed for combined acoustic echo and noise reduction. For instance, the Wiener filter solution to the postfilter problem minimizes the MSE

$$E\{(\tilde{s}(k) - s(k))^2\}$$

under a linear filtering constraint and also provides an interesting perspective on the relation of echo cancellation and residual echo suppression [Hänsler, Schmidt 2000], [Hänsler,
Schmidt 2004], [Enzner et al. 2002].

In the frequency domain, the Wiener filter results in a frequency response for the residual
echo suppression filter

$$C_\mu(\kappa) = \frac{E\{|S_\mu(\kappa)|^2\}}{E\{|\hat{S}_\mu(\kappa)|^2\}} = \frac{E\{|S_\mu(\kappa)|^2\}}{E\{|S_\mu(\kappa)|^2\} + E\{|N_\mu(\kappa)|^2\} + E\{|E_\mu(\kappa)|^2\}}$$

$$C_\mu(\kappa) = \frac{\dfrac{E\{|S_\mu(\kappa)|^2\}}{E\{|N_\mu(\kappa)|^2\} + E\{|E_\mu(\kappa)|^2\}}}{1 + \dfrac{E\{|S_\mu(\kappa)|^2\}}{E\{|N_\mu(\kappa)|^2\} + E\{|E_\mu(\kappa)|^2\}}}, \tag{14.165}$$

provided that $s(k)$, $n(k)$, and $e(k)$ are statistically independent. Compared to the standard
noise reduction case, we now have to account for the disturbing residual echo whose
power spectrum $E\{|E_\mu(\kappa)|^2\}$ must be estimated from the available signals. In analogy to
the Wiener filter for noise reduction, the filter $C_\mu(\kappa)$ can be controlled by the a priori SNR

$$\text{SNR}_\mu = \frac{E\{|S_\mu(\kappa)|^2\}}{E\{|N_\mu(\kappa)|^2\} + E\{|E_\mu(\kappa)|^2\}}.$$

In [Gustafsson et al. 2002] it is proposed to estimate the a priori SNR separately with
respect to $E\{|N_\mu(\kappa)|^2\}$ and $E\{|E_\mu(\kappa)|^2\}$ and to combine it as

$$\frac{E\{|S_\mu(\kappa)|^2\}}{E\{|N_\mu(\kappa)|^2\} + E\{|E_\mu(\kappa)|^2\}} = \frac{1}{\left[\frac{E\{|S_\mu(\kappa)|^2\}}{E\{|N_\mu(\kappa)|^2\}}\right]^{-1} + \left[\frac{E\{|S_\mu(\kappa)|^2\}}{E\{|E_\mu(\kappa)|^2\}}\right]^{-1}}, \tag{14.166}$$

where for each term in the denominator on the right hand side of (14.166) a *decision-directed*
[Ephraim, Malah 1984] estimator can be used.

When the ambient near-end noise is negligible, i.e., $E\{|N_\mu(\kappa)|^2\} = 0$, we may write the
Wiener filter for residual echo suppression as

$$C_\mu(\kappa) = \frac{E\{|\hat{S}_\mu(\kappa)|^2\} - E\{|E_\mu(\kappa)|^2\}}{E\{|\hat{S}_\mu(\kappa)|^2\}}. \tag{14.167}$$

In conjunction with the optimal stepsize parameter of the FDAF (14.164), it is now straightforward to show that

$$\alpha_\mu(\kappa) + C_\mu(\kappa) = 1. \tag{14.168}$$

Thus, it turns out that the control of the FDAF-based echo canceller and of the Wiener postfilter are closely coupled. In both cases the power spectrum of the residual echo is the most
critical control parameter. The residual echo power, however, depends on the magnitude

squared frequency response $|D_\mu(\kappa)|^2$ of the distance vector $\mathbf{d}(k)$ which is not directly mea-surable. However, an efficient statistical approach for the estimation of this quantity is outlined in [Enzner, Vary 2003b], [Enzner, Vary 2006]. In this work, a synergy of FDAF, optimal stepsize control, Wiener postfilter, and convergence state estimation is established on the basis of Kalman filter theory.

### 14.10.5 Initialization with Perfect Sequences

The convergence behavior, in the sense of a fast system equalization, can be improved if a suitable auxiliary signal is applied in the initialization phase of the canceller.

In [Antweiler 1995], [Antweiler 2008], so-called *perfect sequences* [Lüke 1988], [Lüke, Schotten 1995], [Ipatov 1979] are proposed for this purpose. It is shown that, for an undis-turbed adaptation, i.e., for $s(k) = 0$ and $n(k) = 0$, the NLMS algorithm converges in only $m$ steps, and thus exactly identifies the impulse response of the LEM system.

Access to this solution is given by the geometric interpretation of the NLMS algorithm according to Figure 14.17. The adaptation algorithm (14.79f) shortens the system distance vector $\mathbf{d}(k)$ by subtracting the parallel component $\alpha\, \mathbf{d}^{\|}(k)$. With an undisturbed adaptation, the normalized stepsize parameter can be set to $\alpha = 1$. As a result, the component of the distance vector $\mathbf{d}(k)$, which is parallel to the vector $\mathbf{x}(k)$ is completely eliminated.

With the assumption that all $m$ successive state vectors $\mathbf{x}(k), \mathbf{x}(k-1), \ldots, \mathbf{x}(k+m-1)$ are orthogonal in the $m$-dimensional vector space, the complete identification of the unknown LEM system can be achieved in $m$ steps. Periodically applied perfect sequences $p(\kappa)$ ($\kappa = 0, 1, \ldots, m-1$) fulfill the requirements of an optimal excitation signal with

$$x(\lambda \cdot m + \kappa) = p(\kappa); \quad \lambda \in \mathbb{Z}, \tag{14.169}$$

since they are characterized by their periodic auto-correlation function $\tilde{\varphi}_{pp}(i)$, which van-ishes for all out-of-phase values

$$\tilde{\varphi}_{pp}(i) = \varphi_{xp}(i) = \sum_{\kappa=0}^{m-1} p(\kappa)x(k+i) \tag{14.170a}$$

$$= \sum_{\kappa=0}^{m-1} p(\kappa)x(\lambda \cdot m + \kappa + i) \tag{14.170b}$$

$$= \begin{cases} \tilde{\varphi}_{pp}(0) & i \bmod m = 0 \\ 0 & \text{otherwise.} \end{cases} \tag{14.170c}$$

All $m$ phases of the perfect sequences are thus ideally orthogonal in the $m$-dimensional vector space.

The state vector $\mathbf{x}(k)$ meets the orthogonality requirement for $k \geq m$, as it only contains a complete period of the sequence $p(\kappa)$ from this time instant on ($\lambda \geq 1$ in (14.169)).

The convergence behavior for a perfect sequence excitation is illustrated as an example in Figure 14.28, and compared to the adaptation using speech or a white noise signal.

The simulation confirms the behavior to be expected from the geometric interpretation. In the initialization phase, the complete identification of the LEM system (within compu-tational precision) takes $2m$ iterations. After the initialization, during the runtime of the

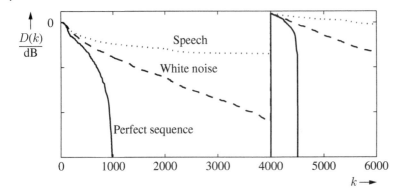

**Figure 14.28** System distance for different excitation signals $x(k)$; undisturbed adaptation, change of the LEM system at $k = 4000$ ($\alpha = 1$; $s(k) = 0$; $n(k) = 0$; $x(k) = p(k_{|\mathrm{mod}\ m}))$; $m = m' = 500$, [Antweiler 1995], odd–perfect sequence $p(\kappa)$ [Lüke, Schotten 1995].

simulation, $m$ iterations are already sufficient for a new equalization; for example, after a sudden change of the time-variant room (see Figure 14.28 at $k = 4000$).

Since white noise fulfills the orthogonality requirement only after statistical averaging, the algorithm converges more slowly under excitation with a white noise signal than with a perfect sequence.

The most frequently used, so-called *odd–perfect* sequences [Lüke, Schotten 1995] are symmetrical, quasi-binary sequences, which, except for a (leading) zero, only take two values $p(\kappa)\in\{+a, -a\}$, $\kappa = 1, 2, \ldots, m - 1$. As the period length must be adapted to the length of the cancellation filter, it is a particular advantage that odd perfect sequences can be generated for every length $m = p^K + 1$ with a prime number $p > 2$, $K \in \mathbb{N}$.

In practice, it is sufficient to apply a few periods of the perfect sequence to the system, only in the initialization phase or following strong changes of the room impulse response. In any case, the power of the perfect sequence must be carefully controlled in order not to disturb the near-end listeners.

## 14.11 Stereophonic Acoustic Echo Control

Multi-channel sound transmission provides spatial realism and is of importance for many applications, such as teleconferencing and multimedia systems. When many different talkers are involved, a realistic rendering of the acoustic scene provides valuable cues about the activity of the various talkers and thus contributes to the naturalness of the presentation. The simplest case of multi-channel reproduction, i.e., stereophonic reproduction, will be discussed in greater detail below. While early work focused on pseudo-stereo systems [Minami 1987], we will consider here sound rendering with two arbitrary loudspeaker signals. In particular, we will discuss the relation between the cross-correlation of the two loudspeaker signals and the performance of the echo canceller. The more general case of a larger number of reproduction and recording channels is treated for example in [Benesty, Morgan 2000b] and [Buchner, Kellermann 2001].

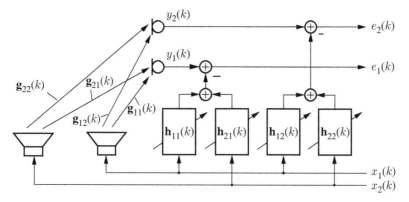

**Figure 14.29** Basic block diagram of stereo acoustic echo cancellation (near-end side).

In the stereophonic transmission set-up the basic echo cancellation model in Figure 14.3 must be extended since each microphone picks up sound from two loudspeakers. Hence, we have to identify two acoustic paths per microphone and will in general need two cancellers per microphone. This is shown in Figure 14.29, where two acoustic paths $g_{11}(k)$ and $g_{21}(k)$ contribute to the microphone signal $y_1(k)$ and two cancellers with impulse responses $h_{11}(k)$ and $h_{21}(k)$ are used.

Typically, the two loudspeaker signals, which are recorded at the far-end side and are transmitted to the near-end side originate from the same sources such as far-end speakers or far-end loudspeakers. In this case, the received loudspeaker signals $x_1(k)$ and $x_2(k)$ in Figure 14.29 can be written as a sum of the convolution of $S$ far-end source signals $s_\ell(k)$, $\ell = 1, \dots, S$, with the impulse responses $f_{1\ell}(k)$ and $f_{2\ell}(k)$ of the acoustic system at the far-end,

$$x_1(k) = \sum_{\ell=1}^{S} \sum_{i=0}^{m_f'-1} f_{1\ell}(i)\, s_\ell(k-i) = \sum_{\ell=1}^{S} f_{1\ell}(k) * s_\ell(k), \tag{14.171}$$

$$x_2(k) = \sum_{\ell=1}^{S} \sum_{i=0}^{m_f'-1} f_{2\ell}(i)\, s_\ell(k-i) = \sum_{\ell=1}^{S} f_{2\ell}(k) * s_\ell(k), \tag{14.172}$$

where $*$ denotes the linear convolution.

When only a single source $s_\ell(k)$ is active we may convolve (14.171) by $f_{2\ell}(k)$ and (14.172) by $f_{1\ell}(k)$ and obtain

$$f_{2\ell}(k) * x_1(k) = f_{2\ell}(k) * f_{1\ell}(k) * s_\ell(k) \tag{14.173}$$

$$f_{1\ell}(k) * x_2(k) = f_{1\ell}(k) * f_{2\ell}(k) * s_\ell(k). \tag{14.174}$$

For time-invariant far-end responses it follows that

$$f_{2\ell}(k) * x_1(k) = f_{1\ell}(k) * x_2(k). \tag{14.175}$$

Hence, $x_1(k)$ and $x_2(k)$ are linearly related. However, as shown below, a large amount of correlation of the loudspeaker signals is detrimental to the fast convergence of a stereophonic echo canceller [Sondhi et al. 1995], [Benesty et al. 1998], [Gänsler, Benesty 2000].

### 14.11.1 The Non-uniqueness Problem

As a consequence of the linear relation of the two loudspeaker signals $x_1(k)$ and $x_2(k)$ in the above scenario, the minimization of the power of the cancellation error signals does not lead to an unambiguous identification of the near-end acoustic system. For the first microphone in Figure 14.29, we obtain the error signal

$$e_1(k) = y_1(k) - x_1(k) * h_{11}(k) - x_2(k) * h_{21}(k). \tag{14.176}$$

However, since $f_{21}(k) * x_1(k) - f_{11}(k) * x_2(k) = 0$ there is a non-uniqueness problem. If the adaptation algorithm would adjust the impulse responses, e.g., as

$$h_{11}(k) = g_{11}(k) - bf_{21}(k) \tag{14.177}$$

$$h_{21}(k) = g_{21}(k) + bf_{11}(k) \tag{14.178}$$

with arbitrary $b \in \mathbb{R}$, we would observe a zero error signal

$$e_1(k) = y_1(k) - x_1(k) * (g_{11}(k) - bf_{21}(k)) - x_2(k) * (g_{21}(k) + bf_{11}(k)),$$

despite the failed system identification. Therefore, the minimization of $E\{e_1^2(k)\}$ cannot result in a unique solution for $h_{11}(k)$ and $h_{21}(k)$.

Moreover, the cancellation error depends on the impulse responses $f_{1\ell}(k)$ and $f_{2\ell}(k)$ of the far-end side. Any change of the far-end source position or alternating far-end speakers will have an immediate effect on the error signal and thus on the convergence of the coefficient vectors. Even in fairly stationary conditions and wideband excitation signals the error signal might be small without proper identification of the near-end acoustic paths. A small error signal, however, does not help in the adaptation of the coefficient vectors.

### 14.11.2 Solutions to the Non-uniqueness Problem

The non-uniqueness can be resolved if the linear relationship between the loudspeaker signals is weakened. In the simplest case, this might be achieved by adding independent white noise to the loudspeaker signals on the near-end side. Improved solutions use spectrally shaped noise to hide the noise below the masked threshold of the audio signal [Gilloire, Turbin 1998].

Another possibility to reduce the correlation of the loudspeaker signals is to use time-varying allpass filters [Ali 1998] or a non-linear processor [Benesty et al. 1998] such as

$$\tilde{x}_1(k) = x_1(k) + \frac{\alpha}{2} \left( x_1(k) + |x_1(k)| \right)$$

$$\tilde{x}_2(k) = x_2(k) + \frac{\alpha}{2} \left( x_2(k) - |x_2(k)| \right). \tag{14.179}$$

Adding a half-wave rectified version of the signal to itself will result in distortions of the signal. Because of the harmonic structure of voiced speech and simultaneous masking effects,

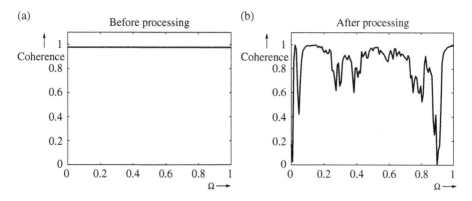

**Figure 14.30** Magnitude squared coherence of two linearly related loudspeaker signals (a) before application (b) after the application of the nonlinear processor (14.179) with $\alpha = 0.5$.

these distortions are hardly noticeable for small values of $\alpha < 0.5$. This method has proven to be very effective without disturbing stereo perception [Gänsler, Benesty 2000].

To illustrate the effect of this non-linear processor on the correlation, we depict the magnitude-squared coherence function

$$C(\Omega) = \frac{|\Phi_{x_1 x_2}(\Omega)|^2}{\Phi_{x_1 x_1}(\Omega)\,\Phi_{x_2 x_2}(\Omega)}, \tag{14.180}$$

of the loudspeaker signals in Figure 14.30 before and after applying the non-linear processor to a monophonic speech signal. For the processed signals, we clearly observe a reduction of the coherence.

Algorithms for the adaptation of the stereophonic echo canceller may be designed by extending the single channel techniques to the multiple channel case [Benesty et al. 1995], [Shimauchi, Makino 1995]. When the loudspeaker signals are independent, any of the extended single-channel echo cancellation algorithms can be successfully applied. As pointed out above, difficulties arise when the two loudspeaker signals are linearly related. Fast convergence is only achieved when the loudspeaker signals are not fully correlated, and when the adaptive algorithms take the correlation of these signals into account. Therefore, multi-channel RLS-type algorithms converge much faster as they mutually decorrelate the input signals. Standard NLMS-type algorithms converge much slower in general since, in most practical situations, the signal covariance matrix is not well conditioned and the amount of additively independent noise or non-linear distortions that can be applied is limited. For the derivation of a multi-channel NLMS algorithm, the gradient computation can be modified to exploit the cross-channel correlation [Benesty et al. 1996]. The multi-channel RLS and FDAF algorithms are especially useful in this context [Benesty, Morgan 2000a]. Efficient implementations of such algorithms have been considered in [Buchner, Kellermann 2001].

## References

Ali, M. (1998). Stereophonic Acoustic Echo Cancellation System Using Time-varying All-pass Filtering for Signal Decorrelation, *Proceedings of the IEEE International Conference on Acoustics, Speech, and Signal Processing (ICASSP)*, vol. 6, pp. 3689–3692.

Antweiler, C. (1995). *Orthogonalizing Algorithms for Digital Compensation of Acoustic Echoes*, PhD thesis. *Aachener Beiträge zu digitalen Nachrichtensystemen*, vol. 1, P. Vary (ed.), RWTH Aachen University (in German).

Antweiler, C. (2008). Multi-Channel System Identification with Perfect Sequences, *in* R. Martin; U. Heute; C. Antweiler (eds.), *Advances in Digital Speech Transmission*, John Wiley & Sons, Ltd., Chichester, pp. 171–198.

Antweiler, C.; Antweiler, M. (1995). System Identification with Perfect Sequences Based on the NLMS Algorithm, *Archiv für Elektronik und Übertragungstechnik*, vol. 49, no. 3, pp. 129–134.

Antweiler, C.; Dörbecker, M. (1994). Perfect Sequence Excitation of the NLMS Algorithm and its Application to Acoustic Echo Control, *Annales des Télécommunication*, vol. 49, no. 7–8, pp. 386–397.

Armbrüster, W. (1988). High Quality Hands-Free Telephony Using Voice Switching Optimized with Echo Cancellation, *Proceedings of the European Signal Processing Conference (EUSIPCO)*, pp. 495–498.

Ayad, B.; Faucon, G.; Le Bouquin-Jeannès, R. (1996). Optimization of a Noise Reduction Preprocessing in an Acoustic Echo and Noise Controller, *Proceedings of the IEEE International Conference on Acoustics, Speech, and Signal Processing (ICASSP)*, pp. 953–956.

Benesty, J.; Morgan, D. R. (2000a). Frequency-Domain Adaptive Filtering Revisited, Generalization to the Multi-Channel Case, and Application to Acoustic Echo Cancellation, *Proceedings of the IEEE International Conference on Acoustics, Speech, and Signal Processing (ICASSP)*, vol. 2, pp. 789–792.

Benesty, J.; Morgan, D. R. (2000b). Multi-Channel Frequency-Domain Adaptive Filtering, *in* S. L. Gay, J. Benesty (eds.), *Acoustic Signal Processing for Telecommunication*, Kluwer Academic, Dordrecht, pp. 121–133.

Benesty, J.; Amand, F.; Gilloire, A.; Grenier, Y. (1995). Adaptive Filtering Algorithms for Stereophonic Acoustic Echo Cancellation, *Proceedings of the IEEE International Conference on Acoustics, Speech, and Signal Processing (ICASSP)*, pp. 3099–3102.

Benesty, J.; Duhamel, P.; Grenier, Y. (1996). Multi-Channel Adaptive Filtering Applied to Multi-Channel Acoustic Echo Cancellation, *Proceedings of the European Signal Processing Conference (EUSIPCO)*, pp. 1405–1408.

Benesty, J.; Morgan, D. R.; Sondhi, M. M. (1998). A Better Understanding and an Improved Solution to the Specific Problems of Stereophonic Acoustic Echo Cancellation, *IEEE Transactions on Speech and Audio Processing*, vol. 6, pp. 156–165.

Benesty, J.; Gänsler, T.; Morgan, D. R.; Sondhi, M. M.; Gay, S. L. (2001). *Advances in Network and Acoustic Echo Cancellation*, Springer-Verlag, Berlin, Heidelberg, New York.

Breining, C.; Dreiseitel, P.; Hänsler, E.; Mader, A.; Nitsch, B.; Puder, H.; Schertler, T.; Schmidt, G.; Tilp, J. (1999). Acoustic Echo Control, *IEEE Signal Processing Magazine*, vol. 16, no. 4, pp. 42–69.

Buchner, H.; Kellermann, W. (2001). Acoustic Echo Cancellation for Two and More Reproduction Channels, *Proceedings of the International Workshop on Acoustic Echo and Noise Control (IWAENC)*, pp. 99–102.

Carini, A. (2001). The Road of an Acoustic Echo Controller for Mobile Telephony from Product Definition till Production, *Proceedings of the International Workshop on Acoustic Echo and Noise Control (IWAENC)*, pp. 5–9.

Claasen, T. A. C. M.; Mecklenbräuker, W. F. G. (1981). Comparison of the Convergence of Two Algorithms for Adaptive FIR Digital Filters, *IEEE Transactions on Acoustics, Speech and Signal Processing*, vol. 29, no. 3, pp. 670–678.

Clark, G. A.; Mitra, S. K.; Parker, S. R. (1981). Block Implementation of Adaptive Digital Filters, *IEEE Transactions on Acoustics, Speech and Signal Processing*, vol. 29, no. 3, pp. 744–752.

Clark, G. A.; Parker, S. R.; Mitra, S. K. (1983). A Unified Approach to Time- and Frequency-Domain Realization of FIR Adaptive Digital Filters, *IEEE Transactions on Acoustics, Speech and Signal Processing*, vol. 31, no. 5, pp. 1073–1083.

Debnath, L.; Mikusinski, P. (1999). *Introduction to Hilbert Spaces with Applications*, 2nd edn, Academic Press, San Diego, California.

Enzner, G. (2006). *A Model-Based Optimum Filtering Approach to Acoustic Echo Control: Theory and Practice*, ISBN = 3-861-30648-4. *Aachener Beiträge zu digitalen Nachrichtensystemen*, vol. 22, P. Vary (ed.), RWTH Aachen University.

Enzner, G. (2008). Kalman Filtering in Acoustic Echo Control: A Smooth Ride on a Rocky Road, *in* R. Martin; U. Heute; C. Antweiler (eds.), *Advances in Digital Speech Transmission*, John Wiley & Sons, Chichester, pp. 79–106.

Enzner, G.; Vary, P. (2003a). A Soft-Partitioned Frequency-Domain Adaptive Filter for Acoustic Echo Cancellation, *Proceedings of the IEEE International Conference on Acoustics, Speech, and Signal Processing (ICASSP)*, vol. V, pp. 393–395.

Enzner, G.; Vary, P. (2003b). Robust and Elegant, Purely Statistical Adaptation of Acoustic Echo Canceler and Postfilter, *Proceedings of the International Workshop on Acoustic Echo and Noise Control (IWAENC)*, pp. 43–46.

Enzner, G.; Vary, P. (2006). Frequency-Domain Adaptive Kalman Filter for Acoustic Echo Control in Handsfree Telephones, *Signal Processing*, vol. 86, no. 6, pp. 1140–1156. Special Issue on Applied Speech and Audio Processing.

Enzner, G.; Martin, R.; Vary, P. (2002). Partitioned Residual Echo Power Estimation for Frequency-Domain Acoustic Echo Cancellation and Postfiltering, *European Transactions on Telecommunications*, vol. 13, no. 2, pp. 103–114.

Enzner, G.; Buchner, H.; Favrot, A.; Kuech, F. (2014). Acoustic Echo Control, Chapter 4, *in* J. Trussell; A. Srivastava; A. K. Roy-Chowdhury; A. Srivastava; P. A. Naylor; R. Chellappa; S. Theodoridis (eds.), *Academic Press Library in Signal Processing*, vol. 4, *Academic Press Library in Signal Processing*, Elsevier, pp. 807–877.

Ephraim, Y.; Malah, D. (1984). Speech Enhancement Using a Minimum Mean-Square Error Short-Time Spectral Amplitude Estimator, *IEEE Transactions on Acoustics, Speech and Signal Processing*, vol. 32, no. 6, pp. 1109–1121.

Fabry, J.; Kühl, S.; Jax, P. (2020). On the Steady State Performance of the Kalman Filter Applied to Acoustical Systems, *IEEE SPL*, vol. 27, no. 1070–9908, pp. 1854–1858.

Faucon, G.; Le Bouquin-Jeannès, R. (1995). Joint System for Acoustic Echo Cancellation and Noise Reduction, *Proceedings of the European Conference on Speech Communication and Technology (EUROSPEECH)*, pp. 1525–1528.

Ferrara, E. R. (1980). Fast Implementation of LMS Adaptive Filters, *IEEE Transactions on Acoustics, Speech and Signal Processing*, vol. 28, no. 4, pp. 474–475.

Ferrara, E. R. (1985). *Frequency-Domain Adaptive Filtering, in* C. F. N Cowan, P. M. Grant (eds.) *Adaptive Filters*, Prentice-Hall, Englewood Cliffs, New Jersey, pp. 145–179.

Gänsler, T.; Benesty, J. (2000). Stereophonic Acoustic Echo Cancellation and Two-Channel Adaptive Filtering: An Overview, *International Journal of Adaptive Control and Signal Processing*, vol. 14, pp. 565–586.

Gay, S. L.; Benesty, J. (eds.) (2000). *Acoustic Signal Processing for Telecommunication*, Kluwer Academic, Dordrecht.

Gay, S.; Tavathia, S. (1995). The Fast Affine Projection Algorithm, *Proceedings of the IEEE International Conference on Acoustics, Speech, and Signal Processing (ICASSP)*, vol. 5, pp. 3023–3026.

Gierlich, H. W. (1992). A Measurement Technique to Determine the Transfer Characteristics of Hands-Free Telephones, *Signal Processing*, vol. 27, pp. 281–300.

Gierlich, H. W.; Kettler, F. (2005). Advanced Speech Quality Testing of Modern Telecommunication Equipment: An Overview, *Signal Processing*, vol. 86, no. 6, pp. 1327–1340. Special Issue on Speech and Audio Processing (to appear).

Gilloire, A. (1994). Performance Evaluation of Acoustic Echo Control: Required Values and Measurement Procedures, *Annales des Télécommunications*, vol. 49, no. 7–8, pp. 368–372.

Gilloire, A.; Turbin, V. (1998). Using Auditory Properties to Improve the Behavior of Stereophonic Acoustic Echo Cancellers, *Proceedings of the IEEE International Conference on Acoustics, Speech, and Signal Processing (ICASSP)*, pp. 3681–3684.

Gilloire, A.; Vetterli, M. (1992). Adaptive Filtering in Subbands with Critical Sampling: Analysis, Experiments, and Application to Acoustic Echo Cancellation, *IEEE Transactions on Signal Processing*, vol. 40, no. 8, pp. 1862–1875.

Gustafsson, S.; Martin, R.; Vary, P. (1998). Combined Acoustic Echo Control and Noise Reduction for Hands-Free Telephony, *Signal Processing*, vol. 64, pp. 21–32.

Gustafsson, S.; Martin, R.; Jax, P.; Vary, P. (2002). A Psychoacoustic Approach to Combined Acoustic Echo Cancellation and Noise Reduction, *IEEE Transactions on Speech and Audio Processing*, vol. 10, no. 5, pp. 245–256.

Hänsler, E.; Schmidt, G. (2000). Hands-Free Telephones – Joint Control of Echo Cancellation and Postfiltering, *Signal Processing*, vol. 80, no. 11, pp. 2295–2305.

Hänsler, E.; Schmidt, G. (2004). *Acoustic Echo and Noise Control – A Practical Approach*, John Wiley & Sons, Ltd., Chichester.

Haubner, T.; Kellermann, W. (2022). Deep Learning-Based Joint Control of Acoustic Echo Cancellation, Beamforming and Postfiltering, *Proceedings of the 30th European Signal Processing Conference (EUSIPCO)*, Belgrade, Serbia, 2022, pp. 752–756.

Haykin, S. (1996). *Adaptive Filter Theory*, 3rd edn, Prentice Hall, Upper Saddle River, New Jersey.

Ipatov, V. P. (1979). Ternary Sequences with Ideal Perfect Autocorrelation Properties, *Radio Engineering Electronics and Physics*, vol. 24, pp. 75–79.

ITU-T G.131 (2003). *Talker Echo and its Control*, International Telecommunication Union (ITU).

ITU-T P.340 (2000). *Transmission Characteristics and Speech Quality Parameters of Hands-free Terminals*, International Telecommunication Union (ITU).

Kalman, R. (1960). A New Approach to Linear Filtering and Prediction Problems, *Journal of Basic Engineering*, vol. 82, no. 1, pp. 35–45.

Kellermann, W. (1985). Kompensation akustischer Echos in Frequenzteilbändern, *Frequenz*, vol. 39, no. 7/8, pp. 209–215 (in German).

Kellermann, W. (1988). Analysis and Design of Multirate Systems for Cancellation of Acoustical Echoes, *Proceedings of the IEEE International Conference on Acoustics, Speech, and Signal Processing (ICASSP)*, pp. 2570–2573.

Kellermann, W. (1989). *Zur Nachbildung physikalischer Systeme durch parallelisierte digitale Ersatzsysteme im Hinblick auf die Kompensation akustischer Echos*, PhD thesis, *Fortschrittsberichte VDI*, Series 10: Informatik/Kommunikationstechnik, Nr. 102, TU Darmstadt (in German).

Kettler, F.; Gierlich, H.-W. (2008). *Evaluation of Hands-Free Terminals*, Springer-Verlag, Berlin, Heidelberg, pp. 339–377.

Kühl, S.; Antweiler, C.; Hübschen, T.; Jax, P. (2017). Kalman filter based system identification exploiting the decorrelation effects of linear prediction, *Proceedings of the IEEE International Conference on Acoustics, Speech, and Signal Processing (ICASSP)*, IEEE, pp. 4790–4794.

Kuttruff, H. (2017). *Room Acoustics*, 6th edn, CRC Press.

Kühl, S. (2022). *Adaptive Algorithms for the Identification of Time-Variant Acoustic Systems*, PhD thesis, RWTH Aachen University.

Le Bouquin-Jeannès, R.; Scalart, P.; Faucon, G.; Beaugeant, C. (2001). Combined Noise and Echo Reduction in Hands-Free Systems: A Survey, *IEEE Transactions on Speech and Audio Processing*, vol. 9, no. 8, pp. 808–820.

Lüke, H. D. (1988). Sequences and Arrays with Perfect Periodic Correlation, *IEEE Transactions on Aerospace and Electronic Systems*, vol. 24, no. 3, pp. 287–294.

Lüke, H. D.; Schotten, H. (1995). Odd-perfect, Almost Binary Correlation Sequences, *IEEE Transactions on Aerospace and Electronic Systems*, vol. 31, pp. 495–498.

Mader, A.; Puder, H.; Schmidt, G. (2000). Step-Size Control for Acoustic Echo Cancellation Filters – An Overview, *Signal Processing*, vol. 80, no. 9, pp. 1697–1719.

Makino, S.; Kaneda, Y. (1992). Exponentially Weighted Step-Size Projection Algorithm for Acoustic Echo Cancellers, *IEICE Transactions on Fundamentals of Electronics, Communications and Computer Sciences*, vol. E75-A, no. 11, pp. 1500–1508.

Mandic, D.; Kanna, S.; Constantinides, A. (2015). On the Intrinsic Relationship Between the Least mean Square and Kalman Filters, *IEEE Signal Processing Magazine*, vol. 32, no. 6, pp. 117–122.

Mansour, D.; Gray, A. H. (1982). Unconstrained Frequency-Domain Adaptive Filters, *IEEE Transactions on Acoustics, Speech and Signal Processing*, vol. 30, pp. 726–734.

Martin, R. (1995). *Hands-free Systems with Multi-channel Echo Cancellation and Noise Reduction*, PhD thesis. *Aachener Beiträge zu digitalen Nachrichtensystemen*, vol. 3, P. Vary (ed.), RWTH Aachen University (in German).

Martin, R.; Altenhöner, J. (1995). Coupled Adaptive Filters for Acoustic Echo Control and Noise Reduction, *Proceedings of the IEEE International Conference on Acoustics, Speech, and Signal Processing (ICASSP)*, pp. 3043–3046.

Martin, R.; Gustafsson, S. (1996). The Echo Shaping Approach to Acoustic Echo Control, *Speech Communication*, vol. 20, pp. 181–190.

Martin, R.; Vary, P. (1996). Combined Acoustic Echo Control and Noise Reduction for Hands-Free Telephony - State of the Art and Perspectives, *Proceedings of the European Signal Processing Conference (EUSIPCO)*, pp. 1107–1110.

Minami, S. (1987). An Acoustic Echo Canceller for Pseudo Stereophonic Voice, *Globecom*, pp. 1355–1360.

Myllylä, V. (2001). Robust Fast Affine Projection Algorithm for Acoustic Echo Cancellation, *Proceedings of the International Workshop on Acoustic Echo and Noise Control (IWAENC)*.

Myllylä, V.; Schmidt, G. (2002). Pseudo-Optimal Regularization for Affine Projection Algorithms, *Proceedings of the IEEE International Conference on Acoustics, Speech, and Signal Processing (ICASSP)*, vol. II, pp. 1917–1920.

Naylor, P. A.; Tanrikulu, O.; Constantinides, A. G. (1998). Subband Adaptive Filtering for Acoustic Echo Control Using Allpass Polyphase IIR Filterbanks, *IEEE Transactions on Speech and Audio Processing*, vol. 6, no. 2, pp. 143–155.

Nitsch, B. (1997). The Partitioned Exact Frequency Domain Block NLMS Algorithm, *Proceedings of the International Workshop on Acoustic Echo and Noise Control (IWAENC)*, pp. 45–48.

Nitsch, B. H. (2000). A Frequency-Selective Stepfactor Control for an Adaptive Filter Algorithm Working in the Frequency Domain, *Signal Processing*, vol. 80, pp. 1733–1745.

Ozeki, K.; Umeda, T. (1984). An Adaptive Filtering Algorithm Using an Orthogonal Projection to an Affine Subspace and its Properties, *Electronics and Communications in Japan*, vol. 67-A, no. 5, pp. 19–27.

Paleologu, C.; Benesty, J.; Ciochin?, S. (2013). Study of the General Kalman Filter for Echo Cancellation, *IEEE Transactions on Audio, Speech, and Language Processing*, vol. 21, no. 8, pp. 1539–1549.

Petillon, T.; Gilloire, A.; Theodoridis, S. (1994). The Fast Newton Transversal Filter: An Efficient Scheme for Acoustic Echo Cancellation in Mobile Radio, *IEEE Transactions on Signal Processing*, vol. 42, no. 3, pp. 509–518.

Rupp, M. (1993). The Behavior of LMS and NLMS Algorithms in the Presence of Spherically Invariant Processes, *IEEE Transactions on Signal Processing*, vol. 41, no. 3, pp. 1149–1160.

Schultheiß, U. (1988). *Über die Adaption eines Kompensators für akustische Echos*, PhD thesis, *Fortschrittsberichte VDI*, Series 10: Informatik/Kommunikationstechnik, Nr. 90, TU Darmstadt (in German).

Shimauchi, S.; Makino, S. (1995). Stereo Projection Echo Canceller with True Echo Path Estimation, *Proceedings of the IEEE International Conference on Acoustics, Speech, and Signal Processing (ICASSP)*, pp. 3059–3062.

Shynk, J. (1992). Frequency-Domain and Multirate Adaptive Filtering, *IEEE Signal Processing Magazine*, vol. 9, no. 1, pp. 14–37.

Slock, D. T. M. (1993). On the Convergence Behavior of the LMS and the Normalized LMS Algorithms, *IEEE Transactions on Signal Processing*, vol. 41, no. 1, pp. 2811–2825.

Slock, D. T. M.; Kailath, T. (1991). Numerically Stable Fast Transversal Filters for Recursive Least Squares Adaptive Filtering, *IEEE Transactions on Signal Processing*, vol. 39, no. 1, pp. 92–113.

Sommen, P. C. W. (1989). Partitioned Frequency Domain Adaptive Filters, *Proceedings of the 23rd Asilomar Conference on Signals, Systems, and Computers*, pp. 677–681.

Sommen, P. C. W.; van Valburg, C. J. (1989). Efficient Realisation of Adaptive Filter Using an Orthogonal Projection Method, *Proceedings of the IEEE International Conference on Acoustics, Speech, and Signal Processing (ICASSP)*, pp. 940–943.

Sondhi, M. M.; Morgan, D. R.; Hall, J. L. (1995). Stereophonic Acoustic Echo Cancellation – An Overview of the Fundamental Problem, *Signal Processing Letters*, vol. 2, pp. 148–151.

Soo, J. S.; Pang, K. K. (1990). Multidelay Block Frequency Domain Adaptive Filter, *IEEE Transactions on Acoustics, Speech and Signal Processing*, vol. 38, no. 2, pp. 373–376.

Tanaka, M.; Kaneda, Y.; Makino, S.; Kojima, J. (1995). Fast Projection Algorithm and Its Step Size Control, *Proceedings of the IEEE International Conference on Acoustics, Speech, and Signal Processing (ICASSP)*, pp. 945–948.

van de Kerkhof, L. M.; Kitzen, W. J. W. (1992). Tracking of a Time-Varying Acoustic Impulse Response by an Adaptive Filter, *IEEE Transactions on Signal Processing*, vol. 40, no. 7, pp. 1285–1294.

Vary, P.; Martin, R. (2006). *Digital Speech Transmission: Enhancement, Coding and Error Concealment*, John Wiley & Sons.

Yamamoto, S.; Kitayama, S. (1982). An Adaptive Echo Canceller with Variable Step Gain Method, *Transactions of the IECE of Japan*, vol. E65, pp. 1–8.

# 15

# Microphone Arrays and Beamforming

In this chapter, we will study the question of how sound pickup in a noisy and reverberant environment can be improved by adding spatial diversity to the signal acquisition front-end. Spatially distributed receivers, i.e., microphone arrays, and multichannel signal processing and machine learning techniques allow the exploitation of spatial, temporal and spectral features of signals. They achieve a performance, which surpasses that of single microphone systems.

## 15.1   Introduction

When more than one microphone is available for sound pickup, the signal enhancement task may be facilitated by exploiting the multivariate deterministic and stochastic properties of these signals. From a deterministic viewpoint, the signals at the various microphones differ in that they arrive via different acoustic paths at the microphones and thus also differ in their spectral amplitudes and phases. From a stochastic perspective, multichannel methods allow the evaluation of the second-order and higher-order statistics of the spatial sound field. Sources, which are close to the array in a reverberant room will generate mostly coherent microphone signals while distant and distributed sources lead to less correlated signals. Thus, the spectral amplitude and phase and the statistics of the signals may be used to differentiate between sources and to perform source separation.

Array technology has been used in radar and sonar systems for quite some time [Monzingo, Miller 1980], [Haykin 1985], [Gabriel 1992]. Frequently, these systems are designed for spectrally narrow signals. The application of array technology to speech signals can thus be more challenging, as speech is a wideband signal spanning several octaves. Furthermore, in many speech processing applications the environment is highly reverberant and a multitude of acoustic sources may be present. As a consequence, the desired signal will arrive not only from one primary source direction but also via reflections from the enclosing walls, and interfering signals may have directional as well as diffuse components.

In this chapter, we first develop the basic scenario and define signal models and performance measures. We will then consider microphone arrays and their properties for stationary environments, i.e. with fixed beam patterns. We explain typical design

*Digital Speech Transmission and Enhancement*, Second Edition. Peter Vary and Rainer Martin.
© 2024 John Wiley & Sons Ltd. Published 2024 by John Wiley & Sons Ltd.

procedures for beamformers, where we assume that the direction of incidence of the source signal is given and that the sound field is stationary. Finally, in Section 15.7, we briefly discuss postfilter techniques and adaptive beamforming approaches. This topic and further extensions toward source separation are treated in more detail in, e.g., Brandstein, Ward [2001], Gannot et al. [2017].

## 15.2 Spatial Sampling of Sound Fields

For our purposes, microphones may be modeled as discrete points in space at whose location the spatial sound pressure field is sampled. Unless explicitly stated, we assume that the microphones are omnidirectional (sound pressure) receivers, i.e. they have the same sensitivity regardless of the direction of the impinging sound.

### 15.2.1 The Near-field Model

Figure 15.1 illustrates the general scenario of a single sound source and a distributed microphone array. The position of the source and the positions of the $N_M$ microphones with respect to a reference coordinate system are denoted by vectors $\mathbf{r}_s$ and $\mathbf{r}_\ell$, $\ell = 1, \ldots, N_M$, respectively. $\mathbf{e}_x$, $\mathbf{e}_y$, and $\mathbf{e}_z$ denote orthogonal unit vectors spanning a Cartesian coordinate system.

In an *anechoic* environment (no reverberation or noise), the analog microphone signals $y_{a,\ell}(t)$, $\ell = 1, \ldots, N_M$, are delayed and attenuated versions of the analog source signal $s_{a,0}(t)$,

$$y_{a,\ell}(t) = \frac{1}{\|\mathbf{r}_\ell - \mathbf{r}_s\|} s_{a,0}(t - \tau_\ell),$$

(15.1)

where $s_0(t)$ originates from a point source at $\mathbf{r}_s$ and $\|\mathbf{r}_s\|$ denotes the norm of vector $\mathbf{r}_s$. The absolute propagation delay $\tau_\ell$ of the $\ell$-th microphone signal is given by

$$\tau_\ell = \frac{\|\mathbf{r}_\ell - \mathbf{r}_s\|}{c}.$$

(15.2)

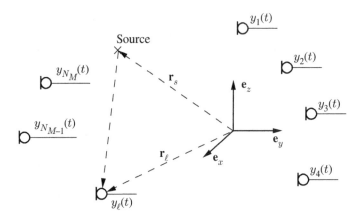

**Figure 15.1** Microphone array in the near-field of a single source. $\mathbf{e}_x$, $\mathbf{e}_y$, and $\mathbf{e}_z$ are orthogonal unit vectors spanning a Cartesian coordinate system.

We denote the signal, which is received at the origin of the coordinate system by $s_a(t)$. $\tau_0$ is the signal delay from the source to the origin. The origin of the coordinate system is henceforth referred to as the *reference point*. Then, we have

$$s_a(t) = \frac{1}{\|\mathbf{r}_s\|} s_{a,0}(t - \tau_0),$$ (15.3)

where the reference point may not coincide with the source location.

The reference point could be the geometric center of the array or, for convenience, the location of one of the microphones. Frequently, we are not interested in the absolute delay of the signal from the source to the microphones but rather in the delay of the signals relative to the signal, which is received at the reference point. The *relative signal delay* $\Delta\tau_\ell$, i.e., the time delay difference between the received signal at the reference point and at the $\ell$-th microphone, is then given by

$$\Delta\tau_\ell = \tau_0 - \tau_\ell = \frac{1}{c}\left(\|\mathbf{r}_s\| - \|\mathbf{r}_\ell - \mathbf{r}_s\|\right).$$ (15.4)

Thus, in an anechoic environment the $\ell$-th microphone signal $y_{a,\ell}(t)$ may now be written as a function of the signal at the reference point,

$$y_{a,\ell}(t) = \frac{\|\mathbf{r}_s\|}{\|\mathbf{r}_\ell - \mathbf{r}_s\|} s_a(t + \Delta\tau_\ell).$$ (15.5)

If the Fourier transform $S(j\omega)$ of $s_a(t)$ exists, the Fourier transform $Y_\ell(j\omega)$ of the microphone signals $y_{a,\ell}(t)$ may be written as

$$
\begin{aligned}
Y_\ell(j\omega) &= \frac{\|\mathbf{r}_s\|}{\|\mathbf{r}_\ell - \mathbf{r}_s\|} S(j\omega)\exp\left(j2\pi f\Delta\tau_\ell\right)\\
&= \frac{\|\mathbf{r}_s\|}{\|\mathbf{r}_\ell - \mathbf{r}_s\|} S(j\omega)\exp\left(j\tilde{\beta}\left(\|\mathbf{r}_s\| - \|\mathbf{r}_\ell - \mathbf{r}_s\|\right)\right),
\end{aligned}
$$ (15.6)

where $\tilde{\beta} = \frac{2\pi}{\lambda} = \frac{2\pi f}{c}$ is the wave number and $\lambda$ is the corresponding wave length. In the above model, no assumptions about the distance between the source and the array were made. The source may be arbitrarily close to the microphones. Therefore, this model is denoted as the *near-field* model.

### 15.2.2 The Far-field Model

We now assume that the distance between the sound source and the microphone array is much larger than the largest dimension (the *aperture*) of the array and much larger than the wavelength, i.e. $\tilde{\beta}\|\mathbf{r}_\ell - \mathbf{r}_s\| \gg 1, \forall\ell$. In this case, the sound waves, which are picked up by the microphones may be modeled as plane waves. This *far-field* scenario is illustrated in Figure 15.2. When the source is far from the array, then

$$\frac{\|\mathbf{r}_s\|}{\|\mathbf{r}_1 - \mathbf{r}_s\|} \approx \frac{\|\mathbf{r}_s\|}{\|\mathbf{r}_2 - \mathbf{r}_s\|} \approx \cdots \approx \frac{\|\mathbf{r}_s\|}{\|\mathbf{r}_{N_M} - \mathbf{r}_s\|}.$$ (15.7)

The absolute and relative attenuation of the source signal at the microphones are approximately the same for all microphones. The phase differences between the microphone signals, however, depend on the distance between the microphones and the wavelength of the impinging wave.

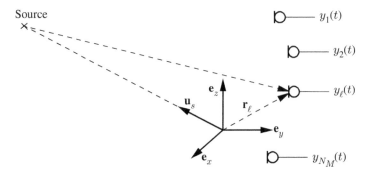

**Figure 15.2** Microphone array in the far-field of a single source.

For the single source far-field scenario, when the reference point is close to the microphones, $\|\mathbf{r}_\ell\| \ll \|\mathbf{r}_s\|$, we may write

$$y_{a,\ell}(t) \approx s_a(t + \Delta\tau_\ell) \approx s_a\left(t + \frac{\langle \mathbf{r}_\ell, \mathbf{u}_s \rangle}{c}\right), \tag{15.8}$$

where $\mathbf{u}_s$ is a unit vector, which points from the reference point toward the source, i.e. $\mathbf{u}_s = \mathbf{r}_s/\|\mathbf{r}_s\|$, and $\langle \mathbf{r}_\ell, \mathbf{u}_s \rangle$ denotes the inner product of vectors $\mathbf{r}_\ell$ and $\mathbf{u}_s$. For $\|\mathbf{r}_s\| < \|\mathbf{r}_\ell - \mathbf{r}_s\|$, this inner product is negative. Then, the reference point is closer to the source than the $\ell$-th microphone, as shown in Figure 15.2. In Cartesian coordinates, we may write the inner product as

$$y_{a,\ell}(t) \approx s_a\left(t + \frac{\mathbf{r}_\ell^T \mathbf{u}_s}{c}\right). \tag{15.9}$$

The relative delays depend on the array geometry and the components of $\mathbf{r}_\ell$ in the direction of $\mathbf{u}_s$.

In the frequency domain, and for the far-field model, the microphone signals can then be expressed as

$$Y_\ell(j\omega) = S(j\omega)\exp(j2\pi f\Delta\tau_\ell), \tag{15.10}$$

where we assume that the Fourier transform of signal $S(j\omega)$ exists. When the microphone signals are sampled with sampling rate $f_s$, we may write the signal spectra of the sampled (and bandlimited) microphone signals $y_\ell(k)$ as a function of the normalized frequency $\Omega$ as[1]

$$Y_\ell(e^{j\Omega}) = S(e^{j\Omega})\exp\left(j\Omega f_s\Delta\tau_\ell\right), \tag{15.11}$$

or, in vector notation,

$$\mathbf{Y}(e^{j\Omega}) = S(e^{j\Omega})\mathbf{a}^*. \tag{15.12}$$

with a vector of signal spectra

$$\mathbf{Y}(e^{j\Omega}) = \left(Y_1(e^{j\Omega}), \dots, Y_{N_M}(e^{j\Omega})\right)^T \tag{15.13}$$

---

1 For ease of notation, we use the same symbols for continuous-time and discrete-time variables in the Fourier domain.

and, in correspondence to the notation in (15.12), the *propagation vector*

$$\mathbf{a}^* = \left( \exp\left(-j\Omega f_s \Delta\tau_1\right), \exp\left(-j\Omega f_s \Delta\tau_2\right), \ldots, \exp\left(-j\Omega f_s \Delta\tau_{N_M}\right) \right)^T. \tag{15.14}$$

Then, in the far-field scenario, $\mathbf{a}^T \mathbf{Y}(e^{j\Omega})$ will yield the sum of perfectly phase aligned microphone signals.

### 15.2.3  Sound Pickup in Reverberant Spaces

In a reverberant space, the microphones will pick up not only the direct sound from the source but also sound from wall reflections and late reverberation. Then, the single-source signal model may be extended toward

$$y_{a,\ell}(t) = \int_0^\infty g_{a,\ell}(\tau) s_{a,0}(t - \tau) \, d\tau + v_{a,\ell}(t), \tag{15.15}$$

where $g_{a,\ell}(\tau)$ denotes the (analog) *room impulse response* (RIR) from the source to the $\ell$-th microphone and $v_{a,\ell}(t)$ is any additional ambient or microphone noise. In the discrete-time domain we then have

$$y_\ell(k) = \sum_{\kappa=0}^\infty g_\ell(\kappa) s_0(k - \kappa) + v_\ell(k), \tag{15.16}$$

and in the frequency domain

$$Y_\ell(e^{j\Omega}) = G_\ell(e^{j\Omega}) S_0(e^{j\Omega}) + V_\ell(e^{j\Omega}). \tag{15.17}$$

Furthermore, as both signals and the acoustic system may be non-stationary, we analyze these in the short-time Fourier transform (STFT) domain. To this end, we take an $M$-point discrete Fourier transform (DFT) on overlapped and windowed signal frames. We thus obtain

$$Y_{\ell,\mu}(\lambda) = G_{\ell,\mu}(\lambda) S_{0,\mu}(\lambda) + V_{\ell,\mu}(\lambda), \tag{15.18}$$

where $\mu$ and $\lambda$ indicate frequency and frame indices. Strictly speaking, the above relation holds true when the transform length $M$ is large enough to accommodate the full extent of the RIR, i.e. a finite impulse response. Usually, this is not given, thus leading to additional disturbance, which may be absorbed in the noise term.

Finally, also in the more general case of reverberant rooms, it is common to relate RIRs and their frequency responses to the response at the reference point. We thus relate the signal model (15.18) to the DFT $G_{0,\mu}(\lambda)$ of the RIR of the reference point. For the signal $s(k)$ at the reference point with $S_\mu(\lambda) = G_{0,\mu}(\lambda) S_{0,\mu}(\lambda)$ we obtain

$$Y_{\ell,\mu}(\lambda) = a_{\ell,\mu}(\lambda) S_\mu(\lambda) + V_{\ell,\mu}(\lambda), \tag{15.19}$$

where the *relative transfer functions*

$$a_{\ell,\mu}(\lambda) = \frac{G_{\ell,\mu}(\lambda)}{G_{0,\mu}(\lambda)} \tag{15.20}$$

may be again summarized in a generalized propagation vector $\mathbf{a}^*$ as in (15.12) and (15.14).

### 15.2.4 Spatial Correlation Properties of Acoustic Signals

Besides the deterministic differences in spectral phase and amplitude, multi-channel signal processing techniques allow to exploit the statistical properties of the microphone signals, especially the spatial correlation of sound fields. In the context of speech enhancement, three cases are of special interest:

1. The microphone signals $y_\ell$ are highly correlated, as considered in the preceding sections. Highly correlated signals occur when the sound source is close to the microphones and the microphone signals contain little noise and reverberation. Then, the direct sound dominates and leads to high and frequency-independent correlation of the microphone signals.
2. The microphone signals $y_\ell$ are uncorrelated. Fully uncorrelated signals originate from microphones and amplifiers as thermal self-noise. Although we may assume that electro-acoustic components are selected such that this noise does not degrade the perceived quality of the microphone signals, we must be aware that self-noise might be amplified by the multi-channel signal processing algorithm.
3. The microphone signals $y_\ell$ exhibit a frequency-dependent correlation. A typical example of frequency-dependent correlation is the ideal diffuse sound field, see Section 13.2. In this sound field, sound energy impinges with equal intensity from all spatial directions onto the microphone array. For omnidirectional receivers, the magnitude squared coherence can then be written as a squared sinc function (13.14). This type of correlation is encountered in reverberant spaces with multiple, distributed noise sources, such as car compartments or noisy offices.

### 15.2.5 Uniform Linear and Circular Arrays

So far, no specific array geometry has been assumed. Figure 15.3 depicts two important special cases: the linear and circular arrays, both with $N_M$ uniformly spaced microphones in the far-field of an acoustic source. The reference point is in the geometric center of the arrays. The microphones of the uniform linear array (ULA) are symmetrically aligned with an inter-microphone distance $d$ along the $z$-axis. Because of the symmetric constellation, it is convenient to introduce polar coordinates. We define an azimuth $\varphi_s$ and an elevation $\theta_s$, which characterize the direction of the desired sound source. Due to the symmetry and orientation of the ULA, the relative signal delays and thus, the array response, do not depend on the azimuth $\varphi_s$.

For the ULA in Cartesian coordinates, the relative delays are given by $\Delta\tau_\ell = \pm \mathbf{e}_z^T \mathbf{u}_s \|\mathbf{r}_\ell\|/c$, or, using just the source elevation $\theta_s$, by

$$\Delta\tau_\ell = \frac{d}{c}\left(\frac{N_M+1}{2}-\ell\right)\cos(\theta_s), \quad \ell = 1, 2, \ldots, N_M. \tag{15.21}$$

When the elevation $\theta_s$ is zero (or $\pi$) we have an *endfire* orientation of the array with respect to the source. Then, the *look direction* of the array coincides with the positive (or negative) $z$-axis and $\Delta\tau_\ell$ attains its maximum absolute value. In *broadside* orientation and a source in the far-field, the look direction is perpendicular to the $z$-axis, i.e. $\theta_s = \pi/2$ and $\Delta\tau_\ell = 0$ for all $\ell$.

(a)                                    (b)

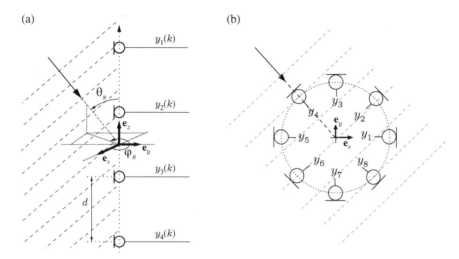

**Figure 15.3** Uniform linear array (ULA) with $N_M = 4$ microphones (a) and uniform circular array (UCA) in the $x - y$ plane with $N_M = 8$ microphones (b).

For the UCA in Figure 15.3, the angular microphone positions (measured counter-clockwise from the positive $x$-axis) are given by $\phi_\ell = 2\pi(\ell - 1)/N_M$ and the propagation vector for a source in the $x - y$ plane by

$$\mathbf{a} = \left( e^{j\Omega f_s r \cos(\varphi_s - \phi_1)/c}, e^{j\Omega f_s r \cos(\varphi_s - \phi_2)/c}, \ldots, e^{j\Omega f_s r \cos(\varphi_s - \phi_{N_M})/c} \right)^H, \tag{15.22}$$

where $r$ denotes the radius of the array and $\varphi_s$ is the angle towards the source (measured counter-clockwise from the positive $x$-axis).

### 15.2.6  Phase Ambiguity in Microphone Signals

The sound pressure field $p(\mathbf{r}, t)$ is a function of space and time. The microphone array samples this wave field at discrete points in space. In analogy to the sampling theorem of discrete time signals, the spatial sampling of the sound field must obey a spatial sampling relation if *spatial aliasing* is to be avoided. The analysis of a sound field by means of a microphone array can lead to ambiguous results if the spatial and temporal sampling constraints are not properly taken into account.

Since the complex exponential in (15.10) is periodic, integer increments in $f\Delta\tau_\ell$ lead to the same value of $\exp\left(j2\pi f\Delta\tau_\ell\right)$. Therefore, for a fixed relative delay $\Delta\tau_\ell$, the signal phase will vary as a function of frequency $f$ and, as a result, the complex exponential may attain the same value for multiple frequencies.

For a harmonic plane wave and two closely spaced microphones located along the $z$-axis at

$$\mathbf{r}_1 = \frac{d}{2}\mathbf{e}_z \quad \text{and} \quad \mathbf{r}_2 = -\frac{d}{2}\mathbf{e}_z, \tag{15.23}$$

where $\mathbf{e}_z$ is the unit vector in $z$-direction, the sampling condition can be derived from the phase term $\exp\left(j2\pi f\Delta\tau_\ell\right)$. If we denote the phase of the harmonic wave at the first and the

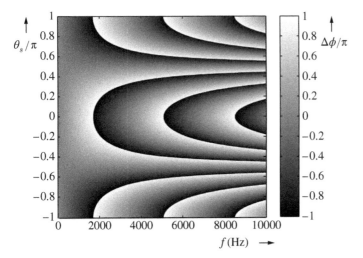

**Figure 15.4** Phase difference $\Delta\phi/\pi$ for $-\pi \leq \theta_s \leq \pi$ of two microphone signals recorded at distance $d = 0.1$ m as a function of frequency $f$ ($c = 340$ m/s).

second microphone by $\phi_1 = \phi + \tilde{\beta} \langle \mathbf{e}_z, \mathbf{u}_s \rangle d/2$ and $\phi_2 = \phi - \tilde{\beta} \langle \mathbf{e}_z, \mathbf{u}_s \rangle d/2$, respectively, the phase difference is given by

$$\Delta\phi = \phi_1 - \phi_2 = \tilde{\beta} d \langle \mathbf{e}_z, \mathbf{u}_s \rangle = \tilde{\beta} d \cos\left(\theta_s\right), \tag{15.24}$$

where $\theta_s$ denotes the angle between the positive $z$-axis and the direction of the sound source. $\tilde{\beta} = 2\pi/\tilde{\lambda}$ is the wave number as before and $\tilde{\lambda}$ is the corresponding wave length. The absolute phase difference is zero when the wave impinges from the broadside direction ($\mathbf{u}_s \perp \mathbf{e}_z$), i.e., $\theta_s = \pm 0.5\pi$. The phase difference is largest when the microphones are in endfire orientation, i.e., $\theta_s \in \{0, \pm\pi\}$ ($\mathbf{u}_s \| \mathbf{e}_z$ or $\mathbf{u}_s \| -\mathbf{e}_z$). Obviously, ambiguous phase values are avoided when, for $\theta_s \in [0, \pi]$, the phase difference $\Delta\phi = \tilde{\beta} d \cos\left(\theta_s\right)$ is in the range $-\pi < \Delta\phi \leq \pi$. Thus, a one-to-one mapping of $\theta_s \in [0, \pi]$ and $\Delta\phi$ is achieved for

$$d \leq \frac{\tilde{\lambda}}{2} \quad \text{or} \quad f \leq \frac{c}{2d}. \tag{15.25}$$

Figure 15.4 plots the phase difference between two microphones as a function of the angle of incidence $-\pi \leq \theta_s \leq \pi$ and frequency $f$. Clearly, for $\theta_s = \pm\pi/2$, the phase difference is zero for all frequencies. Additionally, it can be seen that there is an obvious symmetry for values $-\pi \leq \theta_s \leq 0$ and $\pi \geq \theta_s \geq 0$. A one-to-one mapping of $\theta_s \in [0, \pi]$ and $\Delta\phi$ values is observed below a frequency threshold of $f = c/(2d) \approx 1700$ Hz only.

## 15.3 Beamforming

The task of the beamforming algorithm is to combine the sampled microphone signals such that a desired (and possibly time-varying) spatial selectivity is achieved. In single-source scenarios, this comprises the formation of a beam of high gain in the direction of the desired source and the suppression of all other directions.

### 15.3.1 Delay-and-Sum Beamforming

The simplest method to solve the signal combination problem is to combine the microphone signals such that the desired signal components add up constructively. In an acoustic environment where direct sound is predominant and reverberation plays only a minor role, this phase alignment can be achieved by appropriately delaying the microphone signals. Then, the resulting phase-aligned signals are added to form a single output signal. This is known as the *delay-and-sum* beamformer.

The delay-and-sum beamformer as shown in Figure 15.5 is simple in its implementation and provides for easy steering of the beam toward the desired source. When written as a function of a continuous time argument $t$, the signals picked up by the individual microphones are

$$y_{a,\ell}(t) = s_{a,\ell}(t) + n_{a,\ell}(t) = s_a(t + \Delta\tau_\ell) + n_{a,\ell}(t), \tag{15.26}$$

where $s_a(t)$ is the desired source signal at the reference position and $n_{a,\ell}(t)$ is the noise signal at the $\ell$-th microphone. Again, we assume that the desired source is in the far-field and any signal attenuation is absorbed in $s_a(t)$. We denote the delayed signals by

$$\tilde{y}_{a,\ell}(t) = \tilde{s}_{a,\ell}(t) + \tilde{n}_{a,\ell}(t) = s_a(t + \Delta\tau_\ell - T_\ell) + n_{a,\ell}(t - T_\ell), \tag{15.27}$$

where the delay $T_\ell$ is applied to the $\ell$-th signal. Besides a relative delay, which depends on $\ell$, $T_\ell$ includes a channel-independent delay $T_B$ such that $T_\ell \geq 0$ for any $\Delta\tau_\ell$ and all $\ell$. For a digital implementation and since $T_\ell$ is, in general, not equal to an integer multiple of the sampling period, the accurate compensation of signal delays requires some form of signal interpolation. The interpolation may be implemented by fractional delay filters in the time or the frequency domain [Crochiere, Rabiner 1981], [Crochiere, Rabiner 1983], [Laakso et al. 1996]. In what follows, we will consider only sampled signals and digital beamformer implementations. Therefore, the channel-independent constant delay $T_B$ is chosen to be equal to an integer multiple of the sampling period $1/f_s$.

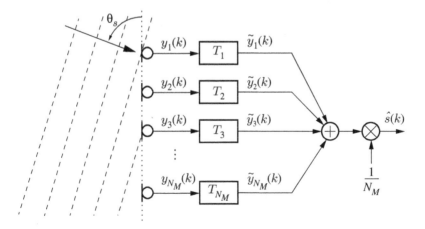

**Figure 15.5** Delay-and-sum beamformer. Note: A/D converters are omitted.

The sampled output signal $\hat{s}(k) = \hat{s}_a(kT)$ of the delay-and-sum beamformer may then be written as t

$$\hat{s}(k) = \frac{1}{N_M}\sum_{\ell=1}^{N_M}\tilde{y}_\ell(k) = \frac{1}{N_M}\sum_{\ell=1}^{N_M}\tilde{s}_\ell(k) + \frac{1}{N_M}\sum_{\ell=1}^{N_M}\tilde{n}_\ell(k), \tag{15.28}$$

where the signals $\tilde{y}_\ell(k)$, $\tilde{s}_\ell(k)$, and $\tilde{n}_\ell(k)$ denote the sampled versions of the delayed microphone, target, and noise signals of the $\ell$-th channel. For a broadside orientation of the array with respect to the source, $T_\ell$ for all $\ell = 1, \ldots, N_M$, and hence $T_B$, are set to zero. Then, the delay-and-sum beamformer comprises a scaled sum of the microphone signals only.

For the far-field scenario and when the delay from the source to the microphones is perfectly equalized, we have $T_\ell = T_B + \Delta\tau_\ell$ and with (15.28)

$$\hat{s}(k) = s(k - T_B f_s) + \frac{1}{N_M}\sum_{\ell=1}^{N_M}\tilde{n}_\ell(k). \tag{15.29}$$

### 15.3.2 Filter-and-Sum Beamforming

A more general processing model is the *filter-and-sum* beamformer, as shown in Figure 15.6 where, before summation, each microphone signal is filtered,

$$\tilde{y}_\ell(k) = \sum_{m=0}^{M}h_\ell(m)y_\ell(k - m). \tag{15.30}$$

We assume that FIR filters of order $M$ are used. To simplify notations, we define a vector of signal samples

$$\mathbf{y}_\ell(k) = (y_\ell(k), y_\ell(k - 1), \ldots, y_\ell(k - M))^T \tag{15.31}$$

and a vector of real-valued filter coefficients

$$\mathbf{h}_\ell = (h_\ell(0), h_\ell(1), \ldots, h_\ell(M))^T. \tag{15.32}$$

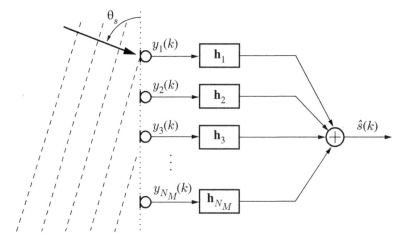

**Figure 15.6** Filter-and-sum beamformer. Note: A/D converters are omitted.

Then, the output signal of the filter-and-sum beamformer may be written as

$$\widehat{s}(k) = \sum_{\ell=1}^{N_M} \mathbf{h}_\ell^T \mathbf{y}_\ell(k).$$
(15.33)

For the signal $s(k)$ of a single (desired) source and the anechoic far-field scenario, we obtain the output signal in the frequency domain (provided all Fourier transforms exist)

$$\widehat{S}(e^{j\Omega}) = \sum_{\ell=1}^{N_M} H_\ell(e^{j\Omega}) Y_\ell(e^{j\Omega})$$

$$= S(e^{j\Omega}) \sum_{\ell=1}^{N_M} H_\ell(e^{j\Omega}) \exp\left(j\Omega f_s \Delta\tau_\ell\right)$$

$$= S(e^{j\Omega}) H(e^{j\Omega}, \mathbf{u}_s).$$
(15.34)

$H_\ell(e^{j\Omega})$ denotes the frequency response of the $\ell$-th FIR filter. Thus, for a fixed source position and fixed microphone positions $\mathbf{r}_\ell$ the array response

$$H(e^{j\Omega}, \mathbf{u}_s) = \sum_{\ell=1}^{N_M} H_\ell(e^{j\Omega}) \exp(j\Omega f_s \Delta\tau_\ell) = \mathbf{a}^H \mathbf{H}(e^{j\Omega})$$
(15.35)

of the filter-and-sum beamformer depends on the propagation vector $\mathbf{a}$ (15.14) of the point source in the far-field and the vector of filter responses

$$\mathbf{H}(e^{j\Omega}) = \left( H_1(e^{j\Omega}), H_2(e^{j\Omega}), \ldots, H_{N_M}(e^{j\Omega}) \right)^T.$$
(15.36)

When a delay-and-sum beamformer is steered towards the source in the far-field, we obtain a special case of the filter-and-sum beamformer with the filter coefficient vector

$$\mathbf{H}_{DSB}(e^{j\Omega}) = \frac{\exp(-j\Omega f_s T_B)}{N_M} \mathbf{a} = \frac{1}{N_M} \mathbf{e},$$
(15.37)

where

$$\mathbf{e} = \left( \exp(-j\Omega f_s T_1), \exp(-j\Omega f_s T_2), \ldots, \exp(-j\Omega f_s T_{N_M}) \right)^T$$
(15.38)

is the *steering vector*.

In the case of random source signals and extending the results of Section 5.8.4, we obtain the power spectrum of the output signal as the Fourier transform of the autocorrelation function

$$\mathrm{E}\left\{ \widehat{s}(k)\widehat{s}(k+k') \right\}$$

$$= \mathrm{E}\left\{ \sum_{\ell=1}^{N_M} \sum_{\ell'=1}^{N_M} \sum_{m=0}^{M} \sum_{m'=0}^{M} h_\ell(m) y_\ell(k-m) y_{\ell'}(k-m'+k') h_{\ell'}(m') \right\}$$

$$= \sum_{\ell=1}^{N_M} \sum_{\ell'=1}^{N_M} \sum_{m=0}^{M} \sum_{m'=0}^{M} h_\ell(m) \varphi_{y_\ell y_{\ell'}}(k'-m'+m) h_{\ell'}(m').$$
(15.39)

Hence,

$$\Phi_{\widehat{s}\widehat{s}}(e^{j\Omega}) = \mathbf{H}^H(e^{j\Omega}) \Phi_{yy}(e^{j\Omega}) \mathbf{H}(e^{j\Omega}),$$
(15.40)

where

$$
\Phi_{yy}(e^{j\Omega}) = \begin{pmatrix} \Phi_{y_1 y_1}(e^{j\Omega}) & \Phi_{y_1 y_2}(e^{j\Omega}) & \cdots & \Phi_{y_1 y_{N_M}}(e^{j\Omega}) \\ \Phi_{y_2 y_1}(e^{j\Omega}) & \Phi_{y_2 y_2}(e^{j\Omega}) & \cdots & \Phi_{y_2 y_{N_M}}(e^{j\Omega}) \\ \vdots & \vdots & \ddots & \vdots \\ \Phi_{y_{N_M} y_1}(e^{j\Omega}) & \Phi_{y_{N_M} y_2}(e^{j\Omega}) & \cdots & \Phi_{y_{N_M} y_{N_M}}(e^{j\Omega}) \end{pmatrix}
\tag{15.41}
$$

is a matrix of cross-power spectra.

When the microphone signals originate from a single point source, we obtain the Fourier transform of the cross-correlation function of the microphone signals

$$
\Phi_{y_\ell y_{\ell'}}(e^{j\Omega}) = \Phi_{ss}(e^{j\Omega}) e^{-j\Omega f_s (\Delta\tau_\ell - \Delta\tau_{\ell'})}
\tag{15.42}
$$

and hence with (15.35)

$$
\Phi_{\hat{s}\hat{s}}(e^{j\Omega}) = \Phi_{ss}(e^{j\Omega}) \mathbf{H}^H(e^{j\Omega}) \mathbf{a}\mathbf{a}^H \mathbf{H}(e^{j\Omega}) = \Phi_{ss}(e^{j\Omega}) |H(e^{j\Omega}, \mathbf{u}_s)|^2.
\tag{15.43}
$$

When the microphones pick up spatially uncorrelated noise and no coherent target signal, the output of the beamformer is given by

$$
\Phi_{\hat{s}\hat{s}}(e^{j\Omega}) = \sum_{\ell=1}^{N_M} |H_\ell(e^{j\Omega})|^2 \Phi_{y_\ell y_\ell}(e^{j\Omega}).
\tag{15.44}
$$

In all the above cases, the design of the filter-and-sum beamformer is reduced to computing filter coefficient vectors $\mathbf{h}_\ell(m)$ such that a performance measure is optimized. Before we investigate these methods, we introduce such measures.

## 15.4 Performance Measures and Spatial Aliasing

### 15.4.1 Array Gain and Array Sensitivity

Microphone arrays are designed to improve the SNR of a desired source signal. The *array gain* characterizes the performance of a microphone array as the ratio of the SNR at the output of the array with respect to the average SNR of the microphone signals. Using the matrix notation of Section 15.3.2 and the trace operator Tr(), the average powers of the desired signals and of the noise signals at the microphones are given by $\mathrm{Tr}(\Phi_{ss}(e^{j\Omega}))/N_M$ and $\mathrm{Tr}(\Phi_{nn}(e^{j\Omega}))/N_M$, respectively, whereas the corresponding powers at the output are given by $\mathbf{H}^H(e^{j\Omega})\Phi_{ss}(e^{j\Omega})\mathbf{H}(e^{j\Omega})$ and $\mathbf{H}^H(e^{j\Omega})\Phi_{nn}(e^{j\Omega})\mathbf{H}(e^{j\Omega})$, respectively. Therefore, the frequency-dependent array gain may be defined as [Herbordt 2005]

$$
G(e^{j\Omega}) = \frac{\mathrm{Tr}(\Phi_{nn}(e^{j\Omega}))}{\mathrm{Tr}(\Phi_{ss}(e^{j\Omega}))} \frac{\mathbf{H}^H(e^{j\Omega})\Phi_{ss}(e^{j\Omega})\mathbf{H}(e^{j\Omega})}{\mathbf{H}^H(e^{j\Omega})\Phi_{nn}(e^{j\Omega})\mathbf{H}(e^{j\Omega})}.
\tag{15.45}
$$

Assuming a far-field scenario with identical speech and identical noise power spectral densities at all microphones and mutually uncorrelated noise signals, we obtain the array gain of the delay-and-sum beamformer with (15.43) and (15.44),

$$
G(e^{j\Omega}) = \frac{N_M \Phi_{nn}(e^{j\Omega})}{N_M \Phi_{ss}(e^{j\Omega})} \frac{\frac{1}{N_M^2} N_M^2 \Phi_{ss}(e^{j\Omega})}{\frac{1}{N_M^2} N_M \Phi_{nn}(e^{j\Omega})} = N_M.
\tag{15.46}
$$

Under the above assumptions, the gain of the delay-and-sum beamformer does not depend on frequency. For partially correlated noise signals $\tilde{n}_\ell(k)$ the improvement can be significantly lower. For example, in the diffuse noise field, the gain is close to zero for frequencies $f < \frac{c}{2d}$ or $\tilde{\lambda} > 2d$, where $d$ is the inter-microphone distance.

Furthermore, we introduce a performance measure, which characterizes the *sensitivity* of the array with respect to spatially and temporally white noise. For instance, this noise can be thermal noise originating from the microphones or the amplifiers. It can also serve as a model for random phase or position errors due to the physical implementation of the array. For a source in the far-field with propagation vector **a** and mutually uncorrelated noise signals with

$$\Phi_{n_\ell n_{\ell'}}(e^{j\Omega}) = \begin{cases} \Phi_{nn}(e^{j\Omega}) & \ell = \ell' \\ 0 & \ell \neq \ell' \end{cases}, \tag{15.47}$$

the array gain is given by

$$G_W(e^{j\Omega}) = \frac{|\mathbf{H}^H(e^{j\Omega})\mathbf{a}|^2}{\mathbf{H}^H(e^{j\Omega})\mathbf{H}(e^{j\Omega})}. \tag{15.48}$$

The inverse of the white noise gain $G_W(e^{j\Omega})$ is called the susceptibility of the array. It characterizes the sensitivity of the array with respect to uncorrelated noise.

### 15.4.2 Directivity Pattern

The spatial selectivity of the array in the far-field of a source is characterized by its *directivity pattern*

$$\Psi(e^{j\Omega}, \mathbf{u}_s) = |H(e^{j\Omega}, \mathbf{u}_s)|^2 = \mathbf{H}^H(e^{j\Omega})\mathbf{a}\mathbf{a}^H\mathbf{H}(e^{j\Omega}) = |\mathbf{H}^H(e^{j\Omega})\mathbf{a}|^2, \tag{15.49}$$

where the vectors $\mathbf{u}_s$ and **a** denote a unit vector in the direction of the source and the propagation vector (15.14) of the impinging sound, respectively. The directivity pattern depicts the power gain from a given direction. It is a useful tool for array performance analysis, especially when sounds propagate coherently and there is no or only little reverberation. Besides frequency and direction of arrival, $\Psi(e^{j\Omega}, \mathbf{u}_s)$ in (15.49) also depends on the inter-microphone distances and the filter coefficients.

As an example, we consider the delay-and-sum beamformer and the far-field scenario. With (15.35) and (15.37), we obtain the power gain directivity pattern of the ULA

$$\Psi(e^{j\Omega}, \mathbf{u}_s) = \begin{cases} \dfrac{\sin^2\left(\pi(\cos(\theta) - \cos(\theta_s))N_M d/\tilde{\lambda}\right)}{N_M^2 \sin^2\left(\pi(\cos(\theta) - \cos(\theta_s))d/\tilde{\lambda}\right)} & \cos(\theta) \neq \cos(\theta_s) \\ 1 & \cos(\theta) = \cos(\theta_s). \end{cases} \tag{15.50}$$

$\theta_s$ denotes the direction of arrival and $\theta$ the look direction of the beamformer. The corresponding directivity pattern of the power gain is shown in Figure 15.7 for the broadside (a) and the endfire (b) orientation. In this example, we use a ULA with $d = 4\,\text{cm}$. The beams in the respective look directions can be clearly recognized. Furthermore, we note that the directivity at low frequencies is not very pronounced, and that for high frequencies, spatial aliasing in the form of side lobes is visible.

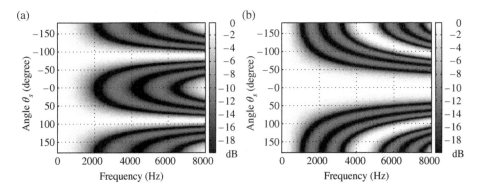

**Figure 15.7** Directivity pattern (power gain) of the ULA with $N_M = 4$ microphones, $d = 4$ cm and delay-and-sum beamformer for the broadside (a) and the endfire (b) orientation.

The directivity pattern (15.50) is a periodic function of the cosine difference of look direction and source direction $\vartheta = (\cos(\theta) - \cos(\theta_s))$ with period $\tilde{\lambda}/d$. For a broadside array $(\cos(\theta) = 0)$, $\vartheta$ attains values between $-1$ and $1$ when $\theta_s$ sweeps from $0$ to $\pi$. Thus, spatial aliasing is avoided when $\tilde{\lambda}/d \geq 1$. For an endfire array, $(\cos(\theta) = 1)$ the same sweep will result in values $0 \leq \vartheta \leq 2$ and therefore in this case $\tilde{\lambda}/d \geq 2$ will avoid spatial aliasing.

In general, $|\cos(\theta) - \cos(\theta_s)| \leq 1 + |\cos(\theta)|$ holds and the above relations can be cast into the more general form [Kummer 1992]:

$$\frac{\tilde{\lambda}}{d} \geq 1 + |\cos(\theta)|, \tag{15.51}$$

describing the range for $\tilde{\lambda}$ without spatial aliasing.

Directivity patterns may also be represented in two- or three-dimensional polar directivity plots. Figure 15.8 depicts two- and three-dimensional polar plots for a filter-and-sum

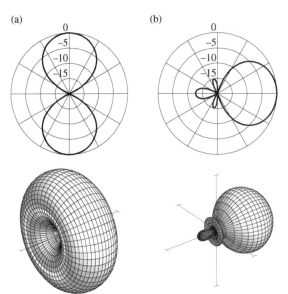

**Figure 15.8** Directivity pattern in dB of a filter-and-sum beamformer and ULA with four omnidirectional microphones, $d = 4.25$ cm and broadside (a) and endfire (b) orientations ($f = 1000$ Hz).

beamformer (see Section 15.5) and a ULA with four microphones. These plots are computed for a frequency of 1000 Hz and for broadside and endfire orientations.

### 15.4.3 Directivity and Directivity Index

The directivity pattern characterizes the performance of the array when directional sources are present. When noise impinges from many different directions, an integral measure is better suited. The *directivity* is defined as the ratio of the directivity pattern in look direction and the directivity pattern averaged over all directions,

$$D(e^{j\Omega}) = \frac{\Psi(e^{j\Omega}, \mathbf{u}_s)}{\frac{1}{4\pi} \int_{A_1} \Psi(e^{j\Omega}, \mathbf{u}) \, da}, \tag{15.52}$$

where $A_1$ denotes the surface of a unit sphere with the array at its center. In polar coordinates, we have $da = \sin(\theta)d\theta d\varphi$ and therefore

$$D(e^{j\Omega}) = \frac{\Psi(e^{j\Omega}, \mathbf{u}_s)}{\frac{1}{4\pi} \int_0^\pi \int_0^{2\pi} \Psi(e^{j\Omega}, \mathbf{u}(\theta, \varphi)) \sin(\theta) \, d\theta d\varphi}. \tag{15.53}$$

The directivity thus characterizes the array's performance for a single desired source and an ideal diffuse noise field. For the far-field scenario, the directivity can be expressed in terms of the array filters

$$D(e^{j\Omega}) = \frac{|\mathbf{H}^H(e^{j\Omega})\mathbf{a}|^2}{\sum_{\ell=1}^{N_M} \sum_{\ell'=1}^{N_M} H_{\ell'}^*(e^{j\Omega}) H_\ell(e^{j\Omega}) \gamma_{n_\ell n_{\ell'}}(e^{j\Omega})}, \tag{15.54}$$

where $\gamma_{n_\ell n_{\ell'}}(e^{j\Omega})$ is the spatial coherence function of the noise signals $n_\ell$ and $n_{\ell'}$ in the ideal diffuse sound field with,

$$\gamma_{n_\ell n_{\ell'}}(e^{j\Omega}) = \frac{1}{4\pi} \int_{A_1} \exp(j\widetilde{\beta} \langle (\mathbf{r}_\ell - \mathbf{r}_{\ell'}), \mathbf{u}(a) \rangle) \, da \tag{15.55}$$

$$= \begin{cases} \dfrac{\sin\left(\widetilde{\beta} d_{\ell,\ell'}\right)}{\widetilde{\beta} d_{\ell,\ell'}} & \ell \neq \ell' \\ 1 & \ell = \ell' \end{cases} = \begin{cases} \dfrac{\sin\left(\Omega f_s d_{\ell,\ell'} c^{-1}\right)}{\left(\Omega f_s d_{\ell,\ell'} c^{-1}\right)} & \ell \neq \ell' \\ 1 & \ell = \ell' \end{cases},$$

where $\mathbf{u}(a)$ is a unit vector pointing towards the infinitesimal area $da$ and $d_{\ell,\ell'}$ represents the distance between the $\ell$-th and the $\ell'$-th microphone.

Figure 15.9 plots the *directivity index* $DI(e^{j\Omega}) = 10\log_{10}\left(D(e^{j\Omega})\right)$ of the same delay-and-sum beamformer as in Figure 15.7 for the ideal diffuse noise field. For low frequencies, the directivity index of the beamformer in endfire orientation is significantly higher than for broadside orientation. For the broadside orientation, the gain does not exceed 3 dB below 2000 Hz.

### 15.4.4 Example: Differential Microphones

Before we discuss more advanced beamformer designs, we consider the simple case of two closely spaced microphones ($d < \widetilde{\lambda}/2$). The two microphone signals $y_1(t)$ and $y_2(t)$ are

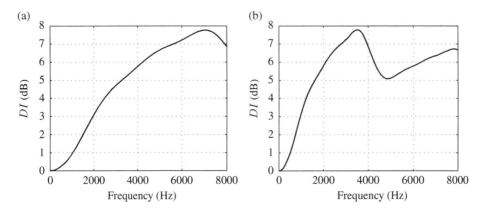

**Figure 15.9** Directivity index of the delay-and-sum beamformer with $N_M = 4$ microphones and $d = 4$ cm for the broadside (a) and endfire (b) orientation.

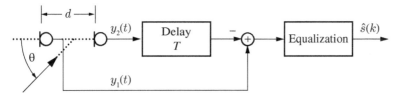

**Figure 15.10** Differential microphones.

combined in a simple delay-and-subtract operation as shown in Figure 15.10. The delay $T$ is sometimes termed *internal delay* as opposed to the external delay $d/c$ of the acoustic path between the two microphones. The array design and the resulting beam patterns are reminiscent of directional (pressure gradient) microphones, where the delay-and-subtract operation is achieved by means of the acoustic design. Typically, the differential microphone array is combined with an equalizer to compensate for undesirable effects of the subtraction on the frequency response.

We now compute the frequency response of the differential microphone array under the far-field assumption. When the reference point is selected to be the mid-point of the line joining the two microphones, the two microphone signals may be written as

$$y_1(t) = s\left(t + \frac{d}{2c}\cos(\theta)\right) \quad \text{and} \tag{15.56}$$

$$y_2(t) = s\left(t - \frac{d}{2c}\cos(\theta)\right). \tag{15.57}$$

Thus, after summation and before equalization we obtain

$$\tilde{s}(t) = s\left(t + \frac{d}{2c}\cos(\theta)\right) - s\left(t - \frac{d}{2c}\cos(\theta) - T\right) \tag{15.58}$$

and in the Fourier transform domain

$$\tilde{S}(j\omega) = S(j\omega) \cdot \left( e^{j\frac{\omega d}{2c}\cos(\theta)} - e^{-j\omega\left(\frac{d}{2c}\cos(\theta) + T\right)} \right) \tag{15.59}$$

$$= S(j\omega) \cdot e^{-j\omega\frac{T}{2}} \cdot \left( e^{j\frac{\omega d}{2c}\left(\cos(\theta) + \frac{cT}{d}\right)} - e^{-j\frac{\omega d}{2c}\left(\cos(\theta) + \frac{cT}{d}\right)} \right). \tag{15.60}$$

The magnitude of the frequency response can be written as

$$|H(j\omega)| = \left| \frac{\tilde{S}(j\omega)}{S(j\omega)} \right| = 2 \cdot \left| \sin\left( \frac{\omega d}{2c}\left(\cos(\theta) + \frac{cT}{d}\right) \right) \right|. \tag{15.61}$$

With the assumption that $\frac{\omega d}{2c} \ll \frac{\pi}{2}$, which is equivalent to $2d \ll \tilde{\lambda} = \frac{c}{f}$, and with $\frac{cT}{d} \lesssim 1$ we may use the approximation $\sin(\theta) \approx \theta$ and therefore,

$$\left| \frac{\tilde{S}(j\omega)}{S(j\omega)} \right| \approx 2 \cdot \left| \frac{\omega d}{2c}\left(\cos(\theta) + \frac{cT}{d}\right) \right| = \left| \frac{\omega d}{c}\left(\cos(\theta) + \frac{cT}{d}\right) \right|. \tag{15.62}$$

With the substitution

$$T = \frac{d}{c}\frac{\beta}{1 - \beta} \quad \text{or} \quad \beta = \frac{T}{T + \frac{d}{c}} = \frac{Tc}{Tc + d} < 1 \tag{15.63}$$

we may write the magnitude response as

$$\left| \frac{\tilde{S}(j\omega)}{S(j\omega)} \right| = \left| \frac{\omega d}{c}\left(\cos(\theta) + \frac{\beta}{1 - \beta}\right) \right| \tag{15.64}$$

$$= \frac{1}{1 - \beta}\frac{d}{c} \cdot \left| \omega \cdot \left( (1 - \beta) \cdot \cos(\theta) + \beta \right) \right|. \tag{15.65}$$

Since the magnitude of the approximate frequency response increases linearly with $\omega$, the frequency response of the array corresponds to a first-order differentiator. Therefore, a first-order lowpass filter may be used for equalization. This, however, might amplify low-frequency noise. Table 15.1 lists the parameter $\beta$ for several directional characteristics and the resulting average directivity index. The resulting polar directivity patterns are illustrated in Figure 15.11.

The differential microphone array can easily be extended into an adaptive null-steering array. Assuming a look direction $\theta \in [-\pi/2, \pi/2]$ and combining, e.g., the omni-directional

**Table 15.1** Parameters and average directivity index $DI_{av}$ of the differential microphones for $d = 0.015$ m.

| Characteristic | $T\frac{c}{d}$ | $\beta$ | $DI_{av}$ / dB |
|---|---|---|---|
| Dipole | 0 | 0 | 4.7 |
| Cardioid | 1 | 0.5 | 4.8 |
| Super-cardioid | 0.57 | 0.3631 | 5.7 |
| Hyper-cardioid | 0.34 | 0.2537 | 6.0 |

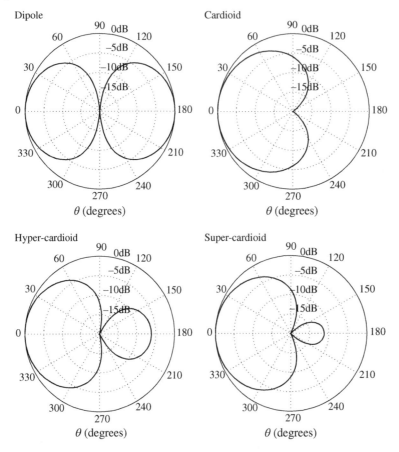

**Figure 15.11** Polar directivity pattern of the differential arrays for $d = 0.015$ m, $f = 1$ kHz and several values of $\beta$.

and dipole patterns, a null of the combined beampattern can be steered towards a disturbing signal source in the range $\theta \in [\pi/2, 3\pi/2]$ [Elko, Nguyen Pong 1997]. Furthermore, differential beamformers have been extended to planar arrays of any geometry and to higher-order differential designs [Huang et al. 2018], [Huang et al. 2020a], [Huang et al. 2020b].

## 15.5 Design of Fixed Beamformers

When the wavelength is much larger than the array aperture, the delay-and-sum beamformer delivers only little directivity. By optimizing the coefficients of a filter-and-sum beamformer, the directivity of a beamformer at low frequencies can be significantly improved. In this section, we will examine frequently used beamformer designs, the *minimum variance distortionless response* (MVDR) beamformer, the linearly-constrained minimum variance beamformer, and the *maximum SNR* beamformer, in greater detail.

## 15.5.1 Minimum Variance Distortionless Response Beamformer

The MVDR beamformer minimizes the output noise variance while constraining the array to have unit gain in look direction. To derive the optimal beamformer coefficients, we write the noise power spectral density at the output of the filter-and-sum beamformer as

$$\Phi_{\hat{n}\hat{n}}(e^{j\Omega}) = \mathbf{H}^H(e^{j\Omega})\Phi_{nn}(e^{j\Omega})\mathbf{H}(e^{j\Omega}) \tag{15.66}$$

where $\Phi_{nn}(e^{j\Omega})$ denotes the power spectral density matrix of the noise signals $n_\ell(k)$, $\ell = 1, \dots, N_M$. The distortionless response constraint is written as

$$\mathbf{a}^H\mathbf{H}(e^{j\Omega}) = 1 \Leftrightarrow \mathbf{H}^H(e^{j\Omega})\mathbf{a} = 1, \tag{15.67}$$

where $\mathbf{a}$ is the propagation vector as before. Note that this (complex-valued) constraint corresponds to two real-valued constraints. To improve the readability of the following derivations, we will drop the dependency on frequency $\Omega$.

Since the source signal is not distorted, the MVDR beamformer also maximizes

$$\tilde{D} = \frac{|\mathbf{H}^H\mathbf{a}|^2}{\mathbf{H}^H\Phi_{nn}\mathbf{H}} \tag{15.68}$$

for a source in the far-field of the array and thus the directivity when $\Phi_{nn}$ is the noise power matrix of the ideal diffuse noise field.

To solve the constrained optimization problem, we use Lagrange's method with a complex multiplier $q = q_1 + jq_2$ and minimize

$$\mathfrak{L}(\mathbf{H}, q) = \mathbf{H}^H\Phi_{nn}\mathbf{H} + \text{Re}\{q^*(\mathbf{H}^H\mathbf{a} - 1)\} \tag{15.69}$$

$$= \mathbf{H}^H\Phi_{nn}\mathbf{H} + q_1\left(\frac{1}{2}(\mathbf{H}^H\mathbf{a} + \mathbf{a}^H\mathbf{H}) - 1\right) + q_2\left(\frac{1}{2j}(\mathbf{H}^H\mathbf{a} - \mathbf{a}^H\mathbf{H})\right).$$

Computing the derivatives with respect to $\mathbf{H}$ and $q^*$ results in the necessary conditions [Kreutz-Delgado 2009]

$$\mathbf{H}^H\Phi_{nn} + \frac{q}{2}\mathbf{a}^H = \mathbf{0} \quad \text{and} \quad \mathbf{H}^H\mathbf{a} = 1. \tag{15.70}$$

Solving the first condition for $\mathbf{H}^H$ and using the result in the second condition we obtain

$$\mathbf{H} = -\frac{1}{2}q\Phi_{nn}^{-1}\mathbf{a} \quad \text{and} \quad q = \frac{-2}{\mathbf{a}^H\Phi_{nn}^{-1}\mathbf{a}}, \tag{15.71}$$

and finally

$$\mathbf{H}_{\text{MVDR}} = \frac{\Phi_{nn}^{-1}\mathbf{a}}{\mathbf{a}^H\Phi_{nn}^{-1}\mathbf{a}}. \tag{15.72}$$

Besides the dependency on frequency, the optimal solution is a function of the spectral correlation matrix of the noise and of the propagation vector (15.14).

Figure 15.12 shows the directivity pattern and the gain of an MVDR beamformer for the broadside (a) and endfire (b) orientations. The array geometry is the same as in Figure 15.7. The array is optimized for ideal diffuse noise field. We observe significantly increased directivity at low frequencies. For high frequencies, the performance of the MVDR beamformer is close to the performance of the delay-and-sum beamformer.

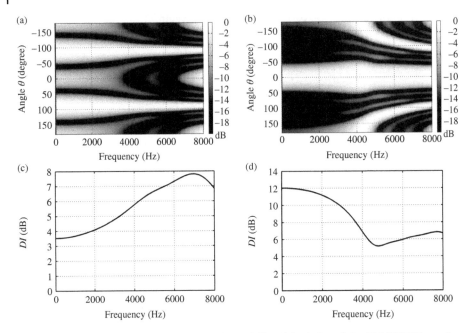

**Figure 15.12** Directivity pattern (power gain) and directivity index of the ULA MVDR beamformer with $N_M = 4$ microphones and $d = 4$ cm for the broadside (a and c) and endfire (b and d) orientation optimized for the ideal diffuse noise field.

As a special case, we consider mutually uncorrelated noise signals with identical power spectra. Then, the spectral noise power matrix is a diagonal matrix, and the optimal filter vector (15.72) evaluates to

$$\mathbf{H} = \frac{\mathbf{a}}{N_M},$$ (15.73)

which is the delay-and-sum beamforming solution for the design condition (15.67) and when the channel-independent delay $T_B$ is neglected.

The high directivity of the MVDR beamformer at low frequencies also results in a high susceptibility to uncorrelated noise. In order to minimize the noise power in each frequency bin, the MVDR beamformer cancels out correlated noise components. The distortionless response, however, can be only maintained if the response of individual filters $H_\ell(e^{j\Omega})$ is much larger than unity for low frequencies. The high gain of the individual filters at low frequencies leads to an amplification of uncorrelated noise in the microphone signals. For high frequencies and in an ideal diffuse noise field, the MVDR beamformer approaches the delay-and-sum solution. For the delay-and-sum beamformer, we obtain (see (15.48))

$$\frac{1}{G_W} = \mathbf{H}^H \mathbf{H} = \frac{1}{N_M}.$$ (15.74)

This is the smallest value that the susceptibility can attain under the distortionless response constraint [Dörbecker 1998].

Furthermore, the propagation vector **a** can be modified to suit the actual transmission conditions. For example, *head-related transfer functions* can be used to model head-related sound transmission effects, which are of importance in hearing aid applications [Lotter, Vary 2006].

### 15.5.2 MVDR Beamformer with Limited Susceptibility

The susceptibility of the MVDR beamformer can be controlled by including an additional constraint in the design procedure [Cox et al. 1986]. We require that the susceptibility attains a fixed value $K_0$,

$$\frac{1}{G_W} = \mathbf{H}^H\mathbf{H} = K_0. \tag{15.75}$$

Using Lagrange's method with an additional real-valued multiplier $q_3$ we now have

$$\mathfrak{L}(\mathbf{H}, q, q_3) = \mathbf{H}^H\mathbf{\Phi}_{nn}\mathbf{H} + \mathrm{Re}\{q^*(\mathbf{H}^H\mathbf{a} - 1)\} + q_3(\mathbf{H}^H\mathbf{H} - K_0). \tag{15.76}$$

Computing the derivatives, we obtain the necessary conditions

$$\mathbf{H}^H\mathbf{\Phi}_{nn} + \frac{q}{2}\mathbf{a}^H + q_3\mathbf{H}^H = 0, \tag{15.77a}$$

$$\mathbf{H}^H\mathbf{a} = 1, \text{ and} \tag{15.77b}$$

$$\mathbf{H}^H\mathbf{H} = K_0. \tag{15.77c}$$

The solution to these equations is given by

$$\mathbf{H} = -\frac{q}{2}\left(\mathbf{\Phi}_{nn} + q_3\mathbf{I}\right)^{-1}\mathbf{a} \quad \text{and} \tag{15.78a}$$

$$q = \frac{-2}{\mathbf{a}^H(\mathbf{\Phi}_{nn} + q_3\mathbf{I})^{-1}\mathbf{a}}, \tag{15.78b}$$

with $q_3$ as an implicit parameter. With these equations, the vector of optimal filters can be written as

$$\mathbf{H}_{\mathrm{MVDR,K0}} = \frac{(\mathbf{\Phi}_{nn} + q_3\mathbf{I})^{-1}\mathbf{a}}{\mathbf{a}^H(\mathbf{\Phi}_{nn} + q_3\mathbf{I})^{-1}\mathbf{a}}. \tag{15.79}$$

The constraint on the susceptibility leads to a *diagonal loading* term and thus improves the condition of the noise power spectral density matrix. The relation between the Lagrange multiplier $q_3$ and the desired susceptibility $K_0$, however, is implicit

$$K_0 = \frac{\mathbf{a}^H(\mathbf{\Phi}_{nn} + q_3\mathbf{I})^{-H}(\mathbf{\Phi}_{nn} + q_3\mathbf{I})^{-1}\mathbf{a}}{\left[\mathbf{a}^H(\mathbf{\Phi}_{nn} + q_3\mathbf{I})^{-1}\mathbf{a}\right]^2}, \tag{15.80}$$

with no closed form solution for $q_3$. For a given $K_0$ the corresponding Lagrange multiplier can be found by using an iterative procedure [Dörbecker 1997]. Using the optimal $q_3$, the coefficient vector $\mathbf{H}$ can then be computed.

The results of the beamformer optimization with the susceptibility limited to $K_0 = 1$ are shown in Figure 15.13 for the same conditions as in the previous figures. As a consequence of the limited susceptibility, the directivity at low frequencies is lower than the directivity without this constraint.

Besides the MVDR design, there are several other design methods for beamformers such as the design for nearfield conditions and constant beamwidth [Goodwin, Elko 1993],

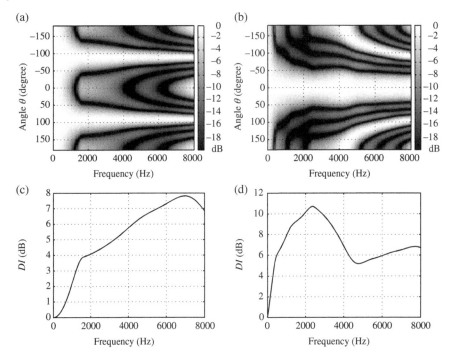

**Figure 15.13** Directivity pattern (power gain) and directivity index of the ULA MVDR beamformer with $N_M = 4$ microphones and $d = 4\,$cm with susceptibility limited to $K_0 = 1$, for the broadside (a and c) and endfire (b and d) orientation optimized for the ideal diffuse noise field.

[Brandstein, Ward 2001]. Furthermore, the design of such arrays may be made robust against amplitude, phase, and microphone position errors by including suitable error distribution functions into a *least-square* design criterion [Doclo, Moonen 2003].

### 15.5.3 Linearly Constrained Minimum Variance Beamformer

The concept of the MVDR beamformer may be further generalized to include a larger number of (complex) constraints in the optimization procedure. This is especially useful, when additional directional sources are present as spatial nulls may be placed in their respecticve directions. The resulting beamformer is known as the *linearly constrained minimum variance* (LCMV) beamformer.

As for the MVDR beamformer, the objective is to minimize undesired interference at the output of the beamformer

$$\Phi_{\hat{n}\hat{n}}(e^{j\Omega}) = \mathbf{H}^H(e^{j\Omega})\Phi_{nn}(e^{j\Omega})\mathbf{H}(e^{j\Omega}),\tag{15.81}$$

subject to additional constraints. For the narrowband LCMV beamformer $N_c$ constraints are summarized into a matrix equation

$$\mathbf{A}^H\mathbf{H} = \mathbf{F}\tag{15.82}$$

where $\mathbf{A}$ is the $L \times N_M$ constraint matrix of (relative) transfer functions (composed of propagation vectors as in (15.20)) and $\mathbf{F}$ denotes the $1 \times L$ response vector. Clearly, the response shall be unity for desired sources and zero for undesired (interfering) sources. Thus, undistorted transmission of the target sources and strong suppression of directional interferers can be achieved.

Using a similar procedure as for the MVDR beamformer, the cost function is formulated via Lagrange's method with the complex-valued $1 \times L$ vector $\mathbf{q}$

$$\mathfrak{L}(\mathbf{H}, \mathbf{q}) = \mathbf{H}^H \Phi_{nn} \mathbf{H} + \mathrm{Re}\{\mathbf{q}^H (\mathbf{A}^H \mathbf{H} - \mathbf{F})\} \tag{15.83}$$

$$= \mathbf{H}^H \Phi_{nn} \mathbf{H} + \mathrm{Re}\{\mathbf{q}\}^T \left( \frac{1}{2}(\mathbf{A}^H \mathbf{H} + \mathbf{H}^H \mathbf{A}) - \mathbf{F} \right)$$

$$+ \mathrm{Im}\{\mathbf{q}\}^T \left( \frac{1}{2j}(\mathbf{A}^H \mathbf{H} - \mathbf{H}^H \mathbf{A}) \right).$$

to obtain the solution

$$\mathbf{H}_{\mathrm{LCMV}} = \Phi_{nn}^{-1} \mathbf{A} \left( \mathbf{A}^H \Phi_{nn}^{-1} \mathbf{A} \right)^{-1} \mathbf{F}. \tag{15.84}$$

### 15.5.4 Max-SNR Beamformer

The MVDR beamformer works well when the propagation vector $\mathbf{a}$ of the target source is known. However, in reverberant environments and when only the direction of the source is known, the use of a simple propagation vector as in (15.14) will result in target signal distortions. An alternative is the Max-SNR beamformer (also known as the generalized eigenvector beamformer [Warsitz, Haeb-Umbach 2006]) that provides a mechanism for dealing with multipath propagation and unknown array geometry from the outset.

The Max-SNR beamformer aims to maximize the output SNR

$$\mathbf{H}_{\mathrm{Max\text{-}SNR}} = \arg\max_{\mathbf{H}} \left( \mathrm{SNR}_{\mathrm{out}} \right), \tag{15.85}$$

where $\mathrm{SNR}_{\mathrm{out}}$ is defined by the generalized Rayleigh quotient (e.g., [Strang 1988])

$$\mathrm{SNR}_{\mathrm{out}} = \frac{\mathbf{H}^H \Phi_{ss} \mathbf{H}}{\mathbf{H}^H \Phi_{nn} \mathbf{H}}, \tag{15.86}$$

and $\Phi_{ss}$ and $\Phi_{nn}$ are the spatial covariance matrices of the target signal and the noise signal, respectively, which absorb all information about the acoustic transmission including the array geometry. Note, that $\mathrm{SNR}_{\mathrm{out}}$ is independent of the scalar gain of $\mathbf{H}$. The derivative of $\mathrm{SNR}_{\mathrm{out}}$ w.r.t. $\mathbf{H}$ leads to the generalized eigenvalue problem

$$\Phi_{ss} \mathbf{H} = b \Phi_{nn} \mathbf{H}, \tag{15.87}$$

where the eigenvalue $b$ is identical to $\mathrm{SNR}_{\mathrm{out}}$. The generalized eigenvalue problem may be converted into an ordinary eigenvalue problem via a multiplication with $\Phi_{nn}^{-1}$ or by using the Cholesky decomposition, since $\Phi_{nn}$ is a positive-definite Hermitian matrix. The Cholesky decomposition of the noise covariance matrix is written as

$$\Phi_{nn} = \Lambda \Lambda^H, \tag{15.88}$$

where $\Lambda$ is a lower triangular matrix. Then, the generalized eigenvalue problem is simplified into an ordinary eigenvalue problem $\mathbf{Ax} = b\mathbf{x}$

$$\left( \Lambda^{-1} \Phi_{ss} \Lambda^{-H} \right) \left( \Lambda^H \mathbf{H} \right) = b \left( \Lambda^H \mathbf{H} \right) \tag{15.89}$$

with $\mathbf{x} = \mathbf{\Lambda}^H \mathbf{H}$. The maximum-SNR solution (up to an arbitrary scalar gain) corresponds to the eigenvector of $\mathbf{A} = \mathbf{\Lambda}^{-1} \mathbf{\Phi}_{ss} \mathbf{\Lambda}^{-H}$ that belongs to the largest eigenvalue, i.e., the largest $\text{SNR}_{\text{out}}$. Since we also have

$$\text{SNR}_{\text{out}} = \frac{\mathbf{H}^H \mathbf{\Phi}_{yy} \mathbf{H}}{\mathbf{H}^H \mathbf{\Phi}_{nn} \mathbf{H}} - 1, \tag{15.90}$$

the spatial covariance matrix of the microphone signals $\mathbf{\Phi}_{yy}$ may be used instead of $\mathbf{\Phi}_{ss}$. For the computation of the Max-SNR beamformer, these spatial correlation matrices must be known, where the formulation in (15.90) only requires the estimation of the noise covariance matrix. This matrix can be estimated using a voice activity detector, as proposed in Warsitz, Haeb-Umbach [2007], Araki et al. [2007], Heymann et al. [2016].

The optimization of the narrow-band SNR in each frequency bin separately as outlined above, leads to distortions of the broad-band signal. These distortions may be mitigated via a single-channel *blind analytical normalization* (BAN) postfilter [Warsitz, Haeb-Umbach 2007],

$$G_{\text{BAN}} = \frac{\sqrt{\mathbf{H}_{\text{Max-SNR}}^H \mathbf{\Phi}_{nn} \mathbf{\Phi}_{nn} \mathbf{H}_{\text{Max-SNR}} / N_M}}{\mathbf{H}_{\text{Max-SNR}}^H \mathbf{\Phi}_{nn} \mathbf{H}_{\text{Max-SNR}}}. \tag{15.91}$$

## 15.6 Multichannel Wiener Filter and Postfilter

As an extension to beamforming methods and the well-known single-channel Wiener filter, we now consider minimum mean squared error (MMSE) estimation of the target signal in the multichannel case. In the optimization of the MMSE, the source signal $s(k)$ is explicitly used and no constraint with respect to the look direction is imposed. However, as shall be seen below, the optimal filter may be decomposed into a distortionless response beamformer and a single channel Wiener-type *postfilter* [Edelblute et al. 1966, Simmer et al. 2001].

The MMSE of the filtered array output with respect to the reference signal $s(k)$ is given by

$$J = E\left\{ (s(k) - \hat{s}(k))^2 \right\}$$

$$= E\left\{ \left( s(k) - \sum_{\ell=1}^{N_M} \sum_{m=0}^{M} h_\ell(m) y_\ell(k-m) \right)^2 \right\}. \tag{15.92}$$

With $E\{s(k)y_{\ell'}(k-i)\} = \varphi_{y_{\ell'} s}(i)$ and $E\{y_\ell(k-m)y_{\ell'}(k-i)\} = \varphi_{y_{\ell'} y_\ell}(i-m)$, the differentiation with respect to the $i$-th coefficient of the $\ell'$-th filter leads to

$$\sum_{\ell=1}^{N_M} \sum_{m=0}^{M} h_\ell(m) \varphi_{y_{\ell'} y_\ell}(i-m) = \varphi_{y_{\ell'} s}(i). \tag{15.93}$$

The Fourier transform of (15.93) yields

$$\sum_{\ell=1}^{N_M} H_\ell(e^{j\Omega}) \Phi_{y_{\ell'} y_\ell}(e^{j\Omega}) = \Phi_{y_{\ell'} s}(e^{j\Omega}) \quad \text{for} \quad \ell' = 1, \dots N_M. \tag{15.94}$$

Therefore, using vector matrix notation, these equations may be stacked to yield

$$\Phi_{yy}\mathbf{H} = \Phi_{ys}, \tag{15.95}$$

and, for invertible $\Phi_{yy}$, the optimal solution

$$\mathbf{H} = \Phi_{yy}^{-1}\Phi_{ys}. \tag{15.96}$$

In the case of additive noise, which is not correlated with the source signal $s$, and identical noise and source power spectral densities at all microphones, we find, for far-field conditions,

$$\Phi_{ys} = \Phi_{ss}\mathbf{a} \tag{15.97}$$

and

$$\Phi_{yy} = \Phi_{ss} + \Phi_{nn} = \Phi_{ss}\mathbf{a}\mathbf{a}^H + \Phi_{nn}. \tag{15.98}$$

Hence,

$$\mathbf{H} = \left(\Phi_{ss}\mathbf{a}\mathbf{a}^H + \Phi_{nn}\right)^{-1}\Phi_{ss}\mathbf{a}. \tag{15.99}$$

Using the matrix inversion lemma (e.g., [Haykin 1996]),

$$\mathbf{A} = \mathbf{B}^{-1} + \mathbf{C}\mathbf{D}^{-1}\mathbf{C}^H \Leftrightarrow \mathbf{A}^{-1} = \mathbf{B} - \mathbf{B}\mathbf{C}(\mathbf{D} + \mathbf{C}^H\mathbf{B}\mathbf{C})^{-1}\mathbf{C}^H\mathbf{B} \tag{15.100}$$

with $\mathbf{B} = \Phi_{nn}^{-1}$, $\mathbf{C} = \sqrt{\Phi_{ss}}\mathbf{a}$, and $\mathbf{D} = 1$ we obtain

$$\mathbf{H} = \left[\Phi_{nn}^{-1} - \frac{\Phi_{ss}\Phi_{nn}^{-1}\mathbf{a}\mathbf{a}^H\Phi_{nn}^{-1}}{1 + \Phi_{ss}\mathbf{a}^H\Phi_{nn}^{-1}\mathbf{a}}\right]\Phi_{ss}\mathbf{a}$$

$$= \left[\Phi_{nn}^{-1}\mathbf{a} - \frac{\Phi_{ss}\Phi_{nn}^{-1}\mathbf{a}}{(\mathbf{a}^H\Phi_{nn}^{-1}\mathbf{a})^{-1} + \Phi_{ss}}\right]\Phi_{ss} \tag{15.101}$$

and the final result

$$\mathbf{H} = \frac{\Phi_{nn}^{-1}\mathbf{a}}{\mathbf{a}^H\Phi_{nn}^{-1}\mathbf{a}} \frac{\Phi_{ss}}{\left(\Phi_{ss} + \left(\mathbf{a}^H\Phi_{nn}^{-1}\mathbf{a}\right)^{-1}\right)}$$

$$= \mathbf{H}_{\mathrm{MVDR}}\frac{\Phi_{ss}}{\Phi_{ss} + \left(\mathbf{a}^H\Phi_{nn}^{-1}\mathbf{a}\right)^{-1}}. \tag{15.102}$$

The optimal multichannel Wiener filter (MWF) solution thus comprises a distortionless minimum-variance beamformer and a single-channel Wiener *postfilter*. Since

$$\mathbf{H}_{\mathrm{MVDR}}^H\Phi_{nn}\mathbf{H}_{\mathrm{MVDR}} = \left(\mathbf{a}^H\Phi_{nn}^{-1}\mathbf{a}\right)^{-1} \tag{15.103}$$

it can be seen that $\left(\mathbf{a}^H\Phi_{nn}^{-1}\mathbf{a}\right)^{-1}$ represents the noise power at the output of the beamformer. Therefore, to compute the single-channel postfilter, the speech power spectral density at the microphones and the noise power spectral density at the output of the beamformer are required. There are several proposals (for a survey, see [Simmer et al. 2001]) for computing these quantities using the available microphone signals. For example, [Zelinski 1988] uses cross-periodograms averaged over all microphone pairs to estimate the power spectral densities of the speech signal and of the noisy signal. This approach rests on the assumption that noise signals are mutually uncorrelated. Due to the large variance of the cross-periodograms and the residual correlation of noise components, additional post-processing of the estimates is necessary to suppress undesired fluctuations in the output of the postfilter. Zelinski [1988] then combines a delay-and-sum beamformer with this postfilter and [Marro et al. 1998] computes a postfilter that accounts for the frequency response of a filter-and-sum beamformer. Furthermore, the MWF formulation has been extended to explicitly control target signal distortions. This *speech distortion-weighted* MWF [Spriet et al. 2004] can balance the level of noise reduction and target signal distortion between the solutions provided by the standard MWF or the MVDR beamformer. Also, the speech distortion-weighted MWF might be implemented in an adaptive GSC-like fashion.

Finally, it is interesting to note, that the above decomposition of the MMSE solution into a linear spatial (MVDR) filter and a spectral postfilter is not optimal when the signal is disturbed by non-Gaussian noise distributions [Hendriks et al. 2009].

## 15.7 Adaptive Beamformers

When the spatial properties of the acoustic noise field are a priori unknown and possibly time varying, the beamformer coefficients cannot be precomputed but must be adapted on-line. While in Frost [1972] the coefficients of a filter-and-sum beamformer are adapted using a constrained version of the LMS algorithm, most practical implementations are based on the *generalized side-lobe canceller* (GSC) or *Griffiths-Jim* beamformer [Griffiths, Jim 1982]. In this section, we discuss both the Frost beamformer and several GSC designs.

### 15.7.1 The Frost Beamformer

Figure 15.14 depicts the block diagram of an adaptive filter-and-sum beamformer as proposed in Frost [1972], where the filter coefficients are now functions of time. The adaptation rule minimizes the noise power at the output while maintaining a constraint on the filter response in look direction. For the development of the Frost beamformer, we stack the coefficients of the $N_M$ FIR filters of length $M + 1$ into a single coefficient vector

$$\tilde{\mathbf{h}} = (h_1(0), h_2(0), \ldots, h_{N_M}(0), h_1(1), \ldots, h_1(M), \ldots, h_{N_M}(M))^T \tag{15.104}$$

and the corresponding microphone signal samples into the vector

$$\tilde{\mathbf{y}}(k) = \Big( y_1(k), y_2(k), \ldots, y_{N_M}(k), y_1(k-1), \ldots,$$
$$\ldots, y_1(k-M), \ldots, y_{N_M}(k-M) \Big)^T. \tag{15.105}$$

For each time lag, a constraint on the sum of the beamformer coefficients is applied. These constraints may be written as

$$\tilde{\mathbf{C}}^T \tilde{\mathbf{h}} = \tilde{\mathbf{F}} \tag{15.106}$$

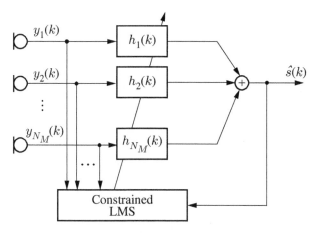

**Figure 15.14** Frost beamformer.

where the constraint matrix of dimensions $(M + 1)N_M \times (M + 1)$ is given by

$$
\tilde{C} = \begin{pmatrix} 1 & 0 & 0 & \cdots & 0 \\ 0 & 1 & 0 & \cdots & 0 \\ \vdots & \vdots & \vdots & \ddots & \vdots \\ 0 & 0 & 0 & \cdots & 1 \end{pmatrix}
\tag{15.107}
$$

and

$$
\mathbf{0} = (0, 0, \ldots, 0)^T \quad \text{and} \quad \mathbf{1} = (1, 1, \ldots, 1)^T
\tag{15.108}
$$

are vectors of dimension $N_M$.

For example, the undistorted and delay-free array response in the broadside look direction implies a constraint

$$
\sum_{\ell=1}^{N_M} \sum_{i=0}^{M} h_\ell(i)\delta(k) = \delta(k) \quad \text{for} \quad k = 0, \ldots, M
\tag{15.109}
$$

on the array response of the filter-and-sum beamformer, which can be also cast into the matrix equation (15.106) with (15.107) and the $M + 1$-dimensional vector

$$
\tilde{F} = (1, 0, \ldots, 0)^T.
\tag{15.110}
$$

An optimal stationary solution is again achieved by minimizing the output power $\tilde{h}^T R_{\tilde{y}\tilde{y}} \tilde{h}$ while maintaining the constraint (15.106). Using the method of the Lagrange multiplier, we obtain

$$
\mathfrak{L}(\tilde{h}, q) = \tilde{h}^T R_{\tilde{y}\tilde{y}} \tilde{h} + q\left(\tilde{C}^T \tilde{h} - \tilde{F}\right).
\tag{15.111}
$$

Differentiating with respect to $\tilde{h}$ and $q$ results in the necessary conditions

$$
2R_{\tilde{y}\tilde{y}} \tilde{h} + q\tilde{C} = 0 \quad \text{and} \quad \tilde{C}^T \tilde{h} = \tilde{F}.
\tag{15.112}
$$

Solving the first condition for $\tilde{\mathbf{h}}$, and, using the result in the second condition, we obtain

$$\tilde{\mathbf{h}} = -\frac{1}{2}q\mathbf{R}_{\tilde{y}\tilde{y}}^{-1}\tilde{\mathbf{C}} \quad \text{and} \quad q = \frac{-2}{\tilde{\mathbf{C}}^T\mathbf{R}_{\tilde{y}\tilde{y}}^{-1}\tilde{\mathbf{C}}}\tilde{\mathbf{F}}. \tag{15.113}$$

Thus, the optimal linearly constrained solution is given in analogy to (15.84) by

$$\tilde{\mathbf{h}}_{LC} = \mathbf{R}_{\tilde{y}\tilde{y}}^{-1}\tilde{\mathbf{C}}\left(\tilde{\mathbf{C}}^T\mathbf{R}_{\tilde{y}\tilde{y}}^{-1}\tilde{\mathbf{C}}\right)^{-1}\tilde{\mathbf{F}}. \tag{15.114}$$

Based on the time domain solution in (15.114), deterministic and stochastic adaptive solutions may be developed [Frost 1972]. From the gradient of (15.111), a deterministic update rule for the coefficient vector may be derived as

$$\tilde{\mathbf{h}}(k+1) = \tilde{\mathbf{h}}(k) - \mu\left(\mathbf{R}_{\tilde{y}\tilde{y}}\tilde{\mathbf{h}} + q\tilde{\mathbf{C}}\right) \tag{15.115}$$

where $\mu$ is a positive stepsize parameter. Applying the constraint (15.106) to (15.115) leads to

$$\tilde{\mathbf{F}} = \tilde{\mathbf{C}}^T\tilde{\mathbf{h}}(k+1) = \tilde{\mathbf{C}}^T\tilde{\mathbf{h}}(k) - \mu\left(\tilde{\mathbf{C}}^T\mathbf{R}_{\tilde{y}\tilde{y}}\tilde{\mathbf{h}} + q\tilde{\mathbf{C}}^T\tilde{\mathbf{C}}\right). \tag{15.116}$$

Solving for $q$ and using the result in (15.115) yields the deterministic update rule

$$\tilde{\mathbf{h}}(k+1) = \tilde{\mathbf{h}}(k) - \mu\left(\mathbf{I} - \tilde{\mathbf{C}}(\tilde{\mathbf{C}}^T\tilde{\mathbf{C}})^{-1}\tilde{\mathbf{C}}^T\right)\mathbf{R}_{\tilde{y}\tilde{y}}\tilde{\mathbf{h}}$$
$$+ \tilde{\mathbf{C}}(\tilde{\mathbf{C}}^T\tilde{\mathbf{C}})^{-1}\left(\tilde{\mathbf{F}} - \tilde{\mathbf{C}}^T\tilde{\mathbf{h}}(k)\right). \tag{15.117}$$

Using the sample covariance $\tilde{\mathbf{y}}(k)\tilde{\mathbf{y}}^T(k)$ as an approximation to $\mathbf{R}_{\tilde{y}\tilde{y}}$, and with

$$\mathbf{P} = \mathbf{I} - \tilde{\mathbf{C}}(\tilde{\mathbf{C}}^T\tilde{\mathbf{C}})^{-1}\tilde{\mathbf{C}}^T, \tag{15.118}$$

$$\tilde{\mathbf{f}} = \tilde{\mathbf{C}}(\tilde{\mathbf{C}}^T\tilde{\mathbf{C}})^{-1}\tilde{\mathbf{F}}, \tag{15.119}$$

$$\hat{s}(k) = \tilde{\mathbf{h}}^T(k)\tilde{\mathbf{y}}(k), \tag{15.120}$$

the *stochastic constrained least-mean-square* (CLMS) algorithm can be written as

$$\tilde{\mathbf{h}}(k+1) = \tilde{\mathbf{h}}(k) + \mathbf{P}\left(\tilde{\mathbf{h}}(k) - \mu\,\hat{s}(k)\,\tilde{\mathbf{y}}(k)\right) + \tilde{\mathbf{f}} \tag{15.121}$$

with the initial condition $\tilde{\mathbf{h}}(0) = \tilde{\mathbf{f}}$. In each iteration, the algorithm updates the coefficient vector such that the constraint is met [Frost 1972].

## 15.7.2 Generalized Side-Lobe Canceller

An improved solution to the constrained adaptive beamforming problem decomposes the adaptive filter-and-sum beamformer into a fixed beamformer and an adaptive multi-channel noise canceller. The resulting system is termed *generalized side-lobe canceller* [Griffiths, Jim 1982], a block diagram of which is shown in Figure 15.15. Here, the constraint of a distortionless response in look direction is established by the fixed beamformer. Since a *blocking matrix* eliminates the target signal in the parallel path, the multi-channel adaptive noise canceller receives noise signals only and can then be

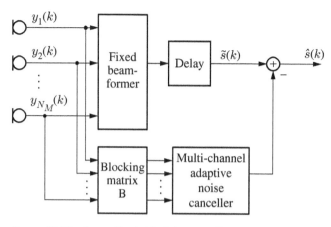

**Figure 15.15** Generalized side-lobe canceller.

adapted without a constraint. The implementation of an adaptive, distortionless response beamformer is therefore greatly facilitated.

The fixed beamformer, blocking matrix, and adaptive noise canceller can be implemented via one of the previously discussed methods in the time or the frequency domain. For the fixed beamformer we may chose, for example, a delay-and-sum beamformer. To avoid distortions of the desired signal, the input to the adaptive noise canceller must not contain the desired signal. Therefore, the *blocking matrix* $\mathbf{B}$ must be designed such that the estimated noise signals

$$
\begin{pmatrix}
\hat{y}_1(k) \\
\hat{y}_2(k) \\
\vdots \\
\hat{y}_{N_M-1}(k)
\end{pmatrix}
= \mathbf{B}
\begin{pmatrix}
y_1(k) \\
y_2(k) \\
\vdots \\
y_{N_M}(k)
\end{pmatrix}
\tag{15.122}
$$

are free of the desired signal. It is generally assumed that the rows of $\mathbf{B}$ are linearly independent. Then, the blocking of a single desired signal by means of a linear operation on the microphone signals reduces the number of independent signal components by one. Therefore, $\mathbf{B}$ is a $(N_M - 1) \times N_M$ matrix.

To illustrate the construction of the blocking matrix, we consider the simplest case: when the desired signal originates from a point source under far-field conditions, and is received from the broadside look direction, blocking of the desired source can be achieved by a pairwise subtraction of the microphone signals. In this case, the blocking matrix (of rank $N_M - 1$) may be written as

$$
\mathbf{B} =
\begin{pmatrix}
1 & -1 & 0 & 0 & \cdots & 0 & 0 \\
0 & 1 & -1 & 0 & \cdots & 0 & 0 \\
\vdots & \vdots & \vdots & \vdots & \ddots & \vdots & \vdots \\
0 & 0 & 0 & 0 & \cdots & 1 & -1
\end{pmatrix}.
\tag{15.123}
$$

When the source signal impinges from other directions, the blocking matrix must implement appropriate delays for target signal cancellation. In the general case with multi-path propagation and reverberation a frequency-domain blocking matrix may be composed of relative transfer functions [Gannot et al. 2001]

$$\tilde{A}_\ell \left( e^{j\Omega} \right) = \frac{A_\ell(e^{j\Omega})}{A_1(e^{j\Omega})} \tag{15.124}$$

such that

$$\mathbf{B}\left( e^{j\Omega} \right) = \begin{pmatrix} -\tilde{A}_2^* \left( e^{j\Omega} \right) & 1 & 0 & \cdots & 0 \\ -\tilde{A}_3^* \left( e^{j\Omega} \right) & 0 & 1 & \cdots & 0 \\ -\tilde{A}_4^* \left( e^{j\Omega} \right) & 0 & 0 & \cdots & 0 \\ \vdots & \vdots & \vdots & \ddots & \vdots \\ -\tilde{A}_{N_M}^* \left( e^{j\Omega} \right) & 0 & 0 & \cdots & 1 \end{pmatrix}. \tag{15.125}$$

It is easily verified that for a single source $S(e^{j\Omega})$ the blocking matrix and the propagation vector are orthogonal, i.e., $\mathbf{B}\left( e^{j\Omega} \right) \mathbf{Y}\left( e^{j\Omega} \right) = \mathbf{B}\left( e^{j\Omega} \right) \mathbf{a}^* S(e^{j\Omega}) = \mathbf{0}$ with $\mathbf{a}^* = \left( A_1(e^{j\Omega}), A_2(e^{j\Omega}), \ldots, A_{N_M}(e^{j\Omega}) \right)^H$.

The output of the blocking matrix is then used in the adaptive noise canceller to estimate and subtract the noise components at the output of the fixed beamformer. Since both the fixed beamformer and the multi-channel noise canceller might delay their respective input signals, a delay in the signal path is required.

In reverberant environments, it is in general difficult to prevent the desired speech signal from leaking into the noise cancellation branch. A number of countermeasures have been proposed (e.g., [Hoshuyama, Sugiyama 2001]), such as

- the use of an adaptive blocking matrix
- improved target tracking
- adaptation-mode control
- coefficient and coefficient-norm constrained adaptive filters.

One of these designs is explained in Section 15.7.3. Furthermore, for hands-free, full-duplex speech communication, it is desirable to combine beamforming microphone arrays with echo cancellation. The GSC and extensions toward combined beamforming and acoustic echo cancellation are discussed, e.g., in Herbordt [2005].

### 15.7.3 Generalized Side-lobe Canceller with Adaptive Blocking Matrix

In reverberant environments, the simple blocking matrix in (15.123) does not deliver sufficient attenuation of the desired signal since it does not account for multi-path propagation effects. The identification of the (relative) transfer functions from the source to the microphones via adaptive filters is necessary. An early implementation of the GSC principle with adaptive filters for blocking the desired signal is proposed in Hoshuyama et al. [1999].

The resulting algorithm is shown in Figure 15.16. The output of the fixed beamformer is fed into adaptive filters, which minimize the error with respect to the microphone signals.

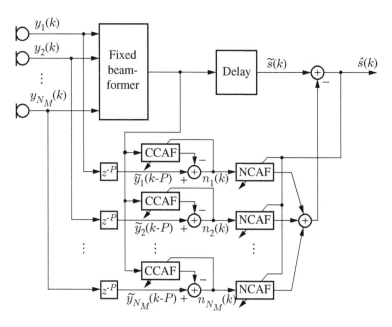

**Figure 15.16** GSC with adaptive blocking matrix. Source: Adapted from Hoshuyama et al. (1999).

Thus, any coherent signal originating from the look direction of the fixed beamformer can be reduced by the adaptive blocking method. For the adaptive blocking filters, a coefficient constrained adaptive filter (CCAF) is used. The coefficients of the adaptive filter are constrained to produce the dominant peak for those taps, which correspond to the look direction with some tolerance interval (typically ±20 taps).

The blocked microphone signals are then fed into a multi-channel noise canceller, which is implemented as a norm-constrained adaptive filter (NCAF). The norm constraint on the coefficient vector prevents excessive growth of the filter coefficients and thus leakage of the desired signal into the output of the multi-channel noise canceller. For adapting both the blocking and the noise cancellation filters, the NLMS algorithm with the above modifications is used.

### 15.7.4 Model-Based Parsimonious-Excitation-Based GSC

In many real-world applications, we need to consider multiple and moving sources. To this end, a versatile GSC implementation has been proposed in [Madhu, Martin 2011], utilizing the spectral sparseness of speech sources [Rickard, Yilmaz 2002]. All components of this GSC are selectively adapted via frequency-dependent source-presence probabilities. The method, known as *parsimonious-excitation* GSC (PEG), comprises four main steps: the estimation of the spatial source distribution, the estimation of source presence probabilities, steering the fixed beamformer towards the desired source, and the adaptation of the blocking matrix and the multi-channel noise canceller. In what follows, we briefly outline these steps.

The spatial distribution of active acoustic sources is determined via the *steered response power* (SRP) approach [DiBiase et al. 2001]. As described in Section 15.3.2, the power spectrum at the output of a filter-and-sum beamformer is given by

$$\Phi_{\hat{s}\hat{s}}(\Omega) = \mathbf{H}^H(e^{j\Omega})\Phi_{yy}(\Omega)\mathbf{H}(e^{j\Omega}), \tag{15.126}$$

where the spatial covariance matrix of the microphone signals is defined as in (15.41). For a single source in the far-field, anechoic conditions and additive noise, we find the cross-power spectrum of the $\ell$-th and $\ell'$-th microphone as

$$\Phi_{y_\ell y_{\ell'}}(\Omega) = \Phi_{ss}(\Omega)e^{-j\Omega f_s(\Delta\tau_\ell - \Delta\tau_{\ell'})} + \Phi_{n_\ell n_{\ell'}}(\Omega).$$

The power spectrum is now rewritten as

$$\Phi_{\hat{s}\hat{s}}(\Omega) = \sum_{\ell=1}^{N_M}\sum_{\ell'=1}^{N_M} H_\ell^*(e^{j\Omega})H_{\ell'}(e^{j\Omega})\Phi_{y_\ell y_{\ell'}}(\Omega) = \sum_{\ell=1}^{N_M}\sum_{\ell'=1}^{N_M}\Psi_{\ell\ell'}(\Omega)\Phi_{y_\ell y_{\ell'}}(\Omega), \tag{15.127}$$

where the spectral weighting function $\Psi_{\ell\ell'}(\Omega)$ is introduced to represent the filters $H_\ell(e^{j\Omega})$. A variety of weighting functions has been proposed [Knapp, Carter 1976], [Madhu, Martin 2008]. The weighting function depends on channel indices $\ell$ and $\ell'$, on steering parameters, and may include an additional normalization. For a single source in additive noise we thus obtain

$$\Phi_{\hat{s}\hat{s}}(\Omega) = \sum_{\ell=1}^{N_M}\sum_{\ell'=1}^{N_M}\Psi_{\ell\ell'}(\Omega)\left(\Phi_{ss}(\Omega)e^{-j\Omega f_s(\Delta\tau_\ell - \Delta\tau_{\ell'})} + \Phi_{n_\ell n_{\ell'}}(\Omega)\right). \tag{15.128}$$

Note, that the SRP method [DiBiase et al. 2001] computes the total output power. It equals the auto-correlation $\varphi(0)$ at lag $\tau = 0$ and corresponds to an average across frequency

$$\varphi_{\hat{s}\hat{s}}(0) = \frac{1}{2\pi}\int_{-\pi}^{\pi}\sum_{\ell=1}^{N_M}\sum_{\ell'=1}^{N_M}\Psi_{\ell\ell'}(\Omega)\Phi_{y_\ell y_{\ell'}}(\Omega)d\Omega. \tag{15.129}$$

However, the total steered power is not further considered here as we strive to exploit spectral sparsity of speech sources in a multi-source scenario.

For the localization of wide-band acoustic sources using cross-correlation or cross-power spectra, it is common to select the lumped spectral weighting function $\Psi_{\ell\ell'}(\Omega)$ such that localization performance is improved. For speech signals with their specific sloping power spectrum, the phase-transform (PHAT) as been shown to provide robustness as it equalizes the amplitudes across frequency and thus enables contributions across a wide range of frequencies [DiBiase et al. 2001]. The SRP-PHAT cost function at frequency $\Omega$ and steering direction $\theta$ results in a cost function

$$J(\Omega, \theta) = \sum_{\ell=1}^{N_M}\sum_{\ell'=1}^{N_M}\frac{e^{j\Omega f_s(\Delta\tau_\ell(\theta) - \Delta\tau_{\ell'}(\theta))}}{\left|Y_\ell^*(e^{j\Omega})Y_{\ell'}(e^{j\Omega})\right|}Y_\ell^*(e^{j\Omega})Y_{\ell'}(e^{j\Omega}). \tag{15.130}$$

Using the DFT instead of the discrete-time Fourier transform, we obtain for each time frame $\lambda$ and discrete frequency bin $\mu$,

$$J_\mu(\lambda, \theta) = \sum_{\ell=1}^{N_M}\sum_{\ell'=1}^{N_M}\frac{e^{j\Omega_\mu f_s(\Delta\tau_\ell(\theta) - \Delta\tau_{\ell'}(\theta))}}{\left|Y_{\ell,\mu}^*(\lambda)Y_{\ell',\mu}(\lambda)\right|}Y_{\ell,\mu}^*(\lambda)Y_{\ell',\mu}(\lambda). \tag{15.131}$$

The method then evaluates the SRP-PHAT cost function

$$\hat{\theta}_\mu(\lambda) = \underset{\theta}{\text{argmax}} \left( J_\mu(\lambda, \theta) \right) \tag{15.132}$$

to identify the direction $\hat{\theta}_\mu(\lambda)$ of the dominant source in each time-frequency bin. Hereby, we make implicitly use of the spectral sparsity assumption for concurrent speech sources [Rickard, Yilmaz 2002].

In the next step, the estimated source directions $\hat{\theta}_\mu(\lambda)$ are collected into a histogram for each frame $\lambda$ and approximated via a Gaussian mixture model (GMM, See Section 5.5.2)

$$\hat{\theta}_\mu(\lambda) \sim \sum_{q=1}^{Q(\lambda)} P^{\langle q\rangle}(\lambda) \mathcal{N}(\hat{\theta}_\mu(\lambda); \theta^{\langle q\rangle}(\lambda), (\sigma^{\langle q\rangle})^2(\lambda)), \tag{15.133}$$

where $Q(\lambda)$ is the number of active sources in this frame and the means $\theta^{\langle q\rangle}(\lambda)$ are estimates of their location. Then, given this model we assume that each component of the GMM corresponds to one source and compute the posterior probability for the $q$-th source in frequency bin $\mu$ as

$$P_\mu^{\langle q|\hat{\theta}\rangle}(\lambda) = \frac{\dfrac{1}{\sigma^{\langle q\rangle}(\lambda)} P^{\langle q\rangle}(\lambda) \exp - \left( \dfrac{(\hat{\theta}_\mu(\lambda) - \theta^{\langle q\rangle})^2}{2(\sigma^{\langle q\rangle})^2(\lambda)} \right)}{\sum_{q'=1}^{Q(\lambda)} \dfrac{1}{\sigma^{\langle q'\rangle}(\lambda)} P^{\langle q'\rangle}(\lambda) \exp - \left( \dfrac{(\hat{\theta}_\mu(\lambda) - \theta^{\langle q'\rangle})^2}{2(\sigma^{\langle q'\rangle})^2(\lambda)} \right)}. \tag{15.134}$$

The final step comprises the signal enhancement following one of two options.

**Extraction of the Target Source Using a Spectral Mask** The extraction of the signal of the $q$-th source can be implemented using the posterior source presence probabilities as a spectral mask $M_\mu^{\langle q\rangle}(\lambda) = P_\mu^{\langle q|\hat{\theta}\rangle}(\lambda)$. The estimated source spectra are then given by

$$\hat{S}_\mu^{\langle q\rangle}(\lambda) = M_\mu^{\langle q\rangle}(\lambda) \left( \mathbf{H}_\mu^{\langle q\rangle}(\lambda) \right)^H \mathbf{Y}_\mu(\lambda). \tag{15.135}$$

Here, $\mathbf{H}_\mu^{\langle q\rangle}(\lambda)$ denotes the filter vectors of the fixed beamformer steered towards the $q$-th source. No further noise cancellation is applied.

**Computation of Target and Noise Signal Sub-spaces** The second option makes full use of the GSC structure and comprises the computation of an adaptive blocking matrix and an adaptive noise canceller. The computation of the adaptive blocking matrix uses the posterior probabilities for each source and proceeds along the following steps:

- recursive estimation of source signal projector

$$\mathbf{L}_\mu^{\langle q\rangle}(\lambda) = (1 - P_\mu^{\langle q|\hat{\theta}\rangle}(\lambda))\mathbf{L}_\mu^{\langle q\rangle}(\lambda - 1) + P_\mu^{\langle q|\hat{\theta}\rangle}(\lambda) \frac{\mathbf{Y}_\mu(\lambda)\mathbf{Y}_\mu^H(\lambda)}{||\mathbf{Y}_\mu(\lambda)||^2} \tag{15.136}$$

- computation of complementary (interference) space projection matrix

$$\mathbf{C}_\mu^{\langle q\rangle}(\lambda) = \mathbf{I} - \mathbf{L}_\mu^{\langle q\rangle}(\lambda) \tag{15.137}$$

- computation of the adaptive blocking matrix

$$\mathbf{B}_{\mu}^{\langle q \rangle}(\lambda) = D_{(M-1),M}\left(\mathbf{C}_{\mu}^{\langle q \rangle}(\lambda)\right) \tag{15.138}$$

where $D_{a,b}(\cdot)$ selects the first $a$ rows and $b$ columns of the matrix argument.

The update rule of the noise canceller is derived from the MMSE criterion

$$\mathfrak{L}_{\mathbf{W}} = \mathrm{E}\left\{\left(\left(\mathbf{H}_{\mu}^{\langle q \rangle}(\lambda)\right)^{H}\mathbf{Y}_{\mu}(\lambda) - \left(\mathbf{W}_{\mu}^{\langle q \rangle}(\lambda)\right)^{H}\mathbf{B}_{\mu}^{\langle q \rangle}(\lambda)\mathbf{Y}_{\mu}(\lambda)\right)^{2}\right\} \tag{15.139}$$

and uses an LMS-type update, which includes a power normalization

$$\mathbf{W}_{\mu}^{\langle q \rangle}(\lambda+1) = \mathbf{W}_{\mu}^{\langle q \rangle}(\lambda)$$

$$+ \zeta_{\mu}(\lambda)\left(\left(\mathbf{H}_{\mu}^{\langle q \rangle}(\lambda)\right)^{H}\mathbf{Y}_{\mu}(\lambda) - \left(\mathbf{W}_{\mu}^{\langle q \rangle}(\lambda)\right)^{H}\mathbf{B}_{\mu}^{\langle q \rangle}(\lambda)\mathbf{Y}_{\mu}(\lambda)\right)^{*}$$

$$\cdot \frac{\mathbf{B}_{\mu}^{\langle q \rangle}(\lambda)\mathbf{Y}_{\mu}(\lambda)}{||\mathbf{B}_{\mu}^{\langle q \rangle}(\lambda)\mathbf{Y}_{\mu}(\lambda)||^{2}}, \tag{15.140}$$

with the step size $\zeta_{\mu}(\lambda) = (1 - P_{\mu}^{\langle q|\hat{\theta} \rangle}(\lambda))\zeta_{c}$ and the fixed constant $0 < \zeta_{c} < 2$.

Finally, the spectrum of the $q$-th source is extracted via adaptive blocking and noise cancellation

$$\hat{S}_{\mu}^{\langle q \rangle}(\lambda) = (\mathbf{H}_{\mu}^{\langle q \rangle}(\lambda))^{H}\mathbf{Y}_{\mu}(\lambda) - (\mathbf{W}_{\mu}^{\langle q \rangle}(\lambda))^{H}\mathbf{B}_{\mu}^{\langle q \rangle}(\lambda)\mathbf{Y}_{\mu}(\lambda). \tag{15.141}$$

This approach has also been extended towards an application in binaural hearing aids [Zohourian et al. 2018].

## 15.8 Non-linear Multi-channel Noise Reduction

In analogy to the single-channel case, non-linear optimal estimators may be derived which account for the probability distribution of the acoustic signals. For instance, the short-term spectral amplitude of the desired signal can be estimated using information about the spectral coefficients of all microphone signals.

Thus, as discussed in Chapter 12, MMSE and MAP estimators may be developed for the short-term spectral amplitude in the $\mu$-th frequency bin of the desired signal. The MMSE solution is again given by the conditional expectation

$$\hat{A}_{\mu} = \mathrm{E}\left\{A_{\mu} \mid Y_{\mu,1}, \dots, Y_{\mu,N_{M}}\right\}, \tag{15.142}$$

which is now conditioned on the complex spectral amplitudes $Y_{\mu,\ell}$ of the noisy signal. This expectation yields a closed form solution for Gaussian distributed signals [Balan, Rosca 2002], which also provides a decomposition into a spatial filter and a postfilter. The Gaussian assumption is, as discussed in Section 5.10.3, not valid for short analysis frames.

Hence, multi-channel noise reduction solutions have been extended to data driven (DNN-based) methods which use the multi-channel information either to estimate the parameters of the beamformer as in Heymann et al. [2016] and Gu et al. [2021] or to implement an entirely non-linear processing model that delivers a spectral amplitude mask [Chakrabarty et al. 2018] or the target signal directly [Tan et al. 2022]. The distinctive feature of these approaches resides in the estimation and utilization of the joint temporal, spatial, and spectral statistics of the noisy microphone signals, which clearly goes beyond the simplifying assumptions of traditional methods using independent frequency bins and second-order statistics only. Furthermore, in order to exploit properly the phase relationships between the microphone signals, recent methods use DNN layers with complex-valued computations [Gu et al. 2021] and the complex-ratio mask [Williamson et al. 2016]. The contributions of spatial, spectral and temporal statistics to the overall quality has been systematically investigated in Tesch, Gerkmann [2023], where it is shown that non-linear multi-channel DNNs are successful in exploiting spatial-spectral relationships and thus outperform the traditional MVDR-plus-postfilter beamformer.

# References

Araki, S.; Sawada, H.; Makino, S. (2007). Blind Speech Separation in a Meeting Situation with Maximum SNR Beamformers, *Proceedings of the IEEE International Conference on Acoustics, Speech and Signal Processing (ICASSP)*, Hawaii, USA, vol. 1, pp. I–41–I–44.

Balan, R.; Rosca, J. (2002). Microphone Array Speech Enhancement by Bayesian Estimation of Spectral Amplitude and Phase, *Proceedings of the IEEE Sensor Array and Multichannel Signal Processing Workshop*, Rosslyn, Virginia, USA, pp. 209–213.

Brandstein, M.; Ward, D. (eds.) (2001). *Microphone Arrays*, Springer-Verlag, Berlin, Heidelberg, New York.

Chakrabarty, S.; Wang, D.; Habets, E. A. P. (2018). Time-Frequency Masking Based Online Speech Enhancement with Multi-Channel Data Using Convolutional Neural Networks, *2018 16th International Workshop on Acoustic Signal Enhancement (IWAENC)*, pp. 476–480.

Cox, H.; Zeskind, R. M.; Kooij, T. (1986). Practical Supergain, *IEEE Transactions on Acoustics, Speech and Signal Processing*, vol. 34, no. 3, pp. 393–398.

Crochiere, R. E.; Rabiner, L. R. (1981). Interpolation and Decimation of Digital Signals – A Tutorial Review, *Proceedings of the IEEE*, vol. 69, no. 3, pp. 300–331.

Crochiere, R. E.; Rabiner, L. R. (1983). *Multirate Digital Signal Processing*, Prentice Hall, Englewood Cliffs, New Jersey.

DiBiase, J.; Silverman, H.; Brandstein, M. (2001). Robust Localization in Reverberant Rooms, *in* M. Brandstein; D. Ward (eds.), *Microphone Arrays: Signal Processing Techniques and Applications*, Springer-Verlag, Berlin.

Doclo, S.; Moonen, M. (2003). Design of Broadband Beamformers Robust Against Microphone Position Errors, *Proceedings of the International Workshop on Acoustic Echo and Noise Control (IWAENC)*, Kyoto, Japan, pp. 267–270.

Dörbecker, M. (1997). Small Microphone Arrays with Optimized Directivity for Speech Enhancement, *Proceedings of the European Conference on Speech Communication and Technology (EUROSPEECH)*, Rhodes, Greece, pp. 327–330.

Dörbecker, M. (1998). *Mehrkanalige Signalverarbeitung zur Verbesserung akustisch gestörter Sprachsignale am Beispiel elektronischer Hörhilfen*, PhD thesis. *Aachener Beiträge zu digitalen Nachrichtensystemen*, vol. 10, P. Vary (ed.), RWTH Aachen University (in German).

Edelblute, D. J.; Fisk, J. M.; Kinnison, G. L. (1966). Criteria for Optimum-Signal-Detection Theory for Arrays, *Journal of the Acoustical Society of America*, vol. 41, no. 1, pp. 199–205.

Elko, G. W.; Pong, A.-T.N. (1997). A Steerable and Variable First-Order Differential Microphone Array, *Proceedings of the IEEE International Conference on Acoustics, Speech, and Signal Processing (ICASSP)*, Munich, Germany, pp. 223–226.

Frost, O. L. (1972). An Algorithm for Linearly Constrained Adaptive Array Processing, *Proceedings of the IEEE*, vol. 60, no. 8, pp. 926–935.

Gabriel, W. F. (1992). Adaptive Processing Array Systems, *Proceedings of the IEEE*, vol. 80, no. 1, pp. 152–162.

Gannot, S.; Burshtein, D.; Weinstein, E. (2001). Signal Enhancement Using Beamforming and Nonstationarity with Applications to Speech, *IEEE Transactions on Signal Processing*, vol. 49, no. 8, pp. 1614–1626.

Gannot, S.; Vincent, E.; Markovich-Golan, S.; Ozerov, A. (2017). A Consolidated Perspective on Multimicrophone Speech Enhancement and Source Separation, *IEEE/ACM Transactions on Audio, Speech, and Language Processing*, vol. 25, no. 4, pp. 692–730.

Goodwin, M. M.; Elko, G. W. (1993). Constant Beamwidth Beamforming, *Proceedings of the IEEE International Conference on Acoustics, Speech, and Signal Processing (ICASSP)*, Minneapolis, Minnesota, USA, vol. I, pp. 169–172.

Griffiths, L. J.; Jim, C. W. (1982). An Alternative Approach to Linearly Constrained Adaptive Beamforming, *IEEE Transactions on Antennas and Propagation*, vol. 30, no. 1, pp. 27–34.

Gu, R.; Zhang, S.-X.; Zou, Y.; Yu, D. (2021). Complex Neural Spatial Filter: Enhancing Multi-Channel Target Speech Separation in Complex Domain, *IEEE Signal Processing Letters*, vol. 28, pp. 1370–1374.

Haykin, S. (ed.) (1985). *Array Signal Processing*, Prentice Hall, Englewood Cliffs, New Jersey.

Haykin, S. (1996). *Adaptive Filter Theory*, 3rd edn, Prentice Hall, Englewood Cliffs, New Jersey.

Hendriks, R. C.; Heusdens, R.; Kjems, U.; Jensen, J. (2009). On Optimal Multichannel Mean-Squared Error Estimators for Speech Enhancement, *IEEE Signal Processing Letters*, vol. 16, no. 10, pp. 885–888.

Herbordt, W. (2005). *Sound Capture for Human/machine Interfaces - Practical Aspects of Microphone Array Signal Processing*, vol. 315 of *Lecture Notes in Control and Information Sciences*, Springer-Verlag, Berlin, Heidelberg, New York.

Heymann, J.; Drude, L.; Haeb-Umbach, R. (2016). Neural Network Based Spectral Mask Estimation for Acoustic Beamforming, *Proceedings of the IEEE International Conference on Acoustics, Speech and Signal Processing (ICASSP)*, Shanghai, China, pp. 196–200.

Hoshuyama, O.; Sugiyama, A. (2001). Robust Adaptive Beamforming, *in* M. Brandstein; D. Ward (eds.), *Microphone Arrays*, Springer-Verlag, Berlin, Heidelberg, New York.

Hoshuyama, O.; Sugiyama, A.; Hirano, A. (1999). A Robust Adaptive Beamformer for Microphone Arrays with a Blocking Matrix Using Constrained Adaptive Filters, *IEEE Transactions on Signal Processing*, vol. 47, no. 10, pp. 2677–2683.

Huang, G.; Chen, J.; Benesty, J. (2018). On the Design of Differential Beamformers with Arbitrary Planar Microphone Array Geometry, *The Journal of the Acoustical Society of America*, vol. 144, no. 1, pp. EL66–EL70.

Huang, G.; Benesty, J.; Cohen, I.; Chen, J. (2020a). A Simple Theory and New Method of Differential Beamforming With Uniform Linear Microphone Arrays, *IEEE/ACM Transactions on Audio, Speech, and Language Processing*, vol. 28, pp. 1079–1093.

Huang, G.; Cohen, I.; Chen, J.; Benesty, J. (2020b). Continuously Steerable Differential Beamformers With Null Constraints for Circular Microphone Arrays, *The Journal of the Acoustical Society of America*, vol. 148, no. 3, pp. 1248–1258.

Knapp, C. H.; Carter, G. C. (1976). The Generalized Correlation Method for Estimation of Time Delay, *IEEE Transactions on Acoustics, Speech and Signal Processing*, vol. 24, pp. 320–327.

Kreutz-Delgado, K. (2009). The Complex Gradient Operator and the CR-Calculus, arXiv, https://arxiv.org/abs/0906.4835.

Kummer, W. H. (1992). Basic Array Theory, *Proceedings of the IEEE*, vol. 80, no. 1, pp. 127–140.

Laakso, T. I.; Välimaki, V.; Karjalainen, M.; Laine, U. K. (1996). Splitting the Unit Delay - Tools for Fractional Filter Design, *IEEE Signal Processing Magazine*, vol. 13, no. 1, pp. 30–60.

Lotter, T.; Vary, P. (2006). Dual Channel Speech Enhancement by Superdirective Beamforming, *EURASIP Journal on Applied Signal Processing, Special Issue on Advances in Multimicrophone Speech Enhancement*, vol. 2006, pp. 1–14. Article ID 063297.

Madhu, N.; Martin, R. (2008). Acoustic Source Localization with Microphone Arrays, *in* R. Martin; U. Heute; C. Antweiler (eds.), *Advances in Digital Speech Transmission*, John Wiley & Sons, Inc.

Madhu, N.; Martin, R. (2011). A Versatile Framework for Speaker Separation Using a Model-Based Speaker Localization Approach, *IEEE Transactions on Audio, Speech, and Language Processing*, vol. 19, no. 7, pp. 1900–1912.

Marro, C.; Mahieux, Y.; Simmer, K. U. (1998). Analysis of Noise Reduction and Dereverberation Techniques Based on Microphone Arrays with Postfiltering, *IEEE Transactions on Speech and Audio Processing*, vol. 6, no. 3, pp. 240–259.

Monzingo, R. A.; Miller, T. W. (1980). *Introduction to Adaptive Arrays*, John Wiley & Sons, Inc., New York.

Rickard, S.; Yilmaz, O. (2002). On the Approximate W-Disjoint Orthogonality of Speech, *Proceedings of the IEEE International Conference on Acoustics, Speech, and Signal Processing (ICASSP)*, Orlando, USA, vol. 1, pp. I–529–I–532.

Simmer, K. U.; Bitzer, J.; Marro, C. (2001). Post-Filtering Techniques, *in* M. Brandstein; D. Ward (eds.), *Microphone Arrays*, Springer-Verlag, Berlin, Heidelberg, New York.

Spriet, A.; Moonen, M.; Wouters, J. (2004). Spatially Pre-Processed Speech Distortion Weighted Multi-Channel Wiener Filtering for Noise Reduction, *Signal Processing*, vol. 84, no. 12, pp. 2367–2387.

Strang, G. (1988). *Linear Algebra and Its Applications*, 3rd edn, Thomson Learning Inc.

Tan, K.; Wang, Z.-Q.; Wang, D. (2022). Neural Spectrospatial Filtering, *IEEE/ACM Transactions on Audio, Speech, and Language Processing*, vol. 30, pp. 605–621.

Tesch, K.; Gerkmann, T. (2023). Insights Into Deep Non-Linear Filters for Improved Multi-Channel Speech Enhancement, *IEEE/ACM Transactions on Audio, Speech, and Language Processing*, vol. 31, pp. 563–575.

Warsitz, E.; Haeb-Umbach, R. (2006). Controlling Speech Distortion in Adaptive Frequency-domain Principal Eigenvector Beamforming, *Proceedings of the International Workshop on Acoustic Echo and Noise Control (IWAENC)*, Paris, France, pp. 1–4.

Warsitz, E.; Haeb-Umbach, R. (2007). Blind Acoustic Beamforming Based on Generalized Eigenvalue Decomposition, *IEEE Transactions on Audio, Speech, and Language Processing*, vol. 15, no. 5, pp. 1529–1539.

Williamson, D. S.; Wang, Y.; Wang, D. (2016). Complex Ratio Masking for Joint Enhancement of Magnitude and Phase, *Proceedings of the IEEE International Conference on Acoustics, Speech and Signal Processing (ICASSP)*, Shanghai, China, pp. 5220–5224.

Zelinski, R. (1988). A Microphone Array with Adaptive Post-Filtering for Noise Reduction in Reverberant Rooms, *Proceedings of the IEEE International Conference on Acoustics, Speech, and Signal Processing (ICASSP)*, New York City, USA, pp. 2578–2581.

Zohourian, M.; Enzner, G.; Martin, R. (2018). Binaural Speaker Localization Integrated Into an Adaptive Beamformer for Hearing Aids, *IEEE/ACM Transactions on Audio, Speech, and Language Processing*, vol. 26, no. 3, pp. 515–528.

# Index

*Digital Speech Transmission and Enhancement*, Second Edition. Peter Vary and Rainer Martin.
© 2024 John Wiley & Sons Ltd. Published 2024 by John Wiley & Sons Ltd.